BIOGEOGRAPHY AND ECOLOGY OF TURKMENISTAN

MONOGRAPHIAE BIOLOGICAE

VOLUME 72

Series Editors
H.J. Dumont and M.J.A. Werger

The titles published in this series are listed at the end of this volume.

Biogeography and Ecology of Turkmenistan

Edited by

VICTOR FET

Dept. of Biological Sciences, Loyola University, New Orleans, Louisiana, USA

and

KHABIBULLA I. ATAMURADOV

Natural Conservation Society, Ashgabat, Turkmenistan

Springer-Science+Business Media, B.V.

Library of Congress Cataloging-in-Publication Data

```
Biogeography and ecology of Turkmenistan / edited by Victor Fet and
  Khabibulla Atamuradov.
      p.    cm. -- (Monographiae biologicae ; v. 72)
    Includes index.
    ISBN 978-94-010-4487-5    ISBN 978-94-011-1116-4 (eBook)
    DOI 10.1007/978-94-011-1116-4
    1. Biogeography--Turkmenistan.  2. Ecology--Turkmenistan.
  I. Fet, Victor.  II. Atamuradov, Kh. I. (Khabibulla Ishchanovich)
  III. Series.
  QP1.P37  vol. 72
  [QH191]
  574 s--dc20
  [574.958'5]                                                   94-6952
```

ISBN 978-94-010-4487-5

Printed on acid-free paper

All Rights Reserved
© 1994 Springer Science+Business Media Dordrecht
Originally published by Kluwer Academic Publishers in 1994
Softcover reprint of the hardcover 1st edition 1994
No part of the material protected by this copyright notice may be reproduced or utilized in any form or by any means, electronic or mechanical, including photocopying, recording or by any information storage and retrieval system, without written permission from the copyright owner.

Table of Contents

List of contributors   vii
1. Introduction: One hundred years of natural history in Turkmenistan   1
   V. Fet
2. Landscapes of Turkmenistan   5
   A.G. Babaev
3. Climate of Turkmenistan   23
   N.S. Orlovsky
4. Paleogeography of Turkmenistan   49
   K.I. Atamuradov
5. Desertification of the arid lands of Turkmenistan   65
   N.G. Kharin
6. Vegetation of the deserts of Turkmenistan   77
   I.G. Rustamov
7. Flora of Kopetdagh   105
   D. Kurbanov
8. Kopetdagh-Khorassan flora: Regional features of Central Kopetdagh   129
   G.L. Kamakhina
9. Vegetation of Southwest Kopetdagh   149
   G.N. Fet
10. Trees, shrubs, and semishrubs in the mountains of Turkmenistan   173
    K.P. Popov
11. Ecosystem structure of subtropical arid pistachio woodlands in Southern Turkmenistan   187
    R.I. Zlotin
12. Biogeographic position of Khorassan-Kopetdagh   197
    V. Fet
13. Vertebrates in the Red Data Book of Turkmenistan   205
    A.K. Rustamov & O. Sopyev
14. Ecology of the bearded goat (*Capra aegagrus* Ersleben, 1777) in Turkmenistan   231
    V.M. Korshunov
15. Ecology of birds in the Karakum Desert   247
    A.K. Rustamov

16. Ecological structure of the bird population in the Transcaspian region: Cartographic analysis and problems of conservation   265
E.A. Rustamov
17. Kidney structure and its role in osmoregulation in desert birds   281
M.A. Amanova
18. On the evolution of the pheasant (*Phasianus colchicus* L.) in Middle Asia   295
A.V. Solokha
19. Zoogeographic analysis of the reptiles of Turkmenistan   307
N.N. Shcherbak
20. Reptiles of Kopetdagh   329
C. Ataev, A.K. Rustamov & S. Shammakov
21. Geographic variability of *Phrynocephalus rossikowi* Nik. (Reptilia: Agamidae) in Turkmenistan and adjacent regions   351
M.L. Golubev, V.V. Manilo & A.A. Tokar
22. Formation of the fish population in the artificial hydrographic network of Turkmenistan (the Amudarya River basin)   365
V.B. Salnikov
23. Arthropods inhabiting rodent burrows in the Karakum Desert   389
V.A. Krivokhatsky
24. Zoogeography of Coleoptera in Turkmenistan   403
O.L. Kryzhanovsky & K.I. Atamuradov
25. Buprestid beetles (Coleoptera: Buprestidae) from Kopetdagh and the adjacent regions of Southern Turkmenistan   419
M.G. Volkovich & A.V. Alexeev
26. Fauna, zoogeography, and ecology of Orthoptera in Turkmenistan   451
T. Tokgaev
27. Encyrtid wasps of Turkmenistan (Hymenoptera: Encyrtidae)   467
S.N. Myartseva
28. Zoogeography and ecological aspects of the formation of horse fly fauna (Diptera: Tabanidae) in Turkmenistan   481
R.V. Andreeva
29. Ant-lions (Neuroptera: Myrmeleontidae) in Turkmenistan   495
V.A. Krivokhatsky
30. Fauna and zoogeography of spiders (Aranei) of Turkmenistan   499
K.G. Mikhailov & V. Fet
31. Fauna and zoogeography of scorpions (Arachnida: Scorpions) in Turkmenistan   525
V. Fet
32. Fauna and zoogeography of molluscs of Turkmenistan   535
Y.I. Starobogatov
Bibliography   545
Index of Taxa   605
Index of Subjects   637

List of Contributors

AMANOVA, M.B., Department of Biology, Turkmen State University, Ashgabat, Turkmenistan.
ANDREEVA, R.V., I.I. Schmalhausen Institute of Zoology, Ukrainian Academy of Sciences, Kiev, Ukraine.
ALEXEEV, A.V., Pedagogical Institute, Orekhovo-Zuevo, Russia.
ATAEV, Ch., Institute of Zoology, Turkmen Academy of Sciences, Ashgabat, Turkmenistan.
ATAMURADOV, K.I., Natural Conservation Society, Ashgabat, Turkmenistan.
BABAEV, A.G., Desert Institute, Turkmen Academy of Sciences, Ashgabat, Turkmenistan.
FET, G.N., Department of Biological Sciences, Loyola University, New Orleans, Louisiana, USA.
FET, V., Department of Biological Sciences, Loyola University, New Orleans, Louisiana, USA.
GOLUBEV, M.L., Seattle, Washington, USA.
KAMAKHINA, G.L., Institute of Botany, Turkmen Academy of Sciences, Ashgabat, Turkmenistan.
KHARIN, N.G., Desert Institute, Turkmen Academy of Sciences, Ashgabat, Turkmenistan.
KORSHUNOV, V.M., Ecocenter, Ashgabat, Turkmenistan.
KRIVOKHATSKY, V.A., Zoological Institute, Russian Academy of Sciences, St. Petersburg, Russia.
KRYZHANOVSKY, O.L., Zoological Institute, Russian Academy of Sciences, St. Petersburg, Russia.
KURBANOV, Dz., Department of Biology, Turkmen State University, Ashgabat, Turkmenistan.
MANILO, V.V., I.I. Schmalhausen Institute of Zoology, Ukrainian Academy of Sciences, Kiev, Ukraine.
MIKHAILOV, K.G., Zoological Museum, M.V. Lomonosov Moscow State University, Moscow, Russia.

MYARTSEVA, S.N., Institute of Zoology, Turkmen Academy of Sciences, Ashgabat, Turkmenistan.
ORLOVSKY, N.S., Desert Institute, Turkmen Academy of Sciences, Ashgabat, Turkmenistan.
POPOV, K.P., Desert Institute, Turkmen Academy of Sciences, Ashgabat, Turkmenistan.
RUSTAMOV, A.K., Turkmen Institute of Agriculture, Ashgabat, Turkmenistan.
RUSTAMOV, E.A., Department of Biology, Turkmen State University, Ashgabat, Turkmenistan.
RUSTAMOV, I.K., Department of Biology, Turkmen State University, Ashgabat, Turkmenistan.
SALNIKOV, V.G., Institute of Zoology, Turkmen Academy of Sciences, Ashgabat, Turkmenistan.
SHAMMAKOV, S., Institute of Zoology, Turkmen Academy of Sciences, Ashgabat, Turkmenistan.
SHCHERBAK, N.N., Zoological Museum, I.I. Schmalhausen Institute of Zoology, Ukrainian Academy of Sciences, Kiev, Ukraine.
SOLOKHA, A.V., Turkmen Institute of Agriculture, Ashgabat, Turkmenistan.
SOPYEV, O.S., Turkmen Institute of Agriculture, Ashgabat, Turkmenistan.
STAROBOGATOV, Ya.I., Zoological Institute, Russian Academy of Sciences, St. Petersburg, Russia.
TOKAR, A.A., I.I. Schmalhausen Institute of Zoology, Ukrainian Academy of Sciences, Kiev, Ukraine.
TOKGAEV, T.B., Institute of Zoology, Turkmen Academy of Sciences, Ashgabat, Turkmenistan.
VOLKOVICH, M.G., Zoological Institute, Russian Academy of Sciences, St. Petersburg, Russia.
ZLOTIN, R.I., Department of Biogeography, Institute of Geography, Russian Academy of Sciences, Moscow, Russia.

1. Introduction: One Hundred Years of Natural History in Turkmenistan

VICTOR FET

> *Brotherhood is our custom,*
> *Friendship is our law.*
>
> Makhtumkuli,
> Turkmen national poet (18th century)

As part of the famous "Great Game" between the Russian and British Empires in Central Asia, Turkmenistan was the last colonial prize of the Russian tsars; its delineation from Afghanistan was completed only in the 1890s. The Russian Empire's Transcaspian Region (*Zakaspiiskaya Oblast*) was roughly what Turkmenistan is today; its neighbors were the semi-independent emirate of Bokhara to the east and khanate of Khiva to the north, both remnants of medieval Muslim empires.

A stunning rate of technological, educational, and cultural progress in this desert land of nomads was achieved in less than 30 years of imperial Russian rule (Pahlen 1963). The famous Transcaspian railroad ran from Krasnovodsk to Tashkent. Scientific research, which had never touched this remote corner of Asia before, went in pace with advances in road building, industry, and irrigation. Traditional interests of nineteenth century Russian naturalists in Central and Middle Asia, so lively portrayed in *The Gift* by Vladimir Nabokov (1952), naturally extended to the newly colonized territories of Transcaspia. Since the 1880s, naturalists have attempted to describe the rich and peculiar flora and fauna of the magnificent sand deserts of Turkmenistan.

Early notes described the rich natural resources in deserts and mountains as well as the severe deforestation. Logging of juniper in the mountains, pistachio trees in the foothills, and saksaul shrubs in sand desert began as early as in the Neolithic Age, when early farming settlements emerged in the foothills of Kopetdagh (Shishkin 1981). It continued through the era of the ancient Parthian Empire, whose capital, Nisa, now lies in ruins a few miles from Ashkhabad, the capital of Turkmenistan. Timber was used in construction, as firewood, and also as a charcoal supply for smelting of metals. Green and populous oases, with such centers of culture and education as Khwarazm (Khiva), thrived in the Transcaspia in the times of the magnificent empires of Alexander the Great and his followers, only to be destroyed in the next millenium by Genghis Khan, Tamerlane, and other warriors. Human-influenced desertification expanded in these times; extensive grazing of sheep and camels by Huns and, later, Turkic tribes, contributed to soil deflation and erosion by desert winds and rare, but intensive rains.

An early naturalist of the 1900s, coming by ferry across the Caspian Sea from well-established Russian settlements in the Caucasus, was able to see herds of large game animals such as gazelles, onagers, and wild sheep. Hyenas, leopards, cheetahs, and even Turanian tigers preyed on a variety of wild game. The Transcaspian Region was immediately recognized as an important area for scientific studies. The world-famous Repetek Sand Desert Station was established in 1912 to study the geology of the Karakum sand desert. Biological stations and museums followed; the first extensive collections were made from the 1890s through the first decade of the twentieth century for major Russian natural history museums in Moscow, St. Petersburg, and Tiflis.

With the establishment of the Soviet regime after 1917, Russian science was artficially severed from European scientific thought. Original, mandatorily isolated Russian schools of theory in ecology and biogeography developed in the 1920s and 1930s. Primary data for this development flowed from many geographical areas of the Soviet Union, including Middle Asia; the deserts and mountains of Turkmenistan continued to be an important site of basic field research (Laptev 1934; Pavlovsky 1934; Kryzhanovsky 1965).

Limited in their abilities to travel abroad, Soviet scientists traveled to exotic, "colonial" domestic places. The deserts and mountains of Turkmenistan were a favorite "spring vacation" site for many Russian entomologists, herpetologists, and bird watchers. As a result, the rich faunas of this republic are extremely well-known as compared to many other areas in Middle East or Central Asia (Kryzhanovsky and Atamuradov this volume; Shcherbak this volume; Rustamov this volume). The well-known volumes of the "Fauna of the USSR" and even the more comprehensive "Flora of the USSR," published since the 1930s, were important landmarks in the scientific development for Middle Asia, similar to the work of British naturalists in India. And, as was true of English for the former subjects of the British Empire, Russian become the only scientific and educational language for all Middle Asian republics. Scientific works were published almost exclusively in Russian. This, on the one hand, prevented the Turkmen language from becoming the tongue of learned people, as Arabic or Farsi had been in the past; on the other hand, it allowed free communication among scientists. (I remember being amused many times by listening to lively conversations in Russian between local Turkmen ornithologists and visiting bird-watchers from Estonia or Lithuania. There, the Russian language performed a communication role among subjects of the Empire, with conversation otherwise hardly possible.)

Study of the natural resourses of Turkmenistan accompanied attempts to preserve its biodiversity, even under the strongest political pressure of the epoch. The famous Russian geneticist Nikolai Vavilov, who perished in 1940 under Stalin's terror, established the first plant breeding station in Kara-Kala (Southwest Kopetdagh) to study the tremendous biodiversity of wild ancestors of domestic plants in the mountains of Turkmenistan. Collection and selection work on hundreds of strains of wild grape, apple, pear, pomegranate, almond, walnut, pistachio, barley, and oat allowed future geneticists to explore the last remnants of gene pools of these species. Badghyz Natural Reserve, established in 1941, became a refuge for the last existing population of the Turkmen onager (*Equus hemionus onager*) and a unique pistachio woodland.

A new generation of local Turkmen scientists, many of whom were trained by the Russian researchers in the graduate schools of Moscow and Leningrad arose from the 1930s through the 1950s. The Turkmen Academy of Sciences and its journal, *Proceedings* (including the monthly biological series), served to record the results of diverse biological studies in the republic.

While basic science in the Middle Asian republics rather gained from the Russian "colonial" influence, natural resources, in contrast, were severely damaged by the Soviet way of handling the economy and social issues. Severe environmental problems have been inherited by the now independent Turkmenistan, including overgrazed desert pastures, deforested mountains, depleted water resources, accumulated pesticides in cotton fields, declining populations of endangered species of animals and plants, and – worst of all – progressing, human-caused desertification (Kharin this volume). In order to approach a solution to these problems, scientists and officials in the republic will need the close attention and help of the international scientific community.

A so-called ecotourism, currently practiced in countries rich in biodiversity (e.g., Costa Rica and Belize), might be one way for Turkmenistan to finance the conservation of its natural protected areas, so vulnerable under the continuing aridization. There is enough to see in Turkmenistan: herds of bighorns and onagers in the wilderness of pistachio forests of Badghyz; a breathtaking view of ancient basaltic volcanoes in the midst of the pink-salt Lake Yeroyulanduz; the magnificent sand dunes of the Karakum Desert; flocks of flamingoes on the Caspian seashore; and the Kopetdagh Mountain valleys in early spring, blooming with almonds and hyacinths. Ecotourist facilities, as well as joint scientific environmental projects, could be based in the eight existing Natural Reserves (Krasnovodsk, Kaplankyr, Syunt-Khasardagh, Kopetdagh, Badghyz, Repetek, Amudarya, and Kugitang) which represent all major landscapes of the republic. Although, in the past, these reserves have never achieved the tourist attendance level or financial security of Western national parks, they have traditionally played the role of biological field stations, housing each year dozens of field researchers and university students.

The newly independent republics of Middle Asia are economically likely to stay under the strong influence of Russia. Culturally and linguistically, however, Turkmenistan belongs to the Turkic-speaking part of the Islamic world. Today, it is important that the positive legacy of the last hundred years of Russian and Soviet influence, as well as the gained momentum in the scientific development of Turkmenistan, not be lost. Critical for the scientific community of the republic is its openness to international cooperation; combined with the solid level of existing research, such cooperation is bound to yield progress.

This book combines the results of basic scientific research in biogeography and ecology; its purpose is to give a fairly comprehensive account of the nature of Turkmenistan. It is also the authors' common desire to see its human population living in balance with this diverse nature, and the state of Turkmenistan being peaceful and prosperous.

Acknowledgements

This monograph could not have been completed without the tremendous help of Elsa Galbraith (Midwestern State University, Wichita Falls, Texas, USA) who volunteered to undertake the painstaking task of editing its English translation.

Michael Golubev (Seattle, Washington, USA) advised on aspects of reptile taxonomy. Robert Vezzetti and Dana Pelligrin assisted with manuscript typing.

One of the first expeditions to Badghyz, ca. 1935. Photo by Mikhail P. Rozanov. Restored by Vladimir M. Potapov, 1976.

2. Landscapes of Turkmenistan

AGADZHAN G. BABAEV

Abstract

Lowland and mountainous desert landscapes of Turkmenistan are described, including the following regions of the republic: the Kopetdagh Mouintains with their northern submontane plain, the isolated Bolshoi Balkhan and Maly Balkhan Mountains, the Ustyurt Plateau (its part within Turkmenistan), the Krasnovodsk Peninsula, the Cis-Caspian Lowland, western sands of Chilmamed and Uchtagan, the Sarykamysh-Khwarazm Lowland, the Uzboi dry valley, the Karakum Desert, the area between the Tedzhen and Murghab Rivers, the Badghyz and Karabil Plateaus, the Amudarya River Valley, and the Kugitangtau Mountains. Landscape structure is determined by a complicated geological history and modern, extremely arid climate. The description of natural hydrographic network is given; the most important rivers are the Amudarya, Tedzhen, and Murghab. Ten lithoedaphic types of deserts are identified within Turkmenistan. Thirteen ecological regions are separated according to their physical-geographical features, climate, and potential for agricultural development.

Introduction

Turkmenistan is located in western Middle Asia between latitudes of 35° 08' N and 42° 48' N and longitudes of 52° 27' E and 66° 41' E. It extends from west to east for 1,100 km, and from north to south, for 650 km, and constitutes 488,100 km^2.

Boundaries of the republic reflect historic and geographic features of migrations of the Turkmen people in Middle Asia. Turkmenistan borders Kazakhstan to the north, Uzbekistan to the east and northeast, Iran to the south and southwest, and Afghanistan to the southeast. To the west, Turkmenistan is limited by the Caspian Sea.

The physical geography of Turkmenistan is determined primarily by the sharply continental climate with its extreme shortage of moisture and high summer temperatures. All landscapes of the republic – whether the Kopetdagh Mountains, the Ustyurt Plateau, or the Karakum sands – bear a desert image. Rivers which flow through the Turkmen lands do not change the desert appearance of surrounding plains and alter only the narrow strips of land in their valleys. Even the influence of the Caspian Sea is minute on its eastern shore, where landscapes are as dry, severe, and desert-like as inland. All green oases with dense populations, vast gardens, and fields are man-made. There are only a few relatively small natural oases in places where ground water reaches the surface; today these are mostly incorporated into larger artificial oases.

Turkmenistan lies within the temperate desert zone, which is south of the semidesert zone. Within the Turanian Lowland, which includes the lowland part of Turkmenistan, Berg (1938) distinguished three desert subzones: northern Tertiary plateaus which intergrade in the north with the semidesert zone; a subzone of sand deserts; and a subzone of submontane loess plains. The southernmost part of the republic bordering Iran and Afghanistan is occupied by mountains which also have desert character as well as altitudinal zonality. Within Turkmenistan, the first desert subzone includes the southern promontories of the Ustyurt and Mangyshlak Plateaus. The subzone of sand deserts includes the great sand desert of Karakum. Finally, loess plains are well developed along the northern foothills of Kopetdagh.

Relief of Turkmenistan

More than 75% of Turkmenistan territory consists of lowlands. The mountains and plateaus occupy only the southern border and small isolated areas in the west (Fig. 1); these elevated areas do not reach snowline and were not glaciated even in the Ice Ages. The maximal altitude in Turkmenistan is 3,137 m in the Kugitang Mountains; the lowest point of the republic is −92 m in the Akhchakaya Depression (the northwestern part of the Trans-Unguz Karakum); thus, the gypsometric amplitude is 3,229 m. Landscapes, however, do not exhibit significant altitudinal changes since maximal elevations are expressed only in the south and east of the state where the influence of increasing aridity and temperatures prevents the distinct expression of altitudinal differences.

The lowlands of Turkmenistan lie primarily between 50 and 200 m above sea level. Only the Caspian Lowland and some depressions are located below this mark while the eastern and southeastern Karakum and parts of the Krasnovodsk and Ustyurt Plateaus are higher. In general, the lowlands of Turkmenistan are tilted from south toward north and from east toward west.

Major mountain ranges of Turkmenistan exceed 1,500 m; e.g., Bolshoi Balkhan rises to 1,800 m; Kopetdagh, to 2,872 m (Mt. Chopandagh, southwest of Ashkhabad); and Kugitang (or Kugiutangtau), to 3,137 m. Other mountains and plateaus (such as Maly Balkhan, Badghyz, and Karabil) usually do not reach 1,000 m. Therefore, mountains of Turkmenistan can be classified as of medium height.

Fig. 1. A schematic map of Turkmenistan. 1 – stabilized sands, 2 – semi-stabilized sands, 3 – drift sands, 4 – the Karakum Canal.

The Kopetdagh Mountains are the northern range of the Turkmeno-Khorassan mountain system. Kopetdagh stretches along the southern border of the republic for ca. 500 km between the meridians of the cities of Kazandzhik and Tedzhen. Turkmenistan includes only the northern portion of Kopetdagh, which is as narrow as 10 km in the east, 25 to 50 km in its central part, and 100 to 120 km in the west. Southern and almost all of eastern Kopetdagh lie in Iran. Within Turkmenistan, Kopetdagh is formed by a number of separate ranges which have their own names. Most of these ranges are anticlinal and are formed from Cretaceous sediments; exposure of maternal rocks is always well expressed.

In the north, Kopetdagh is limited by the submontane plain, inclined from 300 to 50 m. Here are located main settlements of the republic, including its capital, Ashkhabad (altitude 255 m), and numerous agricultural oases irrigated by small Kopetdagh rivers and spring waters. The submontane plain is also inclined from east toward west and slightly dissected. Gullies, river beds, and depressions are well expressed next to the Kopetdagh piedmont, but flatten out farther from the mountains.

The submontane plain is formed from alluvial fan deposits with coarse material concentrated next to the mountains, and fine fractions covering the main plain area. These deposits are usually covered by loess. The width of the submontane plain is 16 or 17 km next to the city of Kazandzhik, 40 km next to the Iskander Station, and 10 to 12 km next to Ashkhabad. In the north, the submontane plain is limited by the sands of the Karakum; this boundary is not straight but forms a row of capes, bays, and semiclosed depressions among sands. Mudslide (*sel*) waters from the plain which reach these depressions form temporary lakes. Against the background of slightly inclined plain one can distinguish separate scattered low hills and clusters of dune or hill sands with scant vegetation.

The isolated mountain ranges of Bolshoi and Maly Balkhan are located northwest of Kopetdagh. Maly Balkhan is a small (ca. 30 km) anticlinal range stretched from west-southwest to east-northeast, and is formed from Cretaceous and – partially – Tertiary sediments. It reaches 955 m and is asymmetric: the highest part of the range rises sharply above the northern submontane inclined plain. On this plain, at some distance from Maly Balkhan, stands a small ridge formed from Upper Tertiary sediments. The southern slope of Maly Balkhan is more extended and consists of several monoclinal ridges parallel to the main range. Both northern and southern slope are dissected by a network of ravines. From both sides, Maly Balkhan is limited by depressions ("corridors") which connect the Caspian Lowland and the Central Karakum.

Bolshoi Balkhan is a more complex mountain system. Geologically, it is a diffuse anticline with its core comprised of Jurassic sediments, and its limbs, of Cretaceous ones. Bolshoi Balkhan proper reaches 1,880 m; it is stretched latitudinally and is bordered in the north by a rocky, sometimes vertical, cliff, while in the south it forms a steep slope, and in the west gives two offshoots. The southern-inclined surface of the Bolshoi Balkhan is dissected by numerous ravines. The eastern part of the range is lower (highest point 1,376 m) and its northern slope, dissected by short ravines, is less steep than the southern one.

The Caspian, or West Turkmen, Lowland lies to the west of Kopetdagh and to the west and southwest of Krasnovodsk and Balkhan Bays of the Caspian Sea and of Bolshoi Balkhan. Its elevation is between −27.6 and 100 m. The Caspian shelf adjacent to the lowland is very shallow: the 10 m isobathe extends out 10 to 20 km from the shore. The shoreline forms a number of bays, peninsulas, islands, and alluvial sand banks. Most of the lowland is occupied by sand ridges and dunes, solonchaks, and takyrs (clay desert). In the north, the plateau-like summit of Nebit-Dagh (or Nefte-Dagh, 45 m) rises from the giant solonchak Kyolkor; farther to the east lie the mountain ranges of Monzhukly and Boyadagh (134 m). In the south of the Caspian Lowland lies the delta of the Atrek River. In the west, the lowland intergrades into the foothills of Kopetdagh, forming a submontane inclined plain tens of kilometers wide and dissected by dry mudslide beds.

The Krasnovodsk Peninsula is bordered by the Caspian Sea to the west, by Krasnovodsk and Balkhan Bays to the south, and by the Kara-Bogaz-Gol Bay to the north. Most of the peninsula is occupied by the Krasnovodsk Plateau, which bears depressions (50 to 70 m in depth, with accumulated sand) and buttes. Average height of this plateau is 200 m; highest points in the south reach 320 m. In the south, southwest, and north, the plateau ends in high cliffs exhibiting outcrops of Tertiary rocks. In the northwest, the plateau gradually descends toward the seashore plain of the western part of the Krasnovodsk Peninsula. Most of this plain is covered by the sand massif, Oktum, formed by intermitting sand ridges and depressions; a strip of solonchaks stretches along the seashore. In the south, the plain turns into the long and narrow Krasnovodsk sand bank which separates Krasnovodsk Bay from the Caspian Sea, and in the north, it turns into a similar Karabogaz sand bank which separates Kara-Bogaz-Gol Bay from the Caspian.

Between the Krasnovodsk Peninsula and Kara-Bogaz-Gol Bay in the west, the Ustyurt Plateau in the north, and Bolshoi Balkhan and the Uzboi dry bed in the south and southeast, lies an orographically diverse area which comprises a mountainous system of Tuarkyr. This territory, composed of dislocated Cretaceous and Jurassic beds, has narrow monoclinal ridges, steep cliffs, closed depressions, and buttes. The mountains of Irsarybaba and Tuarkyr are ca. 300 m high; mountains Begi-Arslan and Akkyr reach 400 m and higher. In depressions, the elevation drops to 100 m and even lower. Sands and solonchaks occupy small areas in the bottoms of closed depressions. Outcropping maternal rocks can be seen almost everywhere.

To the south and east of the Tuarkyr area lie the sands of Chilmamed (or Chilmamedkum) and Uchtagan. The Chilmamed sands are located between Tuarkyr and Bolshoi Balkhan, increasing in altitude from 0 to 200 m toward the northwest. This massif includes sand ridges and interridge depressions stretching from northwest to southeast. The ridges are from 30 to 35 m, and sometimes even to 50 m high. Takyrs are absent; small clay desert areas appear only in the eastern part of the Chilmamed sands. The second sand massif, Uchtagan, lies eastward of Tuarkyr. Elevations here are from 22 to 120 m, rising toward the northwest. Large valley-like depressions filled by small ridged sands are intersected by high sand ridges. Main ridges are oriented 20 to 25° from northwest toward southeast. In some places, rocks of the Trans-Unguz continental formation and of Miocene age outcrop from under the sand.

Between the Uchtagan sands and the Kaplankyr Plateau (which is a southern offshoot of the Ustyurt Plateau) lies a deep depression extending from northwest toward southeast. Its bottom lies at a level of −19 to +20 m and is occupied by the giant (ca. 100 km long) solonchak Karashor. In the west, Karashor is bordered by a terrace ca. 10 m high.

Of the Ustyurt and South Mangyshlak Plateaus, only the southern parts belong to Turkmenistan. The South Mangyshlak Plateau borders the Kara-Bogaz-Gol Bay from the north as a cliff with a good outcropping of Tertiary sediments. Its average altitude is from 100 to 130 m; some points are elevated from 5 to 20 m above this surface. In the southwest, the plateau declines and becomes a sand bank which separates the northern part of Kara-Bogaz-Gol Bay from the Caspian Sea. Small salt lakes and solonchaks can be found within this sand bank, especially in the transition zone from plateau to sand bank.

The Ustyurt Plateau outlines Kara-Bogaz-Gol Bay from the east, and yields two offspurs to the south known as Chelyungkyr and Kaplankyr. These two plateaus, which reach the Uzboi dry bed at 40° N, are separated by the abovementioned Karashor Depression and Uchtagan sand massif. In the west and south, Ustyurt often forms high cliffs (*chinks*) known under a variety of names. Especially impressive (from 300 to 320 m high) are the Kulandagh chink on the shore of Kara-Bogaz-Gol Bay and the chinks of Kaplankyr above the Karashor Depression, all of which exhibit excellent outcrops of Tertiary and Mesozoic sediments. Here, Ustyurt reaches its maximal absolute elevations (330 m in Kulandagh and 302 m in Kaplankyr); average elevation of the Ustyurt surface within Turkmenistan equals 200 to 250 m. This surface lacks large depressions (which appear farther north, in the Kazakh and Karakalpak portions of the plateau) but often possesses deep pan-like depressions, eroded hills and buttes next to the chinks. Since the Ustyurt surface is inclined toward the east and northeast, i.e., toward the direction opposite from the chinks, water runoff is directed toward the inside of the plateau, and the network of ravines along the Ustyurt chinks is, therefore, sparse.

The Sarykamysh-Khwarazm Lowland includes the Sarykamysh Depression in the west and the alluvial plain of the Amudarya River in the east. The lowest point in the Sarykamysh Depression is -45 m.

The bottom of the depression (exposed before flooding by the discharged irrigation waters which formed modern Lake Sarykamysh) was covered by solonchaks and sand areas; its southern and eastern periphery included buttes, dry river beds, takyrs, and sand deposits. The Sarykamysh Depression is bordered southeasterly by the sands of the Karakum Desert. Outcrops of the maternal Tertiary rocks can be found in the Ustyurt escarps, in the buttes, and in the bottom of the Sarykamysh.

The alluvial plain of the Amudarya River within Turkmenistan is inclined from the river westward. Within the plain, elevation falls from 70–80 to 50–55 m. The plain is dissected by numerous natural river beds and artificial canals and contains sparse, table-shaped buttes formed from maternal rocks.

The Uzboi dry bed formerly was a river which carried surplus water from the ancient Lake Sarykamysh to the Caspian Sea. The Uzboi Valley divides two geologically different areas: a so-called Trans-Uzboi folded region, and the Karakum Desert. Most of the valley is excellently preserved; only locally are some beds eroded and smothered by sand due to recent denudation. Terraces of the former river are also quite well preserved. The Uzboi is 550 km long, and its valley is 2 to 3 km wide, with maximal depth of 40 m. The total gradient of the river is 75 m. The Uzboi Valley reaches the Caspian Lowland through the so-called Balkhan Corridor between the mountains of Bolshoi and Maly Balkhan, and it disappears in the Kyuolkor solonchack. The extension of the Uzboi is the Aktam dry bed, which stretches from Kyuolkor to Balkhan Bay of the Caspian Sea.

The Karakum sand desert occupies a giant territory of 350,000 km between the Uzboi in the west, the Amudarya in the east, the Kopetdagh and Paropamiz mountains in the south, and the Kwarazm (or Khiva) Oasis in the north. This vast territory is divided into the Trans-Unguz and Lowland Karakum; the latter, in turn, is divided into the Central and Southeast Karakum.

The Central Karakum lies northward from the submontane plain of Kopetdagh. Its absolute elevations vary from 20 m in the west to 200 m in the east. Sand ridges appear immediately at its commencement from the submontane plain and often contain hard clay takyrs in the interridge depressions. Especially stable is the takyr belt in the central and western parts (eastward to the Tedzhen delta), where it is from 30 to 80 km wide. Sand ridges which separate takyrs can rise 15 to 20 m, and sometimes (e.g., in the lower part of the Tedzhen Valley) can be more than 10 km long. Takyrs are of great importance in the desert since they are watersheds in which precipitation collects; wells and settlements (*auls*) are often located next to takyrs in the Karakum.

In the southern part of the Central Karakum lies a latitudinal belt of solonchaks (*shors*) which increase in width from 10 to 40–45 km from east to west. The shor depth, commonly from 8 to 15 m, can reach 40 m. Northward from the shor belt ridge, sands reappear which stretch to the Unguz area. Ridges here are low, dense, and separated by interridge depressions, creating a ridge-depression relief. Many depressions are ocupied by takyrs. Primary ridges in the Karakum Desert are usually meridional or submeridional in their direction.

The depressed area known as Unguz separates the Central and Trans-Unguz Karakum; it is a linear chain of depressions which lie at the same level. The Unguz can be traced from the Amudarya to the Uzboi Valley. Depressions are usually two to four km wide; their bottoms are often occupied by shors. Some depresssions are divided by massifs of maternal rocks, up to 40 m high. The Unguz is limited in the north by Trans-Unguz *kyrs* (flat-topped ridges) which are elevated from 60 to 80 m above depressions.

The Trans-Unguz Karakum lies between the Unguz area and the Sarykamysh-Khwarazm Lowland. Its relief is highly dissected, with long meridional buttes, or *kyrs*, 20 to 30 m high (rarely 40 m in the western part), formed from Upper Tertiary rocks. Kyrs are tens of kilometers long and are separated by depressions one to three km wide which are usually filled by sands or occupied by takyrs. Facing the Unguz, these depressions form dissected chinks with deep "bays." About 50 km north of the Unguz, the maternal kyrs disappear under sands, and the landscape transforms into one of sand ridges with sparse takyrs in depressions. Absolute elevations of the Trans-Unguz Karakum vary from 220 m in the southeast to 100 in the north; some takyrs in the west lie at 50 to 75 m.

A belt of dune sands from 10 to 50 km wide stretches along the eastern edge of the Karakum Desert, parallel to the Amudarya Valley between the cities of Kerki and Deinau. Some dune (*barkhan*) ridges here can reach 25 m.

The Tedzhen and Murghab Rivers end blindly in the Karakum, forming wide subaerial deltas with branching dry beds which stretch far into the desert. Delta areas are rich in sand, and contain patches of takyrs along the dry river beds. These takyrs sometimes form flat valley-like depressions with low edges turning into sand ridges.

The Southeast Karakum is formally separated from the Central Karakum by the Chardzhou – Ashkhabad Railroad. This part of the Karakum lies higher than the rest of the desert, with elevation from 190 or 200 m next to the railroad to 300 to 350 m in the south. The desert continues to the south without any natural barriers. Between the lower parts of the Tedzhen and Murghab Rivers, which flow in the terraced valleys, stretch the uniform clay plains, rarely interrupted by sands or small hills. High and stable sand ridges appear farther southward between the Tedzhen and Murghab and at the right bank of the Murghab. Farther eastward lies the sand steppe (Obruchev Steppe) which is very slightly dissected by wind or water erosion. Here, the Southeast Karakum is penetrated by the so-called Kelif Uzboi, a linear strip of shors extending northwesterly.

The foothills of the Paropamiz Mountains rise eastward from the Tedzhen River along the Afghanistan border. These foothills are separated from the Southeast Karakum by a wave-like plain with sparse buttes. The part of these foothills between the Tedzhen and Murghab is called Badghyz; the part to the east of Murghab, Karabil. These are desert plateaus with smooth relief. Badghyz rises up to 1,255 m; Karabil, to 950 m. These plateaus lack the extended network of rivers or ravines which appear farther south, in Afghanistan. Only the Tedzhen and Murghab Valleys branch into steep but deep ravines. Hills (*bairs*) in Badghyz reach sometimes 200 m of relative height; depressions between bairs often contain solonchaks, takyrs, and small lakes. Very characteristic of Badghyz are closed depressions: the largest, Yeroilan (or Yeroyulanduz) lies at 273 m, contains two salt lakes, and is distinctly expressed in the relief by its northern cliffs. Karabil is wider and lower than Badghyz, with uniform hilly relief. There are no rivers, and dry beds and depressions are rare.

The relief of the right bank of the Amudarya within Turkmenistan below the city of Kerki is not significantly different from the Karakum Desert. It includes plain desert, mostly occupied by the Sundukli sand massif; dune chains are expressed next to the valley (as well as in the left bank). The Sundukly sands descend to the Amudarya as a low but steep escarp. A few closed depressions with salt lakes or solonchaks in their bottoms are present. Across the Amudarya from the city of Chardzhou ends the dry bed of the Zeravshan River. There are numerous groove-like dry depressions, closed depressions, and dry beds separated by narrow plateau-like ridges of maternal rocks. Clay plain is predominant eastward from Kerki. Absolute elevations of the right bank of the Amudarya within Turkmenistan fall from 400 m in the southeast to 200 m in the northwest; bottoms of closed depressions lie at yet lower elevations.

The mountains of the Gaurdak-Kugitang region, which belong to the Ghissar mountain system, rise in the easternmost part of Turkmenistan on the right bank of the Amudarya. The highest range (up to 3,137 m) is Kugitangtau (or Kugitang), which is comprised of Jurassic and Paleozoic rocks. It forms steep slope eastward toward Uzbekistan. The western slope of Kugitangtau, facing the valley of the Kugitang-Darya River, is less steep but dissected by deep ravines. To the west and south of Kugitangtau lie lower plateaus formed of Cretaceous and, partly, Jurassic rocks. These gradually decrease toward the southwest and approach the Amudarya Valley as ridges separated by wide takyr plains.

Natural Hydrographic Network of Turkmenistan

The natural hydrographic network in Turkmenistan is extremely weakly expressed. There are no significant rivers arising within the republic. Only small rivers originate from the mountains of Kugitangtau and Kopetdagh, and their water is spent for irrigation. Only the mighty Amudarya and, sometimes, Atrek, reach their base level of erosion. The dry beds which cross lowland Turkmenistan for tens and hundreds of kilometers emphasize the scarcity of an active river network.

The Atrek River is the only river in Turkmenistan that belongs to the Caspian Sea basin. It originates from Iran and, west of the mouth of its tributary Sumbar, delineates the state border between Turkmenistan and Iran. The entire delta of the Atrek lies within Turkmenistan. The Atrek is 495 km long, of which 145 km flow within Turkmenistan; its drainage is ca. 40,000 km^2; its average annual debit is 10.4 m^3/sec. The river gradient is 1,265 m from Kuchan in Iran to the mouth; within Turkmenistan, where the Atrek flows along the Caspian Lowland, its gradient is only 84 m. In Turkmenistan, the Atrek is only 10 to 15 m wide and not more than 0.5 m deep; its water is completely spent for irrigation, and it reaches the Caspian Sea only during floods. The long-time deposits of the Atrek form a vast ancient delta.

The largest tributary of the Atrek is the Sumbar River (203 km long), which, together with its tributaries Chandyr and Tersakan, forms the drainage of West and, partly, Central Kopetdagh. In its upper part, the Sumbar is a mountain river; its middle part flows across the wide and flat plain sparsely dotted with hills. The Sumbar normally does not reach the Atrek since its water is taken for irrigation.

The Amudarya, the largest river of Middle Asia (2,287 km long), flows into the Aral Sea. It enters Turkmenistan from Uzbekistan below the mouth of the Surkhan-Darya and leaves the republic via Tyuamuyun Reservoir for Karakalpakistan. Unlike other rivers, the Amudarya has two flood periods: in spring and summer. The summer flood is the result of the thawing of snow and glaciers in the Pamir Mountains. The average annual debit of the Amudarya is 1,700–2,000 m^3/sec. Within Turkmenistan, the river flows in a wide but depressed valley; its bed is from 300 m to 5 km wide. Due to its fast flow, the Amudarya erodes banks in many places. The summer flood level is one to three meters higher than the low water bed; ramparts are constructed for protection against high floods. An enormous amount of deposits is carried by the river in the summer; these deposits form banks and islands and accumulate in river beds and canals. The valley is covered by tugai vegetation and is developed as oases.

In Turkmenistan, the Amudarya has only one tributary, a small river named Kugitang-Darya. It is 75 km long, and collects water from the Kugitangtau

Mountains. This water, however, is primarily used for irrigation and only a small portion of it reaches the Amudarya. The remaining hydrographic network associated with the Amudarya includes dry beds and gullies that are filled by water only during the rare heavy rains.

Only the lower portions of the Tedzhen and Murghab Rivers belong to Turkmenistan; these rivers' origins lie in Afghanistan. The Tedzhen (or Harirud) enters Turkmenistan at the juncture of the state borders of Turkmenistan, Afghanistan, and Iran. Above the town of Serakhs, the Tedzhen forms the state border between Turkmenistan and Iran. Most water of the Tedzhen remains in Afghanistan where it is used for irrigation of the Gerat Valley; the remainder is used for irrigation in Turkmenistan. The drainage of the Tedzhen constitutes 77,700 km^2; its length within Turkmenistan (including the border region with Iran) is 320 km. The average annual debit of the Tedzhen is 25 m^3/sec. During the spring floods (March-April) the debit doubles or triples; in extraordinary cases, it can increase ten-fold for a short time.

The Murghab River enters Turkmenistan from Afghanistan between the plateaus of Badghyz and Karabil and crosses the Southeast Karakum Desert from south to north. Commonly, the Murghab flows only 30 to 40 km north from the city of Mary, but in years rich in precipitation, flood waters of the Murghab extend 140 km north of the Chardzhou – Ashkhabad railroad. The drainage of the Murghab is 62,700 km^2. Its average annual debit is 49 m^3/sec, with maximum volume from April to May; the debit can fluctuate three-fold during the year. Below the city of Takhta-Bazar, the Murghab is up to 70 m wide; its flow rate there is normally up to 1 m/sec but can reach 4 m/sec during the flood.

The Murghab accepts two tributaries in Turkmenistan, the Kash (or Kashan) and the Kushka Rivers. The Kash, within Turkmenistan, is 70 km long and contains water only in spring or during heavy rains. The Kushka River, 117 km long within Turkmenistan, holds a small amount of mineralized water which is used for irrigation. This river, usually shallow and quiet, can carry water with great speed during flooding, eroding its bottom and banks.

The rivers of Kopetdagh are relatively numerous (ca. 80) but their debit is unequal. In summer many of them dry out or are spent for irrigation. The debit of Kopetdagh rivers is highest in spring when, during the heavy rains, they turn into large and threatening streams. Among the largest, we can list the following: the Arvaz, Kurkulab, Firyuzinka (or Firyuza), Artyk, Keshi, Lainsu, Archinyansu, Dushak, Kelatachai, Chaachachai (or Chaacha), Meanachai (or Meana), and Kazganchai Rivers. Kopetdagh rivers are used for irrigation and water supply of cities and settlements of the submontane plain.

There are very few natural lakes in Turkmenistan. Several lakes around the Khwarazm Oasis are located in depressions between sand ridges bordering the Karakum Desert and are fed by discharged irrigation waters. Small and medium lakes are also found in the Amudarya Valley and in the bed of the western Uzboi where groundwater comes to the surface. Some closed depressions which are now dry were occupied by lakes in the recent geological past. During rains, many closed takyrs in the Karakum are filled by water and become shallow, temporary lakes.

Swamps, in the precise sense, are absent from Turkmenistan (if one does not include swampy areas in the deltas of the Atrek and Amudarya). Under existing physiogeographical conditions, solonchaks (shors) are formed instead of swamps and lakes. There are numerous shors along the Uzboi, Unguz, Kelif Uzboi, under chinks of the Ustyurt, and in the Central Karakum. The largest shors are the Karashor and Kumsebshen north of the Uzboi, and the Kyolkor in the Caspian Lowland.

Lithoedaphic Types of Deserts in Turkmenistan

The following types of deserts are distinguished in Turkmenistan according to the lithology of maternal rocks and soils (Petrov 1973; Babaev and Orlovsky 1981): sand, sand-clay, sand-stony, stony submontane, clay-stony, gypsum, clay and loam, loess, salt deserts, and desert valley landscapes (Fig. 2). Below, we give a brief characteristic of each desert type.

1. *Sand deserts* (including drift, semi-stabilized, and stabilized dunes) differ from all other types in mobility of the substrate, lowest carbonate and salt content, and (with deep groundwater position), in the presence of the hanging moisture horizon which is 20 to 120 cm deep. Due to sand mobility, soils are weakly developed or lacking. Sandy sierozems are formed only in areas of stabilized by vegatation. The existence of the hanging moisture horizon allows for growth of psammophytes. Plants and animals of sand desert are adapted to living on a drifting substrate. Sand deserts have a characteristic, highly dissected ridge relief which differs sharply from the even surfaces of takyrs (clay deserts), solonchaks, or clay-stony plateaus.
2. *Sand-clay deserts* are regions with intermittent sand massifs and clay areas (usually takyrs). This desert type occupies the alluvial plain of the Central Karakum, northwest of the modern delta of the Tedzhen River. This area represents the ancient delta of the Tedzhen; strata of groundwater lie relatively close to the surface. Such landscape of alternating sand and clay desert is observed also in the ancient delta of the Murghab.
3. *Sand-stony deserts* are regions of sand deserts developed on the maternal rocks which often outcrop at the surface. These deserts are present, e.g., on the Trans-Unguz plateau with its numerous outcrops of Tertiary sandstone and kyrs, and on in the Badghyz and Karabil plateaus. These areas vary in the depth of the groundwater as well as in the mechanical composition of substrates.
4. *Stony submontane deserts* are formed on alluvial fan deposits in the piedmont area of Kopetdagh, Bolshoi Balkhan, and Maly Balkhan; they accumulate rubble and gravel and have low levels of substrate salinization. Soils here are desert sierozems (gray desert soils) covered by sagebrush or ephemerous vegetation. Groundwater is usually scattered and does not form strata.

5. *Clay-stony deserts* are the second most widespread type (after sand desert). They have hard, usually salinized, substrates developed on maternal rocks. Presence of rubble facilitates leaching of the surface soil horizon; at a certain depth, however, gypsum crystals and crusts of calcium carbonate are formed. Presence of gypsum at some depth is characteristic for almost all stony deserts. Highly solonets-like brown and grey-brown soils develop on the stony loams. Plant and animal life here is impoverished.
6. *Gypsum deserts* are a variety of clay-stony desert with frequent outcrops of gypsum. In some places, especially in the Ustyurt Plateau, gypsum forms a layer between the soil and maternal rock, from 8 to 60 cm thick (sometimes up to 100 cm and more). In the middle and southern Ustyurt, gypsum is usually present at a depth from 5 to 30 cm, and it outcrops at small, elevated areas. Especially characteristic for the gypsum desert are grey-brown solonchak soils, or gypsum-bearing sierozems. Due to the high concentration of salts, vegetation here is very scarce and consists of a special group of gypsophytes.
7. *Clay and loam deserts* are developed at the sites of ancient river valleys, lakes, in the mouths of rivers (such as the Tedhzen and Murghab), and in piedmont areas, e.g., that of Kopetdagh. Heavy loam sierozems and takyr-like soils are formed on alluvial or, sometimes, alluvial fan deposits of clay and loam. Clay soils are weakly permeable, have low aeration, and are rich in nutrients; plants growing here usually have shallow, weakly branched root systems. Groundwater lies close to the surface, and heavy soils are often salinized; therefore, complicated melioration is required for their development. Clay deserts include also takyrs, vast areas with a smooth clay surface covered by a characteristic cracked hard crust and almost devoid of higher plants.
8. *Loess (and gravel-loess) deserts* are widespread in piedmont areas and are the transitional zone from the plain to the low mountain belts. Extremely fertile soils, typical sierozems, are formed on loess and are similar to loess in their mechanical and chemical composition. Most of these desert areas are used for agriculture with artificial irrigation. Characteristic for the loess desert is the seasonality in soil development, plant, and animal life due to the appearance of ephemerous vegetation during the short period of spring rains. Combination of high temperature and maximal precipitation in spring facilitates biochemical processes in soils.

    Loesses, and sierozems developing on them, are rich in carbonates, but salinization is low due to permeability and leaching. Since groundwater usually lies deep, soils are not solonchak-like. Sierozems have rich soil fauna (earthworms, insect larvae) which creates a "perforated" soil horizon. The air in loess desert landscape has a whitish haze due to the extremely fine loess dust carried by wind.
9. *Salt deserts* are found as large areas or as smaller solonchaks (*shors*, or *sors*) among different types of deserts. As a rule, they lie within river terraces, on coastal plains, and in the bottoms of depressions with close groundwater. Salt deserts are widespread due to the dry climate, presence of highly

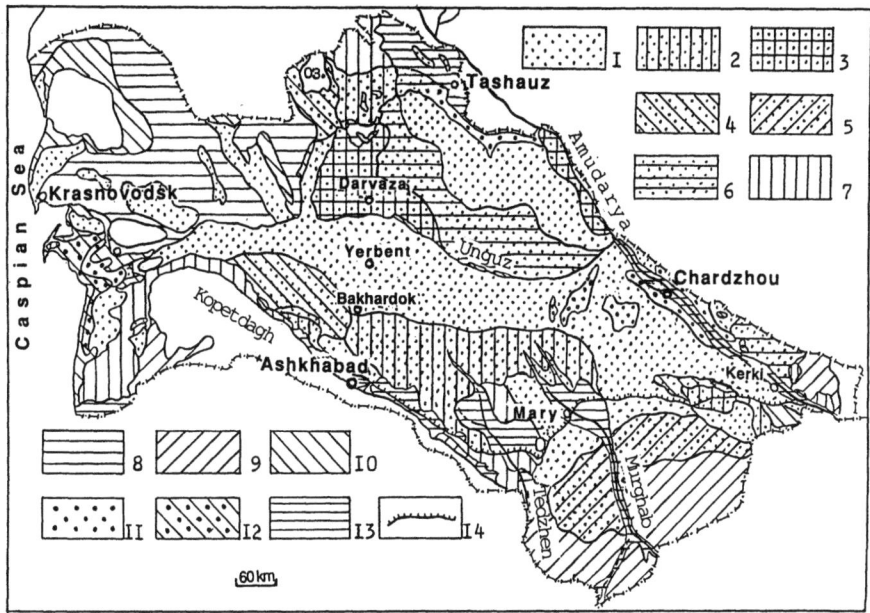

Fig. 2. Lithoedaphic types of deserts and their complexes in Turkmenistan. 1 – sand deserts, 2 – complex of sand and clay deserts, 3 – complex of sand, stony, clay, and salt deserts, 4 – complex of sand and salt deserts, 5 – complex of sand and loess deserts, 6 – complex of sand and stony deserts, 7 – clay deserts, 8 – stony (gypsum) deserts, 9 – loess deserts, 10 – salt deserts, 11 – sand dunes, 12 – complex of sand dunes and salt deserts, 13 – oases, 14 – the Karakum Canal.

salinized Tertiary and Quaternary deposits, and presence of numerous closed depressions and lowlands. High salt concentration prevents agricultural development of salt desert, although they are sometimes used as pastures. Vegetation here is impoverished and consists of highly specialized halophytes. Shors, which are 30 to 40 cm thick, often are completely devoid of vegetation and exhibit smooth salt-covered surfaces, which brilliantly gleam in the sunlight.

10. *Desert valley (and delta) landscapes* are formed on recent sand and clay alluvial deposits. Groundwater here lies close to the surface, and soil formation is constantly interrupted. A specific vegetation of riparian forests (*tugais*), meadows, and oases is present. Meadows, and especially tugais, have the most diverse and abundant plant and animal life of all the desert habitats.

Ecological Regions of Turkmenistan

Thirteen ecological regions are distinguished in Turkmenistan according to the climatic, lithological, and soil characteristics. These include: Cis-Ustyurt, West

Turkmen (or Coastal), Sarykamysh, Trans-Unguz, Karakum, Amudarya, Sundukli, Atrek-Sumbar, Kopetdagh, Kopetdagh Submontane, Murghab-Tedzhen, Karabil-Badghyz, and Kugitang Regions (Fig. 3).

1. *Cis-Ustyurt Region* (including the Krasnovodsk Plateau) is located in the far northwest of Turkmenistan. It includes narrow plateaus separated by narrow valleys, ravines, and depressions. The bottoms of depressions are occupied by solonchaks. Flat-topped buttes, or kyrs, reach 430 m. Large sand massifs within this region are, as a rule, stabilized by vegetation. Average annual temperature is 12 to 15 °C, absolute maximum is 43 °C, absolute minimum is −30 °C; length of a frostless period is 210 to 225 days. Annual precipitation is ca. 100 mm. Vegetation is dominated by *Salsola* and *Artemisia*, sometimes combined with ephemers; the region's feed value for domestic animals is low.
2. *West Turkmen (or Coastal) Region* is a lowland plain with sand-salt and clay-salt desert, recently exposed by the Caspian Sea. Unstabilized dune sands are well developed here; like solonchaks, they lack vegetation. Groundwater (usually highly mineralized) lies from 0.3 to 2.0 m deep. The climate is influenced by the Caspian Sea. Average annual temperature is 15.4 °C, absolute maximum is 44 °C, absolute minimum is −18 °C; length of a frostless period is 260 days. Annual precipitation is ca. 150 mm. The impoverished vegetation is comprised mostly of halophytes which do not form a continuous cover and have low feed value.
3. *Sarykamysh Region* lies between the Tashauz Oasis and the Ustyurt Plateau, including the Sarykamysh Depression. The surface of this area is formed by alluvial deltaic deposits; its relief is a lowland inclined westward. Ancient river beds (e.g., Daryalyk, Daudan, and Akdarya) can be distinctly traced. Takyrs and sands are widespread; sand massifs usually lie next to dry beds. Solonchaks are found within the Sarykamysh Depression. Groundwater lies at a depth of ca. 20 m. Average annual temperature is 12 °C, absolute maximum is 43 °C, absolute minimum is −32 °C; length of a frostless period is 222 days. Annual precipitation is 100 mm. Vegetation is dominated by *Salsola richteri, S. orientalis, Haloxylon aphyllum, Anabasis salsa*, and various ephemers. Stands of *Tamarix* are preserved only in the Sarykamysh Depression.
4. *Trans-Unguz Region* is an elevated, mostly ancient alluvial plain dissected by large (30 to 60 m high) submeridional ridges. The surface of many sand ridges is stabilized by the ancient waste mantle (*kyrs*) which includes carbonates and gypsum. Such ridges have a rather complicated origin; their formation is due to combined action of eolic processes, erosion, and other mechanisms. Between the ridges, in depressions, are developed sand desert soils and, rarely, takyrs. In the south the Trans-Unguz Karakum is limited by the shor depressions of the Unguz. Average annual temperature is 15.4 °C, absolute maximum is 45 °C, absolute minimum is −30 °C; length of a frostless period is 233 days. Annual precipitation is 110 mm. Groundwater is mostly

mineralized and lies at a depth of 15 to 40 m. Pastures of this region are rarely used due to the absence of fresh water sources. Vegetation is dominated by psammophytes and ephemers (*Carex physodes*); typical shrubs include *Haloxylon spp.*, *Calligonum spp.*, *Ephedra strobilacea*, and *Salsola richteri*.

5. *Karakum Region* occupies the largest part of the Lowland and Southeast Karakum Desert. Sands, which are well developed here, represent alluvial deposits of the ancient Amudarya as well as of the Murghab and Tedzhen. These deposits were subject to ancient river erosion and subsequent wind erosion, which led to the creation of various forms of eolic relief. Various sand ridges extend predominantly from northeast to southwest. The eastern part of this region is occupied by sand dunes (barkhans) known as the Amudarya barkhan belt.

   Average annual temperature here is 15.8 °C, absolute maximum is 45 °C in the north of the region and 50 °C in the south, absolute minimum is -33 °C in the north and -28 °C in the south; length of a frostless period is ca. 230 days. Annual precipitation is 115 mm in the north and 130 mm in the south.

   Vegetation includes psammophyte trees, shrubs, semi-shrubs, and herbaceous plants (e.g., *Ammodendron conollyi*, *Haloxylon spp.*, *Calligonum spp.*, *Ephedra strobilacea*, *Salsola richteri*, *Stipagrostis spp.*, and *Carex physodes*). The richest vegetation is present in the eastern part of the Karakum region where are found large massifs of *Haloxylon aphyllum* and *H. persicum*.

6. *Amudarya Region* is an area of well developed Quaternary and modern deposits. It includes a narrow strip from northwest to southeast in the middle part of the Amudarya Valley. Average annual temperature is 12 °C in the north and 16.7 °C in the south, absolute maximum is 43 °C in the north and 47 °C in the south, absolute minimum is -32 °C in the north and -24 °C in the south. Length of a frostless period is ca. 200 days. Annual precipitation is 110 mm in the delta of the Amudarya, and 170 mm in the southern part of the region. The landscape is almost entirely transformed by human activity: it is an important agricultural region for cotton, rice, kenaf, melons, watermelons, vegetables, fruits, and grapes.

7. *Sundukli Region* lies on the right bank of the Amudarya and is a southern offshoot of the Kizylkum Desert which lies within adjacent Uzbekistan. Its complicated relief includes ridge-like hills and buttes with altitudes from 275 to 280 m, separated by wide depressions which often include shors or salt lakes. Fresh and weakly mineralized groundwater lies ca. 20 m deep. Average annual temperature is 16 °C, absolute maximum is 45 °C, absolute minimum is -30 °C; length of a frostless period is 220 days. Annual precipitation is 120 mm. Sand massifs possess psammophyte vegetation with predominance of *Calligonum spp.* and *Ammodendron conollyi*.

8. *Atrek-Sumbar Region* is located at the very southwest of Turkmenistan and includes only the lowland portion of these rivers' drainages. The climate is mild and subtropical. Average annual temperature is 17.1 °C, absolute maximum is 48 °C, absolute minimum is -16 °C. Annual precipitation is 187

mm. High amounts of sunlight and a long frostless period (271 days) allows cultivatinon of suchvaluable subtropical crops as olive, fig, pomegranate, and date palm. The groundwater is highly mineralized and lies close to the surface. Less salinized territories are used for agriculture; irrigated areas are small and scattered.

9. *Kopetdagh Region* includes parallel; mountain ranges comprised of Cretaceous and Paleogene sediments (sandstone, limestone, clay, and marl) which in the foothills are covered by younger Quaternary loess deposits. This region has high seismic activity. The relief is highly dissected by erosion; slopes, especially northern ones, are usually steep and have cliffs. These slopes are dissected by a dense network of deep transverse ravines and gorges. Cuestas are often developed in southern slopes.

    Average annual temperature is ca. 10 °C, absolute maximum is 35 °C, absolute minimum is −24 °C; length of a frostless period is 190 days. Annual precipitation is 300 mm, and at certain elevations, non-irrigated agriculture (*bogara*) is possible. There are many small rivers, whose water is completely spent for irrigation. Vegetation is extremely diverse; in higher belts, shrubs and trees (juniper, maple, and others) are present as well as herbaceous plants. Mountain grasses such as species of *Elytrigia*, *Stipa*, and *Festuca* have high feed quality.

10. *Kopetdagh Submontane Region* includes the narrow inclined submontane plain next to the northern slope of Kopetdagh. It is formed from alluvial fan deposits and loess deposits represented by heavy and light loams.

    Climate is similar to that of the Karakum Desert but is somewhat softened by the influence of the Kopetdagh Mountains. Average annual temperature is 16 °C, absolute maximum is 48 °C, absolute minimum is −26 °C; length of a frostless period is 230 days. Annual precipitation is 228 mm. The plain includes a developed agricultural zone (grapes, vegetables, fruits, and cotton).

11. *Murghab-Tedzhen Region* embraces the valleys and deltas of the Murghab and Tedzhen Rivers which are separated by the Karakum Desert. Sand and clay deposits are developed here. In the deltas lie irrigated lands; sand massifs often surround the oases. This region is a typical arid zone landscape transformed by human culture. Average annual temperature is 16.5 °C, absolute maximum is 48 °C, absolute minimum is −26 °C; length of a frostless period varies from 210 to 248 days. Annual precipitation is 130 mm. Climatic conditions allow cultivation of thin-fiber cotton strains with significant yield. Groundwater in the irrigated part of the region lies from 1 to 3 m deep, and at its periphery, from 3 to 8 m deep. Construction of the Karakum Canal has connected the formerly separate Murghab, Tedzhen, and Kopetdagh submontane oases and created conditions for their concerted development.

12. *Karabil-Badghyz Region* is formed from thick continental deposits of fine-grained clay sandstone, loam, and loamy sand. Wide ancient valleys and depressions represent erosion forms of relief. Generally, the relief is soft and

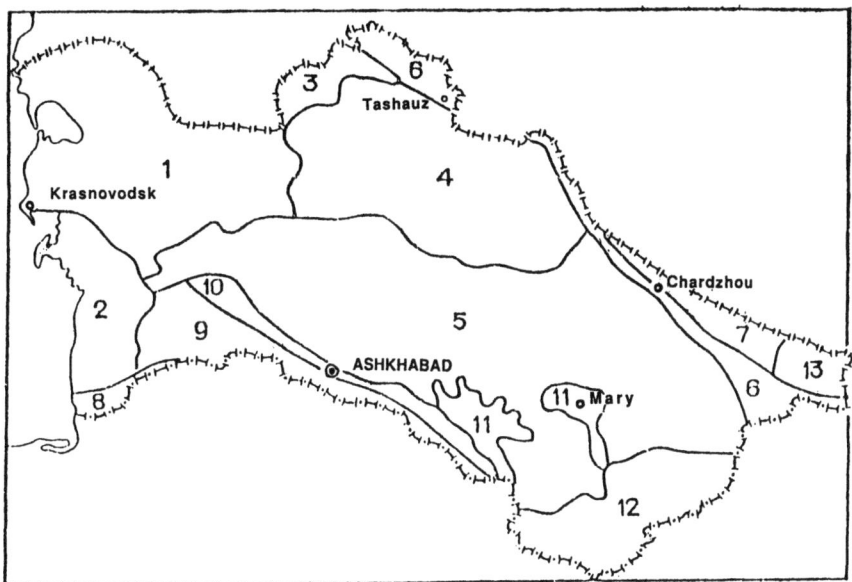

Fig. 3. Ecological regions of Turkmenistan. 1 – Cis-Ustyurt, 2 – West Turkmen (or Coastal), 3 – Sarykamysh, 4 – Trans-Unguz, 5 – Karakum, 6 – Amudarya, 7 – Sundukli, 8 – Atrek-Sumbar, 9 – Kopetdagh, 10 – Kopetdagh Submontane, 11 – Murghab-Tedzhen, 12 – Karabil-Badghyz, 13 – Kugitang Region.

hilly with a semidesert-steppe landscape. Average annual temperature is 16.8 °C, absolute maximum is 47 °C, absolute minimum is -32 °C; length of a frostless period is ca. 230 days. Annual precipitation varies from 200 to 240 mm. Vegetation is dominated by ephemers and ephemeroids (annual and perennial herbaceous plants with winter-spring growth). The western part of Badghyz is especially rich in herbaceous vegetation and is used for pastures throughout the year.

13. *Kugitang Region* is located in the very southeast of Turkmenistan. It has desert landscapes on mountainous/valley relief, highly dissected by ravines. Southward and westward the relief turns into foothills with ridges and cuestas, and then, into alluvial fan plain. Karst processes are developed in areas containing leaching carbonate rocks. The only river in this region, the Kugitang-Darya, has a low debit and does not reach the Amudarya. Average annual temperature is 17 °C; average temperature in January is above 0 °C; average temperature in July is 31 °C; length of a frostless period is 233 days. Annual precipitation is ca. 150 mm. Climatic conditions allow non-irrigated (bogara) cultivation of feed crops.

In conclusion, we should note that the geographic zonality does not strictly determine formation of certain lithoecological types of deserts in Turkmenistan. The characteristics of the ecological regions listed above are defined by local geological formations, geomorphology, and climatic conditions rather than by general zonality. Therefore, the landscapes of Turkmenistan have a complex structure. In lowland areas, the most common type of complex landscape is a combination of sand, salt, and clay (takyr-like) deserts in the ancient dry river valleys or in dry lake depressions.

Walnut (*Juglans regia*) forest along the Aidere River Valley, Southwest Kopetdagh. Photo by I.A. Mukhin.

3. Climate of Turkmenistan

NIKOLAI S. ORLOVSKY

Abstract

Detailed characteristics of the climate of Turkmenistan are given, including data on climate-forming factors, distribution of separate meteorological elements throughout the republic of Turkmenistan, and climatic features of seasonality. Turkmenistan has a very continental and exceptionally dry climate. It is determined by the low latitude position of this area, its significant distance from the oceans, features of atmospheric circulation, character of the underlying surface, and presence of mountain ranges in the southwest, south, and southeast. The continentality of climate in Turkmenistan is expressed by the sharp daily and annual changes of meteorological elements, the contrast transition between seasons, and high probability of dust storms, strong frosts, and late spring and early fall cold spells. Dryness of the climate is expressed by the very low precipitation, low air humidity, low cloudiness, high evaporation, and frequent droughts and dry winds. Ecological conditions in Turkmenistan are favorable for the development of natural vegetation only in the cold period of the year, when wet and humid winter-spring periods facilitate growth of ephemers and ephemeroids. This type of vegetation dries up in the hot and dry summer period. Growth of agricultural crops in Turkmenistan is possible only under artificial irrigation.

Introduction

Climate is one of the most significant factors influencing human activity and environmental conditions. It determines development of vegetation and soils, defines the image of landscapes, and creates a background for agriculture. Climate of a territory depends on its geographic location and underlying surface. Turkmenistan is located in the center of the Asian continent and is neighbored by the Mediterranean Region, Indostan, Central Asia, and Siberia. This geographic location determines four climatic features of the republic: significant sunshine duration, high temperatures of air and soil, sharp continentality, and extreme dryness.

The meteorology of Turkmenistan is well known. The first meteorological stations were established in 1869 in Krasnnovodsk and in 1876 in Kizyl-Arvat. By 1917, the territory of Turkmenistan possessed 24 active meteorological stations. Today, there are 56 stations and 44 posts conducting meteorological observations. Table 1 gives a list of representative stations and their altitudinal position. We should note that only lowland Turkmenistan is well characterized climatically. In the mountains, climate is more complex, and the number of existing stations is not sufficient for thorough monitoring: there are only four stations located at altitudes from 500 to 1,000 m (Kushka, Germab, Firyuza, and Kuitan), two stations located from 1,000 to 2,000 m (Saivan and Gaudan), and only one above 2,000 m (Kheirabad).

Table 1. A list of basic meteorological stations in Turkmenistan.

Station	Altitude (m)	Station	Altitude (m)	Station	Altitude (m)
Tashauz	87	Kazandzhik	33	Kerki	241
Kunya-Urgench	80	Ogurchinsky Island	−26	Chaskak	235
Shakhsenem	62	Chardzhou	188	Gaurdak	482
Bekdash	−26	Kizyl-Arvat	98	Gaudan	1,486
Danisher-Kala	137	Sayat	200	Kizyl-Atrek	32
Kara-Bogaz-Gol	−23	Bakhardok	87	Mary	222
Yekedzhe	59	Cheshme	147	Bairam-Ali	240
Chagyl	115	Bekibent	208	Nichka	232
Kizyl-Kun	−17	Repetek	185	Charshanga	265
Dargan-Ata	142	Bugdaili	−1	Gasan-Kuli	−25
Koshoba	104	Archman	157	Tedzhen	187
Kuuli-Mayak	−22	Saivan	1,036	Kaakhka	308
Darvaza	94	Burdalyk	211	Iolotan	259
Davali	46	Kara-Kala	312	Dushak	248
Krasnovodsk	−13	Bakharden	159	Tedzhenstroi	215
Zeagli	139	Geok-Tepe	204	Ata	235
Yaskhan	−9	Uch-Adzhi	185	Serakhs	275
Dzhebel	−10	Chat	90	Sary-Yazy	306
Akmolla	108	Germab	988	Pulikhatum	395
Ilchik	175	Ashkhabad	227	Takhta-Bazar	349
Cheleken	−14	Firyuza	660	Kushka	625
Aidin	−16	Chashkent	200		
Yerbent	87	Kuitan	790		
Deinau	181	Kheirabad	2,028		

We used data published in five issues of the Reference Book on the USSR Climate (*Spravochnik po klimatu SSSR*, 1964–1969, 144 tables) and in the Scientific and Applied Reference Book on the USSR Climate (*Nauchno-prikladnoi spravochnik po klimatu SSSR*, 1989, which contains 147 climatic criteria). We also used data from published literature, including the author's publications.

Climate-forming Factors

The climate of lowland Turkmenistan is very continental and extremely dry. These climatic features are due to the geographic location of this territory at low latitudes, its significant distance from the oceans, atmospheric circulation, the character of its underlying surface, and presence of mountain systems in the southwest, south, and southeast.

Solar Radiation. Turkmenistan's southern location provides for a high position of the sun. During the winter solstice, its height at noon is from 26° to 32°, and during the summer solstice, it is from 72° to 75°. High noon position of the sun and low cloudiness in the warm period of the year determine long sunshine duration. The probability of clear sky is 90 to 95%; there are only 25 to 30 days in a year without sun. As a result, annual sunshine duration varies from 2,600 to 3,100 hours or even more (Spravochnik 1966; Yurin and Myagkov 1959). This is comparable to sunshine amount in such southwestern American states as Utah, California, and New Mexico (Babushkin 1981). Areas of maximal sunshine duration include the eastern part of the Lowland Karakum, southern part of the low Amudarya, and the entire middle part of the Amudarya. Sunshine duration decreases north, west, and southwest of these areas.

The republic's southern location and long sunshine duration also provide high sun radiation. The annual direct sun radiation varies from 63 to 73% and equals ca. 4,000 to 4,600 Mj/m^2 (Table 2). Global radiation varies from 6,000 to 6,800 Mj/m^2 (Table 2), which is the maximum observed in Middle Asia (Orlovsky and Shlikhter 1975). About 25 to 30% of the global radiation is reflected from the earth; the absorbed portion varies in Turkmenistan from 4,300 to 5,200 Mj/m^2. In winter, due to the distribution of snow cover with its high albedo, 62 to 75% of the global radiation is absorbed; in summer, the absorbed portion is from 70 to 80% (Table 2). The annual radiation balance is rather low; in the Lowland Karakum, it reaches 2,000 Mj/m^2; in oases, it exceeds 2,500 Mj/m^2 and reaches 2,940 Mj/m^2 in the middle part of the Amudarya Valley.

The radiation balance is spent in the turbulent heat exchange between the atmosphere and surface, the heat flow to the soil, and evaporation. The structure of the heat balance determines the heat regime of the air and its humidification.

Fig. 1 demonstrates the annual dynamics of the components of heat balance for different natural regions of Turkmenistan. It illustrates the relationship between physical/geographical conditions and components of the heat balance and their changes under hydromeliorative transformations. Especially characteristic is the structure of heat balance in the warm period of the year. In summer, minimal heat (4 to 7%) is spent in evaporation from the soil in deserts, while from 72 to 88% of heat balance is spent in heating of the air. Therefore, in summer the Karakum Desert is the center of the formation of overheated air. Average July temperature here is 32° C and can maximally reach 50° C.

Table 2. Solar radiation at horizontal surface, in Mj/m² (Nauchno-prikladnoi..., 1989)

Period	Meteorological station						
	Kara-Bogaz-Gol	Bekibent	Gasan-Kuli	Yaskhan	Akmolla	Chardzhou	Ashkhabad
	Direct radiation						
Annual	3,976	3,963	3,761	4,207	4,511	4,610	3,958
December	101	141	139	88	110	107	101
July	587	517	520	603	668	704	592
	Diffuse radiation						
Annual	2,091	2,316	2,467	2,120	2,206	2,206	2,183
December	109	103	106	96	102	105	94
July	251	273	302	239	229	216	248
	Global radiation						
Annual	6,067	6,279	6,226	6,327	6,715	6,816	6,139
December	210	247	244	184	212	213	195
July	838	790	774	842	897	920	839
	Absorbed radiation						
Annual	4,324	4,303	4,538	4,659	4,546	5,250	4,584
December	155	163	188	134	147	159	147
July	612	557	570	640	612	700	608
	Radiation balance						
Annual	n/a	2,002	2,317	n/a	1,898	2,914	2,147
December	n/a	30	44	n/a	20	34	22
July	n/a	317	332	n/a	313	450	332

Atmospheric Circulation. A year in Turkmenistan is distinctly divided into two periods: a very dry warm period with stable hot weather, and a relatively humid cold period with extremely unstable weather. During the cold period, the republic is influenced by the southwestern periphery of the Siberian anticyclone as well as by air mass inbreaks from the northwest and north (Table 3). Cyclonic inbreaks from the south also play a significant role during the cold period. The frequent repetitiveness of cyclones produces unstable winter weather, increased cloudiness, shifts from dry weather to rain and snow, and sharp changes in air temperature and humidity. Inbreaks of cyclones from the south of the Caspian Sea, from the upper parts of the Murghab, Tedzhen, and, more rarely, from the upper Amudarya River carry tropical air; thus, sudden cold attacks in winter often are followed by short periods of warming. As a result, snow cover in Turkmenistan is not formed every year. Winter weather varies depending on the prevalence of certain atmospheric processes. If cyclones carrying warm air prevail, anomalous warm winters occur. In contrast, the predominance of cold inbreaks results in very severe winters with long frost periods, especially if the Siberian anticyclone develops significantly.

Summer brings hot dry weather, and the role of radiation increases. The intensity of atmospheric processes weakens, and cyclonic activity almost ceases. Local tropical air forms which is very similar to the tropical air carried from

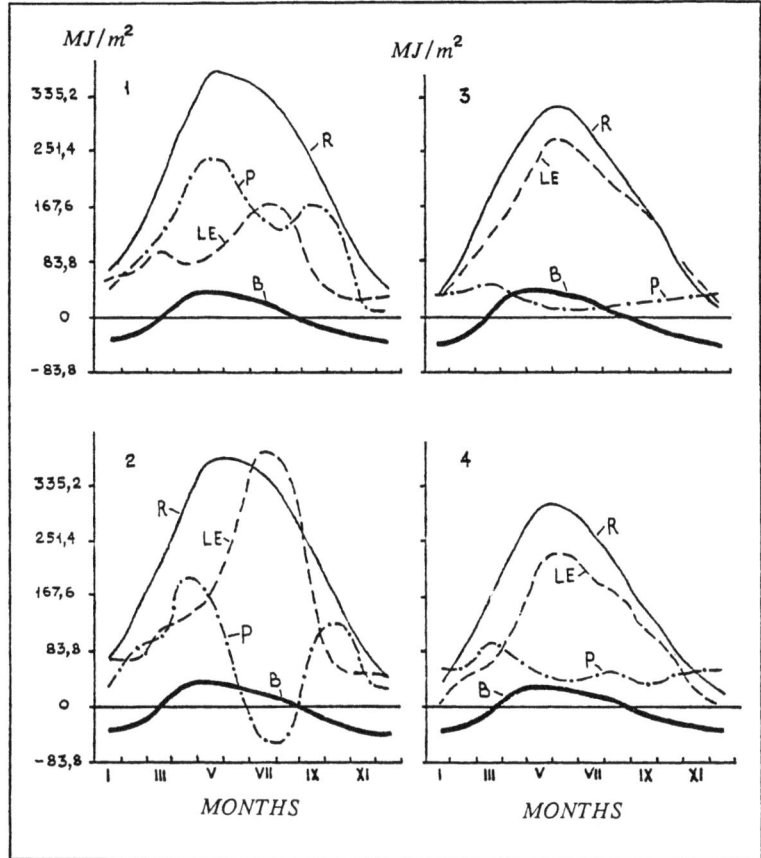

Fig. 1. Annual dynamics of the components of heat balance : (1) Tedzhen Oasis, (2) Murghab Oasis, (3) Central Karakum, (4) Southwest Turkmenistan. P – heat balance, P -- turbulent heat exchange with the air, E – heat spent on evaporation, B – heat flow to the soil.

Asia Minor and the eastern part of the Mediterranean Sea. Intensive heating of the underlying surface of this tropical air leads to the formation of thermal lows above southeastern Turkmenistan, resulting in cloudless skies, dusty hazes, very high temperatures, and low relative air humidity.

During the warm period, cold air invasions may osccur from northwest and north, ususally under cloudless skies and accompanied by strong wind, dust storms, a temperature drop of 4 to 6°, and increased humidity. West and northwest invasions cause rainstorms and heavy rains on the eastern shore of the Caspian Sea and in northwestern Kopetdagh, but clouds rarely reach central Turkmenistan. Significant cooling occurs rarely, and only during multiple invasions of cold air.

Therefore, during the year a successive change of air masses takes place: air masses of temperate latitudes prevail in winter, whereas continental tropical air

Table 3. Probability of basic circulation types (number of years) (Bugaev et al. 1957).

Month	Inbreaks from			Invasions from			Southwest periphery of the Siberian anticyclone	Thermal low
	South of the Caspian Sea	Murghab and Tedzhen	Upper Amudarya	northwest	north	west		
January	12	7	4	16	5	10	27	0
February	11	11	5	17	5	12	24	0
March	10	10	3	17	9	17	23	0
April	11	6	6	13	9	20	26	0
May	11	3	4	22	10	27	19	0
June	3	0	2	32	9	33	0	5
July	0	0	0	23	15	31	0	22
August	1	1	0	19	16	22	0	17
September	2	1	2	21	11	19	40	2
October	0	7	3	21	8	18	38	0
November	12	10	2	16	8	12	35	0
December	10	7	4	16	11	12	34	0

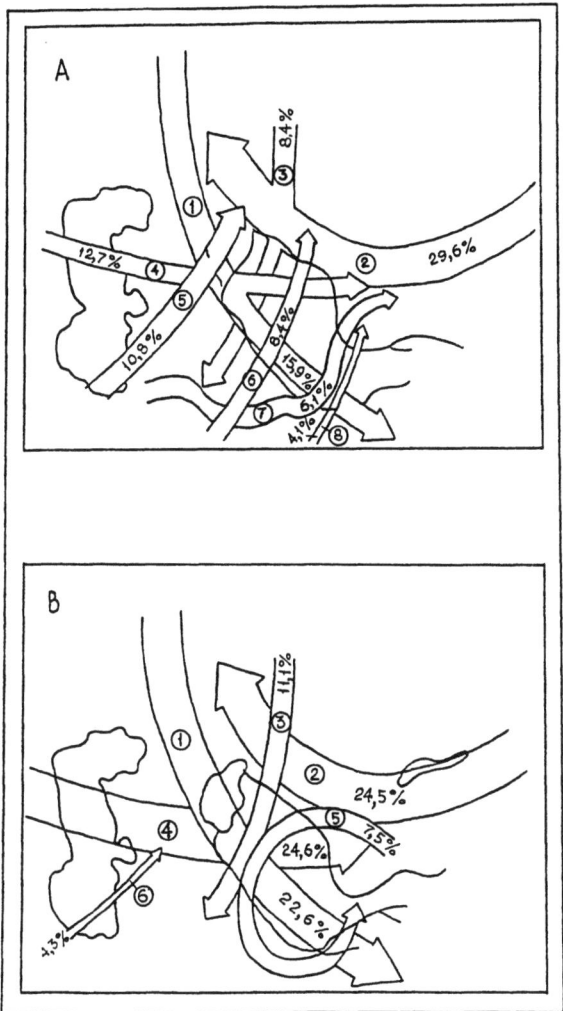

Fig. 2. Dynamics of the formation of the climate of Middle Asia (after Bugaev *et al.* 1957). (A) in the cold period of year: 1 – northwestern cold invasion, 2 – southwestern periphery of the anticyclone, 3 – northern invasion, 4 – western invasion, 5 – South Caspian cyclone, 5 – Murghab cyclone, 7 – wave activity, 8 – Upper Amudarya cyclone. (B) in the warm period of year: 1 – northwestern cold invasion, 2 – southwestern periphery of the anticyclone, 3 – northern invasion, 4 – western invasion, 5 – thermal depression, 6 – South Caspian cyclone.

prevails in summer. Fig. 2 shows the general routes of air masses during the year.

Role of the Underlying Surface. Along with the radiation regime and atmospheric circulation, formation of the climate depends on orographic features (e.g., relief character, altitude, slope exposure, and location of mountain ranges); on presence of, and distance from, water bodies; and on the character of soil and vegetation.

Lowland Turkmenistan is occupied by deserts; desert also influences the mountainous part of the republic. In the warm period, enormous sand desert areas facilitate significant transformation of incoming Atlantic air masses. Thus, the climate of Turkmenistan acquires its extreme dryness.

A specific lowland regime of atmospheric circulation is somewhat modified in the foothills and mountains. Changes in altitude, slope orientation, and steepness are followed by changes in temperature, humidity, cloudiness, and precipitation. In winter, temperature inversions are often observed in the mountains. Mountains cause formation of foehn winds and mountain-valley circulation. Elevated areas form barriers against air masses and change wind direction and speed, and sometimes also serve as a barrier against cold air invasions. For example, the Kopetdagh mountain range prevents invasions of cold air masses from the north and northwest to the southwest areas of Turkmenistan; these southwestern areas are, therefore, much warmer in winter than southern and southeastern parts of the republic.

The Caspian Sea affects only the shore area where, compared to other lowlands, humidity increases, the annual maximum humidity occurs in August, and breezes circulate. Existing large oases also affect climate, especially during warm periods in calm and clear weather. In oases, temperature in summer is lower, and humidity, higher. Temperature inversion forms above the wide irrigated areas, wind speed decreases, and a specific microclimate forms.

Geographic Location. Due to its southern location, Turkmenistan receives a great deal of heat from the sun. Summer is long, hot, and very dry, and winter is short and generally has a non-stable temperature regime. The climate of Turkmenistan belongs to the warm climates of the earth. However, the immediate proximity of temperate areas with continental climates, exposure of the territory in the north, and the republic's separation from the subtropical zone by large mountain systems in the south determine some temperate features of the climate.

Turkmenistan's position close to the center of the giant Eurasian continent and far from the oceans provides sharp continentality of its climate, expressed in the large annual amplitude of air temperature, significant and often sudden changes of meteorological elements and their very sharp fluctuations from year to year, as well as significant daily changes in weather. Continentality is also expressed in sharp contrasts during the transition between seasons.

General Characteristics of the Climate

Air Temperature. In lowland Turkmenistan, daily and annual air temperature varies significantly. Average annual air temperature varies between 12 to 13 °C in the north (12.4 °C in Tashauz) and 17 to 18 °C in the Central and Southeast

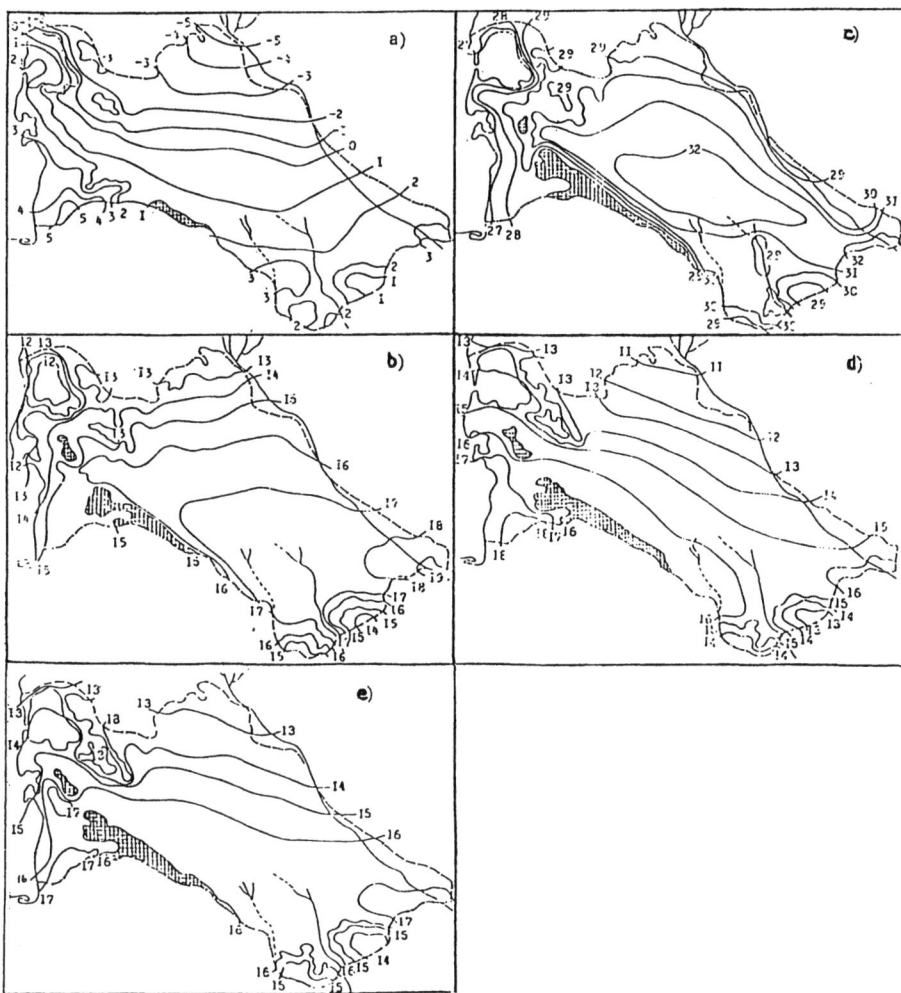

Fig. 3. Air temperature distribution. (a) January, (b) April, (c) July, (d) October, (e) annual.

Karakum (18 °C in Charshanga). Air temperature is lower in the mountains (Gaudan, Lekker) and on the Caspian Sea shore (Krasnovodsk, Gasan-Kuli, Kara-Bogaz-Gol); there, the rule of zonal increase of annual air temperature from north toward south is broken (Fig. 3). The sharpest decrease in average annual air temperature toward the south is observed in the Trans-Unguz Karakum, whereas in the Central and Southeast Karakum the thermal regime is relatively homogeneous. Average amplitude of annual air temperature is 30 to 34 °C in the northeastern Turkmenistan, and 23 to 25 °C in the southwest and on the Caspian shore.

The lowest average monthly air temperatures are recorded from December to February when, in northern Turkmenistan, the temperature falls below 0 °C.

Average maximal monthly temperature during these months in the daytime is from 0 to 5 °C in the north, and from 5 to 10 °C within the remaining lowland part of Turkmenistan. Average minimal monthly temperature in Turkmenistan (except in the southwest) is below 0 °C from December to February; it varies from − 5 to −9 °C in the north, and from 0 to −2 °C in the southeast. The average absolute maximum varies from 1 to −5 °C in the north, and from 3 to 10 °C in the southeast. Average amplitude of daily temperature is from 7 to 11 °C in the most areas.

From March to May, average monthly air temperatures sharply increase, from 7 to 10 °C every month throughout the republic. This period is optimal for vegetation. Average monthly air temperature in March is 5–6 °C in the north, and 9–10 °C in the southeast. In May, average temperature in the Central and Southeast Karakum reaches 24 to 25 °C; in the rest of the lowland territory, it is above 25 °C. Average amplitude of daily temperature varies from 9 to 12 °C in the coastal areas, and from 12 to 15 °C in the rest of Turkmenistan.

The highest air temperatures occur from June to August. The hottest month is July (except in the Caspian coastal zone, where it is August), but in June and August the average monthly air temperatures are only 1 to 2 °C lower than in July. Such a high thermal background is formed due to the stability of the radiation balance of the surface, which also is responsible for the lowest average amplitude of daily temperatures. Generally, June and July are characterized by monotonous hot weather, especially in the Central and Southeast Karakum. Average maximal monthly temperatures during these months reach from 35 to 40 °C (from 31 to 34 °C on the Caspian shore). Average amplitude of daily temperature is from 15 to 20 °C (from 8 to 12 °C on the Caspian shore). The absolute maximum (50 °C) temperatures were recorded from Repetek and Uch-Adzhi in 1915, 1925, and 1937.

In fall (September and October), a sharp decrease of the average, maximal, and miminal monthly air temperatures occurs everywhere at a rate similar to their increase in spring. Average monthly temperatures fall from 21 to 26 °C to 4 to 5 °C in the north, and from 23 to 24 °C to 8 to 10 °C in the rest of the republic. Average maximal air temperature falls from 26 to 28 °C to 12 to 18 °C and from 31 to 35 °C to 16 to 18 °C, respectively. Generally, this period of the year is also favorable for the development of vegetation. However, precipitation is much lower in fall than in spring, and many desert plant species resume their growth after the summer dormant period only with the onset of the first fall rains.

Soil Temperature. The thermal regime of soils is an important ecological element. Due to the dry climate of Karakum, almost all incoming solar radiation is spent on soil warming; therefore, soil temperature during the warm period is high everywhere. Average monthly soil surface temperature in July varies from 32 to 38 °C (Table 4) and, on certain days, can reach a maximum of 76 to 78 °C. Average annual temperature of soil surface varies from 14 to 15 °C in the northern Karakum, and from 18 to 20 °C in the Central and Southeast

Table 4. A temperature regime of soils in Turkmenistan.

Station	Soil surface temperature (°C)			Sums of temperatures higher than 10 °C			Days with temperatures higher than 10 °			H-index
	annual	January	July	surface	at 10 cm	at 20 cm	surface	at 10 cm	at 20 cm	
Kunya-Urgench	14	−5	34	5143	5123	5034	209	218	222	1.27
Leninsk	14	−5	32	4943	4774	4683	207	212	216	1.16
Tashauz	15	−4	34	5268	n/a	n/a	213	n/a	n/a	n/a
Yekedzhe	15	−4	35	5441	5306	5206	215	222	226	1.13
Chagyl	18	−2	36	5832	5759	5619	224	232	238	1.21
Zeagli	18	−1	37	6133	5989	5956	235	244	252	1.15
Yaskhan	18	0	38	6320	6093	6128	238	250	256	1.15
Dzhebel	20	3	37	6470	6554	6617	250	266	278	1.21
Aklmolla	19	0	38	6316	6160	6178	235	248	256	1.14
Chardzhou	18	1	34	5846	5654	5821	237	248	256	1.12
Kizyl-Arvat	19	0	37	6214	6149	6159	237	249	254	1.16
Bekibent	19	3	35	6194	6079	6088	250	259	265	1.18
Repetek	19	1	38	6596	6745	6736	246	259	269	1.27
Kara-Kala	19	3	36	6409	6238	6147	257	263	265	1.22
Ashkhabad	19	1	37	6317	6031	6098	241	251	259	1.13
Kerki	19	2	35	6198	6232	6252	251	268	274	1.18
Kizyl-Atrek	21	5	36	6918	6918	6928	274	282	290	1.26
Bairam-Ali	19	2	37	6269	6103	6075	242	255	263	1.16
Tedzhen	19	2	36	6490	6416	6160	255	263	269	1.19
Kaakhka	19	1	37	6280	6231	6231	246	254	261	1.18
Iolotan	18	2	35	6144	6049	5885	246	258	266	1.20
Serakhs	20	3	36	6460	6550	6554	255	268	275	1.21
Lekker	17	0	34	5508	5458	5349	228	240	246	1.28
Takhta-Bazar	19	3	37	6502	6468	6849	251	259	270	1.20
Kushka	16	2	32	5356	5197	5134	233	249	255	1.14

Karakum. The sum of temperatures varies from 4,900 to 6,900 °C, both at the soil surface and at a depth of 20 cm (Table 4). Dimo (1968) proposed an index of soil heatability (H) which is a ratio between the sum of soil temperatures at a depth of 10 cm and the sum of air temperatures. This index characterizes the relationship between the climate of the surface air layer and climate of the soil. The H-index values for Turkmenistan (Table 4) indicate a high ability of soils to absorb heat.

Cloudiness. The length of sunshine duration is related to cloudiness. Average annual cloudiness in Turkmenistan is from 3 to 4 points. The lowest cloudiness is recorded from June to August (1 point); it increases toward winter (6 to 7 points in January). Total number of cloudy days varies from 45 to 80 a year. The maximum is observed on the Caspian shore (from 60 to 80 days) and in the foothills (60 to 70 days). The minimal number of cloudy days is recorded for the Trans-Unguz and Southeast Karakum.

The average annual number of cloudless days is maximal in the Southeast Karakum (from 166 to 185); the minimal average annual number of cloudless days (100) is recorded from the Caspian shore.

Precipitation. Distribution of precipitation in the territory of Turkmenistan reflects both zonality and local features (relief, underlying surface, large water bodies, and industrial activity). Average annual precipitation varies from 110 mm (Kara-Bogaz-Gol Bay in the northeast of the republic) to 398 mm (Koine-Kesir in the Kopetdagh Mountains). Precipitation is minimal in the Trans-Unguz Karakum; it increases to the south, southeast, and west of this area. Four regions can be distinguished within Turkmenistan according to the level of annual precipitation: (1) the northeasterly portion of the republic (Trans-Unguz Karakum and Kara-Bogaz-Gol Bay, with precipitation less than 110 mm); (2) the Lowland Karakum, with precipitation from 110 to 150 mm; (3) the foothills of the south and southeast, with precipitation from 150 to 200–250 mm; and (4) mountains, with precipitation more than 250 mm (Fig. 4).

Low precipitation in the areas of the Trans-Unguz Karakum and Kara-Bogaz-Gol Bay is due to the lesser significance of such precipitation-forming synoptic processes as outbreaks of southern cyclones and wave activity at the mountain-based atmospheric front. In these areas, precipitation is primarily formed only during western or northwestern invasions of temperate air or during slow-moving high cyclones over the lower part of the Amudarya River.

Another feature of the precipitation regime in Turkmenistan is its great fluctuation in time and significant variation of annual and monthly averages compared to a multi-year average (Table 5), especially in the warm period of the year. Throughout the territory, precipitation occurs predominantly from October to May, with monthly maximums from March to April (Fig. 4). From June to September, occasional precipitation may reach the surface in the west and north of the republic, due to the invasion of western or northwestern temperate or wet Mediterranean air masses which contain high humidity. In

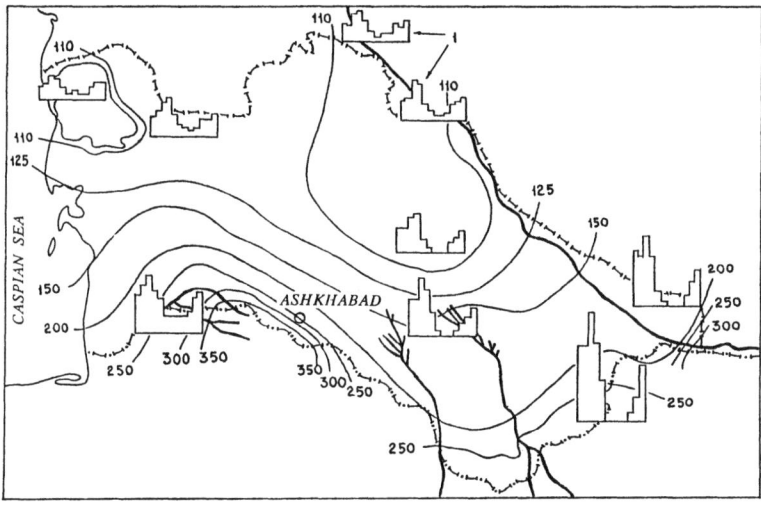

Fig. 4. Average annual and monthly precipitation (mm).

summer, the effect of heated surface on these air masses is their rapid tranformation into tropical ones, with a high level of condensation of the water vapor. In the west and north, such transformation is less intensive, but occasional precipitation does occur; in the east and southeast, however, water either does not precipitate from the completely transformed air, or precipitation does not reach the surface. A prolonged summer drought, from June to September, is common for eastern Turkmenistan. For instance, in Bairam-Ali a total of only 23 mm of precipitation was recorded from June to August during a 10 year period; in Iolotan, only 6 mm was recorded in the same months during 30 years.

These features of precipitation are reflected also in its daily distribution, which is determined by the combination of moisture content in the tropospheric air and the transformational influence of the underlying surface. In the east of the republic, where transformation of the invading air masses is the most significant, daily maximum precipitation is observed from March to April (Balashova *et al.* 1960; Chelpanova 1963) although maximal moisture content in the air throughout the republic is observed during July and August (Kuznetsova 1983). The daily maximum precipitation in the Caspian coastal areas is observed from July through August due to the moisture content of the invading air masses. Throughout the rest of lowland Turkmenistan, the daily maximum precipitation is observed from May through July. In the east, the stationary cyclone in the lower part of the Amudarya River can cause up to 60 to 75 mm of daily precipitation.

In western Turkmenistan, the daily maximum precipitation depends on western and northwestern air invasions and usually occurs in short, heavy bursts. Its distribution, affected by the relief, varies from 70 to 80 mm on the

Table 5. Precipitation in Turkmenistan

Station	Average precipitation (mm)			Maximal monthly precipitation (mm)	Maximal daily precipitation (mm)
	annual	warm period (from April to October)	cold period (from November to March)		
Tashauz	90	37	53	64	38
Chagyl	102	51	51	n/a	n/a
Darvaza	97	39	58	n/a	43
Bakhardok	124	49	75	n/a	n/a
Cheshme	97	38	59	n/a	n/a
Chardzhou	116	38	78	115	63
Kerki	172	40	132	89	40
Charshanga	149	34	115	n/a	n/a
Uch-Adzhi	118	35	83	n/a	n/a
Bairam-Ali	135	40	95	109	44
Tedzhen	139	47	92	97	41
Takhta-Bazar	241	53	188	122	58
Lekker	255	54	201	n/a	n/a
Kushka	260	57	203	167	73
Kaakhka	200	74	126	n/a	n/a
Ashkhabad	230	97	33	128	56
Bakharden	190	82	108	n/a	n/a
Kazandzhik	148	60	88	92	40
Kizyl-Arvat	205	85	120	132	77
Krasnovodsk	103	40	63	n/a	77
Bekibent	165	83	82	n/a	n/a
Gasan-Kuli	196	91	105	107	79
Kizyl-Atrek	188	81	107	110	91

Caspian shore to 90 to 100 mm in the foothills and mountains. For example, in July, 1928, Kizyl-Arvat had 110 mm of precipitation, whereas the multi-year average was 5 mm for July and 77 mm for the entire warm period of the year; of the 110 mm, 91 mm (or 18 times the monthly average) was recorded on July 5. The highest daily maximum of 123.8 mm was recorded in September, 1963, in Khodzha-Kala (West Kopetdagh).

Air Humidity. The regime of air humidity in Turkmenistan is defined by high summer temperatures, shortage of precipitation, and absence of large water bodies. Average annual absolute air humidity in the Central Karakum varies from 6 to 7 mb (8 to 9 mb in oases). On the Caspian shore, it reaches 11 to 14 mb; and in the Amudarya Valley, 7 mb in the north (Danisher-Kala) and 9 mb in the south (Kerki). In winter, the absolute air humidity is less dependent on the underlying surface and is distributed more evenly, increasing from north to south from 3 to 5 mb. The highest maximal absolute air humidity in summer is recorded from the Caspian shore (21 to 26 mb); it ranges from 14 to 16 mb in the Amudarya Valley, and 10 mb in the center of the Karakum Desert.

The relative air humidity reaches its maximum in January when it is distributed rather evenly throughout the republic (although varying from 75 to 78% in the Trans-Unguz Karakum, the foothills of Kopetdagh, and on the Caspian shore). In the driest period, from June to September, the relative air humidity in the Karakum Desert is from 20 to 30%. It is rather high in summer in the coastal zone (69%, Gasan-Kuli) and in oases (30 to 35%); in the Amudarya Valley relative air humidity in summer is from 37 to 41%, which is higher than in the adjacent non-irrigated areas. On certain days, minimal air humidity in the Central Karakum can fall to 2 to 3%. On the Caspian shore, days with 100% relative air humidity may occur, but they are extremely rare (0.3% of all days). No days with humidity less than 10% have been recorded in summer on the shore, where humidity often can reach 50 to 70% at noon. In the Kopetdagh Mountains, daytime humidity in July usually varies from 20 to 40%.

The lowest values of air humidity deficit are recorded in January, when it is as low as 1.0 to 1.5 mb in northern Turkmenistan, increasing to 3.0 to 3.8 mb toward the south. It sharply increases beginning in February and reaches its maximum in July (32 to 33 mb in the Trans-Unguz Karakum, 37 to 41 mb in the Central Karakum, and 40 to 42 mb in the Southeast Karakum). In certain years, the monthly humidity deficit in the Southeast Karakum can reach 70 mb (absolute maximum was recorded as 73 mb for Uch-Adzhi). In the coastal areas, the maximal monthly humidity deficit is observed in August when air temperature reaches its maximum.

Evaporative Capacity. High values of air humidity deficit facilitate intensive evaporation from the water surface. Annual evaporation from the water bodies of lowland Turkmenistan varies from 1,000 to 2,300 mm and is highly dependent on physical-geographical conditions, air humidity, and air temperature. It reaches the maximal value (from 2,000 to 2,300 mm) recorded for Middle Asia in the Central Karakum. Westerly and easterly from this area, evaporation decreases; it is 1,000 mm in the narrow coastal zone of the Caspian Sea, 1,600 mm in the Amudarya Valley, and from 1,400 to 1,600 mm in the Murghab and Tedzhen oases. Annual evaporation in the foothills of Kopetdagh and in Karabil varies from 1,500 to 1,600 mm; in the Tashauz oasis, from 1,200 to 1,400 mm. In Badghyz and in the submontane plain of Kugitangtau, increased humidity deficit and wind speed lead to an increase of evaporation up to 2,000 to 2,200 mm annually (Durdyev and Orlovsky 1984).

Wind. During the year, northeasterly winds prevail in northern Turkmenistan; easterly winds, in the central part of the republic and along the the submontane plain of Kopetdagh; and northerly winds, in the Southeast Karakum. In some regions in the southeast, northern winds become northwesterly due to the relief.

Average annual wind speed in lowland Turkmenistan is from 3.2 to 4.2 m/sec. In oases with their high trees, wind speed does not exceed 3.1 m/sec. Wind speed changes discernably during the year, with maximal average monthly values usually occuring in spring and summer; only on the Caspian shore and in

Kopetdagh is maximal wind speed recorded in winter. Minimal wind speed is observed in fall.

Weak and moderate winds (from 0 to 5 m/sec) prevail throughout the republic (75 to 85% of all speed records). Only on the Caspian shore and on the northern slopes of Kopetdagh are wind speeds from 6 to 9 m/sec. The number of days in a year when wind speed exceeds 15 m/sec is from 5 to 10 in the Central Karakum, and from 3 to 8 in the Southeast Karakum; it increases on the northern Caspian shore (40 days), and in the eastern foothills of Kopetdagh and in the Amudarya Valley (54 days). Strong winds are commonly recorded from March to April, and only in the southeast is maximal occurence of strong winds shifted to the summer. Wind speed in the Central Karakum and in the deltas of the Murghab and Tedzhen can reach 16 to 18 m/sec; in the southwest and north of the republic, 20 to 21 m/sec; and on the Caspian shore and in the Amudarya Valley, 22 to 25 m/sec (Semenova 1961).

Atmospheric Events. During strong winds, dust storms appear in lowland Turkmenistan, especially in spring and summer. They occur more often during cold invasions from the west, northwest, and north. The maximal annual number of days with dust storms is recorded in Nebit-Dagh (60 days); the number ranges from 30 to 40 days in the Central Karakum and in the Southeast Karakum. Dust storms are rare in the mountains and oases.

Unfavorable weather events include fogs, hail, glaze, and rime deposit. Fogs are most common on the Caspian shore (from 20 to 30 days a year); annual number of foggy days is 16 to 17 in the north of the republic, and from 8 to 10 in the sourhern and southeastern parts of the Karakum Desert. In the lower Amudarya, number of foggy days increases to 20 to 25. Fog there forms primarily from November to March; fogs are very rare in the lowlands from April to September except in the middle part of the Caspian shore, where fogs in summer occur three to four times more often than in winter.

Hail is very rare event in Turkmenistan. In the lowlands, an average of one to five days with hail occur per 10 years. Maximum hail is recorded in the mountains of Kopetdagh.

Glazed frost and rime deposits also are rare weather events. Glaze is observed from November to March (in mountains, from October to May), with maximum of glaze days in December or, sometimes, in January. In the lowlands, there are, on an average, up to three days a year with glaze. In certain years, the number of days with glaze increases to 9 to 13 in the lowlands, and to 17 in Kopetdagh. Rime deposit forms more often than glaze and is observed in the lowlands from November to March, with a maximum of 13 to 20 days a year. Rime is recorded from one to two days every year in Central Karakum, from 4 to 7 times every 10 years in the southern Turkmenistan, and even more rarely on the Caspian shore (two to four times every 10 years).

Climatic Seasons in Turkmenistan

Winter. Climatic seasons are defined by the thermal regime, humidity, and features of the development of pasture vegetation. In Turkmenistan, the climatic seasons do not coincide with the calendar (astronomical) seasons. Winter season in the lowlands begins when the stable average daily air temperatures fall lower than 5 °C (Balashova *et al.* 1960; Babushkin 1964). This time is marked by the beginning of a relatively dormant phase in many shrubs and semishrubs, and with massive falling of branches and fruits of saksaul (*Haloxylon* spp.).

In 30 to 35% of all years, winter period begins simultaneously throughout the republic except in the southwest, which is influenced by the non-freezing Caspian Sea from the west and protected by Kopetdagh from the southeast. In 40 to 45% of all years, winter first begins in the north (early November) and northwest (mid-November). From late November to early December, winter begins in the Krasnovodsk Plateau, in the Central Karakum, and in the foothills of Kopetdagh. Somewhat later (mid- and late December) winter begins on the southern Caspian shore, and finally (from late December to early January) it comes to Southwest Turkmenistan. Time of the winter arrival varies from year to year. For example, if an average arrival of winter in the Central Karakum is late November, it may begin from early November (once every 20 years) or even in late October; at other times, it may be delayed until mid-December.

Duration of the winter decreases from north to south. It lasts from 101 to 130 days in the north, averaging 90 days in the Central Karakum (from 74 in Bairam-Ali to 105 in Zeagli), and decreases to 83 in Badghyz, 76 in the foothills of Kopetdagh, and 53, in the Southeast Karakum. Especially short is the winter in Southwest Turkmenistan (from 10 to 30 days).

Within the winter, so-called "real" winter periods exist which are characterized by average daily air temperatures lower than 0 °C and by the complete dormancy of vegetation (Babushkin 1964). Such periods are almost absent in the southwest, in most parts of the submontane plain of the Kopetdagh, in the south of the Central Karakum, in the Southeast Karakum, in Badghyz, and in Karabil. Farther to the north, "real" winter periods occur more often; e.g., in the northern part of the Central Karakum 30 to 60 days a year (40% of the winter season) have an average daily air temperature of lower than 0 °C. These periods increase from 60 to 102 days in northern Turkmenistan and include from 49 to 54% of all winter days in northwestern Turkmenistan, and from 65 to 74% in the Trans-Unguz Karakum (Orlovsky and Volosuyk 1974).

In some years, the "real" winter spreads throughout the entire republic (e.g., 1929–1930, 1932–1933, 1933–1934, 1934–1935, 1936–1937, 1944–1945, 1948–1949, 1968–1969; Babushkin 1964). Sharp winter cold spells are due either to invasions of Arctic or temperate air masses from the northwest, north, or, sometimes, from the northeast, or by radiation cooling inside an air mass. During especially strong cold invasions, temperatures can drop to a range of −26 to −35 °C even in the south. Only Southwest Turkmenistan, protected by

Kopetdagh, has higher absolute temperature minimums (from −15 to −20 °C).

Warm winters can occur as well as severe cold ones. Cold invasions during warm winters are rare and not intensive; relatively warm air masses (temperate Turanian and South European air) prevail in these years (Bugaev et al. 1957). In warm periods some pasture plant species, usually ephemeroids and ephemers, start growing.

Winter precipitation brought by cold air masses from the north and west, can occur as rain or as snow. In lowland Turkmenistan, the snow cover is not stable and may form and thaw several times. Usually, snow cover forms in the lowlands beginning in late December; on the Caspian shore, snow forms from late December to early January; and on the submontane plain of the Kopetdagh, in Badghyz, and in Karabil, it appears usually in mid-December. In early winters, snow sometimes appears 45 days earlier than average.

Thaws occur almost everywhere in late February, except in the northwest (average March 2). The latest thaw is usually near the end of March, but in Badghyz and Karabil snow may remain until early April (record date, April 13).

In Turkmenistan, there are no areas without occasional snow cover. Even in the warmest southwestern part of the republic (Kizyl-Atrek) there are 5 to 6 days a year with snow (maximum, in 1969, was 20 days in Kizyl-Atrek and 25 in Kara-Kala). Maximal number of days with snow is recorded for the high mountain belt of Kopetdagh, the areas of Badghyz and Karabil, and the northwest of the republic. In Badghyz and Karabil snow cover stays from 16 to 18 days on average (with maximum 61 to 64 days); in the northwest, the average is 21 days of snow cover, and maximum, 73 days. In the rest of northern Turkmenistan, the average number of days with snow cover varies from 11 to 13 (maximum, 51 days, in Kunya-Urgench).

The highest daily snow cover (excepting Kopetdagh) has been recorded in Badghyz and Karabil areas and is, on average, 10 cm (with maximums from 37 to 43 cm). The highest snow height, 68 cm, has been recorded in Akar-Cheshme (the Badghyz Reserve). Average height of the snow cover in the southwest, northwest, and in part of the Central Karakum, is 4 to 5 cm; in some winters, however, it has reached 19 cm (Kizyl-Atrek) or 20 cm (Kara-Kala). In the southern and central parts of the Central Karakum and in the Southeast Karakum, average snow height is from 6 to 8 cm. Maximums recorded are 37 cm (Bairam-Ali), 41 cm (Tedzhen), 43 cm (Zeagli), and 56 cm (Darvaza) (Balakirev 1972).

Spring. In the beginning of March in the lowlands (or in February in the south), air temperature rises and begins to exceed 5 °C. Stable average daily temperature higher than 5 °C marks the beginning of spring; it corresponds to the start of active plant growth, and appearance of herbaceous vegetation and flower buds on fruit trees.

Spring begins early, on average before February 10, in the far southwest (Kizyl-Atrek) and southeast (Charshanga); by February 20, it comes to the rest of the southwest, to the south of the Central Karakum, to the Southeast

Karakum, and to Badghyz. By March 1, spring begins in most of the Central Karakum and in Karabil and is established in the north by March 20. Therefore, for the entire republic the transition from winter to spring extends between 50 to 60 days. Duration of the spring increases from north to south from 48 to 80 days. The southwest has the longest spring (from 90 to 105 days); the Caspian Sea shore, from 80 to 87 days.

During the spring, temperature rises and precipitation increases. Already in March, maximal daytime temperature can reach 30 to 39 °C. However, the weather is still unstable due to the intensification of cyclonic activity. Warm periods may change to sudden colds. Sometimes, near-winter weather appears, with short but significant frosts and snow formation. For example, in March of 1959 and 1960, snow cover was observed in Badghyz and Karabil for 16 days; its maximal height reached 32 cm. During this time, air temperature fell to -15 °C; average maximum wind speed was 6 m/sec. In the Central Karakum and in the northwest of Turkmenistan, the number of the days with snow in March in certain years has varied from 8 to 10, with maximal snow height 12 to 15 cm. Temperature during these periods fell to -21 to -24 °C; average maximum wind speed was 9 to 11 m/sec (Balakirev 1972).

Returning cold weather is very characteristic for the spring. On average, final cold spells occur commonly in early March in the southwest of the republic and in mid-March on the submontane plain of Kopetdagh and in the Central Karakum. Final cold spells in the north and in Karabil occur as late as early April. During warm springs, colds occur from February in the south to early March in the north; during cold springs, however, last colds have been recorded as late as April 1 in the southwest, and from April 17 to 21 in the Central Karakum.

Late spring colds are dangerous for the growth and development of trees and shrubs, most of which by this time have fully developed leaves, and some of which are in bloom. One-year shoots of such desert plant species as *Haloxylon aphyllum, H. persicum, Calligonum spp., Salsola richteri, Ammodendron conollyi,* and *Astragalus spp.*, are completely killed by late spring frosts (Nechavea 1958; Dubyansky and Nardina 1963). Spring colds also cause the decrease of pasture phytomass and seed production, and kill germinating wild and agricultural plants.

A warm spring period with sufficient precipitation facilitates growth of pasture vegetation, which develops earlier in wet and warm springs than in dry and cold ones. Warmth-loving plants begin to vegetate during average daily air temperatures from 8 to 14 °C. For example, the growth of white saksaul (*Haloxylon persicum*) begins when average daily air temperatures reach 8 to 9 °C (Kharin 1966; Gringof 1967) (Fig. 5). The growth of black saksaul (*Haloxylon aphyllum*) begins 5 days later; growth of *Ammodendron conollyi*, 18 days later; and growth of *Calligonum spp.* and *Salsola richteri*, 22 days later than that of white saksaul (Kharin 1975).

In April, increase of air temperature and dryness and decrease in precipitation cause increased evaporation; soil starts to lose moisture, and soil

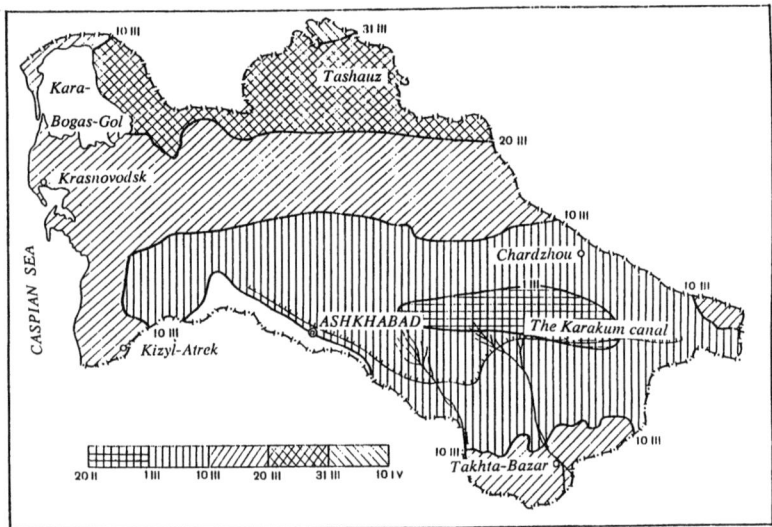

Fig. 5. Beginning of the growth of white saksaul (*Haloxylon persicum*) (after Kharin 1975).

drought develops by the end of spring period. Vegetation begins to dry out. On average, soil drought occurs in most parts of Turkmenistan before April 15; on the submontane plain of Kopetdagh, by the end of April; and in Badghyz and Karabil, by early May (Babushkin 1964).

Summer. Transition from spring to summer is marked by stable increased average daily temperature above 20 °C. Then, increase of the air temperature slows down, late spring colds disappear completely, and dry weather begins (Balashova *et al.* 1960). In northern areas, in the southwest, and on the submontane plain, summer begins in early May. It comes somewhat later to the coastal areas (late May to early July in Kara-Bogaz-Gol). The earliest summer (late April) is observed in the Central and Southeast Karakum. At this time, pasture vegetation dries up extensively (Fig. 6).

Summer is the longest season in Turkmenistan. It lasts everywhere more than 100 days: 110 on the Caspian shore; 120 to 130 in Badghyz, Karabil, and in the north of the republic; 140 in the Krasnovodsk Plateau; 150 in the southwest and in the submontane plain of Kopetdagh; and slightly more than 150 in the Central and Southeast Karakum.

In summer, precipitation sharply decreases. In the Southeast Karakum, practically no precipitation occurs from June to September; in Badghyz and Karabil summer precipitation constitutes about 1% of the annual sum, and occurs primarily in June. From July to August, 4% of the annual precipitation occurs in the Central Karakum; about 7%, in the Trans-Unguz Karakum; and 8%, on the submontane plain. In the west and southwest of the republic the amount of summer precipitation is higher; from June through August 10 to 14% of the annual precipitation occurs in this area.

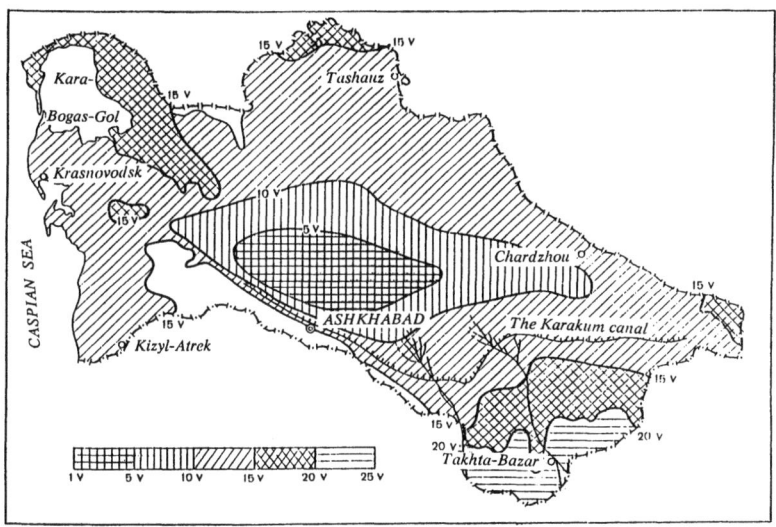

Fig. 6. Dates of the drying of sand sedge (*Carex physodes*) (after Kharin 1975).

Decreased precipitation in summer is accompanied by a rapid increase in temperature. Air temperature reaches its maximum in July (Fig. 3); only in the coastal area is maximal air temperature observed in August. In the daytime, summer air temperature exceeds 40 °C (maximum 46 to 48 °C in the Southeast Karakum, and 49 °C in the Central Karakum). Accompanying high temperatures is low humidity. Relative humidity in desert areas in summer falls to 22 to 25%, with a minimum of 2 to 5%. It is higher in the coastal areas due to the influence of the Caspian Sea (60% in Gasan-Kuli) and in the mountains (47% in Kheirabad). In the Amudarya Valley, humidity is 37 to 41%, which is higher than in adjacent non-irrigated areas.

Probability of drought is from 50 to 75% during summers with extremely high temperatures and low humidity (Table 6). Yield of non-irrigated crops (bogara) in drought years decreases from 40 to 60%; even in irrigated areas, dry winds cause a 30%-decrease of yield. Drying of pasture vegetation occurs 15 to 20 days earlier under drought, and phytomass decreases from 30 to 65%. Droughts, which facilitate degradation of soil and plant communities, act as catalysts of desertification.

According to a humidity deficit measured at 13:00 hours, five levels of air drought are distinguished in Turkmenistan: weak (humidity deficit 50 to 60 mb), medium-level (60 to 70 mb), strong (70 to 80 mb), very strong (80 to 90 mb), and severe drought (more than 90 mb). Weak droughts are observed throughout the southern Karakum Desert. Medium-level droughts are recorded every summer (except in Southwest Turkmenistan, where probability of medium-level drought is from 60 to 93%, and the Amudarya Valley). Strong droughts happen almost every year in the southeast; their annual probability in the well-irrigated areas

Table 6. Droughts in Turkmenistan.

Station	Years of observation	Years with drought	Category of drought				Duration of drought (years)				
			weak	medium	strong	very strong	2	3	4	5	25
Cheleken	72	41	32	1	3	5	4	1	0	1	2
Ashkhabad	93	45	32	9	1	3	8	2	0	1	2
Bairam-Ali	95	37	44	7	2	4	5	2	2	2	2
Kushka	57	33	23	7	3	0	9	2	1	0	0
Repetek	60	30	22	3	2	3	8	1	2	0	2
Chardzhou	90	60	48	6	3	3	4	1	4	0	2

(Murghab Oasis, the Amudarya Valley) is lower (from 20 to 27%). In the drier Tedzhen Oasis and on the submontane plain, probability of a strong drought is from 60 to 80% (Orlovsky and Volosyuk 1974; Orlovsky 1981). Very strong droughts occur often only in the southeast (80% in Nichka, 67% in Takhta-Bazar, and 33% in the far southeast); in the Tedzhen Oasis and on the submontane plain, the probability of very strong drought is from 7 to 20%. Very strong droughts have not been recorded in the Murghab Oasis, but in the Amudarya Valley they occur in 7% of total years recorded, and in the southwest, in 13 to 15%. Severe droughts during the last 20 years have been recorded only in the southeast (10 to 13% of total years). In the irrigated areas, strong and very strong droughts are very rare, and severe droughts have never been recorded (Orlovsky 1981).

Fall. Stable dry and hot weather begins to change in late August or, usually, in September, and the air drought decreases. The beginning of fall corresponds to a stable decrease of average daily air temperature to below 20 °C. First signs of fall appear in early September in Badghyz and Karabil, and in mid-September in the northern regions of Turkmenistan. In late September, fall comes to the Southeast and Central Karakum and to the submontane plain of Kopetdagh, and in early October, to the southwest. Fall in Turkmenistan is short, averaging of 52 days. It lasts 54 to 56 days in the northern Central Karakum; 62 to 69 days in the southeast of this area; a week or two longer on the submontane plain, in the Southeast Karakum, and in the far northwest; and more than 70 days in Badghyz, Karabil, and in the southwest (95 days in Kara-Kala).

The first half of the fall season is characterized by stable warm weather. By October, the thermal depression completely disappears; western and northern cold invasions become less frequent but more significant. Gradual decrease of the air temperature is interrupted by incoming cold air masses which cause frosts. Clouds appear, and precipitation begins. In November, cooling of the air progresses. The temperate air front shifts to the south of Turkmenistan, causing more frequent cyclonic activity. Precipitation becomes more frequent and more intensive.

Due to frequent fall precipitation, soil humidity increases, which facilitates completion of development in some shrubs (*Haloxylon* spp., *Salsola richeri*, and *S. arbuscula*) and semishrubs (*Artemisia* spp.); during this time, their fruits and seeds are formed (Nechaeva 1958). In warm and humid years, *Carex physodes* and some annual spring ephemers renew their growth in the fall. The more frequently such years occur, the more favorable is the local climate for desert pasture vegetation (Nechaeva 1958). Babushkin (1971) estimated the duration of the humid fall period except for the areas of northwest Turkmenistan and northern Central Karakum (Table 7); maximal duration of the humid fall period was recorded as 68 days in the southwest. Fall begins early (late October) in Kopetdagh and the Sumbar Valley; in early November, on the submontane plain of Kopetdagh; in mid-November, in the north, in the Central Karakum, and in the Murghab and Tedzhen Oases; and in late November, in the Southeast

Table 7. The humid fall period in Turkmenistan.

Station	Beginning	Duration (days)	% of years without the humid fall period
Krasnovodsk	Nov. 13	36	12
Chagyl	Nov. 13	2	45
Yekedzhe	Nov. 19	6	55
Tashauz	Nov. 19	8	57
Zeagli	Nov. 19	4	42
Kerki	Nov. 19	25	22
Repetek	Nov. 21	9	35
Tedzhen	Nov. 19	19	27
Bairam-Ali	Nov. 13	22	24
Kazandzhik	Nov. 7	26	21
Ashkhabad	Nov. 3	34	14
Firyuza	Oct. 27	38	10
Germab	Oct. 24	32	16
Gaudan	Oct. 27	32	16
Kheirabad	Oct. 24	13	32
Kizyl-Atrek	Nov. 3	68	2

Karakum. Very rarely, the humid fall period begins in the lowlands in mid-October, and in some places even in early October (Babushkin and Kogai 1971).

Lowland Turkmenistan can be divided in five regions, according to the favorability of climate for fall-winter vegetation of desert pastures: (1) the submontane plain of Kopetdagh, Badghyz, and Karabil, where the humid fall period occurs and fall-winter vegetation develops 80% (or more) of the time; (2) the central part of the Lowland Karakum and the Southeast Karakum (from 50 to 70% of the time); (3) the northern and western parts of the Lowland Karakum (40 to 45% of the time); (4) the northern part of the Trans-Unguz Karakum (10 to 15% of the time); and (5) the north of the republic, where the humid period on average begins after air temperature drops below 5 °C and, thus, fall-winter vegetation is absent (Nechaeva 1960).

Natural Changes of the Climate

Fluctuations of climate significantly affect many natural processes, such as river debit, evaporation, productivity of vegetation, and distribution of animals. Only a few meteorological stations in Turkmenistan maintain observations long enough to study climatic fluctuations. We analyzed the secular variations of average annual air temperatures and annual sums of precipitation using data from these stations (Fig. 7).

Alteration of the following climatic rhythms was revealed: warm and dry, cold and humid, cold and dry, and warm and humid. In the west of the republic (Cheleken), the warm and humid period during the 1930s was followed by a cold

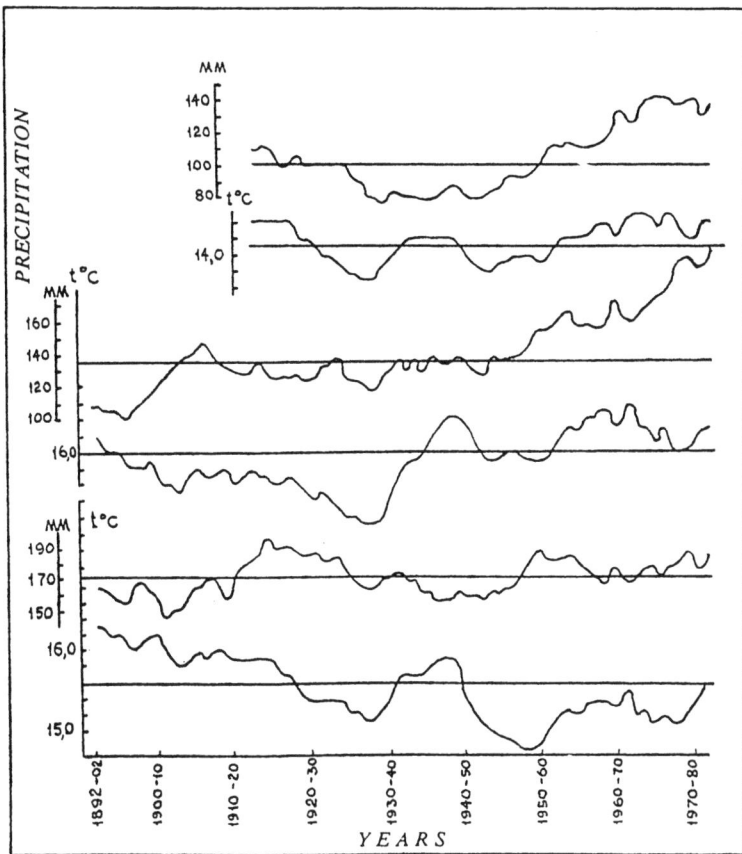

Fig. 7. Sliding 11-year average air temperatures and precipitation. (a) Cheleken, (b) Bairam-Ali, (c) Chardzhou.

and dry one which lasted until the early 1960s. Since the 1960s, a warm and humid climatic rhythm has again been observed. In the Kazandzhik area, two climatic rhythms have been observed within last 50 years: a warm and dry period (from the late 1930s to the late 1960s), and warm and humid (since the late 1960s).

In the Ashkhabad area, five climatic rhythms were observed from 1892 to 1981. The cold and dry rhythm of 1892–1912 was followed by a warm and dry one which lasted until the early 1930s. In the 1930s, a cold and humid rhythm replaced it and, in the 1940s, was itself replaced by a prolonged warm and dry period. Since the late 1970s, the current cold and humid rhythm has been established.

In Bairam-Ali, four rhythms were recorded from 1892 to 1981. The warm and dry period of 1892–1912 was followed by a long cold and dry period, which changed to a warm and dry rhythm in 1944. That was followed by a warm and humid rhythm in the mid-1950s.

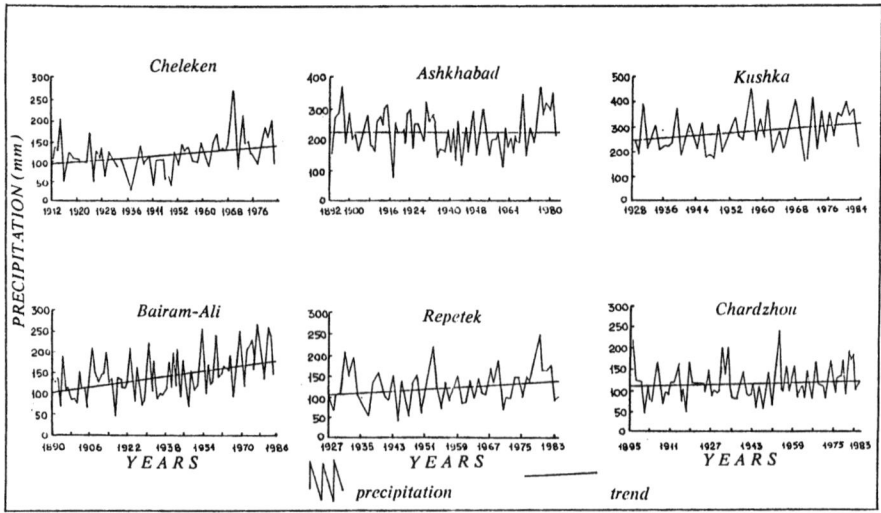

Fig. 8. Variation of annual sums of precipitation. (1) annual sums of precipitation, (2) trend of variation.

In southernmost Turkmenistan (Kushka) within the last 50 years, three climatic rhythms have been observed: cold and dry before the early 1940s, warm and dry from the 1940s to the late 1950s, and warm and humid since 1956.

In the Southeast Karakum (Repetek, Chardzhou, and Kerki stations) the amplitude of the annual sum of precipitation is not so high; the air temperature, however, experiences more clear-cut rhythms. For example, five climatic rhythms have been recorded in Chardzhou from 1895 to 1981.

In general, since the 1950s and 1960s the southern part of Turkmenistan has been characterized by a warm and humid climatic rhythm (Fig. 8). Exceptions are Ashkhabad and Kerki, where the humid climatic rhythm has been expressed by air temperatures close to average multi-year values.

4. Paleogeography of Turkmenistan

KHABIBULLA I. ATAMURADOV

Abstract

The paleogeographic history of Turkmenistan since the Upper Cretaceous period is reviewed, with emphasis on climatic changes and development of the biota. In the Cretaceous, climate become more differentiated; seasonal temperature changes and latitudinal zonality of vegetation appeared; and the desert climatic regime was established. In the Paleogene, the remnants of the Tethys Sea covered only the lowest portions of lowlands; in the early Eocene, the sea spread over a significant area of the modern Karakum Desert. Most of lowland Turkmenistan from the Paleocene to the Lower Oligocene was covered by tropical savanna with sparse vegetation. The Neogene was a period of intensive tectonic movements and fluctuations of the Proto-Caspian Sea. Since the late Miocene and early Pliocene, most of Turkmenistan has been a continental land. By the Middle Pliocene, only two small lakes, Khachmas and Lenkoran, remained as relicts of the Pontic Sea; they were divided by the Kilyazi-Krasnovodsk mountains, which connected the Greater Caucasus and Bolshoi Balkhan and disappeared by the end of the Balakhanian age. In the Upper Pliocene, the so-called Akchagylian Sea expanded over Turkmenistan, and it receded again by the very end of the Pliocene. The aridization and continentalization of the climate of Turkmenistan continued in the Pliocene; vegetation differentiated into lowland and mountain types, and active plant speciation took place. New land surfaces emerged where the littoral flora could give rise to various types of xerophytes. The fauna of open arid landscapes existed in Turkmenistan in the Neogene.In the Quaternary, the tectonic activity in the plains ceased, and the eolic relief of the Karakum Desert was formed. The climate remained within a desert or semidesert regime. The Turkmenistan lowlands experienced four sea transgressions, the Amudarya River turned northward, and deltas of the Murghab and Tedzhen were formed. By the middle of the Quaternary, plant communities of the sand desert were completely formed; shiblyak and forest became reduced in the Kopetdagh Mountains by the middle of the Holocene. The continuing mountain uplift resulted in further, ongoing formation of young, endemic mountain biota of Kopetdagh.

Introduction

This review is based on known data on the paleogeography (Luppov 1956; Sinytsin 1962), paleobotany (Krishtofovich 1936; Vasilevskaya 1949, 1957; Vakhrameev 1964), paleoclimatology (Sinytsin 1965, 1967, 1980; Yasamanov 1978), paleoecology (Korovin 1934a), paleontology of mammals (Vereshchagin and Batyrov 1967; Ishunin and Tetyukhin 1989), and geology (Geologiya Turkmenistana 1958, 1984) of Turkmenistan and adjacent regions of Middle, Southwest, and Central Asia. Following Kryzhanovsky (1965), we begin detailed discussion of the paleogeography of Middle Asia from the Upper Cretaceous. However, we first will give a brief survey of the Jurassic paleogeography.

Jurassic Period

In the Lower and Middle Jurassic, modern Middle Asia and Kazakhstan were a dry land, with well pronounced differences between its western and eastern portions. The western part was a lowland (which it remains today) with sparse hilly areas in the areas of Balkhan, Tuarkyr, Mangyshlak, and Sultan-Uizdagh; the eastern part was occupied by elevated areas of Tien Shan and by the denudational plateau of Kazakhstan.

In the Lower Jurassic, all of Middle Asia and Kazakhstan possessed a humid climate. Elevated areas were occupied by coniferous and ginkgo forest; high lowlands, by mixed forests with sago palms and bennetites; and coastal plains, by fern vegetation. Uplift of the Middle Asian lowlands began in the second half of the Middle Jurassic. The receding sea formed groups of evaporating lagoons and, facilitated by the humid climate, coal deposits formed: e.g., in Tuarkyr in the western Turkmenistan sedimentation had already ceased in the late Jurassic, and erosion of sediments had started. Further south, in Bolshoi Balkhan and Kubadagh, gradually drying lagoons appeared, and the processes of sedimentation coninued.

A large enclosed lagoon appeared in the Gaurdak-Kugitang area and farther to the east (within modern southern Uzbekistan and Tajikistan). The Kopetdagh geosyncline emerged in the Lower and Middle Jurassic and existed throughout the Jurassic, Cretaceous, and Paleogene.

In the Upper Jurassic, climate become drier; the lowlands of Middle Asia were occupied by arid woodlands with tree-like ferns and conifers (Sinytsin 1966) and savannas. Aridization of climate, however, facilitated deforestation: first, ferns, club mosses, and horsetails disappeared; then, mesophylic cycads. The estimated average annual temperature in the arid Middle Asia-Kazakhstan area in the Jurassic was 12° to 15 °C, and annual sum of precipitation was from 500 to 800 mm (Sinytsin 1966).

Cretaceous Period

The western, relatively low portion of Middle Asia-Kazakhstan (including the Turan plateau, Cis-Aral area, and southwestern partt of the Tajik Depression) in the Cretaceous was dominated by coastal landscapes with lagoons and river deltas. In the Valanginian, sea water covered only Mangyshlak and parts of Kopetdagh and adjacent areas.

In the Barremian, the sea expanded over the Krasnovodsk Peninsula and Tuarkyr; there, thin sand and clay deposits accumulated in vast shallow waters. The sea also expanded toward the east, where it filled the Gaurdak-Kugitang area and penetrated far into the Tajik Depression.

Throughout the Aptian time of the Cretaceous period, thick sand and clay sediments were formed in Kopetdagh. The northern areas of Turkmenistan and adjacent parts of Ustyurt and Mangyshlak were further covered by the sea in the early Aptian; for the first time the sea which covered Turkmenistan was connected with the sea of the Russian Platform to the north. In the Upper Aptian and Lower Albian, more land was submerged, and the coastal line moved farther east. Marine basins of southwestern Turkmenistan had a homogeneous, humid climate. In the Lower Albian, islands of this sea were covered by tropical vegetation, and, in the Upper Albian, by the vegetation of a temperate and warm climate.

In the Cenomanian and Turonian times of the Upper Cretaceous, climate in Middle Asia differentiated: it was hot in the south, and warmer and more temperate in the north. Korovin (1961) concluded that climatic conditions during this time were not favorable for dispersal and exchange of paleofloras. Also, the existence of extensive water bodies and numerous islands should have facilitated regional differentiation of the floras, thus creating known Cenomanian floristic types.

The end of the Cretaceous period was characterized by uplifts which resulted in sea retreating westward. In the Danian, the sea liberated areas adjacent to the Amudarya River, and probably most of the Karakum Desert. In Badghyz and East Kopetdagh, large lagoons were formed with accumulation of gypsum and red-colored, gypsum-bearing deposits. Farther to the west, the sea remained during the Danian.

Coastal vegetation included Taxodiaceae, other conifers, ferns, and some palms. The inland lowlands of Middle Asia during this time had dry savanna-type landscapes, with solitary oases and riparian forests along the rivers. In the Lower Cretaceous, these forests included xerophile conifers, ginkgo, and sago, and, in the Upper Cretaceous, angiosperms such as laurels and myrtles. In the north (Mangyshlak and Ustyurt areas), climate was colder, with an annual temperature in the Maestrichtian from 11° to 18 °C (Yasamanov 1978). In general, during the Cretaceous period climate became more differentiated; seasonal temperature changes occured; latitudinal zonality of vegetation evolved; and the desert climatic regime appeared in Middle Asia.

Paleogene

Middle Asian relief in the Paleogene partially resembled the modern one, with a primarily lowland western portion and alternating mountain ranges and depressions in the east. The sea covered only the lowest parts of the lowlands (Southwest Turkmenistan and edges of the Karakum). Mangyshlak, Tuarkyr, and the Balkhans were marine islands. In the early Eocene, sea spread over a significant area of the Karakum and penetrated to the Kizylkum Desert. This transgression continued in the middle Eocene and waters completely covered the Karakum, Kizylkum, and the northern Cis-Aral area. In the Lower Oligocene, large shoals with island groups emerged, and, by the mid-Oligocene, the Karakum, Kizylkum, Cis-Aral area, and Tashkent area became exposed. By the Upper Oligocene, the sea again receded and covered only western Turkmenistan. No Oligocene marine sediments are known from East Kopetdagh, Badghyz, or the Gaurdak-Kugitang area.

Volcanic activity in Badghyz is dated by the mid-Paleogene; a large cover of effusive rocks (andesites and basalts) formed here during three separate eruption events. Volcanic formations of approximately the same age are found in Bolshoi Balkhan.

Most of lowland Middle Asia during the Paleocene, Eocene, and Lower Oligocene was covered by sparse tropical savanna, although riparian forests and a few oases flourished in the river valleys. A well-studied fossil flora of Badghyz (Korovin 1934a, 1934b, 1958; Vasilevskaya 1949, 1957; Abuzyarova 1956; Sikstel and Khudaiberdyev 1968; Pulatova 1971) is commonly dated by the Eocene (although some authors ascribe it to the Lower Oligocene). This flora includes 36 species belonging to 14 families (e.g., Proteaceae, Anacardiaceae, Myrtaceae, Rhamnaceae, Sapindaceae, Myricaceae, Melastomaceae, Araliaceae, and Lauraceae) which now either have entirely tropical distribution or are found both in the tropics and within the Ancient Mediterranean region. Recent palynological data (Pulatova 1971) show dominance of such plants as palms, *Ephedra*, *Myrica*, and *Rhus*, in the Eocene floras of Badghyz. General composition of flora indicates climatic conditions of high temperature and periodic dryness. Vasilevskaya (1957) suggested that the Eocene flora of Badghyz existed in a climate with an average annual temperature of 15° to 20 °C, and annual precipitation from 250 to 1,000 mm, with most precipitation occurring in winter. Korovin (1934a) estimated that the Paleogene climate of Badghyz had an average annual temperature of 16 °C, and annual precipitation of 500 mm. The modern climate of Badghyz (average annual temperature of 15 °C, and annual precipitation 250 mm) does not allow for growth of Eocene-type plant species. The Eocene climate was definitely warmer and more humid than now, probably resembling the modern climate of the southern Mediterranean. A record of palms in the Eocene flora of Badghyz, and especially of mangrove palms (Gladkova 1962) that currently grow only in the tropics, indicates that, in the Upper Eocene, average annual temperature in Badghyz could have been about 20 °C. In general, the Eocene vegetation of southern Turkmenistan may have been similar to that of modern tropical or subtropical savannas.

Krishtofovich (1936) demonstrated that the territory of the former USSR in the Eocene-Oligocene was shared by the Poltava, Turgai, and Greenland paleofloristic provinces (Fig. 1). Of these, the Poltava Province was an area of combined tropical and European floras, while the Turgai Province was an area of temperate forests. The territory of Turkmenistan lay within the tropical Poltava Province. The southern border of the Turgai flora was between 47° and 50° N, and its southernmost known record is the north shore of the Aral Sea. Therefore, two distinct botanico-geographic provinces existed within Middle Asia in the early Paleogene (Korovin 1958). A significant part of Tien Shan and a number of islands had mesophytic mixed forests, whereas the southern mainland (including parts of modern Turkmenistan) had xerophyte woodlands. Among the modern families representing the legacy of the xerophyte Oligocene flora, are Zygophyllaceae, Capparidaceae, Asparagaceae, Tamaricaceae, Rutaceae, Pedaliaceae, Chenopodiaceae, Fabaceae, Asteraceae, Lamiaceae, and Plumbaginaceae.

The Tertiary flora of southern Middle Asia possibly gave rise to xerophyte tree and shrub communities of so-called protoshiblyak, and, later, to the true shiblyak (Kamelin 1965, 1973; Kurbanov 1992). Among modern representatives of these communities found in the valleys of Southwest Kopetdagh are *Ziziphus jujuba*, *Rhus coriaria*, *Celtis caucasica*, *Punica granatum*, *Ficus carica*, *Jasminum fruticans*, and *Euonymus velutina* (Kurbanov 1992). Along with these plant species, ancient floras of Kopetdagh and Badghyz included those of broadleaf forests; e.g., fossils from Akarcheshme in Badghyz

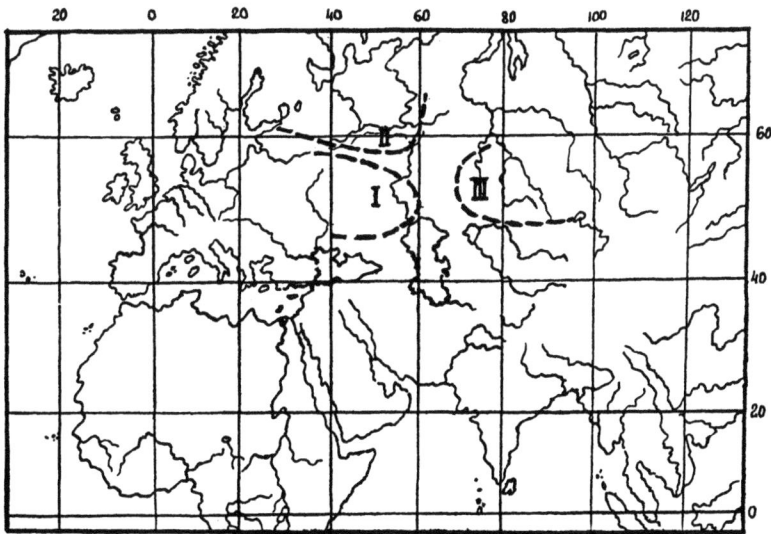

Fig. 1. Ecological provinces in the Paleogene (after Korovin 1934a):
I - Poltava Province, II - Greenland Province, III - Turgai Province.

contained *Carya typica* as well as pollen of *Juglans, Alnus, Betula, Quercus*, and species of Taxodiaceae and Cupressaceae.

A number of ancient sand desert genera such as *Eremosparton, Ammothamnus,* and *Ammodendron*, probably emerged in the late Oligocene in Kopetdagh and adjacent sand semideserts and foothills. Korovin (1961) suggested that these genera are the direct descendants of the Eocene Badghyz flora. Continental plains of Turan were inhabited by the ancestral species of *Haloxylon, Halothamnus,* and *Salsola* as early as the Tertiary (Kurbanov 1992).

In the late Eocene, the climate became colder, and vegetation differentiated according to the thermic regime. The arid belt became drier, and the zone of true deserts was established.

Detailed fossils of the Paleogene terrestrial mammals in Eurasia are known first from the Upper Eocene; no Paleogene mammals are known from Turkmenistan. In surrounding areas, the second half of the Paleogene witnessed the change from a so-called brontotherium fauna of Upper Eocene and Lower Oligocene to an indricotherium fauna of Middle and Upper Oligocene. Brontotherium fauna (named after an elephant-size ungulate) was connected to humid, swampy habitats. It ranged from England and France to Japan; most fossils of this fauna are concentrated in the periphery of the arid zone; similar fossils have been found in Europe, Kazakhstan, Kyrghyzstan, Ferghana, Mongolia, China, and the Far East (Sinytsin 1965). The so-called indricotherium fauna comprised two ecological complexes: animals of riparian forests and swamps, and animals of savannas. Typical savanna species included giant rhinoceroses, burrowing rodents, and tortoises. Indricotherium mammalian fauna of the Middle Oligocene is known from Transcaucasia, Kazakhstan, Mongolia, and China. In the Late Oligocene, it included a variety of lagomorphs, rodents, tapirs, swine, and deer found in the Caucasus, Kazakhstan, Central Asia, and Pakistan (Zoogeography of the Paleogene of Asia 1974).

In general, the primary difference of the Paleogene climate and the modern one was an absence of a defined cold season. However, beginning in the middle of the Lower Oligocene, the climate became colder, and a temperate warm climatic zone began to move southward. Climate in Middle Asia became more differentiated and continental due to the drying of the sea. Well-developed denudational surfaces appeared, and the relief became more pronounced and elevated due to tectonic movements of large blocks (Agakhanyants 1981). New climatic conditions should have stimulated evolutionary processes. Therefore, the inner arid regions of Asia could have been an important center of formation of xerophile flora and fauna as early as the Paleogene (Kryzhanovsky 1965).

Neogene

In Turkmenistan, the Neogene was a period of intensive tectonic movement which drastically changed the appearance of this territory and defined basic features of the modern landscapes. Since that time, the history of this area has been associated with the development of the Caspian Sea basin, which sometimes had a broad connection to the basin of the Black Sea. Lowlands of Middle Asia underwent uplift in the Oligocene, and the sea gradually receded westward. In the very beginning of the Miocene, modern Turkmenistan was completely liberated from water. In the Early Miocene, the sea existed only northward to the modern Ustyurt Plateau. In the Late Miocene, the so-called Sarmatian transgression flooded the Ustyurt and West Turkmenistan. In East Turkmenistan, the continental climate persisted, with inner lakes containing accumulating painted clay sediments. Mangyshlak, Bolshoi Balkhan, Kubadagh, Tuarkyr, and West Kopetdagh were islands in the Sarmatian sea. Later, the sea subsided, and by this post-Sarmatian time, the mountain climatic regime was formed in Kopetdagh and other mountains.

In the Upper Pliocene (the Akchagylian), regression stopped, and the sea expanded, again flooding West Turkmenistan and the western part of the Karakum Desert. Later, it retreated again and, by the end of the Pliocene, the sea existed only within the Caspian Depression, where, by the beginning of the Quaternary period, it was enclosed. In the Late Miocene – Early Pliocene, the folded structure of Bolshoi Balkhan and Kubadagh was completely formed. Orogenesis was also expressed in the areas adjacent to Turkmenistan. In the Miocene, due to the uplift of the Iranian Plateau, the island arch of Elburz-Paropamiz was replaced by a mountain range. Zagroz (still surrounded by the sea from both sides, to the west from Mesopotamia, and to the east, from the Central Iran) also became a mountain range. Other high mountain chains, such as Pamiro-Alai and Mekran, were created as well.

At the same time, most continental alluvial deposits in the Trans-Unguz and Southwest Karakum were formed due to the Paleo-Amudarya River (Babaev and Fedorovich 1970) which originated in the high mountains of Pamiro-Alai. Other rivers participating in the formation of the Neogene continental deposits in the Karakum were the Paleo-Murghab and Paleo-Tedzhen, which originated in the Paropamiz Mountains, as well as – in the east – rivers originating from the western offspurs of Tien Shan. Formation of the closed depressions in Turkmenistan is dated to the post-Sarmatian time (Sidorenko 1952).

Therefore, in the Late Miocene – Early Pliocene the sea either completely liberated the territory of Turkmenistan or covered only small portions next to the modern shore of the Caspian Sea. Since that time, most of Turkmenistan has been a continental land, allowing free dispersal of lowland flora and fauna from the northeast and southwest.

In the Middle Pliocene, a tectonic depression involved significant portions of Turkmenistan, including the Caspian Lowland. One of the reasons for the sea regression in the Pontian was a deep tectonic depression in the area of the southern Caspian Sea. At the end of the Pontian, due to large ascending tectonic fluctuations, two lakes (North Caspian and South Caspian) were formed in the Caspian area, separated by land with a central strait (Fig. 2). Continuing tectonic movements closed this strait, connecting by land the Krasnovodsk and

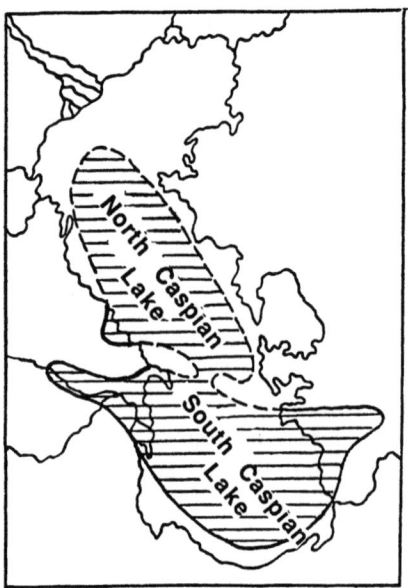 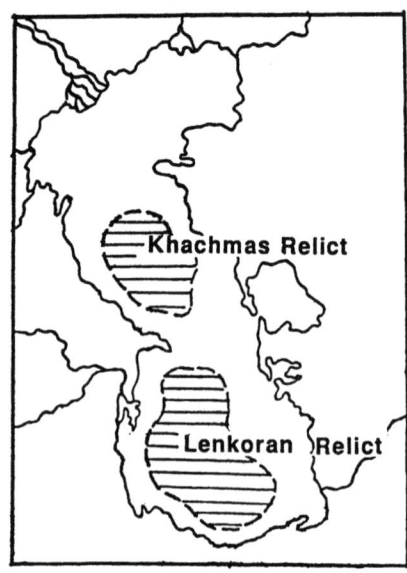

Fig. 2. Upper Pontian basin of the Caspian Region (after Ali-Zade 1961).

Fig. 3. Relicts of the Pontian basin of the Caspian Region (after Ali-Zade 1961).

Kilyazi Peninsulas (Fig. 3). Only two relatively small lakes, Khachmas and Lenkoran, remained by this time as relicts of the Pontian Sea.

Ali-Zade (1961) concluded that by the beginning of the Balakhanian (Middle Pliocene), the relict Lake Lenkoran was bordered to the north by a low Kilyazi-Krasnovodsk mountain chain, which connected the Greater Caucasus and Bolshoi Balkhan. To the south, however, the relict Lake Lenkoran was bordered by high mountains (Greater Caucasus, Lesser Caucasus, Talysh, Elburz, Bolshoi Balkhan, and Kopetdagh), neighbored by vast depressed lowlands. The relict Lake Khachmas was bordered to the east by the Krasnovodsk, Ustyurt, and Mangyshlak Plateaus, and to the north by the Cis-Caspian Lowland with a well-defined hydrographic network (e.g., Paleo-Volga, Paleo-Ural, and Paleo-Emba Rivers). All these rivers probably reached the relict Lake Khachmas. In its turn, the Cis-Caspian Lowland was bordered to the northeast, north, and northwest by the Mugodzhary Mountains, the southern offshoots of the Urals, the Common Syrt, and the Volga Plateau (Fig. 4).

The Paleo-Uzboi, Paleo-Atrek, and other rivers eroded Kopetdagh, Bolshoi and Maly Balkhan, Kubadagh, the Krasnovodsk Mountains, Karakum Plateau, and Elburz, and deposited material which formed the Balakhanian stage of the Caspian Lowland westward to West Kopetdagh. The main source of the material for the Balakhanian deposits of the Apsheron Peninsula could only have been located in the central part of the Kilyazi-Krasnovodsk Mountains (Ali-Zade 1961). These mountains were destroyed and leveled by the

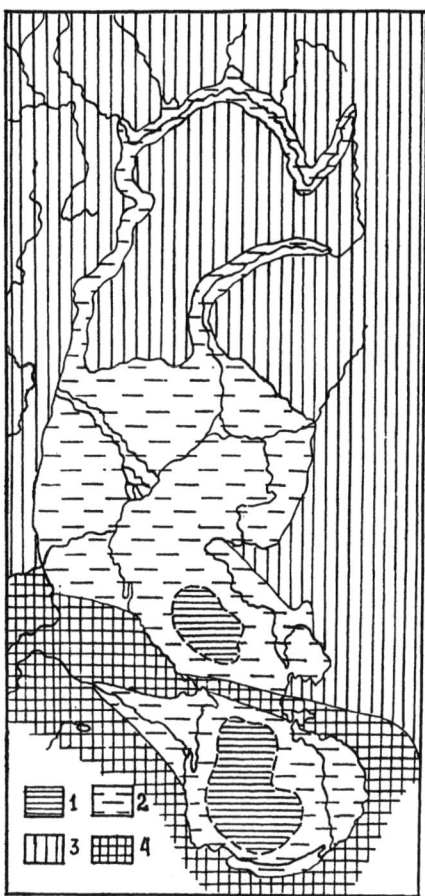

Fig. 4. Paleomorphology of the Caspian Region before the Balakhanian Age (after Ali-Zade 1961): 1 – relicts of the Upper Pontian basin; 2 – lowlands, plains, and valleys; 3 – plateaus; 4 – mountains.

end of the Balakhanian, and their final descent under water effectively liquidated the Middle Caspian land mass.

The Middle Pliocene tectonic depression involved partially the Central Karakum, therefore beginning its separation from the Trans-Unguz area. At this time, the Paleo-Amudarya and its tributaries, Paleo-Tedzhen and Paleo-Murghab, eroded a deep valley to the Caspian Sea. Therefore, the expressed dissection of the surface of Turkmenistan occurred in the Middle Pliocene, before the Akchagylian transgression.

In the Upper Pliocene, the so-called Akchagylian Sea expanded over Turkmenistan eastward to Uchadzhi, and northward (via the Uzboi corridor) to Karakalpakistan (Fig. 5). The non-flooded areas were Kopetdagh, Balkhans, Badghyz, Karabil, portions of the Ustyurt, and the Trans-Unguz and Southeast Karakum. The Akchagylian Sea was an inland sea which was only temporarily connected via a narrow strait to the Black Sea basin.

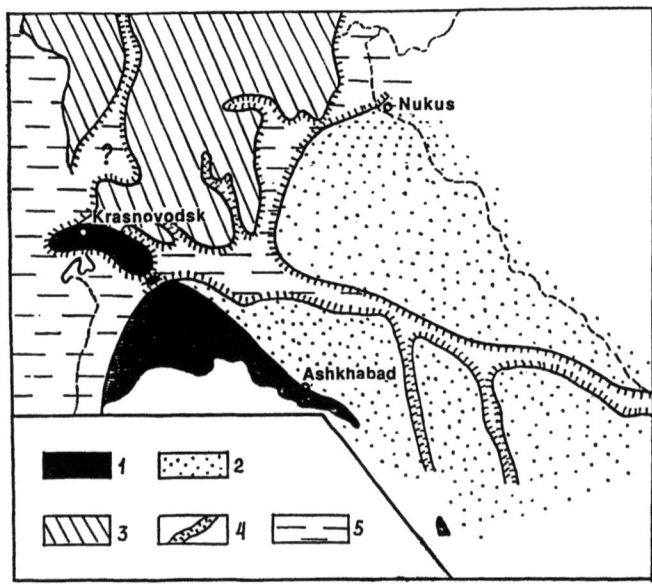

Fig. 5. Geomorphology of Turkmenistan in the Middle Pliocene (after Yurevich 1966; Gorelov 1972): 1 – mountains, 2 – lowland swamps and primarily sand desert foothills, 3 – desert plateaus, 4 – river valleys and erosion valleys, 5 – water bodies.

By the very end of the Pliocene, the sea receded again until it covered only westernmost Turkmenistan. At this time, new tectonic activity involved the Turkmeno-Khorassan Mountains, including Kopetdagh (Kalugin 1977) as well as Gaurdak-Kugitang. In contrast, the Caspian Lowland at the end of the Pliocene continued tectonic depression and became covered by the waters of the Apsheronian Sea. This sea also covered the western part of the Krasnovodsk Peninsula, approached the western offshoots of Kopetdagh, and penetrated the western Karakum. The Apsheronian Sea, like the Akchagylian Sea, was closed. Lowland Karakum was also tectonically depressed at the end of the Pliocene. In the Trans-Unguz Karakum, formation of eolic sand ridges commenced (Luppov 1956; Fedorovich 1960). The first known inland water bodies in the depressions of Sarykamysh and Aral (without any Caspian connection and fed by rivers) also are dated by the Upper Pliocene.

The aridization of the climate of Turkmenistan continued in the late Pliocene. Sinytsin (1967) notes that the annual precipitation in the area between the Caspian Sea and Amudarya was 200 to 300 mm, and average monthly temperatures were from 0° to 5 °C in the coldest month and 25° to 30 °C during the hottest month. Therefore, temperate deserts (although somewhat drier than modern ones) could have existed there in the late Pliocene. The Middle-Asian vegetation differentiated into lowland and mountain types in the Pliocene (Korovin 1958), along with the mountain uplift. Palynological data show that the Middle Asian lowlands in the Pliocene already possessed flora similar to that of modern deserts and included species of *Haloxylon* and *Ephedra* (Fedorovich 1946). Petrosyants (1956) analyzed the early Akchagylian palynological samples from the Central and Southwest Karakum. Xerophytic plants (species of *Haloxylon, Anabasis, Salsola, Calligonum, Artemisia,* and *Ephedra*) were not abundant, whereas a high number of mesophytes and even hydrophytes (e.g., *Crepis, Matricaria, Centaurea,* and other non-identified Asteraceae) was present. Such diversity of pollen can be explained by the existence of mesophile vegetation along the rivers and lake shores; the xerophyte pollen could easily have been brought by wind from the surrounding deserts. In the Repetek deposits, remnants of such hydrophytes as *Carex* and *Scirpus* have been found (Raevsky 1969). In West Kopetdagh, elements of tropical and subtropical flora have been found as fossils (*Cinnamommum polymorpha*; Ali-Zade 1961) as well as pollen of grasses and poplars, which suggest a tugai landscape (Ushko and Isaeva-Petrova 1959). Tugai fossil flora (*Phragmites, Populus, Cercis,* and *Periploca*) (Kara-Murza et al. 1953) is known from the Pliocene deposits of Cheleken as well as pollen of desert plant species (Gladkova 1957). Large amount of xerophyte pollen was also discovered in the Upper Pliocene deposits of the Balkhan Corridor near Yaskhan (Malgina 1958). All these sites reveal similarity to typical modern plant communities of desert Middle Asia. The Akchagylian fossils of Turkmenistan include also the typical Meditrerranean algae *Acucularia italica, Ovulites renata,* and *Chara meriane*. A number of fossil insect species associated with forest-type vegetation was found in the Akchagylian deposits of Syrtlanli, Boyadagh, and Monzhukly, including Trypaniedae, Dolichopodidae, Fungivoridae, Limoniidae (Diptera), and a mesophile forest beetle *Palandra* sp. (Coleoptera: Prinoidae) (Ali-Zade 1961).

Korovin (1961) derived the forest communities of Middle Asia from the Miocene flora; the Miocene and Pliocene forests were dominated by such large trees as species of *Platanus* and *Populus*. With progressing aridization and cooling of the climate in the Oligocene and Miocene, local vegetation differentiated into two major complexes. The first one occupied the ancient continental surface (e.g., species of *Pistacia, Ficus, Sageretia, Cissus, Rhamnus, Ammothamnus, Ammodendron, Eremoparton,* and *Smirnowia*). Plants belonging to the second complex dispersed along the beds of erosion which dissected the Tertiary plateaus and occupied accumulative valleys with high humidity. Some of these plants (e.g., *Erianthus, Alhagi,* and *Halimodendron*) still inhabit riparian communities (tugais) of the river valleys in Middle Asia (Ovchinnikov 1940; Korovin 1958).

Modern ranges of many plant genera, e.g., *Cousinia, Verbascum, Amygdalus, Aegilops, Onobrychis, Acantholimon, Medicago,* and *Tragacantha,* confirm the active process of speciation in arid Asia (Agakhanyants 1981). Kamelin (1979) suggested that florocoenotypes of Turkmenistan *Salsola* species (*S. botschantzevii, S. kopetdaghensis, S. iljinii,* and *S. bungeana*) originated in the Neogene on the lowlands of Turan, Iran, and Central Asia from the Ancient Mediterranean ancestors of the temperate warm zone. Due to the constant aridization and continentalization of climate in Middle Asia, evolution of xerophyte flora here is considered one of the main florogenetic trends (Ilyin 1950; Korovin 1961; Ovchinnikov 1948; Agakhanyants 1981).

Due to the multiple sea regressions, new land surfaces emerged; these new substrates were colonized by species derived from the Paleogene flora (Korovin 1958) or by the salt-tolerant flora of the subtidal zone. Ilyin (1947) demonstrated that tidal (littoral) flora could have given rise to various types of xerophytes. Radiation of the continental Chenopodiaceae (e.g., of *Anabasis*, *Arhrophytum*, *Nanophyton*, *Hammada*, and *Nitraria*) around the Ancient Mediterranean Sea (the Tethys Sea) is dated by the Miocene but probably not later (Grubov 1966; Yemelyanov 1972). Plants belonging to the so-called "hammada" vegetation of stony or gypsum deserts (Popov 1927), including species of *Zygophyllum*, *Limonium*, *Goniolimon*, *Reaumuria*, *Cleome*, *Thesium*, *Haplophyllum*, *Ferula*, *Trichanthemis*, and *Aristida*, became adapted to the new salt-bearing substrates. Endemic solonchak species of *Salsola* and *Suaeda* were formed on the shores of drying coastal salt lakes of the Aralo-Caspian basin (Korovin 1958).

In general, by the Neogene the following plant communities could have existed in Middle Asia: broad-leaf forests, riparian forests (tugais), solonchak vegetation, gypsum hammada, and xerophyte shrub and tree community (shiblyak) (Kamelin 1979). Desert climatic regime was established by the Pliocene, since the first stages of the Cenozoic orogenesis. Accompanying mountain buildup, a new vegetation type of mountain xerophytes evolved, as well as savanna-type herbaceous vegetation and mountain forests.

Neogene fauna of Middle Asia was highly diverse. Due to the climatic changes since the second half of the Miocene, swamp subtropical forests in the valleys and lowlands of the arid zone were transformed into dry broad-leaf forests; interfluvial woodlands became savannas and then turned into steppes and semideserts. Cooling and aridization of the climate caused the dispersal of the so-called Hipparion fauna of the Upper Miocene and Pliocene.

In Turkmenistan, there is fossil evidence of the fauna of open arid landscapes existing at this time. The fossil record of a giraffe is known from the sands of the so-called Esenbai stratum of the Upper Miocene and Pliocene in Badghyz (Godina and Dubyansky 1963) and a record of a tooth of a fossil horse, *Equus caballus fossilis*, near Kara-Bogaz (45 km north from Kizyl-Arvat; Amursky 1961). A small bear, cheetah, and a small gazelle are known from the Pliocene of the Trans-Unguz Karakum (the Aktash Well) (Vereshchagin 1956). In the Pliocene deposits of West Kopetdagh are found camels, gazelles, wild sheep, and wild cat (Amanniyazov *et al.* 1979). The sandstones of the Gokcha stratum (the Upper Pliocene) yielded remnants of monitors, agamids, and other arid lizards (Ananyeva and Gorelov 1981). Fossils of the Upper Pliocene and the very beginning of the Pleistocene contain animals connected with steppe landscapes with shrubs and pockets of forest vegetation; these are hipparions (a small three-toed horse), saber-tooth cats, bears, wolves, camels, and ostriches (Vereshchagin and Batyrov 1967; Gorelov 1969; Babaev and Fedorovich 1970;

Dubrovo and Nigarov 1990). An interesting fossil elephant, *Palaeoloxodon turkmenicus* Dubrovo, has been found in the Upper Pliocene pebble strata near Krasnovodsk (Dubrovo 1960). This animal could have inhabited savanna or steppe landscapes and fed on branches, leaves, and grass. Remnants of other ancient elephants have been also found in Uzboi area (Okladnikov 1956) and between Khudaidagh and Monzhukly (Fedorov 1946).

Quaternary Period

In the Quaternary, the territory of Turkmenistan experienced decreasing tectonic activity, fluctuation of the Caspian Sea level, wandering of the Amudarya River, and formation of the eolic relief of the Karakum Desert. Alteration of pluvial (more humid) and xerothermic (more arid) climatic periods occurred within the Quaternary time, with corresponding movements of the biotas. In Middle Asia the climate remained within a desert or semidesert regime (Fedorovich 1952; Luppov 1956). It differed from the dry Pliocene climate in that cold winters appeared, and rainy seasons disappeared (Fedorovich 1952). In the Pleistocene, the lowlands of Middle Asia were more affected by the cooling of the climate (during the glaciation in Tien Shan and Pamir) than the more southern deserts of Iran, which were protected from the northern air masses by the mountain chains of Elburz and Paropamiz.

In the Quaternary period, lowlands bordering the Caspian Depression experienced four sea transgressions (Baku, Khazarian, Khvalynian, and New Caspian). In the first half of the Quaternary (during the Baku and Khazarian), the sea expanded primarily over the Caspian Lowland, sometimes covered the western edge of the Krasnovodsk Peninsula, and penetrated slightly to the western part of the Central Karakum. The sea fluctuated back from and forth toward West Kopetdagh, creating the lowland coastal plain which to the south contacted the delta of the Proto-Atrek, and to the north, the delta of the Proto-Amudarya.

The Lowland Karakum in the late Pliocene – early Quaternary was occupied by a vast alluvial plain of the Proto-Amudarya. This river was especially large in the Khazarian, when most of the alluvial material was deposited which later was transformed by the wind. The end of the Baku and the Khazarian experienced a more humid climate which might have been responsible for the formation of large mountain slides along the chinks of Ustyurt (Fedorovich 1946). Around this time, large mammals such as elephants (Fedorov 1946) penetrated to the east shore of the Caspian Sea along the Proto-Amudarya Valley.

The eroding activity of the Proto-Amudarya and its left tributaries in the east and southeast during the first half of the Quaternary period resulted in the dissection of the Neogene plain and the formation of the Murghab and Tedzhen Valleys and the Obruchev Steppe. In the north, the modern and ancient (or Kunyadarya) deltas of the Amudarya were filled with sediments, due to the activity of a river network beginning in Kazakhstan and connected to the Amudarya near the southern end of the Upper Uzboi corridor (Yamnov and Kunin 1953).

In the middle of the Quaternary period, two major geological events changed the paleogeography of the plains of Turkmenistan. First, the Amudarya turned toward the north. It left the Lowland Karakum, cutting through the Neogene land between the Trans-Unguz Karakum and Kizylkum, and flowed toward the Aral Depression. That event led to the transformation of the alluvial plain of the Lowland Karakum by the wind, resulting in the creation of the modern sand desert.

The second major event was the Khvalynian transgression of the Caspian Sea, which rose 75 m higher than its modern level. The Khvalynian Sea covered most of the Caspian Lowland and the western part of the Krasnovodsk Peninsula, and it penetrated deep into the Karakum, reaching the meridian of the city of Kizyl-Arvat.

At the end of the Khvalynian, the Amudarya turned toward the Sarykamysh Depression and filled it as well as the Assake-Audan Depression and most of the Upper Uzboi corridor. Thus, the Uzboi River was created, which originated in the Sarykamysh and flowed into the Caspian Sea. Later, when the Amudarya turned entirely toward the Aral Sea, Lake Sarykamysh started drying out, and the Uzboi River disappeared.

The latest (New Caspian) stage of the geological history of the Caspian basin continues today. It is characterized by a new increase of the sea level (with the highest mark during the New Caspian period, however, only 7 m higher above the modern one).

In the second half of the Quaternary period, subaeral deltas of the Murghab and Tedzhen were formed which ended blindly in the Karakum Desert; four subsequent deltas, partially overlapping, have been found in the lower reaches of the Murghab (Fedorovich and Kes 1934).

At this time, the tectonic activity in the plains of Turkmenistan practically stopped. Mountains continued their uplift, but in Kopetdagh this process was interrupted; up to six terraces were created in Kopetdagh river valleys during the Quaternary time (the highest is now elevated 100 m or more above riverbeds). In the Gaurdak-Kugitang area, Quaternary tectonic movements are also well expressed; in Badghyz and Karabil, Quaternary geological history reflects the uplift of the Paropamiz Mountains, which resulted in deep erosion of the Murghab and Tedzhen Rivers and creation of several terraces. On the other hand, no Quaternary uplift has been found in Bolshoi Balkhan and Kubadagh; terraces which exist there in mountain valleys were probably formed during the fluctuation of the Caspian Sea.

The Quaternary vegetation of lowland Middle Asia was similar to that of modern deserts, and in the river valleys, to that of tugais (Fedorovich 1946; Korovin 1958). Its development was influenced by the arid centers of speciation in Central Asia. By the middle of the Quaternary period, the plant communities of the sand desert had been completely formed. At this time, psammophiles of

Tertiary origin (e.g., species of *Calligonum* and *Astragalus* sect. *ammodendron*) experienced radiation. The sand desert flora became enriched by modified hammada and sublittoral solonchak species (Kultiasov 1946; Korovin 1958).

Formation of the loess deposits allowed a new avenue for speciation of plants and accounted for emerging dry herbaceous steppes or semi-savannas on the periphery of the mountain ranges (Ovchinnikov 1940; Kultiasov 1946; Korovin 1958). In the Pliocene-Pleistocene, modern endemic species of sagebrush (e.g., *Artemisia balchanorum*, *A. turcomanica*, and *A. deserti*) as well as such steppe grasses as species of *Stipa*, *Festuca*, and *Poa* became dominant in the mountain vegetation (Kurbanov 1992). In the Quaternary, steppes in the mountains of Middle Asia expanded, and the area occupied by mesophile plant communities (including broad-leaf forests and mountain meadows) decreased. The Turkmen juniper (*Juniperus turcomanica*) replaced broad-leaf trees in Kopetdagh mountains, forming complexes with maple (*Acer turcomanicum*) and various shrubs, and in the foothills, semisavannas, sagebrush desert, and salt desert plant communities replaced communities of shiblyak (Kamelin 1979). Due to general aridization, shiblyak and forest communities became reduced in the middle and lower belts of Kopetdagh by the middle of the Holocene; however, fragments of forest vegetation (e.g., *Juglans regia*, *Allium paradoxum*, and *Jasminum fruticans*) were preserved in refugial deep valleys.

The fossil record of the Quaternary fauna of Middle Asian plains is not rich, which confirms its desert character at this time. Fossil data on small desert and semi-desert mammals are known from Ustyukov (Nastyukov 1976), Mangyshlak (Gromov and Fokanov 1961), and Badghyz (Fokanov 1961; Gorelov 1972). Knyazev (1976) demonstrated that the fauna of small mammals in Badghyz did not significantly change during the middle and late Holocene. Since the formation of sand and clay deserts in the late Pliocene, constant complexes of desert Turanian species have emerged. Such complexes could have been formed via a combined dispersal of desert fauna from other regions as well as by autochthonous evolution; some forms could have dispersed from the southern deserts of Iran and North Africa (Kashkarov *et al.* 1929). Finally, the continuing mountain uplift and dissection resulted in further differentiation of mountain faunas and formation of a wide array of young endemic species in the mountains of Turkmenistan (Kryzhanovsky 1965).

Dinosaur footprints on the Jurassic limestone, Kugitangtau Mountains. Photo by K.I. Atamuradov.

5. Desertification of the Arid Lands of Turkmenistan

NIKOLAI G. KHARIN

Abstract

Desertification in Turkmenistan is described, with various processes indicating desertification level on different scales. Local, regional, and zonal criteria of desertification characterizing this process can include, e.g., data on species composition, vegetation, and productivity in specific plant communities. Such criteria were used for map construction, including the desertification map of Turkmenistan on the scale 1:4,000,000. The basic types of desertification include degradation of vegetation, deflation, water erosion, pasture swamping, salinization of irrigated lands, and formation of solonchaks.

Introduction

Desertification is one of the global problems acknowledged by the United Nations. Experts define desertification as the degradation of lands in arid, semiarid, and subhumid areas caused by the destructive activities of man (Odingo 1990). In this definition, "lands" include soil, local water resources, land surface, natural vegetation, and agricultural crops. The process of desertification includes the degradation of vegetation, wind and water erosion, technogenic desertification, and soil salinization.

The global estimate of desertification performed by UNEP shows that, by 1985, in arid zones worldwide, 80% of all pastures (3,100 million ha), 30% of irrigated arable land (40 million ha), and 60% of non-irrigated arable land (80 million ha), had been subjected to desertification. Annual losses of agricultural production due to desertification are estimated as $ 25 billion.

The definition of desertification given above characterizes the degradation of land as an anthropogenic process due to such activities as an increase of cattle grazing, road construction, population growth, or mining in the desert. However, if these activities were combined with protective measures, it would be possible to stop desertification at its current level or even to restore degraded areas. At the same time, many natural factors influence the rate of desertification, and these factors should be analyzed for an understanding of this process.

Approaches to the Study of Desertification

Our long-term studies in the Desert Institute of the Academy of Sciences of Turkmenistan have resulted in compilation of a database on desertification (Appendix 1) and construction of mathemathical models. The size of an area under desertification is defined as

$$y = (x_1, x_2, x_3, x_4),$$

where x_1 is grazing load, x_2 is population density, x_3 is rate of desertification, and x_4 is internal danger of desertification defined by stability of an arid ecosystem. The grazing load in desert pastures is the major factor in natural ecosystems. In Turkmenistan, of the 39 administrative districts studied, 16 had a high grazing overload, 10 had moderate overload, and only 13, slight overload (Kharin et al. 1989). Table 1 presents the regression equations showing the relationship between the grazing overload (expressed in per cent of a regular grazing load) and desertified area. These data show high statistical significance for three regions of Turkmenistan.

Table 1. Relationship between the desertified area (y) and grazing load (x) in Turkmenistan

Regions[a]	Number of districts[a]	Regression equation	R[b]
Krasnovodsk	6	y = 30.095 + 6.198x	0.674
Ashkhabad	8	y = 25.431 + 0.341x	0.808
Mary	8	y = 30.851 + 0.240x	0.364
Chardzhou	10	y = 26.014 + 0.179x	0.485
Tashauz	7	y = 17.831 + 0.541x	0.954

[a] According to the administrative division of Turkmenistan in 1985
[b] Coefficient of correlation

To estimate desertification, we must know its background level corresponding to natural conditions in arid ecosystems. In Turkmenistan, as well as in most arid regions of the world, there are virtually no ecoytems undisturbed by humans. Fig. 1 shows our basic approaches in dealing with the problem of background (natural) level of desertification. These approaches include:

1) Scale consisting of five classes. Using this approach, one has no knowledge of the background level of desertification; instead, a given condition of ecosystems in a certain time period is used. For example, during creation of the desertification map (25 km in 1 cm) for the arid territories of the former

USSR, we (Kharin et al. 1988) used 1965 desertification as a background level. The map demosntrated changes that occured from 1965 to 1985.
2) Scale consisting of four classes. This approach is used in the rare case when background level is known or can be calculated; it can be used for studies of long-existing protected territories (Natural Reserves).
3) Scale consisting of three classes. Background level is not known; a scale is used for schematic desertification maps of small scale. This approach was used for the creation of the desertification map of Turkmenistan given below.

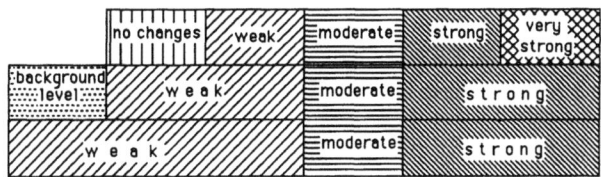

Fig. 1. Scales of desertification.

Indicators and Criteria of Desertification

Various types of processes characterizing the degradation of arid geosystems can be used as indicators of desertification (Nechaeva 1973). Diagnostics and monitoring of desertification are based upon criteria which allow qualitative and quantitative estimation of these processes. Such criteria may be local, regional, and zonal.

Local criteria of desertification characterize this process within a separate geosystem. They can include, e.g., data on species composition and vegetative cover in specific plant communities. We have used such criteria for construction of large-scale desertification maps of separate key plots.

Regional criteria of desertification are established for geographic regions (such as administrative districts or regions of new land development) and include generalized data on anthropogenic environmental changes. For example, during the creation of the desertification maps of Turkmenistan we used criteria characterizing Karakum Desert desertification such as the decrease in productivity of vegetation (in kg/ha) (Kharin et al. 1983).

Zonal criteria of desertification are even more generalized: for example, to characterize the degradation of vegetation, we measured decreases in productivity of vegetation not in kg/ha but in percentages. This approach allows one to compare desertification among different types of deserts where the absolute productivity can vary.

The estimation of desertification is complicated by the presence of various factors. For example, in sand desert such as the Karakum, decrease of area covered by vegetation and development of wind erosion are caused by shrub logging for fuel, by overgrazing, and by automobile movement. These details can be reflected separately only on large-scale desertification maps (0.1 km in 1 cm to 0.25 km in 1 cm), whereas small-scale maps should either reflect the predominant process or characterize desertification as a complex of estimates.

In Turkmenistan, the basic types of desertification include degradation of vegetative cover, deflation, water erosion, soil swamping in pastures, salinization of irrigated lands, and formation of solonchaks due to the closure of Kara-Bogaz-Gol Bay of the Caspian Sea. We estimated desertification using a scale consisting of three classes defined by the degree of degradation of geosystems: low desertification, when geosystems are rated from non-disturbed to slightly disturbed; moderate, when geosystems are moderately disturbed; and high, when geosystems are rated from seriously disturbed to having completely lost bioproductivity (Fig. 2). Rates of desertification were estimated by comparison of theme maps created during 1965 to 1990. We have also used remote sensing materials, statistical data, population surveys, and field observations.

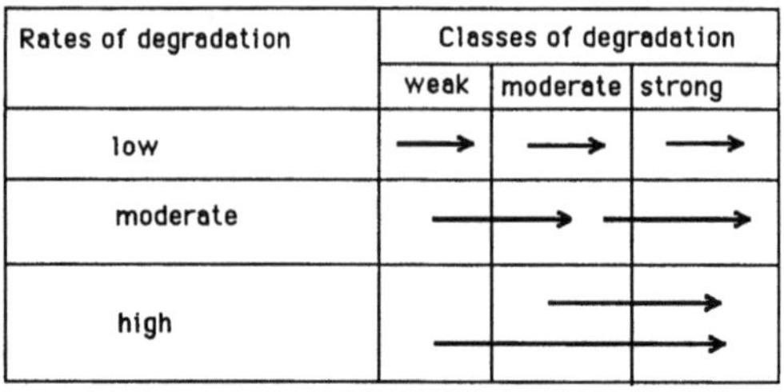

Rates of degradation	Classes of degradation		
	weak	moderate	strong
low	→	→	→
moderate	→	→	
high	→	→	

Fig. 2. Estimation of the degradation rates of geosystems.

Degradation of vegetative cover caused by human activities is the most common type of desertification in Turkmenistan (Tables 2, 3, and 4; Kharin et al. 1983). An important qualitative indicator of such degradation is a succession of dominant plant species. Succession effects qualitative changes in community productivity as well as amount and quality of food plants in desert pastures.

A special case of vegetation degradation in the Karakum Desert is "moss formation," in which soil surface becomes covered by a thick layer of a desert moss (*Tortula desertorum*). In some desert areas, where there is no grazing due to the absence of water sources, this moss covers up to 40% of the surface (Fig. 3) (Kalyonov 1977). Moss cover suppresses shrubs and herbaceous vegetation and reduces reproduction of pasture plants and bioproductivity of desert ecosystems. Moderate grazing, on the other hand, supports plant reproduction because animals break the moss cover and loosen the soil.

A separate type of the degradation of vegetative cover is technogenic desertification (Table 5). Common in all Turkmenistan deserts, it is due to the construction of canals, roads, and gas pipelines, as well as to random automobile movement, etc. An especially barbaric degradation is caused by the dragging of the oil drilling equipment from rig to rig. These ulcers on a desert's face are easily detected on satellite pictures. Technogenic desertification is especially hard to rehabilitate.

Desertification of the Arid Lands of Turkmenistan 69

Fig. 3. Distribution of the desert moss *Tortula desertorum* in the western Trans-Unguz Karakum Desert based on airplane remote photo (1:28,000). 1 – sands, 2 – moss, 3 – solonchaks, 4 – takyr (after Kalyonov 1977).

Table 2. Criteria of degradation of vegetation

Criteria	Classes of desertification		
Plant communities	Low Climax, or slightly modified	Moderate Long-time derived	High Short-time derived
Decrease in productivity (%)			
Sand desert	<15	15–35	>35
Gypsum desert	<20	20–40	>40
Clay desert	<35	35–70	>70
Foothill loess desert	<20	20–40	>40
Mountains:			
Low belt	<20	20–40	>40
Middle and high belts	<15	15–30	>30
Decrease in area covered by climax vegetation (%)			
Sand desert	<10	10–35	>35
Gypsum desert	<10	10–50	>50
Clay desert	<20	20–45	>45
Foothill loess desert	<20	20–40	>40
Mountains:			
Low belt	<10	10–50	>50
Middle and high belts	<5	5–40	>40

Table 3. Dynamics of plant communities under desertification in different types of deserts in Turkmenistan

Classes of desertification	Plant community	Number of plant species
	1. Sand desert	
Background	Haloxylon persicum – Carex physodes	42
Low	Haloxylon persicum – Stipagrostis pennata + Carex physodes	36
Moderate	Salsola richteri – Stipagrostis pennata	17
High	Stipagrostis karelinii – Bromus tectorum	8
Extremely high	Stipagrostis karelinii	3
	2. Gypsum desert	
Background	Artemisia badhysi + Salsola orientalis – Carex pachystylis	31
Low	Artemisia badhysi + Salsola orientalis – Carex pachystylis	29
Moderate	Ephedra distachya + Artemisia badhysi	19
High	Ephedra distachya + Horaninovia anomala	14
Extremely high	Peganum harmala	3
	3. Clay desert	
Background	Salsola gemmascens + Artemisia kemrudica – Gamanthus gamocarpus	35
Low	Salsola gemmascens + Artemisia kemrudica – Gamanthus gamocarpus	25
Moderate	Artemisia kemrudica + Salsola gemmascens	11
High	Artemisia kemrudica + Climacoptera lanata	8
Extremely high	Peganum harmala	3
	4. Foothill loess desert	
Background	Carex pachystylis + Poa bulbosa	50
Low	Carex pachystylis + Poa bulbosa	47
Moderate	Poa bulbosa + Carex pachystylis	30
High	Poa bulbosa + Iris songarica	26
Extremely high	Peganum harmala	3

Soil deflation in sand desert was studied at the Desert Instiute of Turkmenistan by Znamensky (1958), Dobrin (1964), and Ivanov (1972). Criteria for deflation are given in Table 6. Sand particles smaller than 0.04 mm can be carried great distances by the wind. Particles from 0.04 to 2.0 mm in size, however, are transported by wind in a suspended condition along the sand surface. Sand can form eolic deposits, i.e., wind-born sediments capable of surface movements. These movable eolian landforms can be either formed in situ or transported from elsewhere; the thickness of the deflated non-sand layer may be partially compensated for later by dust sedimentation. The surface layer of the atmosphere contains layers with various whirlpool structures; 97% of the sand is transported by the wind within 15 cm from the surface (Znamensky 1958). In the sand desert, deflation is not caused exclusively by human disturbance. Eolic landforms there occupy about 15% of the area (data of G.S.

Table 4. Changes in food value in sand desert pastures (food units/ha)

Classes of desertification	Seasons			
	Spring	Summer	Fall	Winter
1. Ridge-hill sands; *Haloxylon persicum* – *Carex physodes* community				
Background	88	134	119	80
Low	92	132	91	63
Moderate	103	76	17	7
High	73	36	16	6
2. High ridge sands; *Calligonum rubens* – *Mausolea eriocarpa* – *Carex physodes* community				
Background	195	112	61	30
Low	219	136	83	35
Moderate	100	83	45	20
High	65	60	34	12
3. Small hill sands; *Salsola arbuscula* – *Artemisia kemrudica* – *Carex physodes* community				
Background	170	160	71	30
Low	193	117	73	36
Moderate	109	50	38	18
High	70	45	28	14

Table 5. Criteria of technogenic desertification

Criteria	Classes of desertification		
	Low	Moderate	High
Disturbance of vegetation a) logging of trees and shrubs (% of total area)	<25	25–50	>50
b) destruction of turf (% of turfed area)	<25	25–50	>50
Erosion due to the irregular movement of cars and mechanisms (% of total area)	<10	10–25	>25
Area occupied by technogenic sands (% of total area)	<10	10–25	>2
Roads (km/100 sq. km)	<40	40–80	>80

Kalyonov) and include sand dunes (barkhans), deflation hollows, and other disturbances of vegetation caused by natural factors.

Tables 7 to 9 characterize other desertification processes. Water erosion (Table 7) is expressed primarily in mountainous Turkmenistan. Soil swamping in pastures (Table 8) occurs due to discharge of irrigation waters from oases to desert. Productivity in this case increases; however, plant succession takes place, and cattle-forage species commonly are replaced by non-edible ones. Finally,

Table 6. Criteria of deflation in the sand desert

Criteria	Classes of desertification		
	Low	Moderate	High
Area of drift sands (%)	15–30	30–70	>70
Turfness (%)	30–50	10–30	<10
Coverage of vegetation (%):			
a) shrubs	10–15	5–10	<5
b) herbaceous vegetation	40–65	10–40	<10

Table 7. Criteria of water erosion

Criteria – Type of water erosion	Classes of desertification		
	Low – Sheet erosion (single cavities)	Moderate – Sheet erosion (up to 10 cavities/km, formation of single gullies)	High – Gully erosion (more than 10 cavities/km)
Area of drift sands (%)	15–30	30–70	>70
Ablation of surface soil layer (cm)	<5	5–20	>20
Coverage of vegetation (%):			
a) trees and shrubs	<20	20–50	>50
b) herbaceous vegetation	<20	20–50	>50

Table 8. Criteria of pasture swamping

Criteria	Classes of swamping		
	Low	Moderate	High
Coverage of hygrophilous vegetation (%):			
a) *Tamarix ramosissima, Alhagi persarum*	<30	–	–
b) *Tamarix ramosissima, Alhagi persarum, Karelinia caspia*	–	30–70	–
c) *Phragmites australis, Glycyrrhiza glabra, Alhagi persarum*	–	–	>70
Depth of fresh or low mineralized ground water (m)	5–10	2–5	<2
Soil humidification regime	automorphic	semi-hydromorphic	hydromorphic

Table 9. Criteria of salinization in the irrigated lands

Criteria	Classes of desertification		
	Low	Moderate	High
Degree of salinization			
total solid residue, %	0.210–0.400	0.410–0.600	>0.610
Cl^-, %	0.001–0.030	0.030–0.100	>0.101
Na^+, %	0.023–0.046	0.047–0.092	>0.093
Mineralization of ground water (g/l)	3–6	6–10	10–30
Mineralization of irrigation water (g/l)	0.5–1.0	1.0–1.5	<1.5
Decrease in raw cotton yield (% of background level)	<15	15–40	40–80
Seasonal salt accumulation			
a) %	0.11–0.20	0.21–0.30	0.31–0.60
b) t/ha	16–30	30–45	45–90
Degree of pollution of irrigation water (ratio of content of toxic chemicals to their allowed concentration)	1.0–6.0 (weak)	6.0–11.0 (moderate)	>11.0 (high)

great economic damage is caused by the salinization of irrigated lands (Table 9), due to incorrect use of water resources for irrigation and insufficient melioration.

Desertification Map of Turkmenistan

A desertification map of Turkmenistan based on remote satellite pictures (Fig. 4) was created on the scale 1:4,000,000 (40 km in 1 cm) (Fig. 5). Although such a small scale does not allow for depiction of all details of degradation of the arid ecosystems in Turkmenistan, it gives a general picture of the distribution of desertification in this republic. Especially strong human influence is manifested in oases and adjacent territories. Extremely high salinization, for example, is demonstrated for the Tedzhen oasis. Another ecological danger zone is the dried-up Kara-Bogaz-Gol Bay of the Caspian Sea. A vast solonchack (salt pan), from which salt is now carried by the wind to adjacent areas, has been formed here due to the closure of the bay by an artificial dam in 1983. This decision was made in order to "save" the Caspian Sea, whose level was predicted by some to decrease in the near future. This forecast has been proved incorrect: the Caspian Sea level, on the contrary, has been increasing from 1988 to 1992, flooding settlements, roads, and industrial structures.

The lowest levels of desertification are recorded along the Turkmenistan border with Iran and Afghanistan (Fig. 5: 11). This protected territory is close to the background level of desertification, due to restrained development of this area.

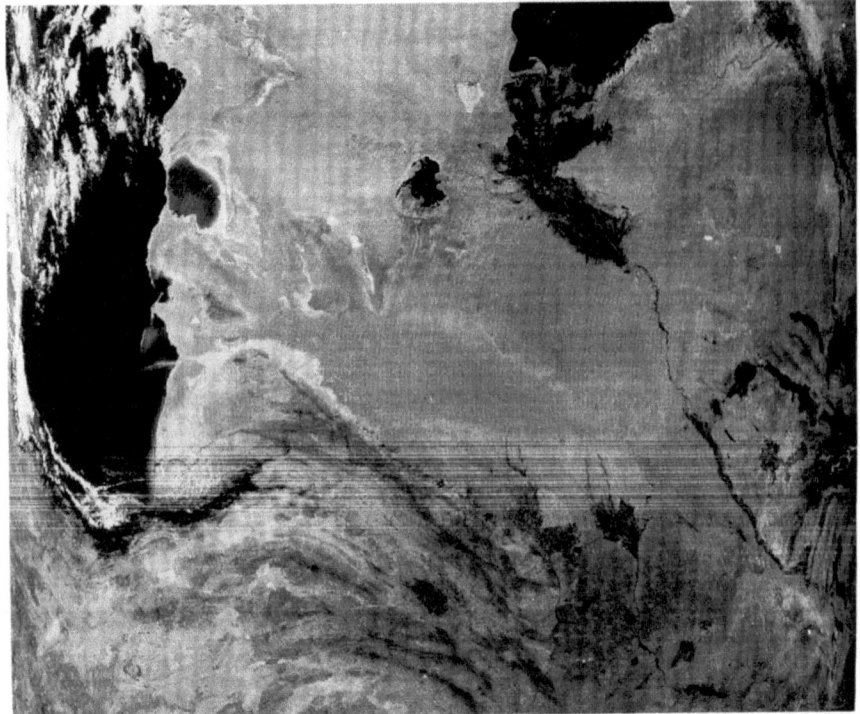

Fig. 4. Territory of Turkmenistan (after Meteor satellite remote photo). Darker areas represent oases, lighter areas represent deserts and solonchaks.

Acknowledgements

The author thanks G.S. Kalyonov, A.A. Kiriltseva, and P. Esenov (Remote Sensing Laboratory of the Desert Institute, Academy of Sciences of Turkmenistan) for their help in creation of the desertification map of Turkmenistan and in establishment of the criteria of desertification.

Structure of the Database of Desertification Data (maintained in the Desert Institute, Academy of Sciences of Turkmenistan, Ashgabat)

I. Causes of desertification (C). CN – natural causes; CNa – air temperature, CNb – albedo, CNc – precipitation, CNd – air humidity, CNe – wind speed, CNf – dust storms. CH – anthropogenic causes; CHa – centralized norms of planned economy, CHb – control figures of economic development by separate regions, CHc – reported data on fulfilment of economic plan by separate regions, CHd – population density, CHe – area of deforestation, CHf – grazing load on pastures, CHg – areas not used for

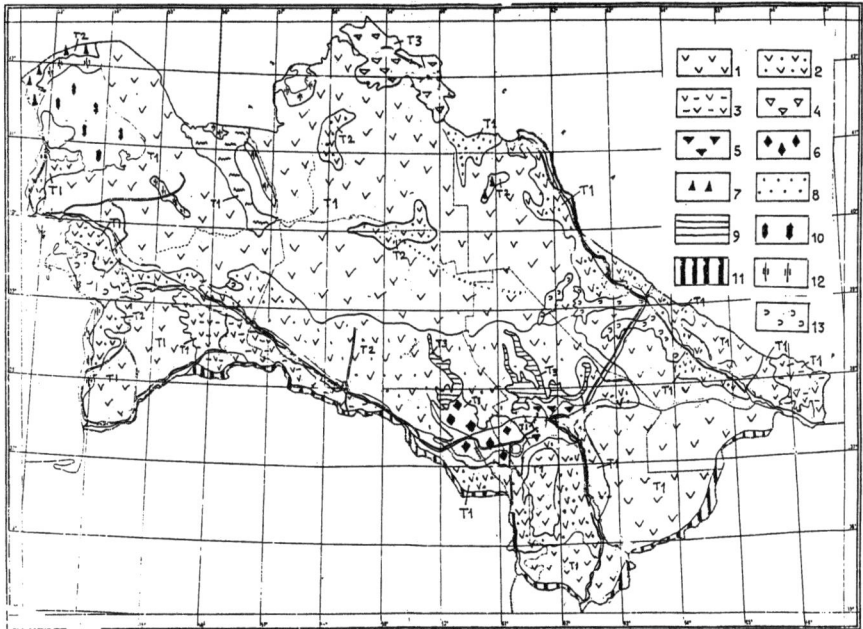

Fig. 5. Desertification Map of the Arid Territories of Turkmenistan
(1:4,000,000). 1 – low degradation of vegetation, 2 – moderate degradation of vegetation combined with deflation, 3 – degradation of mountain vegetation (70%) combined with water erosion (30%), 4 – low salinization of irrigated lands, 5 – moderate salinization of irrigated lands, 6 – high salinization of irrigated lands, 7 – technogenic desertification, 8 – deflation, 9 – pasture swamping, 10 – the solonchak formed in place of the former Kara-Bogaz-Gol Bay, 11 – protected border territories, 12 – automorphic solonchacks, 13 – drift sands of natural origin, T1 – low desertification rate, T2 – moderate desertification rate, T3 – high desertification rate.

crops, CHg1 – area used for roads, CHg2 – area used for residence and construction, CHg3 – area used for industry, CHh – ploughed area, CHi – irrigated area.

II. The process of desertification (D). DV – degradation of vegetation; DVa – species composition of vegetation, DVb – bioproductivity, DVc – forested areas separated by classes of productivity, DVd – forest deposits, DVe – area under pastures, DVf – pasture productivity, DVg – area of dried forests, Dvh – disappeared plant species. DS – soil degradation; DSa – areas under wind erosion, DSb – areas under water erosion, DSc – areas with decreasing humus content, DSd – areas with hardened soil, DSe – areas with salinization and swamping. DE – deterioration of water resources; DEa – river water debit, DEb – lake water volume, DEc – amount of surface runoff, DEd – level of ground water, DEe – quality of surface and ground waters. DA – changes in animal populations; DAb – animal species composition, DAc – disappeared animal species, DAd – diseases of animals.

III. Consequences of desertification (A). AS – social consequences; ASa – number of nomadic population, ASb – migrations of population, ASc – number of new settlements, ASd – sanitary conditions, ASe – birth rate, ASf – death rate, ASg – specific human diseases, ASh – life span. AE – economic consequences; AEa – crop structure, AEb – yield of agricultural crops, AEc – number of cattle, AEd – food consumption, AEe – water consumption, AEf – fuel consumption, AEg – per capita income.

IV. Measures against desertification (L). LS – monitoring of desertification; LSa – criteria of desertification, LSa1 – criteria of cxurrent conditions, LSa2 – rate criteria, LSa3 – internal danger, LSa4 – grazing load, LSa5 – density of population, LSa6 – methodology. LM – measures against desertification; LMa – areas where measures against wind erosion were undertaken, LMb – areas where measures against soil salinization were undertaken, LMc – areas where measures against soil hardening were undertaken, LMd – areas of meliorated pastures, LMe – areas of low productive arable lands where melioration was undertaken, LMf – meliorated forested areas, LMh – usage of mineralized water, LMh1 – deposit of mineralized water, LMh2 – area of irrigated crops, LMh3 – yield of irrigated crops, LMi – areas where complex measures against desertification were undertaken. LP – forecast of desertification; LPa – models of desertification, LPb – climate forecast, LPc – population density forecast, LPd – economic development forecast.

V. Archive materials (AR). ARa – topographic maps, ARb – theme maps, ARc – airplane remote photographs, ARd – satellite remote photographs, ARe – field descriptions, ARf – bibliography.

Sand dunes in the Repetek Reserve, Karakum Desert (1 of 2) Photo by H.R. Levenshtein.

6. Vegetation of the Deserts of Turkmenistan

IGOR G. RUSTAMOV

Abstract

The desert vegetation in Turkmenistan consists predominantly of semishrub sagebrush-halophyte and psammophyte communities, with dominant formations of *Haloxyleta, Salsoleta, Calligoneta,* and *Artemiseta.* We have separated plant communtites into 26 formations and 28 groups of associations with a relatively simple phytocoenological structure and homogeneous species composition. The community structure is usually determined by a few dominant species of semishrubs which also provide most of the phytomass. Other sinusia, such as herbaceous cover, are built mainly by annual ephemerous species which do not play a significant role in the community structure.

In deserts, which occupy more than 80% of the territory of Turmenistan, vegetation is a valuable feed resource for the livestock industry. Desert areas are used throughout the year for sheep and camel grazing. Desert vegetation provides an important ecological role by stabilizing the sand. Several natural reserves and other protected areas have been established to preserve desert vegetation and landscapes in Turkmenistan.

Introduction

The desert vegetation of Turkmenistan is relatively well studied. All basic plant communities have been described and characterized and a number of studies conducted on the dynamics and productivity of many desert plant communities as well as on the role of natural and anthropogenic factors. This review is based on data published by other authors as well as on our original data gleaned from many years of studies in the various desert regions of the republic.

Desert vegetation strongly depends on climatic conditions. In the deserts of Turkmenistan, the severe water deficiency (annual precipitation does not exceed 120 to 130 mm, and in some regions is even as low as 75 mm), an unequal seasonal distribution of this scarce precipitation, and high summer temperatures have resulted in many specific evolutionary adaptations in desert plants. For instance, desert plants may partially lose leaves (or lose assimilating branches that replace leaves) in the summer; leaves may possess protective hairs or wax cover; fewer stomata may be present than in non-desert species, and their size may be smaller or they may become embedded in the leaf tissue; the root system can be extremely developed and reach the level of ground water or the horizon of capillary moisture; and a system of accessory roots may develop in plants growing on sand dunes.

Characteristics of Desert Vegetation

Deserts of Turkmenistan present various types of habitats and plant communities. The vegetation in the sand desert of Karakum is dominated by such shrub species as saksaul, cherkez, and kandym, with an herbaceous cover of sand sedge (*Carex physodes*) and ephemerous plants. The black saksaul (*Haloxylon aphyllum*) sometimes forms peculiar "desert forests." Clay deserts (*takyrs*) are almost devoid of vascular plants but possess specific communities of algae and lichens. Vast clay and gravel plateaus of West Turkmenistan are dominated by semishrub sagebrush-halophyte communities. Below, we present a classification scheme for the desert vegetation in Turkmenistan (Table 1), and give characteristics of its formations and associations. Each subdivision in this classification has a number code referred to in the text.

1. Euxerophyte Desert Vegetation

This vegetation type embraces the most common plant communities found both in the sand desert and low plateaus. Within this type, we distinguish between two classes of formations: desert semishrub and small shrub vegetation (1.1) and desert shrub and large shrub vegetation (1.2). The most common small semishrub communities are characterized by formations of sagebrushes and halophytes. Shrub and large shrub vegetation is found in sand deserts and on thin *kyr* sands underlaid by maternal rocks. In the Karakum and Chilmamedkum sand deserts, the most characteristic communities are those of saksaul and psammophite shrubs (the latter dominated by cherkez, dzhuzgun, or syuzen).

1.1. Desert Semishrub and Small Semishrub Vegetation

This class of formations includes communities dominated by a typical desert ecobiomorph of small semishrubs; these are predominantly various sagebrush (*Artemisia*) species of the subgenus *Seriphidium*, species of *Salsola* (*S. gemmascens* and *S. orientalis*), biyurgun (*Anabasis salsa* and *A. ramosissimum*), and sarsazan (*Halocnemum strobilaceum*). Desert small semishrub vegetation is diverse and dominates the desert vegetation. Within this class of formations, we

Table 1. Classification of desert vegetation of Turkmenistan

Vegetation type	Class of formations	Group of formations	Formation	Groups of associations
1. Euxerophyte desert vegetation	1.1. Desert semishrub vegetation	1.1.1. Sagebrush deserts	1.1.1.1. Kemrud sagebrush formation (*Artemisia kemrudica*)	typical Kemrud sagebrush associations; ephemerous and ephemeroid Kemrud sagebrush associations; psammophyte Kemrud sagebrush associations
			1.1.1.2. Badghyz sagebrush formation (*Artemisia badhysi*)	typical Badghyz sagebrush associations; ephemerous and ephemeroid sagebrush associations
		1.1.2. Halophyte semishrub deserts	1.1.2.1. Tetyr formation (*Salsola gemmascens*)	typical tetyr associations; ephemerous and ephemeroid tetyr associations
			1.1.2.2. Kevreik formation (*Salsola orientalis*)	typical kevreik associations
			1.1.2.3. Biyurgun formation (*Anabasis salsa*)	typical biyurgun associations
		1.1.3. Succulent-halophyte deserts	1.1.3.1. Sarsazan formation (*Halocnemum strobilaceum*)	typical sarsazan associations
		1.1.4. Small halophyte shrub deserts	1.1.4.1. Boyalych formation (*Salsola arbuscula*)	typical boyalych associations; ephemerous and ephemeroid boyalych associations
			1.1.4.2. *Reaumuria* formation (*Reaumuria* spp.)	typical *Reaumuria* associations; ephemerous and ephemeroid *Reaumuria* associations

Table 1. Continued

Vegetation type	Class of formations	Group of formations	Formation	Groups of associations
	1.2. Desert shrub and large shrub vegetation	1.2.1. Saksaul deserts	1.2.1.1. White saksaul formation (*Haloxylon persicum*)	typical (shrub) white saksaul associations
			1.2.1.2. Black saksaul formation (*Haloxylon aphyllum*)	typical (shrub) black saksaul associations; ephemerous and ephemeroid black saksaul associations; moss/black saksaul associations
			1.2.1.3. Mixed saksaul formation (*Haloxylon persicum* and *Haloxylon aphyllum*)	typical mixed saksaul associations
		1.2.2. Psammophyte shrub deserts	1.2.2.1. Syuzen formation (*Ammodendron conollyi*)	typical syuzen associations
			1.2.2.2. Dzhuzgun formation (anthropogenic) (*Calligonum* spp.)	typical dzhuzgun associations
			1.2.2.3. Cherkez formation (*Salsola arbuscula*)	typical cherkez associations
			1.2.2.4. Bordzhok formation (*Ephedra strobilacea*)	typical bordzhok associations
2. Mesoxerophyte desert vegetation	2.1. Desert herbaceous vegetation	2.1.1. Large perennial herbaceous vegetation	2.1.1.1. Selin formation (*Aristida pennata*)	
			2.1.1.2. Yuzarlik formation (*Peganum harmala*)	

Table 1. Continued

Vegetation type	Class of formations	Group of formations	Formation	Groups of associations
			2.1.1.3. *Agropyron* formation (*Agropyron fragile*)	
		2.1.2. Ephemerous-grass vegetation	2.1.2.1. Yepelek formation (*Anisantha tectorum*)	
			2.1.2.2. Arpagan formation (*Eremopyrum orientale*)	
		2.1.3. Annual halophyte vegetation	2.1.3.1. Ebelek formation (*Ceratocarpus utriculosus*)	
			2.1.3.2. Ketgen formation (*Salsola paulsenii*)	
3. Psychroxerophyte desert vegetation	3.1. Desert thallomous vegetation	3.1.1. Algal vegetation (on takyrs)		
		3.1.2. Lichen vegetation (on takyrs)		
		3.1.3. Moss vegetation		

separate three groups: sagebrush deserts, halophyte small semishrub deserts, and succulent-halophyte deserts.

1.1.1. Sagebrush Deserts

Sagebrush communities are typical in the deserts of Turkmenistan, where they are widespread and well-studied (Prozorovsky 1940; Rodin 1940, 1963; Korovin and Granitov 1949; Momotov 1953; Kogan 1954; Nechaeva 1956; Rodin and Rubtsov 1956; Rachkovskaya 1957; Korovin 1961; Rustamov 1962; Granitov 1967). Sagebrush communities are present on the vast lowlands of the Krasnovodsk and Ustyurt Plateaus, in the modern and old deltas of the Amudarya River, and in the Trans-Unguz Karakum. In the Karakum Desert proper, relatively homogeneous sagebrush communities are found in lowland and small-dune sands. Formations of sagebrush are represented also in the underhill lowland and foothills of Kopetdagh (Rodin 1963). Soils under sagebrush communities are loams, sandy loams, grey-brown soils, and sometimes light serozyoms.

Dominant in sagebrush communities are *Artemisia kemrudica*, *A. badhysi*, *A. badhysi* var. *arenicola*, *A. halophida*, and *A. santolina*. The most characteristic, widespread, and prevailing formation is that of *Artemisia kemrudica* (Table 2).

Table 2. The vegetation of the formation *Artemisieta kemrudicae*

Species	Height (cm)	Abundance (Drude scale)	Coverage (%)	Density (plants/ha)
Shrubs:				
Salsola richteri	70	Sol	1–2	200
Calligonum alatum	40–60	Sol	1–2	200
Salsola arbuscula	30–80	Sp-Cop^1	5–15	1,400–3,800
Small shrubs:				
Ephedra distachya	10–30	Sol	1	200
Semishrubs:				
Astragalus turcomanicus	35–40	Sol	1	200
Small semishrubs:				
Halothamnus subaphyllus	25	Sol	<1	100–200
Salsola orientalis	20–30	Sp	1–2	100–700
Salsola gemmascens	10–15	Sp	1–2	1,400–3,800
Artemisia kemrudica	20–45	Cop$^{1-3}$	20–55	6,000–22,000
Perennial herbaceous species:				
Stipagrostis pennata	20–40	Sol	1–2	–
Iris longiscapa	10–20	Sol	<1	–
Carex pachystylis	10–20	Cop$^{1-3}$	2–5	–
Carex physodes	15–20	Sp	2–5	–
Gagea reticulata	8–15	Sol-Sp	<1	–
Allium fibrosum	15–20	Sol	<1	–
Tulipa sogdiana	10–12	Sol	<1	1
Ferula foetida	35–40	Sp	2–3	500
Annual herbaceous species:				
Ceratocarpus utriculosus	5–10	Sol	<1	–

Table 2. Continued

Species	Height (cm)	Abundance (Drude scale)	Coverage (%)	Density (plants/ha)
Climacoptera lanata	15–20	Sol	<1	–
Ceratocephala falcata	2–5	Sol	<1	–
Hypecoum pendulum	10–20	Sol	<1	–
Roemeria hybrida	10	Sol	<1	–
Goldbachia laevigata	8–10	Sol	<1	–
Strigosella grandiflora	25–30	Sp	1	–
Strigosella sp.	10–15	Sol	<1	–
Tetracme quadricornis	10–12	Sol	<1	–
Isatis minuta	30–40	Sp	1	–
Leptaleum filifolium	4–5	Cop^1	1	–
Astragalus oxyglottis	5–6	Sol	<1	–
Erodium oxyrrhynchum	5–6	Sol	<1	–
Lappula semiglabra	20	Sol	1	–
Nonea caspica	5–10	Sol	<1	–
Arnebia decumbens	5–20	Sol-Sp	<1	–
Koelpinia linearis	5–20	Sol	1	–
Senecio subdentatus	8–12	Sol	<1	–
Amberboa turanica	5–10	Sol	<1	–
Eremopyrum orientale	10–15	Sp-Cop^2	1–2	–
Anisantha tectorum	15–20	Sol-Sp	1	–
Atriplex dimorphostegia	2–3	Sol	<1	–
Lallemantia royleana	3–4	Sol	<1	–

The Kemrud sagebrush communities occupy large areas in Northwest Turkmenistan, on the Krasnovodsk, Ustyurt, and Beltau Plateaus, and in the modern and old deltas of the Amudarya River, and are found on grey-brown sandy loams and light loams, covered by gravel, with gypsum-bearing rock located at a depth of 50 to 70 cm. These communities are rather homogeneous in structure: the upper layer, when present, is 60 to 100 cm high and consists of sparse shrubs of low *Haloxylon aphyllum* and *Calligonum* spp.; sometimes, on sandy soils, it includes sparse *Salsola richteri* and *Halothamnus subaphyllum* var. *arenaria*.

The major layer is the second one, built by *Artemisia kemrudica*, with participation of *Salsola arbuscula*, *S. gemmascens*, and occasional *Astragalus turcomanicus*. This layer is usually 20 to 40 cm; however, such species as *Salsola arbuscula* and *Astragalus turcomanicus* often form a separate superlayer up to 60 cm or higher. The dominant species, *Artemisia kemrudica*, is always outstanding and conspicuous. This plant is 30 to 60 cm high; its perennial portion is up to 15 cm long, and annual one is 15 to 30–35 cm long. During wet years, the annual shoots of this small semishrub can reach maximal length of 43 cm (Nechaeva 1956), whereas in dry years their length is 6 to 20 cm long. Lifespan of the Kemrud sagebrush is 15 to 20 years.

The vegetative period of *Artemisia kemrudica* begins in late February to early

March; growth occurs in April and May; flowering begins in late August and continues until September; seed production occurs in October. The root system of sagebrush changes significantly with age. In young plants, the main root is well developed, going down to 60 cm and producing a number of side roots. Later, the sagebrush root system develops primarily as side roots, and the main root almost completely disappears; this developmental peculiarity is a result of the differentiation of branches of an original plant which receives nutrition through side roots. The diameter of a sagebrush root system is about two meters, and the system can penetrate to a depth of 70 to 80 cm.

Also typical for *Artemisia kemrudica* growth is the change in aboveground to underground biomass ratio: juvenile plants can have two to three times more biomass above ground than underground (Nechaeva *et al.* 1973), whereas in adult plants these biomasses are almost equal.

Among other species that can subdominate in sagebrush communities are *Salsola arbuscula*, *S. gemmascens*, and *S. orientalis*; their abundance, however, is insignificant in ephemerous and ephemeroid Kemrud sagebrush communities. About one to two per cent of coverage in the main sagebrush layer can be contributed by *Ephedra distachya*, *Convolvulus* sp., and *Haplophyllum ramosissimum*. Rodin (1963) gives a total list of 20 species of semishrubs found in the Kemrud sagebrush communities; on the description plots this number does not exceed eight species.

The lowest (third) layer in the Kemrud sagebrush communities consists of herbaceous, primarily ephemerous and ephemeroid vegetation. A total list of 20 to 25 species is given for sagebrush pastures of Northwest Turkmenistan by Nechayeva (1956). On concrete plots there are five to ten, rarely up to fifteen, species of ephemers and ephemeroids. They develop primarily in the spring although growth is highly dependent on annual weather conditions.

Within the Kemrud sagebrush formation, four groups of associations can be found: typical, semisavanna, ephemerous and ephemeroid, and psammophyte. Species diversity is highest in typical and ephemerous and ephemeroid Kemrud sagebrush associations (up to 50 species of plants and more). More than 50% of this list consists of annual species, primarily ephemerous ones; plants with a short vegetative period (ephemers and ephemeroids combined) represent 55 to 66% of the species list in these communities. There are six species of semishrubs and small semishrubs, and five to six species of shrubs and small shrubs (altogether 22 to 24% of the species list). Less diverse are psammophyte Kemrud sagebrush communities, but here also more than half (54%) of the species are plants with a short vegetative period. In total, 60% of species present in the Kemrud sagebrush formation have a short vegetative period, and 22% are shrubs and semishrubs.

1.1.2. Halophyte Small Semishrub Deserts
This group of formations includes the typical desert zone small semishrubs tetyr (*Salsola gemmascens*), biyurgun (*Anabasis salsa*), and kevreik (*Salsola orientalis*).

1.1.2.1. Tetyr Formation (Salsola gemmascens) (table 3). Although tetyr formation is not the most widespread, it is one of the most typical desert formations of vegetation. Communities of *Salsola gemmascens* are represented in the Uzboi dry bed, in the southern Ustyurt and Krasnovodsk Plateaus next to the Karabogazgol Bay of the Caspian, and in the western Trans-Unguz area. Patches of tetyr communities are also found throughout the Karakum Desert in takyrs and takyr-like habitats among sand dunes.

S. gemmascens is found on grey-brown *solonets* loams and sandy soils bearing gypsum as well as on grey-brown primitive soils of takyrs (the latter may have a certain percentage of gravel and gypsum).

The dominant species of this formation, *Salsola gemmascens*, is a small (30 to 50 cm) shrub which is a typical xerophyte and, to a certain degree, a halophyte. Individual plants live for 10 to 15, and sometimes to 20 or 25 years, and can

Table 3. The vegetation of the formation *Salsoleta gemmascentes*

Species	Height (cm)	Abundance (Drude scale)	Coverage (%)	Density (plants/ha)
Shrubs:				
Salsola arbuscula	20–40	Sp	2–3	600
Small semishrubs:				
Salsola orientalis	20–30	Sol-Sp	2–3	600–800
Salsola gemmascens	10–15	Cop^{2-3}	10–25	7,600–28,100
Artemisia kemrudica	20–30	Sol-Sp	1–2	300–500
Perennial herbaceous species:				
Heliotropium sp.	n/a	Sp	<1	–
Astragalus xiphioides	5–10	Sol	<1	–
Carex pachystylis	10–20	Cop^{1-2}	1–2	–
Gagea reticulata	5–8	Sp	<1	–
Allium sp.	15–20	Sol	<1	–
Tulipa sogdiana	5–10	Sol	<1	–
Annual herbaceous species:				
Halimocnemis villosa	5–6	Sp	<1	–
Ceratocephala falcata	2–4	Cop^1	1	–
Hypecoum pendulum	8–15	Sol	<1	–
Roemeria hybrida	8–10	Sol	<1	–
Strigosella grandiflora	20–30	$Sp-Cop^2$	1–2	–
Strigosella sp.	5–7	Sol	<1	–
Leptaleum filifolium	3–15	$Sp-Cop^2$	1–2	–
Astragalus arpilobus	3–5	Sp	1	–
Astragalus oxyglottis	3–6	Sol	<1	–
Lappula semiglabra	5–12	Sol	<1	–
Nonea caspica	5–15	Sol	<1	–
Arnebia decumbens	8–10	Sol	<1	–
Koelpinia linearis	10–15	Sol	<1	–
Senecio subdentatus	20–25	Sp	<1	–
Eremopyrum orientale	7–20	$Sol-Cop^1$	1–2	–
Epilasia hemilasia	10	Sol	<1	–

annually produce up to 600 seeds. Seven- to ten-year old plants form an expressed crown and root system (which is superficial and reaches 70 to 100–125 cm; Nechaeva *et al.* 1973).

Habitus and condition of *Salsola gemmascens* vary with soil conditions. On the gravel- and gypsum-bearing soils of Krasnovodsk Plateau and areas south from Ustyurt, tetyr plants are depressed and small-sized (7 to 15 cm); their crown is poorly developed. On the other hand, on grey-brown soils with low content of gypsum and disrupted surface *S. gemmascens* grows more vigorously, develops a normal crown, and reaches a height of 20 to 40 cm. Depending on the habitat, coverage of tetyr can vary from 10 to 25%, and its density varies from 12,000–16,000 to 24,000–28,000 plants/ha (our data; Pelt 1956).

Floristically, this formation is relatively poor (the total list has 30 to 35 species; the concrete plots house 12 to 17 species). As in other desert formations, most of these species (especially ephemerous ones) are annual herbaceous plants (62 to 64% of the total list). The combined number of annual and perennial herbaceous species with spring and fall vegetative period (i.e., ephemers and ephemeroids) represent 57 to 75% of all species in the tetyr formation. Semishrubs and small semishrubs contribute 11 to 13%, and shrubs and small shrubs, only 4 to 8% of the species list. Certain shrubs that participate in this formation, such as *Calligonum setosum*, *Salsola arbuscula*, and *Haloxylon aphyllum*, are so depressed there that they rarely reach a height of 50 or 60 cm.

Vertical structure in tetyr communities is weakly expressed; it basically comprises two layers, but the lower layer is expressed only in the spring period, as in the sagebrush communities. The upper layer (20 to 40 cm high) is composed of *S. gemmascens*, with some participation of *Artemisia kemrudica*, *Salsola orientalis*, and *S. arbuscula*. Plants of the latter species can be taller than *S. gemmascens* but are sparse. The lower (herbaceous) layer, formed by 20 to 23 species of ephemers and ephemeroids, is no higher than 20 cm. The dominant species here is the desert sedge *Carex pachystylis*; other co-dominant ephemers are *Eremopyrum orientale*, *Ceratocephala falcata*, and the species *Leptaleum filifolium*, which is specific for some tetyr communities. In summer and fall, there is significant participation of annual plants with extended vegetative periods, such as *Climacoptera lanata*, *Salsola sclerantra*, *Halimocnemus villosa*, and *H. karelinii*. Soil in tetyr communities is covered by clusters of specific flour-white lichens.

In total, the tetyr formation is one of the most "desert" kinds, judging from the sparse coverage, poor species diversity, and low biomass/ha values. Within this formation, four goups of associations can be separated: typical, petrophyte, psammophyte, takyr, and halophyte communities of *S. gemmacsens*. Of these, typical tetyr associations are the most common ones.

*1.1.2.2. Biyurgun Formation (*Anabasis Salsa*) (table 4).* Biyurgun formation is widespread in the deserts of Middle Asia and Kazakhstan, especially in the subzone of northern deserts, and is fairly well described and studied. Kuznetsov (1959, 1966) conducted a study of biyurgun formation for the entire arid zone

Table 4. The vegetation of the formation *Anabaseta salsae*.

Species	Height (cm)	Abundance (Drude scale)	Coverage (%)	Density (plants/ha)
Shrubs:				
Atraphaxis spinosa	40–50	Sol	1	–
Semishrubs:				
Nanophyton erinaceum	5–10	Cop^1	4–5	2,500–5,600
Salsola gemmascens	15–20	Sp-Cop^1	2–3	3,000–5.000
Salsola orientalis	25–40	Sol-Sp	1	300–600
Artemisia kemrudica	20–25	Cop$^{1-2}$	2–3	2,000–3,500
Anabasis salsa	15–30 (40)	Cop^3	30–35	12,000–50,000
Anabasis eriopoda	10–15	Cop$^{1-3}$	20–55	6,000–22,000
Annual herbaceous species:				
Ceratocarpus utriculosus	5–10	Sol	<1	–
Suaeda arcuata	20–30	Sol	<1	–
Climacoptera lanata	10–12	Sol	<1	–
Halimocnemis karelinii	5–7	Sol	<1	–
Ceratocephala falcata	2–3	Sol	<1	–
Strigosella africana	5–6	Cop^1	1	–
Leptoleum filifolium	3–5	Sol	<1	–
Goldbachia laevigata	5–10	Sol-Sp	<1	–
Tetracme quadricornis	5–10	Sol	<1	–
Arnebia decumbens	5–10	Sol	<1	–
Nonea caspica	5–10	Sol-Sp	<1	–
Senecio subdentatus	5–8	Sol	<1	–
Amberboa turanica	5–10	Sol	<1	–
Eremopyrum orientale	5–7	Sol	<1	–
Lepidium perfoliatum	5–15	Sol	<1	–

the former USSR; in separate deserts, this formation was studied in detail by Korovin and Granitov (1949), Kogan (1954), Kubanskaya (1956) and Rodin (1963).

The communities of *Anabasis salsa* are especially characteristic for the Ustyurt Plateau and takyrs of the ancient delta of the Amudarya; small patches of biyurgun are found also within the sand dunes of the Trans-Unguz Karakum. Biyurgun formation is found on grey-brown, solonchaks, loams and grey-brown primitive (takyr) soils with a compressed surface covered by a film of algae.

The dominant species, *Anabasis salsa*, is a small shrub or semishrub, 10–15 to 20–40 cm high; pure *A. salsa* communities cover 20 to 30% of the soil surface, with the density 40,000 to 50,000 plants/ha. This number can drop to 14,000 to 18,000 plants/ha in the communities when other species co-dominate (e.g., *Artemisia kemrudica* or *Salsola gemmascens*; Nechaeva 1956); generally, the density can vary from 10,000 to 80,000 plants/ha (Korovin and Granitov 1949).

Communities of *A. salsa* usually appear to have only one (upper) layer (15 to 30 cm high) formed primarily by biyurgun, rarely in combination with *Artemisia kemrudica*, *Salsola gemmascens*, or *S. orientalis*. Another species typical of this layer (but somewhat smaller, 5 to 15 cm high) is *Nanophyton erinaceum*.

A lower, herbaceous layer may sometimes be expressed. It is homogeneous and includes such ephemers as *Leptaleum filifolium*, *Strigosella africana*, *Lepidium perfoliatum*, *Goldbachia laevigata*, *Ceratocephala falcata*, *Arnebia decumbens*, and *Eremopyrum orientale*, as well as annual plants with summer and fall vegetative periods such as *Climacoptera lanata*, *Suaeda arcuata*, and *Halimocnemis karelini*. The diversity of species within this formation in Turkmenistan deserts is low (20 to 25); interestingly, Kubanskaya (1956) found 122 species for biyurgun formation in the Betpakdlala Desert (Kazakhstan), and Kuznetsov (1959) listed 218 species found within this formation throughout the arid zone of the former Soviet Union.

Annual herbaceous species constitute 40 to 90% of the species list, with annual ephemers representing 30 to 60%. The absence of perennial herbaceous species (including ephemeroids) is notable, and there are one to five (10 to 50%) species of semishrubs. Four groups of associations are separated within the biyurgun formation: typical, petrophyte, and takyr associations.

*1.1.2.3. Kevreik Formation (*Salsola Orientalis*) (table 5).* These communities are not as well studied as others, probably due to their limited distribution. Granitov (1967) found that *S. orientalis* is relatively common in the Southeast Kizylkum Desert (Uzbekistan) but rarely dominates plant communities. The kevreik formation was studied by Korovin and Granitov (1949), Nechaeva (1956), Rodin (1963), and Granitov (1967). In Turkmenistan, kevreik communities are common in the ancient alluvial plain of the Kunyadarya River and in the Meshed-Messerian Plain (Rodin 1963). Large areas covered by kevreik communites were recorded on the *kyr* plateau which lies between Koimat and the Uchtagan Sands (Nechaeva 1966), as well as in the Tashauz Region of Northwest Turkmenistan, near Edikhauz and Butentau. These communites grow mainly on the lands of ancient irrigation (Rodin 1963), on grey-brown, slightly loamy or sandy soils (rarely on sands).

The dominant species, *Salsola orientalis*, is a small semishrub, 30 to 50 cm high; it covers 20 to 30% of the soil surface and has a density of 5,000–7,000 to 10,000–12,000/ha. Community structure is similar to that of *S. gemmascens*; the upper layer is formed by *Salsola orientalis*, with sparse *Artemisia kemrudica*, *Salsola arbuscula*, and *S. gemmascens*. Rare *Haloxylon aphyllum* can form a sparse superlayer. The second (lower) layer is built of ephemers (15 species), ephemeroids, and certain annual Chenopodiaceae with a summer – fall vegetative period. In spring, the most conspicuous plants are *Carex pachystylis*, *Ceratocephala falcata*, *Leptaleum filifolium* and *Eremopyrum orientale*. Among less abundant species are *Strigosella grandiflora*, *S. circinata*, *Astragalus oxyglottis*, *Lappula semiglabra*, *Arnebia decumbens*, and *Koelpinia linearis*. In summer and fall, the annual species of Chenopodiaceae found in these communities are *Climacoptera lanata*, *Salsola sclerantha*, *Halimocnemis karelinii*, *Ceratocarpus utriculosus*, and *Girgensonnia oppositiflora*. Floristically,

Table 5. The vegetation of the formation *Salsoleta orientales*

Species	Height (cm)	Abundance (Drude scale)	Coverage (%)	Density (plants/ha)
Trees:				
Haloxylon aphyllum	80–100	Sol	1	–
Shrubs:				
Salsola arbuscula	40	Un	<1	–
Small semishrubs:				
Salsola orientalis	35–40	Cop²	20–25	5,000–12,000
Salsola gemmascens	10–20	Sp	2–3	2,300–3,400
Artemisia kemrudica	20–30	Sp	1–2	1,400–2,000
Anabasis eriopoda	10–15	Sol-Sp	1	–
Perennial herbaceous species:				
Carex pachystylis	10–150	Cop¹	1–2	–
Annual herbaceous species:				
Ceratocarpus utriculosus	5–10	Sol	<1	–
Climacoptera lanata	10–15	Sol	<1	–
Salsola sclerantha	10–12	Sol	<1	–
Halimocnemis karelinii	5–6	Sol	<1	–
Ceratocephala falcata	2–3	Sp-Cop¹	1	–
Hypecoum pendulum	10–15	Sol	<1	–
Strigosella grandiflora	20–25	Sp	1–2	–
S. circinata	10–15	Sp	<1	–
Leptaleum filifolium	5–10	Sp-Cop¹	1	–
Tetracme quadricorris	5–10	Sol	<1	–
Astragalus oxyglottis	5–6	Sol-Sp	<1	–
Arnebia decumbens	8–10	Sol	<1	–
Lappula semiglabra	10–12	Sol	<1	–
Nonea caspica	5–10	Sol	<1	–
Koelpinia linearis	10–12	Sol	<1	–
Epilasia hemilasia	10–15	Sol	<1	–
Amberboa turanica	5–10	Sol-Sp	<1	–
Eremopyrum orientale	5–6	Sol	<1	–
Girgensonnia oppositiflora	10	Sol-Cop¹	1	–

this formation is not very different from that of *S. gemmascens*; the list of species for kevreik formation includes 20 to 60 species (Rodin 1963; Rustamov 1973). Of these, 66 to 75% are annual plants (mostly ephemers), and 13 to 27 species are small semishrubs. Perennials are represented here only by *Carex pachystylis*.

Within the kevreik formation, we separate typical and psammophyte kevreik groups of associations.

1.1.3. Succulent-Halophyte Deserts
Communities belonging to this group of formations are widespread on salt areas (solonchaks) and solonchak soils. The most typical formation is that of

*1.1.3.1. Sarsazan (*Halocnemum Strobilaceum*) (table 6).* Characteristics of these communities can be found in many sources (Prozorovsky 1940; Korovin

Table 6. The vegetation of the formation *Halocnemeta strobilaceae*

Species	Height (cm)	Abundance (Drude scale)	Coverage (%)	Density (plants/ha)
Trees:				
Haloxylon aphyllum	40–70	Sp	2–3	100–300
Shrubs:				
Tamarix sp.	40–60	Sol	1–2	200–400
Halostachys caspica	60–80	Sol	1–2	200–300
Nitraria schoberi	60–70	Sol	1	200
Lycium ruthenicum	30–40	Sol	1–2	200–300
Reaumuria fruticosa	40–50	Sol	1–2	200
Salsola arbuscula	40	Sol	1–2	100–300
Small shrubs:				
Limonium subfruticosum	35–40	Sol	1–2	200–400
Small semishrubs:				
Halocnemum strobilaceum	20–40	Cop$^{1-3}$	10–20 (25)	700–4,500
Salsola gemmascens	10–15	Sol	1	300
Perennial herbaceous species:				
Frankenia hirsuta	20–30	Sol-Sp	<1	–
Alhagi persarum	20–30	Sol	<1	–
Annual herbaceous species:				
Climacoptera lanata	10–20	Sol-Sp	<1	–
Salsola sclerantha	10–15	Sol	<1	–
Halimocnemis longifolia	10–15	Sol-Sp	<1	–
Petrosimonia glauca	10	Sol	<1	–

Granitov 1949; Kogan 1954; Kubanskaya 1956; Korovin 1961; Rustamov 1962; Rodin 1963; Granitov 1967); however, only a few of these authors published descriptions of concrete plots and floristic lists.

Sarsazan communities are found on typical solonchaks and in solonchak depressions covered by specific small salt hills (*chokalaks*). The largest areas occupied by *H. strobilaceum* are located in Southwest Turkmenistan around the ancient delta of the Atrek River, along the Kelkor solonchak, and on the shores of Karabogazgol Bay. These communities can be also found in combinations with other succulent-halophyte desert vegetation. In the Karakum Desert, small areas occupied by *H. strobilaceum* are typical for the solonchak depressions with close ground waters.

The dominant species, *Halocnemum strobilaceum*, is a stem succulent, leafless, small semishrub, 20 to 40 cm high. Its growth is not depressed even with high salt concentration in the soil. Roots of sarsazan can penetrate to the depth from 40–50 cm (Rumyantseva 1953) to 130–140 cm if ground water lies deep (Rustamov 1962). Sarsazan plants cover 10 to 20% of the soil surface, very rarely up to 25%. Its density varies from 700–1,200 to 2,200–4,500 plants/ha.

Typical sarsazan communities have only one layer, often formed exclusively by *H. strobilaceum*. Its characteristic flat crowns can be seen on small (50 to 100 cm high) hills of salt (*chokalaks*) standing two to three meters apart; there is virtually no vegetation between chokalaks. When mineralized groundwater lies at deeper levels, sarsazan communities can include a number of species less tolerant to salt concentrations, such as *Halostachys caspica, Limonium subfruticosum, Nitraria schoberi,* and *Frankenia hirsuta.* Very rarely are found depressed shrubs of *Tamarix hispida, Haloxylon aphyllum,* and *Lycium ruthenicum.* Under lower soil salt content, the herbaceous cover is built mostly of annual Chenopodiaceae: *Climacoptera lanata, Salsola sclerantha, Halimocnemis longifolia,* and *Petrosimonia glauca*; sometimes, *Alhagi persarum* is present. Usually, there are no herbaceous species with winter-spring development. Diversity in sarsazan communities in Turkmenistan is very low, totalling about 15 species. Granitov (1967) described fifteen sarsazan associations from the Kizylkum Desert (Uzbekistan); of these, thirteen had one to sixteen species, and only two were unusually rich (31 and 39 species, including 5 or 6 species of ephemers). Kubanskaya (1956), however, listed 59 species for the sarsazan formation in the Betdpakdala Desert (Kazakhstan), and 53 species for the sarsazan association proper (including five species of ephemers and ephemeroids); interestingly, the coverage on these plots reached 50 to 65% which is a very significant figure for the communites of *H. strobilaceum.*

About 50% of all species in sarsazan communites are arboreal plants (mostly shrubs and semishrubs), about 25% are annual non-ephemerous species, and there are no ephemers. Even fewer species (13%) are small semishrubs and perennial herbaceous species.

Within the sarsazan formation, we separate two groups of associations: typical and meadow sarsazan associations.

1.2. Desert Shrub and Large Shrub Vegetation

This class of formations is represented by two groups: saksaul deserts (1.2.1) and psammophyte shrub deserts (1.2.2.). Desert shrub and large shrub vegetation is widespread in sand deserts, in clay desert (takyr) lowlands, and in modern and ancient river deltas.

1.2.1. Saksaul Deserts
This group includes formations of white saksaul (*Haloxylon persicum*), black saksaul (*Haloxylon aphyllum*), and a mixed formation with both *Haloxylon persicum* and *H. aphyllum.*

1.2.1.1. White Saksaul Formation (Haloxylon persicum) (table 7). This is the most characteristic formation of the sand dunes of the Karakum Desert. Within Turkmenistan, the white saksaul communities are widespread in the Trans-Unguz, Lowland and Southeast Karakum, as well as in the sand massifs of Uchtagan, Kumsebshen, and Chilmamedkum; they are found typically in sand dunes, more rarely in lowlands, depressions, and intradune depressions. These communities are found not only on sands proper, but also on the thick sand deposits covering maternal rocks. *Haloxylon persicum* also grows on weakly

Table 7. The vegetation of the formation *Haloxyleta persica*

Species	Height (cm)	Abundance (Drude scale)	Coverage (%)	Density (plants/ha)
Shrubs:				
Haloxylon persicum	140–250	Cop$^{1-3}$	15–25	400–900
Calligonum caput-medusae	140–200	Sol	2–3	100–200
Calligonum setosum	80–110	Sol-Sp	2–4	100–300
Salsola richteri	140	Sol	1–2	100
Ephedra strobilacea	50–100	Sol-Sp	2–3	200–400
Perennial herbaceous species:				
Stipagrostis pennata	40–50	Sp	2–3	100–300
Astragalus flexus	25–30	Sol	<1	–
Rheum turkestanicum	20–30	Sol-Sp	1–2	–
Eremurus anisopterus	30–40	Sol	1	–
Carex physodes	15–20	Cop$^{1-3}$	10–15	–
Gagea divaricata	5–7	Sol-Sp	<1	–
Tulipa sogdiana	10–15	Sol-Sp	<1	–
Annual herbaceous species:				
Ceratocephala falcata	5–6	Sp-Cop^1	1	–
Consolida rugulosa	30–35	Sol	<1	–
Hypecoum pendulum	20–30	Sol-Sp	1	–
Roemeria hybrida	25–40	Sol	<1	–
Streptoloma desertorum	10–20	Sol	<1	–
Isatis minima	50–60	Sol	<1	–
Strigosella circinnata	30–40	Sol-Sp	<1	–
Strigosella grandiflora	40–55	Sol	<1	–
Tetracme recurvata	20–25	Sol	<1	
Astragalus arpilobus	10–15	Sp-Cop^1	2–3	–
Erodium oxyrrhynchum	15–25	Sol-Sp	<1	–
Arnebia decumbens	20–25	Sol-Sp	<1	–
Lappula semiglabra	20	Sol	<1	–
Nonea caspica	15–20	Sol	<1	–
Koelpinia linearis	25	Sol	<1	–
Microcephala lamellata	10–12	Sol	<1	–
Senecio subdentatus	20–25	Sol-Sp	<1	–
Amberboa turanica	25–30	Sol	<1	–
Epilasia hemilasia	25–30	Sol	<1	–
Eremopyrum orientale	15–20	Sol	<1	–
Anisantha tectorum	25–30	Sp	1	–
Cutandia memphitica	20–30	Sol-Sp	<1	–

developed sandy soils of grey-brown type, which are humus-poor and sometimes low in salt content.

The communities of *H. persicum* have probably the most complex structure of all the desert plant communities of Turkmenistan; they contain several layers – at least two or three. The upper (first) layer is 1.5 to 2 meters high and consists of *H. persicum* and other large shrubs such as *Salsola richteri, Calligonum caput-medusae, C. setosum, C. eriopodum,* and *Ephedra strobilacea.* The second layer is represented by shrubs and semishrubs which are up to one meter high:

Ephedra intermedia, Artemisia kelleri, and *Astragalus spp.* A special sublayer can be formed by small semishrubs such as *Convolvulus divaricatus, C. korolkovii,* and *Acanthophyllum* sp.

The third (herbaceous) layer is formed by a large cespitose grass, *Stipagrostis pennata,* as well as by other perennials (*Heliotropium argusioides, Tournefortia sibirica,* and *Astragalus chivensis* and biennial *Cousinia oxiana*). Among perennial species with short vegetative periods (ephemeroids), a significant role is played by a desert sedge, *Carex physodes,* which creates thick turf; also found are *Rheum turkestanicum* and *Eremurus anisopterus*. Most herbaceous species are ephemerous (especially in years with high precipitation, when up to 30 or more species can be detected).

The white saksaul formation, therefore, is floristically diverse and may include representatives of almost all desert ecobiomorphs. Rodin (1963) listed more than 150 species for this formation in West Turkmenistan; within the associations, this number varies from 30 to 80, and on concrete plots there are usually 30 to 35 species.

In Northwest Turkmenistan, white saksaul communities include about 70 species (our data); in Southwest Kizylkum (Uzbekistan), from 28 to 48 species (Granitov 1967); and in the sand deserts of Kazakhstan, white saksaul communities also include several dozen species (Kurochkina 1966). Annual herbaceous plants prevail (51 to 64% of all species), especially ephemers (42 to 64%). The combined share of perennial and annual herbaceous plants is often more than 50% of the species list, and in ephemerous or ephemeroid white saksaul associations this share reaches 79%. Shrubs constitute 10 to 15% of all species.

The dominant species, *Haloxylon persicum,* is a large shrub, 3 to 5 meters high. It forms a short trunk (10–20 cm) and can produce six to seven levels of branches (Nechaeva *et al.* 1973). White saksaul lives to 30 years; its vegetative propagation and seed production do not occur every year. It has a root system of universal type which penetrates down to a depth of four to six meters (Petrov 1935; Nechaeva *et al.* 1973). The ratio of aboveground to underground dry biomass is 1:0.6.

Haloxylon persicum does not form dense thickets but rather grows as solitary bushes, with a density 100–200 to 400–700 plants/ha, and coverage of 10 to 30%. White saksalul communites are valued as pastures although they produce low edible biomass (0.3 to 0.5 ton/ha); they are used in all seasons (although primarily in winter). Size of white saksaul shrubs varies significantly with ecological conditions, as do number and abundance of species in these communities. Extensive grazing and woodcutting for many years in certain areas has led to the replacement of white saksaul communities by those of kandym (*Calligonum* sp. div.). Rodin (1963) separated three groups of associations within the white saksaul formation in West Turkmenistan: typical, ephemerous and ephemeroid, and moss/white saksaul associations. In the West Uzboi area we (Rustamov 1962) found three associations: *Haloxylon persicum – Stipagrostis pennata + Carex physodes* ass.; *Haloxylon persicum – Carex physodes* ass.; and *Haloxylon persicum* + *Calligonum* sp. div. - *Stipagrostis pennata* + *Carex physodes* ass. For the Karakum Desert, Rodin (1963) listed seven common associations; the most widespread there are communities of *Haloxylon persicum* and *Carex physodes*. Within these communities, *H. persicum* is usually abundant and well developed.

1.2.1.2. Black Saksaul Formation (Haloxylon aphyllum) (table 8). These communities are found primarily in modern and ancient river deltas, as well as in depressions within the sand deserts of Middle Asia. Within Turkmenistan, the major areas covered by black saksaul are found in the lowland with ancient alluvial deposits along the Amudarya River, around the Sarykamysh Depression, in the Assake-Audan Depression (Rodin 1963), and within the sand desert of Southeast Karakum. Small areas of *H. aphyllum* can be found in the deserts of West and Northwest Turkmenistan, and also within the sands of the Trans-Unguz and Lowland Karakum, Uchtagan, and Kumsebshen.

Table 8. The vegetation of the formation *Haloxyleta aphylla*

Species	Height (cm)	Abundance (Drude scale)	Coverage (%)	Density (plants/ha)
Trees:				
Haloxylon aphyllum	60–220	Sol	1	–
Shrubs:				
Calligonum setosum	80–100	Sol	1	–
Salsola richteri	40	Un	<1	–
Ephedra strobilacea	80–100	Sol	1	–
Smirnowia turkestana	80–100	Sol	1	–
Astragalus excedens	80–100	Sol	1	–
Tamarix ramosissimum	80–100	Sol	1	–
Salsola arbuscula	80–100	Sol	1	–
Small semishrubs:				
Salsola gemmascens	40	Un	<1	–
Artemisia badhysi	40	Un	<1	–
Salsola orientalis	40	Un	<1	–
Perennial herbaceous species:				
Stipagrostis pennata	10–150	Cop[1]	1–2	–
Carex physodes	10–150	Cop[1]	1–2	–
Eremurus anisopterus	10–150	Cop[1]	1–2	–
Tulipa sogdiana	10–150	Cop[1]	1–2	–
Alhagi persarum	10–150	Cop[1]	1–2	–
Heliotropium dasycarpum	10–150	Cop[1]	1–2	–
H. argusioides	10–150	Cop[1]	1–2	–
Allium fibrosum	10–150	Cop[1]	1–2	–
Annual herbaceous species:				
Cutandia memphitica	10–15	Sol	<1	–
Eremopyrum orientale	5–6	Sol	<1	–
Halimocnemis longifolia	5–6	Sol	<1	–
Salsola incanescens	5–6	Sol	<1	–
S. sclerantha	5–6	Sol	<1	–

Table 8. Continued

Species	Height (cm)	Abundance (Drude scale)	Coverage (%)	Density (plants/ha)
Climacoptera lanata	5–6	Sol	<1	–
Ceratocephala falcata	2–3	Sp-Cop[1]	1	–
Consolida rugulosa	2–3	Sp-Cop[1]	1	–
Hypecoum pendulum	10–15	Sol	<1	–
Streptoloma desertorum	10–15	Sol	<1	–
Papaver pavoninum	10–15	Sol	<1	–
Atriplex dimorphostegia	10–15	Sol	<1	–
Senecio subdentatus	10–15	Sol	<1	–
Roemeria hybrida	10–15	Sol	<1	–
Isatis minima	10–15	Sol	<1	–
Strigosella circinnata	10–15	Sp	<1	–
Strigosella grandiflora	20–25	Sp	1–2	–
Tetracme recurvata				
Astragalus arpilobus	5–6	Sol-Sp	<1	–
Erodium oxyrrhynchum	10–15	Sp	<1	–
Arnebia decumbens	8–10	Sol	<1	–
Lappula semiglabra	10–12	Sol	<1	–
Nonea caspica	5–10	Sol	<1	–
Amberboa turanica	5–10	Sol-Sp	<1	–
Epilasia hemilasia	10–15	Sol	<1	–
Koelpinia linearis	10–12	Sol	<1	–
Matricaria lamellata	10–15	Sp	<1	–

In the modern river valleys and temporary (seasonal) river beds, communities of *Haloxylon aphyllum* form an ecological transition to riparian forests (tugais), e.g., in the lower portion of the Kunyadarya, Murghab, and Tedzhen deltas, along the Uzboi, in the Sarykamysh Depression, and in the ancient delta of the Amudarya. Small patches of *H. aphyllum* are found in the large salt depression of Eroyulanduz in Badghyz.

Communities of black saksaul are found mainly in places with close ground water and additional surface water supply, on primitive (often takyr) soils such as serozyom and grey-brown, usually solonchak or slightly solonchak, sometimes gypsum-bearing, and rarely sandy soils. *H. aphyllum* has high ecological plasticity and can form different ecobiomorphs. In valley-like depressions with close ground water, black saksaul can form peculiar forests, in which it grows as a large shrub or even a tree, seven to nine meters high (Nechaeva et al. 1973). In small depressions, *H. aphyllum* is four to five meters high and grows sparsely, forming mixed saksaul communities with *H. persicum*. A specific ecobiomorph of black saksaul, a small, one-to-two meter tall tree, grows on grey-brown soils of kyrs in Northwest Turkmenistan. In the upper portion of the Tedzhen and Murghab deltas, *H. aphyllum* grows sparsely as a small-sized (1 to 1.5 m) shrub.

The total number of species in *H. aphyllum* formation has been estimated as

132 for all communites in West Turkmenistan (Rodin 1963); in some local communities, such as those in Repetek (Southeast Karakum), species diversity can be more than 140 species (Mikhelson 1955). Our data (Rustamov 1962) showed 30 to 35 species in *H. aphyllum* communities on description plots along the western part of Uzboi. Of 132 species found in black saksaul formation listed by Rodin (1963), 40 species were shrubs and semishrubs, and 10 species were perennial herbaceous plants with prolonged vegetation; there were also 13 species of ephemeroids, 20 summer – fall annual herbaceous species, and 49 species of ephemers.

The concrete plots in West Turkmenistan usually have only 7 to 18 species (Rodin 1963). Black saksaul communities have a simple structure: the upper layer is built primarily of *H. aphyllum*, sometimes together with *Salsola richteri, Calligonum* sp. div., and *Reaumuria oxiana*. In some cases there is a sparse synusium of small shrubs, usually of *Salsola arbuscula*, and rarely of *Reaumuria fruticosa* and *Atraphaxis spinosa*. The second layer (30–40 cm) is built predominantly by small semishrubs: *Artemisia kemrudica, Salsola orientalis*, and (less often) *Salsola gemmascens*. Their density and abundance are usually low.

The third (lowest) layer is formed mainly by annual herbaceous plants; ephemeroids (except *Carex physodes*) and perennial species with a long vegetative period are nearly absent. Among the common ephemers are *Eremopyrum orientale, Anisantha tectorum, Ceratocephala falcata, Leptaleum filifolium, Arnebia decumbens, Lappula semiglabra*, and *Amberboa turanica*. Ephemeroids are represented by *Carex physodes* and *C. pachystylis*; rarely found are *Ferula assafoetida, Tulipa sogdiana*, and *Allium sabulosum*. The desert moss *Tortula desertorum* often completely covers the soil in black saksaul communities, sometimes together with lichens. The moss saksaul communities are especially typical of the interdune depressions in the Trans-Unguz Karakum. There, the black saksaul and other shrubs are sparse, small and suppressed in growth, and there are many dried branches and dead shrubs.

The dominant species of the formation, *Haloxylon aphyllum*, has a peculiar growth form. It can reach 5 to 8 meters high, with a trunk of 50 to 70 cm long and 25 to 40 cm in diameter; it produces seven to eight levels of branches. Black saksaul lives 50 to 70 years, and some individual plants have been found to be about 100 years old. It matures by age 25 or 30; seed production starts at 5 to 6 years but does not occur every year. A well-developed root system penetrates down to four to seven meters (Petrov 1935; Prikhodko 1968). The ratio of aboveground to underground dry biomass is 1:0.5, often with 100 to 150 kg of aboveground dry biomass. Under favorable conditions for seed propagation, black saksaul communities can restore themselves in the place of cuttings in about ten years. In places where *H. aphyllum* has not been cut or removed, there is usually a great amount of dried saksaul (often more than live shrubs).

Within the formation of *H. aphyllum* we (Rustamov 1973) distinguished typical, ephemer-ephemeroid, and moss associations. Rodin (1963) separated five groups of associations for West Turkmenistan: typical, tugai, petrophyte, semisavanna, and moss black saksaul associations. Of these, the typical black saksaul communities are primarily concentrated in the ancient alluvial deposits along the Amudarya River and around the Sarykamysh Depression. Tugai communities of *H. aphyllum* are widespread in the Uzboi Valley, along the dry beds of Daryalyk and Daudan, and in the Kunyadarya lowland; small patches of tugai black saksaul communities are found also on the kyrs of the Trans-Unguz area and on the plateaus of Eshek-Ankren-Kyr and Tarimkaya. The semisavanna (ephemer-ephemeroid) black saksaul communities are common in the Trans-Unguz Karakum, around the Sarykamysh, in the upper portion of the Uzboi, and in the sands of Uchtagan. Most widespread in Turkmenistan are the moss communities of *H. aphyllum* which are found in the Trans-Unguz Karakum, around the Sarykamysh, on the plateaus of Eshek-Ankren-Kyr and Tarimkaya, in the upper portion of the Uzboi, and in the sands of Uchtagan.

1.2.1.3. Mixed Saksaul Formation (Haloxylon persicum and Haloxylon aphyllum). The communities of this formation are far less common than pure white or black saksaul ones. Sometimes, however, they occupy significant areas, e.g., in the Uzboi Valley, in the eastern part of the Trans-Unguz Karakum, and in the extreme southeastern Karakum desert, between the Amudarya and Murghab Rivers. Floristically, these communities are also mixed and probably closer to those of white saksaul.

Haloxylon persicum and *Haloxylon aphyllum* co-dominate in these communities and form the upper layer; other shrubs present include *Salsola richteri*, *Ephedra strobilacea*, and various *Calligonum* species (usually *C. setosum*). Small semishrubs such as *Convolvulus korolkovii* and *Artemisia santolina* are found solitarily. Perennial long-vegetating herbaceous plants are represented by sparse *Stipagrostis pennata*; the most common are ephemeroid *Carex physodes* and various ephemers. There are also typical "patches" of halophyte species around black saksaul plants; in spring, these patches are distinguished by the presence of ephemerous species.

The number of species in mixed saksaul communities varies from 15 to 55; there are usually 5 to 20 species of ephemers and 7 to 19 species of shrubs and semishrubs. Three groups of associations are distinguished: typical, petrophyte, and moss/mixed saksaul associations.

1.2.2. Psammophyte Shrub Deserts
This group of formations is characteristic for sand dunes, including both stabilized and non-stabilized sand (barkhans). Typical formations are dominated by syuzen (*Ammodendron conollyi*), dzhuzgun (*Calligonum* sp. div.), cherkez (*Salsola richteri*), and bordzhok (*Ephedra strobilacea*).

1.2.2.1. Syuzen Formation (Ammodendron conollyi). These communities are found on non-stabilized sands in the Karakum Desert. Syuzen *(Ammodendron conollyi)*, the endemic species of Middle Asian sand deserts (Karakum and Kizylkum), dominates these communities. It is a tree, 4 to 10 meters high and 8 to 10 cm in trunk diameter, with a lifespan of 50 to 60 years. *A. connolyi* produces each year up to 900 seeds which may retain the ability to germinate for several decades; Perskaya (1963) reported syuzen seeds that gave shoots after 60 years of storage in a herbarium. Vegetative propagation of syuzen is achieved by the formation of offshoots from horizontal roots which are exposed from under the sand by wind action.

In the southern part of the Central Karakum Desert, abundance of syuzen can reach 75 plants/ha, and biomass 570 kg/ha (Nechaeva *et al.* 1973). The root system is of the universal type, which develops vertically (main root penetrating 2 to 3 m deep) as well as horizontally (side roots which lie in the two upper meters of the soil and spread sidewards to 5 to 8 m).

Floristically, this formation is very poor and is built primarily of perennial herbaceous psammophytes. Of all trees and large shrubs, only *Ephedra strobilacea* can be found here; shrubs and small shrubs are represented by *Calligonum caput-medusae*, *C. arborescens*, *C. microcarpum*, *Eremosparton flaccidum*, *Smirnowia turkestana*, and *Convolvulus divaricatus*. Only about five perennial herbaceous non-ephemeroid species are found here, with *Stipagrostis pennata* dominant; ephemeroids are absent. Common summer – fall annuals include *Agriophyllum minus*, *A. latifolium*, *Corispermum papillosum*, *C. lehmannianum*, *Horaninovia ulicina*, and *Chrozophora gracilis*. Compared to other sand communities, very few ephemers are recorded here: *Isatis emarginata*, *Chartoloma platycarpum*, *Cithareloma vernum*, *C. lehmanni*, and *Spirorhynchus sabulosus*.

1.2.2.2. Dzhuzgun Formation (Calligonum sp. div.). These communities are dominated by species of the extremely polymorphic genus *Calligonum*; they are typical of non-stabilized and stabilized sands alhough they occupy small areas and have highly variable composition of dominant and subdominant species. More than ten *Calligonum* species are found, including *C. eriopodum*, *C. caput-medusae*, *C. arborescens*, *C. rubens*, *C. leucocladum*, *C. setosum*, and *C. microcarpum*. Other shrubs are *Salsola richteri*, *Halothamnus subaphyllus*, *Smirnowia turkestana*, and sometimes solitary *Haloxylon persicum*; semishrubs *Mausolea eriocarpa*, *Convolvulus divaricatus*, and *Acanthophylum elatus* are also found. Common herbaceous plants are *Stipagrostis pennata*, *Heliotorpium dasycarpum*, *Ferula assafoetida*, *F. litwinowiana*, and *Senecio subdentatus*.

Dominant species, dzhuzguns (*Calligonum* sp. div.), often participate in other desert plant communities, particularly ones of *Haloxylon persicum*. Among *Calligonum* species, there are tree-like forms and large shrubs (3 to 6 m high), as well as medium and small shrubs (0.6 to 2.5 m high). Depending on growth conditions and especially on level of ground water, the same species can acquire different growth forms, e.g., *Calligonum eriopodum* and *C. caput-medusae*. The root system of dzhuzgun is universal and specialized; numerous roots are developed horizontally next to the sand surface and can stretch 15 to 30 m in length. The main root develops within the first year, and side roots form later. On the portions of the *Calligonum* stem that become covered by sand, additional roots may develop which can extract atmospheric and condensed moisture from the surface layer of sand (Petrov 1933, 1935). In nature, *Calligonum* propagates only by fruits and does not form root offshoots.

The abundance of *Calligonum* varies with habitat, from 50–100 plants/ha on non-stabilized (barkhan) sands to 100–300 plants/ha on stabilized sands. Annual productivity of green biomass varies from 50 to 150 kg/ha (Nechaeva et al. 1973).

The communities of *Calligonum* are one of the stages in the natural succession of sand vegetation from non-stabilized to stabilized sands. Most of present *Calligonum* communities are believed to be replacements of former white saksaul communities which were destroyed by human activities.

1.2.2.3. Cherkez Formation (Salsola richteri). This is widespread on all types of sands along the Amudarya, in Southwest Turkmenistan, and in the Lowland Karakum. These communities are also found around old, rarely-used wells along the West Uzboi.

The dominant species is cherkez (*Salsola richteri*), a constant subdominant of many other psammophyte communities. It is a large shrub, 1.5 to 2 m high; its lifespan is 25 to 30 years.

The root system of cherkez is of a universal type, developed both in vertical and horizontal directions. The main root goes down to 120 cm, the side roots stretch 7 to 9 m and give vertical extensions which go down 3 to 4 m and reach the ground water level (Petrov 1933, 1935). On sands where ground water level is deep (20 to 25 m), the roots of cherkez grow mostly horizontally, whereas on sands with close water level (8 to 10 m) they are developed primarily vertically. As do many other shrubs, *S. richteri* forms additional roots on stems covered by sand, but it does not produce shoots from roots which are exposed by wind.

The vegetative period of cherkez is extended from late March to October or November; it flowers from May to July and forms fruit in September to October. One plant can produce up to 1300 seeds. The abundance of *S. richteri* in the Karakum desert varies from 15 to 100 plants/ha, and the annual productivity of green biomass is 30 to 50 kg/ha (Nechaeva et al. 1973).

From 12 to 20 species of plants have been recorded in the *S. richteri* formation; all are adopted to life in non-stabilized sands (Rodin 1963). From seven to ten of these species are large shrubs (*Calligonum* sp. div., *Eremosparton flaccidum*, and *Smirnowia turkestana*), and small semishrubs (*Artemisia santolina*, *A. dimcana*, *Convolvulus divaricatus*, and *Acanthophyllum elatus*). There are only five herbaceous species with a long vegetative period: *Stipagrostis karelinii*, *S. pennata*, *Heliotropium argusioides*, *H. dasycarpum*, and *Cistanche flava*; the common ephemeroids are *Prangos diduma* and *Carex physodes*. There are few annual herbaceous species: summer – fall annual species are represented by *Agriophyllum minus*, *Horaninovia ulicina*, and *Salsola leptoclada*; and ephemers, by *Euphorbia turkomanica*, *Lappula semiglabra*, *Anisantha tectorum*, *Senecio subdentatus*, and *Cutandia memphitica*.

Salsola richteri and related *S. paletzkiana* are used as important species for sand-stabilizing melioration in the desert zone of Middle Asia. They also have importance for the livestock industry both as erosion inhibitors of pasturelands and as feed (young shoots, leaves, and seeds of cherkez are consumed by grazing sheep).

1.2.2.4. Bordzhok Formation (Ephedra strobilacea). Vegetation belonging to this formation is found in barkhan sands of the northern part of Central Karakum and on solonchaks of this area; it also occupies small territories around the sands of Chilmamedkum, Khanbaagykum, and the sands of the Dardzha Peninsula of the Caspian (Rodin 1963).

The dominant species, bordzhok (*Ephedra strobilacea*), is a gymnosperm shrub, one to two m high; it can participate as a subdominant in *Haloxylon persicum* and many other psammophyte communities. Bordzhok is found in the deserts of Middle Asia (Karakum and Kizylkum) on thick dune sands but also is an indicator of gypsum-bearing soils containing chloride and sulphate salts (Mukhammedov 1972, 1979).

Ephedra strobilacea is a dioecious species; the female plants bear about 11,000 seeds. It is one of the longest-lived desert shrubs, with a lifespan of 50 to 70 or even 100 years (Mukhammedov 1972). The root system of *E. strobilacea* is universal and well-developed; its tap root penetrates 0.5 to 4 m; numerous side roots and rhizomes stretch to 3 to 4 m; horizontal roots are more strongly expressed in plants growing on thin sands with close maternal rock than elsewhere.

The abundance of *E. strobilacea* in the Karakum varies from 100 to 400 plants/ha. It dominates the floristically poor formation, with rare *Salsola richteri, Haloxylon persicum, Calligonum caput-medusae*, and *C. setosum* found in the shrub layer; the lower layer is represented by solitary *Artemisia santolina, Salsola arbuscula,* and *Convolvulus erinaceus.* Perennial herbaceous plants include only *Stipagrostis karelinii* and *S. pennata.* The cover of ephemers and ephemeroids is represented by *Carex physodes, Eremopyrum orientale, Koelpinia lineata,* and *Hordeum leporinum. Agriophyllum minus* and *Salsola leptoclada* are typical summer – fall annual species.

2. Mesoxerophyte Desert Vegetation

Communities of this type are fragmentary in deserts and occupy patches usually less than one hectare in size. Many of the herbaceous communiites described here are common as synusia in shrub and semishrub associations. Three groups of formations belong here: large perennial herbaceous vegetation, ephemerous-grass vegetation, and annual halophyte vegetation.

2.1.1. High Perennial Herbaceous Vegetation
This vegetation is represented by formations with a dominance of selin (*Stipagrostis karelinii* and *S. pennata*), yuzarlik (*Peganum harmala*), and

Agropyron fragile. We will give a brief characteristic of the selin formation (2.1.1.1.)

2.1.1.1. Selin Formation. The large grasses *Stipagrostis karelinii* and *S. pennata*, widespread in the Karakum Desert, rarely dominate plant communities. *Stipagrostis karelinii* is a perennial grass 1 to 1.2 m tall; it grows solitarily or in groups on barkhan sands. Sparse groups of selin are not in contact with each other and threfore cannot be defined as true communities (Korovin 1961). However, due to the ability of its root system to form accessory roots and stabilize the sand, *S. karelinii* serves as a pioneer species, which is followed by other perennial species in the process of succession occurring on barkhan sands. Another selin species, *Stipagrostis pennata*, forms denser communities (covering 10 to 25%) and inhabits more or less stabilized sand dunes. *S. pennata* is 60–70 cm high, and its density varies from 1,200 to 2,600 plants/ha.

Shrubs and small shrubs are found only occasionally in these communities; these include *Calligonum* sp. div., *Salsola richteri*, *S. arbuscula*, *Halothamnus subaphyllus*, and *Ephedra strobilacea*. Semi-shrubs and small semishrubs, including *Astragalus ammodendron*, *A. transcaspicus*, *Convolvulus erinaceus*, *Mausolea eriocarpa*, *Artemisia santolina*, and *A. badhysi*, are more frequent but have low abundance. The herbaceous vegetation is represented, first of all, by long-vegetating annual species of Chenopodiaceae such as *Horaninovia* sp., *Agriophyllum latifolium*, *Kochia odontoptera*, and *Ceratocarpus utriculosus*. Ephemers (e.g., *Eremopyrum orientale*, and *Lappula semiglabra*) are rarely found. Among perennial plants, *Heliotropium argusioides* and, sometimes, *Cousinia schistoptera*, *Peganum harmala*, *Carex pohysodes*, and *Ferula assafoetida*, as well as *Stipagrostis pennata*, are present. Total floristic composition of *S. pennata* and *S. karelinii* communities is about 30 species, and on concrete plots thare are rarely more than ten to twelve species. Abundance of virtually all of these plants is low, and no codominant species can be separated.

2.1.2. Ephemerous-grass Vegetation
This group includes formations with two dominant grass species: yepelek (*Anisantha tectorum*; 2.1.2.1) and arpagan (*Eremopyrum orientale*; 2.1.2.2).

2.1.2.1. Yepelek (Anisantha tectorum); and 2.1.2.2. Arpagan (Eremopyrum orientale). Their role in desert vegetation (especially that of *E. orientale*) is rather insignificant; these communities are found mainly in the submontane plain of Kopetdagh.

Communities dominated by *Anisantha tectorum* are found in the depressions between sand dunes where the dune crests are occupied by the ephemerous-grass sagebrush communities, as well as by shrub and semishrub communities (*Calligonum* sp. div., *Ephedra strobilacea*, *Astragalus* sp.). Ephemerous-grass vegetation can form independent communities as well as participate as secondary sinusia in sagebrush vegetation. The dominant species, *Anisantha tectorum*, is an ephemerous grass 15 to 30 cm tall; its coverage is commonly 15 to 20%, and can reach 25 to 30%. *Eremopyrum orientale* often subdominates in these communities.

Floristically, the formation of *Anisantha tectorum* is composed of 22 to 25 mostly ephemerous species with fall – spring and spring developmental cycles, such as *Eremopyrum orientale, Carex physodes, Hypecoum pendulum, Astragalus oxyglottis, Eriodium oxyrrhynchum, Arnebia decumbens, Nonea caspica, Koelpinia linearis, Epilasia hemilasia*, and (sometimes very abundant) halophyte *Salsola carinata*. Of other Chenopodiaceae, solitary *Horaninovia* sp., *Agriophyllum latifolium, Kochia schrenkiana*, and *Salsola sclerantha*. Only rarely these communities include solitary small semishrubs such as sagebrush.

2.1.3. Annual Halophyte Vegetation
Within this group of formations, we separate formations with two dominant species: ebelek (*Ceratocarpus utriculosus*; 2.1.3.1) and ketgen (*Salsola paulsenii*; 2.1.3.2).

2.1.3.1. Ebelek (Ceratocarpus utriculosus); and 2.1.3.2. Ketgen (Salsola paulsenii. Characteristic communities of the ebelek have evident anthropogenic origin and are found on roadsides and degraded pastures. The dominant species, *Ceratocarpus utriculosus*, is a long-vegetating annual plant only 5 to 10 cm high. It has coverage of 10 to 25%, and often grows on gypsum soils with high gravel content. Globular bushes of ebelek are formed toward the fall period; due to the root fragility, plants can be easily broken and dispersed by the wind, which is its means of seed dispersal. Floristic composition of these communities is very poor; there are about 10 to 15 species, with not more than 5 or 6 species on description plots. There is only one (herbaceous) layer with participation of annual Chenopodiaceae such as *Salsola sclerantha, S. carinata*, and *Halimocnemis villosa*, and solitary epehemers (*Eremopyrum orientale, Ceratocephala falcatus, Strigosella* sp., and *Astragalus oxyglottis*).

Conclusion

Desert vegetation of Turkmenistan consists mostly of small semishrub sagebrush-halophyte and shrub psammophyte communities with a relatively simple structure and a homogeneous species composition. The structure of most phytocoenoses is determined by a few dominant species providing most of productivity. Other sinusia, in particular the herbaceous cover, consist mainly of annual ephemerous species.

Deserts occupy more than 80% of the territory of Turmenistan and are used throughout the year for sheep and camel grazing, thus providing a feed resource for a profitable livestock industry. In recent decades negative anthropogenic influence on desert vegetation has increased. Overgrazing, combined with cutting of shrubs for firewood, has led to the decreasing productivity of desert pastures (e.g., those in the southern part of Central Karakum Desert). Several natural reserves and other protected areas have been established in order to preserve and restore desert vegetation and landscapes in Turkmenistan. Extension of this network and other forms of environmental protection is essential to the preservation and optimal management of desert landscapes in Turkmenistan.

Lake Yeroyulanduz, a salt desert depression in Badghyz (1 of 2). Photo by H.R. Levenshtein and I.A. Mukhin.

'Bad lands' near the town of Kara-Kala, Southwest Kopetdagh (outcrops of Cenomanian clays in the foothills). Photo by I.A. Mukhin.

Juniper forest on rocky slopes in Central Kopetdagh. Photo by I.A. Mukhin

7. Flora of Kopetdagh

DZHUMAMURAD KURBANOV

Abstract

The modern flora of Kopetdagh includes 1,942 species, 806 genera, and 104 families of vascular plants, with 372 species of monocotyledons and 1,546 of dicotyledons. A prominent position in this flora is occupied by the typical Ancient Mediterranean families of Asteraceae, Poaceae, Fabaceae, Brassicaceae, and Chenopodiaceae. The desert and low arid mountain character of the flora is expressed by the diversity of Chenopodiaceae, Brassicaceae, and Boraginaceae.

The flora of Kopetdagh is generally an East Mediterranean one, with elements of low mountain subtropical flora (in Southwest Kopetdagh) and xerophylic highland flora. Fifty-five genera of this flora are not found in Middle Asia eastward from Kopetdagh. We have separated 102 types of geographic ranges, among which species of Ancient Mediterranean, Iranian, Irano-Turanian, and Kopetdagh-Khorassan ranges prevail. Our analysis confirms the connection of Kopetdagh flora with the autochthonous, Paleogene-Neogene Iranian center of origin of xerophylic flora; Kopetdagh lies on the periphery of this center.

Within Turkmenistan, 337 species of rare and endemic vascular plants are recorded in Kopetdagh, many of them narrow endemics of West, East, or Central Kopetdagh, Bolshoi or Malyi Balkhan, or the Krasnovodsk Plateau. These endemics are taxonomically related to the Kopetdagh-Khorassan, Iranian, Southwest Asian, Turanian, Transcaucasian, East Mediterranean, or Ancient Mediterranean species. Most endemic plants of Kopetdagh are neoendemic species, no older than the Pliocene.

Flora and vegetation of Kopetdagh have evolved since the Eocene under a constant climatic aridization. This has led to the xerophytizaton of flora through evolution of local xerophytes as well as dispersal of desert species from Iran and Turan. However, in the deep valleys of Southwest Kopetdagh, which have a mild subtropical climate, some representatives of ancient mesophylic flora have survived into the present.

Introduction

Along the southern border of Turkmenistan lies the periphery of a vast mountain system which continues into eastern Iran and western Afghanistan, separates the great deserts of Turan and Iran and connects mountains of Zagroz, the Armenian Plateau, Transcaucasia, and Atropatene (northern Iran) to Pamiro-Alai and Hindu Kush. The western part of this mountain system is known as the Turkmeno-Khorassan, or Kopetdagh-Khorassan Mountains. The system's width varies from about 10 km (in the eastern part) to 50 km (in the central part) to 100–120 km (in the western part).

Within Turkmenistan, the Kopetdagh Mountains consist of several separate ranges which are limited by the great desert of Karakum to the north, and by desert lowlands to the southwest. Kopetdagh stretches from east to west, between the longitudes of Tedzhen and Kazandzhik, for more than 500 km. Northern mountain chains are 700 to 2,000 m high; their steep rocky slopes form terraces which reach 250 to 300 m in height. Northern ranges are dissected by the deep gorges of small rivers flowing toward northern lowland such as Sekizyab, Kelyata, Chuli, and Firyuza. The highest chain of Kopetdagh is the Gaudan range, with its highest peak of 2,895 m (Mt. Riza) located in Iran.

To the east, Kopetdagh is connected to the Paropamiz Mountains (Badghyz Plateau); from the composition of flora and vegetation, we can draw the eastern boundary of Kopetdagh along the ranges of Gyazgyadyk, Kelyatkaya, and Zyulfagar, then from Pulikhatyn to the Zyulfagar Pass, and further through the Pinkhancheshme, Kerlek, and Akarcheshme Springs in western Badghyz.

The western periphery of the Kopetdagh-Khorassan Mountains, closest to the Caspian Sea, includes Maly Balkhan, Bolshoi Balkhan, the Krasnovodsk Plateau, adjacent small isolated mountain massifs (Dagada, Kaifata, Ufra, Monzhukly, and Cheleken) and small plateaus such as Boyadag, Syrtlanly, and Kumdagh.

The Krasnovodsk Plateau is an elevated, slightly concave area comprised of limestone dolomites, sandstone conglomerates, and marly gypsum-bearing sediments. It includes several small plateaus and ranges such as Kubadag, Shakhgadam, Dagada, Kyuryanyn-Kyure, and Ufra. Some ranges (e.g., Sary-Chengarak) are only 300 m high and are composed of the Akchagyl limestones underlain by Miocene-Paleogene clays.

An isolated range of Bolshoi Balkhan (up to 1,883 m) forms the northwestern periphery of Kopetdagh and is surrounded by the sand deserts of Cherkezli to the north and Chilmamedkum to the northeast; this range stretches for 80 km from west to east, with maximal width of about 40 km. Its plateau-like surface is divided by numerous steep, deep valleys. Bolshoi Balkhan is built primarily of Mesozoic and Cenozoic sediments; the most ancient are Lower Jurassic clays and shales which form the Kutlyubil Plateau next to the Kosha-Seira-Porsukh range (up to 700 m). Jurassic (Neocome) limestones are expressed along Bolshoi Balkhan, especially in its northwestern part (Nazarekerem, Yatyk, and Uchgyoz). The Sekidagh range (up to 1,383 m) forms the eastern part of Bolshoi

Balkhan, separated by the dry Umbelmes Valley; here, Upper Jurassic, Upper Cretaceous, and Paleogene sediments can be found.

Eastward from Bolshoi Balkhan lies a small separate range of Malyi Balkhan (ca. 330 sq. km, height up to 820 m), surrounded by clay desert lowland, with a small sand massif, Chetli, adjacent to the north. Malyi Balkhan is formed from Neocome limestones and Cretaceous (Upper Barremian and Aptian) clay and marl sediments. Late Cretaceous and Tertiary clay sediments also outcrop due to strong denudation, especially in the northwest part of this range.

Northwest Kopetdagh proper (i.e., the mountains between the city of Kazandzhik in the west and the Ereushdagh range in the east) consists of many separate small ranges which stretch about 100 km southward from Kazandzhik. The Kazandzhik and Kyurendagh ranges are formed by Lower Pliocene sediments as well as by those of Upper Barremian and Aptian and by Turon-Cenomanian clays. The entire northern system of ranges in Northwest Kopetdagh westward from the longitude of Kizylarvat is comprised of clay aleurolithes layered by clam limestone.

In West Kopetdagh, Upper Cretaceous sediments form the northern as well as southern slopes of such ranges as Danata, Oboi, Chaldzha, Kuilyar, Eishem, Seit-Kerderi, Sindzhou, Trgoi, and Adzhikuyu, as well as the southern slopers of the Peredovoi ("Outpost") northern range. A significant role is also played by Paleogene and Neogene clays (grey, greenish-grey, or marly ones). Marl prevails in the northwestern part of this area (Oboi, Kizylcheshme, and Karagyoz), whereas marl and limestone clays (sometimes with significant aleurolithe admixture) are expressed further to the south (Uilyua and Kulmach). The Peredovoi range of West Kopetdagh is built by the Hauterivian and Barremian limestones, whereas its southwest part is formed by younger Cretaceous clays.

Vast foothills and middle-size mountains of West Kopetdagh include Trgoi Mountain (635 m), built of Upper Cretaceous clays and adjacent to the town of Khodzhakala; the Kulmach range, built of Lower Sarmatian clays and clam limestones layered by sandstone; and separate small ranges such as Aladag, Zirik, and Akoba to the north and west, formed by the Apsheronian sea sediments. Quaternary sediments are represented by the marine, deluvial, and alluvio-proluvial deposits which form most of the northern submontane plain as well as the Khodzhakala Valley, which is filled by clay sandstones and deposits of gravel and pebble.

The extreme southwestern part of Kopetdagh is defined by the Songudagh range (up to 735 m), which is comprised predominantly of Albian and Upper Barremian sands and clays as well as of Lower Cretaceous sandstone and limestone (Kalugin 1977; Babaev and Durdyev 1982).

Climatically, Kopetdagh belongs to a homogeneous Turanian climatic province (Babushkin 1960). Climate here is dry and continental, defined by the adjacent lowlands to the west and southwest, the Karakum Desert to the north, the high ranges of Khorassan to the south, and the Badghyz Plateau to the east. Winter weather is not stable: warm surges occur and snow often covers the upper belts of the mountains; summers are hot, dry, and cloudless. Annual precipitation in Kopetdagh is 320 mm; in the rest of Turkmenistan there is virtually no precipitation from June to September, but in Kopetdagh summer rains, sometimes very intensive, occur almost every year. Summer temperatures normally reach 42° C; minimal temperatures in winter can drop as low as −22 °C. The absolute maximum temperatures in the northern ranges of Kopetdagh (Ashkhabad, Kizylarvat, Kazandzhik, Danata, and Nebitdagh) can reach 46° to 48 °C in June, July, and early August. At the same time, average temperatures in the areas closer to the Caspian Sea (western part of the Krasnovodsk Plateau) might be only 32° to 34 °C; however, even there absolute maximums are sometimes 38° to 40 °C. Due to the close proximity of the Caspian Sea to the Krasnovodsk Plateau, Bolshoi Balkhan, and West Kopetdagh, these areas manifest a milder, wetter climate than do Central and East Kopetdagh.

Well-formed soils in Kopetdagh are mostly fragmentary and appear in small patches among rough gravel substrates, rocky slopes, and outcrops of the maternal rock. Basic soil types in Kopetdagh are dark serozyoms, light serozyoms, loess serozyoms, takyrs and takyr-like soils, and solonchak-meadow soils.

Taxonomic Analysis of the Flora

According to our data, the vascular flora of Kopetdagh is now estimated to comrise 1,942 species belonging to 680 genera and 104 families. Lichens and mosses of Kopetdagh are described separately by Dzhuraeva (1978) and Sirotina (1988), respectively.

The vascular flora of Kopetdagh is relatively well known. Linchevsky (1935) listed 926 species from West (primarily Southwest) Kopetdagh. Gudkova *et al.* (1982) published a list of 1,266 species for West Kopetdagh, including Kyurendag and Maly Balkhan. Kamelin (1973) estimated the flora of Kopetdagh within Turkmenistan to include 1,800 species, with 1,387 species in Central Kopetdagh. For a part of East Kopetdagh, Nikitina (1954) listed 955 species, and the combined flora of Badghyz and the Karabil Plateaus has been recently estimated at 1,118 species (Kamelin 1989).

Such plants as horsetails and gymnosperms, as well as aquatic vegetation, are poorly represented in Kopetdagh. There are only eleven species of ferns, all of which are rare. Although there are few gymnosperm species (one species of juniper and ten species of ephedras), these play a significant role in the formation of plant communitites. The majority of flora belongs to angiosperms, including seven subclasses of Magnoliopsida (dicotyledons) and three subclasses of Liliopsida (monocotyledons) according to Takhtadzhyan's (1987) classification (Table 1).

There are 1,546 species and 531 genera of dicotyledonous plants. Each of three subclasses of Magnoliopsida includes two outstandingly diverse families: Caryophyllaceae and Chenopodiaceae (Caryophyllidae), Fabaceae and Apiaceae (Rosidae), and Boraginaceae and Lamiaceae (Lamidae). Other

Table 1. Floristic spectrum of the Kopetdagh flora

Taxa	Number of genera	Number of species
Angiosperms		
Class Magnoliopsida		
Subclass Ranunculidae	22	62
Of these:		
Ranunculaceae	13	36
Berberidaceae	3	5
Papaveracae	3	10
Hypecoaceae	1	3
Fumariaceae	2	8
Subclass Caryophyllidae	83	244
Of these:		
Aizoaceae	1	1
Portulacaceae	1	1
Caryophyllaceae	23	80
Amaranthaceae	1	3
Chenopodiaceae	34	111
Polygonaceae	6	24
Plumbaginaceae	1	1
Limoniaceae	7	23
Subclass Hamamelidae	2	2
Of these:		
Platanaceae	1	1
Juglandaceae	1	1
Subclass Dilleniideae	110	265
Of these:		
Hypericaceae	1	4
Primulaceae	5	5
Violaceae	1	4
Tamaricaceae	2	21
Frankeniaceae	1	2
Salicaceae	2	6
Cucurbitaceae	1	2
Capparaceae	3	5
Brassicaceae	75	142
Resedaceae	2	5
Cistaceae	2	3
Malvaceae	1	2
Ulmaceae	1	2
Celtidaceae	1	1
Moraceae	1	2
Urticaceae	2	5
Euphorbiaceae	3	29
Thymeleaceae	4	6
Subclass Rosidae	129	428
Of these:		
Saxifragaceae	1	1
Grossulariaceae	1	1

Table 1. Continued

Taxa	Number of genera	Number of species
Crassulaceae	3	6
Rosaceae	19	57
Lythraceae	1	4
Punicaceae	1	1
Onograceae	1	6
Mimosiaceae	1	1
Caesalpinaceae	1	1
Fabaceae	29	186
Aceraceae	1	1
Rutaceae	1	5
Tetradiclidaceae	1	1
Simarubaceae	1	1
Zygophyllaceae	3	13
Nitrariaceae	1	2
Peganaceae	2	2
Anacardiaceae	1	1
Liniaceae	1	2
Biebersteiniaceae	1	1
Geraniaceae	2	13
Celasteraceae	1	1
Santalaceae	1	2
Loranthaceae	1	1
Rhamnaceae	3	3
Elaeagnaceae	1	1
Vitaceae	1	2
Apiaceae	40	77
Caprifoliaceae	2	4
Valerianaceae	2	18
Dipsaceae	4	13
Subclass Lamidae	69	252
Of these:		
Rubiaceae	8	25
Gentianaceae	2	4
Apocrinaceae	1	1
Asclepidaceae	3	3
Oleaceae	1	1
Solanaceae	7	14
Convolvulaceae	3	12
Boraginaceae	25	71
Scrophulariaceae	10	46
Plantaginaceae	1	5
Verbenaceae	2	2
Lamiaceae	31	68
Subclass Asteridae	98	293
Of these:		
Campanulaceae	2	2
Asteraceae	96	291
Total Magnoliopsida	538	1,546

Table 1. Continued

Taxa	Number of genera	Number of species
Class Liliopsida		
Subclass Alismatidae	3	5
Of these:		
Juncaginaceae	1	1
Potamogetonaceae	1	3
Zannicheliaceae	1	1
Subclass Lilidae	123	359
Of these:		
Iridaceae	4	9
Liliaceae	11	43
Hyacinthaceae	1	2
Alliaceae	1	33
Amaryllidaceae	3	4
Ixioliridaceae	1	1
Asparagaceae	1	6
Orchidaceae	6	12
Juncaceae	1	10
Cyperaceae	12	32
Poaceae	82	207
Subclass Arecidae	4	8
Of these:		
Araceae	2	4
Lemnaceae	1	1
Typhaceae	1	3
Total Liliopsida	130	372
Other plants		
Subclass Gnetopsida		
Ephedraceae	1	10
Subclass Pinophyta		
Cupressaceae	1	1
Subclass Polypodiophyta	9	11
Of these:		
Ophioglossaceae	1	1
Athyriaceae	1	1
Aspidiaceae	1	1
Aspleniaceae	3	5
Hemionitidaceae	1	1
Sinopteridaceae	1	1
Adianthaceae	1	1
Subclass Equisetophyta		
Equisetaceae	1	1
Total for Kopetdagh	680	1,942

diverse families are Ranunculaceae (Ranunculidae), Brassicaceae (Dilleniideae), and Asteraceae (Asteridae).

Among Liliopsida, subclasses Alismatidae and Arecidae include only thirteen species and seven genera (Table 1) and are confined to the aquatic communities next to springs and reed swamps. The subclass Lilidae, however, is diverse and includes eleven families; in total, there are 372 species and 130 genera of Liliopsida.

The fourteen largest families (Table 2) include 1,450 species and 482 genera which represent more than 70% of the flora. The richest families are Asteraceae and Poaceae, followed by the Fabaceae, which contain an especially polymorphic genus, *Astragalus*. The fourth and fifth largest families are Brassicaceae and Chenopodiaceae. These five families consist of typical Ancient Mediterranean taxa as well as numerous Iranian and Kopetdagh-Khorassan species. Other important and diverse families include Caryophylaceae, Apiaceae, Boraginaceae, and Lamiaceae (Table 2). The diversity of Rosaceae, Scrophulariaceae, and Liliaceae is achieved primarily through the presence of the highly polymorphic genera *Crataegus*, *Rosa*, *Cerasus*, *Scrophularia*, *Veronica*, *Eremurus*, *Gagea*, and *Tulipa*.

A desert character of the flora is reflected in a significant number of Chenopodiaceae, Brassicaceae, and Boraginaceae; presence of Tamaricaceae, Frankeniaceae, and Nitrariaceae reveals connections between desert and seashore floras.

In the Kopetdagh flora, 34 genera of plants contain more than 10 species each (Table 3); such polymorphism is typical for the Ancient Mediterranean flora.

An outstanding polymorphic genus is *Astragalus* (Fabaceae) which includes 80 species. *Cousinia* and *Artemisia* include numerous endemics, mostly in low and middle mountain belts of Kopetdagh. Other important polymorphic genera

Table 2. Polymorphic families in the flora of Kopetdagh

Family	Number of genera	Number of species
Asteraceae	91	291
Poaceae	82	207
Fabaceae	26	186
Brassicaceae	75	142
Chenopodiaceae	34	111
Caryophyllaceae	23	80
Apiaceae	40	77
Boraginaceae	25	71
Lamiaceae	31	68
Rosaceae	19	57
Scrophulariaceae	10	46
Liliaceae	12	45
Ranunculaceae	13	36
Alliaceae	1	33

Table 3. Polymorphic genera in the flora of Kopetdagh

Genus	Number of species	Genus	Number of species
Astragalus	80	Bromus	11
Cousinia	39	Poa	11
Allium	33	Carex	11
Salsola	24	Polygonum	11
Euphorbia	24	Ranunculus	11
Artemisia	23	Strigosella	11
Veronica	18	Lygophyllum	11
Gagea	16	Alcea	11
Silene	16	Galium	11
Valerianella	16	Juncus	10
Vicia	15	Ephedra	10
Stipa	13	Acanthophyllum	10
Tamarix	13	Ferula	10
Acantholimon	13	Convolvulus	10
Jurinea	13	Lappula	10
Trigonella	12	Scrophularia	10
Medicago	12	Eremurus	8
Centaurea	12	Tulipa	8

are *Allium, Euphorbia, Lappula, Ferula, Veronica, Gagea, Strigosella, Silene, Tamarix, Acantholimon, Jurinea, Trigonella*, and *Scrophularia*. Some of these genera include species endemic for Kopetdagh-Khorassan, Kopetdagh, or certain areas in West or East Kopetdagh. The majority of genera which are polymorphic in Kopetdagh (e.g., *Astragalus, Artemisia, Cousinia, Allium, Veronica, Gagea, Ferula*, and *Silene*) also exhibit polymorphism in other Middle Asian mountains, and in Iran and Afghanistan; *Astragalus, Allium, Polygonum*, and *Silene* are also polymorphic in the southern Transcaucasia. On the other hand, typical desert genera such as *Salsola, Euphorbia, Climacoptera, Tamarix, Zygophyllum, Calligonum*, and *Reaumuria*, as well as steppe and semidesert genus *Stipa* usually exhibit low polymorphism in the mountains of Iran and Middle Asia. Outstanding polymorphic examples of the Ancient Mediterranean flora include *Valerianella* and *Verbascum*; the number of species in these genera decreases eastward from Kopetdagh. Many genera which are poly- or oligomorphic in the Kopetdagh flora (e.g., *Polygonum, Poa, Bromus*, and *Centaurea*), exhibit polymorphism throughout the southern Holarctic. However, certain Holarctic genera which are well represented in the eastern mountains of Middle Asia (*Carex, Potentilla, Festuca*, and *Salix*) are not polymorphic in the Kopetdagh flora, and the genus *Alchemilla* is notably absent from Kopetdagh.

The composition of families and genera described above characterizes the flora of Kopetdagh as an East Mediterranean one, influenced by subtropical flora of low- and middle-height mountains as well as by highland xerophylic flora. These features are reflected in a dominant role played in the floristic spectrum by such polymorphic families as Asteraceae, Poaceae, Fabaceae, Brassicaceae, Chenopodiaceae, and Caryophyllaceae, which include significant number of the Iranian, Kopetdagh-Khorassan, West Kopetdagh, and East Kopetdagh endemics.

Therefore, diversity of plant families in the Kopetdagh flora displays its typical East Mediterranean features, the influence of foothill desert floras (in West and East Kopetdagh), and the presence of rich highland (in Central Kopetdagh) and dry subtropical (in Southwest Kopetdagh) floras.

Several genera which are known from Kopetdagh have not been found in any other mountain system of Middle Asia. Linchevsky (1935) recorded 22 such genera from West Kopetdagh; Kamelin (1979) added 31 genera to this list. In addition, the author discovered five more genera of this type in Kopetdagh (*Anogramma, Peltariopsis, Periploca, Lasiopogon,* and *Urospermum*). Some of the genera in these lists were later recorded for East Kopetdagh or Pamiro-Alai. According to our data, there are now 56 genera (primarily Ancient Mediterranean or East Mediterranean) in Kopetdagh whose ranges do not extend to the eastern mountains of Middle Asia; these include *Anogramma* Link. (Hemiontidaceae); *Tragus* Hall., *Gaudinopsis* Eig, *Arrenatherum* P. Beauv., *Tetrapogon* Desf., *Cynosurus* L. (Poaceae); *Hyacinthus* L. (Hyacinthaceae); *Ophrys* (L.) Svv. (Orchidaceae); *Beta* L. (Chenopodiaceae); *Telephium* L. (Caryophyllaceae); *Ficaria* Guett., *Myosurus* L. (Ranunculaceae); *Aizoon* L. (Aizoaceae); *Auchonium* D.C., *Moriera* Boiss., *Aethionema* R. Br., *Carpoceras* D.C., *Dielsiocharis* Schulz., *Peltaria* Tacq., *Peltariopsis* (Boiss.) N. Busch., *Calepina* Adans. (Brassicaceae); *Homalodiscus* (Boiss.) Boiss. (Resedaceae); *Mespilus* L., *Cydonia* Mill. (Rosaceae); *Coronilla* L., *Hyppocrepis* L. (Fabaceae); *Malvalthaea* Iljin (Malvaceae); *Fumana* Spach (Cistaceae); *Chaerophyllum* L., *Reuteria* Boiss., *Foeniculum* L., *Smyrnium* L., *Ferulago* Koch, *Tordylium* L. (Apiaceae); *Plumbago* L. (Plumbaginaceae); *Molucella* L., *Satureia* L. (Lamiaceae); *Atropa* L., *Mandragora* L. (Solanaceae); *Orthantha* Kern. (Scrophulariaceae); *Sherardia* L. (Rubiaceae); *Sambucus* L. (Caprifoliaceae); *Eupatorium* L., *Codonocephalum* Fenzl, *Varthemia* D.C., *Pallenis* Cass., *Calendula* L., *Gundelia* L., *Siebera* D.C., *Carlina* L., *Perplexia* Iljin, *Callicephalus* C. A. Mey., *Shumeria* Iljin, *Leontodon* L., *Helminthia* , and *Turanophyllum* Poljak. (Asteraceae). Of this list, 23 genera are found in Northwest Kopetdagh, 25 in West and Southwest Kopetdagh, five in Bolshoi Balkhan, West and Northwest Kopetdagh, and two, in the Krasnovodsk Plateau. Although their ranges are limited, species of these genera play a significant role in the formation of xerophylic flora of the low and middle mountain belts, primarily in West, and less in Central and East Kopetdagh. Many of these species participate in phryganoid shiblyak and in complexes of juniper woodlands and vegetation of "painted rocks."

The core flora of Kopetdagh consists of Irano-Turanian species. However, its high endemism, as well as the presence of many species for which Kopetdagh is either the eastern or western boundary of their geographic range, allowed many authors to separate the Turkmeno-Khorassan or Kopetdagh-Khorassan floristic province within the Irano-Turanian subregion of the Ancient Mediterranean (Korovin 1961, 1962; Lavrenko 1965; Kamelin 1973, 1979). Takhtadzhyan (1978) includes Kopetdagh in the Armeno-Iranian botanico-geographic subregion of the Irano-Turanian region of the Ancient Mediterranean. Among the almost 2,000 species of vascular plants inhabiting Kopetdagh one can find various types of distribution: from cosmopolitan to extremely narrow, endemic. Endemics may be limited to a small territory within Kopetdagh, such as the chains and summits of Chopandagh, Aselma, Missinev, Kheirabad, Eureshdagh, Ezetdagh, Trgoi, Songudag, Kyurendag, Maly Balkhan, and Bolshoi Balkhan; some low western plateaus (Dardzha, Monzhukly, Cheleken, and Kumdagh); and the Krasnovodsk Plateau.

We conducted our analysis of the geographic ranges of Kopetdagh plant species according to general principles of botanical geography (Walter 1927; Walter and Alyokhin 1936; Kleopov 1938). We also used a detailed classification of geographic ranges introduced by Kamelin (1971, 1973) for the mountains of Middle Asia, as well as data on geographic distribution recorded in the monographic floras of the USSR, Turkey, Iran, Palestine, and Cyprus and data recorded in the monographic keys to the plants of Central Asia and Middle Asia. We also used treatises on separate groups (Bochantsev 1969; Tsvelev 1976; Grubov 1982; Pratov 1986), and, in some cases, reference monographs on the floras of Europe, Arabia, Algeria, Egypt, and Tunisia.

Among the 1,942 vascular plant species found in Kopetdagh, we have separated 102 types of geographic ranges. About 800 species have a broad Ancient Mediterranean or East Mediterranean type of distribution. However, more than 1,000 species, including the endemics of different levels, belong to the Iranian or Irano-Turanian flora. We recorded 180 narrow endemics, some of which are found only in Kyurendagh, West Kopetdagh, Southwest Kopetdagh, Central Kopetdagh, and East Kopetdagh. Such floristic composition emphasizes the northern Iranian character of this flora, and confirms the role of an autochthonous Iranian center of origin of the xerophytic flora in the Paleogene and Neogene.

Rare, Endemic, and Endangered Species of Plants

The vegetation of Kopetdagh experiences a strong anthropogenic influence: extensive overgrazing has resulted in soil erosion; trees and shrubs have been cut for firewood. Recent observations have revealed the combined effect of anthropogenic changes and unusually dry years on certain plant species: *Glaucium oxylobum, Cleome turkmena, Erysimum kerbabaevii, Lactuca rosularis, Salsola botschantzevii, S. iljinii, S. glabella, Allium transvestiens, A. kopetdaghense, Eremurus kopetdaghensis, Pteropyrum aucheri, Reaumuria botschantzevii, Ferula plurivittata, Siebera nana, Phagnalon androssovii, Cousinia oreoxerophila, C. chaetocephala,* and *Centaurea androssovii* develop extremely slowly and may even not reach reproductive age.

Reliable conservation of the vegetation of Kopetdagh can be realized only within the protected territories, the natural reserves (*zapovedniks*). In addition to the Badghyz Reserve, which has existed since 1941, two more parks have been established recently, the Kopetdagh (est. 1976) and the Syunt-Khassardagh (est. 1979). None of these territories, however, protects the unique "painted rock" vegetation of the arid foothills indigenious to West, Central, and Eastern Kopetdagh.

The International Union for the Conservation of Nature (IUCN) recommends five grades of evaluation of rare and endangered species: (1) extinct; (2) endangered, (3) vulnerable, (4) rare, and (5) indeterminate. These categories define the eligibility of a species for the IUCN Red Data Book (or similar lists of rare and endangered species); they are, however, not very useful for application to a local, regional list. Moreover, the IUCN classification lists under "rare" two principally different things: species rare due to their extremely narrow geographic range (narrow endemics), and species rare due to their low abundance in natural communities. We will use not formal IUCN criteria but, rather, descriptive categories which are more appropriate for small protected territories.

A list of the relict, rare, endangered, and disappearing vascular plants of Kopetdagh within Turkmenistan includes 387 species (Kamelin 1967, 1970, 1973, 1979; Kamelin and Kurbanov 1985, 1987; Kurbanov 1981, 1991; Nikitin and Klyushkin 1971; Nikitin 1978; Nikitin and Krasikova 1978; Nikitin *et al.* 1978). Below we discuss several of the rarest and most relict plant species, which fall in two categories. The first includes species whose very existence in the flora of the former USSR (or even in the world flora) is under direct threat; such species cannot survive without special protective measures. The second includes species which are potentially vulnerable due to their extremely narrow range and very specific ecological requirements.

Category A. Species whose Existence is Under a Direct Threat

Calligonum triste Litv. Soskov (1974) used this name for a unique population of dzhuzgun (*Calligonum*) found only in sand dunes within the isolated mountains of Kumdagh. Only about a dozen of suppressed plants with poorly developed fruits exist. Although the fruits resemble ones of the typical *C. triste* , the two races differ in the extent of rigidity of setae covering the fruit. True status of the Kumdagh dzhuzgun population (or race) has yet to be determined; meanwhile, it should be protected from the mining industry developing nearby since it has perhaps led to the decrease of the Kumdagh dzhuzgun population. To preserve this race, an experimental plantation should be created from seeds of the Kumdagh dzhuzgun in a similar area, most likely in the sand areas surrounding Maly Balkhan (Akhchakuima or Pereval).

Pteropyrum aucheri Jaub. et Spach. A narrow endemic of the northern portion of East Kopetdagh (Meana, next to the Karateken ruins) and Badghyz

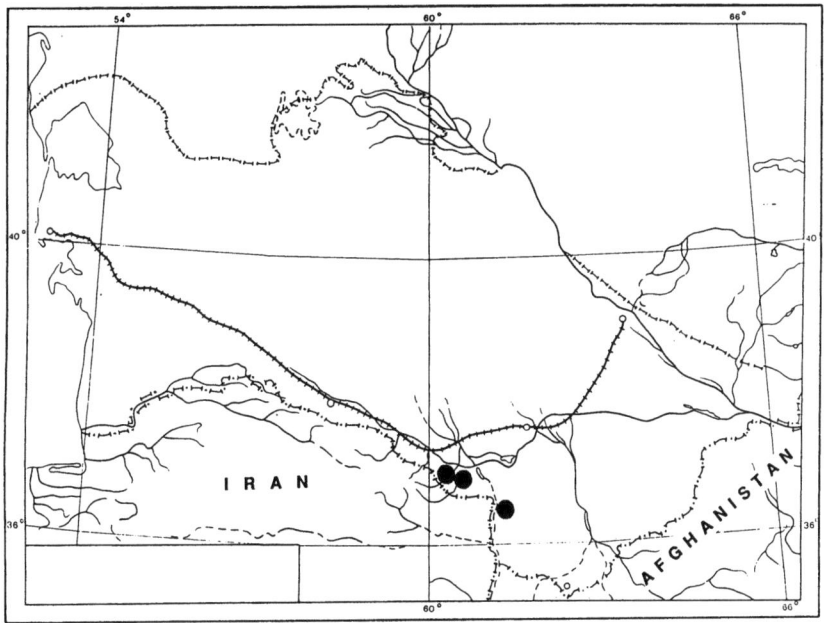

Fig. 1. Range of *Pteropyrum aucheri* Taub. et Spach.

(Pulikhatyn) (Fig. 1). Biology of this species is unknown; it is included in the Red Data Book of Turkmenistan.

Alcea karakalensis Freyn. A narrow endemic of the Songudagh mountain chain. This species is known only from type specimens; recently, it has not been recovered even during a special search at Songudag. *A. karakalensis* is a decorative plant; its biology and ecology are unknown. It is included in a list of species which are candidates for the Red Data Book of Turkmenistan.

Mandragora turcomanica Mizgir – Turkmen Mandrake. A narrow endemic, relict, endangered species. Sparse populations of mandrake are found at the base of southern slopes of the Syunt-Khassardagh range in West Kopetdagh (Fig. 2), within the Syunt-Khassardagh Reserve. Locations include the valleys of Chokhagach, Shevlan, Altybai, Ekechinar, Sarymsakli, and Dagdanli, and specimens are found there among trees and shrubs at 600 m elevation.

This species might have originated in the Miocene – early Pliocene, when Kopetdagh was undergoing major geological and climatic changes. The Mediterranean-Himalayan distribution of the genus could have been disrupted by periodical Pliocene and Quaternary transgressions of the ancient Caspian Sea. These transgressions could have created a number of disjunct, relict ranges of such Kopetdagh species as *Mandragora turcomanica*, *Mespilus germanica*, *Sorbus graeca*, *Evonimus velutina*, *Crataegus androssowi*, *Rhus coriaria*,

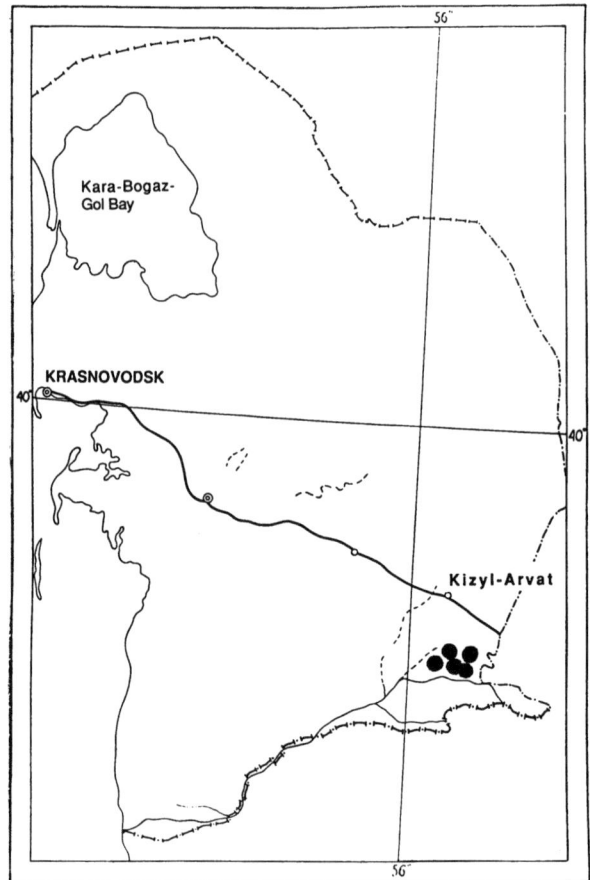

Fig. 2. Range of *Mandragora turcomanica* Mizgir.

Jasminum fruticans, *Malus turkmenorum*, and *Dionysia kossinskyi*. Whereas general mesophytic adaptations of *Mandragora* could have evolved in the Miocene, the aridization of climate which started in Kopetdagh in the Pliocene undoubtedly affected the ecology of *M. turcomanica*, which has a long summer period of xerotermic dormancy. Mandrake is a valuable source of alcaloids (hyoscyamine, atropine, mandragorine) and vitamin C. This species is included in the Red Data Books of IUCN, the USSR, and Turkmenistan.

Reaumuria botschantzevii Zucker. et Kurbanov. A relict endemic of Northwest Kopetdagh. This species is known only from the locality of Akoba (Fig. 3), where only twelve plants now grow on salty clay slopes. A search for new localities as well as study of this extremely rare species is necessary; if no new localities can be found, it will be essential to protect the Akoba site.

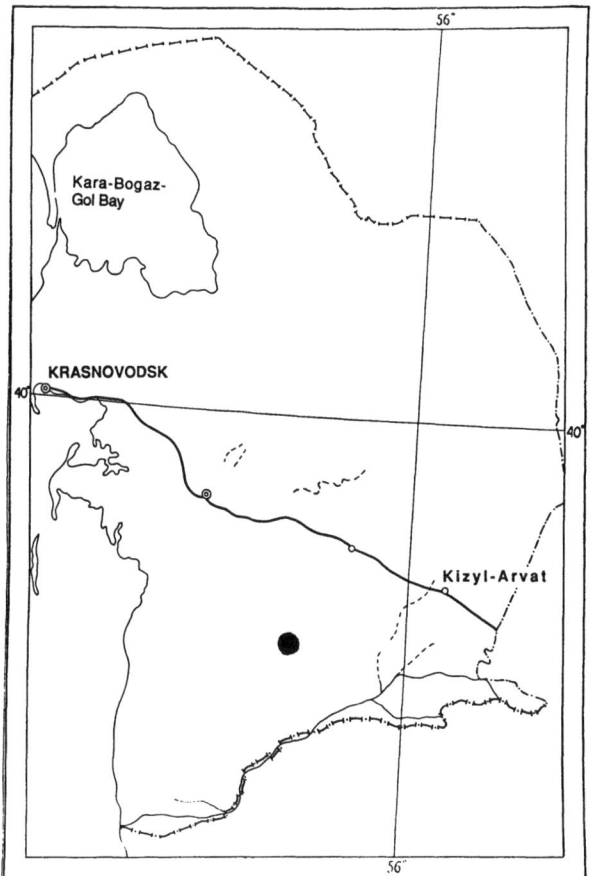

Fig. 3. Range of *Reaumuria botschantzevii* Zucker. et Kurbanov.

Colutea atabajevii B. Fedtsch. A relict, narrow endemic of East Kopetdagh, included in the Red Data Books of the USSR and Turkmenistan. It is found in the localities of Robergovsky and Shamli (Fig. 4) on dry gravel, rocky slopes, and often on outcropping Paleogenic clays. Along with other shrubs, this species suffers from overgrazing and woodcutting.

Category B
Species which are potentially vulnerable due to their extremely narrow range and very specific ecological requirements.

Eremurus kopetdaghensis M. Pop. A narrow endemic of West Kopetdagh. It is known only from the Marghyz (the original site) and Trgoi ranges, and from Kulmachbaba. It grows only on outcropping clays. The species is included in the Red Data Books of the USSR and Turkmenistan; it is a decorative plant cultivated in the Central Botanical Garden of Turkmenistan (Ashkhabad).

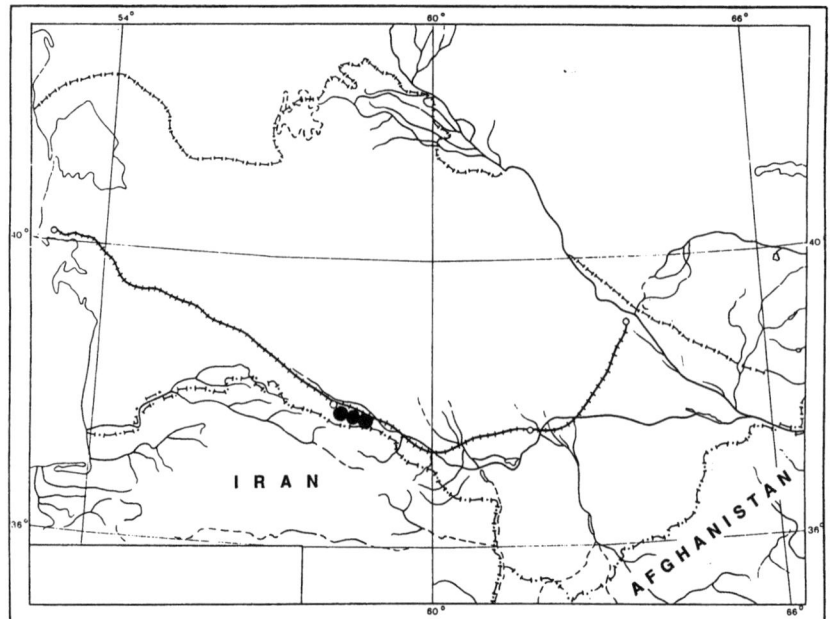

Fig. 4. Range of *Colutea atabajevi* B. Fedtsch.

Salsola iljinii Botsch. A relict, narrow endemic of West Kopetdagh which is known only from the Eureshdagh range, where a single but well-grown population is found next to the Bairamali spring at the locality Zau (Fig. 5). *S. iljinii* belongs to a group of semishrubs of the section *Belanthera* Iljin, which includes also *S. podlechii* Botsch. (South Afghanistan) and *S. titovii* Botsch. (West Tien Shan and West Pamiro-Alai). Bochantsev (1980) placed the ancestral species of this section in a secondary center of speciation for the genus *Salsola* around the Red Sea, with subsequent dispersal in the Pliocene and Lower Quaternary time. Modern ranges of these species are highly disjunct; eastern species of the genus inhabit middle mountain belts where precipitation is higher than in the foothills. Compared to all other species of this section, *S. iljinii* probably grows in the harshest climatic conditions. Differentiation of this species is probably related to the post-Pliocene aridization of climate and formation of modern altitudinal belts in Kopetdagh.

Cleome turkmena Bobr. A narrow endemic of Northwest Kopetdagh (from the Sumbar Valley to Malyi Balkhan). This species is found in many localities (sometimes in great numbers), where it is strictly confined to the gypsum-bearing "painted rocks," and, where in dry years, it can completely disappear from certain localities due to drying of the seedlings. This species, together with its close relative *C. gordjaginii* M. Pop. (which is found on "painted rocks" in Pamiro-Alai), forms a special section *Cleomopsis* Bobr. within the genus

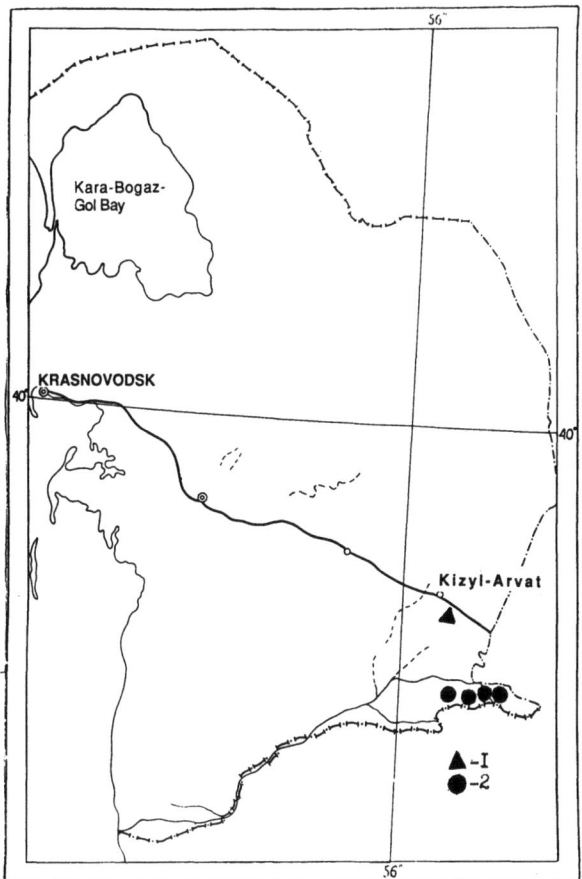

Fig. 5. Ranges of: 1 – *Salsola iljinii* Botsch. 2 – *Atropa komarovii* Blin. et Shal.

Cleome. Popov (1927) related these species to the Australian genus *Roeperia*; this is definitely not accurate: certain similarities are due purely to convergence. The section *Cleomopsis* is related to the more southern subtropical and tropical species groups of this genus and probably differentiated as early as at the boundary of the Paleogene and Neogene, during the colonization of new land exposed by regression of the Tethys. Modern disjunct species in Kopetdagh and Pamiro-Alai were probably formed during Pleistocene aridization.

Endemism of the Kopetdagh Flora

A significant number of species of the Kopetdagh flora are Kopetdagh-Khorassan or even more narrowly endemic. The majority of such species are found in the foothills and middle mountain belts in very specific habitats such as rocks and clays. These endemics could have differentiated at the boundary of the Paleogene and Neogene when the gypsophylic shrub vegetation of the "painted rocks" of Kopetdagh was formed.

The Kopetdagh endemics demonstrate taxonomic relationships to Iranian, Transcaucasian, East Mediterranean, South Pamiro-Alai, Central Asian, and (more rarely) Turanian species. Examples of close relationship between Kopetdagh and Transcaucasian species include *Acantholimon kjurendaghi* (Kopetdagh) and *A. bracteatum* (Transcaucasia), or *Scrophularia kurbanovii* (Kopetdagh) and *S. variegata* (Transcaucasia). Such connections may be ancient and related to the beginning of aridization in the low foothills of both West Kopetdagh and South Transcaucasia in the Upper Pliocene.

An example of an endemic species with Iranian connections is *Eremurus kopetdaghensis* M. Pop ex B. Fedtsch., which is related to the Iranian *E. persicus* Jaub. et Spach; these species could have differentiated autochthonously in the desert foothills of Iran and West Kopetdagh. Iranian connections can also be traced in *Acantholimon korovinii* Czerniak. and *A. balchanicum* Korov.; the first species belongs to a complex species group related to the Iranian *A. bromifolium* Boiss., whereas the second one is related to the Central Iranian *A. tragacanthium* (Jaub. et Spach.) Boiss. and South Iranian *A. scopius* (Jaub. et Spach.) Boiss. Chernyakovskaya (1931) compared *Acantholimon korovinii* to the Baluchistanian species *A. stocksii* Boiss.; it is, however, closer to the Iranian *A. incoptum* Boiss. et Buhse described from Kerman and Yezd. Both *Acantholimon korovinii* and *A. balchanicum* are endemics found in Kyurendagh, Malyi Balkhan, and Bolshoi Balkhan. Another example of the Iranian connections of Kopetdagh endemics can be found in the genus *Scutellaria*, where two Kopetdagh species, *S. litwinowii* Bornm. et Sint. and *S. luteo-coerulea* Bornm. et Sint., form a separate group, *Multicaules* Juz., related to the Iranian species *S. multicaulis* Boiss. Another species, *Onosma longiloba* Bunge, has a close relative, *O. angustilobum* Rech. fil. et Riedl., in Iranian Kurdistan. Certain narrow endemic Chenopodiaceae of Kopetdagh are related to Iranian or Central Asian species; e.g., *Salsola stelullata* Korov. is related to the North Iranian *S. buhseana* (Botsch.) Botsch., and *S. botschantzevii* Kurbanov, to the Central Asian *S. junatovi* Botsch.

A narrow endemic of Kyurendagh, *Tanacetopsis kjurendaghi* Kurbanov, demonstrates relationship with both the Iranian *T. platyrachys* (Boiss.) Kovalevsk. and the Parapamiz endemic *T. paropamisica* (Krasch.) Kovalevsk. Origin of *T. kjurendaghi* can be dated to the late Paleogene-early Neogene, when the relict subtropical xerophytic flora of "painted rocks" was formed in refugial localities of West Kopetdagh. Another narrow endemic of "painted rocks" of the Kyurendagh range, *Scrophularia kjurendaghi* Botsch. et Kurbanov, is also related to an Iranian species, *S. benthamiana* Boiss. (Bochantsev et al. 1983). Its habitats were not completely covered by repeated transgressions of the Caspian Sea in the Pliocene and therefore the species could have differentiated during the Pliocene aridization. This aridization could have led to the differentiation of many other arid endemic species of Iranian connection, such as *Homalodiscus ochradeni, Cephalorrhynchus brassicifolius, Lagochillus balchanicus* and *Reseda dschebeli.*

Many of the Kopetdagh endemics show relationships to the Badghyz, Turanian, and Turanian-Central Asian species: e.g., *Erysimum kerbabaevi* Kurbanov et Gudkova is close to the Badghyz species *E. badchysi* (Korsch.) Lipsky; *Climacoptera czelekenica* Pratov is close to the Turanian *C. sukaczevii* Botsch.; and *Salsola transhyrcanica*, to the Turanian-Central Asian *S. arbuscula* Pall. These connections are probably quite recent compared to the Iranian and East Mediterranean ones. For instance, although the endemic *Climacoptera czelekenica* Pratov is included in the ancient section *Ulotricha* Pratov, it forms the youngest branch of this section and has evolved in the littoral zone of the Caspian Sea, possibly as late as in the Holocene. In the same way the endemic species *Salsola transhyrcanica* could have originated in the western foothills of Kopetdagh in the Pleistocene when the Dardzha Peninsula and the area of Krasnovodsk (the major current habitat of this species) were liberated from sea waters.

Some Kopetdagh-Khorassan endemics such as *Juniperus turcomanica, Acer turcomanicum, Allium christophii*, and *Acantholimon pterostegium* show close relationship not to a single but to several species from adjacent territories. For instance, the Turkmen maple *Acer turcomanicum* is closely related both to the Transcaucasian-North Iranian *A. ibericum* Bieb. and to the South Iranian *A. cinerescens* Boiss. The Turkmen juniper *Juniperus turcomanica* is close to the East Mediterranean *J. excelsa* Bieb., the Transcaucasian-Anatolian *J. polycarpos* C. Koch, and the Pamiro-Alaian *J. seravschanica* Kom. *Allium christophii* Trautv. is related to the Transcaucasian *A. akaka* Gmel., Iranian *A. janischianum* Regel and *A. derderianum* Regel, and the Middle Asian *A. alxeianum* Regel. A narrow endemic of Kopetdagh, *Reaumuria botschantzevii* Zuker. and Kurbanov, exhibits an interesting taxonomic proximity to the Pakistanian *R. pangjgurica* Blatter et Halberg. Differentiation of these species and their narrow disjunct ranges were probably caused by the Neogene aridization over a broad territory.

Our analysis of the endemics in the Kopetdagh flora shows that they belong primarily to large polymorphic genera and are related mostly to Iranian or East Mediterranean species. The number of endemics related to the mountain floras of Middle Asia is negligible.

A significant number of endemic plant species in Kopetdagh indicates that this flora has been authochthonously evolving for a long time (since the Paleogene), primarily from Ancient Mediterranean flora. Xerophylic flora which originated in northern Iran could have dispersed into Kopetdagh during the subsequent Neogene aridization and given rise to local endemics. Thus, analysis of endemism confirms the generally accepted view about the ancient age of the Kopetdagh-Khorassan flora.

Historical Development of Flora and Vegetation

This survey is based upon an extensive literature concerning the origin and evolution of flora and vegetation in Middle and Central Asia (Popov 1927, 1938, 1963, 1983; Korovin 1935, 1961; Ovchinnikov 1936, 1940, 1955, 1971; Ilyin 1936, 1937, 1946, 1958; Nevsky 1937; Kultiasov 1946; Krishtofovich 1936, 1954; Grubov 1959, 1963; Bochantsev 1969a, 1969b; Kamelin 1965, 1973, 1979; and Pratov 1986a, 1986b). Most of these works discuss the formation and development of the desert or littoral flora in the Ancient Mediterranean area based on the example of a certain plant family such as Chenopodiaceae, Cruciferae, Frankeniaceae, and Tamaricaceae. The historical development of the arid mountain flora of Kopetdagh, although touched upon in studies by such prominent botanists as Korovin, Ovchinnikov, and Kamelin, is still poorly reconstructed. Unfortunately, there are no fossil plant data from Kopetdagh. Thus, we can only use fragmentary information from adjacent mountains and lowlands to outline the most general historical patterns.

The formation and development of the Kopetdagh flora has been closely connected to the specific paleogeography and geomorphology of this area, especially to the uplift of the Turkmeno-Khorassan Mountains and fluctuations of the Caspian Sea level. In the early Paleogene (from the Paleocene to Eocene) most of the Ancient Mediterranean, including the Kopetdagh area, was covered by the Tethys Sea. Even then, however, the future major mountain systems of Middle Asia were exposed above the sea level. In the late Eocene to early Oligocene, some ranges of Kopetdagh, together with the Alps, Carpathians, and Caucasus, experienced intensive uplift. In the Eocene, when a warm sea still covered part of the Kopetdagh area, the flora of islands in this sea was probably close to the known Eocene flora of Badghyz. This paleoflora was dominated by now extinct Poltava-type representatives of the Turkmen paleogeographical province as *Dryandra* and *Palibinia* (tentatively placed into Proteaceae) as well as *Rhus turkomanica*, species of *Maytenus*, *Andromeda*, *Amygdalus*, *Pistacia*, evergreen species of *Quercus*, and some Lauraceae (Krishtofovich 1936; Korovin 1936; Vasilevskaya 1957; Kornilova 1960, 1966). It is possible that these floras during the Tertiary period gave rise to the xerophylic tree and shrub communitites of proto-shiblyak, and later to the modern shiblyak (Kamelin 1965, 1967, 1970, 1973). In the valleys of West Kopetdagh, species which could belong to these floras include *Ziziphus jujuba*, *Rhus coriaria*, *Celtis caucasica*, *Cercis griffithii*, *Punica granatum*, *Ficus carica*, *Jasminum fruticans*, and *Euonymus velutina*. Along with these species, ancient floras of Kopetdagh and Badghyz included plants belonging to Proteaceae and Anacardiaceae as well as broadleaf deciduous trees of so-called "*Platanus* forests" (Krishtofovich 1954; Korovin 1961; Kamelin 1965, 1967). Fossils found in Akarcheshme (Badghyz) contain *Carya typica* as well as pollen of walnut, alder, birch, oak, and species of Taxodiaceae and Cupressaceae.

In the late Oligocene, the sea receded significantly, and the aridization of climate increased in the Kopetdagh area. By the end of the Tertiary period

(Pliocene), the outline of Kopetdagh had an almost modern shape. At this time the most ancient sand desert species belonging to the genera *Eremosparton, Ammothamnus,* and *Ammodendron* could have evolved in the sand deserts and foothills surrounding Kopetdagh. Korovin (1961) derives these forms directly from the Eocene Badghyz-type flora. The lowlands of Turan could already have included halophilic and gypsophilic ancestral species of *Haloxylon, Halothamnus,* and *Salsola.* Species of these genera could have dispersed to Kopetdagh as early as in the Miocene and Pliocene.

A comparatively mild climate in the Miocene supported rich flora and vegetation in Kopetdagh, including Maly and Bolshoi Balkhans (Isaeva-Petrova 1959; Malgina 1961; Petrosyants 1961; Surova 1962; Sikstel and Khudaiberdyev 1968). The fossils from Maly and Bolshoi Balkhan include *Phragmites, Populus diversifolia, P. pruinosa, Cercis siliquastrum,* and *Periploca graeca* (Kara-Murza et al. 1953; Gladkova 1957) as well as species of the Turgai flora (*Parrotia* and *Gleditsia*) which inhabited neighboring Kazakhstan (Kornilova 1966). The natural habitats of species of these two latter genera include today the forest valleys of Hyrcania (northern Iran); the vine *Periploca graeca* is still found in West Kopetdagh (Kamelin and Zabelina 1987). The Miocene-Pliocene time in Kopetdagh was characterized by the fluctuations of dry and mild (wet subtropical) climate; for a long time these fluctuations supported coexistence in Kopetdagh of the xerophylic proto-shiblyak Ancient Mediterranean flora and Hyrcanian forest flora. Within these communities evolved such endemics of Kopetdagh-Khorassan, Kopetdagh, and West Kopetdagh as *Mandragora turcomanica* (Fig. 2), *Atropa komarovii* (Fig. 5), *Pyrus boissieriana, P. turcomanica, Crataegus nikitinii, Malus turkmenorum,* and *Cerasus blinovskii.* The subsequent Pliocene aridization of climate and further development of mountain systems led to the reduction of the Hyrcanian forest, to the differentiation of early proto-shiblyak into juniper woodlands and deciduous shiblyak, and to the formation of steppes combined with juniper woodlands.

The aridization in the Oligocene-Miocene facilitated dispersal to Kopetdagh of such ancient xerophylic genera as *Haloxylon, Zygophyllum, Calligonum, Ephedra,* and *Tamarix.* This aridization also prevented many representatives of the Turgai paleoflora (species of *Liquidambar, Fraxinus, Quercus, Ulmus, Populus, Celtis,* and *Ailanthus*) from penetrating Kopetdagh. By the late Miocene the uplift of mountains intensified, and due to the sharp cooling of the climate many warmth-loving subtropical species became extinct. The Sarmatian Sea decreased and retreated far westward in the post-Miocene time.

The late Miocene communities of proto-shiblyak in Kopetdagh could have included tree species widespread at that time throughout the Ancient Mediterranean area such as *Castanea, Acer, Liriodendron, Ulmus, Platanus* (Krishtofovich 1936). Especially prominent may have beeen various oak species (*Quercus ilex* L., *Q. balloot* Griff., and *Q. castaneifolia* C.A. Mey), which dominated the proto-shiblyak of mountainous Middle Asia (Kamelin 1965); similar oak forests still exist in North Iran. It is possible that in the Oligocene-

Miocene various oak species grew in Kopetdagh; fossils of oaks of this age have been found in Kazakhstan (Zaklinskaya 1953, 1958; Abuzyarova 1954; Kornilova 1966). The Oligocene-Miocene proto-shiblyak in Kopetdagh could have been represented by juniper and maple communities including species of *Celtis, Pyrus, Cerasus, Amygdalus, Pistacia, Cercis, Lonicera,* and *Crataegus.* Such hygrophilic elements of the Turgai flora as species of *Castanea, Populus, Salix, Ailanthus, Vitis,* and *Ulmus* could have been also connected in Kopetdagh with proto-shiblyak.

The Neogene was a time of floristic differentiation within proto-shiblyak communities, initially in the foothills and later at middle altitudes. During this period, communities of the Turkmen maple (*Acer turcomanicum*) could have evolved as well as those of other small-leaved maples belonging to the section *Goniocarpa*, e.g., *Acer persicum* and *A. monspessulanum* (which are known as early as from the Oligocene of North Iran). Along with the Turkmen maple, xerophylized proto-shiblyak communities of this time could have included *Cerasus microcarpa, C. pseudoprostrata, C. blinovskii, Pyrus boissieriana, Crataegus nikitinii, Amygdalus scoparia, A. turcomanica, Punica granatum, Rhamnus sintenisii,* and *Lonicera floribunda.* By the middle of the Neogene, a specific arid subtropical Mediterranean flora could have developed in the "painted rock" foothills of Kopetdagh in contact with proto-shiblyak communities, with such representatives as *Homalodiscus ochradeni, Scrophularia kjurendaghi, Gaillonia brugierii, Calligonum triste, Zygophyllum macrophyllum, Euphorbia sclerocyathum, Suchtelenia calycina,* and *Cleome turkmena.* Such Kopetdagh-Khorassan endemics as *Scutellaria litwinowii, S. luteo-coerulea, Andrachne stenophylla,* and *Acantholimon pterostegium,* could have evolved later in communities derived from proto-shiblyak.

In the Pliocene, Kopetdagh experienced further uplift and many mountain summits and ranges reached their present height, including Artlan, Kutlyubil, Kosha, Seira, and Porsukh in the Bolshoi Balkhan; Syunt, Khasar, Kyuren, and Ereushdagh in West Kopetdagh; Markou, Dushak, Chopandagh, and Robergovsky in Central Kopetdagh; and Khiveabad in East Kopetdagh. An almost simultaneous uplift of mountains in Kopetdagh-Khorasan, the Caucasus, and Anatolia gradually led to a separation of the Ponto-Caspian drainage from the Mediterranean. In the Pliocene, the Apsheron-Krasnovodsk barrier divided the northern part of the Caspian from its southern part. At the same time, many warmth-loving xerophytes became extinct in the surrounding lowlands and foothills of Turan due to the increasingly cold winters; some species became relicts.

In the late Pliocene, during the Akchagylian transgression of the Caspian Sea, the climate of Kopetdagh gradually became arid and continental. At this time, communities of shiblyak could have formed at middle and high altitudes while original communities of halophytes (*Salsola botschantzevii, S. stellulata, S. kopetdaghensis, S. iljinii,* and *S. bungeana*) developed in the foothills. Species belonging to the *Salsola* florocoenotype evolved in the Neogene in the lowlands of Turan, Iran, and Central Asia (Kamelin 1979) and could have dispersed to Kopetdagh later.

The end of the Pliocene and beginning of the Pleistocene in Kopetdagh were times of increasingly cold and dry climate. In mountains facing the Caspian, a belt of Hyrcanian forest appeared again, dominated by Turgai elements such as *Parrotia persica*, *Acer laetum*, *Gleditsia caspia*, *Euonymus velutina*, and *Zelkovia carpinifolia*. These elements could have dispersed to the Hyrcanian province as early as in the early Neogene. The Hyrcanian forests included also elm, alder, walnut, and sycamore. In the Pliocene-Pleistocene, areas occupied by Ancient Mediterranean species became reduced, forest florocoenotypes degraded, and areas overtaken by steppes (and also probably by tragacanths) extended. Today, mountain steppes, dominated by such grasses as *Stipa*, *Festuca*, and *Poa*, are well represented in Kopetdagh as well as Balkhans (where also a steppe shrub, *Caragana grandiflora*, is found). Aridization in Pliocene-Pleistocene times could also have facilitated the evolution of endemic Kopetdagh sagebrushes such as *Artemisia balchanorum*, *A. turcomanica*, *A. kopetdaghensis*, and *A. deserti* which later became dominant in many steppe communitites. Steppes also were poplated by many endemic umbelliferous plants such as *Ferula turcomanica* and *Dorema balchanorum* in Bolshoi Balkhan and *Ferula gummosa*, *F. kopetdaghensis*, *F. badrakema*, and *F. ovina* in Kopetdagh. Such endemics as *Ferula karakalensis*, *Cleome turkmena*, *Scrophularia kjurendaghi*, *Tanacetopsis kjurendaghi*, and *Salsola botschanzevii* could have evolved at this time within xerophylized juniper communities of the "painted rock" foothills.

By the late Pleistocene, the Caspian Sea had retreated far west from Kopetdagh, and the climate had become much drier. Mountain xerophytes such as cushion-like *Astragalus piletocladus* and *A. meschedensis* as well as species of *Onobrychis*, *Acantholimon*, *Acanthophyllum*, and *Gypsophila* penetrated the Kopetdagh flora from Iranian plateaus. Many of these species today are narrow local endemics and thus may be of a relatively young (Pleistocene) age, but their ancestors most likely evolved in the Neogene on the vast dry plateaus of Iran or Central Asia. Forests and shiblyak in the late Pleistocene almost completely disappeared from the foothills and, by the middle of the Holocene, took refuge in mountains. For instance, in such deep valleys of West Kopetdagh as Aidere, Ioldere, Pordere, Gyuen, Chakankala, Narli, Naivadai, Khodzhaeken, and Tiamil, one can still find such forest species such as *Juglans regia*, *Vitis sylvestris*, *Ficus carica*, *Pyrus boissieriana*, *Fraxinus sogdiana*, *Cydonia oblonga*, *Jasminum fruticans*, and *Allium paradoxum*. The Pleistocene was also the time of formation of riparian communities (*tugai*), dominated by *Tamarix* and *Populus* with participation of the giant grasses *Erianthus ravennae*, *Arundo donax*, and *Phragmites communis*. Today, tugais are confined to the river valleys (Sumbar and Chandyr) and dry beds (Chalsu, Karadzha, Tersakan, and Adzhidere) of West Kopetdagh.

Mountain mesophylic vegetation experienced further reduction during the Holocene, leading to the formation of modern "black forest" communities devoid of such trees as *Quercus* or *Liquidambar*. With continued mountain uplift, mesophylic forest plants shifted to lower altitudes, and many species became extinct. The Turkmen juniper replaced deciduous forest species on mountain plateaus; in the foothills, areas occupied by juniper woodland, maple forest, and shiblyak continued to shrink. The vacated territory was occupied by semisavanna and sagebrush desert, a process which continues today in Kopetdagh.

In the Holocene, lowland territories westward from Kopetdagh, exposed after the Caspian Sea retreat, became sites of formation of modern desert communities. There, new sand and loess substrates were occupied by typical desert psammophytes such as *Calligonum caput-medusae, C. leucocladum, C. aralense, C. macrocarpum, Astragalus turcomanicus, A. nigricans, A. ammodendron, A. squarrosus, A. ammophilus, Haloxylon aphyllum, Salsola arbuscula, Mausolea eriocarpa, Dendrostellera turkmenorum,* and *Artemisia lobulifolia.* The foothills of West Kopetdagh were inhabited by Turanian desert semishrubs including *Artemisia sieberi, A. diffusa, Kochia prostrata, Salsola orientalis, S. gemmascens, Halothamnus subaphyllus, H. glaucus, H. hispidulus,* and *Reaumuria fruticosa.* The low plateaus of Kopetdagh developed communities of the succulent Turanian semishrubs such as *Anabasis brachiata, A. eriopoda, A. truncata, Zygophyllum turcomanicum, Z. jaxarticum,* and *Z. eichwaldii.* These floristic complexes, although belonging to an ancient euxerophyte core of the hammada-type desert flora, are now recognized as the youngest florocoenotype of the Turanian desert flora (Kamelin 1979); this flora probably appeared in Kopetdagh only in recent (Holocene) time.

A specific halophile vegetation developed along the western foothills of Kopetdagh next to the Caspian Sea. The littoral zone was occupied by communities of *Salicornia europaea, Bienertia cycloptera, Argusia sibirica, Halostachys europaea, Nitraria schoberi, N. komarovii,* and *Tamarix passerinoides.* Along shellstone seashores such species as *Brassica tournefortii, Climacoptera czelekenica,* and *Convolvulus persicus* could be found. Vegetation of isolated mountains was dominated by the narrow endemics, *Salsola transhyrcanica* and *Reaumuria tatarica*; these species may have evolved in degrading shiblyak communities. The vast lowlands adjacent to Kopetdagh were covered by the halophyte communitites of *Suaeda physophora, S. microphylla, Halocnemum strobilaceum, Halostachys belangeriana, Kalidium caspicum, Limonium suffriticosum,* as well as the annual halophytes *Salicornia europaea, Halopeplis pygmaea, Bienertia cycloptera,* and various species of *Climacoptera, Petrosimonia, Halimocnemis,* and *Salsola.* Further into the Kopetdagh foothills, modern communities of ephemerous vegetation developed. Some of the ephemer and ephemeroid dominant species also participated in psammophile communities (e.g., *Carex pachystylis, C. physodes, Smirnowia turkestana,* and *Strigosella grandiflora*).

In summary, the flora of Kopetdagh has evolved from Eocene to Holocene under constant physiographic and climatic changes and in direct contact with the Ancient Mediterranean, Hyrcanian, Khorassan, Iranian, and Turanian floras. The constant gradual aridization has facilitated evolution of an original xerophylic flora; at the same time, remnants of mesophylic flora have survived in deep valleys of West Kopetdagh with their mild dry subtropical climate.

8. Kopetdagh-Khorassan Flora: Regional Features of Central Kopetdagh

GALINA L. KAMAKHINA

Abstract

Four elementary floras are separated within the Central Kopetdagh floristic district. A comparative analysis reveals diversity of the regional flora (1,482 species belonging to 568 genera and 98 families) and its high endemism (249 endemic species, or 16.6%) as well as connections to neighboring floras (37% of species have Ancient Mediterranean or East Mediterranean ranges, and 22% of species have Middle Asian ranges). Rare plant species are discussed in terms of dynamics and evolution of local flora. Degradation of mountain steppe communities have facilitated dispersal of the foothill sagebrush vegetation upwards, thus creating basis for anthropogenic desertification of mountain landscapes. Considerations of conservation strategies for Central Kopetdagh are given.

Introduction

The Kopetdagh Mountains within Turkmenistan are the northeasternmost chain of the Turkmeno-Khorassan mountain range. Kopetdagh defines the southern border of the Karakum desert and lies far from the nearest mountain systems of Hindu Kush and Pamiro-Alai. The chain's geographical position also serves as the northern or northeastern boundary of the geographic range for many Iranian species. Kamelin (1973, 1990), in his discussion on botanical-geographical division of Middle Asia, considered Kopetdagh a natural part of Iran, with Iranian floristic centers, and included it in the Khorassan-Kopetdagh mountain province of the Iranian group of provinces which belongs to the Southwest Asian subregion of the Ancient Mediterranean region.

Central Kopetdagh includes medium-sized arid and subarid mountains (Kamelin 1990) with strongly subdivided terrain. In the foothills of Central Kopetdagh mountain floras combine with elements of northern-type zonal vegetation. Two to three altitudinal belts of vegetation are well expressed in Central Kopetdagh. The vegetation of the middle mountain belt is enriched with trees and shrubs. Certain highland floristic elements are present in the upper belt. The soils of Central Kopetdagh are characterized as medium, typical, and dark serozyoms. In the upper mountain belt, dark serozyoms with fragments of brown carbonate dry steppe soils form complexes with dark-brown soils of thermophile juniper woodlands (Aranbaev 1969).

The highly original flora of Central Kopetdagh has been well studied (e.g., Chernyakovskaya 1927; Fedchenko 1935; Nikitina 1954; Nikitin 1965; Kamelin 1965, 1970, 1973; Kamakhina 1986, 1987, 1989, 1991; Nikitin and Geldykhanov 1988; Kuznetsova 1989), and its analysis is now possible. The flora of northeastern Iran, however, has not been studied completely, which creates difficulties in establishing the southern boundary of our floristic unit. However, the western and eastern boundaries of Central Kopetdagh can be drawn. The eastern boundary (Kamakhina 1989) is not sharp but rather disperse; it is defined by the range of very peculiar flora of "painted rocks" (*pyostrotsvety*). This boundary extends from the Sherlok River to the meridian of Gyaurs Station. The western boundary of Central Kopetdagh is conventional; we follow Nikitin and Geldykhanov (1988), who draw it from Kizyl-Arvat through Khodzhakala along the Tersakan Valley and the Sumbar River to its fusion with the Atrek River next to Chaata. These are approximate boundaries of the vascular flora of Central Kopetdagh (Berezin 1929; Kamelin 1970, 1973).

Floristic Areas of Central Kopetdagh

We have applied the method of concrete floras (Tolmachev 1986; Kamelin 1973; Krasovskaya and Levichev 1986) to divide Central Kopetdagh into four "elementary natural floras" of the Kuruhaudan, Kurtusu-Gaudan, Firyuza and Sekizyab areas (Fig. 1). An elementary natural mountain flora is homogeneous in species composition and is defined on the basis of a drainage system with a common combination of ecological (especially climatic) factors. An additional requirement is the presence of neoendemic species. Within each of these elementary natural floras, we have described detailed "samples of floristic situation" (in the sense of B.E. Yurtsev). Analysis of these samples permits characterization of the taxonomic, geographic, and ecomorphological composition of a single floristic area as well as of the entire Central Kopetdagh.

Kuruhaudan Area

The Kuruhaudan floristic area occupies the high foothills, with maximum elevation 1,006 m (Kamakhina 1989). It lies within the Keltychinar and Sherlok River drainages and includes the mountain ranges of Tekechengasy, Kharvar, Gyaursdagh, and Zirakev. These structures belong to the southern wing of the Manysh syncline and contain such Paleogenic marine sediments as quartz sandstone, aleurolithes with carbonated cements, gypsum clays, and organogenic limestone (Berezin 1927; Kalugin 1977).

The mosaic and relict nature of these Paleogenic rocks is responsible for the peculiar composition of their xerophytic flora, modest in species diversity but very distinctive in morphology and taxonomy. We have recorded 619 species belonging to 354 genera and 74 families within the area of 3,100 ha. Of these, 62% of species belong to the ten most diverse families; the richest genera are *Astragalus*, *Allium*, *Gagea*, *Salsola*, and *Artemisia*. Due to the large areas occupied by the Paleogenic rocks (especially limestones), this flora is enriched with xerophytic elements from the families Chenopodiaceae, Boraginaceae, and Fabaceae.

A number of xerophytic species can be found on aleurolithic slopes covered with fine silica and milk-white quartz gravel. Due to weak cementation of the aleurolithes, the Paleogenic rock is transformed into a pressed sand which is easily eroded by wind and water. Slopes composed of this sand are carved by dry beds of seasonal floodstreams. Desert mosses, lichens, and the desert sedge (*Carex pachystylis*) are absent from this habitat. Only occasionally can be found single plants of *Poa bulbosa*. No soil formation occurs in such "badlands"; plant species do not form dense communities here. We have found 112 species belonging to 33 families which constitute the flora of these Paleogenic painted rocks (*pyostrotsvety*). This flora is composed primarily of perennial (46.4%) and annual (34%) herbaceous plants; xerophytic shrubs and semishrubs contribute 19.1% to the floristic list.

The only tree species that is able to survive in the conditions of severe soil moisture deficiency on the Paleogenic clays is a pistachio (*Pistacia vera*). Shrubs include *Zygophyllum atriplicoides*, *Rubia florida*, and *Cercis griffithi*, and dwarf semishrubs are represented by *Reaumuria turkestanica*, *Hymenocrater bituminosus*, *Colutea atabaevii*, *Artemisia gypsacea*, and *Scariola orientalis*. The relict islets of pistachio communities, which are found only on the water-bearing Paleogenic rocks (Sidorenko and Mikhelson 1948), are considered a result of long-term transformation of xerophytic woodlands of the Ancient Mediterranean (Kamelin 1973). A number of species belonging to the pistachio formation are actively colonizing the newly formed sheath of fine soil along the northern slopes of the Tekechengasy range. The pistachio here grows singly or in groups of 20 to 50 plants, and forms the association of *Pistacia vera* – *Zygophyllum atriplicoides* – *Artemisia kopetdaghensis* + *Poa bulbosa*.

The southern and southeastern slopes of Tekechengasy are occupied by the community of *Cercis griffithii* + *Zygophyllum atriplicoides* – *Artemisia kopetdaghensis* + *Poa bulbosa*, in which the aggregations of *Cercis griffithii* are intermittent with either pure *Zygophyllum atriplicoides* or with *Zygophyllum atriplicoides* + *Pistacia vera* + *Amygdalus turcomanica*.

A very peculiar group of 70 endemic species found here includes such relicts as *Colutea atabaevii* and *Popoviolimon turcomanicum*, which grow only on aleurolithe slopes, as well as the narrow endemics *Leymus nikitinii* and *Centaurea androssowii*. *Chesneya botschantszevii* (also found in East

Kopetdagh) grows on gypsum-clay slopes of Kuruhaudan; this species has also been recorded from the foothills of Mt. Markou. A common species here is a West Kopetdagh sagebrush *Artemisia kopetdaghensis*; relatively rare are *A. olivieriana*, *Ephedra lomatolepis*, and *Hulthemosa blinovskyana*.

Within Kuruhaundan, the major florocoenotypes (in the sense of Kamelin 1973) are semisavanna and East Mediterranean-type deciduous forest ("shiblyak"). The latter exists in combinations with painted rock flora on Paleogenic clays, and with halophyle and riparian communities in the river valleys. Desert semishrubs in the Kuruhaudan area dominate the steppe semishrubs; the vegetation of screes and rocks is very diverse; and the anthropogenic vegetation is actively developing. Hydrophyton (water vegetation) is fairly well represented in non-drying streams. As an eastern outpost of Central Kopetdagh, Kuruhaudan bears the influence of the desert Turan and the neighboring East Kopetdagh combined with the specific ancient flora of painted rocks.

Kurtusu-Gaudan Area

The Kurtusu-Gaudan floristic area occupies approximately 300 sq. km from the high foothills (Yablonovskii village, 580 m) through the intrazonal strip of deciduous tree and shrub vegetation, and up to the juniper-tragacanth and mountain steppe belt; the highest elevation here is 2,534 m (Mt. South Dashtoi). The Gaudan Pass lies at 1,733 m. To the east of this pass is the peak of Mt. Aselma (2,179 m), and to the west, mountains Ludzha (2,497 m) and Sibir (2,423 m). The direct dispersal of Iranian and Southwest Asian mountain xerophytes takes place through the Gaudan Pass and the ancient valley of Kurtusu. This route facilitates an exchange of floristic elements between the areas located to the north and to the south of Central Kopetdagh.

The vegetation here consists predominantly of *juniper woodlands* (*Juniperus turcomanica*) with derived communities of *Ephedra equisetina* and *E. intermedia* and of *shiblyak* (*Celtis caucasica*, *Acer turcomanicum*, *Colutea buhsei*, and *Amygdalus turcomanica*). The florocoenotype of *"black forest"* (chernolesye) is significantly impoverished here; it includes *Berberis densiflora*, *B. turcomanica*, *Ribes melananthum*, *Cotoneaster turcomanicus*, *C. ovatus*, *Lonicera nummularifolia*, and *Rosa* spp. Numerous shrubs and semishrubs form several predominant florocoenotypes. These include *steppe shrubs* (*Cerasus pseudoprostrata* and *Rhamnus sintenisii*); *Irano-Turanian xerophytic semishrubs* (*Perovskia abrotanoides*, *Rubia florida*, and *Jurinea sintenisii*); and an original Kopetdagh cushion plant, or *tragacanth* florocoenotype (*Astragalus piletocladus*, *A. cerasocrenus*, *Acantholimon* spp., *Acanthophyllum* spp., *Gypsophila aretioides*, and *Minuartia litwinovii*). The semisavanna florocoenotype is represented by a *tall herbaceous mountain semisavanna* (*Ferula* spp., *Dorema kopetdaghense*, and *Codonocephalum peacockianum*), and an *Irano-Turanian semisavanna* (*Cousinia congesta*, *Astragalus retamocarpus*,

Haplophyllum versicolor, and *Convolvulus subhirsutus*). Semisavannas are combined with *ephemeretum,* which is distributed from foothills to mountain peaks. The tuft-forming grasses and long-vegetation steppe herbs create a specific *mountain steppe* which can be dominated by different species of *Stipa* as well as by *Elytrigia trichophora* and *Festuca valesiaca,* and has many features in common with South Russian, Iranian, and Armenian steppes.

An intensive exchange of plant species takes place among juniper woodlands with tragacanth communities, tall herbaceous semisavannas, and mountain steppes. The increasing cutting of juniper and other trees and shrubs facilitates the wind erosion of rocks and therefore promotes expansion of the *petrophyton,* including xerophylous rock vegetation (*Cheilanthes persica, Pennisetum orientale, Dianthus turcomanica, Dielsiocharis kotschyi, Rosularia elymaitica, Campanula incanescens,* and *Dionysia tapetoides*) and the vegetation of warm screes (*Melica persica, Allium vavilovii, Gypsophila antoninae, Mesostemma kotschyana,* and *Aethionema kopetdaghi*). In the valley of Kurtusu, where the Paleogenic rocks are covered with deluvial and eluvial sheath, small islets of the ancient "pre-shiblyak" complex (Kamelin 1973) can still be found: the pistachio communities (*Pistacia vera, Artemisia ciniformis,* and *Poa bulbosa*) which cover 4 to 5 hectares and have 120 to 150 pistachio trees) and communities of *Cercis griffithii* with *Zygophyllum atriplicoides* and *Amygdalus turcomanica* which cover 2 to 3 ha and have 300 to 350 *C. griffithii.*

The millenia-long human activities, including mass grazing, fires, and woodcutting, have transformed the foothills and low mountains of Kopetdagh, creating a new type of anthropogenic vegetation. This includes ruderal communities as well as introduced species which have penetrated natural flora such as *Ailanthus altissima* and *Medicago falcata* as well as secondary riparian vegetation, chal (*Apera interrupta, Aeluropus littoralis, Avena barbata, Carduus arabicus,* and *Glycyrrhiza glabra*). The increasing anthropogenic impact has led to the virtual disapearance of the last relict subtropical florocoenotypes such as the riparian mountain forest and subtropical savanna with savannoid grasses (*Erianthus ravennae*) and hygrophilous species (*Phragmites australis, Epilobium tetragonum,* and *Mentha longifolia*).

The vascular flora of the Kurtusu-Gaudan floristic area is comprised of 871 species belonging to 402 genera and 76 families. There are only four species of conifers and Ephedraceae, and five species of ferns. The majority of this flora comprises angiosperm species, with 18 families accounting for 83% of all species. There are four times more dicots than monocots: the leading families are Asteraceae, Poaceae, and Fabaceae (35.8%), which is typical for a mountain flora of eastern Ancient Mediterranean. Among other diverse families are Alliaceae and Rubiaceae; Chenopodiaceae and Cyperaceae are relatively poorly represented; and Orchidaceae are completely absent. The generic composition of Kurtusu-Gaudan flora reveals 247 monotypic genera (62%) and 119 oligotypic genera (29.5%). Only few genera are polytypic (e.g., *Astragalus, Cousinia, Allium, Gagea, Euphorbia,* and *Stipa*). The high percentage of species (67.6%) is contributed by the ten leading families, with the increasing diversity of Brassicaceae (especially ephemerous annual species) and Polygonaceae. Other prominent families are Apiaceae, Lamiaceae, Caryophyllaceae, Liliaceae, Scrophulariaceae, Rosaceae, and Boraginaceae.

Specific ecological conditions in this mountain area are expressed in the numeric dominance of herbaceous polycarpic species (45%) and annual herbs (35.7%). There are seventeen obligate annual herb species with prolonged vegetation periods, 118 ephemerous species, and 37 biennial species. There are few trees (eleven species) and not many shrubs and low shrubs (94 spp., or 10.9%), which is characteristic for the arid low- and middle-height mountains of Middle Asia.

Unique to the Kurtusu-Gaudan flora are numerous rare endemic plant species, most of whoch are of Khorassan origin. We include in this group 157 species; of these, two genera (*Nikitinia* and *Prionotrichon*) and five species (*Rosa fertilis, Astragalus gaudanensis, Cousinia oreoxerophila, Turaniphytum kopetdaghense,* and *Aethionema kopetdaghi*) are local narrow endemics of the Kurtusu-Gaudan area. Two species, *Senecio khorossanica* and *Astragalus chrysostachys*, are probably recent migrants from Khorassan. Our survey (Kamakhina 1986) did not reveal three formerly recorded species (*Dionysia kossinskyi, Cousinia oreoxerophila,* and *Astragalus kucanensis*); these may have disappeared from the Kurtusu-Gaudan flora.

Firyuza Area

The Firyuza floristic area occupies approximately 145 sq. km of the Firyuza River drainage, which starts as small freshwater springs in the northern ranges of the Aladag Mountains; one spring is located in the upper part of the Semansur Valley, and three lie at the foot of Mt. Chopandag. The highest elevations of the Firyuza area are the peaks of Chopandag (2,889 m) and Rizarash (2,980 m); even higher peaks lie within Iranian territory (e.g., Mt. Shakhshakh; Kalugin 1977). These summits are deserted rocky plateaus with occasional quartz veins and slabs and with snow deposits preserved through the entire summer.

The terrain of the Firyuza drainage is a complex combination of rocky crests, steep slopes, and deep canyons and gorges; the elevations are from 1,800 to 2,800 m. The upper Firyuza valley descends as two deep canyons which meet to form one narrow gorge. Further down from the Firyuza resorts (*dachas*), which are located at 343 m, the valley cuts through the Karadag range next to the Vannovskii settlement (600 m) and ascends toward the southwest across the Geabyuldag range (700 to 800 m). At the west, the Firyuza drainage is separated from the Sekizyab area by the Dushakerekdag range; to the east, it borders the Kurtusu-Gaudan floristic area. The Firyuza valley bears outcrops of thick limestone rocks layered by marl, sandstone, and conglomerated quartz pebbles; further down, the alluvial deposits form terraces, each from 0.75 to 3 m thick. The Upper Jurassic limestones and dolomites are visible as outcrops at elevations from 720 to 2,110 m along the Firyuza valley.

High elevations and the very complex terrain of the Firyuza drainage have prevented many botanists from compiling a complete inventory of local flora. The entire known flora consists of 936 species belonging to 448 genera and 74 families (excluding the cultured plants). Since this list does not reflect the true diversity of Firyuza flora, a detailed ecologo-floristic analysis is presently impossible. Some features of this flora, however, can be described and discussed.

Among the ten leading families (65.3%) the top three constitute 35%, namely Asteraceae (129 species), Poaceae (117 spp.), and Fabaceae (82 spp.). That these numbers do not reflect true diversity is evidenced by the numbers of species recorded in such large genera as *Astragalus* (34 species), *Cousinia* (14 spp.), *Euphorbia* (14 spp.), *Gagea* (13 spp.), *Bromus* (11 spp.), and *Ranunculus* (nine spp.).

The summit of Chopandagh is the only locality within Turkmenistan where subalpine meadow species such as *Puccinellia tenuissima*, *Blysmus compressus*, *Nonea pulla*, *Cephalorrhynchus polycladus*, *Cirsium rhizocephalum*, and *Inula rhizocephala* are found. Here is also found the highest diversity of ferns (seven species), as well as Scrophulariaceae (29 spp.), *Euphorbia* (14 spp.), *Poa* (eight spp.), and *Carex* (ten spp.). Among a relatively large group of endemics (142 spp.) found in the subalpine belt are narrow endemic species *Potentilla botschanzeviana* and *Silene czopandagensis* and numerous endemics of Central Kopetdagh and Iranian Khorassan which have been recorded for Turkmenistan (or for the entire Soviet Union) only from the Firyuza area. These endemics include, for instance: *Allium brachyodon*, *Bilegnum bungei* (a representative of an endemic Khorassan genus), *Campanula khorassanica*, *Perplexia microcephala*, and *Erigeron dolichostylus*. An extremely rare Eurasian species, *Phyllitis scolopendrium*, formerly recorded from the Tutly canyon next to the village of Firyuza, has not been found in recent years. There is a possibility that an endemic Khorassan dwarf semishrub, *Pterocephalus khorassanicus*, will be found within Turkmenistan in the upper belt of the mountain range bordering Iran.

Only this area of Kopetdagh, in the upper mountain belt, houses rare cryohumid florocoenotypes. This belt comprises subalpine meadows formed by mesophilic and hygromesophilic perennial grasses and herbs, especially the Holarctic rhizome meadow grasses. Plant communities formed by *Arrhenatherum elatius*, *Bromopsis inermis*, *Bromus popovii*, *B. japonicus*, *Calamagrostis epigeios*, *Catabrosa aquatica*, *Catabrosella parviflora*, *Festuca pratensis*, *F. regeliana*, *Poa trivialis*, and *P. pratensis* are common next to mountain springs. In moist swampy places in river valleys the specific vegetation of *saz* type is formed by *Calamagrostis persica*, *Glyceria plicata*, *Carex diluta*, *C. divulsa*, *C. orbicularis*, *Cyperus glaber*, *C. fuscus*, *Eleocharis meridionalis*, and *Inula rhizocephala*. This florocoenotype is characteristic for warm to temperate mountain floras. Such mesophilic mountain plants as *Geranium regelii*, *Eremopoa oxyglumis*, *Veronica kopetdaghensis*, *V. gaubae*, *Carum carvi*, *Chaerophyllum khorassanica*, *Corydalis chionophylla*, *Ranunculus meyeranus*, *R.*

trichocarpus, *Cirsium rhizocephalum*, and *Taraxacum muricatum* have been recorded next to patches of thawing snow on the mountain summits. The "black forest" flora is relatively rich here; juniper woodlands occupy mountain crests; on the slopes affected by anthropogenic regression, exist the complexes of shiblyak, semisavanna, and ephemeretum. Shrub and steppe communities also are diverse, as well as tragacanth communities (*Astragalus pulvinatus* and *Acanthophyllum* spp.), the cryoxerophylic cushion plant formation (*Acantholimon procumbens*, *Oxytropis czapandaghi*, and *Arenaria insignis*), and diverse petrophilic vegetation (*Elymus longearistatus*, *Leymus kopetdaghensis*, *L. tianschanicus*, *Piptatherum vicarium*, *Allium monophyllum*, and *Arabidopsis wallichii*). Along the outcrops of Paleogene rocks on salt-bearing screes is found the relict vegetation of painted rocks: *Eremurus stenophyllus*, *E. luteus*, *Allium dictyoscordum*, *A. iliense*, *Anthochlamis turcomanica*, *Reaumuria cistoides*, *R. turkestanica*, and *Cuminum setifolum*. In the solonetz and solonchak depressions, meadows are formed of *Carex stenophylloides*, *Merendera sobolifera*, *Bromus sewerzowii*, *Hordeum brevisubulatum*, *Puccinellia distans*, *P. tenuissima*, *Diptychocarpus strictus*, and *Sclerochloa dura*.

The diversity of mountain habitats in the Firyuza drainage provides an existence for a rich and diverse flora even though its true extent is not yet known. This is the only area in Turkmenistan where many rare species exist. High elevations (up to the subalpine zone) of the Firyuza area account for a significant presence of species belonging to cold and temperate floras.

Sekizyab Area

The Sekizyab drainage (ca. 1,000 sq. km) is formed by the small rivers of Mergenulya, Kurkulab, Sekizyab, and Syulyukli, at elevations from 250 to 2,500 m. Within this area, we delineate three independent subareas, or "samples of floristic situation;" these are Dushakerekdagh (230 sq. km), the Germab Depression (100 sq. km), and Missinev (350 sq. km). This last subarea includes part of the territory of Kopetdagh State National Park. The highest elevations are the peaks of Dushak (2,482 m), Missinev (2,480 m), Kheirabad (2,027 m), and Kharlasang (1,643 m). In the foothills of the Dushakerekdagh range (next to Chuli village), the painted rocks terrain is formed on the gypsum-bearing limestone. Here are prevailing marine sediments of the Mesozoic, Paleogenic, Neogenic (sandstone), and Quaternary ages; continental rocks are rare. The maternal rock is covered by a sheath of eluvium, deluvium, and loess deposits with developed soils. The ancient (Paleogenic and Upper Jurassic) sediments do not in most cases reach the surface, and thus do not affect the floristic composition. The arid ecological conditions of soil formation (with ubiquitous carbonate soils) can be best observed in the Mirzadagh range. The anthropogenic effect is enhanced by overgrazing, woodcutting, and fires throughout the Sekizyab area.

At elevations from 400 to 1,500 m, the vegetation of the Sekizyab drainage includes combinations of fragmentary shiblyak communities with steppe shrubs and so-called phryganoid semisavanna (Kamelin 1973) communities, the latter of which are represented by ephemeretum, Irano-Turanian, and mountain semisavanna. At 1,600 to 2,800 m juniper-tragacanth communities and mountain steppes are present; these communities are especially well expressed in the ranges of Missinev and Dushakerekdagh. The contact zone between these principal belts of vegetation along the bottoms of mountain valleys contains combinations of "black forest," shiblyak, and such forest riparian elements as *Malus sieversii, Celtis caucasica,* species of *Cotoneaster, Sorbus, Crataegus, Lonicera, Rosa, Salix,* and *Populus,* entwined by *Clematis orientalis* and *Bryonia dioica,* with thickets of blackberry (*Rubus caesius*) in the undergrowth. The diversity of ferns, however, is low, and the Mediterranean forest elements, while present, are poorly represented due to the general aridity. Ecological originality of the Sekizyab flora is expressed in the presence of meadow plant species (*Bromopsis inermis, Agrostis stolonifera,* and *Poa trivialis*), the mesophilic mountain herbs (*Colchicum bifolium, Gagea dubia, Hyacinthum litwinowii,* and *H. transcaspica*), and riparian species (*Populus euphratica, P. pruinosa, P. alba, Salix acmophylla, Elaeagnus orientalis, Fraxinus lanceolata, Lycium kopetdaghi, Halimodendron holodendron, Tamarix bungei, T. florida, T. hohenackeri, T. ramosissima,* and *Malacocarpus crithmifolium*). Next to the water on terraces subjected to periodic floods are found communities of hygrophilic grasses and herbs such as *Bolboschoenus maritimus, Cladium martii, Orchis pseudolaxiflora, Epilobium parviflora, E. velutinum, Mentha longifolia, Lythrum salicaria,* and species of *Typha* and *Potamogeton.* The salt-bearing areas next to small springs are covered by halophyton (halophyle vegetation) such as *Salsola dendroides, Artemisia gypsacea, Suaeda microphylla, S. acuminata, S. arcuata, Halothamnus glaucus, H. hispidulus, Asparagus brachyphyllus, Frankenia hirsuta, Limonium reniforme, Halostachys caspica,* and *Climacoptera brachiata.* Relict fragments of the mountain riparian forest are preserved at the foot of the Almadzhik range and include thirteen plane trees (*Platanus orientalis*) at the upper portion of the Sekizyab River and 373 walnut trees (*Juglans regia*) in the Karayalchy Valley. The painted rocks flora is expressed in the foothills and includes such species as *Allium iliense, Buhsea coluteoides, Eremurus kopetdaghensis, E. luteus,* and *Convolvulus fruticosus.* The steppe flora is in a constant exchange of species with the neighboring flora of mountain xerophytes. Areas which have been used for grazing for a long time exhibit a characteristic adventitious flora of such pasture weeds as *Peganum harmala* and *Vexibia* (*Sophora*) *pachycarpa.* The anthropogenic regression of desertified semisavannas has promoted an increase of sagebrushes (*Artemisia badhysi, A. ciniformis,* and *A. turcomanica*) in the formation of tall-grass phryganoid communities, especially on the slopes of the Mirzadagh range. Such species as *Amygdalus turcomanica, Cercis griffithi,* and *Pistacia vera* are absent from the shiblyak vegetation in the Sekizyab area.

The mosaic habitat conditions produce different floristic diveristy within three subareas of our "floristic samples": 868 species of flora in Dushakerekdagh, 660 species in Missinev, and only 420 in the Germab subarea. The total flora of Sekizyab includes 1,133 species belonging to 516 genera and 92 families. The floristic saturation index is 8.8 ha per species.

Flora of the Sekizyab area exhibits a diversity pattern which is characteristic for the Ancient Mediterranean. Eighteen families account for 911 species, or 80.5% of the entire flora. The richest families are Asteraceae (12.7%), Poaceae (11.7%), Brassicaceae (8.6%), Fabaceae (8.1%), and Apiaceae (4.8%). However, about 30% of this floristic list is represented by genera with single species.

The mountain floras of Middle Asia and, to a certain degree, of Khorassan, have certain general features: (a) the predominance of the dicotyledonous plants (dicot to monocot species ratio four to one); (b) the high percentage of monotypic (27%) and oligotypic (37.9%) genera; (c) a difference between average number of species and genera in one family for dicotyledons (26 species and 6 genera per family) and monocotyledons (12 species and 6 genera); and (d) numeric predominance of herbaceous polycarpics (544 species, or 48.1%) over annual (365 spp., or 32.3%) and biennial (50 spp., or 4.4.%) plants. There are 31 species of trees, which together with shrubs (141 spp.) comprise only 15.2% of the entire flora. From twelve introduced tree species, only *Ailanthus altissima* and *Acer fraxinifolia* are more or less constant components of the natural flora.

Among 202 endemic species of the Sekizyab drainage (18.2% of its flora) are such West Kopetdagh species as *Astragalus ackerbergensis*, *A. jolderensis*, and *Eremurus kopetdaghensis*. There are numerous narrow endemic species in Sekizyab: *Atraphaxis kopetdaghensis*, *Astragalus fuhsii*, *Helictotrichon turcomanicum*, *Stelleropsis antoninae*, *Hedysarum kopetdaghi*, *Jurinea ludmilae*, *Linum turcomanicum*, *Cousinia cryptadena*, and *C. mucida*. Several rare species have recently disappeared from the flora due to degradation of their habitats; these include *Elytrigia caespitosa*, *Allium kirindicum*, *Lathyrus tuberosus*, *Bidens tripartita*, *Echinops transcaspicus*, *Sonchus palustris*, and *Pseudolinosyris sintenisii*. The author's studies extended the known range of such rare species as *Gypsophila antoninae*, *Linum corymbulosum*, *Allium vavilovii*, *Taraxacum muricatum*, *Malus sieversii var. turkmenorum*, *Pyrus turcomanica*, *Sorbus turkestanica*, and *S. luristanica*.

In summary, the flora of the Sekizyab drainage can be described as typical flora of Central Kopetdagh which is, however, strongly influenced by the neighboring forest communities of West Kopetdagh. It is highly diverse due to its mosaic habitats as well as the anthropogenic factor, which has contributed a number of introduced species and pasture weeds.

Central Kopetdagh as a Floristic District

Our survey demonstrates composition and features of the four elementary natural floras of Central Kopetdagh: Kuruhaudan, Kurtusu-Gaudan, Firyuza, and Sekizyab.

We estimate total floristic diversity of Central Kopetdagh to be 1,482 species belonging to 568 genera and 98 families (Table 1). In comparison, the flora of

Fig. 1. A Scheme of the Central Kopetdagh floristic district.

the Turkmeno-Khorassan Mountains (both in Iran and Turkmenistan) includes approximately 2,500 species, and the flora of Kopetdagh within Turkmenistan comprises 1,900 species (Kurbanov 1987). The former regional estimates assign 1,387 species to Central Kopetdagh (Kamelin 1970), 1,266 spp. to Southwest Kopetdagh (Gudkova *et al.* 1982), and 1,150 spp. to Northwest Kopetdagh (Kurbanov 1987). The flora of Badghyz is estimated as 1,018 species (Kamelin *et al.* 1989), and flora of the Kuhitang Mountains as 1,435 species. Finally, the flora of the mountains of Middle Asia contains approximately 5,500 species of vascular plants (Kamelin 1973). In Central Kopetdagh, natural floristic areas have a higher index of floristic saturation (area to numer of species ratio) than the entire flora of this district.

The angiosperm plant species constitute 98% of the flora of Central Kopetdagh; there are only seventeen species and eleven genera of gymnosperms, ferns, and horsetails. The 4:1 ratio of dicotyledonous to monocotyledonous species is typical for Middle Asian and Iranian floras. Families richest in specific and generic composition are Asteraceae (198 species), Poaceae (180 spp.), Fabaceae (140 spp.), Brassicaceae (108 spp.), Caryophyllaceae (66 spp.), Apiaceae (60 spp.), Lamiaceae (54 spp.), Chenopodiaceae (54 spp.), Boraginaceae (51 spp.), and Rosaceae (46 spp.). Thus, only ten families account for 64% of all species which is typical for Ancient Mediterranean floras, including the mountain floras of Middle Asia. Significant role in Kopetdagh flora is also played by Scrophulariaceae and Liliaceae. Twenty-six families are represented by a single species each.

Table 1. Floristic spectrum of Central Kopetdagh and its natural floristic areas (number of genera/number of species)

Taxa	Floristic areas				Central Kopetdagh
	Kuruhaudan	Kurtusu-Gaudan	Firyuza	Sekizyab	
Equisetophyta	–/–	–/–	1/1	–/–	1/1
Polypodiophyta	–/–	5/5	7/7	4/4	8/9
Pinophyta	2/5	2/4	2/4	2/3	2/7
Subtotal	2/5	7/9	10/12	6/7	11/17
Magnoliophyta	351/614	395/862	438/924	510/1,126	557/1,465
Of these:					
Magnoliopsida	286/490	317/691	363/733	412/897	449/1,159
Liliopsida	65/124	78/171	75/191	97/229	118/306
Total	353/619	402/871	448/936	516/1,133	568/1,482

The difference between average number of species and genera in one family in dicotelydonous (16 species and 6 genera) and monocotyledonous plants (17 species and 7 genera) is insignificant (Table 2). Monotypic (43%) and oligotypic (30%) genera comprise two-thirds of the entire flora. The richest genera are *Astragalus* (64 species), *Cousinia* (28 spp.), *Allium* (26 spp.), *Euphorbia* (18 spp.), and *Veronica* (16 spp.).

The spectrum of different life forms demonstrates adaptive ecological features of the Kopetdagh flora: 636 species (43%) are herbaceous polycarpics, and 67 species (4.5%) are biennial; annual plants contribute 38% of flora (576 species). Shrubs and low shrubs comprise (173 spp., 11.5%), and trees (30 spp., 3%) are not numerous although their appearance increases as aridization and anthropoghenic transformation of habitats increase. The presence of cushion-plant life form is an original feature of Central Kopetdagh flora.

The widespread Ancient Mediterranean and East Mediterranean species, found in Central Kopetdagh as well as in other areas of Middle Asia, contribute 37% of the flora. Kamelin (1973) has estimated that the flora of Kopetdagh includes 15% of Irano-Middle Asian and 7% of Middle Asian species. South Holarctic and cosmopolitan species account for only 7% of the flora. Approximately 16% of species are of Ancient Mediterranean and East Mediterranean origin and are not found eastward from Central Kopetdagh. These are the species that define the modern image of the flora of Central Kopetdagh, together with local endemics (16.6%, or 249 species). Other components of the modern floristic image include the suppressed "black forest" florocoenotype, the expansion of sagebrush communities due to aridization, and the presence of gypsum-bearing areas with the relict flora of "painted rocks." The juniper and tragacanth vegetation belt experiences the highest press of human activities, which facilitates penetration of agressive alien elements of the flora while it decreases ranges of many native highland species.

Dynamics of the Flora

The performed floristic analysis demonstated the general diversity of this flora as well as its significant undersaturation: the absence of certain species, genera, and families which are present in the neighboring territories.

The inventory of flora in Central Kopetdagh by the author, conducted from 1977 to 1991, has revealed new localities for several rare species of plants. The strategy for preservation of a rare species should include inventory of all localities as well as investigation of its biology. The rarity of a species can indicate certain dynamic trends in the local flora. It is possible to incorporate data on anatomy and morphology of rare species in floristic analysis (Krasovskaya and Levichev 1986; Kamakhina 1987): data on seed productivity, for example, can demonstrate the ecological stability of a species in a certain floristic complex. Popov (1963) has emphasized the importance of revealing

Table 2. General features of the flora of Central Kopetdagh.

Taxa	Number of species	% of total number of spp.	Number of genera	% of total number of genera	Number of families	% of total number of families	Floristic ratio (species:genera:families)
Equisetophyta	1	0.07	1	–	1	–	–
Polypodiophyta	9	0.6	8	–	6	–	–
Pinophyta	7	0.47	2	–	2	–	–
Subtotal	17	1.1	11	1.94	9	9.18	2:1:1
Magnoliophyta	1,465	98.86	557	98.06	89	90.81	16:6:1
Of these:							
Magnoliopsida	1,159	78.02	449	79.05	72	73.46	16:6:1
Liliopsida	306	20.84	118	19.01	17	16.35	17:7:1
Total	1,482	100	568	100	98	100	15:6:1

recent trends of speciation in the general floristic picture; such trends can indicate the general patterns of florogenesis.

Within the initial reconstructed mountain flora of Middle Asia, Kamelin (1973) has separated the riparian subtropical and tropical forests, "proto-shiblyak," "black forest," and secondary autochthonous florocoenotypes (shiblyak, shrub steppes, and certain herbaceous complexes). This scheme is useful for providing a detailed analysis of the evolutionary development of different life forms. Among the rare species belonging to the ancient riparian subtropical and tropical forests are *Platanus orientalis*, *Juglans regia*, *Vitis sylvestris*, *Populus* spp., and riparian herbaceous species. Surviving rare species of the "proto-shiblyak" complex include trees *Celtis caucasica* and *Pyrus turcomanica*, shrubs *Cercis griffithii* and *Ficus carica*, low shrubs and semishrubs such as *Colutea atabaevii* and *Gypsophila aretioides*, and perennial herbs such as *Popoviolimon turcomanicum*, *Chesneya astragalina*, and *Mesostemma sp*. The core of the flora of Central Kopetdagh is formed by the following complexes: juniper woodland (*Juniperus turcomanica*); "black forest" (*Sorbus turkestanica*, *Crataegus* spp., *Berberis turcomanica*, and *Malus sieversii*) which is of the same age as "proto-shiblyak;" shiblyak (*Pistacia vera*, *Atraphaxis spinosa*, and *Amygdalus turcomanica*); and perennial herbs (species of *Ixiolirion*, *Eremurus*, and *Tulipa*). Within these core florocoenotypes, different ecological groups of species, i.e., more mesophilic or xerophilic species, can be defined. An especially interesting component of such analysis will be the role of local neoendemics in the dynamics of flora.

We know that all changes in the floristic structure (e.g., its systematic composition) and transformations of plant communities are correlated with certain changes in anatomical and morphological structures of plant species. This evolution of separate floristic elements finally accounts for the evolution of a florocoenotype as a whole. A rare species as an indicator of dynamic trend (e.g., decreasing or extending of a geographic range) is a part of a certain florocoenotype. Therefore, data on the distribution and biology of rare species can confirm the observed dynamic tendencies in a given flora and probably even help to predict future changes.

Rare species are usually vulnerable and require special measures of preservation. The list of rare species of Turkmenistan is constantly increasing due to the human activities, notwithstanding the existence of seven national parks (including Kopetdagh National Park in Central Kopetdagh) which protect many species. The rarity of many plant species in Kopetdagh also has a biogeographic explanation; for these species Kopetdagh represents either the eastern or northern boundary of their range and may be considered a marginal area.

According to our estimates, the flora of Central Kopetdagh includes 249 endemic species which belong to 118 genera and 35 families. Twelve species from this list have been included in the Red Data Books of the former USSR and Turkmenistan. We define an endemic species as a species found only in the Kopetdagh-Khorassan province, that is, in the area from eastern part of the

Elburz Mountains and Bolshoi Balkhan to the Gyazgyadyk range in Badghyz. We distinguish between several groups of endemic species within the flora of Central Kopetdagh. There are endemics of Central Kopetdagh proper (58 species, or 23.3% of all endemics); endemics of Central and East Kopetdagh (5 spp., or 2.0%); and endemics of Kopetdagh within Turkmenistan but excluding Bolshoi Balkhan (65 spp., or 26.1%). Among Kopetdagh-Khorassan species, we separate endemics of Central Kopetdagh and Khorassan (77 spp., or 30.9%) and endemics of northern ranges of Kopetdagh and Khorassan, i.e., from Bolshoi Balkhan to Badghyz and southwards to the Atrek drainage and Turkmeno-Nishapur Mountains in Iran (44 spp., or 17.7%).

The high level of endemism is expressed within the leading families of the flora: Asteraceae (50 species), Fabaceae (50 spp.), Apiaceae (14 spp.), and Lamiaceae (12 spp.), and within the genera *Astragalus* (36 spp.), *Cousinia* (17 spp.), *Allium* (12 spp.), and *Acantholimon* (9 spp.). Analysis of the spectrum of life forms among endemics reveals that these are predominantly perennial herbaceous species (71.1%, with 37.3% being tap root polycarpics) rather than woody plants (endemic shrubs and dwarf semishrubs contribute 28.1%, and two endemic species of trees, omly 0.2%).

From the list of 249 endemics, more than half of them (145 species) qualify as rare ones. According to our knowledge about the distribution of these species, we divide rare endemic Kopetdagh-Khorassan species of Central Kopetdagh into four groups:

(1) *probably extinct* – 5 species: *Astragalus kucanensis, Dionysia kossinskyi, Cousinia oreoxerophila, Pseudolinosyris sintenisii,* and *Allium kirindicum*;

(2) *endangered* – 25 species: *Juniperus turcomanica, Fritillaria raddeana, Ornithogalum arianum, Allium vavilovii, Crocus michelsonii, Epipactis veratrifolia, Ophrys transhyrcana, Gypsophila antoninae, Ranunculus trichocarpus, Corydalis chionophilla, Potentilla botschantzeviana, Rosa fertilis, Aethionema kopetdaghi, Ribes melananthum, Astragalus chrysostachys, A. gaudanensis, Oxytropis czapandaghi, Linum turcomanicum, Marrubium propinquum, Veronica gaubae, V. kopetdaghensis, Senecio khorossanicus, Streptorhamphus linczevskii, Perplexia microcephala,* and *Taraxacum lipskyi*;

(3) *very rare* – 50 species, among them: *Helictotrichon turcomanicum, Leymus nikitinii, Stipa kopetdaghensis, Eremurus kopetdaghensis, Tulipa botschantzevae, Allium brachyodon, Acanthophyllum mikeschianum, Silene czopandagensis, S. litwinowii, Corydalis macrocalyx, Hulthemosa blinovskyana, Rosa karakalensis, Prionotrichon gaudanense, Astragalus bachardeni, A. jarmolenkoi, A. rawlinsianus, A. subdjenarensis, A. xiphidioides, Chesneya astragalina, C. botschantzevii, Colutea atabajevii, Popoviolimon turcomanicum, Bilegnum bungei, Rindera coechinata, Campanula khorassanica, Centaurea androssovii, Cousinia albiflora, C. mucida,* and *Nikitinia leptoclada*;

(4) *rare* – 66 species, among them: *Stipa crassiculmis, S. lingua, Merendera jolantae, Tulipa hoogiana, Hyacinthus transcaspica, H. litwinowii, Allium*

brachyscapum, A. helicophyllum, A. monophyllum, Iris ewbankiana, Atraphaxis kopetdaghensis, Erysimum kerbabaevii, Astragalus ackerbergensis, A. confiniorum, A. curvipes, A. fuhsii, A. kuschkensis, A. nigriceps, A. raddei, Cicer kopetdaghensis, Vicia venulosa, Alcea antoninae, A. kopetdaghensis, Bunium korovinii, Lepechiniella persica ssp. *kopetdaghensis, Cousinia glochidiata, C. hypopolia, Helichrysum kopetdaghense, Jurinea antonovii,* and *J. kultiassovii.*

The rarity of most endemic species can be caused by changes in their habitats. For many steppe, "black forest," or subalpine meadow species, the decrease in humidity which follows general aridization and large-scale cutting of juniper woodlands leads to the reduction of species range and subsequent disappearance. Among endemics, 134 species are found only in the upper mountain belt.

Monitoring of the dynamics of flora based on rare species of different florocoenotypes allows us to determine the stability of the modern "black forest," shiblyak, and juniper woodlands; to predict the future of semisavannas, and petrophylic and anthropogenic florocoenotypes; and to reveal trends of floristic impoverishment in steppes and tall grasslands. This analysis also graphically illustrates high heterogeneity of the discussed floristic elements.

Desertification in Central Kopetdagh

The exploitation of natural resourses is destroying the ecological balance in Central Kopetdagh. Human activity has transformed structure and composition of plant communities, has impoverished biological and genetic diversity of local flora, and has created conditions for the progressing desertification of mountain landscapes.

Paleontologists have demonstrated that the cutting of junipers in the low northern foothills of Kopetdagh started as long ago as the VI to V millenia B.C. During the last 40 to 50 years, areas occupied by juniper woodlands have decreased by 30 to 40%; the total debit of water in mountain rivers has fallen by 50%; and many mountain springs have dried up. As a result, the altitudinal boundary of junipers has been elevated by 500 to 700 m from its ecological optimum, which has transformed many other plant communities. The primary communities have been replaced by secondary ones; the place of exterminated juniper woodlands was first occupied by mountain xerophytes, then by grasslands of *Elytrigia trichophora*, and finally by sagebrush and ephemeroid desert vegetation. The primary mesophilic and xerophilic polycarpics such as species of *Festuca* and *Stipa* have been deleted from steppe flora due to extreme overgrazing, which led to the erosion of alluvial deposits and the outcropping of maternal rocks.

Boundaries of foothill vegetation have been elevated due to the replacement of steppe communities by the sagebrush ones in the course of desertification.

Mikeshin (1940) noticed the expansion of the ephemerous sagebrush communitites from the lowland to the foothills; later, Nikitina (1954) also recorded the formation of a new type of semidesert vegetation, in which sagebrush complexes were enriched by steppe elements and the synusium of *Poa bulbosa-Carex pachystylis*. Today, sagebrush communitites in Central Kopetdagh have expanded from the foothills along the rocky outcrops up to mountain plateaus of such mountains as Dushakerekdagh, Aselma, Chopandagh, Missinev, and Tagarev. Sagebrush communitites in the plateaus are not dense (60 to 150 plants per 100 sq.m); sagebrush occupies the upper layer of the vegetation providing coverage 25 to 30%. Usually sagebrush synusia are combined with sparse juniper woodlands or mountain xerophyte complexes, with an abundance of Sp^1 to Sp^2 by the Drude scale. In replacing steppes, sagebrush complexes incorporate single components of steppe and mountain xerophyte communities combined with diffused tufts of *Poa bulbosa* and *Carex pachystylis*. At the same time, the primary vegetation of preserved steppes (dominated by species of *Stipa* and *Festuca*) undergoes replacement by the secondary one (dominated by *Elytrigia trichophora*).

Thus, only sagebrush species (*Artemisia ciniformis*, *A. turcomanica*, *A. badhysi*, and *A. gypsacea*) have been able to move from desert lowland upward through all altitudinal belts under the conditions of progressing desertification. Due to its high biological, morphological, and ecological plasticity, sagebrush has become a dominant species, successfully creating leading plant formations even at the highest elevations in Central Kopetdagh. The entire landscape of such areas as Mirzadagh has been changed by sagebrush replacement the overgrazed steppe communities and easily occupation of eroded areas. The formation of sagebrush mountain landscapes which replace exterminated juniper woodlands or steppes should, therefore, be considered the first alarm signal of disturbed ecological stability.

Sagebrush "aggressiveness," however, is not unlimited. When grazing is stopped or even temporarily decreased, sagebrush makes a significant retreat. Sagebrush cannot create sufficient layers of soil turf or humus, nor can it be compared to nearly the extent that steppe grasses can accumulate underground phytomass. Usually, sparse sagebrush communities have a very poor representation of annual and perennial herbs (12 to 25 species) and a low degree of soil turfness (up to 30%). Mamedkuliev (1990) demonstrated that low underground productivity usually does not exceed above-ground phytomass of sagebrush. During the maximal productive period (May and September), steppe grasses and ephemeroids have an underground mass exceeding the above-ground one: in *Poa bulbosa* and *Carex pachystylis*, underground mass is 11 to 12 times higher, and in steppe grasses, 6 to 8 times higher in May, and 27 to 50 times higher in September. At the same time, the maximum above-ground phytomass of sagebrush in September was 57% of its underground stock, with only 8.7% of dead phytomass. This is the reason why sagebrush cannot withstand slope erosion during mudslides or wind erosion of soil between single sagebrush plants. The vulnerability of mountain valleys to mudslides is thus enhanced by replacement of the stabilizing woodland and steppe communities by the sagebrush.

The general aridization of habitats, increasing aggression of sagebrush communities, and degradation of juniper woodlands and mountain steppes all have led to the rapid desertification of Kopetdagh, with catastrophic ecological consequenses.

Conservation Measures

An optimization of the natural environment of Kopetdagh should be directed toward facilitating natural processes of soil restoration as well as toward increasing productivity of natural and agricultural communitites.

The first step should be to conduct an ecological survey in order to predict future developments in the upper mountain belts of Kopetdagh, where juniper woodlands are the major stabilizer of ecological situations. Such a survey would provide ecological evaluation of all forests and woodlands. This task, as well as the long-term monitoring of these communities, could be performed by state national parks personnel. The network of protected natural territories in Turkmenistan should be considerably extended, including such areas as Dushakerekdagh and Kuruhaudan in Central Kopetdagh. This would allow not only for preservation of genetic pool of the unique Kopetdagh flora as well as its resource value but also the solution to the problem of agricultural productivity in foothill area. In a broader sense, creation of such a network would also serve as the initial step toward resolution of the severe ecological crisis which threatens Kopetdagh.

A second step in the series of salvaging measures should be the rapid phytomelioration of the dead salt-accumulating lands of ancient irrigation in the underhill lowlands. This would comprise traditional as well as non-traditional agricultural techniques, including protective culturing of phytomeliorative plant species and introduction of drought- and salt-resistant feed species of boreal and tropical origin (e.g., species of sorghum, millet, amaranth, and liquorice). The ecological environment in the Kopetdagh foothills could be also improved through the introduction of active soil-protecting measures, creation of artificial pastures with overhead irrigation, and desalination of abandoned lands.

A third necessary step in the ecological conservation of Central Kopetdagh is the organization of an informational service which would record ecological changes such as the disappearance of a rare plant species, loss of an original, locally selected strain, drying of a mountain spring, transformation of plant communities, and a variety of other direct and indirect events. Analysis of this information would facilitate directed measures of conservation. We must avoid the sad prospect of living between two deserts, the lowland and the mountain one.

Maple forest on Mt. Syunt (Southwest Kopetdagh). Photo by H.R. Levenshtein and I.A. Mukhin.

9. Vegetation of Southwest Kopetdagh

GALINA N. FET

Abstract

A study of the vegetation of Southwest Kopetdagh was conducted from 1978 to 1986 in the Syunt-Khasardagh Natural Reserve and adjacent areas; it resulted in the description of plant communities on 602 plots, with 585 species of higher plants and 14 moss species registered. Fifteen types of vegetation were distinguished and characterized. Spatial structure of vegetation complexes was studied and mapped. Restoration of tree and shrub species in the Reserve, resulting from the protection regime begun in 1978, was also studied.

Introduction

The diverse and unique vegetation of Southwest Kopetdagh has attracted the attention of researchers since the 1900s (Lipsky 1915). It was for the first time described in detail by Linchevsky (1935), who listed 926 higher plant species and outlined the structure of plant communties along an altitudinal gradient. In 1930, the All-Union Plant Breeding Institute (VIR) established a regional Turkmen Experimental Station (TOS) in Kara-Kala primarily to study genetic resources of local wild ancestors of fruit trees and shrubs (Popov 1929; Vavilov 1967). Recent floristic works in this region and adjacent areas have resulted in a list of West Kopetdagh flora (Gudkova *et al.* 1982) which includes 1,226 plant species, and a list of flora for the southernmost range of Southwest Kopetdagh (the Palyzan Range; Proskuryakova 1985). The vegetation of Southwest Kopetdagh remained unstudied after the pioneering works of Chernyakovskaya (1924, 1927), Korovin (1927), and Linchevsky (1935). However, plant communitites of several adjacent areas were described from the 1930s to 1970s; these areas include Bolshoi Balkhan (Bobrov 1931), Kyurendagh (Pyatayeva 1954), Central Kopetdagh (Mikeshin and Alyokhin 1945; Mikeshin 1946a,b; Nikitina 1954), and Iran (Zokhary 1963, 1973).

Flora of Kopetdagh is believed to have originated from the Ancient Mediterranean stock common to all floras of Middle Asia (Kamelin 1973);

however, its unusually high specific endemism (up to 18%; Kamelin 1970) reflects its prolonged isolation in the Neogene from all other mountains of Middle Asia, especially those eastward of Kopetdagh. Arid woodlands (shiblyak) were formed in the lower belts by the early to middle Pliocene, and were impoverished later (Agakhanyants 1981). In the Pliocene/Pleistocene time, the Kopetdagh Range was an important center of origin of mountain xerophyte flora; the mountain steppe communities in Kopetdagh were also formed by the early Pleistocene (Agakhanyants 1981). The mountain steppe communities (in the middle and upper mountain belts) played an important role in the authochthonous evolution of enndemic species. In general, the different ages and origins of flora and vegetation of Southwest Kopetdagh determined the existence here of an extremely diverse set of plant communities. Currently, these communities are profoundly affected by human activity which leads to the rapid degradation of vegetation and desertification of the landscapes.

Altitudinal Belts of Vegetation in Southwest Kopetdagh

Boundaries between mountain belts in Kopetdagh are not distinct, and there is no single opinion on altitudinal zonation. This mountain range does not reach snowline, and only in Central Kopetdagh at the altitudes from 2,500 to 3,000 m, are found subalpine plant communitites (Nikitina 1954). The majority of researchers distinguish several belts of vegetation, with primary separation between the foothills and mountainous portions (the foothills are limited from below by the submontane plain). Chernyakovskaya (1927) distinguished the following altitudinal units of vegetation: submontane plain, sagebrush steppe, grass and herbaceous-grass steppe, juniper forest, highland (including mountain xerophyte formation), canyon vegetation, and intrazonal vegetation of rocky outcrops. In fact, all later schemes of altitudinal zonation have been only variations of this, generally correct division. Linchevsky (1935) distinguished three belts in Southwest Kopetdagh which he named "mountain semidesert" (from 300 to 700 m), "mountain semi-steppe" (from 700 to 1,500–1,800 m), and "mountain steppe" (from 1,600 to 1,800–2,000 m). Lavrenko (1965) characterized the altitudinal zonation of Kopetdagh vegetation as "steppe-tragacanth-woodland-ephemer-sagebrush" type. Korovin (1961, 1962) and Stanyukovich (1955, 1973) classified Kopetdagh as a variant of the "Iranian type of altitudinal zonation," characterized by ephemeroid-sagebrush vegetation in the foothills and lower mountain belt and by steppes in the middle and upper belts. Both Nikitina (1956) and Kamelin (1970) distinguished six belts of vegetation in Kopetdagh although neither clearly defined borders between these belts. Kamelin (1970) discerned that only two belts are truly distinct: "shiblyak and semisavanna belt" (from 400 to 1,500 m) and "juniper and steppe belt" (from 1,600 to 2,800 m). The zone of contact between these two belts contains communitites of deciduous xeromesophilic trees and shrubs.

We accept three altitudinal belts of vegetation for Southwest Kopetdagh; this

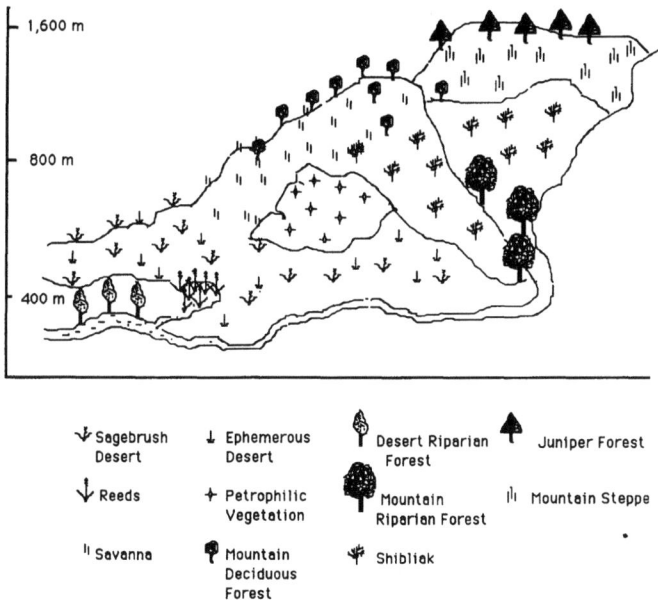

Fig. 1. Altitudinal distribution of vegetation in Southwest Kopetdagh.

division is based on a comparison of all known schemes and generally agrees with those of Nikitina (1954) and Agakhanyants (1981). These three belts are (1) the foothills (where the submontane plain ends at 300 to 400 m, and the foothills proper, or *adyrs*, occupy altitudes from 400 to 800 m); (2) the lower mountain belt (from 800–900 to 1,200–1,400 m); and (3) the middle mountain belt (from 1,200–1,400 to 1,900 m). Fig. 1 represents a general scheme of altitudinal distribution of vegetation in Southwest Kopetdagh.

Methods of Vegetation Survey

Our study of vegetation in Southwest Kopetdagh (a region of approximately 30,000 km²) was conducted from 1978 to 1986 in the Syunt-Khasardagh Natural Reserve (ca. 300 km²) and adjacent areas; it resulted in the description of plant communities on 602 plots, with 585 species of higher plants and 14 species of mosses registered. To study the process of tree and shrub restoration under protection regime of the Syunt-Khasardagh Reserve, the composition of trees and shrubs was described on 204 plots of 100 m². Key plot areas were mapped on a scale of 1:1,000 to 1:5,000. Vegetation maps (1:100,000) were compiled for two areas of the Reserve: the Syunt-Khasardagh (or Central) Area and Aidere Area. Combined, the areas comprise 23,000 ha.

There is no consensus among phytocoenologists on classification of the extremely complex plant communities of mountainous Middle Asia. In arid mountains, plant communities are highly mosaic, polydominant, sometimes lacking in definite dominant species altogether, and often considerably disturbed.

A number of researchers have based their classifications of this vegetation on different principles. The best-known approach (Korovin 1927, 1961, 1962) was based on species ecology and history of community formation. A traditional dominant ecobiomorph approach was used by Prozorovsky (1940), Lavrenko (1940), Rodin (1948, 1958), and Blyumental (1979). Ovchinnikov (1957) and Kamelin (1973) created a florogenetic approach to classification of the Middle Asian mountain vegetation, based on their many years of study in Tajikistan. Zakirov and Zakirov (1978) applied a rather ecological, climatico-edaphic approach. Below, we give a description of the types of vegetation found in Southwest Kopetdagh based on an ecological-phytocoenological analysis of plot descriptions, enhanced by florogenetic interpretation.

Another practical approach toward the classification of plant communities is floristic analysis of plot descriptions of vegetation. A classification built solely on the basis of dominance of certain species creates numerous association units which, in fact, may differ only by the ratio of codominants but which share a common core set of species. For the analysis of vegetation dynamics, for mapping of vegetation, or for ecological comparisons of different habitats, it is essential to take into account not only dominant plant species but entire ecological groups of species that may not be abundant but that indicate a degree of expression of certain ecological factors.

In our first attempt to apply the floristic classification, we used 60 typical descriptions (20 for each mountain belt) to compile three characteristic tables (Tables 1–3). Classification units were tentatively treated as associations and subassociations *sensu* Braun-Blanquet. Currently, we choose not to ascribe a formal syntaxonomic rank to these units; it would be incorrect to place these associations in the classes and orders of vegetation described from the West Europe (Mirkin *et al.* 1984; Mirkin 1985) until further syntaxonomic (floristic) analysis of the vegetation of Middle Asia and Iran is undertaken.

Types of Vegetation in Southwest Kopetdagh

Semishrub (Sagebrush) Deserts. Plant communities of this type in Middle Asia are confined to the submontane plains, foothills, and low mountains. Mountain deserts characteristic for Tien Shan or Pamir are absent from Kopetdagh. Here, dominant plant species are sagebrushes belonging to the subgenus *Seriphiduim*, the *Artemisia herbaalbae* species group (*Artemisia badhysi, A. turcomanica,* and *A. kulbadica*). Sagebrush communities occupy most of the area of the foothills of Southwest Kopetdagh. Most widespread are communities dominated by *Artemisia* spp., *Poa bulbosa*, and desert sedge (*Carex pachystylis*); coverage of *Artemisia* spp. may reach from 5 to 40%; *Poa bulbosa*, from 5 to 20%; and *Carex pachystylis*, from 5 to 10%. These communities include numerous ephemerous and ephemeroid plant species.

Table 1. Vegetation of the foothills of Southwest Kopetdagh. (Per cent of plot area coverage for a plant species is estimated; presence of a species with coverage less than 1% is designated as "+")

Plot Numbers	002	001	004	007	015	005	011	006	022	024	012	025	020	009	018	017	016	010	014	021
Common species of sagebrush association:																				
Artemisia spp.	20	20	10	10	5	10	20	15	30	20	30	15	15	1	5	10	1	5	–	5
Poa bulbosa	5	10	5	5	–	5	+	5	10	20	15	10	5	–	–	+	–	1	–	+
Stachys turcomanica	–	–	–	+	10	–	5	–	+	1	–	–	1	–	1	–	5	1	5	5
Phlomis cancellata	–	–	–	+	5	+	+	–	–	–	–	–	–	–	1	–	–	5	+	–
Lappula barbata	–	–	–	+	+	–	+	+	–	–	–	–	–	+	+	–	–	–	–	–
Ferula gummosa	+	–	+	–	–	–	–	–	5	–	–	–	–	–	–	–	5	–	+	–
Convolvulus fruticosus	–	–	–	+	–	+	–	–	–	–	–	–	–	+	–	–	1	–	–	–
Bongardia chrysogonum	+	–	–	+	+	+	+	+	–	+	–	+	+	+	+	+	–	+	–	+
Callipeltis cucullaris	–	–	–	+	+	+	–	–	–	–	–	–	+	–	+	–	–	+	–	–
Matthiola farinosa	–	–	–	–	–	+	–	–	+	+	–	–	–	+	+	–	–	–	+	+
Astragalus xanthoxiphidium	+	+	–	–	–	–	+	–	–	+	–	+	+	+	–	–	–	–	–	–
Gladiolus italicus	–	–	–	–	–	–	–	–	–	–	–	–	–	–	+	–	–	–	+	–
Bunium longipes	–	–	–	+	+	–	+	–	+	+	–	+	–	+	–	–	–	–	+	+
Nonea turcomanica	–	+	–	–	–	–	–	–	–	–	–	+	–	+	–	–	–	–	+	–
Eremostachys boissieriana	–	–	–	+	–	–	–	–	–	–	–	–	–	+	–	–	–	–	–	–
Characteristic species of ephemerous-sagebrush association:																				
Zygophyllum atriplicoides	+	+	+	+	+	+	+	–	–	–	–	+	+	–	–	–	–	–	–	–
Allium rubellum	1	1	+	+	+	+	+	+	–	+	–	–	+	+	–	+	–	–	–	–
Tulipa micheliana	+	+	+	–	–	–	–	–	–	+	+	+	+	–	+	–	–	–	+	–
Anisantha tectorum	–	–	–	–	+	+	–	–	–	+	1	+	+	–	+	–	–	–	–	–
Eremopyrum orientale	–	+	+	–	+	–	–	+	+	+	+	–	–	–	–	–	–	–	–	–
Helianthemum salicifolium	+	+	+	+	+	–	–	+	–	+	+	–	–	–	–	–	–	–	+	–
Lamium turkestanicum	–	+	–	–	–	–	–	–	–	–	+	–	–	–	–	–	–	–	–	–
Erodium cicutarium	+	+	–	–	–	+	+	+	–	+	+	–	–	+	–	–	–	–	–	–
Gagea gageoides	+	+	–	–	–	+	+	–	–	–	+	+	+	–	+	–	–	–	–	–
Bromus japonicus	+	–	–	–	+	+	+	–	+	–	–	+	+	–	–	+	–	–	–	–
Ziziphora tenuior	–	–	–	–	+	–	+	–	+	–	–	+	+	–	–	–	–	–	–	+

Table 1. Continued

Plot Numbers	002	001	004	007	015	005	011	006	022	024	012	025	020	009	018	017	016	010	014	021
Rostraria cristata	–	+	–	+	–	+	–	–	–	–	–	–	–	–	–	–	–	–	–	–
Arabis auriculata	–	–	–	+	+	–	–	–	+	+	+	–	–	–	–	–	–	–	–	–
Adonis vernalis	–	–	–	+	–	–	+	–	–	–	–	–	+	–	–	–	–	–	–	–
Androsace turczaninovii	–	–	–	–	–	–	–	+	–	+	–	–	+	–	–	–	–	–	–	+
Lithospermum turkestanicum	+	+	–	–	–	–	–	+	+	+	–	–	–	–	–	+	–	–	–	–
Hypecoum trilobum	–	+	–	–	+	+	–	+	–	+	–	–	+	–	–	+	–	–	–	–
Cryptospora falcata	–	–	–	+	–	+	–	+	–	–	–	–	–	–	–	–	–	–	–	–
Phalaris minor	+	+	–	+	–	+	–	+	–	–	–	–	+	–	–	–	–	–	–	–
Iris ewbankiana	–	–	+	–	+	+	–	–	–	–	–	–	+	–	–	–	–	–	+	–
Stipa arabica	+	+	–	+	+	+	+	–	–	–	+	–	+	–	–	–	–	–	–	–
Avena barbata	+	+	–	–	–	+	–	–	–	–	–	–	–	–	–	–	–	–	–	–
Malcolmia africana	–	–	1	–	–	–	–	+	–	–	–	–	–	–	–	–	–	–	–	–
Asparagus verticillatus	+	–	–	+	+	+	+	+	–	–	–	–	–	–	–	–	–	–	–	–
Vicia angustifolia	–	–	+	+	+	–	–	+	–	–	–	–	–	–	–	–	–	–	–	–
Rhinopetalum gibbosum	+	+	–	+	–	+	–	+	–	–	–	–	–	–	–	–	–	–	–	–
Erysimum ischnostylum	+	+	+	+	–	+	–	–	–	–	–	–	–	–	–	–	–	–	–	–
Silene coniflora	–	–	+	+	–	+	–	+	–	–	–	–	–	–	–	–	–	–	–	–
Senetio vernalis	–	–	+	–	–	+	–	–	+	–	–	–	–	–	–	–	–	–	–	–
Differential species of desert sagebrush subassociation:																				
Carex pachystylis	–	–	–	–	–	–	–	–	+	–	+	–	–	–	–	–	–	–	–	–
Differential species of Salsola-Cousinia-sagebrush subassociation:																				
Cousinia albiflora	–	–	–	5	5	+	5	+	–	–	–	15	5	+	–	–	–	–	5	–
Cousinia hypopolia	–	–	–	+	+	–	–	–	–	–	–	–	10	–	–	–	+	–	–	1
Salsola aucheri	+	–	–	+	+	–	+	–	–	–	–	15	5	–	–	–	–	–	–	1
Characteristic species of ecotone communities:																				
Caccinia crassiflora	–	–	–	–	+	+	+	+	–	–	–	–	–	–	–	+	+	–	–	–
Lagochilus kabulicus	–	–	–	–	+	+	–	+	–	–	–	–	–	–	+	+	+	–	–	–
Euphorbia helioscopia	–	–	+	–	+	–	–	–	–	–	–	–	–	+	+	–	+	+	–	–
Ferula oopoda	+	1	–	+	–	–	5	–	–	–	–	–	–	+	–	–	–	–	–	+

Table 1. Continued

Plot Numbers	002	001	004	007	015	005	011	006	022	024	012	025	020	009	018	017	016	010	014	021
Zosima absinthifolia	-	-	-	-	-	-	-	-	-	-	-	-	-	-	-	-	-	-	-	-
Merendera jolantae	+	+	-	-	-	-	-	-	-	-	-	-	-	+	-	+	-	-	-	-
Loliolum subulatum	-	-	-	+	-	-	+	-	-	-	-	-	-	-	+	+	-	-	+	-
Scorzonera pusilla	-	-	+	+	+	+	+	+	-	-	-	-	-	+	+	-	-	-	+	-
Ceratocephala falcata	+	+	-	-	-	+	+	+	-	-	-	-	-	-	+	-	-	-	+	-
Crupina vulgaris	-	+	-	+	-	-	-	+	-	-	-	-	-	+	-	-	-	-	+	-
Galium tricorne	-	-	-	+	-	-	-	+	-	-	-	-	-	-	+	-	-	-	+	-
Galium tenuissimum	-	-	-	+	-	-	-	-	-	-	-	-	-	+	-	+	-	-	-	-
Astragalus ammodendron	-	-	-	-	+	+	+	-	-	-	-	-	-	+	-	-	-	+	+	-
Ziziphora capitata	-	-	-	-	+	-	-	-	-	-	-	-	-	+	-	-	-	+	+	-
Papaver hybridum	-	-	-	-	+	+	-	+	-	-	-	-	-	+	-	+	-	-	-	-
Silene conica	-	-	-	-	-	-	-	+	-	-	-	-	-	+	+	+	-	-	-	-
Silene swertiifolia	-	-	-	-	+	-	+	-	-	-	-	-	-	+	-	-	-	-	-	-
Scutellaria litwinovii	-	-	-	-	-	-	-	-	-	-	-	-	-	-	-	+	+	-	+	-
Convolvulus subhirsutus	-	-	-	-	+	-	-	+	-	-	-	-	-	-	-	-	-	-	+	-
Medicago rigidula	-	-	-	-	+	-	-	-	-	-	-	-	-	-	-	-	-	-	+	-
Chardinia orientalis	-	-	-	-	+	+	+	-	-	+	-	-	-	-	-	-	-	-	-	-
Iris longiscapa	+	-	-	-	-	-	-	-	-	-	-	-	-	+	-	-	-	-	-	-
Differential species of petrophiteephemeroussagebrush subassociation:																				
Rubia florida	-	-	-	-	-	+	-	-	-	-	-	-	+	+	+	+	+	-	+	-
Atraphaxis spinosa	-	-	-	-	-	-	-	-	-	-	-	-	-	+	+	+	-	+	+	-
Teucrium polium	-	-	-	-	-	-	+	-	-	-	-	-	-	+	+	+	-	+	1	-
Prangos latiloba	-	-	-	-	-	-	-	-	-	-	-	-	-	+	-	+	-	-	-	+
Merendera robusta	-	-	-	-	-	-	-	-	+	-	-	-	-	-	+	-	+	-	-	+
Hymenocrater bituminosus	-	-	-	-	-	+	-	-	-	-	-	-	-	-	+	-	+	+	-	+
Acanthophyllum multiflora	-	-	-	-	-	-	-	-	-	-	-	-	-	-	-	-	+	+	-	+

Table 1. Vegetation of the foothills of Southwest Kopetdagh. (Per cent of plot area coverage for a plant species is estimated; presence of a species with coverage less than 1% is designated as "+")

Plot Numbers	002	001	004	007	015	005	011	006	022	024	012	025	020	009	018	017	016	010	014	021
Hieracium procerum	–	–	–	–	–	–	–	–	–	–	–	–	–	–	+	+	+	–	–	–
Eryngium bungei	–	–	–	–	+	–	–	–	–	–	–	–	–	–	+	+	–	–	–	–

Rare species (found in less than 10 % of plot descriptions): *Juno fosterana, Lithospermum officinale, Cerastium glomeratum, Astragalus tribuloides, Holosteum glutinosum* (002), *Arnebia decumbens* (002, 012), *Sisymbrium loeselii, Koelpinia linearis* (004, 022), *Festuca arundinacea, Artemisia serotina, Onobrychis sp., Poa angustifolia, Salsola gemmascens* (007), *Aegylops triuncialis* (007, 015), *Reaumuria reflexa* (007, 005), *Milium vernale* (007, 009), *Adonis parviflora* (007, 006), *Veronica arvensis* (007, 021), *Setaria glauca* (007, 003), *Lactuca serriola, Chesneya kopetdaghensis, Valerianella coronata, V. dufresnia, Cynodon dactylon, Viola occulta, Garhadiolus hedypnois, Onosma dichroanthum, Poterium sanguisorba, Dactylis glomerata* (015), *Haplophyllum acutifolium* (015, 009), *Galium verticillatum, Turgenia latifolia* (015, 011), *Roemeria refracta* (015, 012), *Capparis spinosa, Scrophularia pruinosa, Suchtelenia acmophylla* (015, 014), *Crucianella gilanica* (015, 017), *Tragopogon sp.* (005), *Lallemantia sulphurea* (005, 024), *Nonea caspica* (005, 009), *Hulthemia persica* (005, 011), *Muscari bucharica* (005, 010), *Garhadiolus papposus, Thalictrum isopyroides, Bothriochloa ischaemum* (011), *Bromus danthoniae* (011, 010), *Centaurea squarrosa, Paracaryum turcomanicum* (011,014), *Meniocus linifolius, Thalictrum minus, Lathyrus inconspicuus, Cnicus benedictus* (006), *Lathyrus sphaericus, Valerianella platycarpa* (006, 009), *Eruca sativa, Echinaria capitata, Viola karakalensis* (006, 014), *Phragmites australis, Poa nemoralis, Hordeum spontaneum, Psylliostachys spicata* (022), *Trigonella orthoceras* (022, 009), *Vulpia myuros, Eremostachys labiosiformis, Apera interrupta, Achillea biebersteinii* (024), *Anthemis altissima* (024, 025), *Ixiolirion tataricum* (024, 018), *Malcolmia grandiflora, Astragalus filicaulis, Vulpia persica* (012), *Ranunculus muricatus, Tripleurospermum disciforme, Leptaleum filifolium* (025), *Thlaspi perfoliatum* (025, 009), *Phleum paniculatum, Aegylops cylindrica* (020), *Aetionema transhyrcanum* (009), *Amygdalus scoparia* (009, 014), *Vencetoxicum pumilum, Rhamnus sintenisii* (009,016), *Hypericum perfoliatum, Scrophularia turcomanica, Lycium turcomanicum, Haplophyllum obtusifolium* (018), *Cerasus microcarpa* (018, 016), *Lithospermum arvense* (017), *Thesium arvense, Tragacantha sp., Stipa caucasica, Juniperus turcomanica* (016), *Crucianella sintenisii, Bupleurum gerardii, Clypeola jonthlaspi* (010), *Salvia viridis, Cirsium vulgare, Bromus oxyodon, Delphinium turcomanicum* (004).

Table 2. Vegetation of the lower mountain belt of Southwest Kopetdagh

Plot Numbers	535	409	411	412	404	406	422	407	410	405	408	411	413	424	425	512	423	538	536	537
Common species of the lower mountain belt:																				
Poa bulbosa	+	5	5	5	5	5	10	-	-	5	5	1	1	5	5	+	10	-	-	1
Anthemis altissima	-	-	-	-	+	+	-	-	1	1	+	+	1	1	1	1	-	-	1	1
Crambe kotchyana	-	+	+	+	+	+	-	-	-	+	-	-	+	1	-	-	-	+	-	-
Achillea biebersteinii	-	+	+	+	+	-	-	5	-	-	-	-	-	-	-	-	+	+	-	-
Crupina vulgare	-	-	-	-	+	+	+	+	-	+	+	-	-	-	-	+	-	-	+	-
Meniocus linifolius	-	-	-	-	+	+	+	-	-	+	-	-	+	-	-	+	+	-	-	-
Bunium longipes	+	-	-	-	+	-	+	-	-	-	-	+	-	+	-	+	+	-	-	-
Silene conica	-	-	-	+	+	-	+	-	-	-	-	-	+	-	-	-	+	-	-	-
Scabiosa micrantha	-	+	+	+	-	+	-	-	-	-	+	-	-	+	-	-	-	+	-	-
Eremurus olgae	-	+	1	-	-	+	-	-	-	-	+	-	-	-	-	-	-	+	-	-
Vicia angustifolia	-	-	-	-	-	+	-	-	-	-	+	-	-	-	-	-	-	-	-	-
Galium teniussimum	-	-	-	-	-	+	+	-	-	-	-	-	-	+	+	-	-	-	-	-
Holosteum umbellatum	-	-	-	-	+	+	+	-	-	+	-	-	-	+	+	-	-	-	+	-
Alyssum desertorum	+	+	+	+	+	+	+	-	-	+	-	-	+	+	-	-	-	-	-	-
Cerastium perfoliatum	-	-	-	-	-	+	+	-	-	+	-	-	+	+	-	-	-	-	-	-
Helianthemum salicifolium	+	+	+	+	+	+	+	+	-	+	-	+	+	-	+	+	-	-	-	-
Avena barbata	-	-	-	-	-	-	-	-	-	-	-	-	-	-	-	+	+	-	-	-
Eryngium biebersteinii	+	+	+	+	+	-	-	-	-	-	-	+	+	+	-	-	+	-	-	-
Stachys turcomanica	-	-	5	-	-	-	-	-	-	-	-	+	-	-	5	+	+	1	-	-
Astragalus nephthonensis	1	-	-	-	-	-	-	-	-	-	-	-	-	-	-	-	-	-	-	-
Phlomis kopetdaghensis	+	-	+	-	-	-	-	-	-	-	-	-	-	-	-	+	-	-	-	-
Ephedra equisetina	-	+	-	-	-	-	1	-	-	-	-	+	1	-	-	-	-	-	+	-
Erodium cicutarium	-	-	-	-	-	-	+	-	-	-	-	+	+	-	-	-	-	-	-	-
Ziziphora tenuior	-	+	-	-	-	+	-	-	-	+	+	-	+	-	-	-	-	-	-	-
Characteristic species of the complex of semisavanna and mountain steppe:																				
Elytrigia trichophora	20	40	30	20	-	-	-	-	+	5	5	+	+	-	-	-	-	1	-	-
Festuca valesiaca	20	5	+	-	-	-	-	-	-	-	-	-	-	-	-	-	-	-	-	-

Table 2. Continued

Plot Numbers	535	409	411	412	404	406	422	407	410	405	408	411	413	424	425	512	423	538	536	537
Characteristic species of sagebrush communities:																				
Artemisia ciniformis	+	–	–	–	20	20	25	+	–	–	–	+	+	–	5	20	–	–	10	–
Characteristic species of communities derived from those of semisavanna and sagebrush:																				
Perovskia abrotanoides	1	–	–	+	–	–	15	1	–	25	20	1	+	–	–	–	5	–	10	20
Taeniatherum crinitum	–	+	–	–	–	+	–	–	+	10	–	–	–	–	–	–	–	–	–	–
Aegylops triuncialis	–	–	–	–	1	1	–	10	5	5	1	–	–	–	+	–	–	–	–	–
Differential species of the complex of semisavanna and sagebrush communities:																				
Galium verum	1	+	+	+	+	–	–	–	–	+	–	–	–	–	–	5	1	20	1	1
Convolvulus subhirsutus	1	+	5	5	5	5	10	–	–	5	5	+	–	5	–	–	–	–	–	–
Stipa turkomanica	+	1	5	5	5	5	–	5	1	5	5	+	–	–	1	–	–	+	–	–
Lappula barbata	–	+	+	+	1	+	–	+	+	+	+	–	–	–	+	–	+	–	+	–
Gladiolus italicus	–	–	–	–	–	+	+	+	+	–	–	–	–	–	–	–	–	–	+	–
Teucrium polium	+	+	+	–	–	–	+	+	+	+	+	+	–	–	–	–	–	–	+	–
Allium rubellum	+	–	–	–	–	–	+	+	–	–	+	–	–	–	–	–	–	–	–	–
Eremostachys boissieriana	+	+	–	–	–	+	+	+	+	+	+	–	–	–	–	–	–	–	–	–
Queria hispanica	–	–	–	–	–	–	+	+	1	+	–	–	–	–	–	–	–	–	–	–
Aegylops cylindrica	–	+	+	–	–	+	+	+	–	–	–	+	–	–	–	–	–	–	–	–
Haplophyllum obtusifolium	–	+	1	–	–	–	+	–	–	–	+	+	–	–	–	–	–	–	–	–
Bromus danthoniae	–	–	+	+	–	+	+	–	–	–	–	–	–	–	–	–	–	–	–	–
Erysimum ischnostylum	–	+	+	–	–	–	–	–	–	–	–	–	–	–	–	–	–	–	–	–
Differential species of the complex of forest and meadow communities:																				
Cerasus microcarpa	+	–	–	–	–	–	–	–	–	–	–	+	+	+	5	5	5	10	+	–
Acer turcomanicum	–	–	–	–	–	–	–	–	–	–	+	+	–	–	–	10	5	20	–	–
Salvia virgata	1	–	–	–	–	–	–	–	+	–	–	–	–	+	1	–	+	5	1	5
Dactylis glomerata	–	+	–	–	–	–	–	–	–	–	–	–	+	+	–	+	–	–	10	–
Galium aparine	–	–	–	–	–	–	–	–	–	–	–	+	+	+	+	–	1	–	–	5
Anisantha sterilis	–	–	–	–	–	–	–	–	–	–	–	5	5	–	5	5	5	–	–	–
Colutea gracilis	–	–	–	–	–	–	–	–	–	–	–	–	–	–	5	5	–	–	+	–
Hymenocrater bituminosus	–	–	–	–	–	–	–	–	–	–	–	–	–	–	5	–	1	–	–	5

Table 2. Continued

Plot Numbers	535	409	411	412	404	406	422	407	410	405	408	411	413	424	425	512	423	538	536	537
Ephedra intermedia	-	-	-	-	-	-	-	-	-	-	-	-	-	-	-	-	5	10	-	-
Crataegus spp.	-	-	-	-	-	-	-	-	-	-	-	-	-	-	+	-	1	10	-	5
Melandrium boissieri	-	-	-	-	-	-	-	-	-	-	-	-	+	-	1	-	-	-	-	-
Crucianella sintenisii	-	-	-	-	-	-	-	-	-	-	-	-	-	-	-	+	-	+	+	1
Hieracium procerum	-	-	-	-	-	-	-	-	-	-	-	-	-	-	-	-	+	-	+	+
Scandix stellata	+	-	-	-	-	-	-	-	-	-	+	+	-	-	-	-	+	-	-	-
Potentilla transcaspica	-	-	-	-	-	-	-	-	-	-	+	+	-	-	-	-	+	-	+	+
Medicago sativa	-	-	-	-	-	-	-	-	-	-	-	-	-	5	-	-	+	-	10	5
Variant with xerophylic grasses:																				
Anisantha tectorum	-	-	-	-	-	+	-	-	+	-	-	5	-	1	10	-	-	-	-	-
Hordeum bulbosum	-	-	-	-	-	-	-	-	-	-	-	+	5	25	+	-	-	-	-	-
Ferula ovina	-	-	-	-	-	-	-	-	+	-	-	+	+	+	-	-	-	-	-	-
Variant with Elytrigia repens:																				
Elytrigia repens	-	-	-	-	-	-	-	-	-	-	-	-	-	-	-	-	-	10	10	30

Rare species (found in less than 10% of plot descriptions): *Astragalus retamocarpus, Heteropappus canescens, Gypsophila bicolor* (535), *Imula cordata* (535, 537), *Cotoneaster nummularioides* (535, 421), *Onosma dichroanthum, Delphinium camptocarpum, Scabiosa rotata, Eremopyrum orientale, Thymus transcaspicus, Astragalus vicarius, Daucus carota* (409), *Glycyrrhiza glabra* (409, 411), *Valerianella capitata, Tulipa hoogiana* (409, 406), *Euphorbia bungei* (411, 422), *Verbascum songoricum* (412, 408), *Velezia rigida, Alyssum campestre, Hypericum scabrum* (404), *Ceratocephala falcata* (404, 406), *Malcolmia africana, Scorzonera litvinovii, Milium vernale, Anemone petiolulosa, Trigonella orthoceros, Avena eriantha* (406), *Nonea caspica* (406, 408), *Bongardia chrysogonum, Onobrychis pulchella, Astragalus xanthoxiphidium* (406, 422), *Chardinia orientalis* (406, 405), *Koelpinia linearis* (406, 408), *Camelina rumelica, Ranunculus pinnatisectus, Clypeola jonthlaspi, Crepis sancta, Lallemantia royleana, Muscari bucharica, Gagea stipitata* (422), *Matthiola farinosa, Cryptospora omissa* (405), *Carex pachystylis, Androsace turczaninovii* (408), *Tragacantha sp., Galium tricorne, Prangos latiloba, Euphorbia monostyla, Celtis caucasica, Berberis interregima* (421), *Rhamnus sintenisii* (421, 413), *Rubia florida, Psammogeton setifolium* (421, 425), *Poterium sanguisorba* (421, 536), *Scrophularia turcomanica, Poa trivialis, Juniperus turcomanica* (421), *Araphaxis spinosa* (421, 413), *Amygdalus communis, Medicago rigidula, Cerastium inflatum* (413, 424), *Centaurea vulgare, Calendula persica, Geranium lucidum, Lamium turkestanicum* (413), *Centaurea sintenisii* (413, 423), *Tragopogon sp.* (413, 425), *Cousinia umbrosa, Phleum paniculatum, Cardaria draba, Valerianella coronata, Malva neglecta, Rochelia retorta* (424), *Rumex tuberosus* (424, 425), *Jasminum fruticans, Papaver hybridum* (425), *Echinops ritro* (538, 536), *Ziziphora clinopodioides, Peganum harmala, Allium scabriscapum, Aegylops tauschii, Lactuca serriola, Vulpia myuros, Taraxacum officinalis, Astragalus brachypetalus* (536), *Anthriscus longirostris, Eremostachys labiosiformis* (425).

Table 3. Vegetation of the middle mountain belt of Southwest Kopetdagh

Plot Numbers	528	529	530	531	532	516	517	518	519	515	549	550	551	552	520	521	522	523	533	514
Common species of the middle mountain belt:																				
Galium verum	1	1	1	5	–	1	1	1	10	10	–	5	5	5	5	1	–	5	–	–
Medicago sativa	–	1	+	–	1	–	–	–	–	1	1	1	10	5	1	–	5	10	–	–
Nepeta sintenisii	5	1	1	+	–	–	–	–	–	10	5	–	1	–	+	5	–	+	–	–
Phlomis kopetdaghensis	+	1	1	–	–	+	–	–	5	+	1	1	–	–	+	–	–	–	–	+
Gypsophila bicolor	–	–	–	–	–	–	–	+	+	+	+	+	+	+	–	–	–	–	–	+
Rumex tuberosus	1	+	–	–	–	–	–	–	–	+	+	–	–	+	–	–	–	–	–	–
Lamium turkestanicum	–	+	1	–	–	–	–	–	–	–	–	+	–	+	+	–	–	–	–	–
Salvia virgata	+	–	–	–	–	–	–	+	–	–	–	+	–	+	+	–	+	+	–	–
Verbascum songoricum	–	–	–	–	–	–	–	–	+	+	–	+	–	–	+	–	+	+	+	–
Galium tenuissimum	–	–	–	+	–	–	–	+	+	–	–	–	–	–	–	–	–	–	–	+
Cotoneaster nummularia	–	+	–	–	–	–	–	–	–	–	–	–	–	+	+	–	–	–	–	+
Astragalus brachypetalus	–	–	–	–	–	+	–	–	–	–	–	–	–	–	–	–	–	–	–	–
Hieracium procerum	–	–	+	–	–	–	–	–	–	–	–	–	–	–	+	–	–	–	–	–
Cousinia umbrosa	–	–	–	–	1	–	–	–	–	–	–	–	–	–	–	–	–	–	–	5
Characteristic species of the complex of steppe and semisavanna:																				
Festuca valesiaca	5	10	5	10	10	10	1	+	10	1	–	–	–	–	–	–	–	–	–	–
Elytrigia trichophora	5	5	–	5	1	5	10	10	10	10	–	–	–	–	–	–	–	–	–	–
Crucianella sintenisii	5	–	+	+	–	5	5	5	5	10	–	5	–	–	–	–	–	–	–	–
Hypericum scabrum	5	–	1	1	5	5	10	10	5	5	–	1	–	–	–	–	–	–	–	–
Differential species of the complex of steppe and semisavanna:																				
Poa bulbosa	+	–	1	–	+	+	+	–	–	+	–	–	–	–	–	–	–	–	–	–
Potentilla transcaspica	–	–	1	1	+	–	–	–	–	+	–	–	–	–	–	–	–	–	–	–
Eremostachys boissieriana	+	+	–	–	–	–	–	–	–	+	–	–	–	–	–	–	–	–	–	–
Dactylis glomerata	–	–	–	–	–	+	–	+	+	–	–	–	–	–	–	–	–	–	–	–
Allium rubellum	+	+	–	–	+	–	–	–	–	–	–	–	–	–	–	–	–	–	–	–
Silene bupleuroides	+	–	–	–	–	–	–	–	–	–	–	–	–	–	–	–	–	–	–	–
Ferula ovina	+	–	–	+	–	+	–	–	–	–	–	–	–	–	–	–	–	–	–	–
Helianthemum salicifolium	–	–	+	–	–	+	–	–	–	–	–	–	–	–	–	–	–	–	–	–

Table 3. Continued

Plot Numbers	528	529	530	531	532	516	517	518	519	515	549	550	551	552	520	521	522	523	533	514
Characteristic species of the feathergrass mountain steppe:																				
Stipa hohenackeriana	40	30	10	–	–	–	+	–	–	–	–	–	–	–	–	–	–	–	–	–
Inula cordata	–	+	20	–	–	–	–	–	–	–	–	–	–	–	–	–	–	–	–	–
Characteristic species of the tomillares:																				
Thymus transcaspicus	–	–	–	5	10	–	–	–	–	–	–	–	–	–	–	–	–	–	–	–
Ziziphora clinopodioides	+	+	–	1	5	–	+	+	–	–	–	–	–	–	–	–	–	–	–	–
Acanthophyllum microcephalum	+	–	–	1	1	–	–	–	–	–	–	–	–	–	–	–	–	–	–	–
Differential species of the semisavanna:																				
Lappula barbata	–	–	–	–	–	+	+	+	–	+	–	–	–	–	–	–	–	–	–	–
Zosima absinthifolia	–	–	–	–	–	10	5	–	+	–	–	–	–	–	–	–	–	–	–	–
Scabiosa micrantha	–	–	–	–	–	+	+	+	–	+	–	–	–	–	–	–	–	–	–	–
Centaurea kopetdaghensis	–	–	–	–	–	–	+	+	–	–	–	–	–	–	–	–	–	–	–	–
Euphorbia sequieriana	–	–	–	–	–	+	+	–	–	–	–	–	–	–	–	–	–	–	–	–
Eryngium biebersteinii	–	–	–	–	–	+	+	–	–	+	–	–	–	–	–	–	–	–	–	–
Leontodon asperrimum	–	–	–	–	–	–	+	–	–	+	–	–	–	–	–	–	–	–	–	–
Tragopogon capitatus	–	–	–	–	–	–	+	–	–	–	–	–	–	–	–	–	–	–	–	–
Arenaria serpyllifolia	–	–	–	–	–	+	–	–	+	–	–	–	–	–	–	–	–	–	–	–
Characteristic species of the meadow:																				
Elytrigia repens	–	–	–	–	–	–	–	–	–	10	50	40	30	40	20	50	40	20	1	10
Characteristic species of the maple forest:																				
Delphinium semibarbatus	–	–	–	–	–	–	–	–	–	–	–	–	–	–	–	–	+	–	1	5
Orthurus heterocarpus	–	–	–	–	–	–	–	–	–	–	–	–	–	–	–	–	–	–	5	5
Thelycrania meyeri	–	–	–	–	–	–	–	–	–	–	–	–	–	–	–	–	–	–	5	5
Acer turcomanicum	–	–	–	–	–	–	–	–	–	–	–	–	–	–	–	–	–	10	40	30
Crataegus pseudoambigua	–	–	–	–	–	–	–	–	–	–	–	–	–	–	–	1	–	5	–	5
Galium aparine	–	–	–	–	–	–	–	–	–	–	–	–	–	–	–	–	–	+	+	+
Differential species of the meadow and forest complex:																				
Poa trivialis	–	–	–	–	–	–	–	–	–	–	+	1	1	–	–	–	–	–	–	–

Table 3. Continued

Plot Numbers	528	529	530	531	532	516	517	518	519	515	549	550	551	552	520	521	522	523	533	514
Cardaria draba	–	–	–	–	–	–	–	–	–	–	+	5	+	–	–	–	–	–	–	–
Euphorbia helioscopia	–	–	–	–	–	–	–	–	–	–	1	+	–	+	–	–	–	–	–	–
Convolvulus arvensis	–	–	–	–	–	–	–	–	–	–	+	–	+	+	–	+	–	–	–	–
Peltaria turkmena	–	–	–	–	–	–	–	–	–	–	+	–	–	–	+	–	–	–	–	–
Carex polyphylla	–	–	–	–	–	–	–	–	–	–	+	+	–	–	–	–	–	–	+	–
Allium bodeanum	–	–	–	–	–	–	–	–	+	–	–	–	–	–	+	–	+	+	–	+
Euphorbia falcata	+	–	–	–	–	–	–	–	–	–	–	–	–	–	+	–	+	+	–	–
Alliaria alliacea	–	–	–	–	–	–	–	–	–	–	–	–	–	–	10	–	5	+	+	–
Trisetum flavescens	–	–	–	–	–	–	–	–	–	–	–	–	–	–	+	–	–	+	+	–

Rare species (found in less than 10% of plot descriptions): *Bupleurum gerardii*, *Crupina vulgare*, *Cousinia tenella*, *Anthemis altissima* (528), *Stachys turcomanica*, *Stipa turcomanica*, *Xanthium spinosum* (516), *Thlaspi perfoliatum*, *Taraxacum officinalis*, *Cousinia bipinnata* (517), *Valerianella turkestanica* (518), *Galium verticillatum*, *Milum vernale*, *Psammogeton setifolium* (519), *Onobrychis pulchella*, *Erysimum ischnostylum* (515), *Astragalus retamocarpus* (520), *Dorema hyrcanum* (521), *Anthriscus longirostris*, *Geranium lucidum* (522), *Peucedanum sintenisii* (533), *Euonymus velutina*, *Lonicera bracteolaris*, *Viola suavis* (514).

Taxonomic diversity of species comprising sagebrush desert communities varies according to local soil conditions. The high degree of soil erosion due to cattle grazing in Southwest Kopetdagh leads to the common presence of petrophytes which prefer exposed rock and gravel substrate (e.g., *Stachys turcomanica, Cousinia albiflora, C. hypopodia, Phlomis cancellata,* and *Ephedra intermedia*). Such complexes may completely replace sagebrushes in eroded sites. Under heavy grazing, sagebrush coverage drops to 5 to 10%, and plant communities include such annual grasses as *Eremopyrum orientale, Anisantha tectorum, Avena barbata,* and *Bromus japonicus*. Steep foothill slopes, where erosion may be extreme, are covered by a complex of sparse sagebrushes and *Zygophyllum atriplicoides*. In the Sumbar River Valley and its tributaries, with soils with high salt content communities of *Artemisia olivieriana* and such halophytes as *Salsola dendroides, S. gemmascens,* and *Aellenia subaphylla* are found. In the foothills, sagebrushes form complex with shiblyak shrubs (*Paliurus spinachristii, Colutea gracilis,* and *Hymenocrater bituminosus*).

Another sagebrush species, *Artemisia ciniformis*, is a characteristic dominant species on mountain slopes, where it commonly forms complexes with semisavanna, shiblyak, and petrophites. Mikeshin (1946a) described these communities in Central and West Kopetdagh: *A. ciniformis* is codominant with such mountain grasses as *Festuca valesiaca, Elytrigia trichophora,* and *Stipa turcomanica*, and with shiblyak shrubs (*Ephedra intermedia, E. equisetina,* and *Hymenocrater bituminosus*). Large areas are occupied by a complex of *Artemisia ciniformis* and *Festuca valesiaca*.

Ephemerous Deserts. *Poa bulbosa, Carex pachystylis,* or annual grasses, all highly typical of the foothills of Middle Asia (Korovin 1934), do not occupy significant areas in Southwest Kopetdagh; their absence is undoubtedly due to the high degree of erosion of loess substrate where these communities were originally found (Pavlov 1980). Common to the foothills of Southwest Kopetdagh (up to 1,000–1,200 m) are derived, impoverished communities of annual ephemerous grasses (*Avena barbata, Taeniatherium ornitum, Aegylops triuncialis, Eremopyrum orientale,* and *Anisantha tectorum*). Similar communities dominated by *T. crinitum* have been described from West Tien Shan (Pavlov 1980). Communities dominated by arpagan (*Eremopyrum orientale*) are characteristic for the lowland plains (Rodin 1963; Berdyev 1985) and are practically absent from the mountains.

Semisavanna. This type of vegetation (also called "tall-grass semisavanna" or "*Elytrigia* steppes") includes communities dominated by the perennial grass *Elytrigia trichophora* (with coverage from 50 to 60%) and is characteristic throughout Middle Asia (Demurina 1980). Ecological groups of plants found within these communities can be dominated by ephemeroid species, meadow-steppe herbaceous vegetation, or steppe grasses. Common codominants in the lower portion of the lower mountain belt (from 800 to 1,000 m) include *Poa bulbosa, Helianthemum salicifolium, Convolvulus subhirsutus, Phlomis kopetdaghensis*, and *Perovskia abrotanoides* (the last species is an indicator of a disturbed habitat). Complexes of *E. trichophora* with a steppe grass, *Festuca valesiaca*, also widely found in Uzbekistan (Demurina 1980), spread up to 1,600- -1,700 m and include such characteristic species of meadows and steppes as *Galium verum, Crucianella sinenisii, Thymus transcaspicus*, and *Bunium longipes*. Derived communities of semisavanna may be dominated by *Convolvulus subhirsutus* or *Ferula oopoda*. Other types of derived communities, found in and around former settlements and cattle farms, are dominated by a tall wild barley (*Hordeum bulbosum*).

Meadows. In Southwest Kopetdagh, the meadows are formed primarily by *Elytrigia repens* and are located from 700–800 to 1,500–1,600 m in the depressions of mountain plateaus (e.g., on Mt. Khasardagh or Mt. Gindere) or in the river valleys of the lower mountain belt (e.g., Aidere Valley). *E. repens* covers from 20 to 80 % of these habitats. Most of the described plots are within the territory of the Syunt-Khasardagh Reserve, where hay harvesting is not allowed. Other dominants in meadow communities include *Trisetum flavescens, Dactylis glomerata, Alliaria alliacea, Nepeta sintenisii, Crucianela sintenisii, Anisantha sterilis, Galium verum*, and *G. aparine*. Derived communities are dominated by *Hordeum murinum* and *Eruca sativa*.

Steppe Vegetation. In Southwest Kopetdagh, steppe is developed on mountain plateaus and is similar to the zonal steppes of Asia; dominant grasses here include *Stipa* spp. and *Festuca valesiaca*. The most common are communities of *F. valesiaca* on plateaus from 1,200 to 1,900 m. Common codominant plant species include *Stipa turcomanica, S. hohenackeriana, Galium verum, Thymus transcaspicus, Elytrigia trichophora*, and some species of shrubs. In the foothills, steppe communities may include *Stipa arabica*. Some steppe sites may be dominated by *Hypericum scabrum*.

Shiblyak (Mediterranean Short-Tree Woodland). Diverse tree and shrub communities in Southwest Kopetdagh are classsified within this vegetation type (Popov this volume). In Southwest Kopetdagh, the most commonly found dominant species of shiblyak are the Turkmen maple (*Acer tucomanicum*) and the Christ's thorn (*Paliurus spinachristi*). The Turkmen maple is a small tree or shrub, two to three meters tall; it is widespread in Kopetdagh from 800 to 2,500 m. Maple-shrub communities may cover from 20 to 80% of a given area in which they are found; codominant shrubs found here include *Cerasus microcarpa, Ephedra intermedia, E. equisetina, Cotoneaster nummularia, Colutea gracilis, Paliurus spinachristii, Lonicera bracteolaris, Celtis caucasica, Jasminum fruticans*, and *Amygdalus communis*. Not dominant but very characteristic shiblyak species in Southwest Kopetdagh are pomegranate (*Punica granatum*), jujube (*Zizyphus jujuba*), fig (*Ficus carica*), *Hymenocrater bituminosus*, and *Berberis interregima*. Diverse, usually mesophilic meadow grasses and herbaceous species form the understory, covering from 10–20 to 80–90% of the understory of any given shiblyak stand, with such dominants as *Anisantha sterilis, Trusetum flavescens, Fritillaria raddeana, Allium paradoxum, Orthurus heterocarpus,* and *Lamium turkestanicum.* When disturbed by logging or grazing, maple stands become sparser, and herbaceous cover incorporates derived communities with such dominants as *Elytigia trichophora, Anisantha tectorum, Taeniatherium crinitum,* and *Poa bulbosa.* Complexes of maple and phryganoid (*Thymus transcaspicus* and *Perovskia abrotanoides*) communities are not uncommon. Complexes of maple and juniper communities have features similar to those of juniper woodlands, including the presence of steppe species in the understory. Maple forests in Southwest Kopetdagh are heavily disturbed by uncontrolled logging and grazing of sheep and goats; undisturbed maple stands are found only within the Syunt-Khasardagh Reserve on well-moistened northern slopes (e.g., in the valleys of Aidere and Seutche). The upper altitudinal limit of maple forest is 1,400 to 1,500 m; at its base, maple forest is bordered by communities of the Christ's thorn (*Paliurus spinachristi*).

P. spinachristi, a very characteristic Southwest Kopetdagh shrub, two to five meters tall; it dominates plant communities at altitudes of 300-400 to 700-800 m, and is prevalent on all more-or-less moistened northern slopes and along the beds of temporary streams. The Christ's thorn forms the canopy of its community while the understory is comprised of smaller shrubs such as *Cerasus microcarpa, Colutea gracilis, Jasminum fruticans, Rubia florida, Rhamnus sintenisii,* and *Amygdalus scoparia.* These last two are petrophytes found on rocky or gravel substrates together with herbaceous petrophytes such as *Phlomis kopetdaghensis* and *Cousinia albiflora.* Derived communities are widespread in the foothills and lower mountain belt and include complexes with annual grasses and phryganoid species.

Another common type of shiblyak in the studied region is dominated by hawthorn species (usually *Crataegus pseudoambigua,* but also *C. melanocarpa, C. turkestanica,* and *C. pontica*). Hawthorn communities are similar to those of the Turkmen maple; their complex with maple communities forms the most characteristic tree-shrub vegetation of Southwest Kopetdagh in the lower mountain belt. Only a few shiblyak elements (such as barberry *Berberis interregima* and *Crataegus pontica*) penetrate the middle mountain belt (from 1,600 to 1,900 m) with its juniper woodland and steppe vegetation.

The shiblyak communities of Southwest Kopetdagh house a high number of fruit trees, shrubs, and vines valuable for selection, including pomegranate (*Punica granatum*), wild grapes (*Vitis sylvestris, V. vinifera*), mulberry (*Morus alba*), jujube (*Zizyphus jujuba*), fig (*Ficus carica*), wild apple (*Malus turkmenorum*), wild pear (*Pyrus boissieriana*), wild cherries (*Cerasus microcarpa, C. erythrocarpa, C. blinovskii*), wild plum (*Prunus divaricata*), almonds (*Amygdalus communis* and *A. scoparia*), hawthorns (*Crataegus* spp.,), medlar (*Mespilus germanica*), and quince (*Cydonia oblonga*). Some of these species have been intensively studied, e.g., fig (Petrova 1975), pomegranate (Levin 1985), and almonds (Zaktreger 1966; Denisov 1977; Popov 1981).

Phryganoid Vegetation, or Tomillares. This Mediterranean type of vegetation is represented in Southwest Kopetdagh by communities of the small, fragrant semishrubs belonging to the family Lamiaceae, including *Perovskia abrotanoides, Thymus transcaspicus,* and *Ziziphora clinopodioides.* Phryganoid communities are fragmentarily found from 1,000 to 1,500 m and form complexes with the vegetation of semisavanna and steppe.

Juniper Woodland. Communities of junipers are very common to all arid mountains. There are at least 30 "races" of junipers in the Ancient Mediterranean (Kamelin 1973), and taxonomy of the genus *Juniperus* is controversial. Some authors are confident that only one species, *Juniperus turkomanica* B. Fedtshenko, is found in Kopetdagh and Bolshoi Balkhan (Mukhametshin and Talantsev 1981), while others agree with some reservations (Proskuryakova 1978, 1985); still others think that the population of Bolshoi Balkhan belongs to the Transcaucasian-Iranian "race" *J. polycarpos* C. Koch (Kamelin 1973) or identify all Kopetdagh junipers as *J. excelsa* (Riedl 1960) or *J. polycarpos* (Agakhanyants 1981). Proskuryakova (1985) treats the Turkmen juniper from both Kopetdagh and Bolshoi Balkhan as a separate species (*Juniperus turkomanica*) which differs not only from the Mediterranean *J. excelsa* but also from the adjacent, much less xerophilic form found eastward (Tien Shan and Pamiro-Alai), which commonly is referred to as *J. serawschanica* Kom. in the Russian botanical literature.

Agakhanyants (1981) described complexes that are formed in the arid mountains of Middle Asia by juniper woodlands and steppes, shiblyak, mountain xerophytes, and even mesophilic vegetation. Kopetdagh juniper woodlands have been studied in detail by Zhirin (1974) and Proskuryakova (1978). In Southwest Kopetdagh juniper stands with 30 to 40% coverage can be found only in the easternmost part of the region bordering Central Kopetdagh (watershed plateaus between the Sumbar and Arvaz Rivers). Within the Syunt-Khasardagh Reserve, we found fragments of well-developed juniper stands only on well-moistened northern slopes, e.g., on the eastern and western slopes of the valleys of Kurydere and Kalymkhoz (1,300 to 1,600 m, upper portion of Aidere Valley drainage). Most of the area in these sites is covered by complexes of juniper and maple with mesophilic grass and herbaceous species in the understory. Interestingly, single juniper trees may be found in West Kopetdagh at altitudes as low as 200 or 300 m (in the western, low portion of the Syunt-Khasardagh range or in the Khodzhakala Valley) in shiblyak or petrophyte communities.

Tragacanthoid Vegetation (Tragacanths, or Mountain Xerophytes. Agakhanyants (1981) proposed to narrow the definition of this peculiar ecological group (which, according to various authors, includes a number of shrub, phryganoid, and cushion-plant species) and to limit it in Kopetdagh primarily to the representatives of the genera *Acanthophyllum, Acantholimon,* and *Tragacantha.* In Southwest Kopetdagh, these species are found in the middle and upper mountain belts at the altitudes above 1,500 to 1,600 m in complexes with steppes and juniper woodland. A detailed review is given by Popov and Seifullin (1985).

Deciduous Forest. Well-developed forest communities formed by tall trees such as elm (*Ulmus carpinifolia*), walnut (*Juglans regia*), Syrian ash (*Fraxinus syriaca*, and *Thelycrania meyeri*) are found in Southwest Kopetdagh primarily along the narrow, humid mountain river valleys. Walnut and ash forests are found exclusively in the valleys, whereas elm communities also exist on the mountain plateaus and slopes; their presence there is due to the higher precipitation provided by winter snow and spring rainfall.

Well-developed elm communities include *Thelycrania meyeri*, with *Crataegus* spp. and *Euonymus velutina* as codominant species; the herbaceous layer is formed by *Anisantha sterilis* (on mountain slopes and plateaus) or *Elytrigia repens* (in the valleys, e.g., in the Aidere Valley from 1,000 to 1,200 m).

Walnut riparian forest is confined to a very narrow (50 to 100 m wide) strip along the mountain rivers at altitudes from 1,000 to 1,500 m. This community is formed by *Juglans regia* with codominant trees *Fraxinus syriaca* and *Thelycrania meyeri*, wild plum (*Prunus divaricata*), honeysuckle (*Lonicera floribunda*), blackberry (*Rubus sanguineus*), and wild rose (*Rosa lacerans*). The mesophilic herbaceous layer commonly includes *Anisantha sterilis*, *Elytrigia repens*, *Cousinia umbrosa*, *Anthriscus longirostris*, *Physocaulis nodosus*, and *Allium paradoxum*. Common hygrophilic vegetation of this riparian forest includes horsetail (*Equisetum ramosissimum*) and mint (*Mentha longifolia*).

Walnut riparian forest in Southwest Kopetdagh is represented primarily in the drainage of the Aidere Valley, including the valleys of Aidere, Annakara, Deindere, Karasu, and Kalymkhoz), and also the valleys of Yoldere, Pordere, and Khozly. The total area ocupied by walnut forests does not exceed 80 ha; a detailed description is given by Popov (1985). Distribution of the wild fruit tree species in mountain valleys was studied by Frantskevich (1978).

Fragments of forest communities formed by the Syrian ash (*Fraxinus syriaca*) are represented in such valleys as Aidere and Yoldere. We have described a community at the Annakara Spring (1,200 m) where the understory was formed mainly by *Crataegus* spp., *Rosa lacerans*, *Prunus divaricata*, and *Rubus sanguinoides*, and the mesophilic ground layer was dominated by *Inula oculuschristii*; next to the spring were abundant *Carex* spp., *Equisetum ramosissimum*, *Mentha longifolia*, and *Phragmites communis*. In such highly humid ash and walnut communities of Southwest Kopetdagh we have encountered a number of rare or endemic species of orchids (such as *Ophrys transhyrcana* and *Epipactis veratrifolia*) and ferns (such as *Ophioglossum vulgatum*).

Relict Forest. A magnificent, relict Oriental plane tree (*Platanus orientalis*), included in the Red Data Book of the former USSR, forms complexes with ash and walnut forests and meadow and hydrophilic vegetation along several large valleys. This tree is considered a natural dominant of mesophilic mountain riparian forests in Middle Asia (Zaprygayeva 1976). There are, however, indications that humans may have intentionally spread this shade tree since it is found in virtually all Middle Asian settlements. The major population of plane trees in the region is found in the Aidere Valley, where we surveyed it from 1980 to 1982; no published data exist on this population. It included 204 trees older than 20 years growing on a very narrow strip (10 to 15 m wide) next to the Aidere River; the length of this strip was 3,700 m beginning at the western border of the Aidere Area of the Reserve. The upper altitudinal limit of the plane population was 1,100 m. Within this area, the Aidere Valley contains ruins of several settlements (the most recently abandoned one, Saparbakhar, was populated as late as the 1950s), and dispersal of *Platanus orientalis* may be related to human activity there. We recorded seedlings of the plane tree every year for ten years (1978 to 1987) on the bare alluvial deposits of the river near the water; the density of seedlings varied from 0.1 to 5 per m^2. However, young plane trees (from one to ten years old) are virtually absent from the valley, which may only indicate that young trees are completely destroyed by annual spring or summer mudslides (*sels*). Since adult and old plane trees are also often broken by mudslides, growth from stumps is common. Preservation of this unique population as well as the walnut forest of the Aidere Valley (Popov 1985, this volume) depends entirely on anti-erosion efforts to prevent mudslide formation in the Sumbar River watershed slopes and plateaus.

Relict Tall Grass Vegetation. Florogenetic analysis by Kamelin (1973, 1979) demonstrated that a distinct type of vegetation in Middle Asia is represented in Southwest Kopetdagh by relict mesophilic grasses such as *Imperata cylindrica*, *Arundo donax*, and *Erianthus ravennae*. We found small fragments of this vegetation in complexes with hygrophilic communities in the Sumbar Valley and its tributaries. A characteristic complex of desert riparian forest (tugai), reeds (*Phragmites communis*), and *Erianthus ravennae* is found in the foothill portion of the Sumbar Valley. Communities dominated exclusively by the giant reed *Arundo donax* are found from 400 to 1,000 m next to surfacing groundwater on mountain slopes and in the valleys; this original grass is widely used by local populations for construction purposes.

Desert Riparian Forest, or Tugai. Fragmentary tugais occupy a very narrow strip (10 to 30, rarely 50 to 100 m) along the Sumbar Valley at altitudes from 200 to 700–800 m. Nearly the entire area previously occupied by tugais is now transformed into fields, gardens, and settlements. Communities which may be considered similar to original tugai, are formed by such characteristic trees as the Euphratic poplar (*Populus euphratica*) and, rarely, the Persian willow (*Salix persa*) and *Elaeagnus orientalis*, but especially by salt-tolerant tamarisks (*Tamarix florida* and *T. meyeri*). Tamarisks form communities with annual grasses and halophyte vegetation. Tugai communities in Southwest Kopetdagh are heavily modified (Kulibaba et al. 1982) due to the constant changes in riverbed and in groundwater level as well as erosion, uncontrolled ploughing, cattle grazing, and logging of trees.

Hygrophilic Vegetation. This vegetation is widespread in the region along the rivers and springs in the foothills and lower mountain belt and is represented commonly by the reed (*Phragmites communis*) and also by sedges (*Carex diluta, C. divisa,* and *polyphylla*), *Juncus gerardii, Eleocharis uniglumis,* and hygrophilic moss species. Variants of hygrophilic communities are formed depending on water supply, and swamped areas can be often found locally.

Petrophyte Vegetation. This vegetation is found at all altitudes on rocky and gravel substrates and therefore is treated by some authors as an intrazonal vegetation type (Proskuryakova 1985). It includes trees and shrubs participating in petrophyte shiblyak communities (*Amygdalus scoparia, Rhamnus sintenisii, Celtis caucasica,* and *Colutea gracilis*) and such characteristic herbaceous species as *Stachys turcomanica, Phlomis kopetdaghensis, Cousinia albiflora, Scutellaria litwinovii,* and *Scrophularia turcomanica.*

Complexes of Vegetation in Southwest Kopetdagh

In general, the complex spatial structure of vegetation in the studied area is determined by altitude, humidity, character of the relief, and degree of erosion. The most common complexes of vegetation described and mapped in Southwest Kopetdagh are:

(1) ephemeroid and sagebrush deserts combined with petrophyte vegetation;
(2) maple and juniper woodlands combined with semisavanna;
(3) maple and hawthorn woodlands combined with mesophilic meadows;
(4) juniper woodlands combined with mountain steppe and xerophyte communities;
(5) tree and shrub riparian (tugai) vegetation combined with communities of relict large grasses;
(6) mesophilic deciduous forests and relict forests combined with the hygrophilic herbaceous vegetation.

Examples of spatial structure of vegetation (fragments of large-scale vegetation map of the Syunt-Khasardagh Reserve) are given in Fig. 2.

Restoration of Trees and Shrubs in the Syunt-Khasardagh Reserve

One of the goals of our study was to trace the processes of natural restoration of tree and shrub vegetation under the regime of protection. Since 1978, logging and grazing has been banned in the Syunt-Khasardagh Reserve; adjacent forest

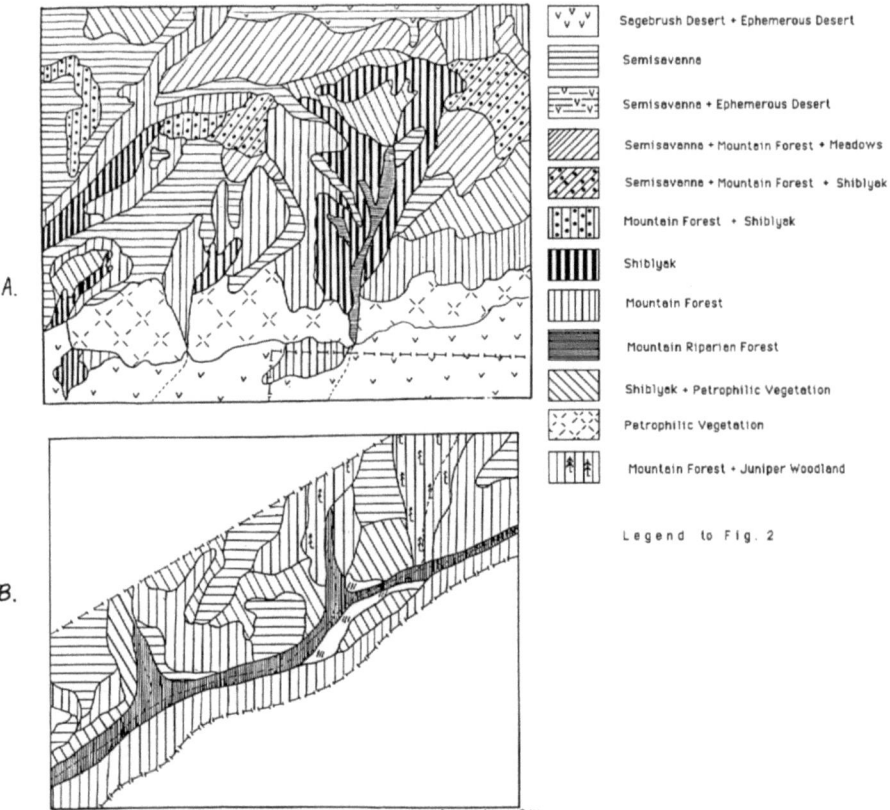

Fig. 2. Fragments of vegetation map of the Syunt-Khasardagh Reserve, Southwest Kopetdagh, Turkmenistan (1 km in 1 cm): (A) central portion of the Central Area, including Mt. Syunt and Mt. Khasardagh; (B) upper portion of the Aidere Area.

communities, however, are still subject to intensive and uncontrolled exploitation. In 1982, the Forest Service surveyed 3,464 ha covered by forest habitats in the Syunt-Khasardagh Reserve (27% of the Reserve territory); most of these habitats qualified as "mountain shrubland" (10% of the territory) or "mountain deciduous forest" (12%).

We have studied the process of natural restoration of tree and shrub vegetation since anthropogenic pressure (cattle grazing and tree logging) was removed in 1978. In 1985 and 1986, we analyzed distribution of different age groups of 30 tree and shrub species along three altitudinal belts.

In the foothills, *Paliurus spinachristi* and *Cerasus microcarpa* were the most successfully reproducing species. In open areas, seedlings of *Paliurus spinachristi* exhibited very successful growth; however, under the canopy of mature *Paliurus spinachristi* as well as in other shaded sites, *Cerasus microcarpa* seedlings were more abundant. Both species possessed a one-year-old seedling density of 30 to 40 per ha, whereas the density of adult plants (ages five and older) was 50 to 60 per ha. Sparse seedlings of *Rubia florida, Colutea gracilis,* and *Jasminum fruticans* were recorded on rocky substrates in the foothills. We found no young seedlings of such shiblyak species as *Zygophyllum atriplicoides, Atraphaxis spinosa, Rhamnus sintenisii,* or *Hymenocrater Bituminosus* although adult shrubs were present.

In the lower mountain belt, trees and shrubs were represented by *Acer tucomanicum, Crataegus* spp., *Prunus divaricata, Ulmus carpinifolia, Lonicera bracteolaris, Cotoneaster nummularia, Cerasus microcarpa, Celtis caucasica, Thelycrania meyeri, Euonymus velutinum, Paliurus spinachristi, Colutea gracilis, Ephedra intermedia, Jasminum fruticans, Rosa* spp., *Juniperus tucomanica, Rubus sanguineus, Ficus carica,* and *Vitis silvestris.*

Dominant species were trees *Acer tucomanicum, Crataegus* spp., *Prunus divaricata,* and *Ulmus carpinifolia,* and shrubs *Lonicera bracteolaris, Rosa* spp., *Cotoneaster nummularia,* and *Cerasus microcarpa.* Communities on slopes and plateaus were formed primarily by *Acer tucomanicum* and *Crataegus* spp., whereas valleys were dominated by *Ulmus carpinifolia.* The dominant tree species in the understory was *Prunus divaricata.* Among these species, one-year-old seedlings were found at densities from 20 per ha (elm and plum) to 60 per ha (maple) to 130 per ha (hawthorn). Densities of trees older than five years of age, however, were higher and varied from 70 to 120 per ha, totaling for all species together 400 to 450 per ha. A competitive relationship may exist between maple and hawthorn; a much higher seedling density in the latter may be secured through its zoochoric dispersal.

Absence of the Christ's thorn in these communities confirms that this tree grows extremely poorly under its own canopy and prefers open spaces. The existence of a large massif of *Paliurus spinachristi* in the Yoldere Valley is related not only to previous fires or cattle grazing; in the last decade before the Syunt-Khassardagh Reserve was established in 1978, this valley housed a children's resort and was a site of intensive construction and recreational activity, which also yielded space in which this species could gain a toehold. Since the altitudinal limits of *P. spinachristi* in Southwest Kopetdagh are from 300–400 to 800–900 m, this characteristic shiblyak species is limited to the upper portion of the foothills and the lower portion of the lower mountain belt.

The middle mountain belt vegetation was surveyed in the mesophilic forest on the plateau of Mt. Khasardagh. Species recorded here included *Acer tucomanicum, Crataegus* spp., *Thelycrania meyeri, Cotoneaster nummularia, Lonicera bracteolaris, Cerasus microcarpa, Euonymus velutinum,* and *Malus turkmenorum.* Dominant species were hawthorn and maple in the canopy, and *Lonicera bracteolaris* and *Cotoneaster nummularia* in the understory.

Density of trees in this belt (83 per ha for hawthorn and 35 per ha for maple) was higher than in the lower mountain belt (37 and 12, respectively) due to higher precipitation. Seedling density was only 2.2 times higher here than in the lower belt for hawthorn (290 versus 133 per ha), but 36 times higher for maple (2,080 versus 58 per ha). This fact is explained by the ability of maple seedlings to grow readily under the canopy of hawthorn trees, with subsequent high survivorship of maple seedlings. Therefore a demutation change of tree species can be expected.

Our survey data in the Syunt-Khasardagh Reserve allow us to make following predictions. In five to ten years, if climatic conditions are favorable and the protection regime persists, further expansion of the Christ's thorn (*Paliurus spinachristi*) will occur in the foothills; the formation of polydominant mesophilic forest combined with shiblyak communitites will continue in the lower mountain belt; and demutation change from hawthorn forest to maple forest will proceed in the middle mountain belt plateaus.

'Bad lands' near the town of Kara-Kala, Southwest Kopetdagh (outcrops of Cenomanian clays in the foothills). Photo by H.R. Levenshtein and I.A. Mukhin.

10. Trees, Shrubs, and Semishrubs in the Mountains of Turkmenistan

KONSTANTIN P. POPOV

Abstract

The tree, shrub, and semi-shrub vegetation of Turkmenistan is discussed. The rapid aridization of climate of southern Turan during the Neogene impoverished forest and savannoid communities and transformed them into sparse xerophitic woodlands. In the historical period, the anthropogenic factor has increasingly affected vegetation, especially woodlands comprised of junipers (*Juniperus turcomanica*) and pistachio (*Pistacia vera*). Structure and current condition of these woodlands are discussed, as well as those of mountainous communities of hawthorn, almond, cushion (tragacanthoid) plants, shiblyak, maple, and mountain riparian forest. By the 1980s, tugais (lowland desert riparian forests) had become practically extinct; the areas occupied by pistachio and juniper woodlands had drastically decreased. Most of the described plant communities are highly disturbed by human activity, and require protection and restoration.

Introduction

Paleobotanical data (Korovin 1934) show that southern Middle Asia in the Tertiary period was covered by subtropical evergreen forests of laurel type as well as by paleosavannas dominated by the tall grasses (*Erianthus ravennae*, *Phragmites communis*) and umbelliferous plants (species of *Ferula*, *Dorema*, and *Prangos*). In the Neogene, however, the forest vegetation was degraded significantly by the rapid aridization of climate. The dendroflora became impoverished; not only evergreen but also many deciduous species and even genera of trees and shrubs became extinct in Middle Asia. Many species of trees and shrubs retreated to the mountains; the lowlands of Turan became dominated by low grasslands which adopted an ephemerous regime of vegetation.

A comparatively rich dendroflora was preserved only in the Middle Asian mountains, in refugial habitats with a relatively high humidity. Zapryagayeva

(1964) listed 78 species of trees and 169 species of shrubs (not counting desert psammophytes) for the extensive mountain range of Pamiro-Alai, whereas Nikitin (1988) lists only 38 tree species and 65 shrub species for the extra arid mountains of Turkmenistan. From this list, only three species of trees play a more or less significant role in the vegetation of foothills and mountains. These are: the pistachio (*Pistacia vera* L.), and two species of junipers (*archa*): *Juniperus turcomanica* B. Fedtsch., and *J. seravschanica* Kom. However, even these trees do not form true forest-type communities but rather rare-standing xerophilic woodlands. Especially sparse are pistachio communities, which are found at the extreme south of Turkmenistan and resemble a low-grass savanna. Mountain shrubs in Turkmenistan do not form the undergrowth characteristic for a forest but rather form fragmentary communities of so-called "shiblyak" (Mediterranean-type communitites of shrubs and small trees). Only some densely branching low shrub species together with spiny semishrubs play a significant role in the mountain vegetation, creating a specific formation of "cushion plants."

Mesophilic forest-type vegetation in Turkmenistan has been preserved only in the Kopetdagh Mountains, mostly in valleys which served as refugia. In West Kopetdagh, fragments of maple forests with shrub undergrowth still can be found. Fragments of riparian, or gallery, forest communities along the non-drying rivers and brooks are severely damaged by human activity. Disrupted fragments of mountain riparian forest can be found along the valleys of Southwest Kopetdagh (such as that of the Sumbar River which is a tributary of Atrek). In lowlands, fragments of former *tugai* riparian communities still exist, comprised of local poplars and willows.

Pistachio Woodlands

Pistachio (*Pistacia vera*) woodland communities of Turkmenistan are concentrated in Badghyz, an area bordering Iran and Afghanistan and located between the Murghab and Tedzhen Rivers. The southern part of Badghyz is characterized by a hilly terrain which transforms southward into a low mountain range with altitudes up to 1,340 m. Pistachio trees here form two large "groves," Kushka and Pulikhatum, with a total area of about 75,000 hectares. These are typical xerophilic woodland landscapes where pistachio is the only tree species (Fig. 1), for no other tree is able to withstand the extreme climatic conditions of Badghyz. Annual precipitation here usually does not exceed 300 mm, and maximal temperature in the shade can reach 45° to 48 °C. From May to September, hot air from the Karakum sand desert brings not only soil drought but also the atmosphere draught to Badghyz. This hilly area has the poorest dendroflora among all foothills in Middle Asia. Apart from the pistachio, only two species of shrublike figs (*Ficus carica* and *F. afghanistanica*) are found here in the valleys of Inzhirli, Turangli, and Pelengovali. This last one, which is also the deepest valley in Badghyz, also has a few *Celtis caucasica*.

Pistachio communities in Badghyz are so sparse that Linchevsky (1935) even classified them as savannas, which is certainly not true. On the other hand, Ovchinnikov (1948) and Zapryagayeva (1964) classify pistachio communities as the forest ones, namely of "shiblyak" type. However, single-species sparse stands of *Pistacia vera*, which are widespread in Middle Asia, have no common characteristics with shiblyak as described by Adamovic (1909) but rather represent a specific vegetative formation, the pistachio semisavanna (Popov 1974; Kamelin and Rodin 1989). Umbrella-shaped crowns of pistachio trees in Badghyz cover only 20 to 30% of the area occupied by its plant association, which includes grasses such as *Poa bulbosa* and sedge *Carex pachystylis*. These ephemeroid plants vegetate in winter and spring and completely dry out in late April. The pistachio tree, which is wind-pollinated, develops its flowers in late March, before the appearance of leaves which cover the trees in the first two weeks of April, just as the grasses dry out. Therefore, pistachio trees and dominant grasses are independent components of vegetation in Badghyz.

All other deciduous species of the genus *Pistacia* are believed to be typical trees; this is not completely true for *P. vera*. This species may naturally develop several trunks and is of a small height (up to 5–6 m), which makes it more similar to a shrub than a tree. It also possesses a huge xylopodium which produces new replacement trunks in the process of aging, a feature also typical of shrubs. We can probably state that *P. vera* is now in the process of evolving from a tree to a shrub. This advancement, as well as high polymorphism, large size of the nuts, and extreme viability, suggest that *P. vera* is a young species (Popov 1979) rather than an ancient one in the genus *Pistacia* (Zochary 1963). *Pistacia vera* is so different from all other pistachios that M.G. Popov (1963) even suggested its hybridogenic origin through spontaneous hybridization (*Juglans* x *Rhus*).

Representatives of the genus *Pistacia* have dominated paleosavannas of Turan in the geological past. Paleobotanical data show that, before the Pleistocene, at least four species of this genus inhabited Middle Asia: *P. mutica*, *P. terebinthus*, *P. chinensis*, and the ancestral form of *P. vera*. Only the last species has large nuts, a feature which undoubtedly facilitated its survival in Middle Asia. *P. vera*.'s possession of large seeds rich in fat and proteins is crucial for the rapid development of deep tap seedling roots which secure the young trees' survival in the first, most critical year of their life. This unusually viable tree normally reproduces even within the extreme conditions of Badghyz and reaches the age of several hundred years (Fig. 2). The pistachio can normally withstand the fires which are fairly common in southern Middle Asia and capable of burning the dry grass over vast areas. After such fires, old trees usually perish but young ones (10 to 15 years old) rapidly develop twig growth from the root neck.

In general, *Pistacia vera* has successfully withstood severe natural trials during the rapid aridization of climate in Turan, but has retreated before humans, who exterminated this remarkable tree in the vast foothill areas of the Middle Asian mountain ranges. Badghyz, which lacks water, has not been populated by humans until recent times, and its pistachio woodlands have not yet suffered significantly. In the Kopetdagh Mountains, however, of all former pistachio communities only one small islet about 500 hectares in size is now left, next to the Iranian border. Single small trees and groups of trees found here and there in Kopetdagh confirm the former broad distribution of the pistachio in Kopetdagh, from its eastern promonitory to the southwestern ranges. We know that the first human settlements in the Kopetdagh foothills emerged as early as in the Paleolithic; a number of highly developed settlements existed here in the Bronze Age. All these cultures consumed wood for the brick and ceramics industry as well as for smelting and treatment of metals.

No pistachio trees or traces of their former existence have been found in western and northwestern ranges of Kopetdagh or in the small ranges of Maly Balkhan and Bolshoi Balkhan, which are surrounded by the desert. The natural range of *P. vera* most likely never extended into this part of the Turkmeno-Khorassan Mountains. In eastern Turkmenistan, small areas covered by pistachio woodlands are found in the foothills of Kugitang, which is a part of the Ghissar range. These communities recently have been completely cut, subsequently producing stump shoots. Unfortunately, the year-round grazing of sheep and goats in this area completely prevents seed reproduction of the pistachio.

Hawthorn and Almond Communities

Besides pistachio, a few species of hawthorns (*Crataegus* spp.) and the common almond (*Amygdalus communis*) form sparse stands in the mountains of Turkmenistan. Among the hawthorns, the most conspicuous are species with large edible fruits, especially the Pontic hawthorn (*Crataegus pontica*). Its communities are located primarily within mountain pastures of West Kopetdagh, where they cover approximately 6,000 hectares. Although local highlanders of the Nokhurli tribe spare these trees in their woodcutting, the seed reproduction of hawthorns is limited due to extensive goat grazing. Because hawthorns exhibit considerable polymorphism, selection for forms bearing large and sweet seeds is potentially possible; such selection, however, has not yet been conducted.

The limited woodlands of the common almond (*Amygdalus communis*) are located in the valleys of Southwest Kopetdagh (the drainage of the Sumbar River). The total area covered by almond woodlands there does not exceed 3,000 hectares. The only other location of *A. communis* in Middle Asia is a small islet in West Tien Shan, at the northern boundary of the species range. Such disjunction is evidence of a relict distribution of *A. communis*. This tree can withstand drought (it sheds leaves in dry periods) and grows fairly well on the dry rocky slopes of the valleys. There, however, the almond periodically suffers from frost attacks during its flowering period, which begins in February. Unregular seed production, nevertheless, does not prevent normal reproduction of almond. The Kopetdagh almond population of almond is relatively stable in its range and abundance. However, the considerable polymorphism of almond nuts, which is observed not only in size, shape, pattern, shell thickness, and shell hardness, but also in seed taste, has changed considerably during the last century. Records reveal that, in the 1880s and 1890s, the Nokhurli collected hundreds of kilograms of sweet nuts of the wild almond. Bogushevsky (1930) still was able to find and describe many trees with sweet and thin-shelled nuts which were introduced into culture for selection. By the 1980s, however, such trees had nearly disappeared from the Aidere Valley (Popov 1981), primarily due to human activity.

Another, mysterious species of almond, *Amygdalus vavilovi* M. Pop., has been described from seeds found in Central Kopetdagh. This population probably become vanished due to natural factors. Outside of Turkmenistan, single trees of *A. vavilovi* are relatively commonly found in the Pamiro-Alai Mountains (Zapryagayeva 1964).

Juniper Woodlands

Fossils of the genus *Juniperus* have been found in the Lower Cretaceous rocks of Middle Asia; it is thus one of the oldest genera in the local flora. In the early geological periods, Kopetdagh formed an island chain in the Tethys Sea which was connected, through the Elburz Mountains, to the Lesser Caucasus range. This bridge made possible the migrations to the east by many Mediterranean and Southwest Asian plant species, which evolved towards xerophytization during the general aridization of climate. Some of these migrants have reached only West Kopetdagh. These are such relicts of the Tertiary dendroflora as walnut (*Juglans regia* ssp. *turcomanica*), pomegranate (*Punica granatum*), *Sorbus graeca*, *Thelycrania meyeri*, and *Evonymus velutina*. Another group of East Mediterranean migrants (*Rhus coriaria*, *Paliurus spina-christii*, *Platanus orientalis*, and *Sorbus persica*) has been able to reach Central Kopetdagh. This group includes also the Turkmen juniper (*Juniperus turcomanica*), which is very close to the Caucasian-Anatolian species *J. polycarpos*. This species dispersed also to the small northern ranges of Maly Balkhan and Bolshoi Balkhan. In the very low range of Maly Balkhan, the Turkmen juniper is close to complete extinction; only single, scattered trees can be found there. In Bolshoi Balkhan, however, this species forms a perfectly viable population and suffers from the lack of moisture only on the narrow terraces and steep slopes, where trees with dry tops can be found. In Bolshoi Balkhan, the lower boundary of juniper communities is 800 to 900 m on the southern slopes, and 400 m on the northern ones. We know, however, that this tree descended until quite recently down into the very desert that surrounds Bolshoi Balkhan, because the old stumps can still be found there. The Turkmen juniper grows under conditions where annual precipitation commonly does not exceed 150 mm, and summer temperature of the extremely dry air can rise up to 46–48 °C. Undoubtedly, *Juniperus turcomanica* can be considered the most drought-resistant species among junipers. Even at low altitudes the trees are well developed and produce numerous fruits which are actively consumed and dispersed by birds (*Mycetobas carnipes*). The age structure of juniper stands of Bolshoi Balkhan demonstrates that *J. turcomanica* here reproduces normally, and can reach the age of several hundred years. It can stand 15 to 18 m high and have a trunk diameter of up to one meter. Unlike many other coniferous trees, the Turkmen juniper can regenerate by stump twig growth after it is cut; however, repeated cuts prevent effective regeneration.

The relative preservation of juniper woodlands in Bolshoi Balkhan is due both to the remote location and difficult access of this mountain range with its steep rocky slopes. Nevertheless, juniper stands in Bolshoi Balkhan, which covered 40,000 hectares in the 1930s, by now have decreased by about 30%, and continue to decrease due to unlimited woodcutting and grazing. Plans to open a Balkhan Reserve in order to protect natural communities of Bolshoi Balkhan have been discussed for almost half a century but so far have had no practical result.

In the Kopetdagh Mountains *Juniperus turcomanica* has had a wide distribution in 19th century, from altitudes of 400 to 600 m up to the mountain plateaus and peaks and covering hundreds of thousands of hectares. Even then, however, juniper woodlands had decreased due to woodcutting, not only for firewood but also for construction purposes: the juniper is the only tree in the Turkmenistan mountains which can provide timber for construction. The estimated 79,000 ha covered by juniper in Kopetdagh in 1930 had decreased to a mere 20,000 ha by late 1960s (Koksharova 1970). Juniper stands have suffered especially in West Kopetdagh, an area densely populated by humans since ancient times. In the lower and easily accessible western ranges of West Kopetdagh juniper now has disappeared (Kurbanov 1988). Juniper woodlands of Central Kopetdagh also have been severely damaged by human activity. Only on scarcely accessible steep slopes of mountain gorges may the juniper tree crowns cover 40 to 50% of the association area, but even these plots cannot qualify as forest-type comunities. Here, magnificent old trees can be found which are 18 to 20 m high and have a trunk diameter of 100–120 cm. At the same time, juniper communities on the mountain plateaus of Central Kopetdagh continue to be degraded, and in many areas they already are practically extinct due to woodcutting and the unlimited grazing of cattle, sheep, and especially goats.

The same fate has befallen juniper communities of Kugitang formed by a different species of Central Asian origin, *Juniperus seravschanica*. Not long ago, tens of thousands hectares of junipers could be found on the slopes of Kugitang from 800 to 3,200 m. Today, due to human impact, the lower boundary of *J. seravschanica* distribution in Kugitang has shifted to as high as 1,300 to 1,700 m, where sparse juniper stands can be found with crown projections covering not more than 30%.

Two new natural reserves, the Kopetdagh Reserve (50,000 ha, in Central Kopetdagh) and the Kugitang Reserve (27,000 ha), have been created in

Turkmenistan, in 1976 and 1986 respectively. The primary goal of both reserves is to protect and preserve relatively intact areas of juniper woodlands and their ecosystems.

Shiblyak Communities

Shiblyak is distributed in Kopetdagh from the foothills to the upper plateaus occupied by the juniper woodlands. Shiblyak does not constitute an altitudinal belt of vegetation but rather is concentrated between 800 and 1,600 m of elevation. It is fragmentarily represented on rocky northern slopes of mountain valleys and also has replaced the exterminated maple forest communities. Shiblyak is a diverse plant community which includes species of *Ephedra, Paliurus, Rosa, Cotoneaster, Cerasus, Amygdalus, Celtis, Ficus, Berberis, Colutea, Lonicera, Rhamnus, Punica, Jasminum, Rhus*, and *Zizyphus*. The most diverse communities of shiblyak are found in Southwest Kopetdagh (the Sumbar Valley), which has a dry, subtropical climate. In this refugium such migrants of the Mediterranean origin have been preserved as jujube (*Zizyphus jujuba*), pomegranate (*Punica granatum*), *Jasminum fruticans, Paliurus spinachristii, Rhus coriaria*, and *Periploca graeca*. This is the only locality in Middle Asia where medlar (*Mespilus germanica*) and quince (*Cydonia oblonga*) are found. In Middle Asia, *Sorbus graeca* is known only from Southwest and Central Kopetdagh; its populations are extremely rare and probably on the verge of extinction. In the dry valleys of Kopetdagh, pomegranate (*Punica granatum*) and the fig tree (*Ficus carica*) are found; the latter is, surprisingly, found also in the dry gorges of Bolshoi Balkhan. Especially conspicuous in the shiblyak communities of Southwest Kopetdagh are jasmine and Christ's spine, which sometimes form dense thickets. Another characteristic species of shiblyak in Southwest Kopetdagh is a leaveless almond (*Amygdalus scoparia*), an endemic of Khorassan flora.

In Central Kopetdagh, shiblyak is predominantly comprised of wild roses, hawthorn trees, wild cherries (*Cerasus* spp.), honeysuckles (*Lonicera* spp.), and the Turkmen almond (*Amygdalus turcomanica*), which is more common in East Kopetdagh.

In Bolshoi Balkhan, with its extraarid climate, fragmentary shiblyak communities are found only along valley bottoms and in the upper portion of the northern slope. Their species composition is impoverished; the dominant species here are the wild cherry (*Cerasus microcarpa*) and the Persian honeysuckle (*Lonicera persica*). They are accompanied by *Ephedra intermedia, Zygophyllum atriplicoides, Atraphaxis spinosa*, and *Cotoneaster* spp. Only in foothills of Bolshoi Balkhan can be found *Berberis turcomanica* and *Caragana grandiflora*; the latter species is also found within Turkmenistan in West Kopetdagh and Ustyurt.

In Kugitang, shiblyak is distributed on mountain terraces and in dry valleys, and is dominated by maple (*Acer pubescens*), the Bokhara almond (*Amygdalus bucharica*), the spiny almond (*A. spinosa*), different species of hawthorns, cherries, roses, honeysuckles, and by species of *Ephedra, Cotoneaster,* and *Atraphaxis.*

Occasionally, *Celtis caucasica* and a peculiar Regel's pear (*Pyrus regelii*) can be found here in shiblyak communities. In some places of Kugitang, shiblyak is combined with juniper stands, but it does not form any kind of undergrowth layer there. Further to the east, within the Ghissar range, shiblyak is enriched by different species of shrubs belonging to the Pamiro-Alai flora, and there in some cases it forms altitudinal belts of the mountain vegetation.

Cushion Plant Formation

This group, represented by several species of densely branching low shrubs and semishrubs with extremely low rates of annual growth, are called "cushion plants" due to their density. Commonly, cushion plants occupy the most hostile habitats and exhibit remarkable ecological plasticity. They grow on dry slopes, in swamps, and in polar and highland deserts. In the Pamir Mountains, cushion plants reach the record elevation for vascular plants which is 5,400 m. Other species are adapted to the extraarid conditions in the Middle Asian foothills and sometimes even in the hot deserts. Cushion plants, with their extremely slow growth, have a very long lifespan, with individual plants of some species reaching more than one hundred years. Cushion plants represent less than 2% of species diversity in the Tien-Shan, Pamiro-Alai, and Turkmeno-Khorassan Mountains, but they play an important role in the formation of vegetation. For instance, cushion plants form a specific vegetation belt of highland cryophytes in the Pamiro-Alai Mountains between elevations of 3,200 to 4,800 m. Cushion plants found in the relatively low mountains of Turkmenistan belong to the group of highland xerophytes. Cushion plants of Turkmenistan, and especially of Kopetdagh, express a high level of endemism; of 38 species of cushion plants found in Turrkmenistan, 28 species are not found elsewhere in Middle Asia. This is undoubtedly due to the migration of cushion plant species from the neighboring arid Iranian Plateau, which is considered a center of speciation of highland xerophytes (Agakhanyants 1985). Kopetdagh has been historically invaded by Khorassan species which have not dispersed further to the north because of the sand desert. In Kopetdagh, cushion plants do not form a pronounced vegetation belt but are widely distributed, mostly within the sparse juniper woodlands between 1,600 to 2,800 m (Nikitina 1958) and in the shiblyak formation, preferring rocky slopes. At these elevations, 21 species of cushion plants are found, whereas only 12 species grow in the middle mountain belt (800 to 1,500 m), and four species, in the foothills. Only two species of cushion plants inhabit sands of the Karakum Desert: *Acanthophyllum adenophorum* and *A. stenostegium.*

Another, not so significant migration flow of cushion plants has occurred from the Lesser Caucasus through the Iranian range of Elburz to the Kopetdagh

Mountains. This route of dispersal probably has been followed by such characteristic cushion plants of Southeast Transcaucasia as *Onobrychis cornuta*, *Tragacantha marshalli*, and *Gypsophila aretioides*. The first two species are found in Bolshoi Balkhan, which has six species of cushion plants (with three species being narrow endemics of this range). *Gypsophila aretioides*, which is found on the rocky slopes of Central Kopetdagh, is an example of the ultimate evolution of a cushion plant: its callouslike "bodies" are so hard that they cannot be destroyed by a gunshot.

Among 26 species of cushion plants found in Kopetdagh, the most common are eleven species of the genus *Acantholimon* (Fam. Plumbaginaceae), seven species of *Acanthophyllum* (Fam. Umbelliferae), and three species of *Tragacantha* (Fam. Fabaceae). All of these species are extremely spiny, typical xerophytes. Only three species of cushion plants lack spines: *Gypsophila aretioides*, *Asperula balchanorum*, and *Dionysia tapetoides*. The last species is the only non-xerophyte cushion plant in Turkmenistan, found on the rocky walls of shadowy gorges in Central Kopetadgh. The relict genus *Dionysia* is represented in Middle Asia only by three species; a fourth one, *D. kossinskyi* Czerniak., has been described from Khorassan (Iran). It has been found only once in Central Kopetdagh, next to the Iranian border, by V.I. Lipsky; later, *D. kossinskyi* has been never recorded and is now considered an extinct species in Turkmenistan.

The cushion plant formation in Kopetdagh undoubtedly belongs to the primary vegetation. In some places, however, the dominance of cushion plants such as species of *Tragacantha* and *Acantholimon* is evidently due to extensive overgrazing of the natural vegetation by goats (e.g., on the mountain pastures around the village of Nokhur in West Kopetdagh). Here, on one hectare we were able to count three to four thousand of these cushion plants, which were unaccessible even to the goat.

Still, the area covered by cushionlike plants has been decreasing since the 1930s when it occupied tens of thousands hectares in Central Kopetdagh (Borisova 1938). After World War II, the species of *Tragacantha* have been intensively dug out for their rubberlike resin, and some species of *Acanthophyllum* used as a "soap root." This exploitation depleted natural cushion plant formations to the extent that management of them became nonprofitable.

Kopetdagh is outstanding among the Middle Asian mountain ranges due to its diversity and the ecological importance of cushion plants. A similar formation is also well expressed in Kugitang, the eastern mountain system of Turkmenistan, which possesses eight species of cushion plants of Pamiro-Alai origin. These species are common in the juniper belt up to the top of the Kugitang Mountains. Dominant cushion plants in Kugitang include *Acantholimon erythraceum*, *Onobrychis echidna*, and an endemic of generic level, *Kuhitangia popovii*.

Cushion plants do not play a significant role in the foothills and low mountains of Badghyz, where four species of *Acanthophyllum* are found between the elevations of 300 to 1,200 m. Three of these species are common for

Badghyz and Kopetdagh, and one (*A. stenostegium*) is common for Badghyz and the sands of Karakum.

Many environmental factors, such as intensive solar radiation, severe winds, drastic temperature changes, dry air, and general aridity of the climate, may promote the evolution of this peculiar life form. Different combinations of these factors should have influenced evolution of cushion plants in different climatic and altitudinal conditions. The only general factor, which always is present, is an insufficient supply of water. Economic spending of scarce moisture should be responsible for the wide convergence of cushion plants, producing this "shrunken" life form equally either as cryophytes in cold highland and Arctic deserts or as xerophytes in xerothermic conditions such as those of Turkmenistan mountains. Both types of cushion plants play an important ecological role in the protection of soils and conservation of precious water.

Maple Forests

Maple forest communities in Kopetdagh, by now near extinction, form only a few remaining islets. In all other mountains of Turkmenistan this type of vegetation disappeared in Post-Pleistocene time due to the rapid aridization of the climate, but in West Kopetdagh remnants of the Turkmen maple (*Acer turcomanicum*) communities are still surviving. They occupy northern slopes between elevations of 800 to 1,700 m and are represented predominantly by shrublike shoots regenerated after numerous cuttings. In places such as the Yoldere Valley and the summit of Mt. Syunt (1,600 m), maple communities are relatively untouched and form a true forest with a high density of canopy (80 to 90%), shrubby undergrowth, and number of shade-requiring forest grasses. The tree layer here is formed only by maple trees, whereas the shrub layer is comprised of *Euonymus velutinum* (the endemic species for Kopetdagh), species of *Rosa, Cotoneaster, Colutea*, and *Jasminum fruticans* (which sometimes forms extremely dense thickets). The famous Russian botanist V.I. Lipsky, who climbed Mt. Syunt in 1912, wrote: "The forest was so dense that it has been extremely hard to get through. Its character resembled a tropical forest." In 1975 we still were able to find this forest, but all maple trees appeared to have grown from stool shoots. The epiphytic lichens hanging from maple trees were so numerous that they, indeed, created an impression of a tropical forest.

Maple communties are preserved also in the Yoldere Valley, where they are, however, significantly suppressed by thickets of the aggressive shrub, *Paliurus spinachristii*, not recorded by Lipsky in 1912. After the Syunt Khossardagh Reserve was established in 1979, the Ioldere Valley was put under a protection regime, and restoration of primary trees and shrubs such as the maple here has been relatively rapid in the subsequent 13 years.

Other than in the described localities, there are no true maple forest communities in Kopetdagh. Shrublike maples which develop as stool shoots after cutting are forming secondary communities where density of canopy does not exceed 40%. These communities can be classified as an anthropogenic shiblyak which covers approximately 6,000 hectares in Kopetdagh. Also, the Turkmen maple can be found in juniper stands throughout Southwest and Central Kopetdagh.

Mountain Riparian Forests

These peculiar communites have a dominant tree layer but lack an expressed layer of shrubs. Instead, the shrubs tend to concentrate at the margins of forest communities. Riparian forests, which are an intrazonal type of vegetation in the arid zone, extend along rivers and brooks. In the past, banks of all lowland rivers in Turkmenistan have been covered by the so-called *tugai* forests formed primarily by two poplar species, *Populus pruinosa* and *P. euphratica*, with participation of *Elaeagnus orientalis* and *Tamarix* spp. Now, the only remnants of *tugai* poplars in in Kopetdagh exist in the lower portion of Sumbar where it joins the Atrek River. Archaeological data (Lisitsyna and Popov 1988) show, however, that *tugai* existed in the Neolithic time up into the northern foothills of Kopetdagh. Moreover, these ancient communites were dominated by the Syrian ash (*Fraxinus syriaca*) rather than by poplar; these ash forests were exterminated by human settlers in the Bronze Age. Ash has been replaced by willows (*Salix* spp.), which still dominate some riparian communities in northern Kopetdagh together with the local species of elm (*Ulmus carpinifolia*). Willow timber is not valued even as a fuel.

The most diverse riparian dendroflora, related to the hygrophilous flora of Southeast Transcaucasia, is found in the moist mountain valleys of Southwest Kopetdagh such as Aidere, Pordere, Annakara, Khozdere, and other branches of the Sumbar River drainage. The most conspicuous trees in these riparian forests are the walnut (*Juglans regia* ssp. *turcomanica*), ash (*Fraxinus syriaca*), elm (*Ulmus carpinifolia*), *Thelycrania meyeri*, and the Eastern sycamore, or *"chinar"* (*Platanus orientalis*). In the same communities can be found wild pear species (*Pyrus communis* and *P. boissieriana*), Turkmen apple (*Malus turkmenorum*), Mahaleb cherry (*Cerasus mahaleb*), plum (*Prunus domestica*), species of willows, maple, and *Elaeagnus orientalis*. Among shrubs, the common species are honeysuckle (*Lonicera floribunda*), *Prunus divaricata*, *Berberis interregima*, and especially blackberry (*Rubus anatolicus*), which sometimes forms entirely unpenetrable thickets. The vines of wild grapes (*Vitis silvestris*) are common here, embracing tree crowns up to the height of eight to ten meters. This species is not found in Middle Asia eastward from Southwest Kopetdagh. The primarily wild condition of grapes here is confirmed by the existence of functionally male and female vines.

Valleys of Southwest Kopetdagh also house another vine species, a Mediterranean-Caucasian *Periploca graeca*; further eastward, this species is found only in cultured conditions. A third peculiar vine species found in Turkmenistan (but not in the rest of Middle Asia) is *Malacocarpus crithmifolium*

(Fam. Peganaceae). This relict plant can be found in several disjunct localities such as the juniper woodlands of Kopetdagh, its northern foothills, shores of the freshwater lakes in the Karakum Desert, and karst depressions of Ustyurt. Growing next to fresh water, *M. crithmifolius* can produce a large amount of seeds with high germinative capacity, but their small size and weight (about one milligram) are probably responsible for a very low seed reproduction. This vine is a prospective species for decorative culturing.

The abundance of wild relatives of fruit trees, shrubs, and vines in the valleys of Southwest Kopetdagh has attracted the attention of the N. I. Vavilov All-Union Institute of Plant Breeding, which has performed genetic studies and selection experiments on these species since the 1930s. Especially detailed studies have been conducted on such valuable species as walnut (*Juglans regia*). In the early 1930s, walnut dominated riparian forests of Southwest Kopetdagh, where 300- to 400-year-old trees could be found, with trunks more than one meter in diameter. The total number of walnut trees in the Sumbar drainage at that time was 3,823, with an average age of 90 years. Of these, about 3,000 were found in the Aidere Valley, where walnut communities formed an uninterrupted strip of riparian forest from 700 m up to the upper portion of this richest refugium of mesophilic flora (Gursky 1932). Even then, the dominance of old and middle-aged walnut trees, with virtual absence of young, seed-reproduced trees, was observed. This phenomenon was explained by the exhaustive collecting of walnut seeds by the local people. The walnut population in the 1930s was also characterized by a high polymorphism of seeds: some trees bore nuts of a quality comparable to that of cultured walnuts.

Fifty years after this survey, the walnut population in the valleys of Southwest Kopetdagh had become drastically reduced. Our survey, conducted in 1980 and 1981, demonstrated that only 1,834 walnut trees, with an average age of 120 years (Popov 1985), were present in the main valleys. About two-thirds of these old trees were already partially dying. Young growth was still absent due to the same reason as manifest in the 1930s: total harvesting of walnut seeds. Later in the 1980s, the valleys of Southwest Kopetdagh were severely damaged by unusually strong mudslides, which exterminated hundreds of old walnut trees. By 1990, the walnut population in the Aidere Valley was less than 800 trees. In general, the Kopetdagh walnut population continues to steadily decrease.

Only one locality of wild walnut is known from the northern slopes of Kopetdagh. About 200 walnut trees of different ages (including some young growth) have been recorded in a small, remote valley of Karayalchi next to the town of Archman. Generally, this population is in better condition than those of Southwest Kopetdagh. Another locality with a small number of large walnut trees is known next to the village of Saivan; this population has not yet been described.

Due to the disappearance of walnut trees, the density of canopy in riparian forests has been disrupted. These canopy gaps have been occupied by species of *Prunus, Cotoneaster, Berberis, Rosa*, and other shrubs, as well as tree species such as ash, maple, *Thelycrania meyeri*, and sometimes, low-sized willows. Along the banks of Sumbar, elm trees are found. However, none of these species is able to form dense forest canopy.

Following the new canopy gaps, the light-loving plane tree (*Platanus orientalis*) has started to disperse along the mountain rivers. This relict Ancient Mediterranean species is extremely hygrophilous and has survived in Middle Asia only in humid mountain habitats. *Platanus orientalis* which for many centuries was a favorite – and often even sacred – tree in the East, is widely distributed in culture. In the valleys of Southwest Kopetdagh this tree undoubtedly grows in wild conditions where it participates in the composition of riparian forest and sometimes even forms small groves. It has a monumental pillar-like trunk and an elevated crown. Unlike the walnut, the sycamore tree can withstand severe mudslides. In the Aidere Valley, *Platanus orientalis* grows as single trees or in small groups and has normal seed reproduction. When the old plane trunk dies, it is usually replaced by shoots, forming up to five replacement trunks.

Another characteristic tree species in the riparian forests of Southwest Kopetdagh is a local species of cornel (*Thelycrania meyeri*), which also is widely distributed in mountain forests of the Caucasus, Elburz, and Khorassan. Its seed reproduction is effectively facilitated by birds. However, as typical of many other plants, this species is not found in Middle Asia beyond the valleys of Southwest Kopetdagh.

In general, the tree and shrub vegetation of Turkmenistan has undergone serious degradation during the Neogene. The rapid aridization of climate of southern Turan impoverished forest and savannoid communities and transformed them into sparse xerophitic woodlands. In the historical period, the anthropogenic factor has increasingly affected vegetation. By the 1980s, tugai communities had become practically extinct, and the areas occupied by pistachio and juniper woodlands had decreased; unfortunately, this process continues.

Cushion-like *Gypsophila aretioides* on the rocks of Central Kopetdagh. Photo by K.P. Popov.

Mixed juniper (*Juniperus tucomanica*) and maple (*Acer turcomanicum*) forest in Southwest Kopetdagh, Deindere Valley (1,400 m). Photo by K.P. Popov.

11. Ecosystem Structure of Subtropical Arid Pistachio Woodlands in Southern Turkmenistan

ROMAN I. ZLOTIN

Abstract

The structure of the pistachio (*Pistacia vera* L.) woodland ecosystem in the Badghyz Reserve was studied and the general pattern of distribution of organic matter determined. Deposits of soil humus and the plant and animal biomass (excluding Protozoa) were measured. Biomass ratio among ecosystem components is 4.8×10^4 (humus):2.4×10^4 (phytomass):1.1×10^1 (microbial biomass):0.5×10^1 (zoomass). Marked differences in soil fauna exist between pistachio sinusia and gaps between pistachio trees. The role of different taxonomic and trophic groups of animals as well as various problems of conservation are analyzed. The ecosystem structure confirms an ancient age for the arid pistachio woodlands in Middle Asia.

Introduction

Within arid Turkmenistan, the four basic types of ecological communities are steppes, semi-deserts, deserts, and woodlands. Vegetation in steppe, semi-desert, and desert communities is formed by herbaceous plants, semi-shrubs, and shrubs, whereas woodland vegetation includes trees. In Turkmenistan, woodlands are represented by white saksaul (*Haloxylon persicum*) ecosystems of the lowland Karakum sand desert, pistachio (*Pistacia vera*) ecosystems in the foothills of Badghyz and Kopetdagh, and juniper (*Juniperus turcomanica*) ecosystems in the middle and high mountain belts of Kopetdagh.

The genus *Pistacia* is believed to have originated in Asia Minor in the Tertiary period (Korovin 1958), when pistachio woodlands were widespread along the modern northern boundary of the subtropics. By the end of Pleistocene, the range of pistachio communities had decreased due to aridization of the climate; since that time, the leading role in these communities belongs to *Pistacia vera* L. In historical times, this tree species was a landscape-forming one in the Irano-Turanian desert subregion. Pistachio woodlands are considered to be "mature" communities which have evolved over a long period of time under conditions of slightly variable aridity.

Pistachio woodlands are distributed among foothills and low mountain belts. In Middle Asia, they are bordered by juniper woodlands in the higher mountain belts, and by desert grass and shrub communities in the lowland deserts. Structure and developmental features of pistachio communities allow them to be classified as subtropical semisavanna (Ovchinnikov 1948; Popov 1979).

The current natural range of pistachio woodlands within the former USSR is about 300,000 hectares; of these, about 30% are well preserved, moderately disturbed communities (crown coverage 0.2). Today, the formerly contiguous range of *P. vera* consists of four separate populations, those of Pamiro-Alai, Tien Shan, Paropamiz, and Turkmeno-Khorassan mountain areas (Popov 1979). These populations are rapidly declining due to logging, fires, cattle and sheep grazing, and nut harvesting, events which prevent natural reproduction of the pistachio tree. The area covered by pistachio woodlands in Tajikistan was reduced to a quarter of its size from the 1930s through the 1970s (Popov 1979), and it continues to decrease.

Study Area

Badghyz is an elevated territory in the extreme south of Turkmenistan between the Tedzhen and Murghab Rivers; it belongs to the foothills of the Paropamiz Range, which is one of the northern mountain ranges of the Hindu Kush mountain system. Turkmen (former Soviet) Badghyz occupies 6,300 sq. km., whereas the rest of geographical Badghyz is located in Iran and Afghanistan.

Pistachio ecosystems are very characteristic for Badghyz although their range does not exceed 6% of its territory. The subtropical pistachio communities of Badghyz are classified as East Mediterranean deciduous xerophyte woodlands (Kamelin 1989). They belong to the Paropamiz population of *P. vera* and constitute about 35% of all pistachio woodlands of the former USSR. One of the largest massifs of pistachio woodland, the so-called Pulikhatum Pistachio Grove, is located within the territory of the Badghyz Reserve and occupies 39,000 hectares, of which 13,000 are occupied by fully developed stands.

Although there are numerous studies on separate components of the arid communities in southern Turkmenistan, the structure of such ecosystems in general is not sufficiently known. This is especially true for pistachio ecosystems. Whereas the flora and vegetation of Badghyz pistachio woodlands have recently been studied extensively (Popov 1979; Kamelin and Rodin 1989; Kamelin 1990), the heterotrophic components of pistachio ecosystems are for the first time described in this study. This article consists of a partial analysis of field data collected during 1978–1986 in the Badghyz Reserve by the researchers of the Department of Biogeography, Institute of Geography, USSR Academy of Sciences (now Laboratory of the General Problems of Biogeography, Institute of Geography, Russian Academy of Sciences).

The basic object of study was the typical pistachio ecosystem in the central part of the Badghyz Reserve, around the Kepele Post (ca. 800 m above sea level). The territory studied was about five square kilometers in size, with hilly relief formed by Paleogene rocks covered by thick deposits of loess.

One of the goals of the biogeographic study was to determine spatial distribution of organic matter in different reservoirs of the ecosystem: soil humus, vegetation, microogranisms, and animals. We estimated the amount of humus deposits, the plant biomass (both woody plants and grasses, underground and aboveground, in dead and living condition), biomass of microogranisms (determined by direct count of live cells of bacteria and fungi), and biomass of all vertebrate and invertebrate animals (excluding only Protozoa). These data allowed us to build a picture of the spatial distribution of organic matter in the pistachio woodlands of Badghyz.

Structure of the Pistachio Ecosystem

Several soil (edaphic) types of pistachio woodlands can be distinguished in Badghyz. The major area is occupied by *Poa bulbosa-Carex pachystylis* pistachio communities on typical and light sandy serozioms (grey soils), which are formed primarily on northern hillsides. The spatial structure of a pistachio woodland depends on the proportion of adult trees to gap areas consisting of herbaceous vegetation. Adult pistachio trees, as opposed to young and medium-aged ones, possess strong environment-forming qualities, and therefore can be separated into a specific subsystem.

The undercrown areas of old pistachio trees occupied about 15% of the total studied area, although up to 28% of all humus deposit and 28% of all soil nitrogen (Table 1) was concentrated here. Pistachio trees represent 58% of the total phytomass of the community (including 11% of the aboveground live phytomass, 18% of the aboveground dead phytomass, and 28% of the root phytomass) (Table 2). In trophic and habitat relationship with this pistachio tree sinusium are more than 20% of the local mammal species, 30% of the bird species (90% of birds nest on the pistachio), 20% of the reptile species, about 22% of the invertebrate biomass, and 25% of the total animal biomass (zoomass).

The structure of organic matter (excluding Protozoa) can be expressed by the following formula: 4.8×10^4 (humus):2.4×10^4 (phytomass):1.1×10^1 (microbial biomass):0.5×10^1 (zoomass). This formula reflects a general ratio among basic fractions of organic matter. It is similar to ratios known for other types of terrestrial ecosystems although the considerably large contribution of living oragnisms (especially plants) is a specifically characteristic feature of arid woodlands. By comparison, the similar formula for a semi-humid steppe (Central Chernozyom Reserve, Kursk Region, Russia) is 6×10^4 (humus):3×10^3 (phytomass):3×10^1 (zoomass), a ratio which reveals a lesser contribution by plants to the organic matter in the ecosystem.

Table 1. Humus and total nitrogen deposits in a pistachio ecosystem

Soil Depth (cm)	0–30	0–50	0–100
Humus (kg/m^2)			
Under old pistachio trees (500 years)	4.65	5.25	6.84
Under middle-age pistachio trees (50 to 100 years)	2.72	3.57	5.04
Under young pistachio trees (10 to 20 years)	2.93	3.65	4.99
In open areas, under herbaceous vegetation	2.38	3.14	4.41
Average	2.75	3.48	4.80
Total Nitrogen (kg/m^2)			
Under old pistachio trees (500 years)	0.338	0.416	0.607
Under middle-age pistachio trees (50 to 100 years)	0.179	0.242	0.369
Under young pistachio trees (10 to 20 years)	0.213	0.289	0.367
In open areas, under herbaceous vegetation			
Average	0.202	0.258	0.357

Table 2. Living and dead phytomass deposits (kg/ha)

	Pistachio trees and undercrown areas	Gaps with herbaceous vegetation	Total
Aboveground Phytomass	7,319.1	477.4	7,796.5
Living phytomass	2,656.9	209.4	2,866.3
herbaceous plants	73.1	209.4	282.5
pistachio leaves and nuts	146.3	–	146.3
pistachio branches and trunks	2,437.5	–	2,437.5
Dead phytomass	4,662.2	268.0	4,930.2
herbaceous litter	24.4	33.5	57.9
pistachio leaves	190.1	25.1	215.2
branches and bark	271.4	–	271.4
mulch in litter	4,176.3	209.4	4,385.7
Underground Phytomass	6,979.4	10,083.5	17,062.9
Living phytomass	6,607.3	8,123.4	14,730.7
herbaceous plants	123.5	7,765.3	7,888.8
pistachio	6,483.8	358.1	6,841.9
Dead phytomass	372.1	1,960.1	2,332.2
herbaceous plants	30.9	1,941.3	1,972.2
pistachio	341.2	18.8	360.0
Total	14,298.5	10,560.9	24,859.4

Table 3. Structure of zoomass deposit in a pistachio ecosystem (Spring Season)

Animal Group	Dry Weight (g/ha)	% of total zoomass	% of group zoomass		
			saprophages	phytophages	zoophages
Invertebrates	49,590	92	22	73	5
dendrobionts	130	0.2	–	73	27
chortobionts	1,910	3.6	–	96	4
pedobionts	47,550	88	23	72	5
Vertebrates	4,430	8	–	94	6
mammals	4,080	7.5	–	99	1
reptiles	210	0.3	–	63	37
birds	140	0.2	–	–	100
Total Zoomass	54,020	100			

The total phytomass in pistachio communities is divided between 31% of the aboveground phytomass (including 20% of dead phytomass) and 69% of roots. The taxonomic and ecological structure of the zoomass is presented in Table 3. The per cent zoomass ratio of three basic trophic groups of animals (saprophages:phytophages:zoophages) is 21:74:5, which indicates a characteristically arid ecosystem dominance of herbivores consuming live plants.

More than 90% of the zoomass in pistachio woodland consists of invertebrates, of which 88% belong to soil fauna (pedobionts). Among soil macrofauna, the major groups in spring under the crowns of pistachio trees include ground beetles (Coleoptera, Carabidae), larvae of Lepidopera, larvae of chafers (Coleoptera, Scarabaeidae), and larvae of crane flies (Diptera, Tipulidae); their total density is 140 animals/m^2, and biomass (dry weight) is 2,840 mg/m^2. Under the herbaceous vegetation of gaps, the major groups of soil macrofauna comprise larvae of chafers, Lepidopera, assassin flies (Diptera, Asilidae), weevils (Coleoptera, Curculionidae), as well as spiders (Aranei) in various stages of development. Here, total density is 300 animals/m^2, and biomass (dry weight) is 3,330 mg/m^2.

The complex of soil mesofauna under the crowns of pistachio trees in the spring time is dominated by Collembola (1,308 mg/m^2, dry weight), larvae of Scyaridae (Diptera) (365 mg/m^2), larvae of beetles (340 mg/m^2), and free-living nematodes (333 mg/m^2). Under the vegetation of gap areas the soil mesofauna is dominated by larvae of weevils (340 mg/m^2), larvae of Boreidae (Neuroptera) (330 mg/m^2), Acari (320 mg/m^2), and nematodes (287 mg/m^2). Presence of Enchytraeidae (Oligochaeta) is characteristic for pistachio woodland soils in the moist, spring season. We determined biomass ratios among the following three size/taxonomic groups of soil fauna: (1) nematodes + Enchytraeidae , (2) microarthropods (soil mesofauna), and (3) soil macrofauna. This ratio (in %) is 9:42:49 under pistachio crowns, and 8:21:71 in open, gap areas. The soil fauna, therefore, is generally dominated by herbivorous inverebrates. A specific complex of saprophagous invertebrates, however, is formed under pistachio crowns in the spring and consists of mesophilic "forest" groups usually active only for a short period of not more than two months.

The biomass of invertebrates inhabiting crowns of pistachio trees, grass layer, and soil surface is approximately two times less than the total biomass of mammals and constitutes ca. 4% of the total zoomass. The total zoomass of dendrobionts is 130 mg/m^2, of which 30% are weevils of the genus *Sitona* (Coleoptera, Curculionidae), 25%, ants of the genus *Crematogaster* (Hymenoptera, Formicidae), and 20%, Cicadoidea (Homoptera). The rest of the dendrobionts are gall aphids (Homoptera, Aphidoidea), moth larvae (Lepidoptera, Noctuidae and Geometridae) and spiders (Aranei). Of dendrobionts, 39% are sucking herbivorous forms and 34% are grazing herbivores; the rest (27%) are predators (including ants, which also can consume sweet secretions of aphids).

The grass-inhabiting invertebrates form two distinct complexes. Under pistachio crowns, the total spring biomass of 600 mg/m^2 was dominated by true bugs *Myrmecophyes* sp. (Hemiptera, Miridae: 60% of total biomass) and larvae of Arctiidae (Lepidoptera: 28%). Total biomass of grass inhabitants in gap areas was 77 mg/m^2, and the dominant groups were larvae of Lepidoptera and larvae of sawflies (Hymenoptera). Very characteristic for open areas were colonies of ant species *Messor intermedius* (biomass ca. 20 mg/m^2).

The structure of the reptile complex in the pistachio ecosystem (Table 4) was determined using data of A. Yu. Tsellarius (pers. comm.) and our original data. About 60% of reptile biomass is contributed by the steppe tortoise (*Testudo horsfieldi*). Snakes contribute 20% (12 species; dominant are viper *Vipera lebetina*, cobra *Naja oxiana*, and *Coluber rhodorhachis*). Among lizards (10

Table 4. Reptiles in a pistachio ecosystem[*]

Species	Density (animals/ha)	Biomass (Dry Weight) (g/ha)
Tortoise (*Testudo horsfieldi*)	0.5	1,330
Snakes, total	1.5	510
Vipera lebetina	0.04	266
Naja oxiana	0.03	100
Coluber rhodorhachis	0.2	67
Coluber ravergieri	0.2	27
Eryx miliaris	0.2	20
Other	0.8	30
Lizards, total	15.0	288
Varanus griseus	0.02	100
Trapelas sanguinolentus	0.8	66
Ablepharus pannonicus	12.0	60
Other	2.2	62

[*] Original data of A.Yu. Tsellarius (pers. comm.) with additional data by R.I. Zlotin.

Table 5. Birds in a pistachio ecosystem (May 1983)

Species	Density (birds/ha)	Biomass (Dry Weight)(g/ha)
Lanius minor	2.4	40.0
Passer domesticus	3.2	28.8
Upupa epops	1.2	28.0
Hippolais pallida	3.8	17.7
Oriolus oriolus	0.8	17.3
Lanius excubitor	0.6	14.0
Streptopelia turtur	0.2	13.9
Emberiza bruniceps	1.6	13.9
Lanius collurio	0.8	12.0
Cercotrichas galactotes	0.6	7.0
Sylvia hortensis	1.2	6.8
Monticola saxatilis	0.4	6.8
Other	2.0	9.8
Total	18.8	216.0

species), the grey monitor (*Varanus griseus*), steppe agama (*Trapelas sanguinolentus*) and Asian ablephar (*Ablepharus pannonicus*) are most prevalent.

The bird population in pistachio woodland varies from year to year: average density is 12 birds/ha, with average biomass (dry weight) 140 g/ha. About 30 bird species nest in pistachio communities. Dominant species belong to families Laniidae, Emberizidae, and Sylviidae (Table 5).

Mammals dominate the total aboveground animal biomass of pistachio communities; these are mostly rodents (gerbils *Rhombomys opimus*, *Meriones libycus*, and *M. tamariscinus*, vole *Microtus afghanus*, *Ellobius fuscocapillus*, and porcupine *Hystrix leucura*) and ungulates (wild sheep *Ovis ammon*, gazelle *Gazella subgutturosa* and, more rarely, onager *Equus onager*). Among carnivores, red fox (*Vulpes vulpes*) is most common; also wolf, hyena, and leopard can be enountered.

The described ratios of different fractions of organic matter in the Badghyz pistachio community refer only to one phenological season (spring – early summer) which lasts for about two months. This season is characterized by maximal biomass and maximal activity of almost all living organisms. Both biomass and activity in all other seasons decrease by two to three times. In spring, various features of humid ecosystems appear in the structure and function of pistachio community: e.g., activity of ephemerous hygrophile soil fauna such as Enchytraeidae, Tipulidae, Sciaridae, Bibionidae, and Collembola; the active process of biogenic destruction; and intensive functioning of the grazing trophic chain. In all other seasons, activity of heterotrophs decreases due to limitations of the hydrothermic regime, and the ecosystem is controlled mostly by abiotic factors.

The average annual total zoomass in the Badghyz pistachio community is about 90 kg/ha. This figure is of the same order as that of the dry steppes with brown soils, and two to three times higher than the zoomass in clay and sand deserts of Middle and Central Asia. The average annual per cent zoomass ratio of saprophages to phytophages to zoophages is 35:57:8. Notwithstanding the significant impact of herbivores, consumption of primary production is only several per cent. This fact is due to the numerous plant adaptations which evolved during the long coevolution of phytophagous animals and their food plants. Such adaptations include fast growth of shoots at the beginning of the vegetative period, effective chemical defense against herbivores, and ability of plants to vegetate in seasons when herbivore activity is suppressed by low humidity and extreme (low or high) temperatures.

Several major features of subtropical pistachio ecosystems demonstrate their ancient age. These include monotonous spatial structure and absence of expressed boundaries within biota; homogenous distribution of sinusia and individual species within the ecosystem; expressed spatial and temporal separation of ecologically similar groups and their minimal interaction; and a dominant role of trophic relationships. These features of "mature" arid ecosystems make them substantially different from younger communities of the temperate zone which exhibit aggregated structure, rigid relationships among their components and a significant regulatory role by heterotrophs.

Problems of Conservation

In the Badghyz Reserve, the general conservation strategy is preservation of the natural functioning of pistachio woodlands as well as fire protection. It is true that fires are necessary for normal functioning of arid ecosystems; however, their frequency and extent should be under control. To preserve the natural age structure of pistachio stands, strong fires should not be allowed in Badghyz more often than once in a century. All pistachio leaf litter, fallen branches, and dead branches in the crown should be left intact. Grazing of domestic animals (given the high density of wild ungulates in the Reserve) and planned harvesting of pistachio nuts must be completely prohibited.

Table 6 demonstrates the basic types of human influences on Badghyz ecosystems and their consequences. Such an overview of anthropogenic transformation of ecosystems can be used to estimate and counter the results of negative human impact in this unique region of Turkmenistan.

Acknowledgements

I thank the members of Arid Ecosystem Field Group of the Institute of Geography, I.V. Gruznova, G.V. Domnikov, Ye.V. Snegireva, G.A. Kryuchkova, and M.P. Fedotov, for their help in field data collection.

Table 6. Basic types of anthropogenic influence on ecosystems of Badghyz and their consequences

Influence	Consequences
1. Grazing of cattle and sheep	Digression of herbaceous vegetation
	Suppression of woody vegetation
	Destruction of young woody plants
	Impoverishment of animal population
	Soil erosion
2. Ploughing (e.g., for barley crops)	Destruction of biota
	Decrease in organic matter deposits and biological productivity
	Soil erosion
	Concentration of wild ungulates feeding on crops
3. Irrigation	Impoverishment of biota
	Sharp increase in biological productivity
4. Logging	Simplification of biota structure
	Decrease in organic matter deposits and biological productivity
5. Fires	Simplification of biota stucture
	Decrease in organic matter deposits
	Increase in biological productivity
6. Nut collection	Breakup of trophic chains and decrease of numbers of carpophages
	Disturbance of natural reproduction and microevolution of pistachio
	Soil modification
7. Pistachio resin collection	Weakening of trees and increase in pressof phytophages
8. Tree planting	Topographic and spatial complexity increase in ecosystems
	Increase in organic matter deposits and biological productivity
	Concentration of ungulates and rodents

Pistachio (*Pistacia vera*) in Badghyz. Photo by K.P. Popov.

12. Biogeographic Position of the Khorassan-Kopetdagh

VICTOR FET

Abstract

The biogeographic position of the Khorassan area, which within Turkmenistan includes the Kopetdagh Mountains, is discussed. Khorassan is believed to be a transitional province included in the Irano-Turanian subregion of the Saharo-Gobian biogeographic region of the Palearctic realm. Specificity of fauna and flora of Khorassan-Kopetdagh is expressed in its mixed character: it includes a combination of western (mostly Mediterranean) and eastern (Turanian) elements, as well as local authochthonous endemics, usually at the level of species. Connections of the Khorassan biota with surrounding deserts and a separate controversial question of Hyrcanian connections are discussed.

Introduction

The majority of biogeographers distinguish the Turkmeno-Khorassan Mountains (including their ranges within Turkmenistan called the Kopetdagh Mountains) as a separate biogeographic unit. Its rank, however, is not firmly established and varies from province to subprovince under different names, including Turkmeno-Iranian Mountainous Province (Lavrenko 1965), Khorassan-Kopetdagh Province (Kamelin 1970, 1973), Khorassan Mountainous Province (Yemelyanov 1974), or Khorassan Province (Takhtadzhan 1978). There is no argument among these authors that Khorassan should be viewed as a separate biogeographic area. There are, however, different opinions regarding the affiliation of this unit with a certain region/subregion of the higher level. The reason for this ambiguity most likely lies in the more or less subjective opinions of certain workers as well as in the general lack of biogeographic studies devoted to this southwestern part of the Palearctic biogeographic realm.

Khorassan-Kopetdagh as a Biogeographic Entity

Our discussion attempts to determine which distinctive biogeographic features of the Khorassan are responsible for its separation from both the mountain systems of Middle and Central Asia (which lie to the east from Khorassan) and from the Armenian and Iranian mountain plateaus (which border the Turkmeno-Khorassan Mountains to the west). The most prominent characteristics of Khorassan flora and fauna are the presence of a large number of European, European-Mediterranean, and East Mediterranean species, and the combination of these elements with those of the Turanian region (the desert, or eremic, element of Middle Asian deserts).

This combination occurs within areas which are physiographically distinctly defined (e.g., the well-studied Sumbar River drainage in Southwest Kopetdagh). The documented patterns of distribution of eremic Turanian species in southern Turkmenistan show that these elements penetrate not only to foothills but also to the high altitudinal belt, or steppe zone, of Kopetdagh. We presume that this kind of combination of elements, different in biogeographic origin, is found within the entire system of the Turkmeno-Khorassan Mountains. These lie predominantly within Northeast Iran, and are very poorly studied compared to the Turkmen Kopetdagh.

The geographical ranges of numerous European-Mediterranean, East Mediterranean and Southwest Asian (or Middle Eastern, or Levantine) species of animals and plants include the Kopetdagh Mountains, which comprise the easternmost limit of their distribution. Examples can be found among many plant species (Kamelin 1970; Takhtadzhan 1978), insects (Kryzhanovsky 1965), and mammals (Rossolimo and Pavlinov 1982).

The similar distribution can be traced in the ranges of representatives of freshwater fauna. For example, Livanov (1951) found that free-living planarian flatworms of the genus *Dugesia* discovered in Kopetdagh belong to a small but distinctly separate species group which is close to European species and represents the easternmost distributional limit for this group. The presence of such forms definitely indicates a similarity between Kopetdagh and Mediterranean biotas (especially regarding East Mediterranean and Southwest Asian elements).

At the same time, the South Turanian species inhabiting desert landscapes also are important contributors to the Kopetdagh biota. For many of these species Kopetdagh is also known to be their limit of distribution (western or southern). Also, Kopetdagh is the northern limit of the geographical range for some Iranian endemics, e.g., the plant genus *Cousinia* (Asteraceae), whose center of diversity lies within the Iranian Plateau (Agakhanyants 1978). A similar range is known for the grasshopper genus *Saxitania* (Acrididae) (Shumakov 1963).

Finally, Kopetdagh (and, probably, the rest of the Khorassan) possesses a number of local, autochthonous endemic species. The flora of Kopetdagh is estimated to include an approximate 18% of endemic species (Kamelin 1970). A list of well-studied darkling beetles (Tenebrionidae) includes 33 endemics to the Khorassan region (Nepesova 1975).

All these features of Khorassan (of which Kopetdagh is the best-studied portion) form a unique combination of floral and faunal elements which includes species of western, eastern, northern, southern, and endemic origin. One can argue for separation of Khorassan into a special biogeographic unit only if the existence of this unique mixture of elements is recognized, and thus, the transitional nature of this combination is fully acknowledged. Defined in this way, Khorassan can be considered a biogeographic province of the same rank as all other provinces included in the Irano-Turanian subregion of the Saharo-Gobian region of the Ancient Mediterranean (Kryzhanovsky 1965). This last unit is considered a subrealm, or subdominion, of the Palearctic Realm (dominion), with some reservations regarding its transitional nature. Another biogeographic classification proposed by Yemelyanov (1974) includes, among others, the following provinces: the Khorassan Mountainous, Southwest Asian, Hyrcanian, Iranian, Afghan, Turkestan, and South Turanian.

Yemelyanov (1974) discerned the transitional character of some of these provinces (e.g., the Southwest Asian and Hyrcanian); he thus includes both in the Irano-Turanian subregion of the Saharo-Gobian ("Sethian" sensu Yemelyanov 1974) region *and* the Mediterranean subregion of the Mediterranean ("Hesperic" sensu Yemelyanov 1974) region. The innovative concept of "transitional" areas proposed by this author eliminates the excessive rigidity of the conventional schemes. This approach as well as Yemelyanov's regionalization of the Irano-Turanian subregion appear to be reasonable. One could only argue against the unnecessary introduction of new names (such as "Sethian" or "Hesperic") for those regions which already have traditional common names.

Turan versus the Mediterranean?

The most prominent of those Russian researchers who considered the problem of biogeographic division of Middle Asia and adjacent areas was O. L. Kryzhanovsky (1965, 1980). He worked out zoogeographical division of the Irano-Turanian subregion using data on Coleoptera and several other animal groups. Kryzhanovsky distinguished a so-called "Irano-Azerbaidzhanian province"; here, the rank of "province" was undoubtedly higher than the corresponding rank used by the majority of other authors (Lavrenko 1965; Kamelin 1970, 1973; Yemelyanov 1974; Takhtadzhan 1978). This is due to the fact that the entire Irano-Turanian area was treated by Kryzhanovsky (1965) as a "superprovince," which included the Irano-Azerbaidzhanian Province as well as the Turanian, Central Iranian, Afghano-Turkestanian, and other provinces. Khorassan was included by Kryzhanovsky (1965) into the limits of the Irano-Azerbaidzhanian province; therefore, its biogeographic rank appeared to be lower than that of other areas of the Irano-Turanian superprovince (or subregion, according to other authors).

Another Russian biogeographer, V.M. Neronov (1976), published the biogeographic division of Iran, using data on the distribution of rodents. His division differs by distinguishing the Southeast Asian ("Peredneaziatskaya") superprovince, which is similar to Kryzhanovsky's Irano-Azerbaidzhanian province. However, this superprovince is included by Neronov (1976) not within the Sahara-Gobian region but within the Mediterranean region. This treatment reflects the traditional understanding of the Mediterranean in the broadest sense which was, in the past, used by many schools of biogeographers, especially plant geographers (Wulff 1944), and which remained popular until recent years (Tolmachev 1974). Such tradition takes its roots in the views of the famous plant geographer, Engler, who included all Southwest Asia, the Caucasus, and part of Iran in the Mediterranean region. The latter, under such treatment, necessarily receives a very vague eastern boundary (Gratsiansky 1971). Most modern plant geographers treat the Mediterranean area according to the tradition of Alfonse De Candolle and Boissier, in a much more narrow sense, within the range of evergreen tree species (such as laurel, oleander, and evergreen oaks). In Asia, the Mediterranean region in this strict sense occupies only narrow coastal zone in Anatolia and the Levant (Lavrenko 1965; Takhtadzhan 1978). We agree with the last delineation of the Mediterranean region. It eliminates the certain ambiguity in definition of this unit, which often triggered criticism from different viewpoints, or – even – proposals to abolish the term "Mediterranean" (Kucheruk 1972).

Taking into account these arguments, we cannot agree with Neronov (1976) in separation of the Southeast Asian superprovince from the Irano-Turanian subregion and its affiliation with the Mediterranean region. We also cannot accept the very rigid boundary between the Mediterranean and Irano-Turanian area (i.e., between the Mediterranean and Saharo-Gobian biogeographic regions) which Neronov (1976) and Neronov and Arsenyeva (1980) draw along the eastern promontory of Kopetdagh, around the central Iranian deserts, and toward the southwest to the southern Zagroz Mountains. Such rigid boundaries are not completely supported by the biogeographic situation in transitional areas.

Transitional Position of Khorassan

The acknowledgment of the essentially transitional character of certain biogeographic areas led Yemelyanov (1974) to the separation of special transitional areas, e.g., his Southwest Asian and Hyrcanian provinces. This approach seems to reflect real situation adequately. It does not require grouping together transitional and non-transitional provinces; therefore, the intermediate category of superprovince is not necessary and can be omitted. One considers only the categories of the region (e.g., Saharo-Gobian), subregion (e.g., Irano-Turanian), and province (e.g., Khorassan). Each province is characterized by a specific set of species and description of its biota, as well as by the combination of abiotic conditions in the history of its development and in present time.

The chorological relationship of the Khorassan with all surrounding provinces, therefore, can be characterized in the following way. The Khorassan province is bordered to the north and east by the South Turanian province; to the southeast, by the Afghan (or Afghano-Turkestan) province; to the south, by the Central Iranian province; to the west, by the Southwest Asian province; and to the northwest, by the Hyrcanian province. The Southwest Asian province is diverse and includes several subprovinces, of which the one directly bordering the Khorassan province is correspondent to the Armenian subprovince of Yemelyanov (1974), or to the Atropatenian subprovince of Takhtadzhan (1978).

This peculiar biogeographic position, combined with the specific historical development (prolonged island isolation in late Paleogene – early Neogene, which led to the formation of numerous endemic species), is the reason why the Khorassan area is separated as a distinct unit by most biogeographers. The fauna and flora of Khorassan (including Kopetdagh within Turkmenistan) present a very characteristic mix of narrow-range species (the endemics of Khorassan, Southwest Asia, South Turan, mountainous Middle Asia, and Iran) and wide-ranging ones (European-Mediterranean, Ancient Mediterranean, and Palearctic). The character of this mixture in Khorassan differs from the combinations of faunas or floras found at the transition between larger biogeographic categories. It should be noted that, where the Mediterranean and Saharo-Gobian regions meet in Asia and in Africa, the boundaries of these two Palearctic regions lie very close to the Paleotropic realm. Moreover, the territories which lie in southern Iran and Pakistan (the Mekran Mountains, Baluchistan, and Sind) were recently affiliated with the Paleotropic, but not the Palearctic realm (Kryzhanovsky 1980). This proximity of Paleotropic desert biotas again emphasizes the transitional character of Khorassan, which effectively separates the southern and nothern deserts of Asia.

Nevertheless, regarding its transitional status, the Khorassan province cannot be classified as transitional area between two realms (as, for example, Baluchistan province is between Palearctic and Paleotropic realms), nor between two regions of the same realm (as are the Southwest Asian or Hyrcanian provinces which separate the Mediterranean and Saharo-Gobian regions), nor even between two subregions of the same region (as the Mesopotamian province is between the Saharan and Irano-Turanian subregions). The mixed character of Khorassan fauna and flora is more local, and it is defined by its location on the crossroad between five different provinces of the Irano-Turanian subregion. This mixture, combined with substantial autochthonous endemism, determines the unmistakable image of the Khorassan province, including the Kopetdagh Mountains within the republic of Turkmenistan.

Contacts Between Mountain and Desert Biotas

Although a mountainous area, the Khorassan province still experiences contact with the surrounding lowland biotas. Since Khorassan is surrounded by a number of vast Asian deserts (which lie within South Turanian, Central Iranian, and, partly, Afghan provinces), it is occupied by xerophile desert species of animals and plants of various origin. These elements often penetrate to the middle and even high (mountain steppe) altitudinal belts, up to 1,800 or 2,000 m. Such penetration may have been enhanced since the last aridization cycle (Agakhanyants 1978). Disjunct ranges of xerophile elements might have been formed in the past by pluvial, humid climatic periods. An example of a species with a disjunct range from a well-studied fauna of reptiles is a lizard species *Chalcides ocellatus* (Scincidae), which recently was found in Kopetdagh; its main range lies in the Mediterranean region and in the south of the Central Iranian province (Darevsky 1981).

The Hyrcanian Connection

Biogeographical analysis of the Khorassan-Kopetdagh necessitates consideration of the often discussed and misunderstood connection between the Khorassan and Hyrcanian provinces. A unique relict biogeographic area, Hyrcania is well defined as the southwestern and southern shores of the Caspian Sea (within the Talysh or Lenkoran areas in Azerbaidzhan, administrative provinces of Ghilan and Mazandaran in Iran, and, possibly, also the far southwestern portion of Turkmenistan in the delta of the Atrek River). This area has distinctive climate and soils; many authors have ascribed to it an even higher rank than province. Takhtadzhan (1978) mentions that Hyrcania is one of the most clearly defined and delineated provinces in the Irano-Turanian region. Kryzhanovsky (1965) classifies Hyrcania within the Mediterranean region; other authors (V.B. Sochava) include it into the European forest province of the Circumboreal subregion of the Palearctic (Physical-Geographical Atlas of the World 1964). These different viewpoints depend on recognition of the relict nature of Hyrcania; nevertheless, in the modern geological epoch, Hyrcania is sharply isolated from adjacent mountain and desert areas. It occupies relict forests of Talysh, northern slopes and foothills of the Elburz Mountains (its southern slope belongs to the Southwest Asian province), and the lower portion of the Atrek River. The relict Hyrcanian forests are located at altitudes from sea level to 500–800 m (Takhtadzhan 1978). In Kopetdagh, this belt is occupied by a completely different landscape of climatically defined semidesert foothills (adyrs). However, the proximity of the Hyrcania and Khorassan, as well as the presence of mesophilic mountain valleys with riparian forests in Southwest Kopetdagh, motivated some authors in the past (Petrov 1945; Rustamov 1945) to characterize certain elements of the fauna and flora of Kopetdagh as "Hyrcanian" ones. This view seems to be based on an misunderstanding. All animals and plants mentioned by these authors as "Hyrcanian" are actually mesophilic species common not only to Hyrcania and Khorassan but distributed much more widely, with European-Mediterranean, Ancient Mediterranean, or European-Caucasian-Khorassan geographic ranges. The relict character of certain species in Southwest Kopetdagh valleys (such as walnut *Juglans regia* or Eastern sycamore tree *Platanus orientalis*) has no connection with the Hyrcanian flora since such Tertiary relicts are present throughout most of the Ancient Mediterranean area, often with a disjunct range. Of the Kopetdagh flora, which is very well studied and includes more than 1,800 species of vascular plants (Kamelin 1970), only very few species are specific Hyrcanian elements. Such characteristic dominants of the Hyrcanian forest as *Quercus castaneifolia* and *Parrotia persica* (the relict genus), are absent from Kopetdagh. There are also very few animal species with the Hyrcanian-Khorassan distribution; these include a ground beetle *Broscus karelini* (Carabidae) and darkling beetle *Metaclisa viridis* (Tenebrionidae), which penetrate from Hyrcania to the humid valleys of Southwest Kopetdagh (Kryzhanovsky 1965). Many genera have vicariant species in Hyrcania and Khorassan; for example, spiders *Dysdera concinna* (Dysderidae) in Hyrcania and *D. pococki* in Kopetdagh (Dunin 1982, 1985).

Hyrcania and Khorassan, therefore, do not have much in common, and there are almost no "Hyrcanian" biogeographic elements in Southwest Kopetdagh. One can only talk about means of penetration of the mesophilic European, Mediterranean, or Caucasian animals and plants to Kopetdagh and further eastward. These ways, naturally, led through the Hyrcanian province because they were limited by the Caspian Sea to the north and the Iranian deserts to the south.

An additional confusion concerning this question probably originated from historical-cultural traditions: ancient Hyrcania included, among other territories, Southwest Kopetdagh, whereas Parthia, adjacent to the east and southeast, included the rest of Turkmeno-Khorassan Mountains and the northern submontane plain (Yefremov 1955).

Walnut (*Juglans regia*) forest along the Aidere River Valley, Southwest Kopetdagh (1,200 m). Photo by K.P. Popov.

13. Vertebrates in the Red Data Book of Turkmenistan

ANVER K. RUSTAMOV AND OVEZ SOPYEV

Abstract

Among the 101 vertebrate species listed in the Red Data Book of Turkmenistan (1985), are 27 mammals, 35 birds, 30 reptiles, one amphibian, and eight fishes. Their distribution in Turkmenistan is given, with comments on their status as rare and endangered species, and with elements of their biology. Turkmenistan is inhabited by about one-third of the animal species listed in the Red Data Book of the former USSR; more than 11% of those species are found, within the territory of the former USSR, only in Turkmenistan. Means of effective conservation of rare species are discussed, including their protection in eight Natural Reserves (*zapovedniki*) of the Republic of Turkmenistan.

Introduction

A loss of any species of plant or animal is the loss of part of a genetic heritage that cannot be recreated. All species are essential for normal functioning of the biosphere, and all plants and animals should be protected and conserved. However, attention to endangered species indicates that some species are increasingly vulnerable and require urgent measures of protection.

The Red Data Book of Turkmenistan was written following the example of the Red Data Book of the International Union for the Conservation of Nature (IUCN). It distinguishes fives categories of rare and endangered species:

I. *Endangered Species*. These are the species whose populations have decreased so sharply or whose habitats have changed so profoundly that their survival is unlikely without special protective measures.
II. *Declining Species*. This category includes species whose populations are still sufficient for survival but which may become endangered if certain factors leading to their decline are not controlled.
III. *Rare Species*. These are species whose existence is not threatened at the

moment but which are found in very low numbers or in very limited areas, and may disappear due to environmental change or human influence.

IV. *Undetermined.* These species are poorly studied and may possibly be rare or even threatened; the lack of data does not allow to classify these species into one of the three aforementioned categories.

V. *Restored.* These are species whose existence is now not threatened due to applied protective measures, and whose abundance is no longer decreasing. These species are not yet eligible for industrial use, and their status should be constantly monitored.

Below, we give an annotated list of vertebrate species included in the Red Data Book of Turkmenistan which was published in two languages (Turkmen and Russian) in 1985. Classification of some species into one of the given categories is tentative, due to incomplete or subjective knowledge.

Mammals (27 species)

I. Endangered Species (eight spp.)

1. Ursus arctos Linnaeus, 1758 – Black Bear. The subspecies *U. a. syriacus* Hemprich et Ehrenberg, 1828, is not a resident of our republic but rather is visits West Kopetdagh from northern Iran. Such records were common in the 1920s (Laptev 1934, 1937); in recent decades, they are much more sporadic. Footprints of two bears were recorded in the winter of 1961–1962 in the Aidere Valley. Bear footprints were registered also in the spring of 1963 and 1964 in the Chandyr Valley (localities Chaili and Narli) and in the Guen Valley. An adult bear was killed in 1967 in Karakala District; another, on Mt. Tagarev in 1967 (Shcherbina 1970). An adult bear looting the nest of wild bees on a juniper was seen in the summer of 1972 in the Chandyr Valley (locality Kelykhalyk). In West Kopetdagh, bears have been observed in remote valleys with thickets of trees and shrubs, up to the juniper belt. The only record outside this area is that of a female bear with cubs from East Kopetdagh (Meana-Chaacha area), next to the Iranian border, in 1980 (V. Fet pers. comm.; information obtained from border patrol). This subspecies is included in the IUCN Red Data Book.

2. Felis lynx Linnaeus, 1758 – Lynx. The subspecies *F. l. isabellina* Blyth, 1847, is rarely found in Kopetdagh (mostly in the more forested western part but also in Central Kopetdagh). Probably, lynx is not resident there but visits from northern Iran. During the 1960s, only ten lynx pelts were bought from hunters by state purchasing commissioners (from Karakala, Bakharden, and Geoktepe Districts).

3. Acinonyx jubatus Schreber, 1775 – Cheetah. In Turkmenistan, the Asian subspecies of cheetah, *A. j. raddei* Hilzh., has been (still is?) found. Before World

War II, there were about 40 cheetahs in Turkmenistan (Sludsky 1973); from 1930 to 1957 in Badghyz, 25 animals were shot or captured, and about 70 encounters recorded. Cheetahs were often found in Badghyz and Karabil until the late 1950s (Dementyev 1956); today they are absent (A. Rustamov 1980). Some still inhabited Northwest Turkmenistan during the 1960s and early 1970s (A. Rustamov 1980). Today the cheetah can be considered extinct in Turkmenistan; even though some non-confirmed records indicate that single cheetahs still are present in the Northwest, the existence of a viable population there is doubtful. We recommend the reintroduction of this unique large predator to the Badghyz and Kaplankyr Reserves (A. Rustamov 1980). It should be done quickly while it is still possible to obtain the Asian subspecies from Iran or Afghanistan; if cheetah in these countries also disappear, we would have to introduce the African subspecies. This species is is included in the IUCN Red Data Book.

4. *Panthera pardus* Linnaeus, 1758 – Leopard. The subspecies *P. p. tullianus* Valenciennes, 1856, was widespread in mountains and foothills of Turkmenistan in the late 19th and early 20th centuries; its range included the entire Kopetdagh, Kyurendagh, Bolshoi and Maly Balkhans, the western Badghyz, the area between the Kushka and Murghab Rivers (Chengurek Mountains), and part of southern Karabil (Dementyev and Rustamov 1956; Kolesnikov 1956; Rustamov and Shcherbina 1957; Heptner and Sludsky 1967; Gorelov 1973; Babaev *et al.* 1978). Today, it is found only in Kopetdagh and the western Badghyz, including ranges of Gyazgyadyk, Keletkaya, Danagermab, the western part of the Pulihatum pistachio grove, Yeroyulanduz Depression, and Kizyldzhar Valley (Heptner and Sludsky 1967; Gorelov and Shcherbina 1971). It is also found in Bolshoi Balkhan (Zarkhidze 1979), and, from there, sometimes visits West Uzboi. Leopard is completely extinct in Maly Balkhan (the last adult animal was killed there in 1966), as well as in the Kyurendagh and Chengurek Mountains (Babaev *et al.* 1978).

The population of Kopetdagh leopards is steadily declining. Estimates from the 1920s gave an average of 0.25 animals/100 sq. km. In Kopetdagh, up to 15 leopards per year were killed from 1940 to 1941; in Badghyz, in 1947 and 1948, fourteen leopards were killed within 500 sq. km. From 1963 to 1970, 46 leopards were killed in Kopetdagh and Gyazgyadyk (western Badghyz). In total, records show that 360 leopards were killed throughout Turkmenistan from 1924 to 1966. Seventy animals were killed within only seven years in the 1960s (Gorelov 1973).

Today, leopard is a very rare and endangered species. Its records are constant but sporadic in Kopetdagh. In the last two decades, leopards were encountered in Central Kopetdagh in Gaudan (February 1974, one female with two cubs); Mt. Dushak (February 16, 1976, footprints of an adult); Pervomaisky village (December 1976, footprints of an adult); Yablonnaya village (December 26, 1976, footprints of two adults and two young animals); Shamli (January 1977, footprints of an adult); Kelat (in the eastern part of Central Kopetdagh;

January 1977, footprints of two adults); Kurkulab (May 1977, footprints of an adult); Firyuza Valley (January 1978, next to the Iranian border; one adult); Mergenulya (May 1978, one adult); Babazo Valley (summer 1979, one adult). In West Kopetdagh in the 1970s, leopards were spotted in the Syulyukli Valley (a tributary of the Chandyr Valley; June 1975, a female with two cubs), next to the village of Makhtumkala in the Sumbar Valley (June 1976, one adult), and on Mt. Khasardagh (June 1976, footprints of an adult). With the establishment of the Kopetdagh Reserve in 1976, and the Syunt-Khasardagh Reserve in 1979, regular observations of leopards were recorded in the 1980s in Central and West Kopetdagh.[1] Other protected areas in Turkmenistan which are inhabited by leopard include the Kopetdagh and Badghyz Reserves. The total number of leopards in Turkmenistan (including those that may visit from adjacent Iranian territory) probably does not exceed 30 to 40 animals. This species is included in the IUCN Red Data Book.

5. Panthera tigris virgata. Illiger, 1815 – Turanian Tiger. Unfortunately, this subspecies is now not only lost from the fauna of Turkmenistan but is probably extinct from the world fauna as well (Potapov 1978). Tiger lived in Kopetdagh (the Atrek, Sumbar, and Chandyr Valleys) as well as in valleys of the Amudarya, Murghab, and Tedzhen, but it was almost completely exterminated in the first decades of the 20th century. The last tiger in the Sumbar Valley was killed in January 1954 (Dementyev and Rustamov 1956).[2] Tigers entering from Iran were recorded in May 1963 (Arvaz Plateau) and in May 1964 (Mt. Khasardagh)[3] (Babaev *et al.* 1978). The last records of tigers from the Amudarya were in summer 1968 and fall 1971, and from Iran, in the late 1960s to the early 1970s[4] (Potapov 1978). The species is included in the IUCN Red Data Book.

6. Cervus elaphus bactrianus Lydekker, 1900 – Bokhara Deer, or Khangul. In the past, this subspecies of deer inhabited the Amudarya Valley and possibly also the Murghab and Tedzhen Valleys (Heptner *et al.* 1967).[5] Within Turkmenistan, the Bokhara deer now inhabits only tugais (riparian forests) of the Amudarya in Darganata and Deinau Districts. In 1941, when the deer reserve was established in the tugais of the Amudarya, the population there was about 60, and by 1947, about 100 (Bannikov 1979; Heptner *et al.* 1967). Abolishment of this reserve led to the decline of species in this area, and, by the middle of the 1960s, there were only twelve or fifteen deer left (Bannikov 1979). Today, although a new Amudarya Reserve has recently been established, there are no more than a total of 20 to 30 deer within Turkmenistan. This population of Bokhara deer, together with those from Uzbekistan and Tajikistan, is invaluable for the preservation of their genetic pool. This subspecies of deer is included in the IUCN Red Data Book.

7. Gazella subgutturosa Güldenstaedt, 1780 – Goiter Gazelle, or Dzheiran. From the mid-19th century up to the 1930s, many researchers recorded herds of thousands of these gazelles. In the early 1940s the population of dzheiran in Turkmenistan was estimated as more than 100,000 animals (Heptner 1949). Today, its range has decreased more than 70%; this gazelle is now found only in isolated herds. Hunting of gazelles has been banned since 1950; however, this ban has not been effectively enforced, and extensive poaching continues. The only place where dzheiran is relatively secure is the Badghyz Reserve, where its population reaches 3,000 to 4,000 animals; the total population in Turkmenistan today is estimated as not more than 5,000 or 6,000. The dzheiran has been reintroduced at the Gyaurs nursery and introduced on Ogurchinsky Island in the Caspian Sea (Rustamov 1979). By mid-1990 there were more than 400 dzheirans on Ogurchinsky Island. This species is included in the Red Data Book of the IUCN.

8. Capra falconeri Wagner, 1839. – Markhor. In Turkmenistan, the subspecies *C. f. heptneri* Zalkin, 1945, is found only on the western slopes of the Kugitang between the villages Khodzhagaskar and Karlyuk (Heptner *et al.* 1967). This population primarily inhabits the highest part of the range next to the village of Khodzheipil (Ishadov and Klyushkin 1978). The number of *C. falconeri* within Turkmenistan is estimated anywhere from 25 to 200 animals; within the entire range of this species in Middle Asia, there are not more than 1,000 goats. This species is included in the Red Data Book of the IUCN and is protected in the Kugitang Reserve.

II. Declining Species (six spp.)

9. Rhinolophus blasii Peters, 1866. This bat is found in Kopetdagh from the Bakharden Cave to the Chandyr Valley. The Bakharden Cave population has declined from 250 to 300 bats in the 1930s (Kuzyakin 1950) to about 100 today.

10. Miniopterus schreibersi Kuhl, 1819. The subspecies *M. s. pallidus* Thomas, 1907, is found in the Bakharden Cave and several other caves in Kopetdagh. In 1937, the number of bats in the Bakharden Cave was estimated as 40,000 (Kuzyakin 1950). In the spring of 1966 and in August, 1968, only 12,000 bats were recorded (Strelkov *et al.* 1979; the second survey counted young bats also). In the spring of 1970 about 19,000 bats were found, and in 1972, 4,000 to 5,000 (Babaev 1974, Strelkov *et al.* 1979). *M. schreibersi* may possibly move from cave to cave: e.g. in May, 1971, when Bakharden Cave population figures were low, several thousand bats were found in a cave to the south of the city of Geok-Tepe (Strelkov *et al.* 1979). This cave was also used as a winter shelter by about 10,000 bats (February 1979). Of 2,000 to 3,000 bats found in May 1967 in the Kelat Cave, approximately 50% were *M. schreibersi* (Strelkov *et al.* 1979). The total population of this species in Turkmenistan is estimated as 10,000 to 20,000, a two- to four-fold decline in the last 50 years.

11. Calomyscus mystax Kashkarov, 1925. This hamster species is an endemic of the Turkmeno-Khorasan Mountains. Within Turkmenistan, it inhabits Kopetdagh (Chakankala, Aidere, Yoldere, Koine-Kesyr, Germab, Firyuza, Dushak, Keltechinar) and adjacent mountains, from the Maly and Bolshoi Balkhans to Gyazgyadyk in Badghyz (Kashkarov 1925; Ognyev and Heptner 1929; Laptev 1934; Heptner 1956; Kolesnikov 1956; Bondar and Zhernovov 1960; Shukurov 1962). Presently it is rare in West Kopetdagh (Rossolimo and Pavlinov 1982) although it was common there in the 1950s (Bondar and Zhernovov 1960). Maintenance of the Turkmenistan population is important for the preservation of this species. The species is protected in the Badghyz, Kopetdagh, and Syunt-Khasardagh Reserves.

12. Hyaena hyaena Linnaeus, 1758 – Striped Hyena. In the early 20th century, hyenas were common in southern Turkmenistan from the Caspian to the Amudarya (mostly in Maly and Bolshoi Balkhans, Kopetdagh, Badghyz, and Karabil). Today, isolated populations of hyenas still exist locally in western Badghyz and in Kopetdagh; single specimens are recorded also from eastern Badghyz and, possibly, from Karabil (Heptner and Sludsky 1967; Shcherbina 1970; Gorelov 1973; Babaev *et al.* 1978; Rossolimo and Pavlinov 1982). It is very rare in Maly Balkhan, the middle portion of Uzboi (Ishadov and Klyushkin 1979), in the Central Karakum (Ishadov 1973a), and Southeast Karakum (Nurgeldyev 1960). The total number of these animals in Turkmenistan is not more than several dozen (Babaev *et al.* 1978). It should be mentioned that only about 200 hyenas survive in the southern parts of Eastern Transcaucasia (Azerbajan), and this species is extremely rare in Tajikistan and Uzbekistan. Hyena is protected within the Badghyz and Kopetdagh Reserves.

13. Capra aegagrus Ersleben, 1777 – Bearded Goat. The Turkmenistan population belongs to the subspecies *C. a. turkmenicus* Zalkin, 1950. It is found in Maly and Bolshoi Balkhans and in Kopetdagh. Its numbers have decreased dramatically in the last decades. The populations on Gyazgyadyk (Gorelov 1959, 1973) and Kubadagh (Heptner *et al.* 1967) Mountains have disappeared; it is almost gone from West Kopetdagh, where it has been recorded in the past (Laptev 1934); and it is on the verge of extinction in Maly Balkhan (Zarkhidze 1980a). Most of the population is concentrated in Central Kopetdagh, where about 2,000 goats are found within the borders of the Kopetdagh Reserve. The Bolshoi Balkhan population is also declining. This mountain range was especially rich in goats in the past (they were found even in the foothills, at 100–200 m above sea level); a survey in 1971 showed 140 goats per 10 km^2 (Zarkhidze 1980a). The total number of bearded goats in Turkmenistan probably does not exceed 2,000. This species is protected in the Kopetdagh and Syunt-Khasardagh Reserves.

14. Ovis ammon Linnaeus, 1758. The Turkmen mountain sheep (*O. a. cycloceros* Hutton, 1842), or arkhar, inhabits mountains and foothills from Bolshoi Balkhan through Kopetdagh to Badghyz and Karabil, as well as isolated mountains and cliffs in lowland Turkmenistan. Another subspecies (*O. a. bucharensis* Nasonov, 1914) is found in Kugitang. Uncontrolled hunting, use of natural pastures by domestic sheep and cattle, and disturbance by human activity has significantly decreased the population of the mountain sheep in Turkmenistan. There are now 8,000 to 10,000 *O. a. cycloceros*, and the Kugitang population of *O. a. bucharensis* does not exceed 15 to 20 animals. This species is included in the IUCN Red Data Book and is protected in several reserves in Turkmenistan.

III. Rare Species (11 spp.)

15. *Rhinolophus euryale* Blasius, 1853. Only several dozen of these bats inhabit the Bakharden Cave. Another record is known from a cave 10 km to the south of Geok-Tepe (Kuzyakin 1950; Strelkov *et al*. 1979).

16. *Rhinolophus hypposideros* Bechstein, 1800. Only seven specimens of this bat species are known from Turkmenistan, mostly from the Sumbar and Chandyr Valleys in West Kopetdagh. It is protected in the Syunt-Khasardagh Reserve.

17. *Nyctalus noctula* Schreber, 1775. This bat has been recorded only twice in Turkmenistan: from the Sumbar Valley (Kyzyldagh, 25 km to the south of Kara-Kala) in West Kopetdagh; and during migration, 200 km to the southeast of the city of Tashauz.

18. *Tadarida teniotis* Rafinesque, 1814. This bat species is rare all over its range; in Turkmenistan, its total number is about several hundred. The colonies have been found in rock crevices and cliffs in Kopetdagh, Badghyz, and in the Kashan River valley, 60 km to the south of the city of Takhta-Bazar (Kuzyakin 1950; Babaev and Dmitrieva 1966; Gorelov 1977a; Babaev *et al*. 1978; Strelkov *et al*. 1979). Each colony contained 15 to 19 bats. It is protected in the Badghyz, Kopetdagh and Syunt-Khasardagh Reserves.

19. *Jaculus turkmenicus* Vinogradov et Bondar, 1949 – Turkmen Jerboa. This rare and subendemic for Middle Asia rodent has been recorded mostly from the lowlands of West Turkmenistan (about 20 records are known). Surveyed density is 0.2–0.25 animals/km for automobile transects and 0.3–0.5 animals/km for foot transect (Babaev and Ataev 1966; Lobachev and Shenbrot 1973).

20. *Vormela peregusna* Güldenstaedt, 1770. – Marbled Polecat. This polytypic mustelid species is represented in Turkmenistan by its subspecies *V. p. koshevnikovi* Satunin, 1910. It is a common inhabitant of lowlands, foothills, and lower mountain belts, where it feeds on gerbils. The Turkmen Anti-Plague Station conducted surveys of density of this species within gerbil colonies for many years (Babaev *et al*. 1978). The density varied from 13 animals per 100 gerbil colonies (Darvaza at Karakum, Spring 1970) to seven per 100 colonies

(submountain plain between the cities of Bakharden and Kaakhka, Fall 1970); 3.2/100 colonies (Dardzha in West Turkmenistan, Spring 1976); 3.0 to 4.6/100 colonies (Sundukli Sands in Southeast Turkmenistan, Spring 1976); and 2.8/100 colonies (Repetek and Unguz, Spring 1976). In Central Karakum (Yerbent) only one animal per 100 gerbil colonies was counted after an especially cold winter (1968-1969) which affected the gerbil density. Low density, also due to another rodent depression, was observed in the Central Karakum (0.3/100 colonies) and Trans-Unguz Karakum (0.2/100 colonies). Average density of marbled polecat in West Turkmenistan, surveyed in 1954 to 1980, was low: 0.3-1.8 animals/100 colonies on the Krasnovodsk Peninsula; 0.3-0.5/100 colonies in Trans-Uzboi area; and 0.1-0.4 animals/100 colonies in South Ustyurt. Preservation of the Turkmenistan population is important because in many other parts of its range, *V. peregusna* is now rare or has disappeared. This species is found and protected in all Turkmenistan reserves.

21. Mellivora capensis Schreber, 1776 – Honey Badger. In Middle Asia, this species is found only in Turkmenistan (represented by the subspecies *M. c. indica* Kerr, 1792). It is recorded sporadically but probably is not as rare as was thought before (Novikov 1956). In 1952-1962, 31 records of honey badger were reported (Sukhinin and Shcherbina 1955; Sapozhenkov *et al.* 1963; Heptner *et al.* 1967), and in 1970 to 1977, 30 records, mostly in the Karakum Desert along the Unguz dry bed and in the Trans-Unguz area (Gorelov *et al.* 1978; Ishadov and Klyushkin 1979). Its numbers are probably neither sharply fluctuating nor increasing, and its range does not seem to expand northward (Heptner *et al.* 1967). Honey badger is protected in several reserves.

22. Lutra lutra Linnaeus, 1758 – Otter. The subspecies *L. l. seistanica* Birula, 1912, inhabits drainages of the Sumbar and Atrek Rivers (including Maloe Delili Lake), as well as the Amudarya, Murghab and Kushka Rivers (Heptner 1956; Heptner *et al.* 1967; Ishadov 1973b; Ishadov and Ishunin 1975, 1976; Babaev *et al.* 1978). Otter may live along the Kashan River (a tributary of the Murghab) but have not been observed in the Tedzhen River drainage in the last 40 years. After the construction of the Karakum Canal, otter dispersed along the canal, and the formerly disjunct populations in the Amudarya and Murghab drainages became connected. Migration of otters into Lake Sarykamysh is also thought possible. There are no more than 200 otters throughout Turkmenistan; even in the past, when otter hunting was permitted, only about 100 pelts were recorded by state fur purchasers during 1924 to 1966.

23. Felis caracal Schreber, 1777 – Caracal, or Sand Lynx. In Turkmenistan, the subspecies *F. c. michaëlis* Heptner, 1945, is found in the Karakum Desert and other lowland and foothill regions (Heptner and Sludsky 1967). Its population now does not exceed 200 to 300 animals. These areas are main part of the species' range in Middle Asia; caracal is on the verge of extinction in the Cis-Aral area of Uzbekistan. A ban on caracal hunting was established in 1937, but its range is decreasing due to the development of desert areas. Some cats die in cold winters with high snow accumulation, or they are attacked by stray dogs. The frequency of records of caracal also depends on density fluctuations of its prey species, the desert hare (*Lepus tolai*). Caracal is listed in the IUCN Red Data Book and is protected in several reserves of Turkmenistan.

24. *Felis manul* Pallas, 1776 – Manul Cat. The subspecies *F. m. ferrugineus* Ognev, 1928, is found in Kopetdagh from the Atrek River to the meridian of Artyk Station, and also in Kyurendagh, Bolshoy Balkhan, western Badghyz (and probably in Karabil), on the southern chink of Ustyurt, and in parts of northern Turkmenistan (Laptev 1934; Heptner 1956; Heptner and Sludsky 1967; Babaev *et al.* 1978). This cat is rare in Kopetdagh and Bolshoi Balkhan, and very rare in Badghyz (Heptner and Sludsky 1967; Zarkhidze 1979; Rossolimo and Pavlinov 1982).

25. *Equus hemionus* Pallas, 1775. – Onager (Wild Ass), or Kulan. The Turkmen subspecies (*E. h. onager* Boddaert, 1785) is currently found only in South Turkmenistan between the Tedzhen and Kushka Rivers, to the south of the line connecting the villages of Shortepe and Kalaimor (Heptner 1948; Shcherbina and Kravchenko 1960; Heptner *et al.* 1967; Babaev *et al.* 1978). Outside of Turkmenistan, the onager is found only in Iran and northern Afghanistan. It was recently introduced to the Meanachaacha and Kalinin Refuges (submountain plain of Central Kopetdagh) and to the Kaplankyr Reserve (northern Turkmenistan).

In the 19th century, thousands of onagers lived in Turkmenistan. By 1935, only about 500 animals were left, all of them in Badghyz. Their number continued to decrease until the Badghyz Reserve was established in 1941, when only 250 onagers were left. In 1969, the Badghyz herd comprised 800 animals and in 1976, 1,254 (Babaev *et al.* 1978). Currently, the Badghyz Reserve has not less than 2,000 onagers. The population growth in Badghyz is limited by the territory of the Reserve and water availability; most water sources and summer pastures are located outside of the Reserve, in valleys of the Kushka and Tedzhen Rivers. The onager is included in the IUCN Red Data Book.

IV. Undetermined

26. *Myotis nattereri* Kuhl, 1818. The endemic subspecies, *M. n. tschuliensis* Kuzyakin, 1935, is described from Turkmenistan. Only seven specimens are known from gorges and abandoned mines of Central and West Kopetdagh (Kuzyakin, 1950; Strelkov *et al.* 1979). This bat species is protected in the Kopetdagh and Syunt-Khasardagh Reserves.

27. *Myomimus personatus* Ognev, 1924. In Middle Asia, this dormouse is found only in Southwest and Central Kopetdagh (Rossolimo and Pavlinov 1982) and Bolshoi Balkhan. Since its discovery (Ognyev 1924), only 10 to 12

specimens have been recorded, including four from North Iran (Nikolsky and Molyukov 1975). Part of the range of *M. personatus* lies within the territories of Kopetdagh and Syunt-Khasardagh Reserves.

Birds (35 species)

I. Endangered Species (six spp.)

1. Anas angustirostris Menetriés, 1832. – Marbled Teal. Until the 1940s, this duck was the commonest waterfowl species on the Amudarya, Murghab, Tedzhen, and the lakes of the western Uzboi and the Atrek River (Shestoperov 1937); part of this population was probably resident. On the lower Atrek, numbers of *A. angustirostris* in the winters of the 1930s reached 17,000 (Laptev *et al.* 1934). Today they are very rarely found nesting in Turkmenistan but are seen during migrations and in winter time. A very large group of ducks (ca. 5,000) was observed in January 1988 on Lake Dengizkul, next to the Uzbekistan border (O. Mitropolsky pers. comm.). A catastrophic decrease of this population has been caused by the development of river valleys as well as by uncontrolled hunting.

2. Haliaëtus leucoryphus Pallas, 1771. – Pallas' Sea Eagle. This species is not resident in Turkmenistan but occasionally is found in winter and summer along the river valleys and on the Caspian shore (Zarudny 1896; Shestoperov 1937; Isakov and Vorobyev 1940; Sukhinin 1971).

3. Gypaëtus barbatus Linnaeus, 1758. – Lammergeier, or Bearded Vulture
Our subspecies, *G. b. aureus* Hablizl, 1783, is resident in Bolshoi Balkhan, Kopetdagh, Badghyz and Kugitang (Zarudny 1896; Dementyev 1952). Today, its total number in Turkmenistan is only 20 to 26 couples. In Kopetdagh, bearded vultures inhabit rocky and stone areas in the upper (juniperous) belt and can forage in lower belts, including human setlements. In Central Kopetdagh, 10 cases of nesting were recorded from 1983 to 1985. One to two eggs are laid as early as February (average 1.7 eggs and one hatchling per couple). This species is protected in the Kopetdagh and Syunt-Khasardagh Reserves.

4. Falco pelegrinoides Temminck, 1829.[6] – Shakhin Falcon. The subspecies *F. p. babylonicus* Sclater, 1861 inhabits Kopetdagh, Badghyz, Karabil and Kugitang. G.P. Dementyev (1952), who was a great authority on falcons, recorded several dozen couples in the republic in 1950s. Not more than 20 couples are currently recorded from Turkmenistan; therefore, the population density is extremely low. This species is protected in the Badghyz, Kugitang, Kopetdagh, and Syunt-Khasardagh Reserves.

5. *Grus leucogeranus* Pallas, 1773. – White Crane, or Sterkh. This species is very rarely found during migration within Turkmenistan. Not more than 20 records are known in the last 100 years: most of them are very old encounters from the 1890s or 1900s in different areas of Turkmenistan such as Sarykamysh (Zarudny 1896), Atrek (Zhitnikov 1900), Uchadzhi (Loudon 1901, 1902) and Mt. Dushak (Dementyev 1952). In 1935, sterkh was observed next to Ashkhabad (Dementyev 1952). On March 10, 1977, five sterkhs were observed in a mixed crane flock at Badghyz, in the Yeroyulanduz Depression (Ataev *et al.* 1978). This crane species is listed in the IUCN Red Data Book; not more than 250 or 300 birds of this species exist.

6. *Picus squamatus* Vigors, 1831. – Scaled Woodpecker. In the 1900s, this species was common in riparian poplar forests (tugais) along the middle portion of the Murghab River. In 1942, A.N. Formozov recorded an average of eight woodpeckers per day in five days of field trips between the villages of Sarychop and Kazyklybent (Dementyev 1952). This Murghab population has most likely disappeared in recent decades due to tugai logging and valley development. The reported observation of a woodpecker couple on May 8, 1986 (Simakin 1991) on the right bank of Murghab (15 km from the Tashkepri Dam) has yet to be confirmed. Reintroduction of this species from Afghanistan to the remaining old tugai forests has been recommended (Gorelov and Gorelova 1976).

II. Declining Species (13 spp.)

7. *Larus ichthyaetus* Pallas, 1773. – Great Black-headed Gull. This seagull is resident from Karabogazgol Bay of the Caspian Sea and the lower Atrek River to Lake Sarykamysh (Chernov 1990); it is also found on the Caspian shore and islands. It is not rare, but its density is decreasing. At Karabogazgol Bay, 270 couples were observed nesting in 1974; 153, in 1975; and 670, in 1976 (Gauzer *et al.* 1976). The island colony at Lake Sarykamysh had 80 to 100 nests in 1987 (Chernov 1990). The species is very rare in winter; in the winters of the 1960s, only six to eight birds were found between Gasankuli and Chikishlyar villages (Dobrokhotov 1962; Orlov 1970). This seagull is protected in the Krasnovodsk Reserve and the Sarykamysh Natural Refuge.

8. *Anthropoides virgo* Linnaeus, 1758. – Demoiselle Crane. This extremely rare migratory bird is found in river valleys and submountain areas (Shestoperov 1937; Dementyev 1952), and is protected in several reserves.

9. *Pelecanus onocrotalus* Linnaeus, 1758. – Pink Pelican. In Turkmenistan, pelicans are found during migrations along the Amudarya, Murghab, Tedzhen, Kushka, and Atrek Rivers, and also along the Karakum Canal and Kopetdagh (Zarudny 1896; Shestoperov 1937; Dementyev 1951b; Rustamov and Sukhinin 1957). Pelicans are resident at Lake Sarykamysh, where from 2 to 457 (average 100) couples were recorded from 1984 to 1989 (Chernov 1990). In the 1890s,

thousands of pelicans could be observed on fall migration at the lower Atrek river; some birds also spent the winter there (Zhitnikov 1900). As early as the 1930s, however, pelicans did not overwinter at the lower Atrek – Gasankuli area and rarely migrated through this region. Currently, only about 150 pelicans are found in winter on the water bodies of Turkmenistan (but not on the Caspian shore). This species is protected in several reserves.

10. Platalea leucorodia Linnaeus, 1758. – White Spoonbill. In the 1930s, this species was commonly observed at the Atrek River (up to 500 birds at the same time; Isakov and Vorobyev 1940). Today it is rarely found nesting in Turkmenistan. In 1971, twelve nests were recorded from the Atrek River (Lake Maloe Delili) (Vengerov 1975); in 1984, two nests were found on Lake Sarykamysh (Chernov 1990). It is protected in the Krasnovodsk and Amudarya Reserves, and in the Kelif and Sarykamysh Natural Refuges.

11. Phoenicopterus roseus Pallas, 1811. – Flamingo. This species is presently not resident in Turkmenistan although nests were recorded in the 1930s from Karabogazgol Bay of the Caspian Sea. The total number of flamingos spending the winter on the eastern Caspian shore (Krasnovodsk Reserve) can reach 15,000 to 16,000 birds. Records for the areas vary from 11,000 to 13,000 in 1950, to 12,300 in 1952, 8,000 in 1953, and 2,000 in 1954 for the Chikishlyar – Gasankuli area; and from 16,000 in 1968 to 6,000–15,300 birds in 1971–1975 in the northern part of the Krasnovodsk Reserve (Rustamov and Vasilyev 1976). A few nesting places are known from Kazakhstan, but the population there is decreasing, and all these birds now spend the winter in the Krasnovodsk Reserve. The Caspian shore of Turkmenistan, therefore, is the most important habitat supporting the flamingo population in Middle Asia.

12. Oxyura leucocephala Scopoli, 1769. – White-headed Duck. Populations of this species are catastrophically decreasing all over the world. In the recent past it was resident in Turkmenistan, probably along the entire Amudarya River, in the lower portions of the Murghab and Tedzhen (Zarudny 1896; Dementyev 1952), and on the Caspian shore (Shestoperov 1937). Today, only a few ducks migrate and spend the winter at the southeast Caspian (Rustamov and Vasilyev 1976); in winter this species has also been recorded from the Kelif Uzboi lakes (Rustamov and Khakyev 1978; Chernov 1990) and Lake Sarykamysh, where a flock of 100 ducks was seen in April 1986, and a flock of 30 birds, on April 7, 1987 (Chernov 1990). The species is protected in the Krasnovodsk and Amudarya Reserves and in the Kelif and Sarykamysh Natural Refuges.

13. Aquila chrysaëtos Linnaeus, 1758. – Golden Eagle, or Berkut. About 120 to 130 couples resident in Turkmenistan belong to the subspecies *A. c. homeyeri* Severtzov, 1888. They are found in Bolshoi and Maly Balkhans, Kopetdagh, Badghyz, and Kugitang, and are protected in several reserves.

14. Haliaëtus albicilla Linnaeus, 1758. – White-tailed Sea Eagle, or Erne. In the 1930s, the species was common on the southeastern Caspian shore; 2,680 birds were recorded there in February 1932 (Laptev *et al.* 1934). The present number of sea eagles observed during migration or winter is very low. On the entire shore from Krasnovodsk Bay to Karabogazgol Bay, only 60 birds were recorded in the winter of 1976 (Vasilyev 1976a,b). In the Chikishlyar area, 20 birds were recorded in January, 1978, and 68 birds on a 30-km transect in February, 1978 (Chernov 1990). More than 300 birds were recorded from an airplane on a 40-km stretch of Daryalyk Canyon (Northeast Turkmenistan) (Chernov 1990). This species is included in the IUCN Red Data Book; it is protected in the Krasnovodsk Reserve and the Sarykamysh Natural Refuge.

15. Falco peregrinus Tunstall, 1771. – Peregrine Falcon, or Sapsan. This species is found during migration and in winter on the southeastern Caspian shore (rarely) and on the lower Atrek (very rarely). It was a quite common bird in these areas in the 1930s (Isakov and Vorobyev 1940).

16. Otis tarda Linnaeus, 1758. – Great Bustard. The typical subspecies *O. t. tarda* L. does not nest in the republic but is found during migration and in winter in the lowlands of Southwest and East Turkmenistan and in the submountain plain of Kopetdagh. Bustard was a common species in the 1940s and 1950s (Zarudny 1896; Shestoperov 1937; Isakov and Vorobyev 1940; Dementyev 1952) but now is quite rare; the total number of birds occuring in Turkmenistan can be as low as several dozen.

17. Otis tetrax Linnaeus, 1758. – Little Bustard. The subspecies *O. t. orientalis* Hartert, 1916 was recorded nesting in Kopetdagh before the 1930s (Zarudny 1896; Shestoperov 1928); today only occasional nests can be found. It is found during migration along the Kopetdagh foothills, the valleys of the Amudarya, Murghab, Tedzhen, Atrek, on the Caspian shore, and locally in the Karakum Desert (Shestoperov 1937; Isakov and Vorobyev 1940; Dementyev 1952). *O. tetrix* was a common migratory bird in Turkmenistan until the 1940s; flocks of 20 to 30, and up to 200–300 (in the Atrek valley) were recorded (Dementyev 1952; Isakov and Vorobyev 1940). Today it is very rare; during field trips in 1983–1986, only three birds were recorded in Central Kopetdagh, on March 28, 1984.

18. Chlamydotis undulata Jacquin, 1784. – Dzhek. The range of this species in the recent past included most of the Turkmenistan lowland, from the Caspian shore deserts and Ustyurt through the Karakum Desert and westward along the foothills of Kopetdagh and Badghyz. In 1948, four birds were recorded on a 15-km transect in Central Karakum (A. Rustamov 1954). Development of desert areas and hunting have steadily reduced dzhek's density. During field trips from 1972 to 1980, dzhek was recorded only once, on July 26, 1979, a few kilometers from Gasankuli village. Only one family was found during several years of field studies in the same area on May 18, 1970, between Bugdaily and Bekibent villages in the Kizyl-Atrek District (Mishchenko and Shcherbak 1980). During automobile surveys totaling 9,940 km in the deserts lying between the Tedzhen and Murghab valleys (spring – summer 1975–1978), 54 birds were recorded, or 0.5 birds/100 km (Rustamov and Makeev 1981).

19. Chettusia gregaria Pallas, 1771. – Sociable Plover. This species in Turkmenistan is found only during migrations and exclusively in the lowlands; it is very rare and only 20 to 30 records are known, mainly from the past (Radde and Walter 1889; Stolzmann 1890; Zarudny 1896; Shestoperov 1936c; Isakov and Vorobyev 1940; Dementyev 1952).

III. Rare Species (14 spp)

20. Pelecanus crispus Bruch, 1832. – Curly Pelican. This species is resident only at Lake Sarykamysh, where 30 colonies have been found, each with 10 to 25 nests (Velikanov and Khokhlov 1979; Chernov 1990). This population increased from 1984 (48 couples) to 1989 (215 couples) (Chernov 1990). *P. crispus* also migrates through the river valleys and along the eastern shore of the Caspian Sea; some pelicans overwinter there. Annually, not less than 350 pelicans spend winter on the water bodies within Turkmenistan (E. Rustamov et al. 1990). This species is protected in the Krasnovodsk and Amydarya Reserves and in the Sarykamysh Natural Refuge.

21. Ciconia nigra Linnaeus, 1758. – Black Stork. A rare migratory and resident bird, the black stork is found in foothills, oases, and, primarily, in river valleys. Nests have been recorded in East and Central Kopetdagh, as well as in Badghyz: next to the Kushka River valley, on the Fistashkovy (Pistachio) Range, and to the northeast of Kalaimor village (Sukhinin 1955). There are only about 10 or 12 nesting couples in Turkmenistan annually, and the number of recorded migrating storks is about 20 or 30.

22. Rufibrenta ruficollis Pallas, 1769. – Red-breasted Goose. This goose was very common in the 1890s in the middle portion of the Atrek River where tens of thousands of birds were recorded with 300 to 500 geese in some flocks (Zhitnikov 1900). Such numbers have never been observed since; only some migratory flocks were recorded from the lower portion of the Atrek in 1938 (Isakov and Vorobyev 1940). No other data are available about this species in Turkmenistan.

23. Cygnus bewickii Yarrell, 1830. – Swan. This very rare species is now found only on the Caspian shores; at the end of the 19th century, it was recorded more often in the lower part of the Atrek River (Zhitnikov 1900).

24. Pandion haliaëtus Linnaeus, 1758. – Osprey. Very sporadic records of this species (during migration and in winter) are known from the Caspian shores, the Amudarya, Murghab, Tedzhen, Atrek, and Karakum Canal. In the past, nests were found in the middle Atrek (Zarudny 1896), in the Murghab Valley, and lakes of the Kelif Uzboi (Tashliev and Yermakova 1961). Nesting in Turkmenistan now needs confirmation. The species is under protection in the Amudarya, Badghyz, and Krasnovodsk Reserves, and in the Kelif Natural Refuge.

25. *Circaëtus gallicus* Gmelin, 1788. – Short-toed or Serpent Eagle. This rare migratory and resident bird (subspecies *C. g. heptneri* Dementiev, 1932) is found in Kopetdagh, Badghyz, and the Karakum Desert (Zarudny 1896; Vorobyev 1955; Heptner 1956; Sukhinin 1957; Ataev 1974). Its nesting areas in the Badghyz Reserve (pistachio woodland) and the Repetek Reserve (saksaul forest) are well protected, and therefore the density of serpent eagles in these areas is constant, about 3 to 5 nesting couples per 10 sq. km. (Ataev 1974).

26. *Aquila rapax* Temminck, 1828. – Steppe Eagle. This rare, non-resident bird can be found during migration and sometimes in the winter (Shestoperov 1937). Its numbers are steadily decreasing over the species' range.

27. *Aquila heliaca* Savigny, 1809. – Imperial Eagle. The nominal subspecies, *A. h. heliaca* Sav., has sporadically nested in Turkmenistan in the past (Zarudny 1896) but now is found only during migration and, mainly, in winter on the Caspian shores, in the Tedzhen and Murghab valleys, and in Badghyz (Dementyev 1952).

28. *Falco cherrug* Gray, 1834. – Saker Falcon, or Baloban. Two subspecies of this rare bird are found in Turkmenistan. A nominal subspecies, *F. c. cherrug* Gray, is found during migration and in winter. Another form (the Turkestan saker falcon *F. c. coatsi* Dementiev, 1945) is resident here in mountains and lowlands, and in rocky and clay cliff habitats (Kopetdagh, Badghyz, Kugitang, Karakum, Ustyurt). Within Turkmenistan, there are 80 to 100 couples. Density of this species over its range is decreasing; it has already disappeared from many areas. The saker falcon is protected in several reserves.

29. *Porphyrio porphyrio* Linnaeus, 1758. – Purple Swamphen. Of many existing subspecies, only one is found in Turkmenistan, *P. p. seistanicus* Zarudny et Haerms, 1911. In the past, it was observed in large numbers in the lower portion of the Atrek River, from Kizyl-Atrek to Chalayuk (Radde and Walter 1889; Zhitnikov 1901; Dementyev 1952). Presently it is recorded only from Lake Maloe Delili (Karavaev 1981), where it lives in dense reeds. It is a resident species there but the total number is not more than 50 birds. This population has declined due to the development of the Atrek Valley. In Middle Asia, *P. porphyrio* has a limited range, and therefore conservation of the Turkmenistan population is very important.

30. Cursorius cursor Latham, 1787. – Cream-colored Courser. In Middle Asia, this bird species (subspecies *C. c. bogolubovi* Zarudny, 1910) is known only from South and Southeast Turkmenistan, where it is both resident and migratory. A few birds spend the winter in Badghyz (Rustamov and Sukhinin 1957). The total number for this species in Turkmenistan is several hundred.

31. Hypocolius ampelinus Bonaparte, 1850. The first record of this species from Turkmenistan was in the early 1960s from Dauletabad and Novruzabad in Serakhs District (Flint 1961, 1962). Later, it was observed several times in the middle portion of the Tedzhen River, and in the southernmost part of Turkmenistan, next to Morgunovsky village in the Kushka River Valley (Chunikhin 1968; Vorobyev 1968; Stepanyan 1978; Eminov 1982). Nests were found in 1979 on the Murghab River, 5 km from the city of Takhta-Bazar (Peklo and Sopyev 1980). Probably, it also nests in the valleys of the Kushka and Tedzhen Rivers. Total number of resident birds is not more than 50 or 60 couples; 400 birds were recorded during the fall migration along the Tedzhen River (Eminov 1982). The Turkmenistan population of *H. ampelinus* is important for the species' conservation at the northern limit of its range.

32. Passer simplex Lichtenstein, 1823. – Desert Sparrow. Not more than 75 to 100 couples belonging to the subspecies *P. s. zarudnyi* Pleske, 1896, are resident in the eastern and central Karakum Desert, from Sundukly Sands, Repetek and Yeradzhi, to Yerbent and Bakhardok in the west (Dal 1936; A. Rustamov 1954, 1958). The species is protected in the Repetek Reserve.

33. Francolinus francolinus Linnaeus, 1758. – Francolin. The nominal subspecies is found in Turkmenistan in two isolated locations: in the lower Atrek (from the town of Kizyl-Atrek to Cheloyuk village), and in the Sumbar River valley (from 12 km above the town of Kara-Kala to 34 km below it) (Polozov 1980, 1981). The total number of francolins in Turkmenistan does not exceed 500. It is protected in the Krasnovodsk Reserve (Gasan-Kuli area) and Syunt-Khasardagh Reserve (Sakharov 1982).

IV. Undetermined (two species)

34. Tetraogallus caspius Gmelin, 1784. – Ular. The range of 300 to 350 birds (nominal subspecies *T. c. caspius* Gmelin) is limited to Central Kopetdagh (between Kheirabad, Firyuza, and Gaudan). It is protected in the Kopetdagh Reserve. In the past, ular was also recorded from the upper part of the Chandyr River in West Kopetdagh (Zarudny 1896).

35. Lobivanellus indicus Boddaert, 1783. – Red-wattled Lapwing. The subspecies *L. i. aigneri* Laubmann, 1913, is found along the Murghab River drainage (from the Egrigek and Kushka Rivers in the south to the oases of Mary in the north), in the Shorgol area, and along the Tedzhen River valley (from its higher part within Badghyz to the Tedzhenstroi II settlement) (Zarudny 1896; Dementyev 1952). It has been quite numerous in the past both in the Murghab and Tedzhen, but its numbers are markedly decreasing due to the development of river valleys.

Reptiles (30 species)

I. Endangered Species (two spp.)

1. Phrynocephalus maculatus Anderson, 1872. An isolated relict population of ca. 2,000 lizards was recently found to the north of Bami Station on the southwest edge of the Karakum Desert (Bogdanov *et al.* 1974; Krasnaya Kniga SSSR 1978; Rustamov and Shammakov 1977, 1979). This population is threatened by extinction due to the projected flooding of its habitat.

2. Ptyas mucosus Linnaeus, 1758. The subspecies *P. m. nigricens* Chernov, 1949 (first described in Terentiev et Chernov 1949), is known from the Murghab and Kushka Valleys of southern Turkmenistan.[7] It is recorded from the vicinity of villages and towns of Childukhtor, Poltavka, Morgunovsky, Cheminibit, Kushka, Takhta-Bazar, Tashkepri, Saryyazy, Imambaba, Krasnoye Znamya, Sultanbent, Iolotan, Talkhatanbaba, Turkmen-Kala, Kaushutbent, Mary, Bairam-Ali, Vekil-Bazar, and Akibai (Bogdanov 1962, 1965; Gorelov and Orlov 1965; Ataev *et al.* 1978; Gorelov 1980; Shammakov 1981). It was a common snake 20 to 25 years ago; up to ten snakes could be seen in a four- to five-hour field trip. The industrial development of river valleys has undermined this population. Currently, its density is not more than 0.2 snakes/ha[8], and the total number of snakes in the Murghab drainage is estimated as 2,000 (Ataev *et al.* 1978, Gorelov 1980). The Turkmenistan population is priceless for the preservation of the genetic diversity of this species. For its protection, a refuge should be established in the Kushka River Valley between Morgunovsky village and Pobeda Sovkhoz (state farm), under the control of the Badghyz Reserve.

II. Declining Species (five spp.)

3. Alsophylax pipiens Pallas, 1813. Presently, this is a common gecko species around Lake Sarykamysh, next to the chink of Ustyurt, and in the Butentau area of the Kunyadarya Plain (Bogdanov 1962; Velikanov 1977; Rustamov and Shammakov 1979); however, its density is decreasing due to the flooding of habitats. The species is protected in the Sarykamysh Refuge.

4. Alsophylax loricatus Strauch, 1887. The subspecies of this gecko, *A. l. szcerbaki* Golubev et Sattorov, 1979, is known from Turkmenistan along the left bank of the Amudarya, between the towns of Dargan-Ata and Kunya-Urgench (Terentyev and Chernov 1949; Bogdanov 1972; Shammakov 1974, 1981;

Rustamov and Shammakov 1979; Golubev et Sattorov 1979; Shcherbak 1979).[9] In the early 1980s it was common (average 28 lizards/ha) but the development of clay desert is destroying its habitats and reducing the population.

5. *Alsophylax laevis* Nikolsky, 1905. This gecko species is relatively common (average 30 lizards/ha) in clay desert habitats from the Messerian Plain in Southwest Turkmenistan, along foothills of Kopetdagh, and eastward to the Tedzhen River. However, in the Karakum Canal area its habitats are intensively developed, and population is reduced. The protection of the Turkmenistan population is important; the species' range is shared between Turkmenistan and Uzbekistan (the Kizylkum Desert).

6. *Varanus griseus* (Daudin, 1803) – Grey Monitor. The monitor's range includes all of lowland and foothill Turkmenistan (Morits 1929; Andrushko 1939; Kartashov 1955; Rustamov 1955; Bogdanov 1962, 1965; Shammakov 1968; Makeev 1979; Makeev et al. 1988).[10] Its density can be high in territories with a high number of rodents: e.g. from 1967 to 1972 in some parts of Southwest Turkmenistan, the monitor's density was 9 to 12 lizards/sq. km. A 1976 spring survey in Southeast Karakum areas next to the Karakum Canal revealed a density of 5 lizards/sq. km. (Ataev et al. 1978); in the area between the Tedzhen and Murghab Rivers, average density from 1975 to 1978 was 0.5 to 0.9 lizards/sq. km. Density in the Badghyz Reserve has not changed significantly from 1955 (0.17 lizards/sq. km) to 1972 (0.1 lizards/sq. km). The grey monitor is protected in all reserves of Turkmenistan. In unprotected territories, its numbers are decreasing due mainly to the development of land and degradation of habitats. This species is included in the IUCN Red Data Book.

7. *Naja oxiana* (Eichwald, 1831) – Middle Asian Cobra. The cobra inhabits both mountains and lowlands (Bolshoi and Maly Balkhan, Kopetdagh, Badghyz, Karabil, Kugitang, Karakum Desert, valleys of the Amudarya, Murghab, Sumbar, and Atrek (Morits 1929; Laptev 1934; Andrushko 1939; Rustamov 1956; Bogdanov 1962; Makeev and Zemlyanova 1968; Makeev 1978,). The entire Middle Asian cobra population is estimated as 300,000 to 500,000 snakes, of which a significant part is found in Turkmenistan.[11] The cobra can locally achieve high density (e.g., up to 50 snakes/sq. km in Southwest Kopetdagh); however, over its range it is a species with decreasing density and is listed in the IUCN Red Data Book. It is protected in all reserves of Turkmenistan.

III. Rare Species (18 spp)

8. *Eublepharis turcmenicus* Darevsky, 1977. This rare lizard is known only by several records from the middle mountain belt (up to 800–900 m) of Central and

West Kopetdagh. It is protected in the Kopetdagh and Syunt-Khasardagh Reserves.

9. Bunopus tuberculatus Blanford, 1876. Only one population of the nominal form *B. t. tuberculatus* Blanf. is known from the Yeroyulanduz Depression in southern Badghyz; the total number of this gecko there is estimated as about 300 lizards. The population is protected in the Badghyz Reserve.

10. Cyrtopodion spinicauda (Strauch, 1887). Small populations of this gecko are found in Central and West Kopetdagh and in the western Badghyz (Ataev *et al.* 1968; Ataev 1985; Shcherbak and Golubev 1978). Recently, one specimen was also found in the Murghab Valley, 10 km south of Tashkepri Station (Bogdanov and Sudarev 1988). It is protected in the Badghyz, Kopetdagh, and Syunt-Khasardagh Reserves.

11. Cyrtopodion longipes (Nikolsky, 1897). The subspecies *C. l. microlepis* Lantz, 1918, is limited to mountain ranges of Danagermab, Gyazgyadyk, Kelatkaya, Akarcheshme, and Zakli, and next to the Pulikhatum Post in Badghyz (Gorelov *et al.* 1974; Bannikov *et al.* 1977; Rustamov and Shammakov 1979). This species is protected in the Badghyz Reserve.

12. Cyrtopodion turkmenicus (Scherbak, 1978). This rare species was recently described from southern Turkmenistan (Morgunovsky village and Chemenibit Station in Badghyz, and Pelengovali gorge in Karabil) and adjacent regions of Afghanistan. Total population is estimated as not exceeding 500 lizards.

13. Phrynocephalus rossikowi Nikolsky, 1899. Most of this species' range lies in Turkmenistan, where two subspecies are found. A nominal form, *P. r. rossikowi* Nik., has sporadically been recorded from the left bank of the Amudarya River (villages of Sheikharyk, Karagyoz and Duyeboyun, Lake Danishor, and towns of Ispas and Neftezavodsk). This population comprises 2,000 to 3,000 lizards. Another subspecies, *P. r. shammakowi* Szerbak et Golubev, 1979, is known only from the Serny Zavod settlement in Central Karakum (Shcherbak and Golubev 1979).

14. Eremias regeli Bedriaga, 1905. In Turkmenistan, this species has been recorded only from the foothills of Kugitang near to Kelif Station and Karlyuk village (Nikolsky 1911; Shcherbak 1974).

15. Eremias arguta Pallas, 1773. A population belonging to the subspecies *E. a. uzbekistanica* Chernov, 1934, is recorded from the Khodzhambas District at the Amudarya River. Four to six lizards per hour of surveying were recorded next to the settlements of Kyrkoili and Amudarya, and Talimardzhan Station in 1964, 1966, and 1970. The population is decreasing due to land development.

16. Eremias nigrocellata Nikolsky, 1896. This species is found only in Southeast Turkmenistan, and the foothills of Kugitang; its density averages 1.7 lizards/ha. The population is decreasing due to land development. The role of the Turkmenistan population, along with those from Uzbekistan and Tajikistan, is important for the species' preservation in the Middle Asian fauna.

17. Lacerta defilippii Camerano, 1877. The main range of this species lies in the Elburz Mountains in North Iran; in Middle Asia, it is found only in the Central Kopetdagh valleys of Firyuza and Bolshie Karanki, and next to Gyaurs village (Morits 1929; Laptev 1937; Ataev 1976; Khomustenko and Ataev 1979). Total size of the population at Bolshie Karanki is estimated as 1,500 lizards on 4 ha. The species is protected in the Kopetdagh Reserve.

18. Eryx elegans (Gray, 1849). In Middle Asia, this rare snake is known only from Central and West Kopetdagh (Boettger 1890; Bogdanov 1962; Nurgeldyev *et al.* 1970; Ataev 1976; Rustamov and Ataev 1976). It is protected in the Kopetdagh and Syunt-Khasardagh Reserves.

19. Eryx miliaris (Pallas, 1773). This is very rare species in Turkmenistan. It is known only from Kugitang, where six snakes were captured next to the Karlyuk settlement (Shukurov 1976).

20. Coluber ataevi N. sp. (in litt.). A rare endemic snake of Central Kopetdagh, formerly identified as *C. najadum* Eichwald, 1831, *C. ataevi* is known from the Sulyukli spring on Mount Dushak, Firyuza Valley, Kuchan highway, and to the south of Ashkhabad (Varentsov 1894; Nikolsky 1905; Ananyeva and Orlov 1977; Bannikov *et al.* 1977). The species is protected in the Kopetdagh Reserve.

21. Coluber jugularis Linnaeus, 1758. The subspecies *C. j. schmidti* Nikolsky, 1909, is rarely found in West Turkmenistan along the valleys of the Sumbar, Chandyr, and Atrek Rivers, and on Lake Maloe Delili (Shammakov 1964, 1981; Ozorovsky 1968; Bogdanov 1970; Ataev 1977).

22. Litorhynchus ridgewayi Boulenger, 1887. The range of this snake lies from the Messerian Plain in the west, along Kopetdagh and Badghyz to the Kugitang foothills, to Unguz and Central Karakum to the north, and to the Amudarya to the east (Andrushko and Mikkau 1964; Shukurov 1973; Bannikov *et al.* 1977; Shammakov 1981). Agricultural development, especially in the Kopetdagh foothill and Khauzkhan Reservoir areas, destroys natural habitats of *L. ridgewayi*. The population of this snake in Turkmenistan does not exceed several hundred. It is protected in the Repetek, Badghyz, Kopetdagh, and Kugitang Reserves.

23. *Elaphe quatuorlineata* Lacepede, 1789. Subspecies *E. q. sautromates* Pallas, 1814, is very rarely found in Turkmenistan, which is the periphery of its range. Only a few records are known from the area of northern Karabogazgol Bay of the Caspian Sea (Vasilyev et al. 1960).

24. *Oligodon taeniolatus* (Jerdon, 1853). The only areas in Middle Asia where this snake species is found are West and Central Kopetdagh and northwestern Badghyz (Nikolsky 1905; Filippov 1947; Bogdanov 1962; Nurgeldyev et al. 1970; Rustamov and Ataev 1976; Bannikov et al. 1977; Ataev et al. 1978; Shcherbak 1979). A total of 23–25 snakes were recorded from these areas. The species is protected in the Badghyz, Kopetdagh, and Syunt-Khasardagh Reserves.

25. *Agkistrodon halys* (Pallas, 1776). Two subspecies of this rare venomous snake are found in the republic: *A. h. caraganus* in the very north (Vasilyev et al. 1960; Bogdanov 1962; Nurgeldyev et al. 1970; Bannikov et al. 1977; Velikanov 1977), and *A. h. caucasicus*, which has been recorded from the Kopetdagh valleys and mountains of Aidere, Parkhai, Shikhindere, Gaudan, Chopandagh, Firyuza, Kheirabad, Mirzadagh, Malye Karanki, Bolshie Karanki, and Chashdepe (Varentsov 1894; Morits 1929; Pestinsky 1939; Bogdanov 1962; Skalon 1982). The species is protected in the Kopetdagh and Syunt-Khasardagh Reserves.

IV. Undetermined (five species)

26. *Chalcides ocellatus* (Forskal, 1775). This very rare lizard (only three specimens have been recorded from Turkmenistan and constitute the only record from Middle Asia) has been found in Badghyz (Akarcheshme) and Central Kopetdagh (Mergenulen). The status of this species needs to be confirmed, but its uniqueness in our fauna is evident.

27. *Ophiomorus chernovi* Anderson et Leviton, 1966. This recently described lizard is known only by several specimens from Badghyz (Pulikhatum village and Byukabaicheshme Spring in Gyazgyadyk Range).[12] This species may also inhabit the territory of the Badghyz Reserve. The conservation of this population is extremely important.

28. *Lycodon striatus* (Shaw, 1802). The subspecies *L. s. bicolor* Nikolsky 1903 is known mainly from West Kopetdagh (Karayelchi, Kara-Kala); Central Kopetdagh (Archman, Bakharden Cave, Karagan, Kurkulab, Firyuza, Berzengi); and Badghyz (Laptev 1934; Filippov 1947; Heptner 1949; Bogdanov 1962, 1965; Rustamov and Ataev 1976). Outside of the Kopetdagh-Badghyz area, *L. striatus* has been recorded in Turkmenistan on the Amudarya River in the Farab and Chardzhou Districts (Shukurov 1973) and in the Kugitang Mountains next to the village of Khodzhapil (Makeev et al. 1988). This snake is protected in the Badghyz, Kopetdagh, and Syunt-Khasardagh Reserves.

29. Telescopus rhynopoma (Blanford, 1874). In Middle Asia, this extremely rare snake is found only in West Kopetdagh (Iskander Station) and Central Kopetdagh (Firyuza Valley, Vannovsky Settlement) (Ataev 1966; Bogdanov and Potopolsky 1956; Rustamov and Ataev 1976; Rustamov and Shammakov 1979). Only three records are known; probably this species has a very low natural population density. It is protected in the Kopetdagh Reserve.

30. Boiga trigonatum (Schneider, 1802). The subspecies *B. t. melanocephala* Annandale, 1904, is found in southern and southeastern Turkmenistan; about 30 records are known. It is considered a rare snake, probably due to insufficient studies. It is protected in the Badghyz and Repetek Reserves and may be found in other reserves of Turkmenistan.

Amphibians (one species)

III. Rare Species

1. Rana macrocnemis Boulenger, 1885. A single record of two brown frogs is known from the Arpaklen Valley, 25 km from Kara-Kala in Southwest Kopetdagh (Borkin 1977). This record, however, should be confirmed.

Fishes (eight species)

I. Endangered Species (one sp.)

1. Aspiolucius esocinus Kessler, 1874. This endemic fish of the Amudarya and Syrdarya was common in the past in the Amudarya River, below the city of Chardzhou, where it was an important fishing species (Berg 1905; Nikolsky 1938). Today, this is a rare and endangered species; more efforts should be undertaken for its conservation and for the control of fishing.

II. Declining Species (two spp.)

2. Pseudoscaphirhynchus kaufmanni Bogdanov, 1882. This endemic fish species of the Amudarya inhabited the entire river in the past (Nikolsky 1938; Shaposhnikova 1950; Tleutov and Sagitov 1973) but now is limited only to its upper portion (above the city of Chardzhou). The population of *P. kaufmanni* has steadily declined (Tleutov and Sagitov 1973). Unconfirmed records of this species exist from the Karakum Canal.

3. Pseudoscaphirhynchus hermanni Kessler, 1877. This very rare species formerly found in the Amudarya around the city of Chardzhou and somewhat higher (Tleutov and Sagitov 1973); remnants of this population possibly still exist. It is probably found in the the Karakum Canal. *P. hermanni* is found in the same habitat as *P. kaufmanni*, but its abundance is much lower (Shaposhnikova 1950; Tleutov and Sagitov 1973).

IV. Undetermined (two spp.)

4. *Nemachilus starostini* Parin, 1982. This unique endemic fish species is the only underground cave fish from the entire territory of the former USSR. It is a small (3-7 cm) blind fish known from a single record (ten specimens; more were observed) from the Kugitang foothills to the east of Karlyuk settlement, 45 km from Charshanga Station (Parin and Dolzhansky 1982).

5. *Discognathichtys rossicus* Nikolsky, 1900. This species was common in the Tedzhen, Kushka, and upper Murghab Rivers (above Tashkepri village) in the 1940s, when schools of 10 to 20 fishes were observed in summer in the Kushka River (Nikolsky 1945). There are no new data on *D. rossicus*.

Conclusions

Turkmenistan is inhabited by a number of endangered, declining, and rare vertebrates. About one-third of the Red Data Book vertebrate species of the former USSR are found in this republic, and more than 11% of those species are found, within this territory, only in Turkmenistan. Their listing in the Red Data Book is merely a first step to the implementation of these species' preservation, as evidence of their endangered status. Additional steps in saving the Red Data Book species should include establishment of new conservation areas, bans on animal capture, breeding programs for animals in captivity, and creation of a live genetic stock of rare species.

The most effective form of species conservation in Turkmenistan has been, and continues to be, their protection in the Natural Reserves (*zapovedniki*). Currently, there are eight such reserves in Turkmenistan, with a total area of 1,109,700 ha, as well as sixteen Natural Refuges (*zakazniki*), with total area of 940,700 ha (Table 1). We should mention, for the sake of comparison, that in Uzbekistan there are thirteen reserves with a total area of 211,300 ha; in Kazakhstan, six reserves (580,500 ha); and in Tadjikistan, only two reserves (63,600 ha). The eight reserves of Turkmenistan (Repetek Biosphere, Krasnovodsk, Badghyz, Kopetdagh, Syunt-Khasardagh, Kaplankyr, Amudarya, and Kugitang Reserves) are unique in many ways. For example, the Turkmen population of onager, or wild ass (*kulan*), threatened to extinction by the 1940s, was saved only by the establishment of the Badghyz reserve. Our reserves play an important role in the preservation of reptile species included in the Red Data Book (e.g., the Asian cobra, *Eublepharis turcmenicus*, as well as *Bunopus tuberculatus* and other species of geckos). In general, Turkmenistan, and Middle Asia as a whole, are among key regions of the world in which rare species conservation is of vital importance.

Table 1. Natural reserves and refuges in Turkmenistan

	Year Established	Area (hectares)
I. Natural Reserves (*zapovedniki*)		
1. Repetek Biosphere	1928	34,600
2. Krasnovodsk:		
southern area	1932	69,700
northern area	1968	192,300
3. Badghyz	1941	86,000
4. Kopetdagh	1976	49,800
5. Syunt-Khasardagh	1978	29,700
6. Kaplankyr	1979	570,000
7. Amudarya	1982	50,500
8. Kugitang	1986	27,100
II. Natural Refuges (*zakazniki*)		
1. Chemenabit	1956	30,000
2. Kizyldzhar	1956	13,000
3. Pulikhatum	1956	15,000
4. Kelif	1970	103,000
5. Meana-Chaacha	1976	60,000
6. Kalininsk	1976	15,000
7. Khodzhakala	1977	40,000
8. Aidere	1977	21,500
9. Sumbar	1977	35,500
10. Yeradzhi	1978	30,000
11. Sarykamysh	1980	210,700
12. Ogurchinsky	1982	7,000
13. Shakhsenem	1983	270,000
14. Karlyuk	1986	40,000
15. Khodzhapil	1986	30,000
16. Khodzhaburdzhibelend	1986	20,000

Notes

1. A detailed survey of the leopard family in the Aidere Valley of Southwest Kopetdagh was conducted by Danov (1985), whose data were based on several years of personal encounters and footprint records. Numerous observations of leopards within the Syunt-Khasardagh Range and Chandyr Valley were made in the 1980s by V. Lukarevsky and Yu. Karpinsky (pers. comm.) – Eds.
2. This specimen is on display in the Ashkhabad Natural History Museum. – Eds.
3. The Khosardagh record is dubious and may refer to the leopard. – Eds.
4. Local hunters reportedly saw a tiger next to the Iranian border in the upper part of the Sumbar Valley (Shalcheklen Plateau) as late as 1980 (M. Mukhammetnurov pers. comm.). The last Turanian tigers might thus still be alive in this remote border region of Kopetdagh – Eds.
5. A local population reported deer hunting in the 1900s along the Chandyr River in West Kopetdagh (now within Iranian territory) (Laptev 1934). There have been no deer in this area at least since the 1940s. – Eds.

6. Sometimes considered a race of *F. peregrinus* (Brown and Amadon 1968) – Eds.
7. Its subspecies status and endemic range need to be confirmed. Terentyev's and Chernov's description (1949: 274) distinguish it on the basis of coloration from Indian and Chinese (i.e., marginal) populations; *P. mucosus* also inhabits North (Anderson *et al.* 1965) and South (Leviton and Anderson 1963) Afghanistan, Pakistan (Minton 1966), and possibly East Iran (Leviton and Anderson 1970) – Eds.
8. Makeev *et al.* (1988) estimate density of *P. mucosus* in the Kashan Valley as 1.33 snakes/ha; near Iolotan as 0.77/ha; and near Krasnoye Znamya, 0.91/ha – Eds.
9. *A. loricatus* is also found in the Chardzhou Region, Geldere area, Kabakly village (left bank of the Amudarya) and next to Eldzhik village (right bank of the Amudarya) (Golubev and Streltsov 1989) – Eds.
10. Makeev *et al.* (1988) estimate the number of monitors in East Turkmenistan about 50,000 (of these, 19,300 are in the Badghyz-Karabil area; 3,100 in Kugitang; 4,900 in Sundukli Sands; and 17,300 in the Karakum sand desert. They estimate that there are not more than 300 monitors left in the Murghab Valley – Eds.
11. Makeev *et al.* (1988) estimate the number of cobras in the Murghab and Kushka Valley as 1,060; in Badghyz and Karabil with adjacent desert areas as 10,130, and in Kugitang as 1,400 snakes. They assume that, in East Turkmenistan, there are not more than 19,600 cobras. – Eds.
12. Although these points are located in mountainous (western) Badghyz, the exact habitat of *O. chernovi* there is alluvial sand with gravel along the bank of the Tedhzen river. – Ed.

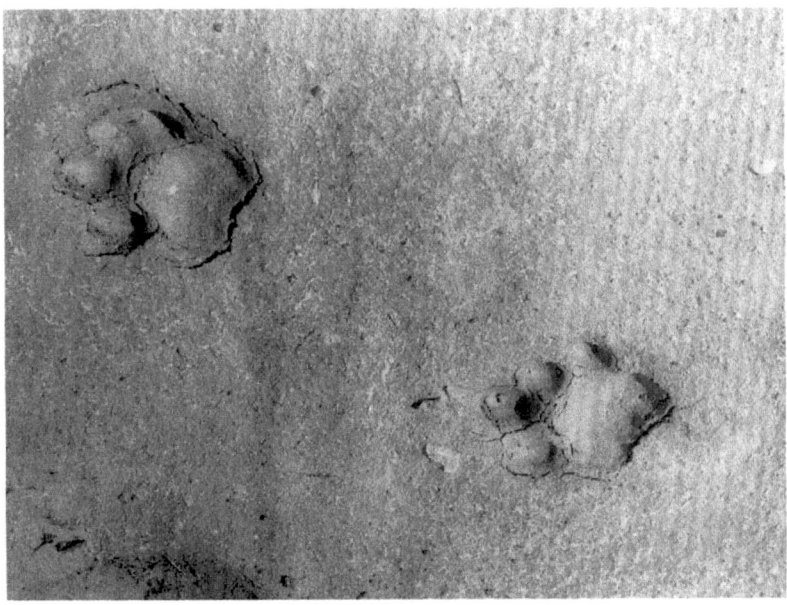

Footprints of leopard. Photo by H.R. Levenshtein.

Eremias arguta. Photo by N.N. Shcherbak.

Dzheirans (*Gazella subgutturosa* in Badghyz. Photo by H.R. Levenshtein.

14. Ecology of the Bearded Goat (*Capra aegagrus* Ersleben 1777) in Turkmenistan

VLADIMIR M. KORSHUNOV

Abstract

Only ca. 8,000 bearded goats (*Capra aegagrus*) are left in Turkmenistan, predominantly in the mountains of Central Kopetdagh. The animal's lives are characterized by four distinct periods: winter/spring period, birthing period, drinking site period, and mating period. Each of these periods has a specific herd structure and hierarchical system in a herd. Average herd size is 15.9 ± 1.1 animals. Average life span of males is 6.23 ± 0.24 years. Herd structure, on average, is 23.9% males, 49.9% females, and 26.2% young goats; male/female ratip is 1:2.1. In response to environmental changes, populations may change in herd structure, size, and social organization. Competition with sympatric mountain sheep is minimized due to habitat separation.

Introduction

Seven species of ungulates are currently known from Turkmenistan: wild boar (*Sus scrofa nigripes* Blanford, 1875), deer (*Cervus elaphus* L., 1758), dzheiran (*Gazella subgutturosa* Gueldenstaedt, 1780), bearded goat (*Capra aegagrus* Ersleben, 1777), Bokhara goat (*Capra falconeri heptneri* Zalkin, 1945), mountain sheep (*Ovis ammon* L., 1758), and kulan (*Equus hemionus* Pallas, 1775). In fall and winter, saiga (*Saiga tatarica* L. 1766) visits republic from adjacent Kazakhstan and Karakalpakistan. The records of the Siberian goat (*Capra sibirica* Pallas, 1776) from Turkmenistan are not confirmed.

These animals and their ecology have been studied to various extents. Here, we present results of the author's original field study of the bearded goat, conducted from 1978 to 1992. We used automobile, horse, and walking surveys, as well as airplane and helicopter surveys of larger territories and direct observation of animals in their pastures and at their drinking sites.

General Characteristics

Bearded goats from Turkmenistan are smaller than related wild goat species. The male is ca. 85 cm high and 155 cm long (measured from upper lip to the root of the tail) (Morits 1930). Legs are relatively short but thick and strong. Average weight of well-fed males in the fall is 55 kg (Morits 1930); maximal weight is 70 to 80 kg (Heptner et al. 1961). Females are about twice as small.

Summer molt in healthy animals ends by May. Winter wool is much longer than summer wool (Morits 1930), but the skin is thicker in summer. The epidermis, particularly in its horny layer, is thinner in the neck skin during summer than winter, and it is more darkly pigmented in summer. Sweat glands are larger and more convolute in summer (Sokolov 1973) whereas the ratio of hair to down changes from 1:6.2 in winter to 1:4.3 in summer. Therefore, loss in insulation capacity of the wool cover in summer is compensated for by body cooling via an increase of sweat gland efficiency. Horns are slender and slightly bent (saber-shaped) in males; their maximum length 127 cm. In females, horns are considerably smaller (20 to 25 cm) and grow until animals are four years old. In males, horns grow throughout their lifetime. Maximal rate of growth is recorded in the second year (Table 1). From the third year, annual horn growth rate decreases. Horn growth rate can thus serve as an individual growth index (M), and the deviation range (r) as a measure of individual variation within an age class (except during the two first years). These indices allow tracing of the changes in a male's status in herd structure with his age; males with well-developed horns occupy higher rank in the herd hierarchy, and those with less well-developed horns, lower rank.

As well as horn length, horn circumference measured at its base in male bearded goats partially correlates with age and herd rank (Korshunov 1988). Horn circumference gradually increases until males approach the mating age;

Table 1. Annual growth rate parameters for bearded goat horns in Central Kopetdagh

Year	M ± m (cm)	P	+	−	r	n
1	11.44 ± 0.26	n/sign	9.6	5.4	14.0	82
2	15.34 ± 0.25	0.001	5.7	5.1	10.8	80
3	13.42 ± 0.24	0.1	5.1	4.4	9.5	74
4	13.42 ± 0.30	n/sign	5.6	6.4	12.0	62
5	12.67 ± 0.37	n/sign	5.3	8.7	14.0	53
6	11.63 ± 0.44	0.1	3.9	9.6	13.5	39
7	9.86 ± 0.47	0.01	4.2	6.6	10.8	26
8	7.97 ± 0.56	0.05	5.0	4.0	9.0	17
9	6.53 ± 0.73	n/sign	3.5	3.5	7.0	8
10	5.83 ± 0.36	n/sign	0.7	0.8	1.5	3

M ± m − average annual growth and standard deviation, P − confidence level of growth rate compared to a previous year (Student's *t*-test), "+" and "−" − maximal individual deviations from average annual growth, r − amplitude of deviation, n − sample size this measure is approximately the same in males older than five years (Table 2). Maximal circumference recorded is 25 cm (eight-year-old male). Average measure of horn circumference is 20.71 ± 0.23 cm, which is somewhat lower than in bearded goats of the Caucasus (22.43 ± 0.20 cm; Tsalkin 1950). Based on this character, bearded goats of Turkmenistan have been described as a separate subspecies, *Capra aegagrus turkmenicus* Zalkin, 1950 (Tsalkin 1950). However, a comparison of our data (Table 2) and data from the Caucasus demonstrates that the sample studied by Tsalkin (1950) did not have any horns with base circumference less than 20 cm. This fact can be explained only by the absence of goats younger than five years from Caucasian samples, which were submitted for study by hunters, and, thus, the largest and longest horns were already selected, as Tsalkin (1950) himself confirms. On the other hand, Tsalkin's sample from Turkmenistan included all available horns (similar to our samples). Comparative analysis of these data is not possible because Tsalkin (1950) does not give age distribution of his material. Therefore, we currently do not support describing the Turkmenistan bearded goat as a separate subspecies.

Table 2. Measure of basal horn circumference in different age groups of male bearded goats

Male death age	2	3	4	5	6	7	8	9	10	11	Mean
Maximum (cm)	16.0	19.0	22.0	23.5	24.0	24.0	25.0	24.0	23.5	24.5	23.0
Minimum (cm)	15.0	16.3	16.5	19.0	18.0	18.5	19.5	18.5	19.7	24.0	18.2
Mean (cm)	15.4	17.9	20.0	21.4	21.1	21.6	21.9	21.5	21.1	24.3	20.7
Stand. deviation	0.24	0.27	0.40	0.60	0.49	0.39	0.54	0.49	0.45	0.25	0.23
Sample size (total, 92)	4	8	13	7	11	17	9	12	9	2	

Species Distribution and Ecological Requirements

The bearded goat is widely distributed in mountains from islands of the Aegean Sea in the west to Baluchistan and western Sind in the east; its range occupies Asia Minor, Iran, mountainous Daghestan, Armenia, Azerbaijan, mountains of southern Turkmenistan, and Afghanistan (Heptner *et al.* 1961). Its ecology is defined by its narrow habitat preference; not mountains themselves, but rocks and cliffs, in any surroundings and at any altitude, are necessary for the presence of the bearded goat (Heptner *et al* 1961). It has been found from sea level to 4,200 m (Mt. Ararat) (Radde 1899).

A major ecological requirement of the bearded goat includes the presence of trees and shrubs for feeding, which must have constant productivity and be available even under snow cover. Mini-springs in the upper mountain belts provide water throughout the year. Due to specific body shape and hoof structure, the bearded goat can easily move along rocks and find shelter in the most secure places free of competitors; of natural predators, leopard (*Panthera pardus* L.) is by far the greatest threat to the goat. Due to their remote habitats, bearded goats do not experience competition with domestic pasture animals such as cows, sheep, and goats, which is a common factor influencing other wild ungulates. Bearded goats, however, are affected by other anthropogenic factors such as illegal hunting, fires, exploitation of water sources, and mere disturbance; all these factors have led to the fragmentation of the animal's range (Korshunov 1988).

Distribution in Turkmenistan

Today, bearded goats in Turkmenistan are found in the southern mountains from Bolshoi Balkhan in the west to the meridian of Chaacha village in the east. A hundred years ago, its range extended from the Kubadagh Mountains near Krasnovodsk (Radde 1899), through Maly and Bolshoi Balkhan and virtually the entire Kopetdagh, to Gyazgyadyk Plateau in Badghyz (Heptner 1949).

Currently, the best-preserved population of bearded goat lives in Central Kopetdagh, where it was much more common 20 to 30 years ago (Ishadov 1965a) than today. We have observed goats in the Tekechingasy Range (Shamli Spring). In 1976 and 1977 goats were observed in the Kishimiri Mountains, and they still may enter this area from the Gyuaursdagh Range, where they are still present. Border patrols have repeatedly recorded goats from the Kizyldagh Range. Goats are common in the Aselma Range, within the Kopetdagh State Reserve (Babazo Area and Firyuza Area), and in the valleys of Sukhaya Balka, Sandagzov, and Zelenaya which lie between these areas. Westward of Firyuza Area goats have been recorded in the Kavalerizou Valley. Horns and other evidence of bearded goats' presence have been found in the Missinev Range. Rare but still frequent are records from Mt. Dushakerekdagh. We have observed goats in the Mirzadagh Range (herd of 54 animals), in the upper portions of Gadameli and Digermendzhik Valleys, and in the Neprokhodimoye Valley.

Goats are still common further west, in the Barsovoe Valley, Mergenchuk Range, and on the northern slope of Mt. Togarev; however, in recent years, due to cattle industry development in the western part of Central Kopetdagh, the range of bearded goats has become fragmented. The animals are not now found between the Missinev and Stolovye Ranges, which created an isolated population in the area of Mt. Togarev (Gyuni, Mergenchuk, Alma, and Stolovye Mountains and Sarzavu, Barsovoe, and Neprokhodimoye Valleys). An isolated group (70 to 80 animals) have formed in the Mirzadagh Range. Due to construction works along the border, part of the Mt. Dushakerekdagh population had also become nearly totally isolated. Bearded goats are not now found east of the Aselma Range, except in small groups in the Tekechingasy and Gyuaursdagh Ranges. These groups are probably maintained by animals migrating from the Zerkau Range (East Kopetdagh), where they are still common. Therefore, the modern range of the bearded goat in Turkmenistan has become fragmented, and survival of certain isolated populations (e.g., those of Dushakerekdagh, Tekechingasy, and Mirzadagh) is now problematic. For example, an isolated but viable (density 1.38 animals/sq. km) population exists on Mt. Togarev, but goats are not found on the Iranian part of this mountain due to agricultural development and elimination of juniper and shrubs. Total area of the bearded goat range in Central Kopetdagh is now 107,200 ha (Korshunov 1986).

Recently (before 1959), bearded goats were found as far east as western Badghyz in Kelet-Kaya, Gyazgyadyk, and Zyulfagar Ranges (Gorelov 1959). Today, the easternmost limit of its range is low ranges (up to 1,200 m) near the village of Meana in East Kopetdagh, where several dozen goats are still found. Sporadic records are currently known from the Kat and Kara-Bulag Ranges. Undoubtedly, these areas served as connections to the western Badghyz population in the past. Also in East Kopetdagh, goats are found from the meridian of Babadurmaz village along the Zerkau Range.

The westernmost limit of the bearded goat range qute recently was the Maly and Bolshoi Balkhan Mountains (Heptner et al. 1961), from where they visited the adjacent Tekedzhik and Begarslan Ranges (Ishadov 1972). Bearded goats were very common 60 to 70 years ago in territories adjacent to Bolshoi Balkhan (Kurtlibil and Koshaseira Ridges), where they now are absent (Shukurov 1962). Where Laptev (1934) recorded herds of 60 to 70 bearded goats in Bolshoi Balkhan in 1925, Shukurov (1962) encountered only sporadic small groups from two to seven animals in the 1950s. Ishadov (1965) quotes local hunters on densities of 5 to 10 goats per 20 km of survey in Maly Balkhan. In the past, bearded goats were also recorded from the Kyurendagh and Karagyoz Ranges of Northwest Kopetdagh (Ishadov 1961).

Very scarce are data on bearded goats from West Kopetdagh. Babaev et al. (1978) published records from the Karagysy and Reush Ranges (near the town of Khodzhakala), Duedzhidagh Range (13 km south of the city of Kizyl-Arvat), and from the mountains near Koine-Kesir and Daina villages in the upper portion of the Sumbar River. We observed goats in West Kopetdagh near the village of Ipai-Kala (i.e., 10 to 15 km west from Mt. Togarev) and five km south of the village of Arpaklen. There are also numerous records of bearded goats from the Khozly Valley (6 km north of the village of Saivan).

In total, there are ca. 8,000 bearded goats now left in Turkmenistan, with the majority concentrated in Central Kopetdagh, where in 1984 we estimated their number as 6,600 animals (Korshunov 1986).

Habitats

In Turkmenistan, bearded goat habitats are different from those in other parts of its range (such as Transcaucasia) (Heptner et al. 1961). It has been found from the Caspian Sea level (Kubadagh) (Radde 1899) to the highest peaks of Kopetdagh (2,600 to 2,889 m). In the Gyazgyadyk Range (West Badghyz), which is only up to 1,100 m high, goats have been recorded in easily accessible, shallow valleys (Heptner et al. 1961); in the Bolshoi Balkhan, they have even descended into the semidesert foothills (Laptev 1934). Our observations from 1979 through 1991 have not confirmed an opinion that females of bearded goat are found at significantly lower mountains levels than are adult males (Ognev and Heptner 1929; Morits 1930); this observation is true only during the birthing period. Favorite habitats of bearded goats are rocks, steep cliffs, and valleys covered by juniper stands and shrubs (Morits 1930). Rich vegetation in such habitats and adjacent plateaus provides goats with food in any season.

We have separated five habitat types within the landscapes of Central Kopetdagh. For bearded goats, distribution of time spent in each habitat is the following: small hills, 5.3%; plateau, 7.2%; grass-covered slopes, 15.3%; rocky slopes, 42.1%; and ledges, 30.1% (Korshunov 1986). These data confirm the preference by bearded goats of highly dissected relief.

Herd Structure

A herd of bearded goats Fig. 1 experiences cyclic changes in its structure. Four periods can be distinguished through the year: winter/spring period, birthing period, drinking site period, and mating period. A winter/spring period starts after the end of mating and ends with the onset of birthing. Stabilized snow cover and reduced pasture area lead to the concentration of goats in relatively small areas, where up to 300 animals may accumulate (without forming a single herd). The largest herd encountered in this period was 99 animals, with an average number of 19.0 ± 2.9 animals. During the winter/spring period herds which contain goats of both sexes and various ages, are recorded more frequently than during other periods (Table 3). Pure male groups are rare; more frequent are female groups; and there are no groups including only adult males and females. Groups of females with young goats are as frequently found as pure female groups.

The birth period starts ca. April 25, when females leave herds in groups of two to four in search for birthing shelters. Adult males also leave herds in small groups or by themselves. The remaining mixed herds (Table 3) contain young

Table 3. Dynamics of bearded goat herd structure in Central Kopetdagh

Periods	Percent of males only	Number of females only	Encountered males and females	Groups consisting of:	
				males, females, and young	females with young
Winter/spring	2.3	16.3	0	65.1	16.3
Birth period	8.5	11.8	13.6	42.4	23.7
Drinking sites	8.3	0	5.6	47.2	38.9
Mating	3.1	12.5	6.3	62.5	15.6

males, barren females, yearlings, and single large males. By the end of the birthing period, with gradual growth of newborns, groups of females with newborn animals become more common. The number of independent, pure male groups with established hierarchy also increases. Pure female groups exist only at the very beginning of the birthing period. In mixed male/female groups, adult males are absent; those remaining are young males which still follow their mothers. Herd size drops to 15.4 ± 2.2 animals.

At the beginning of August, when the air becomes extremely dry, the drinking site period begins; goats, including young growing animals, need more water than they receive from plant food. Concentration of animals around the limited number of mountain springs facilitates formation of large groups. First, several small groups of females with newborns join groups of barren females, females who have lost newborns, and young males. The number of mixed groups increases (Table 3), and pure female groups disappear; they join mixed herds and groups of females with newborn young, which are more frequently found in the drinking site period. Herd size in this period is 15.0 ± 1.7 animals.

The mating period starts in early November. Males and females form stable mixed herds, as numerous as in the winter/spring period. These are full-size herds with established leaders, dominants, hierarchy, and communication. Young and invalid males are expelled from such herds and form separate male groups. Some females with grown young join mixed herds (Table 3). Pure female groups are encountered again, as well as male/female groups. Herd size does not change and is 15.0 ± 2.2 animals; interestingly, average herd size does not change significantly throughout the year.

Our multi-year data show that average herd size from 1979 to 1991 was 15.9 ± 1.1 animals. However, in 1983 density sharply decreased over the entire Central Kopetdagh area (10 to 14 times in separate areas) due to bad ecological conditions; in this year, average herd size decreased significantly (P <0.05), from 18.2 ± 1.7 to 13.3 ± 1.3 animals. At the same time, the share of males increased (by 6.1%), whereas shares of females and young animals decreased (Table 4). Certain differences exist in herd structure within the protected territory of Kopetdagh Reserve as compared to unprotected areas. Herd size is larger within the Reserve (16.6 ± 1.2) as compared to unprotected territory (7.5 ± 1.7) (P <0.001); the share of males within the Reserve is higher (by 4.1%), whereas the share of females is constant and share of young animals lower (by 4.1%) than in unprotected areas (Table 4).

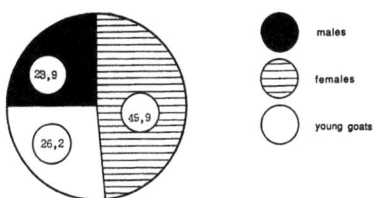

Fig. 1. Age and sex structure of the bearded goat population.

Table 4. Dynamics of population structure of Central Kopetdagh bearded goats dependent on periods of depression and conservation regime

	Population structure (%)		
	males	females	young
Increasing density period	22.1	50.3	27.6
Decreasing density period	28.6	48.2	23.2
Within protected areas (Kopetdagh Reserve)	24.2	50.5	25.3
Outside of protected areas	28.3	50.5	21.2

Behavior

Our observations of bearded goats revealed an established order of following in a herd. An adult female moves in front; then follow females with young goats and young males (up to three or four years old). Following these is the dominant (usually the largest) male, escorted by subdominant males. In the rear move secondary males and females without young. Such order is maintained when the herd is disturbed but the source of disturbance is determined and the way of retreat is generally known.

Usually, leaders and dominants in a goat herd are different animals. In an alarm situation, an old female or young male often accepts the leading role, whereas old males follow at the back of the herd; such behavior may be related to herd protection (Morits 1930). We have often observed cases in which, after a disturbance, a dominant male and his escort move toward its source. Sometimes, this group separates from the herd, positions itself at a suitable observation point, and tries to determine the degree of danger. Then, by their behavior, these males pass the obtained information to the rest of the herd.

The role of a dominant male is revealed in non-standard situations: for instance, when an unexpected obstacle appears in the herd's way, or during a predator's attack. First, a dominant male and his escort take a waiting position to estimate the danger; after this, the male moves in a certain direction, and the herd rapidly follows him. Such movement may also be directed toward the source of danger; its direction depends on the specific situation. Another role of the dominant male is to maintain the herd hierarchy; inobedience in subordination is firmly punished. At the same time, dominant males protect animals that need protection.

Herd members constantly assess their relationships. After splitting into different sex and age groups during the birthing period, the herd again establishes full-size herd hierarchy at the drinking sites, especially when the mating period approaches. The dominant male drinks first, then males of his escort, and then, in subordinate order, the remainder of the herd.

Hierarchy is expressed also in the selection of mating partners. In dense populations during the mating period, herds often consist of dozens of animals. Every herd has several large males; clashes among them become more frequent as the mating period approaches. Under this pressure, young or sick males leave mating herds, and the dominant male spends a great deal of energy to maintain subordination. This dominant male is thus often exhausted and cannot donate his equal share to the population's gene pool; his major role is to remove defective individuals from herd. Mating in the herd is random (promiscuous) without any constant male/female relationship.

When population density decreases (as it did in 1983), groups become smaller and their structure changes. During the mating periods of 1983 and 1984, most herds had six to ten animals; herds with 20 or more goats were very rare. Each of these small herds had only one large male and several adult females with young (including sometimes a young male who had not yet participated in mating). Most groups resembled a typical harem, and thus promiscuous mating was replaced by polygamy.

Feeding

According to Morits (1930), favorite foods of adult bearded goats are juniper and leaves of shrubs and various herbaceous plants, including grasses. Ishadov (1965a) identified 38 species of herbaceous plants eaten by bearded goats but has not published his list. Any such list would probably be incomplete because there are 181 species of herbaceous fodder (eaten by cattle, sheep, and domestic goats) in Kopetdagh (Kamakhina 1986), and most of these would be eaten by wild goats as well. Analysis of stomach contents shows that 61.1% of a bearded goat diet consisted of grasses (Poaceae), and 14.2%, of the Turkmen maple (*Acer turcomanicum*). Juniper and other trees and shrubs were secondary diet items.

However, diet composition can change throughout the year. In summer, the predominant food is grasses (up to 100% in certain goats). Contributions by trees and shrubs in summer does not exceed 30%. In other periods, especially in snowy winters, juniper may constitute 80% of stomach content, and *Ephedra intermedia*, 10%. These percentages confirm the ability of bearded goats to support themselves on trees and shrubs, making them less dependent on seasonal mountain pasture vegetation.

Drinking sites in the arid conditions of Turkmenistan are of great importance for bearded goats, especially from late August through early October. Permanent water sources at that time exist predominantly in rocky areas in the upper mountain belt and in deep gorges as small springs formed by rare summer precipitation, which also fills stony "bowls" available to the goats as they move along the steep cliffs. Such water sources are relatively abundant and evenly distributed among the mountain valleys, which minimizes competition for water.

Goats visit drinking sites several times a day, especially frequently in the morning and more rarely in the daytime. Goats which come to drink in the evening are probably those who did not drink during the day. Goats drink for 20 to 30 minutes, kneeling on their front legs. Drinking follows the hierarchical order described earlier. When goats approach a drinking site, they act extremely carefully; any danger or unusual object can scare them away, in which case they will not return to this site during the same day.

Reproduction and Mortality

We have built a survivorship curve for 98 male bearded goats (Fig. 2) which is based on the age at which these animals died (Table 5). This curve has a shape typical for mammals (Caughley 1966); its maximal curvature is between the fourth and eighth years of life and reflects average life span (6.23 ± 0.24 years). During this age, adult males are especially vulnerable; we relate their increased mortality to their approaching dominant age. Male bearded goats mature by the third year of life and start participating in the struggle for hierarchical status by their fourth year (Heptner et al. 1961). Adult males more often encounter danger than other herd members, and they spend their energy during the mating period. Therefore, unfavorable ecological conditions affect first of all those males which are four to five years old. Starting from the seventh year of life, mortality becomes more evenly distributed (Fig. 2). According to our data, in Central Kopetdagh bearded goats can survive up to 11 years; Heptner et al. (1961) give their natural lifespan from 10 to 12 years. Bearded goats can live even longer in captivity; one male in the Ashgabat Zoo died when he was 14 years old. We have no data about the life span of females because their age cannot be easily determined.

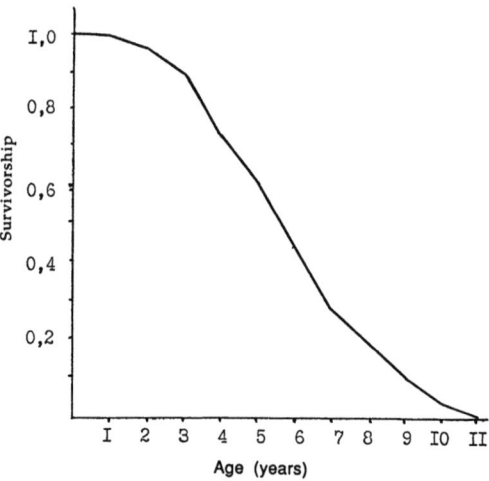

Fig. 2. Survivorship curve of the bearded goat males.

Table 5. Survival of male bearded goats in Central Kopetdagh (n = 98)

Age limits (years)	Number of animals which died within this age limits	Number of animals which survived until this age limits	Survivability (number of survived animals per 1,000 of newborn)
0 to 1	1	98	1,000
1 to 2	3	97	0.990
2 to 3	7	94	0.959
3 to 4	16	87	0.888
4 to 5	11	71	0.724
5 to 6	17	60	0.612
6 to 7	15	43	0.439
7 to 8	9	28	0.286
8 to 9	9	19	0.194
9 to 10	6	10	0.102
10 to 11	4	4	0.041
11 to 12	0	0	0.000

The mating period of bearded goats starts in early November, when males often fight for females, and the sound of their colliding horns can be heard in silent weather far along the mountain valleys (Morits 1930). Oestrus in females begins in late November and ends by the mid-December; during this time, traces of blood can often be seen on rocks. Mating occurs somewhat earlier or later. Pregnancy lasts about five months (Dormidontov and Blokhin 1977). Most young are born in late April or early May (Morits 1930; Ishadov 1964; Korshunov 1983) although sometimes births have occurred in late March or early April (Ishadov 1964), and once, even in July. Usually, adult females have two kids, and young females, only one (Morits 1930; Ishadov 1964, 1965); rarely a female may have three kids (Morits 1930, Heptner et al. 1961). About 20% of females remain barren (Korshunov 1983). Overall maximal amount of kids per female per year is 1.2.

Females take care of their young (Morits 1930), and the number of kids per female per year decreases insignificantly from 0.502 to 0.497 (Table 6). The mother actively drives away even other females approaching her young. However, when ecological conditions (such as drought, disappearing drinking sites, or epizootics) has caused a decrease in goat abundance, this index also has decreased (from 0.549 to 0.480). Even more dramatically, human influence on unprotected territories has caused a decrease in the number of kids per female (from 0.502 to 0.421) as compared to protected areas. Newborn kids grow quickly, and in a week can already follow the mother along cliffs and rocks.

Female bearded goats reach sexual maturity at the age of 19 months (Morits 1930); we observed single two-year-old females with young. Many females, however, give birth only after they are three years old. Young males in stable groups with complex hierarchy start to participate in reproduction after four years of age, and continue throughout their lives.

Table 6. Dynamics of number of young yearded goats per female in Central Kopetdagh*

	Number of females	Number of young	Number of young per female
Birth period to mating period	572	287	0.502
Mating to next birth period	594	295	0.497
Period of increasing density	663	364	0.549
Period of decreasing density	331	159	0.480
Within protected areas	1,111	558	0.502
Outside of protected areas	57	24	0.421

* Mean 0.50

Ecological conditions affect population not only directly via the survival of young goats but also indirectly through the changes in sex ratio. Average multi-year male/female ratio is 1:2.1. During the depression period in 1983 the male ratio increased from 1:2.27 to 1:1.69. Similar changes have been observed under different protection regimes, where the ratio was 1:2.09 within the protected area and 1:1.78 in unprotected territory.

Changes in Density and its Causes

Many authors in the past have attempted estimation of density of the bearded goat in various parts of its range (Silantyev 1898; Laptev 1936, 1937; Tsalkin 1948; Heptner et al. 1961; Shukurov 1962; Ishadov 1964; Gorelov 1973; Babaev et al. 1978). Since these surveys used various accounting approaches, however, it is difficult to compare their results.

According to our estimates, in 1979 about 2,000 bearded goats inhabited the territory of the Kopetdagh Reserve. Since that time, annual surveys in the Reserve have not been conducted; therefore, we can operate only with data from separate years (e.g., 1,780 goats were surveyed in 1983). In 1984, the survey of bearded goats was conducted throughout the entire Central Kopetdagh. West of the Firyuzinka River, in an area with very deep canyons, we recorded 680 goats (density 1.33 animals/sq. km). East of this river, in an area with deep rocky canyons intermittent with plateaus, 5,920 goats were recorded (density 10.54 animals/sq. km).

Dynamics of density were analyzed during the many years of surveys in the Firyuza Area of the Kopetdagh Reserve and in the eastern part of Central Kopetdagh (Fig. 3), where the majority of Turkmenistan population of bearded goats is concentrated. Density changed from year to year; e.g., in the Firyuza Area, the number of goats was 1,610 in 1982, and only 286 in 1983. This decrease cannot be explained solely by migration. In 1983, Firyuza, as well as in the entire Central Kopetdagh, experienced catastrophic ecological changes: maximal air temperatures were present throughout the summer; most springs dried out; and productivity of vegetation was very low. The majority of goats migrated to other

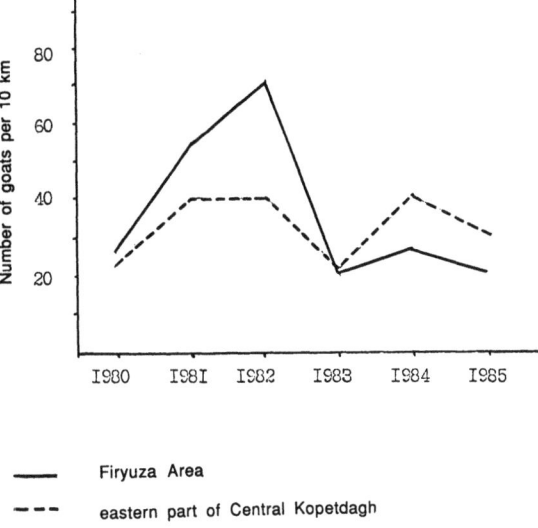

Fig. 3. Data of the surveys of the bearded goat population.

areas or simply died, and it was precisely between 1982 and 1984 when we collected a maximal amount of goat horns. In subsequent years, goat density more or less stabilized. A characteristic dynamic feature of goat population is the maximal decrease in their density in maximal density areas (areas of concentration) when ecological conditions worsen (Fig. 3).

Enemies, Competitors, and Parasites

The major predator of bearded goats in Kopetdagh is the leopard (*Panthera pardus tullianus* Valenc.). In Central Kopetdagh, 85.8% of recorded dead goats have been killed by leopards. A certain selectivity in the leopard's preference was recorded: 67.8% of killed goats were males while females constituted 16.1% and young goats, 14.3%. A much less significant predator of goats may be the lynx (in West Kopetdagh and Bolshoi Balkhan) (Morits 1930); other potential enemies are hyena and fox, and, according to Morits (1930), wild cats, manul, and lammergeier, which are primarily a threat to young kids.

Illegal hunting today is not a limiting factor and has, in the past, been responsible only for 7.1% of the deaths. Current conditions are probably due to better protection enforcement and difficult access to goat habitats.

A potential competitor of the bearded goat might be mountain sheep (bighorn). which inhabits the same area and utilizes similar food and water resources. Competition between these two species, however, is minimal due to their spatial separation. Bearded goats prefer habitats of rocky slopes (42.1%)

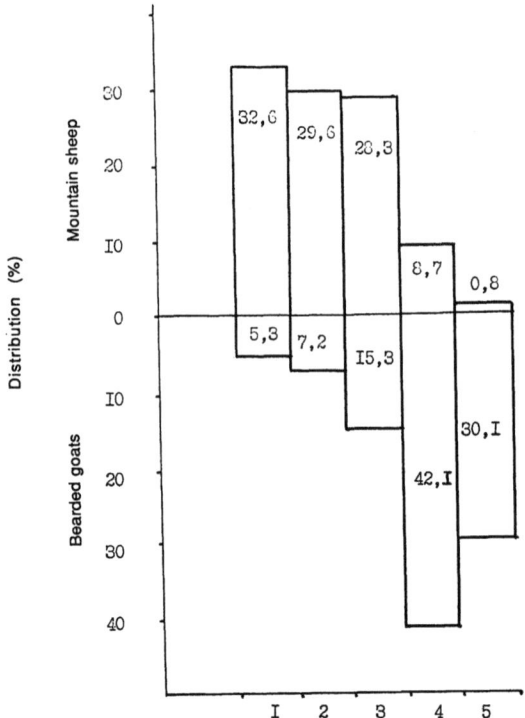

Fig. 4. Habitat distribution of bearded goats and mountain sheep in Central Kopetdagh. Habitats: 1 – small hills, 2 – plateaus, 3 – grassy slopes, 4 – rocky slopes, 5 – ledges of deep canyons.

and ledges of deep canyons (30.1%), whereas mountain sheep prefer small hills (32.6%), plateaus (29.6%), and grassy slopes (28.3%) (Fig. 4).

Among parasites collected on bearded goats in Kopetdagh are six species of ticks (Ixodidae): *Haemaphysalis pospalovashtromae* Hoog, 1966; *H. sulcata* Can. et Fanz., 1877; *H. punctata* Can. et Fanz. , 1877; *H. otophila* P. Sch., 1918; *Dermacentor daghestanicus* Oe., 1929; and *Hyalomma detritum* P. Sch., 1918 (our data; Berdyev 1980). Other ectoparasites of bearded goat include heel fly *Hypoderma silenus* and bot fly *Oestrus ovis* (Diptera) and a chewing louse, *Mallophagus ovinus* (Mallophaga) (Ishadov 1965a). Internal parasites found in bearded goats include a protozoan *Sarcocystis* sp. (Berdyev 1982), and 22 species of helminths (Dobrynin and Korshunov 1991) (Table 7).

Ecological Importance

Bearded goats are a natural part of the ecosystem in mountainous Turkmenistan. Major biotic limiting factors for bearded goat population are

Table 7. Helminths of the Bearded Goat in Turkmenistan

Species	1[a]	2[b]
Moniezia benedeni (Moniez, 1879)	+	−
Taenia hydatigena (Pallas, 1766) (larvae)	+	+
Oesophagostomum columbianum (Curtice, 1879)	+	−
Trichostrongylis axei (Cobbold, 1879)	+	−
T. probolurus (Railliet, 1896)	+	−
T. vitrinus (Looss, 1905)	+	+
Ostertagiella circumcinta (Stadelman, 1894)	−	+
O. trifurcata (Ransom, 1907)	−	+
O. occidentalis (Ransom, 1907)	−	+
O. trifida (Guile, Marotel et Panisset, 1911)	−	+
Camelostrongylus mentulatus (Railliet et Henry, 1909)	+	−
Marshallagia marshalli (Ransom, 1907)	+	+
Haemonchus contortus (Rudolphi, 1803)	+	−
Nematodirus filicollis (Rudolphi, 1802)	+	−
N. abnormalis (May, 1920)	−	+
N. davtiani Grigorian, 1949	−	+
N. dogieli Sokolova, 1948	−	+
N. helvetianus (May, 1920)	+	−
N. oiratianus Raiewskaja, 1929	+	+
N. spathiger (Railliet, 1896)	−	+
Skrjabinema ovis (Skrjabin, 1915)	+	+
Gongylonema pulchrum (Molin, 1857)	−	+

[a] data from other authors
[b] our data
Source: Dobrynin and Korshunov 1991

threat by predators, infection by parasites, and competition with mountainsheep and domestic livestock.

Competition with mountain sheep is moderated by habitat separation. Competition with domestic livestock for pastures today is practically absent because bearded goats have been eliminated from all pasture available for livestock. A major danger now is competition with livestock herds for water sources. The bearded goat, in its turn, influences the vegetation in the Kopetdagh Mountains. Dynamic changes in herd structure, size, and diet, however, can moderate vegetative damage as well as the mechanical load on soil substrate.

Acknowledgements

I am deeply grateful to my late teacher, Professor A.G. Bannikov, for his priceless help in my work; I will always remember his kindness and care.

I thank former staff members of the Kopetdagh Reserve, A.Yu. Belov and N.N. Yefimenko, who helped to collect data; A. Berdyev (Institute of Zoology, Ashgabat), for identification of ticks; A.S. Berdyev, of the same Institute, for identification of protozoan parasites; and M.I. Dobrynin, of the same Institute, for identification of helminths. I also thank G.F. Baryshnikov (Zoological Institute, Academy of Sciences, St. Petersburg) and I.Ya. Pavlinov (Zoological Museum of Moscow State University), who made possible my work with horn collections deposited in these museums.

15. Ecology of Birds in the Karakum Desert

ANVER K. RUSTAMOV

Abstract

The list of birds inhabiting the Karakum Desert includes 238 species. Of 79 nesting bird species, 28 are classified as the "core" fauna, with ancient and profound relationships within the desert ecosystems. We have analyzed the composition, residence status, habitat distribution, and abundance of the core bird species in clay deserts, sand deserts, black saksaul forests, chinks, and micro-oases. Within the core fauna, endemic species are lacking and endemic subspecies are few. This suggests the existence of a young center of origin for the desert avifauna. From the perspective of bird zoogeography, the Karakum and Kyzylkum are both southern-type deserts and can be combined into a single zoogeographic area. Faunal characteristics of bird species in the East, Central, West, and North Karakum are discussed, as well as the ecological, behavioral and morphological adaptations. Among ecological factors, the most important is water. Negative and positive effects of irrigation on bird population are discussed, as well as the human influence on bird habitats. A list of avifauna, with remarks on species abundance and residence status, is given.

Introduction

The birds of Turkmenistan have been studied for more than 120 years. More than 1,000 articles have been published during this time, primarily in Russian, and with only a few in English, German, or French. A detailed bibliography has been given in Dementyev (1952). The major sources of information on the birds of Turkmenistan and the Karakum Desert are books by Radde and Walter (1889), Zarudnyi (1896), Dementyev (1952), and A. Rustamov (1954, 1958).

Composition of the Avifauna

The Karakum Desert is inhabited by 238 bird species (A. Rustamov 1954; new data) including 79 nesting species (33.1%). This fauna can be divided into five groups by the residence status: resident birds (28 species), nesting non-resident (51 spp.), wintering non-resident (26 spp.), migrating (163 spp.), and occasionally observed (34 spp.). Some bird species, however, may be classified within two of these groups; e.g., two local species of *Pterocles* are partially resident in the Karakum (local populations), but members of the same species from populations nesting in the northern parts of Middle Asia migrate through the Karakum Desert. Therefore, these species of *Pterocles* are listed here both as resident and migratory.

The "core" nesting avifauna of the Karakum has ancient and profound relationships with the desert ecosystems, and includes 28 species: *Buteo rufinus rufinus, Aquila chrysaëtus homeyeri, Chlamydotis undulata macqeenii, Burhinus oedicnemus harterti, Charadrius leschenaultii crassirostris, Ch. asiaticus, Cursorius cursor bogolubowi, Pterocles orientalis arenarius, P. alchata caudacutus, Athene noctua bactriana, Caprimulgus aegyptius arenicolor, Dendrocopos leucopterus albipennis, Galerida cristata iwanovi, Calandrella cinerea longipennis, C. rufescens heinei, Lanius excubitor pallidirostris, Podoces panderi, Corvus ruficollis ruficollis, Hippolais rama, H. languida, Sylvia curruca jaxartica, S. nana nana, Scotocerca inquieta platyura, Oenanthe deserti salina, O. isabellina, Parus bokharensis bokharensis, Passer ammodendri ammodendri,* and *P. simplex zarudnyi.*

Of these 28 species, 17 (60.7%) are resident birds: *Buteo rufinus rufinus, Aquila chrysaëtus homeyeri, Charadrius leschenaultii crassirostris, Pterocles orientalis arenarius, Athene noctua bactriana, Dendrocopos leucopterus albipennis, Galerida cristata iwanovi, Calandrella cinerea longipennis, C. rufescens heinei, Podoces panderi, Corvus ruficollis ruficollis, Sylvia curruca jaxartica, S. nana nana, Scotocerca inquieta platyura, Parus bokharensis bokharensis, Passer ammodendri ammodendri,* and *P. simplex zarudnyi.* The status of a resident species does not assume a complete absence of changes in the bird distribution. We consider resident those bird species that are found within their nesting range in seasons beyond their nesting period.

The core fauna is differentially distributed among the desert habitats. Most birds there are found in shrub desert; these include *Buteo rufinus rufinus, Aquila chrysaëtus homeyeri, Chlamydotis undulata macqeenii, Burhinus oedicnemus harterti, Charadrius leschenaultii crassirostris, Ch. asiaticus, Cursorius cursor bogolubowi, Pterocles orientalis arenarius, P. alchata caudacutus, Caprimulgus aegyptius arenicolor, Dendrocopos leucopterus albipennis, Calandrella rufescens heinei, Lanius excubitor pallidirostris, Podoces panderi, Corvus ruficollis ruficollis, Hippolais rama, H. languida, Sylvia curruca jaxartica, S. nana nana, Scotocerca inqueta platyura, Parus bokharensis bokharensis, Passer ammodendri ammodendri,* and *P. simplex zarudnyi.* Four species inhabiting clay desert habitats are *Charadrius leschenaultii crassirostris, Ch. asiaticus, Cursorius cursor bogolubowi*, and *Calandrella cinerea longipennis*. Only two species in the Karakum Desert live in rare habitats which provide vertical surfaces such as isolated rocks and ruins *Buteo rufinus rufinus* and *Aquila chrysaëtus homeyeri* (*B. rufinus* can also nest on high saksaul trees). Finally, six species (*Chlamydotis undulata macqeenii, Pterocles orientalis arenarius, Athene noctua bactriana, Galerida cristata iwanovi, Oenanthe deserti salina*, and *O. isabellina*) are found in different types of habitats, including both sand and clay deserts. Nests of *Athene noctua* can be found in the Karakum in rodent burrows as well as in the crevices of isolated rocks.

The most significant feature of this core avifauna of the Karakum Desert is the *total absence of endemic species*. It is worth adding that there are only two endemic species of birds within the entire desert area of Turan (*Podoces panderi* and *Hippolais languida*). This picture is entirely different from what can be observed in the zoogeography of many groups of insects, reptiles, and mammals in Middle Asia, which have not only numerous endemic species but even endemic genera here.

There are, however, a few endemic subspecies of birds in the Karakum: *Dendrocopos leucopterus albipennis, Sylvia curruca jaxartica, Scotocerca inquieta platyura, Parus bokharensis bokharensis, Passer ammodendri ammodendri*, and *P. simplex zarudnyi*. This list can be supplemented by the list of subspecific forms endemic for the entire Irano-Turanian desert area: *Chlamydotis undulata macqeenii, Cursorius cursor bogolubowi, Pterocles orientalis arenarius, P. alchata caudacutus, Athene noctua bactriana, Caprimulgus aegyptius arenicolor*, and *Galerida cristata iwanovi*. Thus, there is a significant and pronounced subspecific endemism within the avifauna of the arid zone of Southwest and Middle Asia.

The degree of differentiation of Turkestan (or Turan) desert bird fauna, therefore, does not exceed subspecific level, which may indicate that the Turanian center of origin (or speciation) of the eremophile avifauna is not ancient, but rather recent and still active. This hypothesis (A. Rustamov 1950, 1954, 1968) is a logical development of the classical views of M.A. Mensbier (1914), who considered the Turkestan desert avifauna as being of a recent, post-Pliocene age. Such a view is concordant with views of other geographers and botanists on the post-Pliocene origin of modern desert landscapes of Middle Asia, especially those of sand deserts. Fossil pollen analysis of Pliocene and Quaternary sediments in Turkmenistan (Fedorovich 1946) demonstrated that flora very similar to modern saksaul forests of the Karakum existed not only in the Quaternary period but possibly already in the Pliocene. Sinitsyn (1962) argued for the younger age, not only of Middle and Central Asian deserts but also of the Sahara.

Tables 1 and 2 (based on the unpublished data of E. Rustamov) give an idea of bird density and distribution within the habitats of the Karakum desert. The distribution of bird species is heterogeneous, differing among several natural areas. The eastern part of the Karakum differs from the western and northwestern ones. In the east, one finds the best representation of species of the sand shrub desert inhabiting its typical habitat, a saksaul forest (areas of Yeradzhi, Repetek, and Chamchakly). The distinctiveness of the East Karakum is also manifested by the localized range of several species; e.g., *Passer ammodendri* and *P. simplex* do not penetrate further westward than the central part of the Karakum, and the western limit of range of *Dendrocopos leucopterus* is the dry bed of Uzboi. Other species (e.g., *Podoces panderi, Corvus ruficollis, Lanious excubitor, Sylvia nana,* and *Scotocerca inquieta*) can penetrate from the East Karakum westward along the suitable habitats. Still, the majority of the East Karakum species are rather sporadically distributed and are not abundant in the West, Central, or Trans-Unguz Karakum. In the north, a number of typical East Karakum species do not penetrate further than the southern shore of Lake Sarykamysh. The southern limit of the East Karakum bird complex lies between 36°20' and 36°40' N (A. Rustamov 1948).

Table 1. The abundance of the core aviafauna of the Karakum Desert in spring (birds/km^2); average data for many years of observation

Species	Habitat				
	Clay desert	Sand desert	Black saksaul	Chinks	Micro-oases
Buteo rufinus rufinus	<0.1	0.7	<0.1	2.0	–
Aquila chrysaëtus homeyeri	<0.1	<0.1	–	1.3	–
Chlamydotis undulata macqeenii	<0.1	<0.1	–	–	–
Burhinus oedicnemus harterti	0.2	<0.1	–	–	–
Charadrius leschenaultii crassirostris	<0.1	–	–	–	–
Ch. asiaticus	<0.1	–	–	–	–
Cursorius cursor bogolubowi	0.1	–	–	–	–
Pterocles orientalis arenarius	1.0	<0.1	–	–	–
P. alchata caudaculus	1.5	–	–	–	–
Athene noctua bactriana	0.1	0.1	0.2	4.5	2.7
Caprimulgus aegyptius arenicolor	<0.1	0.1	–	–	–
Dendrocopos leucopterus albipennis	–	0.3	0.4	–	0.9
Galerida cristata iwanovi	1.2	3.0	4.5	0.5	0.1
Calandrella cinerea longipennis and *C. rufescens heinei*	100.0	1.0	0.5	–	–
Lanius excubitor pallidirostris	1.4	2.0	0.1	0.1	3.0
Podoces panderi	0.2	3.5	0.1	–	–
Corvus ruficollis ruficollis	0.2	0.2	<0.1	1.6	4.0
Hippolais rama and *H. languida*	1.0	10.8	29.0	<0.1	0.1
Sylvia curruca jaxartica	1.0	11.5	30.0	<0.1	11.0
S. nana nana	1.5	2.0	0.1	<0.1	–
Scotocerca inquieta platyura	0.2	10.2	98.0	<0.1	–
Oenanthe deserti salina	2.0	0.4	0.3	0.4	–
Oe. isabellına	17.0	1.5	0.7	2.3	–
Parus bokharensis bokharensis	–	8.5	44.0	–	1.5
Passer ammodendri ammodendri	–	10.8	70.0	–	46.8
P. simplex zarudnyi	–	0.6	–	–	–
Total	137.8	65.5	278.0	14.9	70.4

Table 2. The spring population of birds in the Karakum Desert (birds/km^2)

Habitats	Abundance	
	Total	The "core" bird species
Clay deserts	163	137.8 (84.54%)
Sand deserts	100	65.45 (65.45%)
Black saksaul forest	532	278.0 (52.24%)
Chinks	173	14.9 (8.61%)
Clay deserts	163	70.4 (12.71%)

The avifauna of the West and Northwest Karakum represents a solid complex which differs from that of the East Karakum. The predominant landscape of the West is clay plains; sand inclusions, when present, are impoverished as compared to the East Karakum sands. The characteristic bird species are *Chlamidotis undulata, Charadrius leschenaultii, Melanocorypha bimaculata,* and *Calandrella rufescens*. Other typical landscape elements in the Northwest Karakum include *chinks* (cliffs) and *ostantsy* (buttes) inhabited by *Aquila chrysaëtus, Falco cherrug, Alectoris chukar, Bubo bubo,* and, to some extent, *Oenanthe deserti*.

The avifauna of the Central Karakum is a transitional one between the described eastern and western complexes. In general, the northern boundary of the Karakum avifauna runs along the dry bed of the West Uzboi (crossing it in two large extensions within the sand massifs Chilmamedkum and Uchtagan), the eastern edge of the Kaplankyr Plateau, and the southern shores of Lake Sarykamysh. The southern limit of the Karakum avifauna comprises the foothills of Kopetdagh.

The Karakum avifauna in total is typical for so-called "southern" deserts (Dementyev and Rustamov 1946). Comparison of this avifauna to that of the Kyzylkum desert in adjacent Uzbekistan reveals 33 common nesting species of birds, usually belonging to the same subspecies. Very few species typical for the Karakum (e.g., *Cursorius cursor* and *Ammomanes deserti*) are absent from the Kyzylkum. On the other hand, several species nesting within the Kyzylkum do not nest in the Karakum; these include *Syrrhaptes paradoxus* and birds nesting on the buttes (*Aquila rapax, Falco pelegrinoides,* and *Emberiza cia*). In general, avifaunas of the Karakum and Kyzylkum deserts can be combined in one zoogeographical area, that of southern deserts. We should note that the avifauna of the southern Ustyurt Plateau is a transition between faunas of southern (sand) and northern (clay) deserts.

Reproduction and Molting of Birds in the Karakum Desert

Table 3 presents the reproductive timetable of Karakum Desert bird species. In general, the reproduction period occurs within the ecological optimum of every

Table 3. The calender of reproduction of the core avifauna of the Karakum Desert

	Late February	March			April			May		
		1 to 10	11 to 20	21 to 31	1 to 10	11 to 20	21 to 30	1 to 10	11 to 20	21 to 31
Buteo rufinus rufinus	+	+	+?	–	–	–	–	–	–	–
Aquila chrysaëtus homeyeri	+	+	+	–	–	–	–	–	–	–
Chlamydotis undulata macqeenii	–	–	–	–	+	+	+	–	–	–
Burhinus oedicnemus harterti	–	–	–	–	–	+?	+	–	–	–
Charadrius leschenaultii crassirostris	–	–	–	–	+?	+	+	+?	–	–
Ch. asiaticus	–	–	–	–	–	–	–	+	+?	–
Cursorius cursor bogolubowi	–	–	–	–	–	–	–	+?	+	–
Pterocles orientalis arenarius	–	–	–	–	–	–	–	+	+?	–
P. alchata caudaculus	–	–	–	–	–	–	–	+	+?	–
Athene noctua bactriana	–	–	–	–	+	+	–	–	–	–
Caprimulgus aegyptius arenicolor	–	–	–	–	–	–	–	–	+	+
Dendrocopos leucopterus albipennis	–	–	–	+	+?	–	–	–	–	–
Galerida cristata iwanovi	–	+	+	–	–	–	–	–	–	–
Calandrella cinerea longipennis	–	–	–	–	–	+	+	+?	+	–
C. rufescens heinei	–	–	–	–	–	+	+	+	–	–
Lanius excubitor pallidirostris	–	–	–	–	+	–	+	–	–	–
Podoces panderi	–	–	–	+	+	+	+	–	–	–
Corvus ruficollis ruficollis	–	–	–	+?	–	+	+	+?	–	–
Hippolais rama	–	–	–	–	–	+	+	–	–	–
H. languida	–	–	–	–	–	+	+?	+	+	–
Sylvia curruca jaxartica	–	–	–	–	–	+?	+	–	–	–
S. nana nana	–	–	–	–	–	+?	+	–	–	–
Scotocerca inquieta platyura	–	–	–	–	–	+?	+	–	–	–
Oenanthe deserti salina	–	–	–	–	–	+?	+	–	–	–
Oe. isabellina	–	–	–	–	–	+?	+	–	–	–
Parus bokharensis bokharensis	–	–	–	–	–	+?	+	–	–	–
Passer ammodendri ammodendri	–	–	–	–	–	+?	+	–	–	–
P. simplex zarudnyi	–	–	–	–	–	+?	+	–	–	–

species. Processes of spermatogenesis and oogenesis, which require high energy expenses, usually occur in March and April (or even February), i.e., a period with maximal development of vegetation and mass abundance of active invertebrates (especially insects). Protective quality of the desert habitat also is optimal in spring. The resident "core" bird species, as a rule, reproduce earlier than others.

Normally, a secondary reproductive cycle is present only among a few eremophilic "core", passerine species. *Galerida cristata*, for example, has two or even three reproductive cycles in the Karakum Desert although these multiple cycles are weakly expressed. Secondary clutches compensate somewhat, however, for the low viability of clutches and hatchlings since 50–60% of these die from spring frosts, heat, smothering by sand, or predation. Many species which possess secondary reproductive cycles elsewhere do not exhibit this feature in the Karakum, however, and in general it may be stated that "core" species have only one cycle, in spite of the southern position of the desert.

This circumstance, along with the low number of eggs in a clutch, is responsible for the relatively low fecundity of such widespread species in the Karakum as *Bubo bubo*, *Athene noctua*, *Corvus ruficollis*, *Galerida cristata*, *Lanius excubitor*, and *Parus bokharensis*. Comparison of the number of eggs in a clutch for species such as *Bubo bubo*, *Athene noctua*, and *Buteo rufinus* between Karakum and temperate populations evidences the low viability for Karakum populations; there are very few exceptions to this rule.

The process of bird molting is also not intensive in Karakum birds: the "core" species, at least, they usually do not molt in the spring. Of these, the spring molt has been observed only in *Pterocles archata*. In *Oenanthe isabellina*, notwithstanding the data from literature, we have never observed a spring molting. Even such species as *Hippolais rama*, *Sylvia nana*, and *Scotocerca inquieta* molt only once (in the fall) in the Karakum whereas related species of Sylviidae usually have both spring and fall molts. Thus, Karakum bird species, as a rule, have only one molt in a year, which starts in June, reaches a peak in July, and ends in September.

It should be mentioned that birds' plumage in the Karakum becomes worn out rapidly and intensively, which, to a certain degree, facilitates its cryptic, camouflaging role within the desert landscape.

Adaptations of Birds to the Arid Climate

Our field data, collected during many years in the Karakum Desert, include numerous examples of convergent and parallel adaptations to the arid climate (A. Rustamov et al. 1974). Such adaptations appear in different systematic groups of birds existing under extreme conditions of water shortage, excessive sun radiation, and sharp temperature changes. A number of morphophysiological, ecological, and behavioral adaptations help desert birds to minimize their energy expenses (A. Rustamov 1954, 1955a,b; Sopyev 1965; A.

Rustamov et al. 1974; Amanova 1979); decreased rates of metabolism are also demonstrated for many other desert animals (e.g., rodents, reptiles, and domestic mammals).

The entire complex of adaptations in arid bird species is often characterized as a specific ecobiomorph ("life form") of a desert bird (A. Rustamov 1955, 1964; Dementyev 1964). Desert communities of birds in general exhibit functional and physiological similarities (Vtorov and Drozdov 1978), and communities inhabiting similar ecosystems reveal maximal similarities even when these ecosystems are geographically remote (A. Rustamov and Drozdov 1984).

The following behavioral adaptations are shared by many desert birds. They usually have two peaks of activity (morning and evening), with morning activity higher; some species (e.g., *Burhinus oedicnemus* and *Caprimulgus aegypteus*) have crepuscular/night activity. Desert birds utilize shade and microclimate for nesting, resting, and drinking sites (summary sun radiation in the shade is four to five times less than under sun rays). Species are able to cover large distances in search of food and water (by fast flight, as in *Pterocles orientalis* and *P. alchata*, or by running, as in *Burhinus oedicnemus, Cursorius cursor*, and *Podoces panderi*). When feeding, desert birds select juicy, moist food and can actively search for salt. They are able to escape predators through a peculiar reaction of "freezing." Competition for nesting sites exists among ecologically similar species as well as within the same species. Adults protect their nests and hatchlings from overheating or sand covering by shifting position along the nest edge, following the movement of sun rays or wind. Those widespread species that normally build nests both on vegetation and on the ground in other geographic zones do not built ground nests in the desert. There is maximal utilization of nesting sites and a resulting absence of "group" nesting. Desert birds search for effective heat-insulation material for their nests. Finally, for migratory (non-resident) desert bird species, early desert arrivals and departures are quite common.

The following features can also be listed as ecological adaptations common to many desert birds. The thickness of nest walls can change depending on reproduction time. In *Podoces panderi*, for example, thickness is 20–47 mm in March versus 10–25 mm in April; in *Scotocerca inquieta*, average nest weight is 67 g in March versus 38.7 g in April. The reproductive cycle can be accelerated by buildup of the nest and wall thickening during egg laying and incubation. When secondary egg layings take place, the nest may be positioned higher due to the increased heat (e.g., in *Scotocerca inquieta* and *Passser simplex*). Generally, eggs that are laid later are smaller than first eggs (e.g., in *Podoces panderi* first eggs are 2–3 mm longer). Depending on temperature conditions and nest position, the rate of incubation and feeding of the hatchlings can change. In the heat, the familiar reflex of "begging for food" in hatchlings disappears in those species that make nests in open places; their hatchlings have much weaker reaction to parents' arrival. This weakening reaction was not observed in species nesting inside tree trunks and thus protected from the heat.

Peculiar adaptations of desert birds related to water balance include the following: accumulation of water in muscles and visceral organs (in *Caprimulgus aegyptius, Galerida cristata, Rhodospiza obsoleta, Podoces panderi, Oenanthe isabellina,* and *Passer simplex*); efficient use of water in digestion and its more intensive reabsorption (in *Caprimulgus aegiptius, Dendrocopos leucopterus, Podoces panderi, Scotocerca inquieta* and *Passer simplex*); absence of skin evaporation; and increased kidney function (e.g., kidney weight in *Passer ammodendri* and *P. simplex* is 12.8% and 11.2% of the total body weight, whereas in non-desert sparrow species *P. montanus* and *P. indicus* it is 9.9% and 9.1%). Frequent visits to drinking sites (an average five times per hour in insectivores, and up to 200 times per hour in granivores) have also been observed.

Both behavioral and ecological adaptations lead to the evolution of parallel morphological adaptations in a number of desert bird species. Examples include long legs (*Cursorius cursor, Caprimulgus aegyptius, Podoces panderi, Lanius excubitor, Scotocerca inquieta* and *Oenanthe isabellina*); presence of sharp "speed" feathers (in *Pterocles*); lengthening of secondary inner long feathers, which facilitates better coverage of the wing and camouflaging of the dorsum (in *Pterocles, Syrrhaptes, Cursorius cursor,* and *Charadrius asiaticus*); existence of spiral-shaped plates on the inner side of the intestine which help to prolonge movement of the food mass and water reabsorption (in *Passer ammodendri* and *P. simplex,* whereas in non-desert sparrow species *P. montanus* and *P. indicus* these plates are straight); and generally, an increased absorbing surface of the intestine.

Drozdov (1977) compared parallel adaptations of the Karakum bird fauna to those of the bird fauna of the Simpson Desert (Australia). Ecological and behavioral parallel features can be traced in *Scotocerca inquieta* (Karakum Desert) and *Stipiturus ruficeps* (Simpson Desert), or *Erythropygia galactotes* (Karakum Desert) and *Rhipidura leucophrus* (Simpson Desert). These species demonstrate parallelism also in nest site selection and nest features. Their density in desert landscapes is also similar; during the nesting period, 200–500 birds per km^2 are found in shrub vegetation of the Karakum Desert, and 200–400, in the Simpson Desert.

Ecological and Geographical Importance of Water for Birds

Irrigation of the Karakum through construction of the Karakum Canal (from the Amudarya River to Kazandzhik) has substantially transformed the natural arid ecosystems. Water shortage is the ecological and physiological norm for strictly desert species. With irrigation, these birds leave their habitats and move deeper into the desert; their density decreases, and habitats shrink. Such modification in structure and density of aboriginal desert bird populations is intrinsically connected to modifications in vegetation, soils, and invertebrate populations; however, the common initial cause of all these changes is artificial irrigation.

A number of less specialized bird species which historically penetrated into desert ecosystems via intrazonal habitats exhibit different attitudes toward irrigation. Before irrigation, these species were connected to caravan routes, wells, and pastures, next to sheep and camels (A. Rustamov and Ptushenko 1948). After irrigation, the intrazonal species, being more ecologically plastic, were incorporated into new transformed ecosystems.

Therefore, water in an irrigated desert is a limiting factor for arid bird species but a beneficial ecological and zoogeographic factor for intrazonal species (A. Rustamov 1976). However, even some arid species settle next to water during hot summer weather (e.g., *Passer, Pterocles*, species of Alaudidae and Columbidae).

The arid lands have been irrigated since ancient times because the very penetration of man into the desert is made possible only in the presence of water. In Middle Asia, people learned to dig wells, to construct *kyarizes* (underground irrigational systems for leading ground water in the foothills to the desert plain) and *sardobas* (reservoirs for preserving winter and spring precipitation). The most ecologically plastic bird species could develop new habitats next to human dwellings although these connections were neither necessarily strong nor constant. The presence of the extensive modern system of canals, reservoirs, settlements, and agricultural areas, however, has had an enormous impact on desert birds.

Irrigation provides animals with water and increases amounts of food. These beneficial ecological conditions expand over vast irrigated areas, and through those areas the animals from already existing anthropogenic landscapes can disperse to new territories. For example, birds inhabiting valleys of the Amudarya, Murghab and Tedzhen Rivers and adjacent desert intrazonal habitats literally besiege settlements and developing irrigated areas along the Karakum Canal. Their dispersal (primarily of the species characteristic for river valleys) rapidly follows the irrigation of new lands.

We should, however, distinguish this type of dispersal from a true dispersal which leads to the expansion of the species' range. Expansion of bird species from the East Turkmenistan river valleys along the Karakum Canal is an example of massive dispersal of animals *within their range*, when more suitable habitats appear. Some of the species, however (e.g., *Acridoteres tristis* which is now nesting along the Karakum Canal) exhibit true dispersal with irrigation.

Irrigation exerts considerable influence on the structure of bird populations during migrations and rewintering. In the past 25 to 30 years, reservoirs and discharging lakes have become mass rewintering sites for birds. In good years, the artificial water bodies of Middle Asia can host from 600,000 to 800,000 ducks, geese and other waterfowl, or 10 to 15% of all birds concentrating in large rewintering sites in the territory of the former Soviet Union. The filtration lakes are less beneficial, and numbers of rewintering birds there are insignificant. Many migrant species of birds during spring and fall migrations are attracted by artificial water bodies, which leads to the breakup of the "wide front" of typical migrations over the deserts. This pattern is replaced by relatively narrow migrating corridors along the artificial water bodies.

Many years of studies (A. Rustamov et al. 1990; A. Rustamov 1993) show that, within the Karakum Desert, up to 500,000 birds belonging to 34 species of waterfowl are rewintering on artificial water bodies. Among them, the dominant species are *Fulica atra* (25%), *Anas platyrhynchos* (24%), *A. crecca* (19%) and *Netta rufina* (20%). The major natural rewintering sites for waterfowl within Turkmenistan are the eastern shore of the Caspian Sea, the Kelif Lakes (southeastern Karakum), and the delta of the Atrek River (southwestern Turkmenistan).

The formation of new "winter quarters" for the waterfowl, and the related formation of new (although temporary) biological diversity can be considered a positive result of irrigation in the desert. However, approximately half of all Amudarya water that is retained for irrigation is sinking into the sand of the drainage network and is causing formation of swampy areas in the desert as well as salinization and degradation of desert ecosystems. These are the negative and even catastrophic results of irrigation (Babaev and Zaletaev 1990; A. Rustamov 1987).

In cases like that of the Aral Sea in neighboring Uzbekistan, the ecological situation is, for all purposes, out of human control. We know that almost all aboriginal freshwater and brackish fauna of the Aral Sea became extinct. Species which were introduced here from the Caspian and Black Seas are on the verge of extinction. The species diversity of mammals and birds has decreased two to three times, as well as the floristic diversity of the Aral area, particularly in the tugai communities (Glazovsky 1990). Birds are definitely changing their aboriginal habitats, their ranges are shrinking, and density is decreasing (A. Rustamov 1991a,b).

The first list of the Karakum birds was published almost four decades ago (A. Rustamov 1954). This is why we considered it necessary to compile a new, updated list (Table 4). The increase in the number of bird species is due to some taxonomic rearrangements at species/subspecies level as well as to new records and enrichment of nesting fauna. The number of recorded nesting bird species is now 79 (in 1954, it was only 60). New species nesting in the Karakum area are, for example, such species of water, bywater, and anthropogenic habitats as *Pelecanus onocrotalus*, *P. crispus*, *Streptopelia senegalensis ermanni*, and *Acridotheres tristis tristis*. Of these, two species of pelicans are found during migration and nesting on the islands of artificial Lake Sarykamysh, which was formed within the last three decades through the discharge of drainage waters to the Sarykamysh Depression. The complete list of the Karakum avifauna now totals 238 species and 249 subspecies of birds (the 1954 list contained 220 species and 249 subspecies).

Table 4. A list of birds of the Karakum Desert

Species	Residence Status				
	resident	nesting	migratory	rewintering	occasional records
Podiceps ruficollis capensis			4		
P. nigricollis nigricollis			2		
P. auritus auritus			2		
P. grisegena grisegena			2		
P. cristatus cristatus			4		
Pelecanus onocrotalus		2	2		
P. crispus		2	2		
Phalacrocorax carbo sinensis			4	3	
Ph. pygmaeus					2
Botaurus stellaris stellaris			2		
Ixobrychus minutus			2		
Egretta alba			2	3	
Ardea cinerea cinerea		2	3	4	
A. purpurea purpurea			2		
Platalea leucorodia					2
Plegadis falcinellus			2		
Phoenicopterus roseus					2
Anser anser			2		
A. erythropus			2		
Tadorna ferruginea	2	2			
T. tadorna	2	3			
Anas platyrhynchos platyrynchos			4	5	
A. crecca crecca			4	5	
A. strepera			3		
A. penelope penelope			3	3	
A. acuta acuta			2	2	
A. querquedula			4		
A. clypeata clypeata			2		
A. angustirostris					2
A. rufina			4		
Aythya ferina			4		
A. nyroca			4		
A. fuligula			4		
A. marila					1
Clangula hyemalis					1
Bucephala clangula clangula			4		
Oxyura leucocephala			2	2	
Mergus albellus			2	2	
M. serrator			2		
M. merganser merganser			2		
Milvus migrans migrans		4	5		
Circus cyaneus cyaneus			4	4	
C. macrourus			2		
C. pygargus			4		
C. aeruginosus aeruginosus			4	4	
Accipiter nisus nisus			3		

Table 4. Continued

Species	Residence Status				
	resident	nesting	migratory	rewintering	occasional records
A. badius cenchroides			3		
Buteo rufinus rufinus	5				
B. buteo vulpinus			3	2	
B. buteo japonicus				1	
Circaëtus ferox heptneri		2	2		
Aquila rapax orientalis			1	1	
A. r. nipalensis			1	1	
A. heliaca heliaca			1	1	
A. chrysaetos homeyeri	2				
Haliaëtus albicilla			2		
Neophron percnopterus percnopterus		4	4		
Aegypius monachus			2		
Gyps fulvus fulvus			1		
Falco cherrug coatsi					2
F. columbarius subsp.					
F. naumanni					
F. tinnunculus tinnunculus		5	5	3	
Alectoris chukar laptevi	4				
A. chukar dementievi	3				
Coturnix coturnix coturnix			4		
Phasianus colchicus zarudnyi		1			
Grus grus lilfordi			3		
Rallus aquaticus aquaticus			4		
R. aquaticus korejewi					2
Porzana porzana				4	
Crex crex					2
Gallinula chloropus chloropus			2	4	
Fulica atra atra		4	5	3	
Otis tarda tarda			1	1	
O. tetrax					1
O. undulata macqueenii		2	2	1	
Burhinus oedicnemus harterti		3	3		
Pluvialis squatarola			3		
P. dominica					1
Charadrius dubius curonicus			4		
C. leschenaultii		2	4		
C. asiaticus		1	2		
C. alexandrinus alexandrinus		2	4		
Chettusia gregaria			1		
Vanellus vanellus			4		
Vanellochettusia leucura			5		
Himantopus himantopus himantopus		2	5		
Tringa ochropus		2	5		
T. glareola			5		
T. nebularia			2		

Table 4. Continued

Species	Residence Status				
	resident	nesting	migratory	rewintering	occasional records
T. totanus totanus			2		
T. stagnatilis			2		
Actitis hypoleucos			4		
Phalaropus lobatus			5		
Philomachus pugnax					1
Calidris minuta			4		
C. ferruginea					2
C. alpina alpina			5	5	
Limnocryptes minimus			2		
Gallinago gallinago gallinago			4		
Numenius phaeopus phaeopus			2		
Cursorius cursor bogolubovi		2	2		
Glareola pratincola pratincola			2		
Larus ichthyaetus			2		
L. minutus			1		
L. ridibundus			4		
L. genei			4		
L. argentatus heuglini			2		
L. canus heinei			2		
Chlidonias hybrida hybrida		2	4		
Gelochelidon nilotica nilotica			2		
Sterna hirundo hirundo			4		
Pterocles orientalis arenarius		2	2	1	
P. alchata caudacutus		2	2	1	
Syrrhaptes paradoxus					2
Columba oenas oenas			1		
C. eversmanni		2	2		
C. livia neglecta	5				
Streptopelia turtur turtur			2		
S. turtur arenicola		4	4		
S. orientalis meena					1
S. senegalensis ermanni	3				
Cuculus canorus subtelephonus		1	4		
Nyctea scandiaca					1
Bubo bubo omissus		3			
B. bubo turcomanus			3	3	
Asio otus otus			4	4	
A. flammeus flammeus			4	4	
Otus scops scops					2
O. brucei					2
Athene noctua bactriana	4				
Caprimulgus europaeus zarudnyi		3	3		
C. aegyptius arenicolor		3	3		
Apus apus pekinensis		3	4		
A. melba petrensis					2
Coracias garrulus semenovi		2	3		

Table 4. Continued

Species	Residence Status				
	resident	nesting	migratory	rewintering	occasional records
Merops apiaster		2	5		
M. superciliosus persicus		2	5		
Upupa epops epops		4	4		
Iynx torquilla torquilla			2		
Dendrocopos leucopterus albipennis	3				
Riparia riparia riparia			5		
Hirundo rustica rustica			5		
Ammomanes deserti parvirostris	4				
Galerida cristata iwanowi	5				
Calandrella cinerea longipennis	5				
C. rufescens heinei	5				
Melanocorypha calandra psammochroa		4	4		
M. bimaculata torquata		4	4		
Eremophila alpestris albigula			2	2	
Alauda arvensis dulcivox			5	4	
Anthus richardi richardi					1
A. campestris boehmii			4		
A. campestris griseus			4		
A. trivialis trivialis			4		
A. pratensis			5	5	
A. cervinus rufogularis			2	2	
A. spinoletta blakistoni			4	4	
Motacilla flava thunbergi			1		
M. flava beema			5		
M. feldegg melanogrisea			4		
M. citreola werae			5		
M. alba dukhunensis			5	5	
M. personata			4	4	
Lanius isabellinus phaenicuroides		2	3		
L. collurio			4		
L. minor			2		
L. excubitor homeyeri			4	4	
L. excubitor pallidirostris		4	4		1
Oriolus oriolus oriolus			2		
O. oriolus kundoo			2		
Sturnus vulgaris poltaratskyi			5	4	
S. v. porphyronotus			2	2	
Pastor roseus			3		
Acridotheres tristis tristis	3				
Pica pica bactriana		2			
Podoces panderi	4				
Corvus monedula monedula		2			
C. frugilegus frugilegus			5	5	

Table 4. Continued

Species	Residence Status				
	resident	nesting	migratory	rewintering	occasional records
C. corone orientalis			3		
C. cornix sharpii			4		
C. ruficollis ruficollis	4				
Bombycilla garrulus garrulus					2
Prunella atrogularis atrogularis		2	2		
Locustella naevia straminea		1			
Acrocephalus agricola brevipennis		2			
A. dumetorum		3			
A. scirpaceus fuscus		2			
A. stentorius brunnescens		4			
A. arundinaceus		2			
Hippolais caligata caligata		4			
H. rama		5	5		
H. pallida elaeica		4	4		
H. languida		5	5		
Sylvia communis communis			4		
S. c. icterops		4	4		
S. curruca curruca			5		
S. c. blythi			5		
S. c. halimodendri			5		
S. c. jaxartica		5	5		
S. mystacea turcmenica			4		
S. nana nana	1	4	4	1	
Phylloscopus trochilus acredula			4		
Ph. collybita fulvescens			5	2	
Ph. trochiloides viridanus			4		
Scotocerca inquiete platyura	5				
Regulus regulus tristis			2	2	
Ficedula parva parva			5		
Muscicapa striata neumanni			4		
Saxicola torquata maura			2		
S. caprata rossorum			2		
Oenanthe oenanthe oenanthe			3		
Oe. pleschanka pleschanka		4	4		
Oe. picata picata			3		
Oe. finschii		4	4		
Oe. deserti salina		4	4		
Oe. isabellina	1	5	5	1	
Cercotrichas galactotes familiaris		4	4		
Monticola saxatilis saxatilis			2		
Phoenicurus phoenicurus phoenicurus			4		
Ph. ochruros rufiventris			4		
Ph. erythronota			3	2	
Luscinia megarhynchos hafizi			3		

Table 4. Continued

Species	Residence Status				
	resident	nesting	migratory	rewintering	occasional records
L. luscinia			2		
L. svecica svecica			3		
L. svecica pallidogularis			4		
Erithacus rubecula rubecula			2	2	
Turdus ruficollis					1
T. atrogularis			4	4	
T. pilaris			3	3	
T. viscivorus bonapartei			3	3	
Panurus biarmicus russicus			4	4	
Remiz pendulinus jaxarticus			3	3	
R. pendulinus coronatus		4	4	4	
Parus bokharensis bokharensis	5				
Passer indicus bactrianus		5	5		
P. hispaniolensis transcaspius		4	5	5	
P. ammodendri ammodendri	4				
P. montanus dilitus	5				
P. simplex zarudnyi	2				
Fringilla coelebs coelebs			4	4	
F. montifringilla			4	4	
Chloris chloris turcestanicus			3		
Spinus spinus			4		
Acanthis cannabina fringillirostris			4	2	
Rhodopechys sanguinea sanguinea					1
Bucanetes githagineus crassirostris		3			
Rhodospiza obsoleta		4	5	3	
Carpodacus erythrinus erythrinus			3		
Coccothraustes coccothraustes nigricans			2		
Emberiza calandra buturlini			4		
E. citrinella erythrogenis			2	1	
E. leucocephala leucocephala			3	3	
E. schoeniclus passerina			3	3	
E. schoeniclus pallidior			4	4	
E. schoeniclus harmsi			2	2	
E. rustica					1
E. hortulana			4		
E. bruniceps		3	4		

1 – solitary records, 2 – rare, 3 – moderately abundant, 4 – common, 5 – numerous

16. Ecological Structure of the Bird Population in the Transcaspian Region: Cartographic Analysis and Problems of Conservation

ELDAR A. RUSTAMOV

Abstract

This 20-year long study of the bird population in Turkmenistan combines methods of ornithology, zoogeography, ecology, and cartography. We have estimated and mapped ecological characteristics, including diversity, density, and similarity. Our approach to the mapping has allowed us to account for changes in habitats, including human-influenced changes. Maps of diversity, density, and types of population are given. Bird population structure is discussed in detail for a model territory (the area between the Murghab and Tedzhen Rivers in the southern Turkmenistan). The data obtained as a result of this analysis can be used for the long-term monitoring and conservation of ecosystems.

Introduction

An analysis of spatial and temporal patterns of ecological structure in the bird population of the Transcaspian Region was performed using the cartographic methods established in Soviet zoogeography (Vernander *et al*. 1959; Voronov and Cheltsov-Bebutov 1962; Tupikova 1969, 1976; Cheltsov-Bebutov 1964, 1966, 1970, 1973, 1976; Danilenko 1974; Tupikova and Komarova 1979; Danilenko and Mirutenko 1986; Rustamov 1988, 1991, 1992; Danilenko *et al*. 1991a,b). Strategies in field data collection and analysis were dictated by our object, a "bird population/habitat" system. We used maps not only for representation of the initial data but also for analysis.

Field Data Collection

Our data were collected from 1969 to 1989 in the Transcaspian Region, located between 35° and 47° north and 53° and 66° east (Fig. 1). Field surveys were conducted through all seasons to obtain a complete temporal picture of the bird population. In order for the data collected from this enormous area to be comparable and representative, survey techniques had to be standardized, and time of data collection compressed as possible. We therefore often surveyed by automobile to reduce field time to cover large areas and to obtain comparable data from different types of arid habitats. Walking surveys were also conducted in limited areas to avoid underestimation of small birds. Stops were made in areas with shrub and tree vegetation, wells, and other habitats attractive to small birds in arid landscapes. We also performed year-round surveys on standard routes near the city of Ashkhabad, in the Tedzhen Valey, and in the Repetek Reserve.

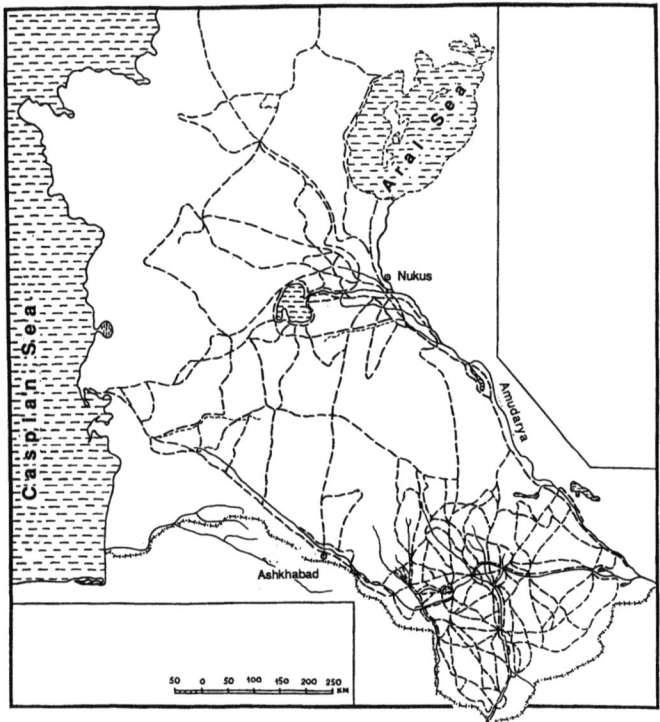

Fig. 1. Field data collection routes of the author.

Observations included characteristics of the landscape, soil, and vegetation, bird species, and their density. We also registered weather conditions and the degree and kind of human activity on the territory. Details of the biology of an observed species (e.g., flocks, feeding) were also registered; nests, when found, were measured and described. We surveyed birds both to the left and the right of the car along three transects of fixed width parallel to the automobile's direction: the closest lane was 25 m; the middle lane, 25 to 100 m; and the farthest lane, 100 to 300 m wide. Birds were registered from these lanes along the route outlined on a topographic map. Automobile speed was 40–45 km/hr on paved highways, 30–35 km/hr on non-paved roads, and 20–25 km/hr on the areas without roads. Mileage readings were registered for all encounters of rare species, and also for all species of birds found next to such landmarks as wells, *cosharas* (sheep corrals), drilling towers, bridges, buildings, artificial reservoirs.

Our data were collected in a territory of 556,800 km^2 (Fig. 2). This area stretches 1,300 km from south (Badghyz) to north (northern chink of Ustyurt); from west to east it covers 1,000 km in its southern part (Chilmamedkum to Kelif Uzboi), and 400 km in its northern part (between western and eastern chinks of Ustyurt). Within this area, we originally attempted to perform a survey of all bird species in all seasons of the year. This, however, proved to be possible only for the spring period, during which we were able to conduct comparable surveys of the entire area. We selected a smaller model territory within the Transcaspian region for characterization of the spatial structure of bird population in other seasons. This model territory covered 87,300 km^2, and included the area between the Tedzhen and Murghab Rivers, and parts of the Karakum Desert adjacent to the north and east.

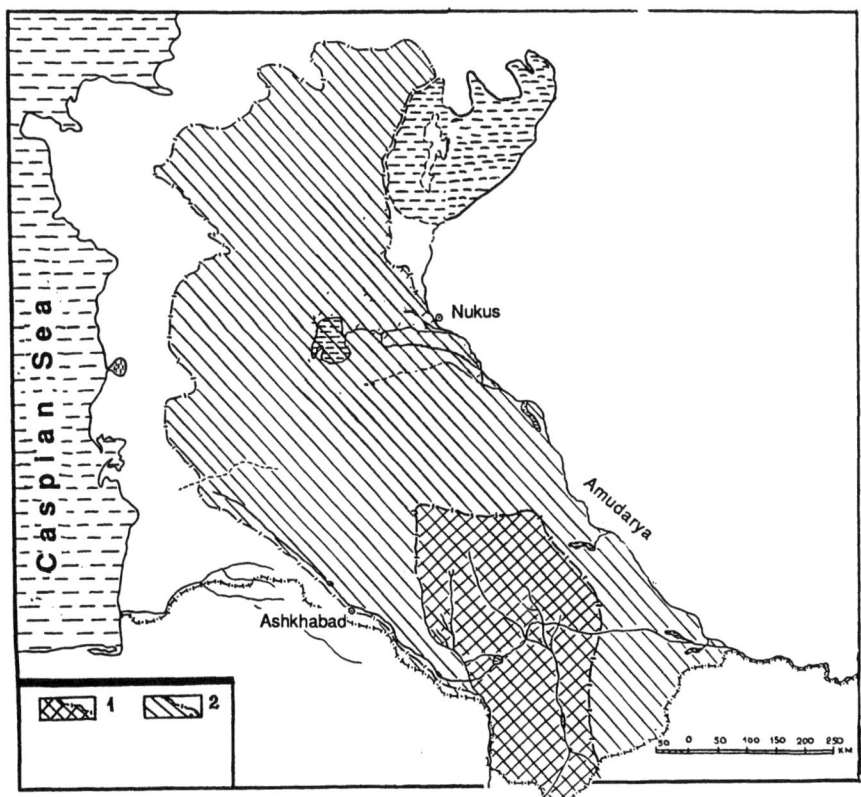

Fig. 2. Areas of data collection throughout the year (1) and in spring (2).

We conducted about 60 field trips totaling 68,600 km of automobile surveys. A complete zoogeographic survey was performed on 43,600 km and accounted for a total of more than one million individual birds belonging to 376 species. In total, 17,979 km were surveyed and 425,800 birds were recorded in spring; 5,756 km and 155,500 birds in summer; 11,227 km and 322,500 birds in fall; and 8,656 km and 14,200 birds in winter time. More than 1,000 days were spent in walking surveys.

Analysis of the field data included: compilation of a habitat map; compilation of maps of seasonal aspects of population; classification and typological division of population; statistical analysis of surveys; and generalization and extrapolation of data from the created habitat map.

Map of Bird Habitats

The map of bird habitats was created primarily using spring survey data; this permitted distinction among habitats which can look similar in other seasons due to the seasonal shifts in bird feeding strategies. The creation of a habitat map started from analysis of survey outlines from the topographic map (scale 1:300,000 or 1:500,000). Characteristics of a territory from field records were aligned with its features on the topographic map, including isolines showing elevations of the terrain. An elementary area was outlined and considered homogeneous if within it both physiographic structure and characteristics of bird population (diversity and density) appeared to be similar. Such homogeneous areas on the topographic map were further aligned with various natural maps (e.g., of landscapes, vegetation). When anthropogenic habitats were recorded in field survey but were not found on maps, we corrected boundaries of habitats according to remote sensing photographs (airplane or satellite data). Corrected boundaries of habitats reflected, therefore, both natural and human-affected territories (e.g., effects of desertification, irrigated areas, fields, construction, gas pipelines, electric lines). Every outline was assigned its own index according to a classification based on certain ecological requirements of bird species.

The habitat map was then generalized by its transformation to a scale of 1:1,000,000. Individual small habitat outlines did not disappear with this generalization but rather were incorporated into a new, complex outline which was given a new index. We attempted to provide maximal reflection of the initial field map data in a final habitat map; this map then served as a basis for subsequent analysis of seasonal changes in bird population.

We also classified habitats to create the map's legend. Within the Transcaspian Region, we distinguished 60 different bird habitat types. These types were hierarchically classified and represented by 82 outlines on the map (Fig. 3; Table 1; the part of the legend with the description of all habitats of the Transcaspian Region is omitted here due to its length). Certain similarities found among these types of habitats are due to both natural and anthropogenic causes.

Table 1. A typological classification of bird habitats

Types	Classes	Automorphic Natural	Automorphic Anthropogenic	Hydromorphic Natural	Hydromorphic Anthropogenic: oases	Hydromorphic Anthropogenic: lakeshores
Northern deserts	solonchak	1				
	clay	2, 3, 4				
	sand	7	43			
	gravel	8				
Transitional deserts	solonchak	9				
	clay	10a, 10b		49, 50a	54a	
	sand-clay	11	44a			
	sand	12, 13				59, 60a
	gravel	14, 15, 16				
Southern deserts	solonchak	17				
	clay	18		50b, 50c, 50d	54b, 54c, 54d, 54e	
	sand-clay	19, 20, 21	45, 46	44b, 44c, 44d	55a, 51b, 52	60b, 60c
	sand	22a, 22b, 23, 24, 25, 26a, 26b, 27, 28, 30, 31a, 31b, 31c, 32, 33, 34, 35, 35a, 36				
Submontane deserts	solonchak	37		53		
	clay	38a, 38b	47		56, 57	60d
	sand-clay		48			
	sand-loess	39, 39a, 40, 41, 42			58	

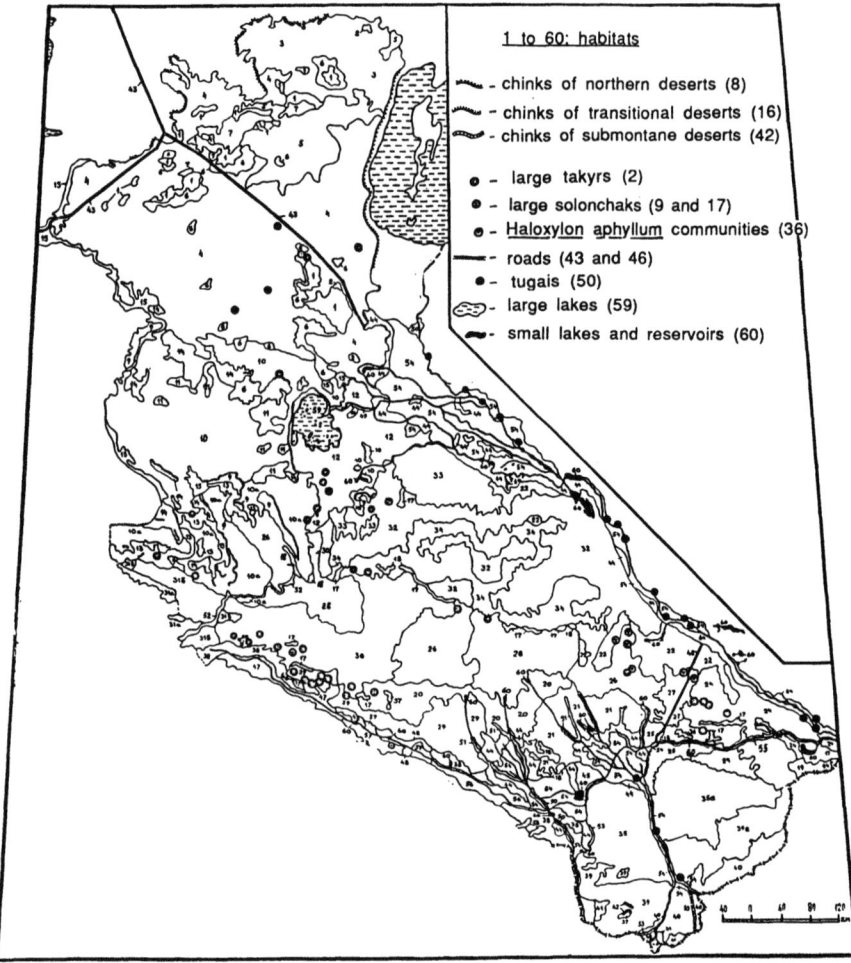

Fig. 3. Bird habitats in the Transcaspian Region (see Table 1 for the typological classification of habitats).

The order of the rows in the legend to Fig. 3 corresponds to the increasing ecological diversity of habitats. For example, the row of habitats for southern desert automorphic habitats starts with solonchaks (17) and ends with more productive black saksaul deserts (36). For desert foothill habitats, the foothills with *Poa bulbosa-Carex pachystylis* vegetation (40) are followed by more diverse and productive foothills with *Poa bulbosa-Carex pachystylis-Pistacia vera* (41). The main text of the legend (omitted here) is a description of the geomorphology and vegetation of every habitat type; these descriptions allow assessment of concrete ecological factors important for the existence of birds.

The completed habitat map was further used for extrapolation of survey data to build maps of seasonal changes in bird population, i.e., species distribution, density, and ecological structure in different seasons.

Statistical Analysis of Survey Data

Statistical analysis of field surveys, using a domestic FORTRAN package, included calculation of total density as well as share of different bird species by habitat and season. We utilized outlines (portions of field routes) with data on surveyed bird species, outlines with characteristics of a given habitat, composition, and density of birds within outlines with certain given characteristics as well as graphs of distribution of selected characters. Thus, we standardized the process of generalization of a zoogeographical survey, i.e., selection and combining of outlines by similar characteristics of a landscape and bird population. These possibilities are important especially for cartographic analysis of birds where the indices of density should not be calculated independently in every outlined area but should be combined according to the factors important for bird ecology. For estimation of density (number of birds of a certain species per square km) we calculated index $\Pi = (A_1 \times 20 + A_2 \times 5 + A_3 \times 1.7)/P$, where A_1, A_2, A_3 are the numbers of birds registered in the first, second, and third lanes of the survey, and P is the length of the survey.

To classify the obtained data on density of species, we built a scale based on the evaluation of number of species and their density for each habitat and season, generally similar to a decimal point scale; we used this scale to build the maps of seasonal changes in bird population.

Results and Discussion

General Considerations

A complex classification of bird population should rely not only on numeric estimation but also include criteria which reveal the similarity of ecological groupings of birds. The typological territorial groups based on dominance of certain species (E. Rustamov 1988) are not units of classification in the full sense. In this work, we attempted to combine numerical and ecological criteria. Birds, as a sensitive component of ecosystems, react to changes in environment, and these reactions are reflected in the territorial distribution of birds.

We selected three ecological parameters for our analysis: feeding preference for certain kind of food, area of feeding, and position of nests. We thus generalized ecological parameters and use them in combination with quantitative indices such as density, saturation of ecological niches, and taxonomic structure. In general, we can state that trophic preferences of a species reflect the realization of its energetic requirements; an above-species taxonomic rank (genus, family) reflects the evolutionary, morpho-physiological basis of a species' ecology; and position of nests and feeding of a certain species reflect the success of realization of all adaptations.

The basic characteristic used to compare units of bird population is still density, calculated in number of birds/km^2 and reflected in relative indices. The density of a certain territorial group of birds, in a sense, also reflects how successfully it utilizes the ecological potential of a habitat. Generally (i.e., without estimation of limiting factors), where density is high, conditions will be more favorable. On the other hand, a comparison of density of ecologically similar species (those similar in food preference or those who overlap territorially in feeding or nesting grounds) helps to reveal a limiting factor in a certain habitat. Therefore, a prediction of ecological distribution becomes possible (especially within a single taxon) which usually is consistent with a principle of competitive exclusion (Odum 1971; Ricklefs 1976). Finally, the total density of ecologically similar species serves as an estimate of the capacity of an ecological niche: the higher the density of a certain ecological group of birds is, the more capacity this given niche has, the more favorable conditions it offers, and the better representation it has in a given territory.

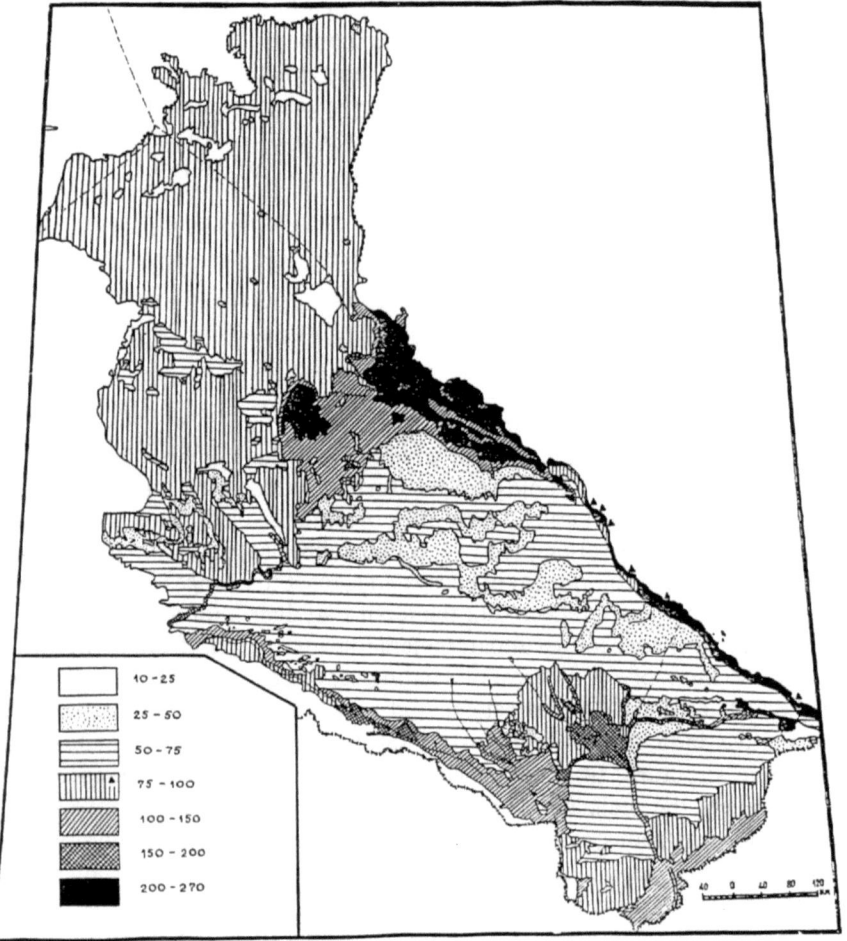

Fig. 4. Index of diversity of the bird population of the Transcaspian Region (in spring).

Cartographic Analysis of the Structure of Bird Population

We distinguished *territorial groups of bird population* (TGBP) which were further combined in 14 *types of bird population*. We calculated the *indices of diversity* (number of species in a habitat multiplied into numer of families in a habitat) and *density*, and built three maps: the first reflects the total index of diversity for TGBP (Fig. 4); the second reflects the relative density of birds (Fig. 5); and the third map reflects types of bird population (Fig. 6).

The first map (Fig. 4) reveals the high diversity in foothill lowlands (index of diversity 100–150), in the valleys of the Tedzhen (100–150), Murghab (150–200) and Amudarya (200–270) Rivers, and in the eastern part of the transitional zone

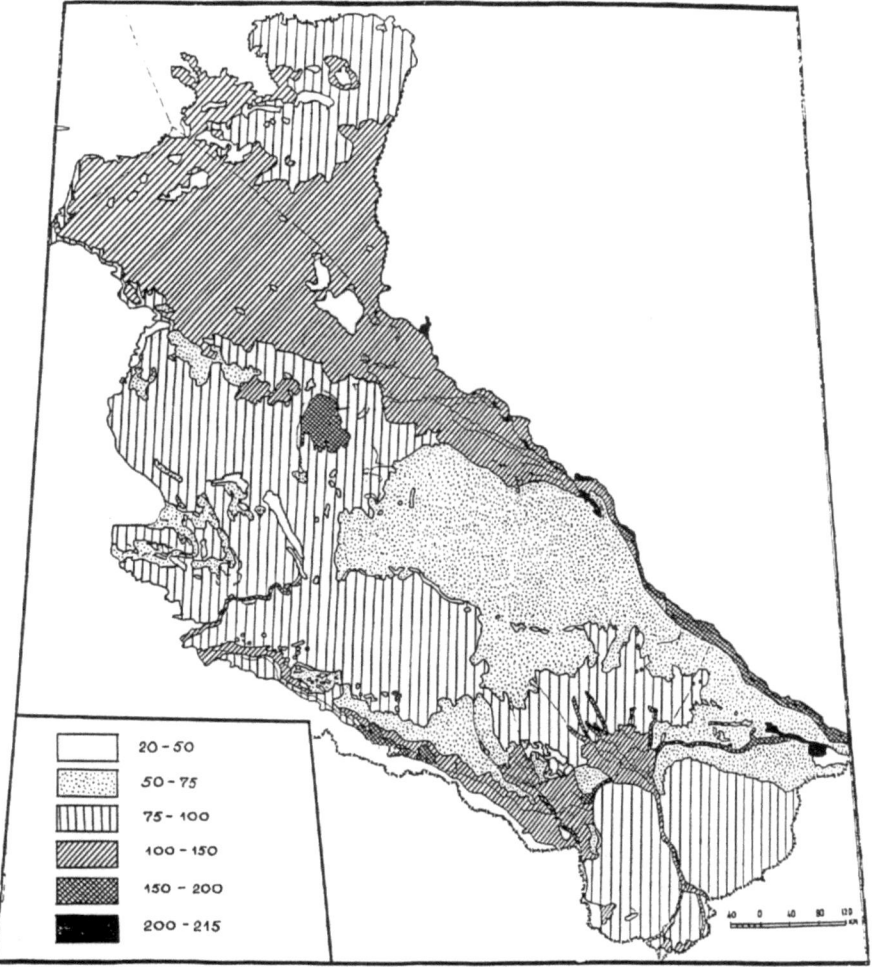

Fig. 5. Bird density in the Transcaspian Region (in spring).

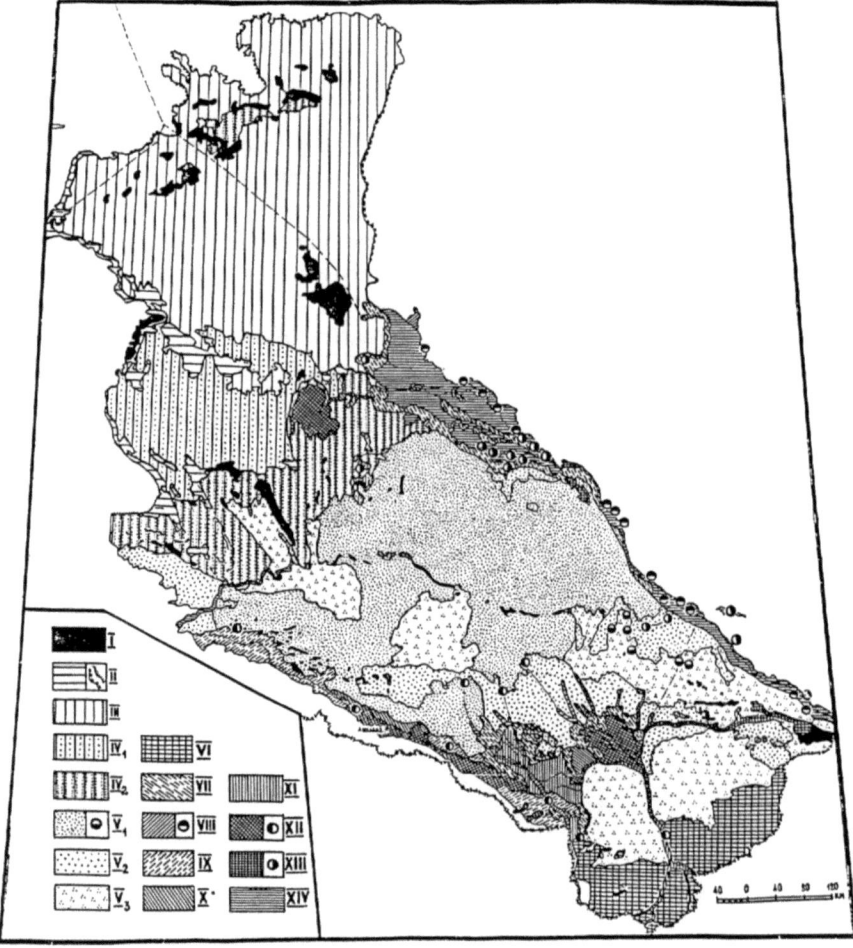

Fig. 6. Types of the bird population in the Transcaspian Region (in spring).

between the northern and southern desers (100–150), including the Sarykamysh Depression (200–270). The vast desert lowlands have average diversity of bird population; their northern part (including all Ustyurt and the western portion of the transitional zone) have higher diversity (75–100) than the southern sand deserts (25–75).

The second map (Fig. 5) shows the distribution of total density (estimated by a six-point scale) in TGBP. A very high density (200–215) was found exclusively in the very small areas of tugais in the Amudarya Valley; the density was also high in the oases of the Amudarya and Sarykamysh, and long the irrigation canals and in oases along the northern foothills of Kopetdagh (150–200). Higher-than-average index of density (100–150) is also characteristic for clay foothill lowlands and the Khorezm Oasis; the same density level is also

Table 2. Similarity between the types of bird population

	I	II	III	IV	IV	V	V	V	VI	VII	IX	X	XI	XII	XIII
II	55.6														
III	58.6	64.7													
IV	60.1	69.4	82.1												
IV	56.4	55.9	76.7	93.3											
V	60.0	60.0	70.4	71.0	69.7										
V	60.1	69.5	70.0	75.0	75.7	80.0									
V	58.2	64.1	63.6	68.8	74.3	75.0	81.2								
VI	52.8	74.3	56.7	65.8	66.7	66.7	70.3	50.0							
VII	52.1	54.8	61.8	65.7	85.7	58.3	66.7	61.5	53.4						
VIII	43.8	43.7	47.5	45.4	50.0	52.5	52.4	52.3	57.8	61.9					
IX	48.8	66.7	57.1	66.7	72.2	58.3	66.7	61.5	76.3	64.1	54.5				
X	52.1	60.0	61.8	62.2	63.1	54.0	62.2	61.5	59.5	73.0	54.5	77.8			
XI	50.8	60.0	41.6	55.8	60.5	48.8	59.5	59.1	60.9	69.0	63.0	77.5	69.8		
XII	46.6	46.3	43.7	48.0	49.0	42.0	51.0	51.0	52.8	69.6	64.0	62.5	69.6	46.2	
XIII	48.8	41.9	48.8	46.7	51.2	43.2	46.7	46.8	51.1	52.5	55.1	55.6	66.7	67.4	82.6
XIV	49.0	50.9	45.8	53.1	54.0	50.0	56.2	56.0	54.7	64.6	69.4	58.0	61.2	79.1	78.8	73.5

expressed in the Central Ustyurt which has, in general, an average diversity index. At the same time, the in all the remaining clay deserts, in all western parts of sand deserts, and within the entire transitional zone (including the part with high diversity index), the density was not higher than 75–100.

This comparison of diversity and density shows that the transformational rows of TGBP are clearly expressed in areas of human influence. The most significant factor for birds appears to be agricultural transformations. TGBP are dramatically modified as a result of irrigation and changes in vegetation (appearance of gardens, crops, and reeds) from the initial clay and clay-sand deserts. The new anthropogenic habitats represent a sharp contrast with the traditional environment of birds. Therefore, they play a significant role in TGBP transformation although these habitats occupy relatively small areas. The TGBP of anthropogenic habitats have higher diversity and density of birds; we may assume that the capacity of ecological niches and the level of realization of these niches by birds in anthropogenic habitats are higher than in natural ones.

The third map (Fig. 6) represents the spatial distribution of all 14 types of bird population by their ecological similarity expressed in Jaccard's index (Table 2), which increases from the northern clay deserts (55 to 82%) to sand and clay transitional deserts (69–93%). Interestingly, the ecological similarity of TGBP in transitional and southern deserts, though high enough (70%), still is lower than that of southern sand deserts (80–81) and northern and transitional deserts (82–93%). As could be expected, the highest similarity (93%) is found between the TGBPs in northern and southern parts of the transitional zone. Very different are southern sand desert and foothill lowland (similarity not higher than 50%). Very specific are TGBP complexes in oases and valleys, including not only valleys of large rivers but also the Karakum Canal, Kelif Uzboi, West Uzboi, and Sarykamysh.

This combined analysis of maps allowed us to built the separate maps of seasonal bird population for the entire region as well as for a smaller model territory. The ecological similarity of TGBP appears to be the most informative index. For example, the similarity of TGBP in the central oases and their periphery is 83%; at the same time, it is quite low (46%) between oasis TGBPs and TGBP of valleys and lakeshores.

Analysis of Spring Bird Population

As one of the results of a comparative cartographic analysis of seasonal patterns of bird population in the enormous territory of the Transcaspian Region, we created the ecological map of the bird population in a spring (nesting) period (the actual map is omitted in this paper due to its complexity). All characteristics of bird population – its diversity, density, types of population with dominant species, and spatial patterns of changes – are especially expressed in the spring season.

In spring, average density decreases from the northern (100–150 birds/km^2) to the southern (75–100 birds/km^2) part of the region; the density clearly increases from lowland deserts (50–75 and 75–100 birds/km^2) to the valleys and lakeshores (150–200 birds/km^2). Indices of diveristy of bird population are higher in the northern and foothill deserts (75–100 and 100–150 birds/km^2) than in southern sand deserts (25–50 and 50–75 birds/km^2). Very high diversity (150–200 and 200–270) and total density (150–200 and 200–215 birds/km^2) are demonstrated by TGBP of lakeshores and river valleys, and especially within oases and tugais. The specificity of spring bird population is well expressed in the transitional zone between northern and southern desert TGBP. Diversity in this zone is closer to northern desert values, but the density (75–100 birds/km^2) is nearer those for southern deserts. In different oases, depending on their size, diversity varies from 50–75 and 100–150 to 150–200, and density can vary from 50–75 to 100–150 birds/km^2). The fluctuation of these indices reflects the influence that an oasis area exerts on the surrounding territories.

Thus, the spring bird population in northern deserts differs sharply from that in southern ones. At the same time, the differences between spring population of southern deserts and of foothill clay deserts are not as sharp and on the map appear diffused.

All these types of bird population are defined by the common history and geography of this territory, which belongs to the desert zone of the Palearctic region. Our results demonstrate the profound unity of all TGBPs of the studied area and the rather local level of their differences. There are many ecological niches which are common for all or most of the TGBPs. The clear dominance of the ecologically flexible and widespread bird species, however, is an evidence of a high level of the anthropogenic transformation of the arid ecosystems of the Transcaspian Region.

Dynamics of Bird Population on a Model Territory

Our "model" territory included the area between the Tedzhen and Murghab Rivers, and parts of the Karakum Desert adjacent to the north and east. This area is highly affected by human influence; it is also an important route of migrations of bird species which may be very different in their ecology. Some species of birds also spend winter here.

The map of seasonal patterns of bird population in the model territory (Fig. 7) was built on the basis of indices of density but without the special evaluation of ecological structure (E. Rustamov 1988). In spring, the highly diverse (50–60 species) bird population is formed in the northern part of this territory, but the density is low (30–70 birds/km^2). The central part of the model territory includes primarily the clay, loess, and sand deserts around oases; here, the diversity is rather high (51–76 species) but density is low (45–95 birds/km^2) or average (100–180 birds/km^2). Anthropogenic habitats (oases), due to the diverse ecological conditions there, have high bird diversity (more than 100 species), average density (100–200 birds/km^2), and considerable variety of TGBPs.

In summer, in the northern part of the model territory the diversity decreases to 24 species, and density, to 25 birds/km^2. In the central part of this area, TGBPs of oases become sharply differentiated, with diversity from 42–46 to 52 species, and density as high as 216–227, and sometimes to 670 birds/km^2. On the contrary, in the rest of this territory diversity drops to 31 species, and density can be from 60 to 151 birds/km^2. The summer population is generally impoverished, and the density increases drastically due to synanthropic and – partially – to desert birds (both resident and nesting migratory species).

In fall, the diversity is comparable to the spring values (50–70 species), and density is intermediate between spring and summer (100–200 birds/km^2). In winter, the northern part of our model territory has low (28–30 birds/km^2) and sometimes even very low (10 birds/km^2) density, and diversity of 25 species. The central part of the area has average density (100–125 birds/km^2) and rather high diversity (58–63 species). Finally, the southern part has low winter density (77–98 birds/km^2) and diversity of 16 to 25 species. The bird diversity in oases is lower than in summer though still high (58–63 species), with density of 102–125 birds/km^2.

Conservation of the Bird Population

Ecological studies constitute a theoretical basis of natural conservation (Shvarts 1960, 1969, 1975); the very survival of humankind depends on the survival of biodiversity. Birds are one of the important components of ecosystems and a common object of conservation (Odum 1983; Temple 1989; A. Rustamov 1991a,b; Flint 1991). According to the different levels of diversity distinguished by ecologists (alpha, beta, and gamma diversity; Whittaker 1960, 1965) we can

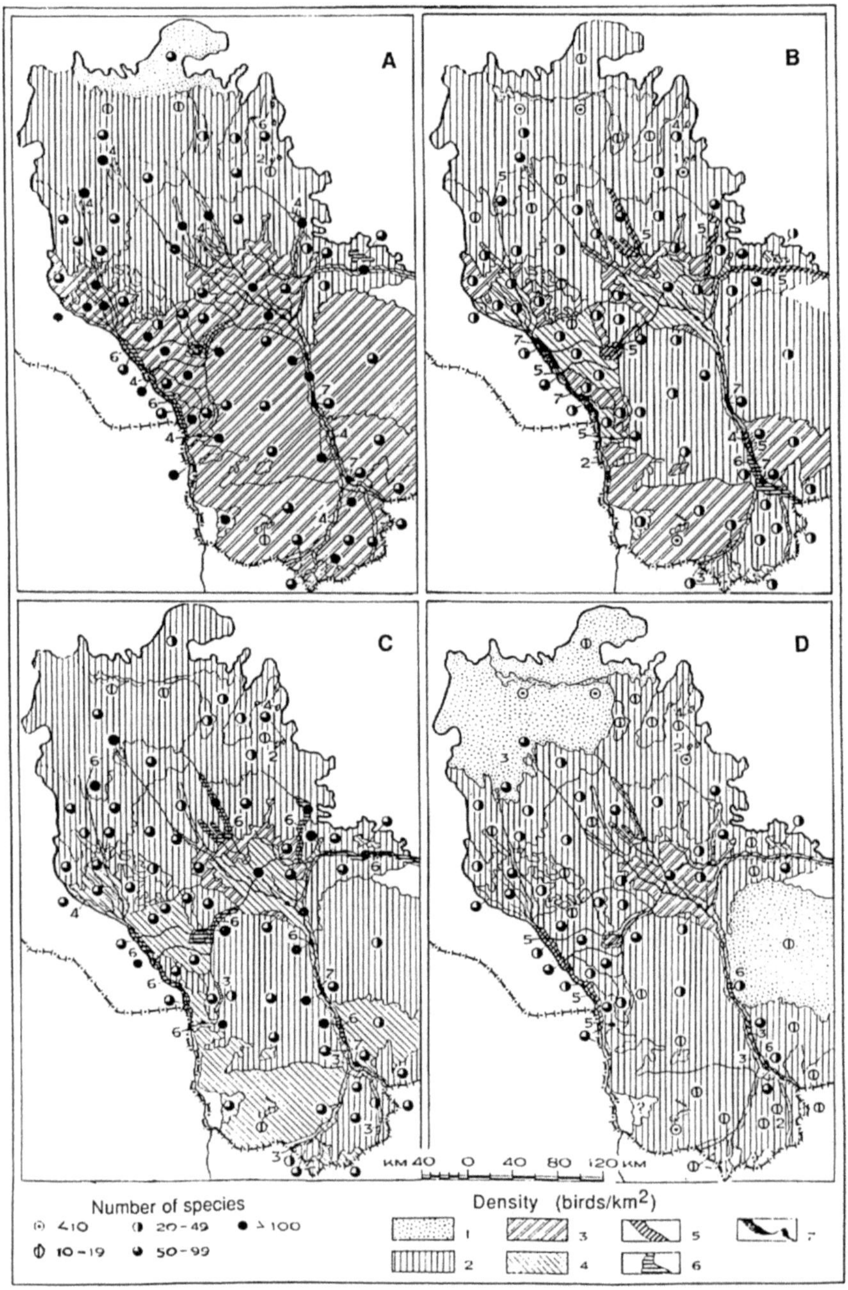

Fig. 7. Seasonal aspects of the bird population; (A) – spring, (B) – summer, (C) – fall, and (D) – winter. Density: 1 – less than 10 birds/km^2, 2 – from 10 to 99 birds/km^2, 3 – from 100 to 199 birds/km^2, 4 – from 200 to 399 birds/km^2, 5 – from 400 to 499 birds/km^2, 6 – from 500 to 999 birds/km^2, 7 – more than 1,000 birds/km^2.

estimate diversity within a certain habitat, among the habitats in a geographical area, or within the entire biome.

Our study was relevant to all three levels of diversity. At the alpha level, we found the specific features of TGBPs in every habitat. At the beta level, we compared similarity of TGBPs among northern, central, and southern parts of the model territory between the Murghab and Tedzhen Rivers. Finally, at the gamma level, we studied the spatial and temporal patterns of diversity within the vast territory of Palearctic deserts in the Transcaspian Region, from Badghyz to Ustyurt.

Our cartographic analysis of seasonal changes in bird population, for the deserts of the Transcaspian Region as a whole and in their particular habitats, is relevant to conservation practice. The transformation patterns of TGBPs (from those minimally influenced by human activity to those which maintain an active balance with anthropogenic habitats) that we described and mapped can be used for the planning of conservation strategies. Every season provides additional information about diversity, density, and structure of bird populations; all can be reflected on maps and used for long-term monitoring of bird populations.

Bird conservation planning in arid landscapes should be directed toward those territories and habitats which are characterized by certain unique features important for bird species. These include both dominant and non-dominant habitats. Comparison of the requirements for the ecological monitoring network (Fig. 8) to the existing system of natural reserves in the Transcaspian Region shows that this system is not sufficient for the conservation of bird diversity and needs to be extended. We attempted to show (Fig. 8) the position of key areas which combine maximal habitat diversity and can provide the maximal realization of ecological potential for birds. This ecological key area network includes, first of all, now existing natural reserves (Fig. 8): (I) in northern deserts, Ustyurt Reserve*[1] (established 1984, area 223.3 km^2); (II) in transitional deserts, Kaplankyr Reserve (established 1979, area 570 km^2); (III) in foothills, Badghyz Reserve (established 1941, area 87.6 km^2); (IV) in southern deserts, Repetek Reserve (established 1928, area 34.6 km^2); and (V) in tugais, Amudarya Reserve (established 1982, area 50.5 km^2), Kizylkum Reserve[2] (established 1971, area 10.1 km^2), and Badaitugai Reserve[2] (established 1971, area 6.5 km^2). Bird surveys should be performed in all seasons in order to create a data bank on bird phenology and density for the long-term monitoring at alpha and beta diversity levels.

We also propose to establish additional nine key points for the monitoring of bird population (Fig. 8): in northern deserts, Samkum (A); in transitional deserts, Akchakaya (B); in foothills, Gyaurs (H), Ashgabat (K) and Meana-Chaacha (D); in southern deserts, Gatygyzyl (D); and in tugais, Sarykamysh (G), Uzboi (F), and Sultandagh (E). The tugais, lakeshores, black saksaul forests, and other key habitats where birds spend the less favorable seasons in or around deserts, should assume an important place in this system of ecological monitoring. The mosaic of habitats in oases also supports the diversity of birds;

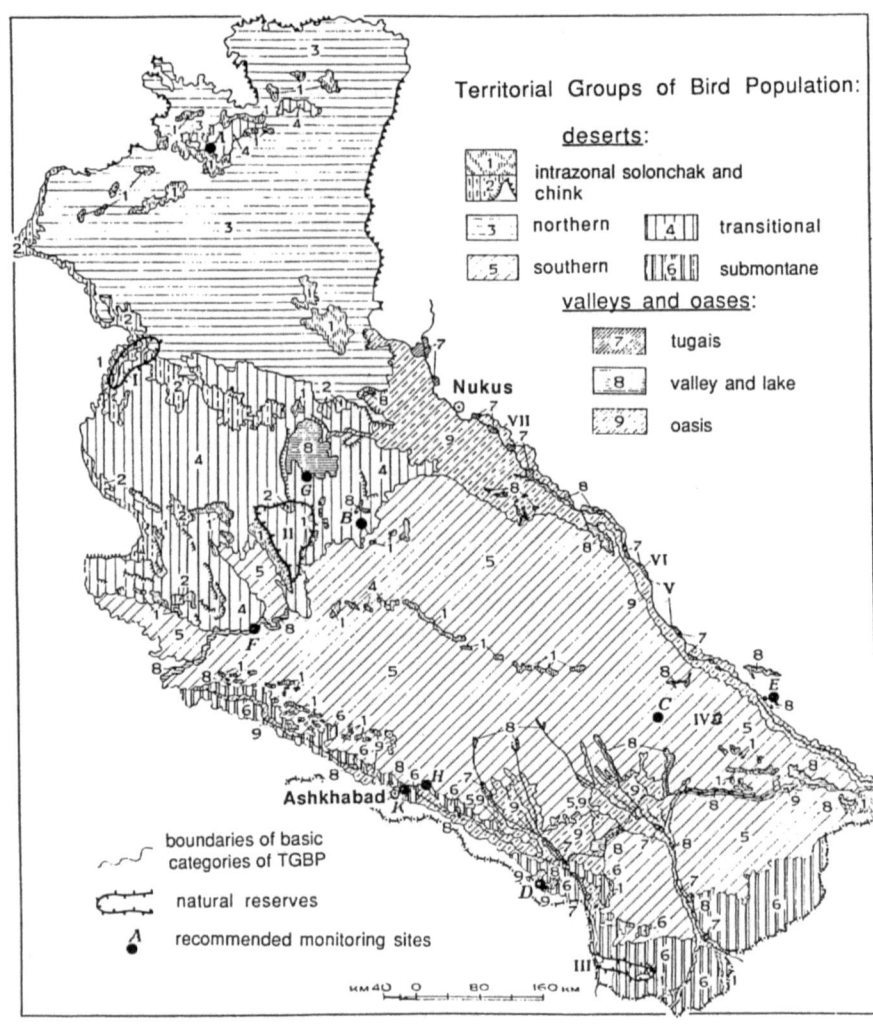

Fig. 8. Protected territories and the key monitoring sites.

therefore some points of monitoring also should be selected in these anthropogenic landscapes.

Notes

1. Outside of Turkmenistan (the reserve lies in adjacent Kazakhstan) – Ed.
2. Outside of Turkmenistan (the reserve lies in adjacent Uzbekistan) – Ed.

17. Kidney Structure and its Role in Osmoregulation in Desert Birds

MARAL B. AMANOVA

Abstract

A number of ecological, physiological, and morphological adaptations allow birds to survive in the extreme climate of the Karakum Desert. Kidney structure of nine desert and oasis bird species (*Streptopelia turtur, S. senegalensis, Galerida cristata, Scotocerca inquieta, Podoces panderi, Passer ammodendri, P. simplex, P. montanus* and *P. domesticus*) was studied using standard microscopic and histochemical techniques. Relative kidney size and, sometimes but not always, size of the medullary zone increase in desert birds but not oasis species. The number of renal medullary cones per unit of kidney weight is specific for a given taxonomic group of birds and is directly related to the rate of water retention. There was no difference between desert and oasis birds in the length of Henle's loops. Adaptation to arid climate (production of concentrated urine) is achieved through an increase either in the number of medullary cones or in the number of Henle's loops per cone. Details of microscopic structure and histochemical analysis of various components of a nephron in desert birds are presented and discussed.

Introduction

Desert birds have evolved very specific adaptations for survival under unfavorable climatic conditions (e.g., high temperature, significant sunlight, and shortage of water). A number of studies have been devoted to ecological, physiological, and morphological adaptations of desert birds (Rustamov 1964; MacMillen and Trost 1966; Willoughby 1966; Hainworth 1969; Johnson 1968, 1970, 1972, 1974; McNabb 1969; Brown and Dantzler 1972; Havold 1983; Davies 1984; Dawson 1984; Thomas 1984a,b,c; Kirkley 1990). Many data on bird adaptations are controversial. The goal of our work was to study details of ecologico-morphological and physiological adaptations of the bird kidney to the arid conditions of the Karakum Desert.

Materials and Methods

Kidneys from the following Karakum bird species were studied: *Streptopelia turtur, S. senegalensis, Galerida cristata, Scotocerca inquieta, Podoces panderi, Passer ammodendri, P. simplex, P. montanus* and *P. domesticus*. Several techniques were used. The number of renal cones and nephrons was detemined by microdissection. Heart blood samples were obtained utilizing a siliconized glass pipet. Urea content in blood plasma and fresh tissue samples was determined by diacetylmonoxime reaction. Concentration of Na^+, Ca^{++}, and Mg^{++} in tissue and plasma was determined using a Leiss-III flame photometer and Hitachi atomic absorbtiometer.

Kidney tissue was fixed with 10% neutral formaldehyde, glutaraldehyde (phosphate buffer, pH 7.2), a mixture of formaldehyde, ethyl alcohol and acetic acid, and with Carnoy's and Shabadash solutions. Sections (5 to 7 mm thick) were stained by hematoxylin and eosin, azure-eosin, methyl green pyronin, van Gieson picrofuchsin, and Mallory stains. For histochemical analysis, we used Feulgen-Schiff and PAS reactions, Hayle staining (colloid iron hydroxide and 0.1% alcian blue with 3% acetic acid and 0.2–0.4 mM $MgCl_2$), toluidine blue (pH 2 to 6), and the reaction of "hidden metachromasy" (Vinogradov and Potapova 1964). We also performed Altman reaction and Heidenhain's iron hematoxylin staining. Kidney sections obtained by freezing microtome were analyzed to determine concentration of acid phosphatase and succinate dehydrogenase. Alkaline phosphatase content was determined on paraffin-embedded sections.

General Kidney Structure

Studies of the bird kidney have been performed by a number of researchers (Hudson et al. 1971; Shoemaker 1972; Liswiller and Farner 1972; Skadhauge 1973, 1975; Amanova 1978; Thomas 1984a,b). The kidney is the organ most sensitive to changes of environmental factors affecting animal metabolism. Variation in the kidney weight to body weight ratio is believed to be the best estimate of variation in ecological conditions of compared species (Shvarts 1954, Shvarts et al. 1958). Smaller birds with higher metabolic activity have larger kidneys (King and Farner 1961); desert mammals are known to have larger kidneys than non-desert ones (Sperber 1944). Kidney weight in our study species varied from 4.9% to 13.9% of the body weight; strictly desert birds had a larger ratio compared to oasis species (Amanova 1968), reflecting intensification of osmoregulation in the desert species.

The kidney is composed of medullary and cortical parts forming lobes (Sperber 1960). The cortex possesses a group of branched central veins (Spanner 1925); the cortical mass is concentrically positioned around each central vein, forming a cortical lobe. Cortical lobes vary in size. Smaller lobes surround distal central veins, whereas the largest ones (two to three times the size of small medullary lobes) are located proximally along the vein network. Most non-passerine birds have cortical lobes larger than one millimeter in diameter; passerines always have smaller cortical lobes (Johnson 1968).

Cortical lobes contain proximal and distal portions of long-looped and short-looped nephrons. Proximal portions of collecting tubules, located in the center of a cortical lobe, further connect to larger tubules which diverge radially in the periphery. These larger tubules connect to even larger ducts which surround a cortical lobe as circular tubules. Several groups of these tubules join together at the peripheries of neighboring cortical lobes and connect to a distal (wide) end of a medullary lobe. Tubules of cortical lobes which lie next to the proximal (narrow) end of a medullary lobe often enter it without joining the peripheral collecting tubule.

Medullary lobes contain loops of Henle and collecting tubules; each medullary lobe receives loops of Henle and tubules from several cortical lobes. Each medullary lobe is enclosed in a cone-shaped layer of connective tissue (Sperber 1960); the top of this cone extends into a connective tissue sheath of the ureter. The conical shape of a medullary lobe is due to the different length of Henle's loops (their number decreases from the wide distal end toward the narrow proximal end of the lobe) and to the branching fusion of collecting tubules. Therefore, the proximal end of a medullary lobe has only one large collecting tubule, which can be called a branch of the ureter. Several such branches from neighboring medullary lobes join together to form a larger branch of the ureter; several levels of such joining take place before tubules reach the ureter proper. This continuous branching of collecting tubules is characteristic for birds, in contrast to mammals (Johnson and Mugaas 1970).

Only the kidney of birds and mammals has a concentrating function; this most important adaptation of the excretory system allows for production of urine which is more concentrated than blood plasma.

The degree to which the renal medulla is developed is directly related to the ability of birds and mammals to live in conditions of water shortage (Sperber 1944; Pernia 1989; Meshchersky 1990a,b; Serebryakov et al. 1990). In mammals, the thickness of the medulla is proportionate to the concentration ability of the kidney (Johnson and Mugaas 1970; Johnson 1974; Johnson and Skadhauge 1975). In birds, the length of the medullary cones is considered to be a measure analogous to the thickness of the medulla in mammals. However, this is not always true (Johnson and Skadhauge 1975): for example, desert birds have relatively thinner renal medulla than do desert rodents, a feature which is probably balaced by the existence of salt glands in birds which osmoregulate by excreting excess salts. Other osmoregulatory parameters of desert species should be searched for.

Results and Discussion

Renal Medullary Cones. Our data did not show constant correlation between the thickness of medulla and the degree of arid adaptation in a given bird species. In three Karakum sparrow species, the average lengths of medullary cones were the following: *Passer ammodendri*, 3.11 mm; *P. domesticus*, 4.56 mm; *P. simplex*, 4.82 mm. In contrast, average cone length in the oasis species, *Passer montanus*, was 5.55 mm (Table 1). Similar results were obtained for doves (Table 1): desert species *Streptopelia turtur* had shorter medullary cones (2.2 mm) than oasis dove *S. senegalensis* (3.6 mm). The same measure in a specialized desert bird, a saksaul jay (*Podoces panderi*), was 3.9 mm (Table 1).

The bird kidney contains two different types of nephrons, those of the so-called mammalian type (with Henle's loops) and reptilian type (without Henle's loops). Brown and Dantzler (1972) studied function of these two nephron types in a desert quail (*Lephothryx gambelii*) and found that under osmotic stress, most reptilian-type nephrons stop functioning whereas mammalian nephrons remain active. This mechanism may be common to all birds; it facilitates water retention and also maintains a concentrated gradient in the medullary cones.

Because of the concentration of Henle's loops in medullary cones, the number of cones per kidney reflects the percentage of long-loop nephrons which, in turn, can increase nephron efficiency (Paulson 1965; Johnson and Mugaas 1970). A comparative study of two Karakum doves, *Streptopelia turtur* and *S. senegalensis*, which live in habitats with different water regimes, confirms Paulson's idea (1965). The desert-inhabiting *Streptopelia turtur* had 28-32 renal medullary cones, with an average of about 370 loops of Henle in each cone (Table 1), whereas oasis dove *S. senegalensis* had 25-27 cones, with an average of only 290 loops of Henle per cone.

Sparrows (*Passer* spp.) have a higher number of cones than do doves. Oasis species *P. montanus* has the lowest number of renal medullary cones (39-41), each with 93-95 Henle's loops. Desert sparrows *Passer ammodendri* and *P. simplex*, which usually have no access to water sources, have 46-47 and 49-52 cones correspondingly (with 92-95 and 94-96 Henle's loops each) (Table 1).

The number of medullary cones per gram of kidney tissue reflects the degree of development of the medulla and can be an indicator of the concentrating ability of the kidney. The highest number of medullary cones was discovered in sparrows *P. domesticus* (212 cones/g) and *Passer ammodendri* (202 cones/g), whereas the oasis sparow *P. montanus* had 155 cones/g (Table 1). The higher number in the desert species may be an indication of an increase of medulla and concentrating ability of a kidney as an evolutionary adaptation to desert conditions. Doves had much lower number of medullary cones per gram of kidney tissue than sparrows (Table 1).

The shape of medullary cones in birds varies within the kidney. Commonly, straight cones are found in the cranial portion of the kidney, whereas intermediate and caudal portions contain mostly convoluted cones. Our studies show that desert sparrows (*Passer ammodendri* , *P. simplex*) had highly

Table 1. Structure of renal medullary cones (MC) in desert birds (number of measurements is given in parentheses)

Species	Kidney weight (g)	MC number		MC length (mm)	MC/renal lobe length ratio	Number of Henle's loops in a MC
		per kidney	per gram of kidney weight			
Podoces panderi	0.498 ± 0.02 (3)	54–56 (6)	112 ± 4.3 (6)	3.9 ± 0.1 (60)	71.0 ± 0.1 (60)	190–310 (7)
Streptopelia turtur	0.401 ± 0.01 (3)	28–32 (14)	75 ± 1.5 (14)	2.2 ± 0.1 (60)	58.0 ± 0.8 (60)	360–575 (7)
S. senegalensis	0.350 ± 0.21 (3)	25–27 (14)	71 ± 1.52 (14)	3.0 ± 0.15 (60)	54.0 ± 0.2 (60)	290–360 (7)
Passer ammodendri	0.228 ± 0.21 (5)	46–47 (14)	202 ± 1.52 (14)	3.11 ± 0.13 (60)	59.0 ± 0.32 (60)	92–95 (7)
P. domesticus	0.207 ± 0.25 (4)	44–46 (14)	212 ± 1.33 (10)	4.56 ± 0.15 (60)	68.2 ± 0.34 (60)	90–93 (7)
P. simplex	0.258 ± 0.23 (6)	39–41 (14)	155 ± 1.58 (14)	5.55 ± 0.13 (60)	73.0 ± 0.32 (60)	93–95 (7)
P. montanus	0.263 ± 0.27 (2)	49–52 (6)	190 ± 1.38 (6)	4.82 ± 0.13 (18)	67.6 ± 0.32 (18)	94–96 (3)

convoluted renal medullary cones in contrast to oasis sparrow species. This feature is also known in desert rodents, whose kidneys have high concentrating ability. Desert birds also have longer medullary cones and thicker interlobe connections than do non-desert species.

The number of mammalian-type nephrons was determined from the number of Henle's loops in desert bird kidneys. The highest number of mammalian-type nephrons was found in the desert dove *Streptopelia turtur* (average 14,160), which has limited access to water sources. Kidneys of oasis dove *S. senegalensis* had only 8,450 mammalian-type nephrons. Sparrow kidney contains fewer mammalian-type nephrons than does dove kidney. The highest number was found in the desert saksaul sparrow, *Passer ammodendri* (4,371); in *P. domesticus* this number was 4,095, and in *P. montanus*, 3,760. Therefore, the number of mammalian-type nephrons can also be an indicator of adaptation to arid climate and increased concentrating ability of the kidney.

Osmoregulation. Birds, like mammals, produce urine that is more concentrated than blood plasma; commonly, it is about two times more concentrated (Dantzler 1969; Goncharevskaya et al. 1978). The renal medulla is less developed in birds than in mammals, and thus the concentrating ability in birds cannot be as significant. However, desert birds have additional osmoregulatory mechanisms: not only do they possess salt glands but, in addition, a modified cloaca and rectum with increased absorptive surface for increased water reabsorption efficiency (our data).

The bird kidney also has concentrating mechanisnms analogous to those of mammals (Johnson 1967). As in mammals, the functional load of different nephron types in birds is regulated under extreme water regimes. The rate of filtration through the renal glomerulus was found to be higher in desert bird species compared to domestic ones (Brown and Dantzler 1972). Under a hydration regime, up to 71% of the proximal tubules in reptilian-type nephrons remained open, whereas under a salt-loaded regime 84% of the same tubules were closed (Brown and Dantzler 1972). This indicates that under a salt-loaded regime most of the reptilian-type nephrons do not participate in glomerular filtration, which is an adaptation to water shortage. Our microscopic data also show that, in desert species of sparrows (*Passer ammodendri* and *P. simplex*) and saksaul jay (*Podoces panderi*), glomeruli of reptilian-type nephrons are excluded from filtration. In contrast, in oasis birds (*Passer montanus* and *Streptopelia senegalensis*), most reptilian-type nephrons have functional glomeruli and open proximal tubules. Therefore, under water shortage the blood flow in birds is redistributed by exclusion of most of the reptilian-type nephrons from the process of filtration; most of the blood is filtered through the deep glomeruli which give rise to mammalian-type nephrons, and most filtered blood plasma is concentrated in mammalian-type nephrons.

The discovered variation in number and length of renal medullary cones and number of Henle's loops in studied desert birds allowed us to study the comparative role of ions and urea in the osmotic gradient between cortex and

Table 2. Concentration of cations and urea in the blood plasma, cortex, and medulla of kidney of *Podoces panderi* and *Streptopelia turtur*[a,b,c]

	Cations			Total Na^+, K^+, Mg^{++}	Urea (mM/kg of wet weight)
	Na^+	K^+	Mg^{++}		
Podoces panderi					
Blood plasma	143.4 ± 2.5 (14)	7.7 ± 0.5 (13)	0.73 ± 0.03 (14)	152.4 ± 0.03 (14)	6.3 ± 0.4 (12)
Cortex	423.7 ± 2.5 (14)	360.8 ± 4.6 (11)	350.0 ± 4.6 (11)	840.0 ± 54.9 (9)	15.0 ± 0.9 (16)
Medulla	567.0 ± 56.3 (10)	475.8 ± 25.6 (13)	76.2 ± 5.7 (11)	107.3 ± 88.7 (9)	21.2 ± 2.2 (16)
Cortical/medullary gradient	1.3 ± 0.07 (10)	1.3 ± 0.04 (13)	1.7 ± 0.22 (11)	1.3 ± 0.04 (9)	1.4 ± 0.1 (14)
Streptopelia turtur					
Blood Plasma	142.7 ± 1.9 (8)	5.7 ± 0.24 (9)	0.71 ± 0.06 (9)	149.3 ± 1.9 (8)	5.8 ± 0.4 (8)
Cortex	390.0 ± 29.0 (6)	371.6 ± 27.7 (6)	64.7 ± 8.5 (5)	827.0 ± 65.7 (5)	10.3 ± 1.8 (8)
Medulla	509.3 ± 64.8 (6)	520.8 ± 26.6 (6)	89.2 ± 10.0 (5)	117.6 ± 108.6 (5)	11.8 ± 1.9 (8)
Cortical/Medullary Gradient	1.4 ± 0.24 (6)	1.4 ± 0.1 (6)	2.1 ± 0.7 (5)	1.5 ± 0.22 (5)	1.2 ± 0.1 (8)

[a] Concentration of cations and urea in blood plasma is given in mM/l, in tissue, in mM/kg of dry weight
[b] Cortical/medullary gradient is a ratio of cation concentration between cortex and medulla
[c] Number of samples is given in parentheses

medulla. Many experiments have revealed a difference in the osmotic concentration of sodium, chlorine, and potassium between medulla and cortex, the medulla being hypertonic (Paulson and Bartholomew 1962; Paulson 1965; Dantzler 1966; Skadhauge and Schmiedt-Nielsen 1967; Skadhauge 1974). Our study was performed on two desert bird species. Saksaul jay (*Podoces panderi*) kidney possesses the longest medullary cones and the highest number of them among studied species. Dove species *Streptopelia turtur* is not a native all-year desert species but, because it nests here, it is exposed to the desert climate. In *S. turtur*, the medullary cones are comparatively short, but the number of Henle's loops is very high, which should be considered a functional compensation.

The concentration of ions and urea in the blood plasma was similar in *Podoces panderi* and *Streptopelia turtur* (Table 2). The concentration gradient of cations (sodium, potassium and magnesium) and urea was found to increase from cortical substance to medullary one (Table 2). Sodium and potassium increased 1.3–1.4 times; magnesium, 1.7–2.1 times; and urea, 1.2–1.4 times. Therefore, these ions and urea create an osmotic gradient of concentrated urea in the kidney of *P. panderi* and *S. turtur*, with hypertonic medulla. This can be achieved either by the increase in medullary cone length (i.e., length of Henle's loops), as in *Podoces panderi*, or by the increase in number of Henle's loops, as in *Streptopelia turtur*.

Nephron Structure. The details of nephron structure have been studied by a number of researchers (Sperber 1960; Kuroda 1963; Paulson 1965; Berger 1966; Johnson 1968, 1970, 1974; Johnson *et al.* 1972; Brown and Dantzler 1972; Goncharevskaya 1975, 1976). Reptilian-type (cortical) nephrons are similar to those of amphibians and reptiles, whereas mammalian-type (medullary) nephrons resemble superficial short-looped nephrons of mammals and also the nephrons of lamprey.

Reptilian-type (long-looped) nephrons are located in the cortical lobes and consist of proximal, distal, and very short intermediate thin tubules (Fig. 1). The

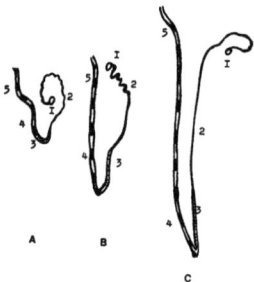

Fig. 1. Nephron structure in *Passer ammodendri*: (A) cortical nephron, (B and C) medullary nephrons; 1 – renal corpuscle, 2 – proximal portion, 3 – thin portion, 4 – distal portion, 5 – intermediate portion.

Table 3. Length of nephrons and their segments in desert bird kidney.

Species and zone	Nephron[a]		Segment (% of nephron length)				
	(mm)	Range	Proximal	Thin	Distal	Connective	Henle's loop[b]
Streptopelia turtur							
Cortical	2.59 ± 0.5 (5)	2.32–2.83	55.1 ± 0.5	4.4 ± 0.02	32.9 ± 0.1	5.7 ± 0.3	Absent
Medullary	6.25 ± 0.7 (10)	2.86–8.72	42.8 ± 0.4	11.5 ± 0.02	44.3 ± 0.4	3.3 ± 0.3	58.4 ± 0.5
Passer ammodendri							
Cortical	2.25 ± 0.2 (5)	1.43–2.33	53.8 ± 0.7	2.9 ± 0.03	33.3 ± 0.4	7.3 ± 0.2	Absent
Medullary	5.21 ± 0.7 (10)	2.46–7.22	36.7 ± 0.4	12.1 ± 0.02	44.3 ± 0.5	3.4 ± 0.3	52.3±0.4
Passer montanus							
Cortical	2.71 ± 0.6 (5)	1.43–2.28	53.2 ± 0.5	2.9 ± 0.03	33.2 ± 0.3	7.1 ± 0.1	Absent
Medullary	4.53 ± 0.5 (10)	2.42–7.00	36.7 ± 0.3	12.1 ± 0.02	44.3 ± 0.2	3.4 ± 0.1	52.3±0.3
Podoces panderi							
Cortical	2.66 ± 0.11 (5)	1.46–3.74	58.7 ± 0.6	3.8 ± 0.2	31.3 ± 0.7	6.3 ± 0.3	Absent
Medullary	7.00 ± 0.6 (10)	2.99–10.02	36.3 ± 2.9	13.8 ± 0.2	46.4 ± 3.0	3.1 ± 0.4	48.0 ± 4.6

[a] Number of isolated nephrons is given in parentheses
[b] Henle's loop includes descending part of the proximal segment, thin segment, and ascending part of the distal segment

length of reptilian-type nephrons in studied species ranged from 2.59 mm in *Streptopelia turtur* to 2.01 mm in *Passer montanus*; the proximal segment occupied about 50% of the nephron length. Details of reptilian-type nephron structure are listed in Table 3.

Glomeruli of mammalian-type nephrons are located adjacent to the medulla, with loops of Henle entering it to various extent. Nephron length varied from 6.25 mm (in *Streptopelia turtur*) to 4.53 mm (in *Passer montanus*) (Table 3). No difference was found between the length of reptilian-type and mammalian-type nephrons in desert and oasis bird species. In both nephron types, considerable length is occupied by a proximal tubule. In mammalian-type nephrons, the thin segment is more developed than in reptilian-type one and constitutes up to 13.8% of the nephron's length (in *Podoces panderi*). Henle's loop consists of the descending part of the proximal tubule, the thin segment, and the ascending part of the distal tubule. It occupies most of a mammalian-type nephron length, and up to 58% in doves (Table 3); it consists of the descending part of the proximal tubule, the thin segment, and the ascending part of the distal tubule. We detected no differences in the length of loop of Henle and its thin segment among desert birds with different water regimes, although the length of thin segment was variable. It is known that the thin segment in bird kidney is extremely short as compared to that of mammals. Details of mammalian-type nephron structure are listed in Table 3.

In general, adaptations to the arid climate have developed by an increase in the number of mammalian-type nephrons rather than by the lengthening of Henle's loop. The fraction of mammalian-type nephrons provides an initial step in concentration of provisory urine; its further concentration and formation is completed in the rectum and cloaca. The predominant final nitrogenous waste is not urea but uric acid, which is present in bird's urine as a colloid suspension and does not create osmotic pressure (Paulson 1965).

Structure of Glomeruli. In all studied species, the cortex contains glomeruli of variable size (from 18.0 to 61.4 mm in diameter). Burachinsky (1970) suggested that the size of glomeruli directly reflects the bird's body weight. *Podoces panderi* and *Streptopelia turtur* have larger glomeruli than do sparrows; however, the size of glomeruli in *P. panderi* was larger than in *S. turtur*, whereas the body weight of the latter species is much higher. In sparrows, the desert species have larger glomeruli than the oasis ones. Therefore, the increase in the size of glomeruli can be considered to be an additional adaptation to arid conditions. The largest glomeruli are located in the cortical lobes that lie next to medullary lobes. These large glomeruli give rise to mammalian-type nephrons. Large glomeruli are characterized by an extensive, branching capillary network, whereas small glomeruli have a simple network comprised of several capillary loops. The visceral layer of the renal corpuscle (capsule of Bowman) lies directly next to the capillaries; the parietal layer is formed by flat epithelium with small oval nuclei. In larger glomeruli the outer capsular layer appears thicker, and the subcapsular slits are well defined.

Most glomeruli in observed desert birds were completely filled with blood; capillaries were wide open, with the blood cells and gaps filled by plasma visible in their lumina. As blood flow decreased, lumina of capillaries appeared narrow, with few visible blood cells and very narrow gaps; the capsular cavity often could not be detected. Such glomeruli were temporarily excluded from filtration.

Glomerular cells were weakly stained by pyroninophilic reaction which reveals nucleoli inside the nuclei of podocytes. The capillary glomerulus was intensely stained by PAS reaction and by Hayle's technique. Under staining with toluidine blue (pH 6), glomerular membranes gave a metachromatic staining which intensified after treatment by 1% acetic acid. Histochemical analysis demonstrated that glomerular membranes contained carbohydrates which included acidic mucoproteins with sialic acids (sialomucoproteins).

Structure of Proximal Tubules. The diameter of proximal tubules varied from 16.22 to 44.2 μm. Among sparrows, proximal tubules in desert species tended to be wider; in *Passer simplex* the diameter of a proximal tubule often was larger than the diameter of a glomerulus. The highest epithelial cells in tubules were found in *P. simplex*, cells were relatively high in *P. domesticus*, and low in *P. ammodendri* and *P. montanus*. Epithelial cells were higher in *Streptopelia senegalensis* than in *S. turtur*; in *Podoces panderi*, *Galerida cristata*, and *Sylvia inquieta* epithelial cells were rather low.

The proximal tubule could be divided into initial, convoluted, and straight portions. The initial portion of the tubule was lined by cuboidal epithelium with non-distinctive intercellular boundaries; large nuclei had a characteristic, round nucleolus. In the basal part of the cells, a rod-like pattern was formed by orderly positioned mitochondria. The middle, convoluted part of the tubule was lined by epithelium which often appeared cylindrical, with non-distinctive intercellular boundaries; nuclei were often shifted to the apical surface of the cells; a rod-like pattern of the basal cytoplasm was distinct, and granules of glycogen were present in the cytoplasm. The straight, descending portion of the tubule was lined by cuboidal epithelium of dome-shaped cells with distinct boundaries. Nuclei were primarily shifted to the basal surface, and a rod-like pattern was not distinct. Pinocytic vesicles were numerous here; it is possible thus that pinocytosis plays an important role in desert birds for filtrate transport from the lumina of tubules. The brush border was intensively stained with PAS reaction, but weakly by Hayle's technique and toluidine blue. Histochemical analysis revealed acidic mucopolysaccharides and high activity of alkaline phosphatase, which provides indirect evidence of glucose reabsorption in this portion of the nephron. Acidic phosphatase was practically absent. Succinate dehydrogenase activity was higher in the proximal tubules than elsewhere in nephron, which suggests the presence of a large number of mitochondria in the epithelium for generation of energy for transport.

The cortex contained a high amount of total protein; the brush border of the proximal tubule was especially intensively stained. Thin argyrophilic fibers and neutral and acidic mucopolysaccharides were found in the basal membrane of tubules. Acidic mucopolysaccharides containing sialomucoproteins were present in all tubular membranes and gave metachromatic staining with toluidine blue (pH 6.0). This staining ("hidden metachromasy") increased after treatment by a weak solution of acetic acid (pH 1.6 to 3.0), which is a specific reaction of sialomucoproteins (Vinogradov and Potapova 1963).

Structure of Henle's Loops. Diameter of the thin portion in birds is generally larger than in mammals. In the studied bird species it varied from 9.3 to 23.8 μm. The largest thin portion diameter was found in *Streptopelia senegalensis*, *Podoces panderi*, *Galerida cristata* and *Scotocerca inquieta*. Among sparrows, significant variations in the diameter of the thin portion of Henle's loop were present only in the *Passer montanus* population from Repetek. The thin portion was lined by flat-celled epithelium with slightly visible cell boundaries; portions of cells containing nuclei extended to the lumens of tubules. Flattened nuclei contained granules of chromatin; cells were rich in cytoplasm which lacked visible structures. A thin layer of Hayle-positive material (uromucoid) was present on the surface of cells facing the lumen. Uromucoid includes acidic mucopolysaccharides related to sialomucoproteins which are stained with Hayle's alcian blue and with PAS reaction and which give "hidden metachromatic" staining with toluidine blue.

Along Henle's loop, the thin segment gradually changes into the thick one. The diameter of the thick portion of Henle's loop in studied species varied from 18 to 35 μm. The height of epithelium varied among different species and populations. Among sparrows, maximal height of epithelial cells was observed in *Passer ammodendri*, medium height, in the *P. montanus* population from Repetek, *P. simplex* and *P. domesticus*, and minimal height, in the *P. montanus* population from Ashkhabad. Epithelial cells were cuboidal; large nuclei included small granules of chromatin and a dense nucleolus. The basal rod-shaped pattern indicated concentration of mitochondria, the presence of which was confirmed by high activity of succinate dehydrogenase. Apical portions of these cells often appeared "empty" and relatively highly vacuolized. Basal membranes of Henle's loop, as opposed to those of proximal tubules, were more finely structured and included single thin argyrophylic fibers and neutral mucopolysaccharides.

Structure of Distal Tubules. A gradual transition from the ascending portion of Henle's loop to the distal tubule was accompanied by a decrease in epithelial cell size and increase in nuclear density. Nuclei contained granules of chromatin and a dense nucleolus; apical portions of cells were slightly vacuolized, whereas basal portions had a dense, rod-shaped pattern. Distal tubules contacted the vascular pole of their glomeruli; in this area, the "dense spot" (*macula densa*) appeared in the wall of a tubule comprised of a cluster of cuboidal cells with dense basal nuclei; a rod-shaped pattern was not distinct. Ogawa and Sokabe (1971) consider cells of the *macula densa* as intermediate between regular cells of distal tubules and cells of the mammalian *macula densa*. It appears that, anatomically and functionally, the bird kidney is intermediate between that of reptiles and mammals in many ways (Sokabe et al. 1962; Leyssac 1967; Siller 1971).

Structure of Collecting Tubules. In all studied species, the diameter of collecting tubules increased from cortical to medullary portion of the tubule, varying from 24.8 to 74.5 μm. The epithelium of the cortical portion of a collecting tubule was cuboidal, with non-distinct cell boundaries; large nuclei contained small granules of chromatin and small nucleoli. In some cells ("dark cells"), the cytoplasm was stained more intensely than in others; their apical cytoplasm extended slightly toward the lumen of a tubule in a "flame" pattern. The epithelium of the medullary portion of a collecting tubule was flatter than that of the cortical portion and of tubules in the proximal end of the medullary cone. Nucleus-containing cytoplasm extended slightly toward the lumen of a tubule. Toward the proximal end of the medullary cone, height of the epithelial cells gradually increased, and, at the very beginning of the cone, acquired cylindrical shape. Their cytoplasm extended slightly toward the lumen, cell boundaries were distinct, and an apical row of microvilli was present.

Along the entire collecting tubule, significant deposits of Hayle-positive uromucoid on the cell surface were revealed by PAS-reaction with alcian and toluidine blue (pH 6.0); these deposits were several times more numerous in tubules of the medullary zone. Uromucoid was also found in the lumina of tubules, its amount increasing in the proximal part of medullary cones. In *Passer montanus*, intracellular deposits of uromucoid were revealed around nuclei of epithelial cells of the medullary collecting tubules. Similar deposits are known from the epithelium of collecting tubules in rodents (Batenko 1972, 1973), where they can perhaps act as ion-exchange material for the capture and reabsorption of cations (Vinogradov 1977).

Conclusions

The concentrating ability of the kidney is an important physiological feature of desert birds which reduces renal water loss and produces urine more concentrated than blood plasma. Water is effectively reabsorbed in nephrons due to their specific structure. Relative kidney size and size of medullary zone are larger in desert birds than in oasis species. The number of renal medullary cones per unit of kidney weight is specific for a given taxonomic group of birds and is related to the efficiency of water retention.

Decreased interstitial space and close contact of all elements in the medullary zone (i.e., Henle's loops, collecting tubules, and capillaries) are characteristic for a desert bird kidney, features which apparently facilitate diffusion of water and cations from tubules to capillaries. Deposits of acidic mucopolysaccharides which contain hialuronic acid and form a "gel filter" to regulate tissue permeability are found in the interstitial tissue of distal portions of medullary cones.

The smaller size of renal medullary cones in some desert bird species is compensated for by a high number of Henle's loops in the cones. It is possible that an increase either in number of cones or in number of Henle's loops per cone might be adaptations to arid climate. No difference was found in the length of Henle's loops between bird species adapted to temporary or permanent water shortage, and the oasis species. Therefore, the adaptive osmotic concentration of urine in desert birds probably evolved via development of medullary nephrons rather than via lengthening of Henle's loops. However, our results show that high level of adaptation to desert conditions is not always accompanied by development of medulla.

Additional adaptations to arid climate in many bird species include limitation of water loss through lungs, increase of osmotic concentration of blood plasma under dehydration, and postrenal (rectal and cloacal) reabsorption of water from feces and urine, accompanied by anatomical modification of rectum, coprodeum, and urodeum. Therefore, a complex of ecological, physiological, and morphological adaptations exists which allows birds to survive extreme desert climatic conditions.

18. On the Evolution of Pheasant (*Phasianus colchicus* L.) in Middle Asia

ALEXANDER V. SOLOKHA

Abstract

Geographic variation of the plumage coloration patterns in the pheasant (*Phasianus colchicus* L.) in Middle Asia and the adjacent territories is discussed. Seven Middle Asian subspecies belong to three geographic races: a semi-collared race (*Ph. c. mongolicus* and *Ph. c. turkestanicus*); a black-gold race (*Ph. c. bianchii* and *Ph. c. chrysomelas*); and a red-belly race (*Ph. c. zerafschanicus*, *Ph. c. zarudnyi*, and *Ph. c. principalis*). The eighth, West Kopetdagh subspecies *Ph. c. persicus* is closer to the Caucasian form *Ph. c. talischensis* than to the three Middle Asian races. Reconstruction of intraspecific differentiation and dispersal of the pheasant in Middle Asia and the Caucasus is proposed, and possible intersubspecific hybridization is discussed.

Introduction

Natural range of the pheasant (*Phasianus colchicus* L.) stretches from the Black Sea shores in the west to the Pacific Ocean in the east, and from northern Iran, deserts and steppes of Central Asia, northern Laos, northern Vietnam, and Taiwan in the south to the Volga River delta, Aral Sea, and Lake Balkhash in the north (Stepanyan 1975). Within this vast range, thirty subspecies are separated, differing primarily in male plumage coloration (Vaurie 1965). Two general groups can be recognized, the eastern (subspecies of *P. c. pallasii*/*P. c. torquatus* group) and the western one (subspecies of *P. c. colchicus*/*P. c. mongolicus* group). Subspecies *P. c. tarimensis* is considered an intermediate form between the eastern and western groups (Delacour 1951). Kozlova (1970) lists general differences between the two groups of subspecies as following: "in males of the eastern group, rear dorsal feathers and upper tail cover feathers differ in color from the rest of the upper body, and are grey-blue with greenish tone and shiny green spots; upper wing cover feathers are grey-blue; scapular feathers always with a distinct colored pattern. In males of the western group, rear dorsal feathers and upper tail cover feathers do not significantly differ in color from the front dorsal feathers and are not grey-blue; upper wing cover feathers are white, whitish or brownish but not grey-blue; a distinct colored pattern in the central part of scapular feathers is not always visible" (Kozlova 1970).

Formation of phenotypic variation in the pheasant is still a subject of discussion; different authors argue about the role of dispersal and age of geographically isolated populations and subspecies (Kozlova 1970; Potapov and Sapozhnikov 1976; Bidos 1985; Potapov 1987a). Here, we attempt a phenotypic and arealogical analysis of the western group of subspecies, with emphasis on pheasant differentiation in Middle Asia.

Materials and Methods

We analyzed the plumage coloration in about 200 pheasants belonging to Caucasian and Middle Asian subspecies deposited in the collections of the Zoological Institute of the Russian Academy of Sciences (St. Petersburg, Russia), Zoological Museum of the Moscow State University (Moscow, Russia), Institute of Zoology and Parasitology of the Academy of Sciences of Uzbekistan (Tashkent, Uzbekistan), as well as in the author's collection. Color descriptions were standardized according to Bondartsev's scale (Bondartsev 1954). To reconstruct dispersal routes, we used published paleogeographic data (Sinytsin 1962; Fyodorov 1972; Smolko et al. 1972).

Male pheasants of the western group of subspecies are well described in the literature although descriptions are sometimes controversial (Zarudny 1896; Buturlin 1935; Delacour 1951). To compare phenotypes, we distinguished three key characters: coloration of the plumage of the front part of the chest; the white "collar" area; and the upper wing cover feathers. Plumage of the front part of the chest forms a subspecies-specific pattern of distinct transverse bands of various colors and widths. In the crop area, feathers are small and details of the pattern difficult to distinguish, and feathers on the remainder of the chest and the bird's sides do not have a specific pattern. Dorsal plumage is subject to individual variation and also can fade under exposure to the sunlight. Coloration of the upper wing cover feathers is also an important character in distinguishing the eastern, Middle Asian and Caucasian subspecies.

The degree to which the white "collar" area is expressed, although variable, can also serve as a diagnostic character since its geographic variation is higher than its individual variation. To estimate this character quantitatively, we introduced the following 10-point scale:

0 – no white color in neck plumage
1 – white marks found under thorough examination
2 – white marks found on the hidden parts of feathers
3 – white marks seen from external examination and present on the feathertips
4 – white external spots

5 – white interrupted crescent-shaped lines on the sides of the neck
6 – distinct white crescent-shaped lines on the sides of the neck
7 – unequal white collar, interrupted in front
8 – relatively wide white collar, not closed in front
9 – wide white collar, narrower in front
10 – wide entire white collar

Below, we give the descriptions of the coloration of male pheasants of different subspecies recognized in Middle Asia. Brief data on their ranges are obtained from the literature (Potapov 1987b) and from our studies (Solokha 1991).

Results

Semirechye Pheasant (Phasianus colchicus mongolicus Br.). Plumage of the front part of the chest has three colored bands: brownish-red basal area, narrower purple-violet band, and black edge (Fig. 1a). White collar usually is wide but sharply narrows or is interrupted in front (8 to 9 points). Lumbar area, upper tail cover feathers, and, to a lesser extent, all dorsal sides have a significantly greenish color. Upper wing cover feathers are white.

Range: eastern and northeastern Middle Asia, including Semirechye, Issyk-Kul plain, and the foothills of Trans-Ili Alatau.

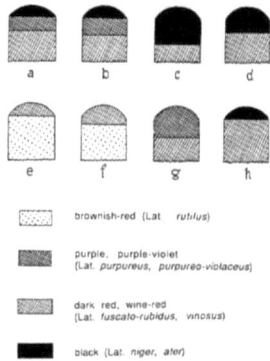

Fig. 1. Color pattern in front chest feathers of pheasant male. Subspecies: a – *Ph. c. mongolicus*; b – *Ph. c. turkestanicus*; c – *Ph. c. bianchii*; d – *Ph. c. chrysomelas*; e – *Ph. c. zerafschanicus*; f – *Ph. c. zarudnyi* (center of the range); g – *Ph. c. zarudnyi* (marginal populations) and *Ph. c. principalis*; h – *Ph. c. persicus*.

Syrdarya Pheasant (Ph. c. turkestanicus Lor.). Similar to the previous subspecies, but the brownish-red coloration on the front part of the chest is less prominently expressed whereas purple is more expressed (Fig. 1b). White collar is narrower and interrupted in front (8 points). Dorsal plumage with weak greenish color.

Range: Syrdarya River drainage, from the foothills of West Tien Shan and Talas Alatau, to the northeastern shore of the Aral Sea.

Tajik Pheasant (Ph. c. bianchii But.). Plumage of the crop area and front part of the chest is black, with green cast. Chest feathertops have two colors: a narrow brownish-red basal band and a wide black distal edge (Fig. 1c). White collar cannot be distinguished externally; average expression of white color in neck area is 1.17 points (range from 0 to 3, n = 12 birds). Collar usually is wide but sharply narrows or interrupted in front (8 to 9 points).

Range: upper part of the Amudarya River drainage including the Amudarya (above the city of Termez), Vakhsh, and Pyandzh Rivers.

Khiva Pheasant (Ph. c. chrysomelas Sew.). Black color with green cast is expressed only in the crop area. Compared to *Ph. c. bianchii*, front chest feathers have wider brownish-red band and narrower black edge (Fig. 1d). White collar is usually incomplete (average 4.61 points, with range from 0 to 7, n = 26 birds).

Range: lower part of the Amudarya River (from the city of Turtkul to the Aral Sea).

Zeravshan Pheasant (Ph. c. zerafschanicus Tarn.). Ventral plumage lighter than in previous forms. Front chest feathers with a wide yellow-red zone and narrow wine-red edge (Fig. 1e). In some birds, black central band or black edge around the feather is present. White collar is usually narrow, widely disjunct in front, or present only as crescent-shaped lateral bands (average 5.5 points, with range from 4 to 7, n = 5 birds).

Range: lower part of the Amudarya River (from the city of Turtkul to the Aral Sea).

Amudarya Pheasant (Ph. c. zarudnyi But.). Plumage coloration exhibits variation within this subspecies. Birds from the central populations of the range (the Amudarya Valley between the cities of Chardzhou and Deinau) resemble *Ph. c. zerafschanicus*: front chest feathers are mostly yellow-red, with relatively narrow wine-red (or purple) edge (Fig. 1f); white collar is well expressed (5 to 6 points, n = 3). Coloration of males from marginal populations (areas of Dargan-Ata and Kerki) is darker; front chest feathers have brownish-red band with rusty tone, and relatively wide purple or purple-violet edge (Fig. 1g); white collar is often inconspicuous (average 2.46 points, range 0 to 6, n = 27).

Range: middle portion of the Amudarya River Valley (from the city of Kelif to the Tyuyamuyun Reservoir) and the Kelif Uzboi Valley.

Murghab Pheasant (Ph. c. principalis Scl.). Plumage coloration in this form is almost identical to that of marginal populations of *Ph. c. zarudnyi*. Front chest feathers are bicolor, with brownish-red basal part and purple-violet edge (Fig.

1f). White collar is inconspicuous (average 1.77 points, range from 0 to 4, n = 32).
Range: valleys of the Murghab and Tedzhen (Harirud) Rivers, and submontane plain of East and Central Kopetdagh.

Persian Pheasant (Ph. c. persicus Sew.). This subspecies differs substantially from all abovementioned forms. Upper wing cover feathers are greyish but not white as in all other Middle Asian subspecies. Front chest feathers have wide wine-red basal band with rusty tone, and narrow black (with blue tone) edge (Fig. 1h). White color in plumage coloration is inconspicuous (average 0.88 points, range 0 to 2, n = 8).
Range: valleys of the rivers in West Kopetdagh (the extreme southwest of Middle Asia).

Discussion

Seven of the eight subspecies of pheasant described from Middle Asia clearly form three basic groups, or races: a so-called semi-collared race (*Phasianus colchicus mongolicus* and *Ph. c. turkestanicus*); a black-gold race (*Ph. c. bianchii* and *Ph. c. chrysomelas*); and a red-belly race (*Ph. c. zerafschanicus, Ph. c. zarudnyi,* and *Ph. c. principalis*). The West Kopetdagh subspecies *Ph. c. persicus* is closer to the Caucasian form *Ph. c. talischensis* than to the three Middle Asian races. We compared plumage coloration of subspecific forms within each of those races as well as their geographic ranges (Fig. 2).

In a semi-collared race, the subspecies *Ph. c. mongolicus*, as compared to *Ph. c. turkestanicus*, has different ratio of brown-red and purple-violet colors in chest feathers, a slightly more developed white collar, and diminished grey or green cast in the dorsal plumage. These two subspecies are closely allopatric and possibly originated from a common ancestral population without major divergence. Both of the subspecies of the semi-collared race (especially *Ph. c. turkestanicus*) have green coloration in the dorsal and tail plumage. This character is shared by several subspecies distributed farther to the east, southeast, and south from *Ph. c. turkestanicus*, namely *Ph. c. satscheuensis, Ph. c. hagenbecki, Ph. c. tarimensis,* and *Ph. c. shawi*.

Ranges of the two subspecies of the black-gold race (*Ph. c. bianchii* and *Ph. c. chrysomelas*) are separated by the range of a red-belly Amudarya pheasant (*Ph. c. zarudnyi*). Of the two black-gold forms, plumage of *Ph. c. chrysomelas* has less black in the crop and chest areas, and more white in the neck area than does *Ph. c. bianchii*. We suggest that these differences stem from hybridization of a black-gold race with a semi-collared *Ph. c. turkestanicus* (or its ancestor), which may have occured in the past on the southeast shore of the Aral Sea. This suggestion is supported by analysis of the Aral subspecies of pheasant (*Ph. c. bergii*), which is extinct today but in the 1900s still inhabited the Aral Sea islands (Uvun-Kair, Uyaly and others) (Zarudny 1915). These localities lie between the

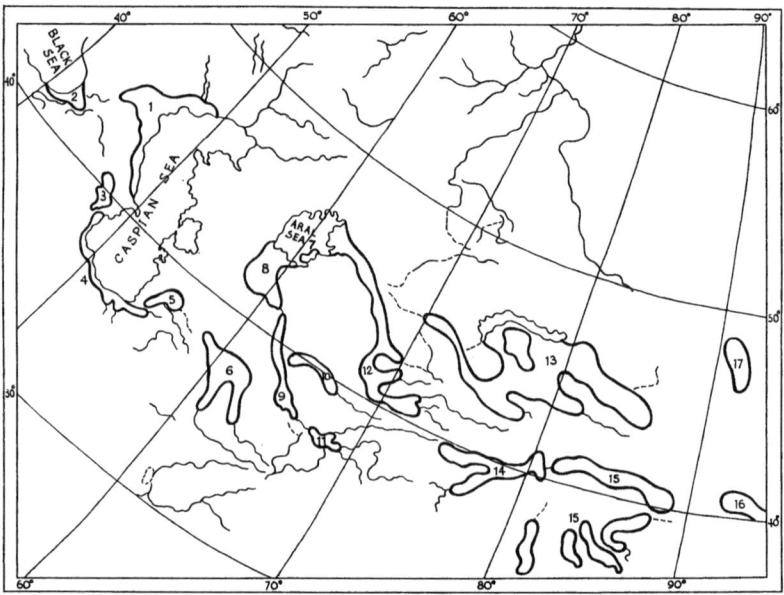

Fig. 2. Ranges of the subspecies of *Phasianus colchicus* belonging to the western group (1 to 13) and neighboring forms (14 to 17): 1 – *Ph. c. septentrionalis*; 2 – *Ph. c. colchicus*; 3 – *Ph. c. lorensi*; 4 – *Ph. c. talischensis*; 5 – *Ph. c. persicus*; 6 – *Ph. c. principalis*; 7 – *Ph. c. bergii*; 8 – *Ph. c. chrysomelas*; 9 – *Ph. c. zarudnyi*; 10 – *Ph. c. zerafschanicus*; 11 – *Ph. c. bianchii*; 12 – *Ph. c. turkestanicus*; 13 – *Ph. c. mongolicus*; 14 – *Ph. c. shawi*; 15 – *Ph. c. tarimensis*; 16 – *Ph. c. satscheuensis*; 17 – *Ph. c. hagenbecki*.

deltas of the Amudarya and Syrdarya, which are the northern parts of the ranges of *Ph. c. chrysomelas* and *Ph. c. turkestanicus*, respectively. Also, the coloration of museum specimens of *Ph. c. bergii* is intermediate between those two subspecies: its chest and crop area feathers have wide black edges (a character of *Ph. c. chrysomelas*), and, at the same time, the white collar is well expressed (a character of *Ph. c. turkestanicus*). *Ph. c. bergii*, isolated on the Aral Sea islands was, probably, a well-defined subspecies. Also, hybrid pheasants were recorded in the 1900s in the contact zone which still existed between *Ph. c. chrysomelas* and *Ph. c. turkestanicus* (Zarudny 1915).

The third, red-belly race includes *Ph. c. zerafschanicus*, *Ph. c. zarudnyi*, and *Ph. c. principalis*. Details of variation within male plumage coloration of *Ph. c. zarudnyi*, as well as its geographic range suggest that this form is a result of hybridization between a red-belly *Ph. c. zerafschanicus* and a black-gold *Ph. c. bianchii*, or between their ancestral forms (Potapov and Sapozhnikov 1976). *Ph. c. zarudnyi* from the center of the range, i.e., from the area adjacent to the range of *Ph. c. zerafschanicus*, resemble the latter form, whereas *Ph. c. zarudnyi* from marginal populations differ by possessing darker coloration and having no white collar. The formation of a new phenotype could have happened gradually, following the dispersal of *Ph. c. zerafschanicus* from the Chardzhou-Deinau area up and down the Amudarya River, with a gradual increase in contribution of characters of the black-gold race: darkening of the plumage (i.e., change of yellow-red color to brown-red, and wine-red to purple-violet; shrinking of the light basal zone; and widening of the dark distal edge) and reduction of the white collar. This variation could have been clinal in the prehistoric time. The advent of human settlers in the Amydarya Valley, however, has disrupted the continuous range of the pheasant and created disjunct populations. This is probably why variation in male coloration of *Ph. c. zarudnyi* today does not appear clinal and why this subspecies appears polymorphic. The Murghab pheasant (*Ph. c. principalis*) is probably just one of these disjunct populations. Its proximity to *Ph. c. zarudnyi* is evident from the comparison of populations from the Tedzhen, Murghab, and middle Amudarya Rivers. There are no significant differences in male plumage coloration; females and young birds also are identical. Contact between the Murghab and middle Amudarya populations could have existed for a long time before these valleys were populated and developed by humans.

The range of the Persian subspecies *Ph. c. persicus* in West Kopetdagh is well separated from all abovementioned forms. Bidos (1985) suggested the origin of this form from the black-gold race; however, the Persian pheasant lacks such common features of all black-gold subspecies as dominant black coloration of the chest, and especially, crop area. We maintain that, of Middle Asia forms, the red-belly Zeravshan pheasant (*Ph. c. zerafschanicus*) is closest to *Ph. c. persicus*. These forms are similar in their egg size and weight, while differing from the Khiva, Amudarya and Murghab pheasants in these parameters (Solokha 1991). The Persian subspecies differs from the Zeravshan one in the following ways: the basal zone of its chest feathers is wine-red, with a rusty cast (whereas in the Zeravshan pheasant it is yellow-red); the chest feather edge is black and narrow (in the Zeravshan pheasant it is wine-red with a purple cast; however, narrow black edges or black quill bands are sometimes also recorded in the Zeravshan pheasant). The unique characters of the Persian subspecies that suggest its ancient differentiation from red-belly Middle Asian forms include the absence of a white collar and, especially, the presence of a grey color of the upper wing cover feathers.

It is important to consider the key characters of male coloration in pheasant subspecies which are found to the west of *Ph. c. persicus*, i.e., in the Caucasus. The Talysh pheasant (*Ph. c. talishensis*) from the southeastern Transcaucasia differs from the Persian form only in darker upper wing cover feathers. The Georgian subspecies *Ph. c. lorensi* has also wider black edges and a brown-red basal zone on the front chest feathers. Finally, the nominal subspecies of the Caucasian pheasant (*Ph. c. colchicus*) has brownish upper wing cover feathers, and its front chest feathers have a purple-violet basal band and even wider black edge. Plumage of all these forms lacks black color. Therefore, the clinal variation in plumage coloration can be traced from east to west, from the range of *Ph. c. zerafschanicus* to that of *Ph. c. colchicus*. It includes substitution, in chest feathers, of a wine-red edge with purple cast by the black edge and its subsequent enlargement; darkening in coloration of these feathers from yellow-red to purple-violet; darkening in coloration of upper wing cover feathers (which become grey and, finally, brown); and disappearance of the white color from neck plumage.

Another form, the North Caucasian pheasant (*Ph. c. septentrionalis*) is found to the north and northeast of the nominal Caucasian subspecies (*Ph. c. colchicus*). The North Caucasian form has lighter plumage coloration, and its chest feather basal band is brownish-red rather than purple-violet. This feature does not fit into the pattern clinal variation, and could have been a result of contact between *Ph. c. septentrionalis* and *Ph. c. talishensis* or *Ph. c. lorensi* along the western shore of the Caspian Sea. Given the modern range of the North Caucasian pheasant (which almost reaches the Apsheron Peninsula in the south), such intergradation could have been quite possible in the past.

We attempted to reconstruct dispersal and differentiation of the pheasant in Middle Asia and the Caucasus on the basis of phenotypic variation and paleogeographic data (Fig. 3). *Phasianus colchicus* most probably evolved in the forests of Southeast Asia (Delacour 1951) with subsequent dispersal to Central Asia (Kozlova 1970). The pheasant could have dispersed from Central Asia to Middle Asia in the Late Pliocene – Middle Pleistocene, following the route from the Tarim Depression to the South Tajik Depression. This route was open until the Middle Pleistocene when the Pamiro-Alai mountains had not yet become a

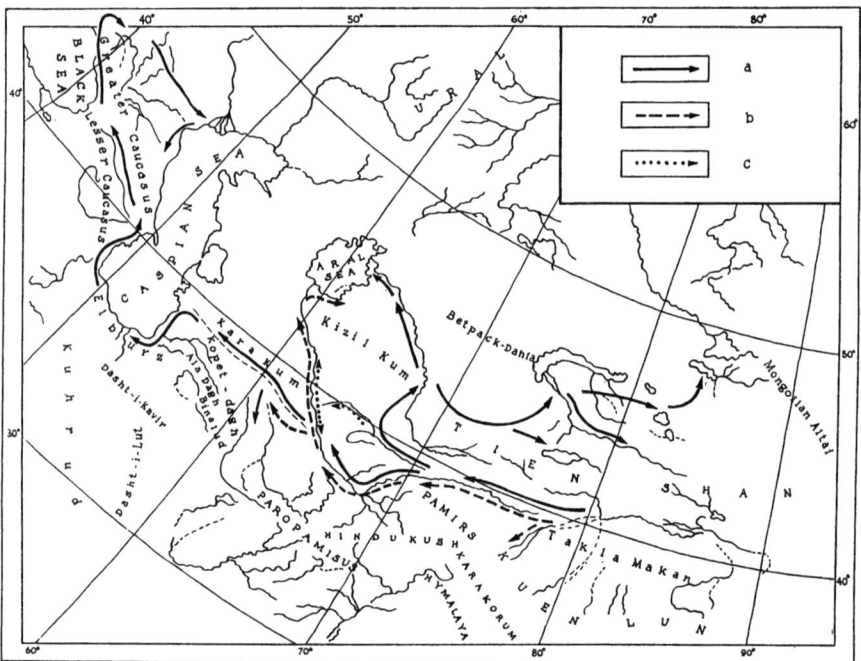

Fig. 3. Suggested dispersal routes of *Phasianus colchicus* in the western part of its range. a – Proto-Zeravshan form; b – black-gold form; c – *Ph. c. zerafschanicus*.

major barrier for species dispersal (Bidos 1985). At the same time, the already extant mountain ranges of Kuen Lun and Karakorum could have barred dispersal of the pheasant from the Tarim River drainage toward the southwest, to India.

The range of the pheasant after its dispersal to Middle Asia could have been continuous from the northern foothills of East Tien Shan and the Zaisan Depression to the Black Sea. Such dispersal would have been facilitated by the Proto-Amudarya which flowed to the Caspian Sea, with its drainage occupying the lowland Karakum Desert. The pheasant dispersed to the Caucasus along the southern shore of the Caspian Sea and the foothills of the Elburz Mountains. Absence of this species in central and southern Afghanistan and Iran (Potapov 1978a) probably can be explained by unfavorable local climatic conditions within the inner ranges of the Paropamiz and Turkmeno-Khorassan Mountains, and the Iranian Plateau, which prevented dispersal of the pheasant to the south of Middle Asia. The arid landscapes there probably already existed in the early Pleistocene.

The dispersal of the ancestors of the western group of subspecies (which includes all Middle Asian forms) easterly was probably limited by the ancient Dzhungar Desert (Kozlova 1970); however, some features in the coloration of *Ph. c. mongolicus* and *Ph. c. turkestanicus* suggest probable contacts between the western and eastern groups in the past.

The ancestral Middle Asian form could have resembled a modern red-belly form such as *Ph. c. zerafschanicus*. The aridization of Middle Asia, which began at the end of the Neogene, gradually confined the pheasant to the river valleys and humid foothills; by this time the ancestral populations of the semi-collared race (*Ph. c. mongolicus* and *Ph. c. turkestanicus*) could have already differentiated.

In Late Khazarian age (ca. 100,000 years ago), the Amudarya turned toward the Aral-Sarykamysh Depression. This event, along with the progressing aridization (including formation of the Karakum Desert), disrupted the range of the pheasant. Therefore, populations to the west of the Karakum became isolated from those to the east in the Amudarya Valley. Subsequent disruption of the western range and evolution of isolated populations led to the differentiation of *Ph. c. persicus, Ph. c. talishensis, Ph. c. lorensi, Ph. c. colchicus,* and *Ph. c. septentrionalis* (formation of the latter could be due to hybridization along the Caspian shores).

Following the change in course of the Amudarya and climatic changes (aridization and cooling), the local pheasant populations probably became rare or extinct in the new valley. The newly formed ecosystems of the Amudarya Valley were then colonized by pheasants of the black-gold race, which dispersed from the upper portion of the Amudarya and resembled modern *Ph. c. bianchii*. This black-gold form penetrated valleys of the Murghab, Tedzhen, and the foothills of East Kopetdagh, following the deltas of the Balkh and other former tributaries of the Amudarya which have formed the Kelif Uzboi. The hybridization of this black-gold ancestor with the Proto-Zeravshan form could have led to the formation of the Murghab subspecies.

Dispersal of the already modern red-belly Zeravshan subspecies (*Ph. c.*

zerafschanicus) probably took place in the Upper Pleistocene. These birds migrated throughout the middle Amudarya, hybridizing there with the black-gold form. A likely cause of such a migration might have been the Valdai glaciation of the Pamiro-Alai mountains: this event should have driven pheasants from the foothills and low mountain belt, facilitating their dispersal along the Zeravshan and Kashkadarya, which became large rivers and reached the Amudarya during this period.

In the lower portion of the Amudarya, the hybridization of black-gold pheasants with a semi-collared *Ph. c. turkestanicus* could have produced the black-gold Khiva subspecies (*Ph. c. chrysomelas*). In a zone of close contact, hybridization between *Ph. c. turkestanicus* and *Ph. c. chrysomelas* produced the Aral subspecies *Ph. c. bergii*, which later became isolated on the Aral Sea islands. Finally, the geographic isolation of a black-gold population in the upper Amudarya produced the Tajik subspecies, *Ph. c. bianchii*.

Our modification of the system of subspecific relationships within *Phasianus colchicus* (Delacur 1951) is presented in Fig. 4. We suggest that the Proto-Zeravshan form dispersed to Middle Asia before the dispersal of the black-gold race. Nevertheless, these two races had a common ancestor, which also gave rise to subspecies *Ph. c. tarimensis* and *Ph. c. vlangalii*.

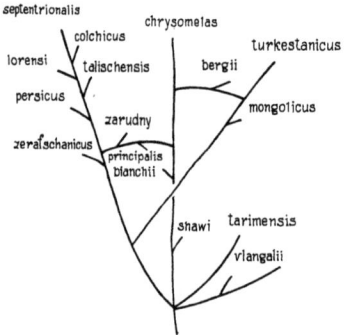

Fig. 4. Suggested phylogeny of the *Phasianus colchicus* subspecies belonging to the western group.

The formation of isolated populations and their subsequent differentiation into separate subspecies in the western part of the species range did not occur simultaneously. There are still cases when contact exists or secondary contact is possible, e.g., between *Ph. c. persicus* and *Ph. c. talishensis* in the Gorgan River Valley (Potapov 1987b); between *Ph. c. zarudnyi* and *Ph. c. bianchii* in the Amudarya Valley between the cities of Kelif and Termez; between *Ph. c. zarudnyi* and *Ph. c. chrysomelas* in the Amudarya Valley around the Tyuyamuyun Reservoir; and between *Ph. c. zarudnyi* and *Ph. c. zerafschanicus* along the Amu-Bokhara Canal. These contact zones, however, are narrow; the gene flow there is not intensive, and these subspecies currently are well-defined.

Dispersal and change in the ranges of various subspecies of the pheasant still occur, primarily due to human activity. Irrigation, through the Karakum Canal, of the formerly arid lands in the deltas of the Murghab and Tedzhen and in the Gyaurs Valley, have increased the range of the Murghab pheasant (*Ph. c. principalis*) two-fold in the last 30 years without any artificial breeding attempts; this range has extended toward the northwest. In the lower Amudarya, the dynamics of irrigation control the range fluctuation of of the Khiva pheasant (*Ph. c. chrysomelas*), which currently is dispersing toward the west. On the other hand, the destruction of natural habitats has led to the shrinking of the Persian subspecies range in its southeastern part (Solokha 1991).

Ophiomorus chernovi. Photo by N.N. Shcherbak.

Eumeces taeniolatus. Photo by N.N. Shcherbak.

19. Zoogeographic Analysis of the Reptiles of Turkmenistan

NIKOLAI N. SHCHERBAK

Abstract

The reptilian fauna of Turkmenistan includes 80 species (3 spp. of turtles, 27 spp. of lizards, and 50 spp. of snakes). Rooted in the Neogene, and formed under desert conditions since the Pliocene, this fauna approached its modern composition by the Holocene. Turkmenistan was the place of endemic origin for the Turanian sand desert fauna and, partially, for the Asian mountain fauna, with important connections to the Steppe, Caucasian, and Irano-Afghan faunal centers. Modern distribution of the herpetofauna is strongly influenced by human activity. The habitat diversity in Turkmenistan promoted the evolution of several ecological complexes of reptiles: psammophiles (10 spp.) and sclerobionts (7 spp.) in the lowlands, a large mountain complex (35 spp.), the species found next to the water (5 spp.), as well as widespread reptiles (6 spp.). Other species can occupy various combinations of different habitats. The reptiles of Turkmenistan can be divided into three groups according to their activity patterns: diurnal (51 spp.), nocturnal (16 spp.), and species with mixed or changing activity (13 spp.). Although the majority of reptiles feed on insects, some snakes feed on vertebrates. The reproductive period of reptiles in Turkmenistan is longer than that of the species from higher latitudes. Clay desert species begin reproduction earlier than do sand desert ones. Compared to the lowland species, mountain reptiles tend to have a longer maturation period, a higher longevity, a lesser number of eggs and clutches, and a shorter period of reproduction. The fauna of reptiles of Turkmenistan is highly endemic (10 endemic species and subspecies). Twenty-three species found here are subendemic to Middle Asia. The core fauna consists of the Turanian species (36 spp.) as well as of the Irano-Afghan/Southwest Asian ones (17 spp.). Other groups are comprised of the Saharo-Sindian, Indian, Caucasian-Anatolian, European-Mediterranean, Kazakhstan-Mongolian, Mountainous Asian, and East Palearctic species. Twenty species of reptiles from Turkmenistan have been included in the Red Data Book of the former USSR; current efforts toward the conservation of reptiles in Turkmenistan are not sufficient.

V. Fet & K.I. Atamuradov (eds.), Biogeography and Ecology of Turkmenistan, 307–328.
© 1994 *Kluwer Academic Publishers.*

Introduction

Desert landscapes of Turkmenistan constitute a true realm of reptiles. It is not by chance, therefore, that all prominent herpetologists of pre-revolutionary Russia and the former Soviet Union, from A.M. Nikolsky (1905, 1908, 1911, 1915, 1916) and S.A. Chernov (1934) to I.S. Darevsky (1955) and O.P. Bogdanov (1960), have visited this area and conducted research there. The author has also devoted many years (1959 to 1991) to the study of reptiles in Middle Asia, especially in Turkmenistan. Also, the reptiles in Turkmenistan have been actively studied in the last decades by Turkmen zoologists, namely S. Shammakov and Ch. Ataev.

Nevertheless, reptiles of Turkmenistan still are incompletely described. Within the past fifteen years, the author has described a new species of gecko (*Tenuidactylus turcmenicus*; Shcherbak 1978); recorded two species of geckos and skinks for the first time (Shcherbak 1974, 1985); described several new subspecies of lizards (Shcherbak 1972, 1979, 1990); and collected numerous new data on taxonomy, distribution, and ecology of many species of reptiles. An initial, faunistic stage of herpetological studies is now, however, completed in Turkmenistan.

The Origin of the Herpetofauna of Turkmenistan

A detailed analysis of the origin of the herpetofauna of Turkmenistan should await accumulation of more paleontological data than are now available. This fauna is generally considered to be ancient, a hypothesis enhanced by the fossil record of lizard bones in Badghyz (Ananyeva and Gorelov 1981). These authors propose that the Miocene species of *Agama* and *Eremias* are closely related to extant ones.

A general scenario of the origin of the herpetofauna in this area can be reconstructed from the phylogenetic analysis of reptiles, their zoogeography, and paleogeographic data. Nikolsky (1916) was the first to discuss this subject. He believed that Middle Asian fauna is very young and was formed in the post-glacial time from those species that migrated here, primarily from deserts of Central Asia. Later, Chernov (1959) reviewed new data incorporating new information obtained by botanists. He concluded that the Middle Asian herpetofauna is a quite ancient one (of Miocene origin) and relatively independent from the Central Asian herpetofauna. Chernov (1959) observed a high level of specific endemism in Middle Asia and hypothesized that a separate center of origin of psammophilous herpetofauna existed here. He also suggested that the desert faunas of North Africa and Middle Asia evolved independently. Shcherbak (1971) demonstrated that the lizard genus *Mesalina* is an element of the North African center of origin of desert fauna, whereas the genus *Eremias* is its counterpart in the Asian centers. The species composition of the latter genus differs for the Middle and Central Asian deserts. Therefore, a special

Central Asian center of origin of the desert herpetofauna can be defined as well as a Middle Asian one (the latter one is heterogeneous and includes, in turn, several centers of origin of lower ranks).

The two most ancient centers of origin, which include specialized, polymorphic and widely distributed species, are: the *Steppe*, or *Kazakhstan Center* (Heptner 1945; represented by the lizards *Eremias arguta* and *E. velox*), and the *Turanian Center* of origin of sand desert fauna (*Eremias grammica, E. scripta, E. lineolata, E. intermedia*), which principally occupies the Karakum Desert in Turkmenistan. A derivative of these two centers is the younger *Irano-Afghan* (*Mountain Steppe*) *Center*, or the center of origin of the "ephemerous fauna." The representative species of this center (e.g., species of lizards related to *Eremias velox*: *E. strauchi, E. persica, E. afghanistanica*, as well as *E. nigrocellata* and *E. arida*) inhabit foothills of Kopetdagh, Badghyz, and Karabil within Turkmenistan, and are also found in Iran and Afghanistan. The *Tien Shan Mountainous Center* was formed at the boundary between Middle and Central Asia; its representative species could be considered to be *Eremias nikolskii*.

Interpretation of the age of the Middle Asian desert landscapes is debatable (Kryzhanovsky 1965; Murzaev 1966). A comparative study of *Eremias* lizards from Middle and Central Asia provides evidence of an approximately equal age for both most ancient centers. At the same time, however, the specific composition of *Eremias* in Central Asia is rather heterogeneous and appears to be a remnant of a formerly considerably more widespread fauna. Several species of *Eremias* here bear certain ancestral characters which may indicate that the Central Asian center is older than the Middle Asian one. In any case, the genus *Eremias* undoubtedly evolved in Asia. If we take into account the paleogeographic data and the timing of orogenic processes which isolated different species groups, we could suggest that this genus probably was formed in the Neogene somewhere between the eastern shores of the Tethys Sea and the Gobi Desert.

The structure of ecological complexes within the genus *Eremias* in Central Asia allows us to separate there two centers of origin: the *Mongolian Steppe* and *Forest-Steppe Center*, (containing such representatives as *Eremias multiocellata, E. argus*, and *E. brenchlei*), and its derivative, the *Kashgarian Mountain Steppe* (or *Ephemerous*) *Center* (with its representative *Eremias buechneri*). Such species as *E. przewalskii, E. vermiculata* and possibly *E. quadrifrons* are related to the *Alashan Sand Desert Center* of origin. A certain symmetry in the structure of the centers of origin in Middle and Central Asia can be observed (Fig. 1).

The relationships between herpetofaunas of Middle Asia and the Caucasus have been recently discussed by Rustamov (1981). He noticed that, of 80 species of reptiles inhabiting Turkmenistan (Table 1), 27 species are also found in Transcaucasia. Darevsky (1957) demonstrated that Turanian forms have dispersed to Transcaucasia from the south (through Iran) as well as through a land bridge that once connected the Apsheron and Krasnovodsk Peninsulas of the Caspian Sea. The dispersal of the Turanian species was also possible by

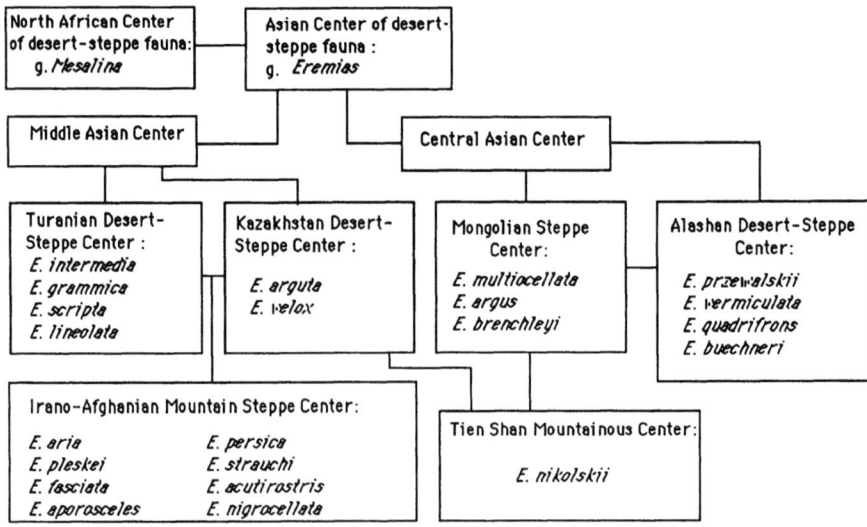

Fig. 1. Relationship among the centers of origin of the Asian desert-steppe fauna based on the example of a lizard genus *Eremias*.

Table 1. Species list of the reptiles of Turkmenistan and their habitat distribution

Species/subspecies	Habitats					
	Sand Desert	Salt Desert	Stony Desert	Clay Desert	Mountains & Foothills	Humid Habitats
Emys orbicularis orbicularis	−	−	−	−	−	+
Mauremys caspica caspica	−	−	−	−	−	+
Agrionemys horsfieldi rustamovi	+	−	+	+	+	−
Alsophylax laevis	−	−	+	−	−	−
A. loricatus szczerbaki	−	−	−	+	−	−
A. pipiens	−	−	−	+	−	−
Bunopus tuberculatus	−	+	+	−	−	−
Tenuidactylus caspius caspius	+	−	−	+	+	−
T. fedtschenkoi	−	−	−	−	+	−
T. turcmenicus	−	−	−	−	+	−
T. longipes microlepis	−	−	−	−	+	−
Mediodactylus spinicaudus	−	−	−	−	+	−
M. russowi russowi	+	−	−	−	−	−
Crossobamon eversmanni eversmanni	+	−	−	−	−	−
Teratoscincus scincus scincus	+	−	−	−	−	−
Eublepharis turkmenicus	−	−	−	−	+	−
Phrynocephalus helioscopus helioscopus	−	−	−	+	−	−
Ph. rossikowi rossikowi	−	−	+	−	−	−
Ph. rossikowi schammakovi	−	−	+	−	−	−
Ph. ocellatus ocellatus (= *Ph. reticulatus*)	−	−	+	−	−	−

Table 1. Continued

Species/subspecies	Habitats					
	Sand Desert	Salt Desert	Stony Desert	Clay Desert	Mountains & Foothills	Humid Habitats
Ph. ocellatus bannikovi	−	−	+	−	−	−
Ph. raddei raddei	−	−	+	+	−	−
Ph. guttatus guttatus	+	−	−	−	−	−
Ph. maculatus	−	+	−	−	−	−
Ph. interscapularis interscapularis	+	−	−	−	−	−
Ph. mystaceus galli	+	−	−	−	−	−
Stellio chernovi	−	−	−	−	+	−
S. caucasius caucasius	−	−	−	−	+	−
S. caucasius triannulatus	−	−	−	−	+	−
S. erythrogaster	−	−	−	−	+	−
S. lehmani	−	−	−	−	+	−
Trapelas sanguinolentus ?aralensis	+	+	+	+	+	−
Ophisaurus apodus	−	−	−	−	+	−
Eremias arguta uzbekistanica	−	−	−	+	−	−
E. grammica	+	−	−	−	−	−
E. intermedia	+	+	−	+	−	−
E. lineolata	−	+	−	+	−	−
E. nigrocellata	−	−	−	+	−	−
E. persica	+	−	+	−	−	−
E. regeli	−	−	−	+	−	−
E. scripta scripta	+	−	−	−	−	−
E. strauchi kopetdaghica	−	−	−	−	+	−
E. velox velox	−	−	−	+	−	−
Mesalina guttulata watsonana	−	−	−	+	−	−
Lacerta strigata	−	−	−	−	+	+
L. defilippii	−	−	−	−	+	+
Ablepharus pannonicus	−	−	−	−	+	−
A. bivittatus (?)	−	−	−	−	+	−
A. deserti	−	−	−	−	+	+
Chalcides ocellatus	−	−	−	−	+	−
Eumeces schneideri princeps	−	−	−	−	+	−
E. taeniolatus parthianicus	−	−	−	−	+	−
Mabuya aurata affinis	−	−	−	−	+	−
Ophiomorus chernovi	+	−	−	−	−	−
Varanus griseus caspius	+	+	+	+	+	−
Eryx elegans	−	−	−	−	+	−
E. miliaris miliaris	+	−	−	−	−	−
E. miliaris tataricus	−	−	+	+	−	−
Natrix natrix persa	−	−	−	−	−	+
N. tessellata	−	−	−	−	−	+
Coluber schmidtii	−	−	+	+	−	−
C. karelini karelini	+	+	+	−	+	−
C. rhodorhachis rhodorhachis	−	−	−	+	+	−
C. rhodorhachis ladacensis	−	−	−	+	+	−
C. najadum	−	−	−	−	+	−
C. ravergieri ravergieri	−	−	−	−	+	−
C. nummifer	−	−	−	+	+	−

Table 1. Continued

Species/subspecies	Habitats					
	Sand Desert	Salt Desert	Stony Desert	Clay Desert	Mountains & Foothills	Humid Habitats
Elaphe dione	−	−	+	+	−	−
Elaphe quatuorlineatus sauromates	−	−	+	+	−	−
Eirenis meda	−	−	−	−	+	−
Pseudocyclophis persicus persicus	−	−	−	−	+	−
Boiga trigonatum melanocephala	+	−	−	+	+	−
Lycodon striatus bicolor	−	−	+	−	+	−
Lytorhynchus ridgewayi	−	−	−	+	−	−
Oligodon taeniolatus	−	−	−	−	+	−
Psammophis lineolatum	+	+	+	+	+	−
P. schokari	−	−	−	−	+	−
Ptyas mucosus	−	−	−	−	−	+
Spalerosophis diadema	+	−	+	+	+	−
Telescopus rhynopoma	−	−	−	−	+	−
Agkistrodon halys caraganus	−	−	−	+	−	−
A. h. caucasicus	−	−	−	−	+	−
Naja oxiana	+	−	−	−	+	−
Typhlops vermicularis	−	−	−	−	+	−
Echis carinatus multisquamatus	−	−	−	+	−	−
Daboia lebetina turanica	−	−	−	−	+	−

detouring the Caspian Sea, to the northeast of the Caucasus, during the migrations of the Proto-Volga River and the Caspian Sea transgressions.

The modern herpetofauna of the Ancient Mediterranean has been formed during the Pliocene and Pleistocene (Shcherbak 1966). A high level of endemism of Turanian fauna suggests its ancient age. It was apparently formed in the Neogene when the vast areas of dry land were exposed due to the drying of the Tethys Sea (the modern Black, Caspian, and Aral Seas are mere remnants of this ancient sea). The sand desert of Karakum was formed thereafter, due to the alluvial accumulation of sand by the Amudarya River that was constantly changing its way from the Aral Sea to the Caspian Sea and back again. The formation of this desert was also facilitated by the erosion of uplifting mountain ranges of Kopetdagh, Hindu Kush, Pamiro-Alai, and Tien Shan. Similar processes led to the formation of sand deserts in the Kizylkum and Balkhash areas, where sand was deposited by the Syrdarya and Ili Rivers, respectively. The desert climate of Middle Asia was already well expressed in the Pliocene (Fedorovich 1946). The Kopetdagh Mountains are relatively young, and the uplift of this mountain range could have caused, for instance, a range disjunction in a lizard, *Phrynocephalus maculatus*. A relict population of this species was found in the northern foothills of Central Kopetdagh (Bami Station), whereas its main range lies within Iran, Iraq and Saudi Arabia. This lizard has not been able to disperse from the west, through the Messerian Plain, because this area until a very recent time was covered by the Caspian Sea (sea shells still are scattered over the surface of this plain). The ranges of Maly and Bolshoi Balkhan were islands at this time. Our study of the isolated population of *P. maculatus* from Bami showed no variation when compared to the nominal subspecies although generally species of *Phrynocephalus* are highly variable.[1]

Strong habitat restrictions have not allowed the Turanian psammophilous species of reptiles to disperse beyond Middle Asia. Many of these species are not able to move even ten meters beyond sands (e.g., lizards *Phrynocephalus mystaceus*, *Eremias scripta*, and *E. grammica*). In the Pliocene, the Circum-Euxinian distribution of species belonging to the core of Ancient Mediterranean fauna (Shcherbak 1966) has been responsible for the dispersal to the modern territory of Turkmenistan of such species as the glass lizard *Ophisaurus apodus*, snakes *Natrix tessellata* and *Coluber schmidti*, and the turtle *Emys orbicularis*.

In the Pleistocene, the mountain uplift disrupted geographic ranges and gave rise to the endemic faunas of the Caucasus and Irano-Afghanian center of origin. At the same time, several species dispersed to Kopetdagh through the Elburz Mountains in northern Iran: these include snakes of the genus *Eirenis* and *Agkistrodon*, and lizards *Lacerta defilippi* and *L. strigata*. At this time, a lizard species, *Eremias strauchi*, probably was separated from the ancestral form, *E. velox*, and an endemic subspecies, *E. s. kopetdaghica*, was formed.

The general aridization of climate in the Holocene, accompanied by river drying and the decrease of the Caspian Sea level, probably was responsible for the dispersal to the North Caucasus of the following reptile species: *Phrynocephalus mystaceus*, *Ph. guttatus*, *Trapelas sanguinolentus*, *Eremias arguta*, *E. velox*, *Eryx miliaris*, and *Elaphe dione*. Several species, such as *Eremias arguta* and *Elaphe dione*, penetrated much further westward. Following this period, a certain humidization of the climate occured. Later, however, fluctuations of aridity caused the formation of a complex "lace" of zoogeographic ranges of reptiles.

The first settlements of shepherds, hunters and farmers have emerged ca. 8,000 years BP, in the foothills of Kopetdagh (southern Turkmenistan), and reached their peak in III-II millenia BC (Sarianidi 1967). Since that time, a new powerful factor – the influence of human species – has influenced the distribution of reptile species. Cutting of Kopetdagh forests has led to the drying of mountain springs and further aridization of the climate. Rustamov (1981) suggested that this recent aridization has caused general shrinking of ranges and a decrease in abundance of Caucasian reptile species in Turkmenistan and their replacement by the Middle Asian species in vacated ecological niches.

Notes on the Systematics of the Reptiles of Turkmenistan

The Kiev school of herpetology has made a significant impact on taxonomic studies of reptiles in Middle Asia. We have accumulated extensive collections (over 40,000 units) and maintained the exchange of collection materials with the largest museums in the USA, Great Britain, Germany, Switzerland, and other countries. Thus, we have been able to conduct taxonomic revisions of many groups of reptiles. Our studies have been based not only on morphological methods but have also involved modern techniques in statistics, karyology, protein electrophoresis, and ecology. Results of these studies have been presented in numerous books, dissertations, and articles, e.g., revisions of the lizard genus *Eremias* (Shcherbak 1974), geckonid lizards (Shcherbak and Golubev 1986), and ablepharid lizards (Eremchenko and Shcherbak 1986); the study of the intraspecific variation in the glass lizard *Ophisaurus apodus* (Shcherbak and Tertyshnikov 1989); revisions of snake genera *Eirenis* (Dotsenko 1986) and *Eryx* (Tokar 1990); and revisions of lizard genera *Eumeces* and *Mabuya* within the USSR (Shcherbak and Akhmedov 1990; Shcherbak 1990). Currently, M.L. Golubev is conducting a revision of the lizard genus *Phrynocephalus*, and T. Fentisova, of the snake genus *Agkistrodon* in the northern Palearctic realm.

The most important results of taxonomic studies of the reptile fauna of Turkmenistan are summarized below.

1. The gecko *Bunopus tuberculatus* Blanf. 1874 is a monotypic species. Arnold (1980) correctly synonymized with this species both *B. blanfordi* Str. and *B. abudhabi* Leviton et Anderson. The only relict population of *Bunopus tuberculatus* in Turkmenistan is known from the Eroyulanduz Depression; no records of this gecko have been published in recent years.
2. We have demonstrated that a certain parallelism exists between the evolution of thin-toed geckos and their zoogeographic distribution. For example, ancestral genera *Gymnodactylus* and *Cyrtodactylus* are found in the Neotropical and Oriental zoogeographic regions, whereas younger generic groups inhabit different provinces of the Palearctic region, e.g., *Tenuidactylus* (the Turanian Province), *Mediodactylus* (the Mediterranean Province), and *Cyrtopodion* (the Irano-Afghan Province) (Shcherbak 1986, 1988, 1991). The most advanced and well-defined Palearctic genus is *Tenuidactylus*. Its representatives (*T. caspius, T. fedtschenkoi, T. turcmenicus* and *T. longipes*) are well characterized by their scale counts, morphometric ratios and coloration as well as by the karyological patterns. The genus *Mediodactylus* includes such species as *M. kotschyi, M. spinicaudus* and *M. russowi* (the latter penetrating far eastward in Middle Asia). The placement of thin-toed geckos from Middle Asia into the genus *Tenuidactylus* has been recently criticized by Bauer (1987).[2]
3. The lizard species *Trapelas sanguinolentus* (= *Agama sanguinolenta*) was previously considered to be monotypic. Recently, Ananyeva and Tsaruk

(1986) made an attempt to revalidate a subspecific form from Middle Asia, *T. s. aralensis* (the nominal subspecies is described from the European part of the species range), based on certain differences in coloration and average scale counts. In this case, however, the presence of clinal variation cannot be excluded, and the validity of this subspecies is doubtful.

4. The nominal subspecies of the lizard species *Stellio caucasius* (= *Agama caucasica*) is known from the Caucasus and most part of Turkmenistan mountains. Ananyeva and Ataev (1984) described a new subspecies, *S. c. triannulatus* from the loess canyons of Madau (western Turkmenistan). The new subspecies has groups of three rings of scales per tail segment whereas the nominal subspecies has two rings per segment. We possess a good sample of specimens from this area and confirm the existence of this peculiar population.

5. Ananyeva (1986) placed the lizard species *Phrynocephalus mystaceus* in a separate monotypic genus *Megalochilus*. We do not share this opinion. One subspecies of this species, *P. m. galli*, described from the Asian part of the species range, was characterized by longer legs and tail; a possibility exists, however, of clinal variation with a gradual increase in size from west to east. We do not discuss the subspecific subdivision of *Phrynocephalus mystaceus*. Golubev (1989) and Golubev and Sattorov (1992) believe that separation of this species into subspecies is not reasonable.

6. *Phrynocephalus ocellatus* is a valid senior synonym for *Ph. reticulatus*, and its population from Turkmenistan should be called *P. ocellatus bannikowi* (Golubev 1991) We should note here that the taxonomy of the genus *Phrynocephalus* is the most complicated one among all Palearctic lizards. New forms are constantly described within this genus, new synonymies are established, subspecies become species and vice versa. This situation probably can be clarified only through the application of new, non-traditional techniques of taxonomy.

7. Tertyshnikov and Shcherbak (1989) demonstrated that only the nominal subspecies of the glass lizard, *Ophisaurus apodus apodus*, is found within Middle Asia (as well as within the entire territory of the former USSR).

8. The revision of *Mabuya aurata* (Akhmedov and Shcherbak 1987) showed that a subspecies of this lizard found in Turkmenistan is *M. a. affinis*, not *M. a. septemtaeniata* as was previously thought (Bannikov *et al.* 1977).

9. Shcherbak and Akhmedov (1990) analyzed the geographic variation of *Eumeces schneideri* and discovered that populations from Transcaucasia and Middle Asia, although different, do not qualify for a level of subspecific forms and thus should be both placed in *E. s. princeps*.

10. Shcherbak (1990) studied the intraspecific variation in *Eumeces taeniolatus* and found that specimens from Turkmenistan form a separate subspecies, *E. t. parthianicus*.

11. Shcherbak (1974) demonstrated that the lizard species formerly known as *Eremias guttulata* belongs to the genus *Mesalina*; in Turkmenistan, the Asian subspecies *M. g. watsonana* is found.[3]

12. Shcherbak (1971) found that the lizard species *Eremias strauchi* is represented in Turkmenistan and northeastern Iran by a subspecies, *E. s. kopetdaghica*, whereas the nominal subspecies inhabits Transcaucasia.
13. Borkin and Darevsky (1987) elevated the subspecific form *Lacerta saxicola defilippii*, known from Kopetdagh, to a specific level.
14. Tokar (1990) demonstrated that the snake species *Eryx miliaris* from Turkmenistan belongs to a nominal form.
15. Until recently, the snake species *Coluber schmidti* and *C. caspius* have been both considered to be subspecies of *C. jugularis* (which is probably correct). Baran (1987), however, elevated their status, arguing that both forms are allegedly sympatric in Turkey. Shätti (1988) established a new genus, *Hierophis*, for *C. caspius*, *C. schmidti*, and a Chinese species, *C. spinalis*; we do not share this opinion.
16. There are two known forms of *Coluber rhodorhachis*. The nominal form, found in southern Turkmenistan, has a narrow red dorsal stripe. The other subspecies, *C. r. ladacensis*, is found sympatrically and also in the western part of the range (southern Uzbekistan to western Kyrghyzstan); it is characterized by a pattern of narrow transverse stripes formed by dark brown spots. Shammakov (1988) suggested that these are two different species; this opinion was uncritically accepted by some researchers (Borkin et al. 1987). However, numerous records of hybrid individuals between these two forms in Turkmenistan contradict their proposed specific status.
17. The intraspecific systematics of *Coluber ravergieri* are also complicated. Shätti and Agasian (1985) elevated the subspecies *C. r. numifer* (which, as well as the nominal form *C. r. ravergieri*, is found in Turkmenistan) to the rank of a separate species. This case requires further detailed study.
18. Dotsenko (1985) revalidated the snake genus *Pseudocyclophis*, which includes the species *P. persicus*, previously placed in the genus *Eirenis*. In Turkmenistan, two subspecies have been previously recorded, *P. p. persicus* and *P. p. walteri* (Bannikov et al. 1977), which differed primarily in their color pattern. Dotsenko (1985) demonstrated that a transverse stripe pattern (*"P. p. walteri"*) is found only in females, and individuals without this pattern (*"P. p. persicus"*) are males. Thus, the variation in pattern represents sexual dimorphism, and *P. p. walteri* is not a valid subspecies.
19. The systematics of the snake genus *Echis* is extremely complicated. Cherlin (1981, 1983, 1990) and Cherlin and Borkin (1990) formally elevated many subspecific forms to a specific rank; the descriptions of new forms sometimes have been based on a single specimen, without any discussion of individual or geographic variation. Stemler (1969) was the first to separate the Middle Asian *Echis* species from the related African and Indian ones. Stemler's work was apparently not noticed by Cherlin (1981), who assigned the name *E. multisquamatus* to the Middle Asian populations. However, the difference between *E. carinatus sochureki* and *E. mutisquamatus* is not more than subspecific (with 27–37 and 34–40 scales around the middle of the

body, respectively). Cherlin and Borkin (1990) even included these two forms in the same subgenus (sensu auct.). The correct name for the form inhabiting Turkmenistan is *Echis carinatus multisquamatus*.

20. The snake subspecies *Agkistrodon halys caucasicus* (which in Turkmenistan is found in Kopetdagh) was recently placed into *A. intermedius*, a species known from Siberia and the Far East (Gloyd and Conant 1990). We believe that this form is closer to *A. halys*.

Ecological Features of the Reptiles of Turkmenistan

A great diversity of landscapes and climates characterizes the territory of Turkmenistan. The distribution of reptile species is very specific, and different communities of reptiles can be found in areas similar in their landscape, soil, or climate. Some species of reptiles occupy very narrow niches, especially psammophiles who rarely come out of the sands and cannot be found even ten meters from a sand dune. An example of such a narrow psammophile is *Phrynocephalus mystaceus*, which is sensitive to a particular type and composition of sand. This species (as well as several others) lives exclusively in wind-driven, movable "pure" sands without any loess fraction and usually is found on the top of a dune.

Moving north from Ashkhabad, one crosses an approximately 10 km-wide belt of clay desert and approaches an area of sand dunes, 30 to 70 km wide and extending from north to south. Here, the sand has an admixture of clay, and *Phrynocephalus mystaceus* is practically absent. To the north of this area, "pure" sand dunes are found; here, this lizard species is very abundant. Its highest density, however, is recorded in completely naked sands next to wells, roads or settlements eroded by sheep grazing.

Certain Turkmenian psammophiles such as the gecko *Teratoscincus scincus* are found further east, in the Kizylkum Desert, where they may inhabit salt areas and even clay patches (*takyr*). Along with *Teratoscincus scincus*, another gecko species, *Crossobamon eversmanni*, is found in the sands (this last species can climb bushes quite well). The density ratio of these two species is 8:1 in "pure" sands (Central Karakum) and 1:1 in salty sands (the Caspian Sea shores). A habitat shift across the geographic range is very characteristic for the geckos. In Turkmenistan, *Mediodactylus russowi* lives on tree trunks of saxaul (*Haloxylon spp.*) and kandym (*Calligonum spp.*), whereas in Uzbekistan it can be found on loess cliffs and in human buildings. The Caspian gecko (*Tenuidactylus caspius*) inhabits rocks and buildings on the shores of the Caspian Sea; in western Turkmenistan, it is also found in buildings, on rocky cliffs of the Uzboi, Unguz, and Ustyurt, and also in the wells (*kyariz*) in the foothills of Kopetdagh. However, in the lowlands of East Karakum and Badkhyz, this species commonly inhabits rodent burrows, and even is found in the sand desert. We (Shcherbak and Golubev 1986) demonstrated that the distribution of the Caspian gecko is limited by moisture. It inhabits rocks in regions where the average humidity in May is not less than 40% and the average precipitation exceeds 100 mm (with about 100 mm of the total occuring during the warm period of the year). A significant decrease in humidity and precipitation east of the longitude of Ashkhabad corresponds to the general aridization of climate; it is exactly here where *T. caspius* shifts into the rodent burrows (Shcherbak and Golubev 1986).

Competition with closely related species can also be an important factor affecting the distribution of geckos. For instance, along the Murghab River *T. caspius* inhabits loess foothills, whereas a larger species, *T. turkmenicus*, is found on the rocks nearby. Several gecko species exhibit vicariant patterns of distribution. *T. caspius* is a common species in Turkmenistan eastward to the Amudarya River; further to the east it is replaced by another common species, *T. fedtschenkoi*. Similar patterns of vicariance are also known among other Middle Asian lizards, such as species of *Eremias*, *Stellio*, and *Phrynocephalus* (Bogdanov 1965).

To conclude our survey of sand desert reptiles, we should mention *Phrynocephalus interscapularis* (the smallest of all *Phrynocephalus* species) as well as the smallest and the largest of all *Eremias* lizards, *E. grammica* and *E. scripta*, respectively. *Eremias lineolata* is also commonly found in the sands, preferring salty areas.

The so-called group of *sclerobiotic* reptiles which inhabit clay and stone deserts is a prominent ecological component of the Turkmenistan herpetofauna. It includes the smallest of all geckos, *Alsophylax laevis*, and the related species, *A. pipiens*, which penetrates to northern Turkmenistan from adjacent steppes. The majority of *Phrynocephalus* species also belong to this group (*Ph. helioscopus, Ph. raddei, Ph. rossikowi, Ph. guttatus, Ph. ocellatus*) as well as *Eremias intermedia, E. nigrocellata*, and *Mesalina guttulata*. A characteristic inhabitant of the salt desert is *Phrynocephalus maculatus*.

The reptiles inhabiting mountains comprise a separate ecological group and the most numerous one (35 spp., Table 1). This group includes numerous geckos, other lizards (species of *Lacerta, Ablepharus, Eumeces*, and all species of *Stellio*), and several species of snakes (*Typhlops vermicularis, Eryx elegans, Coluber najadum*, and species of *Eirenis* and *Pseudocyclophis*). Mountain habitats are diverse enough to provide optimal conditions for very different reptiles, from a mesophile *Lacerta defilippii* to a xerophile *Eublepharis turkmenicus*, and from a soil dweller *Typhlops vermicularis* to the lizards clinging to rock cliffs (species of *Stellio* and many geckos). The majority of mountain reptiles belong to the families Colubridae (8 species), Gekkonidae (7 spp.), Scincidae (6 spp.), Agamidae (4 spp.), and Lacertidae (4 spp.). Their distribution among the different mountain ranges and plateaus is unequal: Kopetdagh has 25 species; Badghyz and Karabil, 21 spp.; Kugitang, 8 spp.; Bolshoi Balkhan, 3 spp.; and Malyi Balkhan, 2 spp. (Ataev 1985). The mountain plateaus of Badghyz and Karabil have 15 species of reptiles in common with Kopetdagh, and only 5 with the Kugitang Mountains. The diversity of herpetofauna, therefore, decreases from west to east, following the general gradient of increasing aridity.

Ataev (1985) demonstrated that the mountain reptiles have larger clutches and more eggs per clutch, longer reproductive cycles, and later maturation times than the reptiles of the lowland deserts in Turkmenistan. The lifespan of the mountain reptiles also is longer than that of the lowland species. Bogdanov (1965) observed that the reproductive season in mountain species is more predictable than in the lowland ones. Our observations of many years on the rare and endemic species of reptiles (more than 50% of all the mountain reptiles in Turkmenistan have narrow, localized ranges) show that they are not subject to the drastic fluctuations of density which characterize wide-ranging, and, especially, lowland species. Density fluctuations in mountain species have not been found even under harsh conditions such as severe winters or the return of cold weather in the spring, after reptiles have already emerged from their winter shelters.

Reptiles attracted to the water comprise the smallest ecological group (five species), a fact which can be expected in the desert landscapes. Many species, however, both in mountains and lowlands, can be found in the river valleys. Some snakes, such as the Asian cobra (*Naja oxiana*), feed here on amphibians. Others use river bank cliffs in the same way they use similar mountain habitats. The viper (*Daboia lebetina*) penetrates far eastward to Chardzhou along the Amudarya River Valley.

Finally, some species of reptiles are widely distributed both in mountains and in the desert (both sand and clay ones) of Turkmenistan. These are the Middle Asian tortoise (*Agrionemys horsfieldi*), lizards *Trapelas sanguinolentus* and *Varanus griseus*, and some snakes. Their life habits vary considerably within their wide ranges. For example, the Middle Asian tortoise has a larger body size in the mountains (average 286 mm) than in the lowland deserts (180 to 200 mm). Its period of surface activity is also longer in the mountains, where hibernation does not begin until late June or July, whereas in the lowland it starts in May. The tortoises also live longer in the mountains (up to 30 years) than in the lowlands (12 to 13 years).

Vegetation density is an important factor in reptile distribution. Lizards inhabiting sand and clay deserts (such as Geckonidae, Agamidae, and *Varanus griseus*) avoid areas where the vegetation covers more than 5 to 15% of the surface. Another ecological group of reptiles prefers the dense vegetative cover: this group includes *Opisaurus apodus*, *Daboia lebetina*, and lizard species of the genera *Ablepharus*, *Mabuya*, *Eumeces*. Some lizards and snakes require intermediate conditions between these two ecological extremes.

The sand and clay desert reptiles almost always build burrows, whereas the rock cliff dwellers use natural crevices and do not construct any shelters. There is a certain pattern in the order in which different age and sex groups of reptiles move down to the burrows for rewintering. First, the adult males move, then females, and eventually (usually well later), juvenile reptiles. In spring, the juveniles emerge from the winter shelters first; then, in a few days, adult males appear, and finally, females.

The reptiles of Turkmenistan can be divided into three groups according to their patterns of activity: (1) nocturnal species (all geckos, *Lycodon striatus, Eirenis meda, Pseudocyclophis persicus,* and *Lytorhynchus ridgewayi*); (2) diurnal species (turtles and tortoises, the majority of lizards other than geckos, many snakes), and (3) species which shift to nocturnal activity during the summer heat (snakes such as *Natrix* species, *Coluber ravergieri, C. karelini, C. rhodorhachis,* and *Naja oxiana*). Among the nocturnal reptiles, not every species engages strictly in night activity. Depending on a season, a species can be active at twilight or even after sunset or before sunrise. For example, geckos (*Tenuidactylus caspius, T. fedtschenkoi,* and *Mediodactylus russowi*) and snakes *Eryx* spp. can be observed to be active in the daylight in spring. The activity of reptiles depends primarily on the outside temperature; we found that the limiting temperature increases from spring to fall and depends on the age of an individual (young reptiles have a lower limiting temperature). The body temperature of an actively moving reptile is somewhat higher than the temperature of the environment. Also, behavioral regulation of body temperature is quite common and is maintained at the optimum of 34° to 37 °C. The lizards, *Phrynocephalus* spp. protect themselves from excessive heat by digging in the sand; the agamid *Trapelas sanguinolentus* climbs into bushes; and turtles (even the Middle Asian tortoise) bathe in water puddles.

The majority of reptiles in Turkmenistan are predators and feed on invertebrates, predominantly insects. Lizards feed primarily on the Coleoptera and Hymenoptera. Only the Middle Asian tortoise is usually considered a herbivore, although it has also been observed feeding on insects and scavenging on small dead animals. Some lizards (species of *Stellio, Phrynocephalus, Eremias*) also consume plants. Curiously, the presence of insects in snake stomachs is usually due to their consumption by a toad or a lizard which was eaten by this snake. However, such snake species as *Eirenis meda* and *Pseudocyclophis persicus* often eat insects. The only entirely insectivorous snake in the Turkmenistan fauna is *Typhlops vermicularis*. The majority of snakes prey on vertebrates such as rodents, birds, and lizards. Water turtles and the water snake *Natrix tessellata* eat principally fish. Amphibians are common in the diet of *Ptyas mucosus, Natrix tessellata,* and *Natrix natrix*; the Asian cobra (*Naja oxiana*) commonly feeds on toads. Some snakes, such as *Psammophis lineolatus, Coluber rhodorhachis* and *C. karelini,* are specialized herpetophages; many other snakes readily eat reptiles (e.g., spp. of *Eryx, Coluber, Ptyas, Lycodon, Litorhycnchus, Naja, Echis, Boiga,* and *Agkistrodon*). Turtles are consumed by the grey monitor, cobra, and viper; snakes have been found in stomachs of other snakes (such as *Ptyas, Naja, Echis,* and *Agkistrodon*), of the agamid lizard *Stellio erythrogaster,* and of the grey monitor. Twelve species of reptiles feed on birds (spp. of *Varanus, Opisaurus, Eryx, Coluber, Natrix, Daboia, Naja, Echis, Boiga,* and *Agkistrodon*); of these, the viper (*Daboia lebetina*) specializes in small passerine birds which it ambushes at drinking spots. Mammals (predominantly rodents) are a common diet of lizards *Varanus griseus* and *Ophisaurus apodus* and of 14 species of snakes (spp. of *Eryx, Coluber, Ptyas, Spalerosophis, Elaphe, Daboia,*

Echis, and *Agkistrodon*). The composition of the reptilian diet often changes seasonally.

A brief characterization of the reproductive features of reptiles in Turkmenistan reveals that they have a longer reproductive period than the reptiles of more northern territories. Snakes and some lizards (*Varanus griseus*, *Ophisaurus apodus*, *Mabuya aurata*, and *Eumeces* spp.) have only one clutch per season, whereas turtles and the majority of lizards lay eggs several times. The geckos lay one to two eggs at a time and also several lizards may lay eggs in the same place. Among the species of *Phrynocephalus*, those that live in clay desert reproduce earlier and lay more eggs than the sand desert species. Viviparity is known in the lizard *Mabuya aurata* as well as in the snakes *Echis carinatus* and *Eryx* spp.; it must be considered the most progressive reproductive adaptation to the dry and hot climate.

Well-nourished individuals of desert reptiles undergo molting in spring, summer or fall. We (Shcherbak and Golubev 1986) found that the first molt in some gecko species occurs during their hatching from the egg.

The enemies of reptiles are usually other vertebrates. Among the invertebrates, only scorpions (*Mesobuthus* spp.) and solpugids have been found occasionally feeding on small geckos. The water snakes (*Natrix* spp.) have been attacked by a wels (*Silurus glanis*). Herpetophagous snakes and lizards have been discussed above. Birds of prey, especially falcons (*Falco tinunculus*, *F. naumanni*) and eagles (*Circaetus ferox*), are also very important predators of reptiles. The list of herpetophagous birds in Turkmenistan includes 66 species. Reptiles are also consumed by many mammals, including the hedgehogs (*Hemiechinus auritus* and *Paraechinus hypomelas*), the shrew *Diplomesodon pulchellum*, the badger (*Meles meles*) and the honey badger (*Mellivora indica*). This last predator is capable of tearing tortoise shells apart with its mighty claws. Other mustelids (*Mustela eversmanni*, *Vormela peregusna*, and *Martes foina*), the wolf (*Canis lupus*), the jackal (*Canis aureus*), the desert fox (*Vulpes corcsac*), and the red fox (*Vulpes vulpes*) are also herpetophagous. One late fall in West Turkmenistan the author found a part of viper (*Daboia lebetina*) head with a poisonous fang in a fox's stomach; this snake evidently had been dug out by the fox from its winter shelter. Turtles also are consumed by the striped hyena (*Hyaena hyaena*). Wild cats (*Felis lybica* and *F. chaus*) and some rodents also feed on lizards and snakes. In all, the list of mammals that feed on reptiles in Turkmenistan includes 20 species.

Severe winters without ample snow cover, which are common in Turkmenistan, and onsets of cold weather in the spring can kill reptiles in numbers. Once, after a severe winter, the author found 18 skeletons of the Caspian gecko (*Tenuidactylus caspius*) in an old loess cliff at Kunya Urgench (northern Turkmenistan). Spring mudslides (*sel'*) also damage populations of the reptiles. A few hours after the mudslide near Kushka (southern Turkmenistan), we collected more than 50 dead or half-dead reptiles, including 39 snakes (*Eryx miliaris* and cobras) and lizards *Trapelus sanguinolentus* and *Eremias* spp. Flooding of human-made reservoirs following winter and spring rains also kills reptiles. In 1977, a major flood destroyed a dam on the Tedzhen River and eradicated the type population of *Alsophylax laevis* and many other lizards. A similar event occured in the 1960s when new-built reservoirs were filled on the Murghab River; dead lizards floating in water were consumed in great numbers by birds (Bogdanov 1965). In general, human activity is the major factor now threatening the natural populations of reptiles in Turkmenistan. Irrigation destroys entire populations; vehicles crush individual reptiles on highways. Recently, water from irrigation fields was disposed onto a salt pan (*shor*) next to Bami Station, the only known site in Middle Asia of *Phrynocephalus maculatus*. In spite of scientists' protests, this unique population of a Red Data Book species was destroyed.

Zoogeographical Analysis of the Herpetofauna of Turkmenistan

A comparative study of the reptilian endemism in Middle Asia and the adjacent countries of Iran and Afghanistan (Rustamov 1981) has demonstrated that the reptilian fauna of Middle Asia has highly expressed specific endemism (15.1% of species), whereas Iran has 12.7%, and Afghanistan, only 2.0%. The following ten species and subspecies of reptiles are endemic or subendemic for Turkmenistan: *Alsophylax loricatus szczerbaki*, *Eublepharis turkmenicus*, *Tenuidactylus turkmenicus*, *Phrynocephalus rossikowi schammakowi*, *Ph. ocellatus bannikowi* (= *Ph. reticulatus*), *Stellio caucasius triannulatus*, *Eumeces taeniolatus parthianicus*, and *Eremias strauchi kopetdaghica*. Twenty-three species subendemic for Middle Asia are found in Turkmenistan, including representatives of the genera *Tenuidactylus*, *Teratoscincus*, *Crossobamon*, *Stellio*, *Trapelas*, *Phrynocephalus*, and several species of *Eremias*. Many species are found only in western Turkmenistan (e.g., *Mesalina guttulata*, *Mabuya aurata*, *Stellio caucasius*, *Coluber schmidti*, and *Eirenis meda*); others are found only in the east of the republic (e.g., *Tenuidactylus fedtschenkoi*, *Eremias regeli*, *Stellio chernovi*, and *S. lehmani*). Still others inhabit only northern portions of the republic (e.g., *Alsophylax pipiens*, *Elaphe quatuorlineata*, *E. dione*, and *Agkistrodon halys caraganus*) or southern regions (e.g., *Tenuidactylus longipes*, *T. turkmenicus*, *Stellio erythrogaster*, *Eremias persica*, and *Ptyas mucosus*). These patterns of distribution depend on historical events as well as on the actual landscape and climatic diversity. One can delimit the biogeographic territories which possess a characteristic set of species. Reptiles have rather limited dispersal abilities and are ecologically conservative animals. These considerations allowed us (Shcherbak 1982; Rustamov and Shcherbak 1985, 1986, 1989, 1990) to design a general scheme of the zoogeographic regionalization of Middle Asia based on the distribution patterns of herpetofauna.

The territory of Turkmenistan lies within the so-called Mediterranean-Asian Arid subregion of the Palearctic region. According to the landscape principle of zoogeographic subdivision, this territory includes parts of three major provinces: the Irano-Afghan Transitional province the Turanian Lowland Desert province, and the Asian Mountain province. The indexes (4, 5, or 6) given here for the provinces, subprovinces, districts, and areas within Turkmenistan correspond to the general indexation adopted for all of Middle Asia (Shcherbak 1982) (Fig. 2). Below we list the zoogeographic subdivisions for Turkmenistan and provide a brief characterization of each.

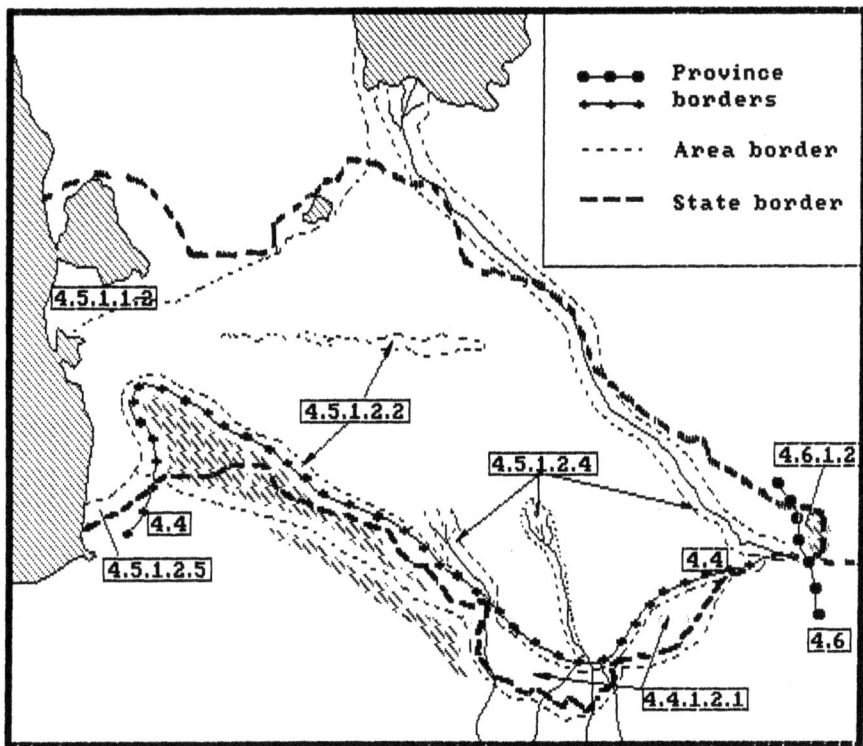

Fig. 2. Zoogeographic regions of Turkmenistan based on the fauna of reptiles (see text for the description of regional units).

Irano-Afghanian Transitional Province

Irano-Afghanian Mountainous Subprovince

Iranian Mountainous District

Khorassan-Kopetdagh Area. Within Turkmenistan, this area includes the Kopetdagh Mountains. The characteristic and endemic reptiles include *Eublepharis turkmenicus, Mediodactylus spinicaudus, Eremias strauchi kopetdaghica*, *Eryx elegans*, *Oligodon taeniolatus*, *Eirenis meda*, and *Pseudocyclophis persicus*.

Northern Afghanian District

Badghyz-Karabil Area. The characteristic and endemic reptiles include *Tenuidactylus turkmenicus*, *T. longipes microlepis*, *Bunopus tuberculatus*, *Ophiomorus chernovi*, *Eremias persica*, and *Stellio erythrogaster*.

Turanian Lowland Desert Province

This province embraces the Middle Asian deserts and semideserts. A large part of Turkmenistan is occupied by the Karakum Desert, which is an important center of origin of the desert herpetofauna, including species of the genera *Eremias*, *Phrynocephalus*, *Teratoscincus*, *Crossobamon*, and *Agrionemys*).

Karakum Subprovince

This subprovince occupies the vast territory from the Caspian Sea to the Amudarya River and includes sand deserts, *takyrs* (clay deserts), semidesert areas of the Ustyurt Plateau, and riparian forests (*tugai*) of the Tedzhen, Murghab, and Amudarya Rivers. It is ecologically diverse and is subdivided into two districts and five areas.

Ustyurt-Caspian District
The district occupies the Ustyurt Plateau and the northeastern shores of the Caspian Sea. The habitats here include rocky cliffs (*chink*), a semidesert plateau with clay desert patches, and shoreline salt desert.

Ustyurt Area. Only the southern part of this area lies within Turkmenistan, where it includes southern cliff (*chink*) of the Ustyurt Plateau, built of limestone and gypsum. This area includes also the Kaplankyr Plateau and adjacent salt depressions (such as Sarykamysh). The characteristic representatives are: *Alsophylax pipiens*, *Phrynocephalus ocellatus bannikovi* (= *Ph. reticulatus*), *Elaphe quatuorlineata sauromates*, and *Eremias arguta*.

Karakum District
This district includes the Karakum Desert with a variety of habitats. Sands with various degrees of vegetative cover are interspersed with clay desert (*takyr*), salt desert (*shor*), and stony-gypsum desert areas.

Karakum Sand Desert Area. This area includes sand desert of various types. Characteristic reptiles include *Phrynocephalus interscapularis*, *Ph. mystaceus*, *Ph. raddei*, *Ph. rossikowi*, *Ph. r. schammakowi*, *Eremias grammica*, *E. scripta*, and *Teratoscincus scincus*.

Karakum Takyr Area. The most prominent *takyr* (clay desert) territory in Turkmenistan forms a latitudinally directed belt to the west and north from the foothills of the Kopetdagh Mountains. *Takyr* patches also can be found within sand desert, but the characteristic *takyr* fauna there is impoverished or absent. Irrigated portions of the clay desert are significantly transformed by human activities. Among the characteristic species are *Alsophylax laevis, Phrynocephalus helioscopus, Eremias intermedia, E. lineolata,* and *E. velox.*

Amudarya Tugai Area. The area includes riparian forests (*tugais*) along the valleys of the Amudarya, Tedzhen, and Murghab Rivers. Here, a number of desert species can be found, including *Tenuidactylus caspius, Trapelus sanguinolentus,* and *Varanus griseus,* and the snakes, *Coluber* spp. Some snakes penetrate along the *tugais* from the north (*Elaphe dione*) and from the south (*Ptyas mucosus* in the Murghab Valley and *Daboia lebetina,* in the Amudarya Valley); the water snake *Natrix tessellata* is also common in the *tugais*. The *tugais* house a relict population of *Ablepharus deserti.* Next to the Amudarya Valley are found endemic forms of the lizards *Alsophylax loricatus szczerbaki* and *Phrynocephalus rossikowi rossikowi.*

Atrek Tugai Area. This includes the lowland portion of the Atrek River. The *tugai* fauna here is quite different from that in the Amudarya Valley. Only in this area can one find species of western origin such as turtles, *Emys orbicularis* and *Mauremys caspica,* and the snake, *Coluber schmidti.*

Asian Mountainous Province

This province includes the mountain ranges of Ghissaro-Alai, Tien Shan, Pamir, and Tibet. Within Turkmenistan, it includes only the mountain system of Kugitang (the easternmost range of Ghissaro-Alai).

Middle Asian Mountainous Subprovince

This occupies the territory from Kugitang in the west to Tien Shan in the east. The landscapes of this subprovince exhibit well-expressed altitudinal zonation. The middle mountain belt is covered by deciduous forest and stands of the treelike juniper, or *archa* (*Juniperus* spp.).

Ghissaro-Alai District
This district includes the mountain ranges lying between the Amu Darya River, Ferghana Valley, and Alai Valley. We treat the altitudinal mountain belts as zoogeographic areas.

Ghissaro-Turkestan Foothill Area. This includes the foothills (*adyrs*) of the Kugitang range (500 to 1,400 m). The characteristic representatives are:

Tenuidactylus fedtschenkoi, Eremias regeli, E. velox, and *E. nigrocellata.* Many Turanian elements, such as *Agrionemys horsfieldi, Eumeces schneideri, Coluber ravergieri, C. rhodorhachis, Eryx miliaris tataricus, Daboia lebetina,* and *Naja oxiana,* penetrate to this area from the west.

Ghissaro-Turkestan Juniper Area. This area occupies the middle belt of the Kugitang range (1,400 to 2,600 m). The only characteristic representative is *Stellio lehmani.* Other reptiles found here include *Agrionemys horsfieldi, Ophisaurus apodus, Varanus griseus, Natrix tessellata, Psammophis lineolatus,* and *Coluber* spp.

Ghissaro-Turkestan Highland Area. This occupies the rock and subalpine belts of Kugitang (2,600 to 3,000 m). The herpetofauna here is impoverished; a characteristic representative is *Stellio chernvoi* (= *S. himalayana*?).

Analysis of geographic ranges shows that the herpetofauna of Turkmenistan is a very heterogeneous one. It includes 9 faunistic groups which differ in their zoogeographic affinities. The core of Turkmenistan reptiles are the *Turanian* species (36 spp.), primarily restricted to the lowland deserts (only fifteen species are found also in the mountains; Ataev 1985). In contrast, the *Irano-Afghanian* and *Southwest Asian* species (17 spp.) are found principally in the mountains. The *Sahara-Sindian* reptiles are represented by seven species: *Mesalina guttulata, Varanus griseus, Mabuya aurata, Eumeces schneideri, Spalerosophis diadema, Echis carinatus,* and *Psammophis schokari.* Five species of snakes comprise the *Indian* group of species; these are *Lycodon striatus, Ptyas mucosus, Oligodon taeniolatus, Boiga trigonatum,* and *Naja oxiana.* The *Caucasian-Anatolian* group includes four species (*Stellio caucasius, Lacerta strigata, L. defilippii, Coluber schmidti*). Another group of four species includes reptiles of the *European-Mediterranean* ranges (*Emys orbicularis, Natrix natrix, N. tessellata,* and *Ophisaurus apodus*). The *Kazakhstan-Mongolian* group of three species includes the lizards *Alsophylax pipiens, Eremias arguta,* and *E. velox,* whereas the *Asian Mountain* group is represented by only two lizard species, *S. chernovi* and *Stellio lehmani.* Finally, two species of the wide *East Palearctic* range are snakes *Elaphe dione* and *Agkistrodon halys.* This rough estimation (several species do not easily fit into these groupings) shows that the majority of reptiles in Turkmenistan are autochthonous elements and that Turkmenistan has strong biogeographic connections with adjacent southern and western areas. Similar conclusions have been reached by other authors (Rustamov 1966; Ataev 1985).

Rare and Endangered Species of Reptiles in Turkmenistan and their Conservation

The problem of conservation of reptiles in Middle Asia has been discussed in numerous papers (Rustamov and Shammakov 1975; Rustamov and Makeev 1981; Rustamov 1983; Rustamov and Shcherbak 1987; Rustamov et al. 1988) and monographs (Shcherbak 1974; Shcherbak and Golubev 1986), as well as in the Red Data Books of the (former) USSR and the republics of Middle Asia. The only effective means to such conservation is the protection of reptiles in the natural reserves (*zapovedniki*). In Turkmenistan, the existing natural reserves play a significant role in the protection of reptiles (Table 2).

Table 2. The number of species of reptiles protected in the natural reserves of Turkmenistan

Natural Reserve	Total number of species of reptiles	Number of species included in the Red Data Book of the USSR
Badghyz	37	11
Syunt-Khasardagh	35	5
Krasnovodsk	35	3
Kopetdagh	34	6
Repetek	21	4
Kaplankyr	13	0

Of 35 reptile species listed in the Red Data Book of the USSR, 20 are found in Turkmenistan. This list includes 12 monotypic species and seven monotypic genera. However, a number of extremely rare and endangered species still remain unprotected, notwithstanding the numerous proposals for protected status submitted to state conservation departments (Rustamov and Shcherbak 1990). These animals include *Phrynocephalus maculatus*, found only in a 6 × 14 km area on the Ulyshor salt desert; the urgency regarding its protection already has been discussed for more than a decade. An area next to the Meshed-Messerian ruins is suitable for protection of the gecko species, *Alsophylax laevis*, and a protected area should be established next to Kunya Urgench including a local medieval cemetery for *Alsophylax loricatus*. *Ptyas mucosus* which could be protected along the Murghab River, and *Phrynocephalus rossikowi*, for which an area should be established in the stony desert next to the Unguz, are also threatened. The natural reserves make only 2.2% of the entire territory of Turkmenistan. Although these reserves constitute two times more than the total territory of the natural reserves in all other Middle Asian republics, they are not sufficient for the effective protection of many species. The desert soils are highly vulnerable to human impact, and they virtually cannot be restored after transformation from such activities as irrigation or dragging of disassembled oil rigs.

Even when conservation efforts are made, they still may not yield the desired effect. Recently, zoologists from Turkmenistan undertook an introduction of about 200 individuals of *Phrynocephalus maculatus* to a *shor* (salt desert patch) surrounded by sands, 50 km north from the original population at the Ulyshor. Four months later, we thoroughly checked this area but did not discover any *Ph. maculatus*. The lizards have evidently been eliminated by predators, most probably by the large agamid lizard, *Trapelas sanguinolentus*, which was absent from their primary habitat.

Notes

1. Shenbrot and Semenov (1990) described this population as a new species, *Ph. golubevi.* – Eds.
2. This viewpoint (see also Szczerbak 1989) was also critisized by Böhme (1985) and Kluge 1985) but accepted by other researchers (Ananyeva *et al.* 1988; Khan 1991). Leviton *et al.* (1992) use the name *Mediodactylus* only as subgenus of the Paleacrtic gecko genus *Cyrtopodion*. – Eds.
3. Arnold (1986) considers *Mesalina watsonana* a separate species. – Eds.

Lycodon striatus. Photo by N.N. Shcherbak.

20. Reptiles of Kopetdagh

CHARY ATAEV, ANVER K. RUSTAMOV AND SAKHAT SHAMMAKOV

Abstract

The northern slopes of Kopetdagh are inhabited by 45 species of reptiles (three species of turtles, 18 spp. of lizards, and 24 spp. of snakes), which corresponds to 57% of the Turkmenistan reptile fauna (79 species). Of these, East Kopetdagh has 30; Central Kopetdagh, 36; Southwest Kopetdagh, 36; and Northwest Kopetdagh, 23 species. Seventeen species of reptiles are found from the foothills to upper altitudinal belts, whereas 23 species are found also in the submontane lowlands. Lower mountain belt (from 100 to 1,000 m) has 22 species (48.8%) of reptiles, middle mountain belt (from 1,000 to 2,000 m) has 14 species (31.1%), and upper mountain belt (from 2,000 to 2,500 m) has 9 species (20.0%). Only four species are highly abundant (more than 10 animals/ha); 15 species are common (from 1 to 10 animals/ha); and 26 species are rare (less than 1 animal/ha). Thirty-nine species are oviparous, and six are ovoviviparous. Ten biological groups are separated: mountain species with extended life cycle, and lowland ones with short life cycle; species with extended life cycle have more diverse reproductive strategies. Eight zoogeographic groups are separated: Irano-Afghanian (21 species), Turanian (7 spp.), Indian (5 spp.), Mediterranean (4 spp.), Saharo-Sindian (4 spp.), Caucasian-Anatolian (2 spp.), European (2 spp.), and East Palaearctic (1 sp.). The Kopetdagh endemics include lizards *Eublepharis turcmenicus* and *Eremias strauchi kopetdaghica*, and snakes *Eryx elegans*, *Coluber atayevi*, and *Eirenis meda*.

Introduction

As compared to the eastern mountains of Middle Asia, the Kopetdagh mountain range contains considerably higher diversity of reptile species; there are 45 species of reptiles known only from northern slopes of Kopetdagh whereas the entire Tien Shan has only 25 species. Kopetdagh's location at the crossroad of different faunistic provinces, its landscape and climatic conditions, and the existence of an ancient center of origin of desert fauna make the herpetofauna of this area unique in its zoogeography and ecology (Rustamov 1966; Darevsky 1981b; Ataev 1985).

Only the northern slopes of Kopetdagh lie within Turkmenistan, whereas the rest of this range (Turkmeno-Khorassan Mountains) is located in Iran. This brief review of reptilian fauna was produced on the basis of our original collections, field observations, and publications as well as on the published data of other authors.

The reference list on the Kopetdagh reptiles includes more than 100 papers. The earliest ones were published in the 1890s and 1900s (Boettger 1890; Boulenger 1891; Zarudnyi 1897, 1903; Varentsov 1894); these were followed by monographic works of Nikolsky (1915, 1916). Numerous later researchers of the Kopetdagh herpetofauna included Morits (1929), Chernov (1934), Terentyev and Chernov (1949), Bogdanov (1962, 1965), Rustamov, Kurbanov and Sopyev (1962), Rustamov (1966, 1981), Shammakov (1968), Shcherbak (1974), Rustamov and Shammakov (1982), Skalon (1982), Makeev, Bozhansky, and Khomustenko (1983, 1986), Bobrov (1986), Shcherbak, Khomustenko, and Golubev (1986), and Rustamov and Shcherbak (1987, 1990). Finally, Ataev (1985) published a monographic study of the mountain reptiles of Turkmenistan, primarily the Kopetdagh fauna. In contrast, reptiles of Iranian Kopetdagh are relatively poorly studied (Zarudnyi 1897; Anderson 1963, 1968, 1974; Latifi 1991).

Annotated List of Species[1]

Order Testudines

Family Emydidae

1. Emys orbicularis (Linnaeus, 1758). Inhabits the Atrek, Chandyr, and Sumbar Rivers and water bodies of their drainages. Density can reach 20–25 turtles/ha (Ataev 1985). They overwinter from October to February, burying themselves into bottom mud 10 to 15 cm deep. Active turtles appear in March and soon start mating (copulation has been recorded both in water and on land). From May to July females can lay eggs up to three times, having two to nine eggs per clutch. Young turtles stay under water until the next spring. Food includes molluscs, frogs (*Rana ridibunda*), and algae (Rustamov *et al.* 1962).

2. Mauremys caspica (Gmelin, 1774). The nominal form, *M. c. caspica*, inhabits the Atrek, Chandyr, and Sumbar Rivers; prefers flowing waters. Density can reach 20 turtles/ha. They spend winter from October to March on the bottom. Females have three clutches from May to July, five to ten eggs in each. Young turtles hatch in September and spend winter in soil. The Caspian turtle feeds in water as well as on land; food includes beetles, molluscs, fish, frogs (*Rana ridibunda*), algae, and reed.

Family Testudinidae

3. *Agrionemys horsfieldi* (Gray, 1844) – Steppe Tortoise. In Turkmenistan, the nominal subspecies (*A. h. horsfieldi*) inhabits Kopetdagh, whereas the adjacent lowland of Southwest Turkmenistan is inhabited by a diffferent subspecies, *A. h. rustamovi* Chkhikvadze *et al.* (Chkhikvadze *et al.* 1990). Steppe tortoise is common in all habitats in Kopetdagh but prefers steppe-like mountain areas. Density is usually 10–15 animals/ha, and in certain favorable habitats it can reach 150–200 animals/ha (Ataev 1985). Tortoises appear from winter shelters in late March and in a few days start mating. Females lay two to five eggs in April or May. Young tortoises hatch after 70–75 days of embryonic development; they stay in soil over winter. Adult tortoises start summer diapause in late May or in June; in late September or October some of them can interrupt hibernation and appear on the surface for a short time to feed on vegetation which appears due to fall rains (Ataev 1974, 1975, 1977, 1979); most, however, stay in their shelters until the next spring. Tortoises feed on dozens of different plant species, including *Astragalus paucijudus*, *Dodartia orientalis*, *Iris kopetdaghensis*, *Bromus tectorum*, *Allium sabulosum*, and *Papaver pavoninum* (Shammakov 1981).

Order Squamata/Suborder Sauria

Family Agamidae

4. *Stellio caucasius* Eichwald, 1831. The nominal subspecies (*S. c. caucasius*) of this agamid lizard is widespread in the Kopetdagh Mountains (Zarudnyi 1897, 1903; Kolesnikov 1956; Kamalova 1969; Bogdanov 1962; Makeev 1972). Another subspecies, *S. c. triannulatus* Ananjeva et Ataev 1984, is found in the Madau desert, 100 km west from West Kopetdagh (Ananyeva and Ataev 1984).

S. c. caucasius inhabits rocky slopes with poor vegetation and other rocky habitats. Density is high, sometimes 30–40 animals/ha; the entiire Kopetdagh population of Caucasian stellions is estimated at 11 million animals (Ataev 1985). It mates in April or May. In May through July, females lay from three to five eggs into a shallow hole under rocks. In approximately 60 days young lizards (40–45 cm long) hatch from the eggs. In late October or November, up to 30 or 40 lizards gather in deep rock crevices for winter shelter (Ataev 1973, 1974, 1977). This species feeds on various insects but also consumes leaves, flowers, and seeds of plants (Shammakov 1963, 1964; Panov and Zykova 1985; Ataev 1985).

5. *Stellio erythrogaster* Nikolsky, 1896. The nominal subspecies is found in Badghyz; in Kopetdagh, a new subspecies was recently discovered and

described as *S. e. nurgeldievi* Tuniev, Atajev et Shammakov, 1991 (Tuniev *et al.* 1991). It was found in the loess foothills of East Kopetdagh next to the village of Khiveabad (elevation 400 to 500 m), in sparse pistachio groves in the Kaptaringa valley (500–600 m, 15–20 km from the Kaakhka Station), and on the right bank of the Meanachai River (Gorelov and Lukarevsky 1990; Ataev *et al.* 1991). The species is common in these habitats; in Khiveabad, nine lizards were found during a five-hour survey, and in Kaptaringa, eleven lizards in four hours.

6. *Trapelas sanguinolentus* (Pallas, 1814). We accept the Kopetdagh population as a nominal subspecies, because the validity of *T. s. aralensis* (Lichtenstein 1823) is problematic.

This agamid species is found everywhere in Kopetdagh; in dry foothills, density reaches 20–30 animals/ha. It lays eggs two or three times from April to July; each clutch has from 3 to 18 eggs, usually 7 or 8. The embryonic development lasts 50 to 60 days; young lizards hatch in June or July. It feeds mostly on Coleoptera, Hymenoptera, Orthoptera, and Lepidoptera (caterpillars) (Bogdanov 1962, 1965a,b; Ataev 1986; Shcherbak *et al.* 1986).

Family Anguidae
7. *Pseudopus apodus* (Pallas, 1775) *(= Ophisaurus apodus)*[2]. The form found in Kopetdagh probably belongs to the western subspecies, *P. a. thracicus* Obst 1978. This "glass lizard" is common and is found everywhere in valleys, on mountain slopes, and on mountain steppes. It appears on the surface in mid-March, but by late June-July begins a hibernation which continues through fall and winter. Glass lizards mate in April or May. Females lay six to ten eggs only once. Young lizards hatch in July through September. *P. apodus* feeds on insects, especially on Coleoptera.

Family Gekkonidae
8. *Cyrtopodion caspius* (Eichwald, 1831) *(= Tenuidactylus caspius)*. The Kopetdagh population belongs to the nominal subspecies. The Caspian gecko is widespread in Kopetdagh in foothills, on mountain slopes, in river valleys, rocky canyons, caves, wells, kyarizes, and buildings. Its density can reach 100–150 animals/ha. The Caspian gecko is usually active from March to October although single active animals can be found in winter. This species has crepuscular and nocturnal activity but can be found in daytime in places that are well protected from sunlight. The Caspian gecko lays eggs two or, possibly, three times in a season (May through August). It feeds on Coleoptera, Orthoptera, Isoptera, and Aranei.

9. *Cyrtopodion spinicauda* (Strauch, 1887) *(= Mediodactylus spinicaudus)*. The spine-tailed gecko is found in all parts of Kopetdagh on rocky slopes with outcrops of limestone and in valleys filled with gravel and pebbles. It is possibly not as rare as it has been considered before; sometimes in Kopetdagh its density

reaches 4–5 animals/ha (Kurilenko 1985; Shcherbak *et al.* 1986). This species is included in the Red Data Book of Turkmenistan (Ataev *et al.* 1985).

10. Eublepharis turcmenicus Darevsky, 1977. Within Turkmenistan, this lizard is found only in Kopetdagh where, in about 100 years of herpetological studies, only about 30 specimens have been captured or recorded (Darevsky 1977; Rustamov *et al.* 1985; Ataev 1978). It is found in rocky foothills and on rocky slopes of the middle altitudinal belt with poor xerophyte vegetation. The Turkmen eublephar is a strictly nocturnal species which hides in the daytime in empty space under rocks. It is active from April to August; females lay two eggs only once in a season. It feeds on various insects (Pashchenko 1964; Shcherbak 1974; Rustamov *et al.* 1985). This species is included in the Red Data Book of Turkmenistan (Ataev *et al.* 1985).

Family Lacertidae
11. Eremias strauchi Kessler, 1878. In Kopetdagh, the endemic subspecies *E. s. kopetdaghica* Szczerbak, 1971, is found. It is common (4–6 animals/ha; Ataev 1985) in Central and West Kopetdagh, where it inhabits mountain steppes, slopes, and rocky valleys with grass, shrub, and tree vegetation. *E. strauchi* is active from April to October; it lays eggs twice in a season and has two to six eggs in a clutch. This lizard feeds primarily on Acrididae, Coleoptera, and caterpillars (Vashetko 1969; Shcherbak 1974; Rustamov and Ataev 1976, 1985; Shcherbak *et al.* 1986).

12. Eremias velox (Pallas, 1771). The nominal subspecies is found everywhere in Kopetdagh (rocky valleys, mountain slopes, and foothills). It is abundant (sometimes 20 to 30 animals/ha). *E. velox* has two to four eggs in a clutch; embryonic development lasts 40 to 50 days. Yound lizards hatch in late June or early July. The common diet of *E. velox* consists of Coleoptera, Formicidae, Acrididae, and Isoptera (Vashetko 1970; Shcherbak 1974; Vashetko and Chernyakhovsky 1972; Shammakov 1964, 1968, 1969, 1981; Ataev 1985).

13. Mesalina guttulata (Lichtenstein, 1823). This lowland lizard is only occasionally found in ephemerous foothills of Kopetdagh; our population belongs to the subspecies *M. g. watsonana* (Stoliczka 1872).[3]

14. Lacerta raddei Boettger, 1892. Two isolated populations belonging to the nominal subspecies *L. r. raddei* are found in the moist, deep, and narrow rocky valleys of Bolshie Karanki and Sushanka in Central Kopetdagh (Khomustenko and Ataev 1979). The entire population is estimated as 3,000 to 4,000 animals. The species is listed in the Red Data Book of Turkmenistan (Ataev *et al.* 1985).

15. Lacerta strigata Eichwald, 1831. This lizard is common in Southwest Kopetdagh (Ataev 1985). It is found in moist places with herbaceous vegetation; its density can reach 15–20 animals/ha.

Family Scincidae
16. Ablepharus pannonicus (Lichtenstein, 1823). This small lizard is common in all parts of Kopetdagh where it prefers habitats with dense vegetation. In Southwest Kopetdagh its density sometimes is extremely high (200–300 animals/ha). *A. pannonicus* is active from April to October. Females lay eggs twice, in late April or May and in June; one clutch has two to six eggs. It feeds on Acrididae and Blattoptera (Bogdanov 1962, 1965a; Ataev 1985; Shcherbak et al. 1986).

17. Chalcides ocellatus (Forskal, 1775). It is known only by two specimens from the Mergenolen Valley in Central Kopetdagh (Darevsky 1981a); identified as a nominal form, *C. o. ocellatus*. This lizard species is included in the Red Data Book of Turkmenistan (Ataev *et al.* 1985).

18. Eumeces schneideri (Daudin, 1802). The subspecies *E. s. princeps* Eichwald 1839 is widespread in Kopetdagh and commonly found in foothills, on mountain slopes, in valleys, and on agricultural lands. It hides in burrows and under rocks. *E. schneideri* is active from April to mid-June; it then goes to shelters for summer hibernation, which lasts through fall and winter (Ataev 1966, 1972, 1985). Skinks lay one clutch of three to six eggs. Their main foods are Coleoptera, Acrididae, Lepidoptera (caterpillars), and Isoptera.

19. Eumeces taeniolatus Blyth, 1854. The Kopetdagh population belongs to the subspecies *E. t. parthianicus* Szczerbak, 1990 (Shcherbak 1990). This skink is common in foothills, mountain slopes and valleys, and in mountain steppe. Its density can reach 10 to 15 animals/ha. *E. taeniolatus* are active from April to July; after that, they hibernate through summer, fall, and winter. From May to June skinks lay one clutch of three to six eggs. Their main foods are Coleoptera and caterpillars (Vasilyev 1905; Bogdanov 1962; Ataev 1970, 1980).

20. Mabuya aurata (Linnaeus, 1758). The Kopetdagh population belongs to the subspecies *M. a. septemtaeniolata* Reus., 1834.[4] This lizard is quite common (up to 10 to 15 animals/ha) from April to September on mountain slopes and in valleys, in foothills, and in ruins. The species is ovoviviparous; females with developed embryos were found in from May to July. The usual foods of *M. aurata* are Coleoptera, Orthoptera, and Aranei (Ataev 1985; Shcherbak *et al.* 1986).

Family Varanidae
21. Varanus griseus (Daudin, 1803). The grey monitor inhabits lowlands of Turkmenistan and very rarely has been found in the mountains. Only 23 individuals have been recorded in a total of 1,082 km surveyed in the mounatins of Turkmenistan (Ataev 1985). This giant lizard is included in the Red Data Book of Turkmenistan (Ataev *et al.* 1985).

Suborder Serpentes

Family Boidae

22. *Eryx elegans* (Gray, 1849). In Turkmenistan, this snake is found only in Kopetdagh. It inhabits valleys, slopes, and mountain plateaus where its density can reach 1.3 animals/ha (Zinchenko and Ritskov 1989). *E. elegans* is active in daytime from April to October. It reproduces once in a season; females with four to eight completely developed embryos were found in August and early September. *E. elegans* actively hunts rodents, mostly mice (*Mus*) and voles (*Microtus*). This species of snakes is included in the Red Data Book of Turkmenistan (Ataev et al. 1985).

23. *Eryx miliaris* (Pallas, 1773). This typical lowland desert snake is known from Kopetdagh only by sporadic records in the Arpaklen Valley, Purnuar and Saivan villages in Southwest Kopetdagh, and from the Dagish Valley in Central Kopetdagh (Babazo Unit of the Kopetdagh Reserve) (Morits 1929; Shcherbak et al. 1986). All these records refer to the nominal subspecies *E. m. miliaris*.

Family Colubridae

24. *Boiga trigonatum* (Schneider, 1802). This snake (subspecies *B. t. melanocephala* Annandale, 1904) is a typical inhabitant of the desert landscape; three records are known from the environs of Kara-Kala (Southwest Kopetdagh; our data). The species is listed in the Red Data Book of Turkmenistan (Ataev et al. 1985).

25. *Coluber caspius* Gmelin, 1789. This species is found in Southwest Kopetdagh (the valleys of Sumbar, Chandyr, and Bami Rivers, and next to the village of Saivan). It prefers valleys and mountain slopes covered with vegetation. The density of *C. caspius* is 0.7–0.9 animals/ha. It is active in the daytime from April to October. It feeds on other reptiles (lizards and *Natrix tessellata*) and rodents. *C. caspius* is included in the Red Data Book of Turkmenistan (Ataev et al. 1985).

26. *Coluber ladacensis* (Anderson 1871). This snake is widespread in Kopetdagh and can be found in rocky canyons, in river valleys, on mountain slopes and foothills covered with sparse vegetation, in cellars, and under roofs of buildings. The density of *C. ladacensis* is 0.6–1 animals/ha. It is active from April to October; sometimes solitary snakes can be found as late as November and December. It is active in daytime in spring and fall, but in summer it shifts to crepuscular and nocturnal activity. Females lay three to nine eggs in June or July; young snakes hatch in early September. *Coluber ladacensis* feeds on small and medium-sized lizards and rodents (Bogdanov 1962, 1965b; Ataev 1985; Shcherbak et al. 1986).

27. Coluber atayevi Tuniev et Shammakov, in litt. This species was previously identified as *Coluber najadum* (Varentsov 1894; Chernov 1949; Bogdanov 1962; Bannikov *et al.* 1977; Ataev 1985). The detailed analysis of *Coluber najadum* from different parts of its range led to the conclusion that a new species existed.

This snake species is found within Turkmenistan only in Central and Southwest Kopetdagh. There, it inhabits rocky slopes of mountain valleys covered with vegetation. *C. atayevi* is active in daytime but is very secretive. Only nine snakes have been found during about 100 years of herpetological studies in Kopetdagh; surprisingly, we detected 30 *C. atayevi* during three-day surveys conducted in May 1990 and 1991 next to the village of Saivan (Southwest Kopetdagh). As a rare species, this snake (under *C. najadum*) has been included in the Red Data Book of Turkmenistan (Ataev *et al.* 1985).

28. Coluber ravergieri Menetriés, 1832. This species is found in foothills, in wide valleys with dense herbaceous vegetation, and in agricultural areas. It is often observed under the roofs of houses. The density of *C. ravergieri* in Kopetdagh is 0.5–0.7 animals/ha. Snakes are active in the daytime from March to November. Females lay four to nine eggs in July; young snakes (20–30 cm long) hatch in September. *C. ravergieri* feed on lizards, rodents, and small passerine birds (Bogdanov 1962, 1965a; Ataev 1985; Shcherbak *et al.* 1986).

29. Coluber rhodorhachis (Jan., 1865). This species is found everywhere in mountain steppe, on slopes and in valleys, and in foothills. Average density is 0.6 animals/ha. Snakes are active for seven to eight months; in spring and fall they are found in daytime but shift to crepuscular and nocturnal activity in summer. Females lay four to nine eggs in June-July. *C. rhodorhachis* feed on lizards, small passerine birds, and rodents (Bogdanov 1962, 1965a,b; Ataev 1985; Shcherbak *et al.* 1986).

30. Coluber karelini Brandt, 1938. This lowland snake species is very rare: it is known from the foothills of Central Kopetdagh (Germab) and East Kopetdagh (Khiveabad). The Kopetdagh population belongs to the nominal subspecies, *C. k. karelini*.

31. Eirenis meda (Cernov, 1949). This snake is found in Central and Southwest Kopetdagh in rocky canyons and on mountain slopes covered with grass, shrubs, and trees. Only 40 individuals have been recorded in the past; however, our surveys in Southwest Kopetdagh in April-May 1989 and 1990 revealed 50 *E. meda* is only ten days (Ataev *et al.* 1991). The species has strictly nocturnal activity and can be found from April to September. Its diet includes insects, spiders, and scorpions (Khomustenko 1981; Ataev 1985; Shcherbak *et al.* 1986; Dotsenko and Shcherbak 1985; Rustamov ans Ataev 1976, 1990).

32. Lycodon striatus (Shaw, 1802). This species of snake is widespread in Kopetdagh but rather rarely found; only about 40 individuals have been recorded. The Kopetdagh population belongs to the subspecies *L. s. bicolor* Nikolsky 1903. It prefers rocky canyons but is found also in mountain steppe; it hunts only at night and spends the day under rocks and in crevices. *L. striatus* is active from April to August; females lay a maximum of five eggs in July. The prey of *L. striatus* includes small lizards, particularly *Ablepharus pannonicus*. This rare snake species is included in the Red Data Book of Turkmenistan (Ataev *et al.* 1985).

33. Litorhynchus ridgewayi Boulenger, 1887. This lowland snake is extremely rare in the mountains. The Kopetdagh records include one in the Nakhduin range (Elpatjewsky and Sabanejew 1907), one in the Arvaz Valley (Mikhailovsky 1904), and one in the lower portion of the Chuli Valley in Central Kopetdagh (our data May 1992).

34. Natrix tessellata (Laurenti, 1768). This water snake is found in the rivers of Kopetdagh, on irrigated areas, and around springs. It is active from March, with the highest density in April. *N. tessellata* feeds on fish, toads (*Bufo viridis*), and frogs (*Rana ridibunda*) (Bogdanov 1962; Skalon 1982; Ataev 1985).

35. Oligodon taeniolatus (Jordan, 1853). This snake is widespread in Kopetdagh on mountain slopes and in valleys with shrub and tree vegetation. The density can reach two to three animals/ha; this strictly nocturnal species is found from April to August. Females lay one or two eggs in May or June (Bogdanov 1962; Dotsenko 1984; Ataev 1985; Shcherbak *et al.* 1986). As a rare species, this snake is included in the Red Data Book of Turkmenistan (Ataev *et al.* 1985).

36. Psammophis lineolatus (Brandt, 1838). A typical lowland species which has occasionally been recorded in the foothills of Central Kopetdagh (Germab) and Southwest Kopetdagh (Kara-Kala) (Bogdanov 1962, 1965a; Skalon 1982; Ataev 1985; Shcherbak *et al.* 1986).

37. Psammophis schokari (Forskal, 1775). This species is found only in East Kopetdagh (nominal subspecies *P. s. schokari*), where it inhabits slopes of hills and valleys, rocky canyons, and various buildings. These snakes are active in daytime from March to October. In some places they are common (three to five snakes detected in one day). *P. schokari* feeds on small lizards and rodents (Bogdanov 1962; Shcherbak 1979; Ataev 1985).

38. Pseudocyclophis persicus (Anderson, 1872). This snake is widespread in Kopetdagh (Boettger 1890; Nikolsky 1903; Bogdanov 1902; Nurgeldyev *et al.* 1970; Sherbak and Golubev 1981; Dotsenko and Shcherbak 1985; Kurilenko 1987; Starkov 1988; Rustamov *et al.* 1990). It is found on mountain slopes covered with gravel and in the foothills with ephemerous vegetation. Only about 60 individuals have been recorded from Kopetdagh; they belong to the nominal form, *P. p. persicus*. This nocturnal species is found from April to August. Females lay one to three eggs in June (Bogdanov 1965a,b; Ataev 1985).

39. Spalerosophis diadema (Schlegel, 1837). This snake is typical for lowland deserts and is rarely recorded in mountains and foothills; it has been found in Central Kopetdagh (Firyuza) and Southwest Kopetdagh (Kara-Kala) (Bogdanov 1962; Skalon 1982). These individuals may belong to the subspecies *S. d. schiraziana* (Jan., 1865).

40. Telescopus rhynopoma (Blanford, 1874). This nocturnal snake is known only by three specimens from valleys and mountain slopes of Central and Southwest Kopetdagh (Bogdanov and Potopolsky 1956; Ataev 1966). As a rare species, it is included in the Red Data Book of Turkmenistan (Ataev *et al.* 1985).

Family Crotalidae

41. Agkistrodon halys (Pallas, 1776). The species is common in valleys, mountain slopes, and steppes of Central and Southwest Kopetdagh. Its density is 0.2 animals/ha on mountain plateaus but reaches 2.3 in valleys and up to 6.5 in areas with rodent colonies (Zinchenko and Ritskov 1989). *A. halys* is active in daytime in the spring and fall and shifts to crepuscular and nocturnal activity in summer; it is found from April to November. This snake feeds on small rodents and lizards (Khomustenko 1982; Ataev 1985; Shcherbak *et al.* 1986). *A. halys* is included in the Red Data Book of Turkmenistan (Ataev *et al.* 1985); the Kopetdagh population belongs to the subspecies *A. h. caucasicus* Nikolsky, 1916.

Family Elapidae

42. Naja oxiana (Eichwald, 1831) – the Middle Asian Cobra. It is found in all habitats, but favors hilly areas with scarce herbaceous cover, river valleys and canyons, and mountain plateaus; it also is common in ruins and cemeteries. Density is low (2 to 5 snakes per 100 ha). With first cold weather (in October) cobras leave for winter shelters; in spring, they come out later than other snakes. Mating takes place in spring; in July, females lay 6 to 19 eggs. Young snakes (28--30 cm long and weighing 10–12 g) hatch in September. Cobras feed mostly on frogs and reptiles, but sometimes on small birds and rodents (Bogdanov 1965a; Makeev and Zemlyanova 1968; Makeev *et al.* 1983; Danov 1984; Ataev 1985; Sopyev *et al.* 1989; Rustamov 1991). This cobra is included in the Red Data Book of Turkmenistan (Ataev *et al.* 1985).

Family Typhlopidae

43. Typhlops vermicularis Merzem, 1820. This small blind snake is widespread in the Kopetdagh Mountains, where it is common (and sometimes abundant) in wide valleys and rocky slopes with scarce vegetation. It density reaches 10–12 animals/ha. *T. vermicularis* is an exclusively nocturnal species and spends daytime sheltered under rocks, in soil crevices, or in burrows. It mates in spring; in June, female lays one or two eggs. Young snakes hatch approximately in a month. It is a specialized predator and feeds only on ants (Boettger 1890; Bogdanov 1965; Ataev 1985; Shcherbak *et al.* 1986).

Family Viperidae

44. *Echis multisquamatus* Cherlin, 1981 *(= Echis carinatus)*. This lowland species is very rarely found in mountains; it has been recorded several times in the Chandyr Valley and around the city of Kara-Kala in Southwest Kopetdagh.

45. *Vipera lebetina* (Linnaeus 1758). This viper is widespread in Kopetdagh where it is represented (as in entire Turkmenistan) by the subspecies *V. l. turanica* Cernov, 1940. This species lives in dry foothills, on mountain slopes, in canyons and river valleys; it is common on irrigated areas and in vineyards. Density is low: 0.05 animals/ha in Southwest Kopetdagh; in other parts of the range up to 0.5 animals/ha. These snakes are active in the daytime in spring and fall but in summer shift to nocturnal activity. They mate in April or May; in July, females lay 18 to 25 eggs (large females can lay up to 40 eggs) (Makeev 1969; Makeev *et al.* 1983, 1986; Danov 1985). The prey of vipers are rodents and other reptiles; in spring and fall they feed on birds of small and medium size (Bogdanov 1962; Ataev 1985; Shcherbak *et al.* 1986; Starkov 1989).

Biology of Reptiles in Kopetdagh

Distribution and Density

In Kopetdagh, there are 45 species of reptiles, including three species of Testudines, 18 species of Sauria, and 24 species of Serpentes. This fauna constitutes 57% of the total herpetofauna of Turkmenistan (79 species). The composition of reptiles varies considerably from the western to eastern parts of Kopetdagh. Of 45 species (3/18/24 species of turtles/lizards/snakes, respectively), 30 are found in East Kopetdagh (1/14/15), 36 in Central Kopetdagh (1/16/19), 36 in Southwest Kopetdagh (3/14/19), and 23, in Northwest Kopetdagh (1/10/12) (Table 1). Of 45 species, 25 live only in the mountains (Table 1); some of these reptiles have narrow niches, while others have more ecological plasticity. For instance, both *Stellio erythrogaster* and *S. caucasius* are mountain species; however, the first one is more flexible in habitat selection (it lives on clay slopes, in gerbil colonies, on rocks, in ruins), whereas the latter is confined to slopes, canyons, and screes. Characteristic reptiles inhabiting rocky substrates are *Coluber atayevi, Sphalerosophis persicus, Lacerta raddei, Typhlops vermicularis, Lycodon striatus*, as well as eurytopic species *Naja oxiana* and *Vipera lebetina*.

Lacerta raddei, Eumeces schneideri, E. taeniolatus, and *Coluber caspius* prefer moist mountain habitats. Snakes commonly found in ruins are *Coluber caspius, C. ladacensis, C. ravergieri,* and *Naja oxiana.*

Table 1. Distribution and activity of reptiles in Kopetdagh

	I	II	III	IV	V	VI	VII	VIII
Testudines								
Emys orbicularis[b]	–	–	+	–	D	323	600	277
Mauremys caspica caspica[b]	–	–	+	–	D	323	400	77
Agrionemys horsfieldi[b]	+	+	+	+	D	232	2,027	1,795
Sauria								
Stellio caucasius[b]	+	+	+	+	D	232	2,027	1,795
Stellio erythrogaster[a]	+	–	–	–	D	293	600	307
Trapelas sanguinolentus[b]	+	+	+	+	D	232	1,000	768
Pseudopus apodus[a]	+	+	+	–	D	232	1,900	1,668
Cyrtopodion caspius[b]	+	+	+	+	N	232	1,400	1,168
C. spinicauda[a]	+	+	+	+	N	330	1,840	1,510
Eublepharis turkmenicus[a]	+	+	+	+	N	330	988	658
Eremias strauchi[a]	–	+	+	+	D	330	2,027	1,697
E. velox[b]	+	+	+	+	D	232	988	756
Lacerta raddei[a]	–	+	–	–	D	1,500	–	–
L. strigata[b]	–	–	+	–	D	400	–	–
Mesalina guttulata[b]	+	–	–	+	D	600	–	–
Ablepharus pannonicus[a]	+	+	+	+	D	300	1,840	1,540
Chalcides ocellatus[a]	–	+	–	–	D	400	988	588
Eumeces schneideri[b]	+	+	+	+	D	232	988	756
E. taeniolatus[b]	+	+	+	–	D	232	988	756
Mabuya aurata[b]	+	+	+	–	D	232	988	756
Varanus griseus[b]	+	+	+	+	D	232	323	91
Serpentes								
Eryx elegans[a]	–	+	+	–	N	1,100	2,027	1,027
E. miliaris[b]	+	+	+	–	N	232	1,000	768
Boiga trigonatum[b]	–	–	+	+	N	232	323	91
Coluber caspius[b]	–	–	+	–	D	323	1,300	977
C. ladacensis[b]	+	+	+	+	D	232	2,027	1,795
C. atayevi[b]	–	+	+	–	D	500	1,840	1,340
C. ravergieri[b]	+	+	+	+	D	223	1,500	1,277
C. rhodorhachis[b]	+	+	–	–	D	223	2,027	1,804
C. karelini[b]	+	+	–	–	D	400	1,000	600
Eirenis meda[a]	–	+	+	–	N	223	1,380	1,157
Lycodon striatus[a]	+	+	+	+	N	250	1,500	1,250
Litorhynchus ridgewayi[b]	–	+	–	–	N	600	1,000	400
Natrix tessellata[b]	+	+	+	–	D	223	988	765
Oligodon taeniolatus[a]	+	+	+	+	N	330	1,500	1,170
Psammophis lineolatus[b]	+	+	+	+	D	223	1,300	1,077

Table 1. Continued

	I	II	III	IV	V	VI	VII	VIII
P. schokari[a]	+	–	–	–	D	293	1,500	1,207
Pseudocyclophis persicus[b]	+	+	+	+	N	223	1,000	777
Spalerosophis diadema[b]	+	–	+	+	N	223	350	127
Telescopus rhynopoma[a]	–	+	–	+	N	600	–	–
Agkistrodon halys[a]	–	+	+	–	N	1,000	2,027	1,027
Naja oxiana[b]	+	+	+	+	D	223	2,027	1,804
Typhlops vermicularis	+	+	+	+	N	223	323	765
Echis multisquamatus[b]	–	–	+	–	N	323	–	–
Vipera lebetina[b]	+	+	+	+	N	223	2,027	1,804

I – East Kopetdagh, II – Central Kopetdagh, III – Southwest Kopetdagh, IV – Northwest Kopetdagh, V – active in daytime (D) or in nighttime (N), VI – minimal altitude (m), VII – maximal altitude (m), VIII – the range of altitudes in which a species is found (m)
[a] species found only in foothills and mountains of Kopetdagh
[b] species found in Kopetdagh as well as in adjacent lowlands

Some species inhabiting both mountains and lowland are eurytopic, e.g., *Agrionemys horsgfieldi, Trapelas sanguinolentus, Eremias velox, Varanus griseus*, and *Eryx miliaris*. Some species, however, can be found both in mountains and in lowland regions of Turkmenistan but are not eurytopic there, e.g., the turtle *Emys orbicularis*, which is found in fresh water bodies of the western Uzboi and in Lake Maloe Delili, and in rivers and thermal springs of Southwest Kopetdagh (Skalon 1982).

Of 45 species of reptiles, 22 (48.8%) are found in the lower mountain belt of Kopetdagh (200–1,000 m), 14 (31.1%) in the middle mountain belt (1,000–2,000 m), and only nine (20%) in the upper belt (2,000–2,500 m). Eight species are found through all mountain belts (*Agrionemys horsfieldi, Stellio caucasius, Ablepharus pannonicus, Eremias strauchi, Coluber rhodorhachis, C. reavergieri, Naja oxiana*, and *Vipera lebetina*).

The altitudinal range of distribution in widespread forms is usually broader than in localized species (Table 1). For example, *Agrionemys horsfieldi, Stellio caucasius, Ablepharus pannonicus*, and *Eremias strauchi* are found up to 2,000–2,500 m (range 1,277–1,804 m). At the same time, highland species (*Eryx elegans, Coluber atayevi*, and *Agkistrodon halys*) have altitudinal ranges from 91 to 1,340 m. Altitudinal zonation is expressed mainly in Central Kopetdagh.

We distinguish three categories of abundance (Rustamov and Shammakov 1982): *abundant* (ten and more animals/ha), *common* (one to ten animals/ha), and *rare* (less than one animal/ha). Only four species of reptiles in Kopetdagh belong to the *abundant* category: *Typhlops vermicularis, Mabyua aurata, Lacerta raddei*, and *Ablepharus pannonicus*; they average from 10.4 to 12.8 animals/ha. Among *common* reptiles (15 species) are all three species of turtles, eleven species of lizards (*Lacerta strigata, Stellio caucasius, S. erythrogaster, Eremias velox, E. strauchi, Trapelus sanguinolentus, Cyrtopodion caspius, C. spinicauda,*

Eumeces taeniolatus, E. schneideri, and *Pseudopus apodus*), and only one snake species (*Natrix tessellata*). The remainder, and majority, of reptilian fauna (26 species, 57.7%) are *rare* species whose density varies from 0.02 to 0.8 animals/ha.

Reproduction

Of 45 species of reptiles discussed here, 39 (86.6%) lay eggs, and only six (13.4%) are ovoviviparous. Below we discuss in detail reproductive strategies and patterns of selected species.

Order Testudines

Steppe tortoises (*Agrionemys horsfieldi*) mature by the age of ten to twelve years (Ataev 1979, 1985); most of the tortoises (78.1%) mate in April. They lay clutches of 3 to 4 eggs twice, in April and May. The embryonic development lasts 70 to 75 days (Ataev 1979). Ovaries examined during the summer diapause contain developed follicules ready for fertilization.

Swamp turtle (*Emys orbicularis*) mating has been recorded in Lake Maloe Delili (Southwest Turkmenistan) in mid-April; a first egg in the Atrek Valley was observed in late April. One clutch has one to eleven eggs. *E. orbicularis* can mate in winter in thermal springs of Southwest Kopetdagh (Skalon 1982).

Order Squamata/Suborder Sauria

Family Gekkonidae
Caspian (*Cyrtopodion caspius*) and spine-tailed (*C. spinicauda*) geckos mature by the age of 10 to 12 months. They mate from late April to July; most animals (61.2%) mate in June or July. They lay eggs twice, in May and in June or July. Each clutch has one to three eggs. The Turkmen eublephare (*Eublepharis turcmenicus*) matures by the age of three and lays a single egg only once. Embryonic development of all these species lasts from 50 to 60 days.

Family Agamidae
In Kopetdagh, *Stellio caucasius* has one clutch per season, and *Trapelas sanguinolentus* has two clutches; the first species matures in its third year of life, and the latter, in its second year. Reproductive capacity of *S. caucasius* changes with season and age. Older females begin reproduction earlier. Of 200 reproducing females of *S. caucasius* surveyed, seasonal distribution of those in the size class 100–115 mm was 16 in April, 15 in May, 26 in June, and 6 in July. Distribution of those in the size class of more than 116 mm was: 3 in March, 35 in April, 48 in May, 44 in June, 6 in July, and 1 in August. Old females of this species produce more eggs than young. In 61 surveyed females, the average number of eggs per clutch was: 7.6 eggs in size class 101–110 mm, 8.5 eggs in size class 111–120 mm, 9.8 eggs in size class 121–130 mm, 10.3 eggs in size class

131–140 mm, and 12.1 eggs in size class 141–150 mm. In spring, females lay more eggs (average 9.9) than in summer (7.5). In *S. caucasius*, 92.0% lay eggs in May to July, whereas 63.8% of *T. sanguinolentus* lay eggs in April to May. We have also recorded adult females with undeveloped ovaries (two in May, five in June, and six in July).

Family Anguidae
Pseudopus apodus matures in Kopetdagh in the fourth year and has one clutch of 7 to 10 eggs.

Family Varanidae
Varanus griseus matures in its fourth year and has one clutch of 15 to 34 eggs; older females have 22–34 eggs, and younger, 15–21 eggs.

Family Scinicidae
Only *Mabuya aurata* and *Chalcides ocellatus* are viviparous species; all other skinks (*Eumeces schneideri, E. taeniolatus*, and *Ablepharus pannonicus*) lay eggs. *M. aurata* has two to seven young from May to August (77.7% in June or July) which mature in 22 or 23 months. Females of *A. pannonicus* mature in 10 or 11 months and lay two to three eggs twice (in April-May and June-July). Skinks *Eumeces schneideri* and *E. taeniolatus* mature in three years and have one clutch of three to six eggs. Their peak of reproduction (92.9%) has been registered in July. Embryonic development lasts about 60 days; females of both species have been found with undeveloped ovaries.

Family Lacertidae
All five species found in Kopetdagh mature in one year and lay eggs twice (*Eremias velox* and *E. strauchi* have two to four eggs in a clutch); not all females reproduce simultaneously, but, rather, egg laying is distributed through the season. Embryonic development lasts 40 to 50 days.

Suborder Serpentes

New hatched young snakes emerge from late July. First matings of *Coluber rhodorhachis* and *Spalerosophis diadema* have been recorded in Kopetdagh on May 24 or 25.

Of 24 species of snakes found in Kopetdagh, 20 lay eggs, and four are viviparous (*Eryx miliaris, E. elegans, Agkistrodon halys*, and *Echis multisquamatus*). Females of *Typhlops vermicularis, Natrix tessellata, Lycodon striatus, Coluber ravergieri, Pseudocyclophis persicus, Oligodon taeniolatus*, and *Vipera lebetina* with developed eggs, ready for laying were recorded from June 1 to August 9; the peak of reproduction (69.0 % of eggs layed) was from June 15 to July 20; egg laying lasts about 60 days. The average number of eggs was 8.7 in *Coluber ravergieri*, 18.3 in *Natrix tessellata*, and 20.4 in *Vipera lebetina*. Small nocturnal species of snakes lay a lesser amount of eggs; for instance, in the oviducts of *Typhlops vermicularis* there was only one egg, in *Oligodon taeniolatus*, two; and in *Lycodon striatus*, five eggs.

All snakes have only one clutch in a season. We have recorded a total of 206 large follicules and eggs ready for laying in 27 females; of these, 42 were found in May, 34 in June, and 139, in July. We also have found undeveloped ovaries in *Vipera lebetina* (one female, size 810 mm long, found May 12, 1964; and three females, size from 970 to 1320 mm, found June 21 and 25, 1969).

Activity

Of 45 species of reptiles found in Kopetdagh, 28 are active in the day, and 17, at night. In the summer time, reptiles have two peaks of activity, and from fall to spring, one peak.

Many day-active species in summer shift their activity to humid or shadowy habitats. For instance, in Firyuza Valley (Central Kopetdagh) in June, July, and August 1964-1967 and 1972-1978, we recorded from 1120 to 1430 hrs total of 19 reptiles belonging to 6 species (*Stellio caucasius, Mabuya aurata, Eumeces schneideri, E. taeniolatus, Coluber ladacensis,* and *Naja oxiana*). Reptiles were active on the top of the range in the daytime. In the juniper belt of Mt. Dushak (Central Kopetdagh) we recorded a total of 75 *Eremias strauchi* in June through September 1969, 1973, and 1976; of these, 68 lizards were active from 1000 to 1400 hrs. The number of active reptile species there varied from fall to spring in the following manner: in November, one species of turtles, seven species of lizards, and four of snakes; December, respectively, three, three, and two; January, two, three, and two; February, two, five, and two species. On the slopes of Novaya Nisa reptiles were active in November for 5.5 hrs, in December, for 2.5 hrs, and in January and February, for 4 hrs. At the same time, strictly nocturnal species of reptiles were active for seven to eight hours, and during the year, only for 5 to 5.5 months.

Feeding

Reptiles of Kopetdagh can be divided into three groups: those feeding on invertebrates ("insectivores" in a broad sense), those feeding on other vertebrates, and herbivores. Invertebrates are found in the diet of 57.7% of all species (23 species of lizards and 4 species of snakes). Insects contributed from 78 to 100% to food items in lizards belonging to families Agamidae (total 1,733 specimens), Scincidae (328), Gekkonidae (110), Lacertidae (258), and Anguidae (*Pseudopus apodus*; 41 specimens). For example, insects represented 95.4% of the diet of *Stellio caucasius* in Kopetdagh in spring, 94.0% in summer, 95.8% in fall, and 91.4% in winter; 92.3% in adult males, 96.9% in adult females, and 96.7% in young lizards. Of all insects, *S. caucasius* preferred Coleoptera (82.7% of the diet in young lizards, 80.9% in males, and 67.3%, in females). Myriapoda and Scorpiones are consumed by *Eublepharis turcmenicus, Eirenis meda,* and *Pseudocyclophis persicus*. Ants and their larvae are the major food (95.2%) of *Typhlops vermicularis*.

Twelve species of snakes (*Eryx elegans, Natrix tessellata, Lycodon striatus, Coluber caspius, C. rhodorhachis, C. ravergieri, Spalerosophis diadema, Psammophis lineolatus, P. schokari, Naja oxiana, Vipera lebetina,* and *Agkistrodon halys*), and also *Varanus griseus* feed predominantly on vertebrates. Some snakes are specialized in their feeding; e.g., *Natrix tessellata* and *Naja oxiana* feed mostly on amphibians; *Psammophis lineolatus* feeds on lizard *Eremias* spp. (90%), and *Lycodon striatus* on lizard *Ablepharus pannonicus* (83.3%). In *Vipera lebetina*, 76.4% of diet consists of rodents and the rest is comprised of other mammals, reptiles, and birds. Lizards, rodents, and small passerine birds are the prey of *Coluber caspius, C. rhodorhachis, C. ravergieri,* and *Spalerosophis diadema*. Seventeen species of reptiles feed on lizards. Five species were found to consume snakes (*Stellio caucasius, S. erythrogaster, Varanus griseus, Naja oxiana,* and *Vipera lebetina*).

The only completely herbivorous reptile is the steppe tortoise *Agrionemys horsfieldi*. In lizards, herbivory is common in *Stellio caucasius* (65.8% of diet items) and *S. erythrogaster* (53.3%), but it is not significant in *Trapelas sanguinolentus* (7.5%) or skinks (*Eumeces schneideri*, 5.9%).

Hibernation

Turtles *Mauremys caspica* and *Emys orbicularis* hibernate for 120 to 160 days; in the thermal springs of Southwest Kopetdagh (Parkhai), however, they are active through the entire winter (Skalon 1982). A steppe tortoise (*Agrionemys horsfieldi*) hibernates for 9–9.5 months. Among lizards, *Cyrtopodion caspius, Trapelas sanguinolentus, Stellio caucasius, Mabuya aurata, Ablepharus pannonicus, Eremias velox,* and *E. persica* leave for winter shelters in late October and reappear generally in early and mid-March; their inactive period, therefore, is only 4 to 4.5 months.

Mabuya aurata has been recorded hibernating from late October to early April; it digs into loose loess soil where one or two (or sometimes three, ten, or even 17) lizards can be found together; where there are two lizards hibernating together, they often curl into rings one above another. Their average body temperature measured in November was 19 °C (measured in two lizards); in December, 10.1 °C (12 lizards); in January, 6.1 °C (11 lizards); and in February, 8.4 °C (12 lizards).

We studied in detail the process of hibernation in *Stellio caucasius*. These agamid lizards can be found in crevices in rocks and slopes, and under rocks at the depth from 5 to 45 cm, either separately or in groups. Of the hibernating groups found, 15 contained two to ten lizards, and six, 12 to 34 lizards. Group contained lizards of different sex and age; e.g., one group contained 25 young and two adult *S. caucasius*; another, ten and two, respectively; and a third group, six and one. The body temperature of these animals during hibernation varied from -0.8 to 9.8 °C (in crevices and cracks, from -1.0 to 8.8 °C). At the body temperature 1 to 4 °C, *S. caucasius* are inactive; at 5 to 8 °C they are able to move slowly; and above 9–11 °C they can move relatively quickly. Young *S.*

caucasius, which hatch in July or August, reach maximum body mass (average 7.0 g) and amount of fat deposit (average 0.195 g) by November. During their first hibernation, they lose up to 20% of body mass and 78.6% of fat deposit.

We believe that we also have recorded summer hibernation in several species of reptiles (*Eumeces schneideri, E. taeniolatus*, and *Varanus griseus*) although this subject is debatable.[5] Summer hibernation begins and continues when the temperature of the air increases. For example, of 243 *E. schneideri*, 171 (70.4%) were observed in April and May; 56 (23.4%), in June; 11 (4.5%) in the first half of July; 3 (1.0%) in the second half of July; and 2 (0.8%) in August. The same data for *E. taeniolatus* are 244 (84.7%), 23 (8.0%), 11 (3.8%), 4 (1.3%), and 6 (2.0%).

Biological Groups of Reptiles

The study of ecological characteristics, including life span, reproduction, feeding, embryonic and postembryopnic development and growth of reptiles in Kopetdagh allowed to distinguish two biological groups (Table 2): mountain (with a long cycle of development) and lowland (with a short cycle) species.

Table 2. Biological groups of lizards in Middle Asia

Characteristic	Life Cycle	
	long cycle (mountain group)	short cycle (lowland group)
Adult body size	Large	Small
Age of maturity	3–4 years	9–10 months
Growth rate	Slow	Fast
Reproduction	For 3–5 seasons	For one season
Clutches per year	One	Two or three
Embryonic development	40–90 days	30–50 days
Life span (years)	5–8	1–2
Population type	Differentiated	Nondifferentiated

The mountain lizard species are usually larger than lowland ones. For instance, *Stellio caucasius* and *S. erythrogaster* have average body length 121.5–132.6 mm; *Eumeces schneideri* and *E. taeniolatus*, 107.0–123.7 mm; *Eublepharis turcmenicus*, 127.9 mm; and *Pseudopus apodus*, 348,3 mm. When young lizards of these species hatch, they are quite small (35.8 to 55.1% of the length of an adult lizard), grow slowly, and reach adult size in three to four years. Since these species live seven to eight years, one can always find individuals of different ages. Their reproductive strategies vary; significant energy is spent for the maintenance of their individual existence; reproduction is extended for several years with decreasing number of clutches (down to once in a season) and number of eggs (Table 2). A clutch of *Pseudopus apodus* in Kopetdagh has an average of 8.5 eggs; in *Mabuya aurata, Eumeces schneideri, E. taeniolatus*, and

Vipera lebetina, from 3.9 to 4.3 eggs; in *Stellio caucasius* and *S. erythrogaster*, and *Trapelas sanguinolentus*, 3.5 to 14.4 eggs; and in *Eublepharis turcmenicus*, 2 eggs. The underdevelopment of ovaries (found in Kopetdagh in *Stellio caucasius*, *Eumeces schneideri*, and snake *Vipera lebetina*) can be a compensatory mechanism for the economy of resources when not every adult female participates in reproduction.

Lowlands in Turkmenistan are occupied mostly by small-size lizard species. The measured average body length of adult *Phrynocephalus helioscopus*, *Ph. rossikowi*, and *Ph. interscapularis* was 33.4 to 53.4 mm; of *Eremias arguta, E. intermedia, E. lineolata, E. nigrocellata*, and *E. scripta*, 42.5 to 67.3 mm, and in *Mesalina guttulata*, 52.0 mm. Due to their rapid development, young lizards in the lowlands reach maturity in their first season (by September or October), or by the time they leave winter shelters (March or April). Most of these lizard species also do not live longer than one year. They have similar reproductive strategy; due to their short lifespan energy is spent rather rapidly through production of multiple progeny (up to three clutches per season; increased number of eggs in a clutch). The embryonic development of small lowland lizards lasts 30 to 60 days in the sand deserts and 30 to 40 in clay desert.

Zoogeography and Faunogenesis

We know very little about the Tertiary and Quaternary faunas of reptiles in Kopetdagh (as well as in entire Middle Asia). The only paleontological data are those of Birman *et al.* (1971) who reported Middle and Upper Pliocene turtle fossils (*Agrionemys* sp.) from Keletchaya Spring (East Kopetdagh). Therefore, zoologists have relied on zoogeographic analysis to reconstruct a tentative history of the Kopetdagh herpetofauna (Nikolsky 1916; Chernvo 1959; Rustamov 1966). Eight zoogeographic groups can be distinguished in the Kopetdagh fauna of reptiles, according to the ranges of species.

Irano-Afghanian Species *(21 spp.)*: *Eublepharis turcmenicus, Stellio caucasius, S. erythrogaster, Cyrtopodion spinicauda, Eremias strauchi, Ablepharus pannonicus, Eumeces schneideri, E. taeniolatus, Mabuya aurata, Varanus griseus, Eryx elegans, Coluber ladacensis, C. atayevi, C. ravergieri, C. rhodorhachis, Eirenis meda, Pseudocyclophis persicus, Naja oxiana, Typhlops vermicularis, Lytorhrhynchus ridgewayi*, and *Echis multisquamatus*. A majority of these species are found in Turkmenistan only in Kopetdagh, and on the Badghyz and Karabil Plateaus; some of them also inhabit Kugitang.

Turanian Species *(7 spp.)*: *Agrionemys horsfieldi, Trapelas sanguinoletus, Cyrtopodion caspius, Eremias velox, Eryx miliaris, Coluber karelini*, and *Psammophis lineolatus*. These reptiles are widespread in lowland Turkmenistan as well as in the mountains of Kopetdagh.

Indian Species *(5 spp.)*: *Boiga trigonatum, Lycodon striatus, Oligodon taeniolatus, Psammophis schokari,* and *Telescopus rhynopoma.* Of these, *Boiga trigonatum* penetrates from the lowlands to the foothills and middle altitudinal belt of Kopetdagh; both *Oligodon taeniolatus* and *Telescopus rhynopoma* are found only within Kopetdagh mountains.

These three groups of species represent the core of the Kopetdagh fauna of reptiles (33 species, or 73.3%). Less significant groups are Mediterranean species (*Mauremys caspica, Pseudopus apodus, Coluber caspicus,* and *Natrix tessellata*), Saharo-Sindian species (*Mesalina guttulata, Chalcides ocellatus, Spalerosophis diadema,* and *Vipera lebetina*), Caucasian-Anatolian species (*Lacerta raddei* and *L. strigata*), one European (*Emys orbicularis*), and one widespread East Palearctic species (*Agkistrodon halys*) (Table 3).

Table 3. Zoogeographic composition of the Kopetdagh reptile fauna

Zoogeographic Group	Number of spp.	% of Total Fauna
Irano-Afghanian	21	46.6
Turanian	7	15.5
Indian	5	11.1
Mediterranean	4	8.8
Saharo-Sindian	4	8.8
Caucasian-Anatolian	2	4.4
European	1	2.2
East Palearctic	1	2.2

The herpetogeographic subdivision of Middle Asia (Rustamov and Shcherbak 1985, 1986; Szczerbak 1982) classifies Kopetdagh as the Khorassan-Kopetdagh area of the Iranian Mountain District of the Irano-Afghan Mountain subprovince. The latter is included in the transitional Irano-Afghan province of the Arid Mediterranean-Asian subregion of the Palearctic Region. Within this scheme, only four species and one subspecies of reptiles are endemic for the Khorassan-Kopetdagh area: lizards *Eublepharis turcmenicus* and *Eremias strauchi kopetdaghica,* and snakes *Eryx elegans, Coluber atayevi,* and *Eirenis meda.* Kopetdagh is a western limit of distribution in Middle Asia of the following reptile species: *Emys orbicularis, Mauremys caspica, Eremias strauchi, Lacerta raddei, L. strigata, Coluber caspius,* and *Telescopus rhynopoma.* Absence of these species in adjacent Badghyz characterizes a zoogeographic boundary between the Kopetdagh and Badghyz-Karabil plateaus. In contrast, some lizard species present in Baghyz (*Bunopus tuberculatus, Cyrtopodion turcmenicus, Ophiomorus chernovi,* and *Eremias persica*) have not been found in Kopetdagh.

The Circum-Euxinian disrtibution of the Ancient Mediterranean core of herpetofauna with presence of European boreal forms probably was responsible for the penetration of many reptile species (*Emys orbicularis, Pseudopus apodus,*

Natrix tessellata, N. natrix, Cyrtopodion caspius, Elaphe quatuorlineata, Lacerta agilis, and *Vipera ursini*) to the Crimea and, around the Caspian Sea, to Middle Asia. At the same time, a non-disrupted range from Anatolia to Middle Asia could have existed in such reptile species as *Mauremys caspica, Mabyua aurata, Eumeces schneideri, Ablepharus pannonicus, A. bivittatus, Stellio caucasius, Typhlops vermicularis, Coluber ravergieri, Psammophis lineolatus, Agkistrodon halys,* and *Vipera lebetina*. We know that land tortoises were already represented by the end of the Pliocene by three separate species in Middle Asia, in the Mediterranean area, and in the Balkans (Rustamov 1981). The mountain uplift during this time could have disrupted geographic ranges and led to the formation of endemic species, such as in the genus *Eirenis* with Caucasian species *E. collarius* and *E. modestus*, and the Middle Asian *E. medus*. Rock lizards of the genus *Lacerta* (*L. raddei* and *L. strigata*), as well as the Transcaucasian subspecies of *Agkistrodon halys*, could have penetrated to Kopetdagh via the Elburz Range (south from the Caspian Sea) from the Caucasus. Interestingly, the Caucasian highland and mountain-steppe species (such as *Vipera kaznakovi* or *V. xanthina*) never penetrated to Middle Asia, probably because by the time of the formation of these species the Elburz was completely covered by mountain forests (Rustamov 1981). These "Hyrcanian" forests have been previously described as an essential route of dispersal for birds from Transcaucasia to Middle Asia (Rustamov 1945, 1961). Our study also confirms their significance for reptile distribution; we can state that the dispersal of the "Western" species of turtles, lizards, and snakes was possible only through the Elburz corridor, and 18 reptilian species found in Kopetdagh or next to it now are represented by populations or local forms descended from reptiles dispersed via this route. These include *Mauremys caspica, Emys orbicularis, Stellio caucasius, Pseudopus apodus, Mabyua aurata, Eumeces schneideri, Eremias strauchi, Lacerta strigata, L. raddei, Typhlops vermicularis, Natrix tessellata, Natrix natrix, Coluber atayevi, C. caspius, C. ravergieri, Elaphe quatorlineata, Vipera lebetina,* and *Agkistrodon halys*.

Since the first agricultural settlements emerged in the Kopetdagh foothills (VIII century B. C.; Sarianidi 1967), the deforestation of Kopetdagh by human activity has added to the general aridization of climate. Deforestation has especially affected juniper forests; the palaces of Nisa, the capital of the ancient Parthian kingdom (II-I centuries B. C.) next to Ashkhabad, include magnificent trunks of *Juniperus turcomanica* which were probably delivered from nearby mountains. Aridization of Kopetdagh has led to the drying of mountain springs, and ranges of the animals of Caucasian origin generally have decreased; these animals have been replaced by the arid Middle Asian forms. These probably are the most recent events which have contributed to the distribution of the modern reptile fauna of the Kopetdagh Mountains.

Notes

1. The systematic position of species and subspecies follows Borkin and Darevsky (1987).
2. Borkin and Darevsky (1987) reinstated generic name *Pseudopus* Merrem, 1820. However, Leviton *et al.* (1992) still use traditional name *Ophisaurus* Daudin 1804 for this species. – Ed.
3. Arnold (1986) treated it as a separate species, *M. watsonana*. – Ed.
4. Akhmedov and Shcherbak (1987) believe that the subspecies found in Turkmenistan is *M. a. affinis*, but not *M. a. septemtaeniolata*. – Ed.
5. Tsellarius *et al.* (1991) believe that there is no evidence of summer hibernation in *Varanus griseus*, which is a predominantly lowland desert lizard, extremely rare in the mountains. – Eds.

Grey monitor *(Varanus griseus)*. Photo by H.R. Levenshtein.

21. Geographic Variability of *Phrynocephalus Rossikowi* Nik. (Reptilia: Agamidae) in Turkmenistan and Adjacent Regions

MICHAEL L. GOLUBEV, VALENTINA V. MANILO AND ANATOLY A. TOKAR

Abstract

We studied karyotypes and morphology of five samples of *Ph. rossikowi* from different parts of the species range. Revealed variability shows that peripheral populations of this species undergo intensive speciation; for further taxonomic conclusions, additional detailed studies of populations from Chardzhou and Lebab are required. Analysis of variability shows that the Amudarya and Syrdarya Rivers played an important role in the formation of modern range of *Ph. rossikowi*. We consider central populations as a relatively stable core as compared to peripheral ones. It is suggested that *Ph. rossikowi* originated through adaptation of ancestral lizards to the local stony semi-desert environment in the West Kizylkum. The most ancient part of the range of *Ph. rossikowi* probably lies within the alluvial fan submontane plain of the Sultanuizdagh Range and the Pitnyak elevation in East Turkmenistan, adjacent to the south. The most ancient part of the submontane plain of the Sultanuizdagh has an early Pleistocene age (Kogai 1957). Therefore, our opinion differs from that of Ananjeva and Tunijev (1992), who suggested that *Ph. rossikowi* is a Pliocene species whose range shifted during the Pleistocene from the middle portion of the Amudarya River to its lower part.

Introduction

The toad agama *Phrynocephalus rossikowi* has been described from "the mountain of Khen-tau near Nukus," on the right bank of the lower Amudarya (Nikolsky 1898: 286). Its karyotype has been described by Sokolovsky (1974, 1975), who studied individuals collected 70 km northwest of the city of Chardzhou (left bank of the Amudarya). A diploid set of 2n = 48 telocentric chromosomes has been determined, of which Sokolovsky tentatively designated the first 12 pairs as macrochromosomes, and the remaining 12 pairs as microchromosomes. Macrochromosomes were subdivided into three size groups separated by considerable size gaps. Satellite chromosomes have not been found.

The type locality of the Turkmen subspecies of *Ph. rossikowi*, *Ph. r. shammakowi* Shcherbak et Golubev, 1979, lies 20 km north from the town of Serny Zavod and ca. 60 km east of the town of Darvaza in one of depressions of the so-called "Unguz bed" in the Central Karakum. Recently, the karyotype of this form has been described (Manilo and Golubev in press) and found to be different from data of Sokolovsky (1974, 1975). Discovery of this difference promoted continuation of this study and analysis of morphology.

Materials and Methods

Table 1 shows size and origin of studied samples. Chromosomal preparations were done according to standard techinques (Ford and Hamerton 1956; McGregor and Varley 1986) with some modifications (Manilo 1986). Slides were stained with 2% Giemsa solution in 0.01 M sodium phosphate buffer (pH = 6.8) for 30 to 40 min. We used standard nomenclature and terminology (Levan et al. 1964) to identify chromosomes. Microphotographs were done under Biolam L-212 light microscope. We studied 30 metaphase plates from each population. On preparations from testes, we studied metaphase plates of spermatogenetic division, diakinetic bivalents, and chromosomes of the meiotic metaphase II.

Table 1. Geographic origin of the studied samples of *Ph. rossikowi*

Population (# on the map)	Number of specimens used for the analysis of:		
	karyotype	proteins	external morphology
Karakalpakstan, Boktybulak Area (Boktybulak Well, Taschukho Well, Bukan Well) (#5)	4 m, 2 f	20 m, 18 f	23 m, 15 f
East Turkmenistan, Village of Lebab (#3)	1 m, 1 f	3 m, 5 f	3 m, 5 f
Turkmenistan, 20 km N of Serny Zavod (#1)	1 m, 4 f	7 m, 7 f	7 m, 7 f
Karakalpakstan, 40 km by road from Nukus to Turtkul (#4)	–	2 m, 1 f	2 m, 1 f
East Turkmenistan, Deinau (#2)	–	3 m, 9 f	3 m, 9 f

For statistical analysis of external morphology, the following 16 parameters were measured in each lizard: L – body length, LCD – tail length, FEM – distance between knee joints of femora positioned at a right angle to the longitudinal body axis, LEG – length of the hind leg from its base through the end of the fourth toe (without the claw), LHD – length of the head from the end of the snout to the posterior edge of temporal shield, CAP1 – maximal width of the "hood," CAP2 – posterior width of the "hood," CAP3 – distance from the posterior edge of the nostril to the base of tubercle behind the eyesocket, INOS – distance between nostrils, HNOS – distance from the nostril to the base of upper lip fold, HHD – height of the head, SQCC – number of scales across the "hood" (without shields above the eyesocket), SQCL – number of scales along the "hood" (from temporal shield to nasal shield), SDL – number of scales in the ventral longitudinal row on the fourth toe of right hind foot, SDR – same, on the right foot, GVA – number of scales along the longitudinal body axis from chin shield to anal opening.

Samples initially were studied to reveal sexual dimorphism; then, for each sample standard statistical values were calculated and averages compared to determine significance of their differences by t-test. To characterize general relationships among the samples, we used principal component, discriminant, and cluster analysis. All calculations were performed on IBM PC/AT 386 computer with standard statistical software (CSS 3/c and Statgraf 3.0).

Chromosomal Analysis

Phrynocephalus rossikowi shammakowi (Unguz Dry Bed, 20 km North from the Town of Serny Zavod, Turkmenistan; Fig. 1, 1). Diploid set includes 46 to 48 chromosomes. Since the number of metaphase plates with 46 and 48 chromosomes is approximately equal (16 and 14, respectively), we could not decide which karyotype is more characteristic for this form. This karyotype is very homogeneous morphologically, a feature which is characteristic for this lizard group (Sokolovsky 1974). All chromosomes are telocentric (except for a few large elements where shoulders can be seen). Since there is a clear size gap between the 12th and 13th pairs, and the centromere in the second size group is practically invisible, we tentatively consider the first 12 pairs as macrochromosomes, and the remaining 12 pairs as microchromosomes. Both versions of chromosome formulas are as follows:

$$2n = 24 \text{ M } (24 \text{ A} + \text{T}) + 22\text{m } (22 \text{ a}) = 46, \text{ NF} = 46 \text{ (Fig. 2b, d)},$$
$$2n = 24 \text{ M } (24 \text{ A} + \text{T}) + 24\text{m } (24 \text{ a}) = 48, \text{ NF} = 48 \text{ (Fig. 2a, c)}.$$

Therefore, the number of macrochromosomes in the Unguz populations is constant, whereas the number of microchromosomes varies from 22 to 24. Metaphase plates with different numbers of chromosomes were observed in all preparations from females and males, including those from testes. In some metaphase plates, the first pair of chromosomes has satellites (Fig. 3).

Ph. rossikowi rossikowi (Lebab Village in east Turkmenistan; Fig. 1, 3). Of 30 metaphase plates studied, 25 (83%) had 46, 3 had 48, and 2 had 44 chromosomes. This population, therefore, has a typical karyotype of 46 chromosomes:

$$2n = 22 \text{ M } (22 \text{ A} + \text{T}) + 24\text{m } (24 \text{ a}) = 46, \text{ NF} = 46 \text{ (Fig. 4, a)}.$$

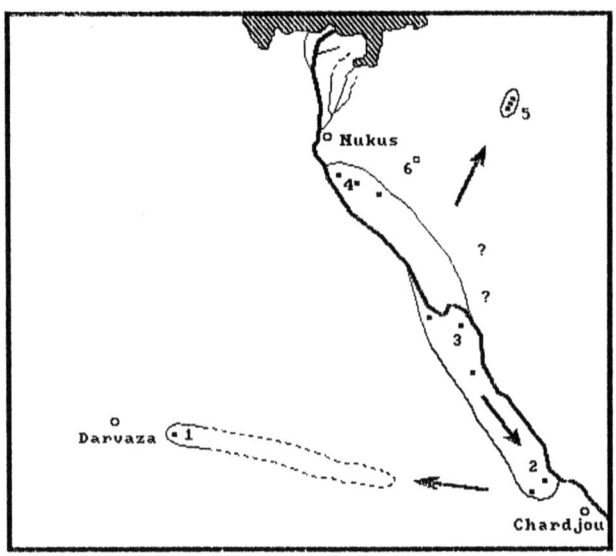

Fig. 1. Distribution of *Phrynocephalus rossikowi*: 1 – Unguz group of populations: Turkmenistan, 20 km north of the town of Serny Zavod (Shcherbak and Golubev 1979), 2 – Chardzhou group of populations: 70 km nothwest of the city of Chardzhou (Sokolovsky 1974); "Radiogram" ford (ZIN collection); village of Deinau (Ananjeva and Tunijev 1992); 3 – Pitnyak-Darganata group of populations: 6 km south of village of Pitnyak (Yevgenov and Parinkin 1955); village of Lebab (ZIK collection); 65 km south of the village of Darganata; east shore of Lake Danisher; village of Sheikharyk in 30 km above village of Khazaraspa; Karagyoz, 7 km south of the village of Khazaraspa; Dyueboyun (Shukurov 1965); 4 – Sultanuizdagh group of populations: Khentau near Nukus (Nikolsky 1898); Lake Khodzhakul near the Sultanuizdagh Mountains (Reshkevich 1958); Kushkanatau (Shukurov 1965); 40 km south of Nukus, west slope of the Sultan-Uvais (= Sultanuizdagh) Range (ZIK collection); 5 – Boktybulak group of populations: Bukan Well (Syroechkovsky 1958); Taschukho Well, Boktybulak Well (ZIK collection).

The number of macrochromosomes here is one pair less than in the Unguz population; the macrochromosome group includes pairs from 1 through 11, and the microchromosome group, pairs from 12 through 22. Satellite chromosomes have been observed in the first pair of chromosomes in the majority of metaphase plates.

Ph. rossikowi rossikowi (Boktybulak Well in Karakalpakstan; Fig. 1, 5). More than 75% of metaphase plates studied had 48 chromosomes. As in the previous cases, the majority of chromosomes are telocentric. The macrochromosome group includes 11 pairs, and has a significant size gap compared to the microchromosome group, which has 13 pairs. Satellite chromosomes in the first pair of chromosomes are present. Karyotype formula is

$$2n = 22 \text{ M } (22 \text{ A} + \text{T}) + 26m (26 \text{ a}) = 48, \text{ NF} = 48 \text{ (Fig. 4, b)}.$$

The testis cells at diakinesis and metaphase II were studied in *Ph . r. shammakowi* and the Boktybulak population of *Ph. r. rossikowi*. No differences

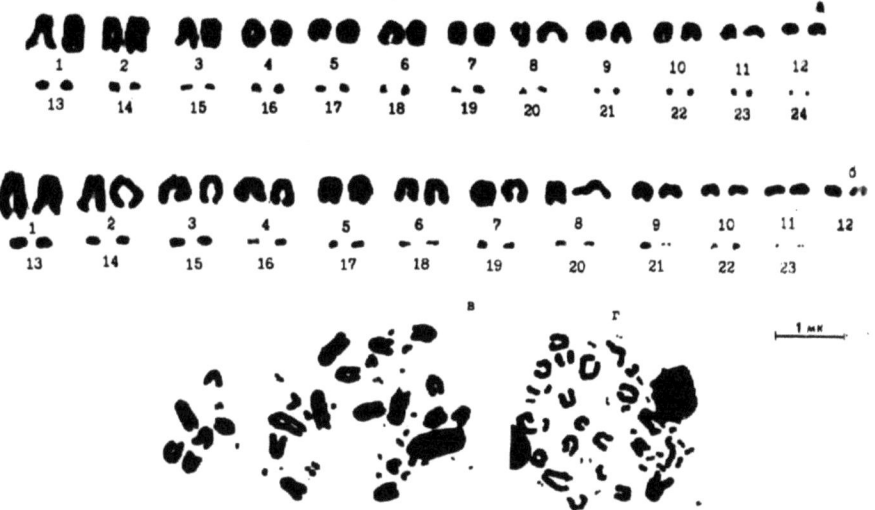

Fig. 2. Karyotype of *P. r. shammakowi* ((a, c) 2n = 48; (b, d) 2n = 46).

Fig. 3. Metaphase plate of *P. r. shammakowi* (arrows show satellite chromosomes).

in the morphology of bivalents were revealed. Differences in the number of bivalents and chromosomes were characteristic for the karyotypes described above.

Our analysis of karyotypes of three populations belonging to two subspecies of *Ph. rossikowi* reveals their similarity in chromosomal morphology; there are two to three acrocentric pairs, and satellite chromosomes are present in the first pair of chromosomes. Differences at various levels are also found. Variation of number of chromosomes in a diploid set (from 46 to 48 chromosomes) within individuals is seen in *Ph. r. shammakowi*. The number of chromosomes of the nominal subspecies, *Ph. r. rossikowi*, is also variable from 46 to 48 chromosomes, but at the level between populations. Finally, the nominal and

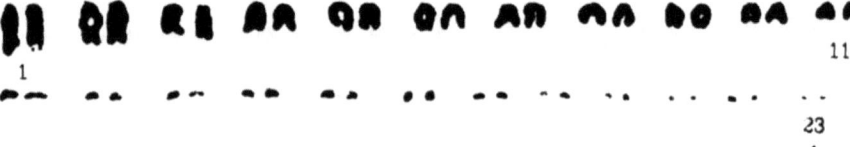

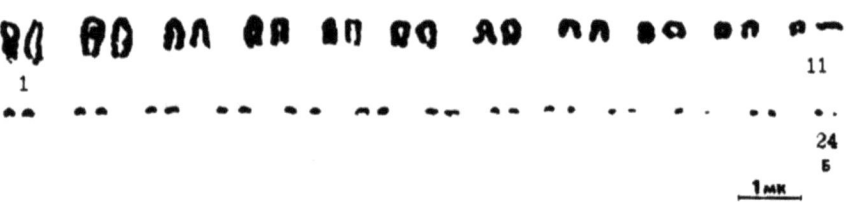

Fig. 4. Karyotype of *P. r. rossikowi* ((a) Lebab population, 2n = 46, (b) Boktybulak population, 2n = 48).

Unguz subspecies differ in their number of macrochromosomes (11 pairs in *Ph. r. rossikowi* and 12 pairs in *Ph. r. shammakowi*). Karyological data show that the existence of these two subspecific forms may indicate splitting of the species into two independent lineages.

Our data, compared to those of Sokolovsky (1974), show the similarity between the Boktybulak and Chardzhou populations. Discovery of satellite chromosomes in the first chromosomal pair, not found by Sokolovsky, can be explained technically (e.g., high spiralization of chromosomes). For further taxonomic conclusions, additional detailed studies of populations from Chardzhou and Lebab are required.

Morphological Analysis

Study of the samples in order to reveal sexual dimorphism shows that only in six meristic characters are there differences between males and females (Table 2). Comparison of correlation matrices (Table 3) suggests that most of the discovered differences are secondary and related to the general differences in size between sexes (L); after we moved from analysis of primary measures to analysis of ratios, sexual dimorphism was observed only in the L/LCD ratio. This ratio was, therefore, studied separately for males and females. Comparison of samples by sexually dimorphic characters L, LCD, FEM, and LEG shows that the second sample of females was of smaller size (Table 4); among male samples no such differences were detected. Other characters reveal differences between the first sample and samples 3, 4, and 5 (Table 5). However, analysis of

Table 2. Student's *t*-criterion comparison for males and females of *Ph. rossikowi*

Character	*t* value	P	Sample size		Mean	
			males	females	males	females
L	4.227	0.000	33	38	40.90	36.33
LCD	6.928	0.000	32	37	46.20	38.53
LLCD	−4.311	0.000	32	37	0.88	0.93
FEM	4.689	0.000	30	34	25.65	22.54
LEG	5.764	0.000	30	35	31.60	27.50
CAP1	2.980	0.004	28	28	8.17	7.69
CAP2	2.829	0.006	32	29	6.70	6.34
LHD	3.842	0.000	32	29	7.36	6.94
CAP3	3.477	0.001	32	29	5.92	5.58
INOS	2.553	0.013	32	29	1.03	0.96
HNOS	3.241	0.002	32	29	0.99	0.91
HHD	1.800	0.077	32	28	5.64	5.42
SQCC	1.548	0.128	29	26	18.24	17.53
SQCL	1.590	0.118	29	27	7.96	7.51
SDL	1.430	0.159	29	27	22.75	22.29
SDR	0.561	0.577	29	27	23.24	23.03
GVA	2.939	0.005	29	26	117.79	113.69

Table 4. Values of *t*-criterion and means for characters significantly different in selected samples of females of *Ph. rossikowi*

	2-tailed test		Groups compared, sample size		Means	
	t value	P level	1	2	1	5
L	9.051	0.000	7	9	39.00000	28.92222
LCD	7.631	0.000	7	9	40.07143	32.61111
FEM	8.529	0.000	7	9	24.80000	18.83333
LEG	8.114	0.000	7	9	29.70000	23.32222
			3	2	3	5
L	8.988	0.000	5	9	38.90000	28.92222
LCD	7.669	0.000	5	9	40.30000	32.61111
FEM	7.610	0.000	4	9	23.72500	18.83333
LEG	7.775	0.000	5	9	29.06000	23.32222
			5	2	4	5
L	7.588	0.000	16	9	38.20625	28.92222
LCD	6.178	0.000	16	9	40.63750	32.61111
FEM	6.232	0.000	13	9	23.30000	18.83333
LEG	6.309	0.000	13	9	28.49231	23.32222

Table 3. Matrices of partial correlations (below the diagonal) and Pearson's linear correlations (above the diagonal) for morphological characters of *Ph. rossikowi*

	L	LCD	FEM	LEG	CAP1	CAP2	LHD	CAP3	INOS	HNOS	HHD	SQCC	SQCL	SDL	GVA
L	1	0.92	0.96	0.95	0.88	0.90	0.84	0.88	0.52	0.52	0.81	−0.20	0.09	0.18	0.11
LCD	−0.03	1	0.93	0.95	0.84	0.83	0.81	0.84	0.53	0.48	0.73	0.23	0.16	0.21	0.22
FEM	0.66	0.14	1	0.97	0.91	0.91	0.88	0.90	0.65	0.65	0.80	0.19	0.07	0.19	0.19
LEG	0.05	0.45	0.47	1	0.87	0.87	0.87	0.88	0.64	0.61	0.75	0.16	0.11	0.19	0.16
CAP1	−0.40	−0.11	0.43	0.00	1	0.93	0.83	0.89	0.54	0.45	0.87	0.17	0.03	0.23	0.20
CAP2	0.09	−0.04	0.18	−0.21	0.36	1	0.88	0.94	0.55	0.52	0.78	0.11	−0.03	0.27	0.08
LHD	−0.14	−0.08	0.10	0.12	−0.05	−0.06	1	0.91	0.61	0.65	0.70	0.15	−0.01	0.28	0.06
CAP3	0.13	0.14	−0.19	0.07	0.11	0.58	0.53	1	0.53	0.57	0.75	0.10	−0.04	0.14	0.11
INOS	−0.04	−0.13	−0.07	0.23	0.01	0.25	0.12	−0.28	1	0.71	0.51	0.09	0.20	0.12	0.08
HNOS	−0.54	−0.19	0.74	−0.20	−0.43	−0.21	0.03	0.25	0.39	1	0.41	−0.11	−0.09	0.11	−0.13
HHD	0.34	0.05	−0.07	−0.21	0.56	−0.17	0.05	−0.05	0.11	0.11	1	0.16	0.08	0.15	−0.17
SQCC	−0.03	0.08	0.28	−0.32	−0.04	−0.11	0.22	−0.05	0.05	−0.28	−0.07	1	0.38	0.09	0.30
SQCL	0.15	0.11	−0.24	0.18	0.04	−0.02	−0.04	−0.05	0.24	0.06	0.04	0.45	1	0.03	0.31
SDL	0.13	0.14	−0.25	0.04	0.08	0.43	0.39	−0.54	−0.17	0.24	−0.06	−0.01	−0.02	1	0.16
GVA	−0.51	−0.01	0.58	−0.20	−0.11	−0.31	−0.21	0.30	0.15	−0.54	0.07	−0.03	0.22	0.34	1

Table 5. Values of t-criterion and means for characters significantly different in selected samples of *Ph. rossikowi*

	2-tailed test		Sample number (size)		Means	
	t value	P level				
SQCL	4.333	0.001	1 (14)	4 (3)	8.2857	6.3333
SQCL	4.359	0.000	1 (14)	4 (3)	8.2857	7.0000
INOS	4.326	0.000	1 (14)	5 (38)	1.1000	0.9500

Table 6. Values of t-criterion and means for characters significantly different in samples 2 and 5 of *Ph. rossikowi*

	2-tailed test		Sample number (size)		Means	
	t value	P level				
FEML	−5.197	0.000	4 (32)	2 (11)	0.387559	0.977884
LEGL	−4.176	0.000			−0.163562	1.078257

indices shows a difference only between samples 2 and 5 in FEM/L and LEG/L (Table 6). Probably most differences in the primary measures listed above are related to age composition of samples, which were randomly collected and were not statistically representative in some characters (e.g., samples 2 and 4).

The following measures and ratios were selected for the comparative analysis of samples: HHD, SQCC, SQCL, SDL, GVA, FEM/L, LEG/L, CAP1/L, CAP2/CAP3, LHD/CAP3, INOS/HNOS, and SDL/SDR. Discriminant analysis (Fig. 5, Tables 7 and 8) allows for identification of each sample and determination of the impact of separate characters. As Table 8 shows, the most important indices were those which reflect body shape (FEM/L and LEG/L); of less importance were indices characterizing the head. Scale counts were the least informative parameters. In general, results of discriminant analysis correspond well to the comparative analysis of samples based on original measures.

To illustrate the structure of relationships among four samples (sample 2 was excluded due to lack of data in many parameters), we calculated the generalized Machalanobis distance (Table 9). Results are presented as a cladogram (Fig. 6).

Results of morphological analysis show that variation of characters in samples is chaotic and, therefore, cannot be used for taxonomic analysis (at least at the intraspecific level). However, proximity in a multidimensional model (Fig. 6) of groups originating from the periphery of species' range suggests that variability of *Ph. rossikowi* can possess a radial pattern. Populations which inhabit areas along the Amudarya River can be considered the most ancient ones, and ancestral for the populations on the periphery.

Table 7. Eigenvalues of discriminant functions

Function	Eigenvalue	% of Variation
1	1.2467049	46.45
2	0.9889400	36.84
3	0.4486023	16.71

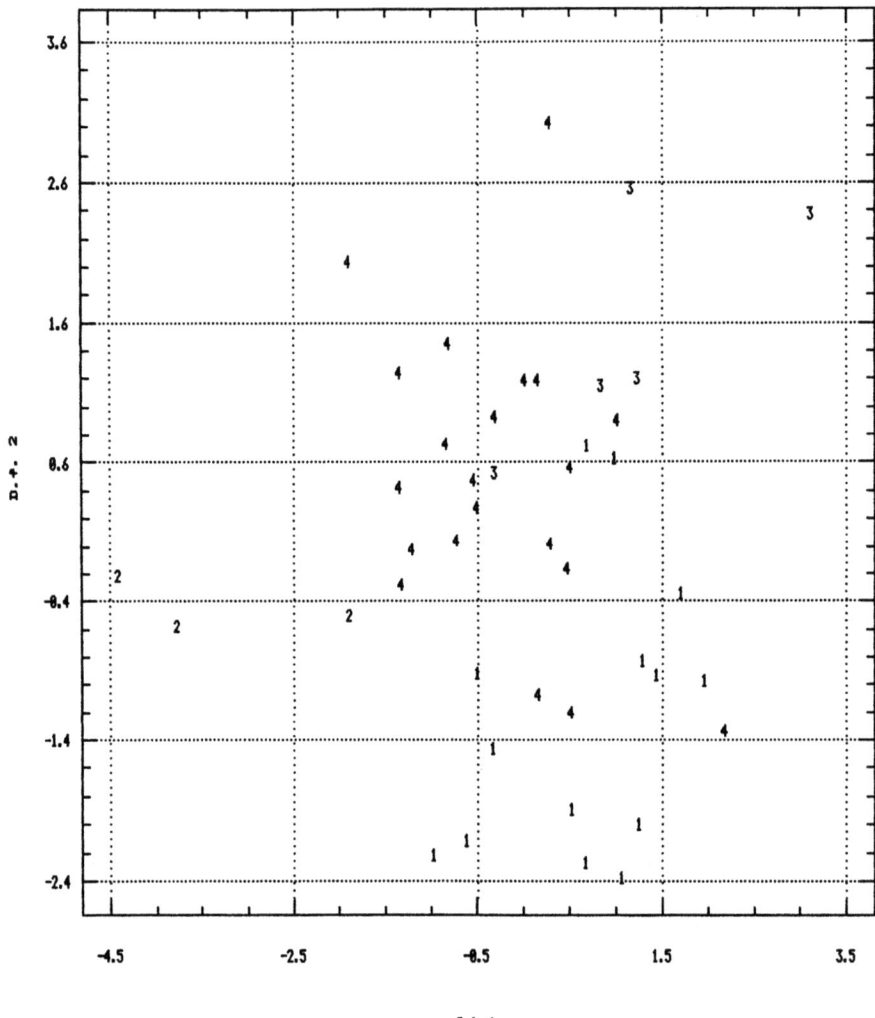

Fig. 5. Position of studied individuals of P. rossikowi in a system of discriminant axes (for sample code see Table 1).

Table 8. Coefficients of discriminant functions

Characters	Coefficients		
	1	2	3
FEML	0.62859	−0.79833	−0.61642
LEGL	0.12631	1.25007	0.51803
CAP1L	−0.20471	−0.08404	−0.34963
CP2CP3	0.29844	0.46828	−0.16728
LHDCP3	0.08933	−1.00528	−0.31238
INSHNS	0.43689	−0.19232	−0.30238
HHD	−0.07197	0.26199	0.12504
SQCC	0.51940	0.36683	0.44321
SQCL	0.38049	−0.55504	0.44054
SDL	−1.43266	−0.40215	0.07042
SDLR	0.98189	0.70780	−0.39385
GVA	−0.33688	−0.16309	0.41983
Constant	0.24489	0.18303	−0.04679

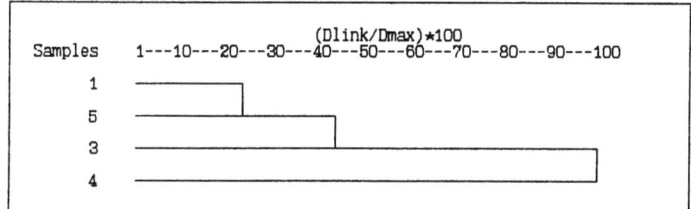

Fig. 6. UPGMA cladogram of relationships among samples (data from Table 9).

Table 9. Machalanobis distance matrix for four samples of *Ph. rossikowi*

	1	3	4
3	8.63		
4	15.39	19.42	
5	3.96	5.62	12.42

Analysis of Selected Non-metric Characters

The original description of the Turkmen subspecies *Ph. r. shammakowi* noted such diagnostic features as carinate tail scales, size of occipital scales, expressed undertail stripes, and peculiar color pattern (Shcherbak and Golubev 1979). This description was based on two lizards which were compared with the nominal population from the vicinity of Nukus (Karakalpakstan). Now, after we have seen numerous samples from different parts of the range, we can state that most of the differences listed above are within the limits of normal individual and intrapopulational variability and do not delineate two subspecies. Only the dorsal color pattern of *Ph. rossikowi* has certain variability (Fig. 7) that can be of diagnostic value.

We divide dorsal pattern into two components: protective and disruptive pattern. The protective pattern, which is direct camouflage, includes general background (dorsal color), small elements of the pattern (dots, small stripes), and "ocelli" with their dark edges. The latter also probably possess a thermoregulative function. Protective components of the dorsal pattern correspond to the color of the substrate.

Disruptive elements of the pattern are dark transverse stripes which are symmetric with respect to the longitudinal body axis. These stripes correspond to large (relative to lizard size), common elements of the environment (such as pebbles, gravel, small stones, and clusters of large dark sand grains). These elements also contrast with the pale desert substrate due to the interplay of light and shade in the desert landscape. Position of dorsal stripes is relatively constant; most Middle Asian *Phrynocephalus* species have shoulder, clavicular, central dorsal, and lumbar stripes. Their shape and darkness may vary considerably; stripes may break into fragments which may be reduced or completely disappear (especially those of the central dorsal stripe). Fragmented stripes become part of the protective pattern. This is precisely the case in *Ph. rossikowi*: its dorsal stripes are reduced; stripe fragments are rotund and resemble pebbles covered by a dark desert tan – the substrate on which lizards usually hide. Size and darkness of dorsal pattern fragments of a lizard correspond to size and darkness of pebbles: e.g., in the gravel-pebble submontane plain of the Sultanuvais Range (Fig. 7, a). Disruptive elements of the dorsal pattern are reduced even more in animals inhabiting clay or loess areas with sparse, light-colored pebbles, grains of sandstone, and occasional sand patches: e.g., areas of Lebab village on the left bank of the Amudarya, or Boktybulak Well in Karakalpakstan (Fig. 7, b). Background dorsal color, which corresponds to the essentially similar light background of desert substrate, does not vary among these populations.

Therefore, we think that the dorsal pattern in *Ph. rossikowi* underwent development to provide camouflage on the surface of the gravel-pebble desert. This suggestion corresponds to the published descriptions of habitats in which this lizard is found (Yevgenov and Parinkin 1955; Rashkevich 1958; Syroechkovsky 1958; Shukurov 1965; Bondarenko 1982). We should note that "pebble-like" fragments of the central dorsal stripe are also expressed in some populations of *Ph. ocellatus* (Fig. 7, c) which is separated from *Ph. rossikowi* only by the sands of the western Kizylkum. Such fragments are completely absent from all other sclerobiotic *Phrynocephalus* of Middle Asia. If *Ph. ocellatus* and *Ph. rossikowi* are sister species, the latter could have been formed in situ within its modern range.

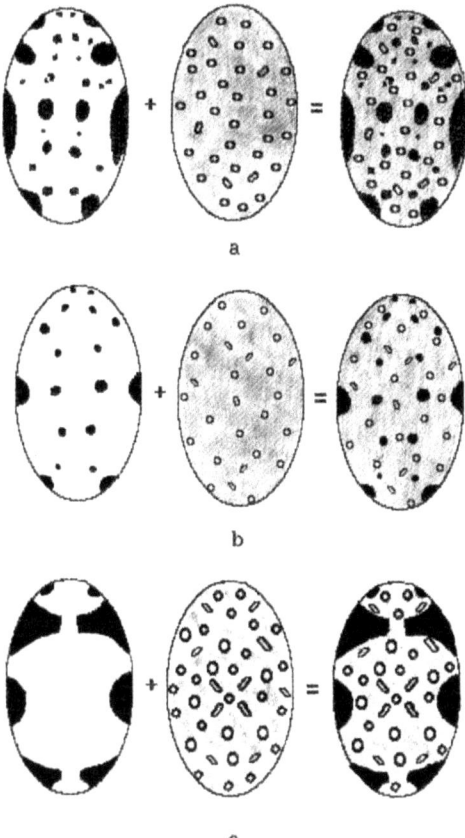

Fig. 7. Schematic dorsal pattern of *P. rossikowi* from different parts of its range ((a) gravel-pebble plain, 40 km south of Nukus, foothills of Sultanuizdagh Mountains; (b) the Unguz bed) and of *P. ocellatus* (southern foothills of the Kuldzhuktau Range, ancient bed of the Zeravshan River, Central Kizylkum).

Discussion

Our results at first glance seem to provide a chaotic picture at all levels. In addition, electrophoretic studies (Mezhzherin and Golubev 1992) have not revealed any differences among the Unguz, Lebab, and Boktybulak populations of *Ph. rossikowi*. It is difficult to expect, however, that relatively young species would exhibit clear intrapopulational differences. Although well-expressed karyological patterns confirm that intraspecific differentiation takes place, these data alone cannot guarantee certain taxonomic conclusions.

Bondarenko (1984) suggested that the Amudarya River affected distribution of *Ph. rossikowi*. We propose our own model for such a process. The Amudarya is known to have changed its course in the Upper Pleistocene from west to north. It cut through the Pitnyak elevation and flowed into Lake Khwarazm, and later, via Lake Sarykamysh, again started to flow into the Caspian Sea. At this time, the waters of the Amudarya filled the Unguz dry bed, which became its left tributary; it became dry again in the second half of the Late Pleistocene (Gorelov et al. 1989). We suggest that *Ph. rossikowi*, with the southern part of its range probably reaching the right bank of the Proto-Amudarya (approximately the site of the modern city of Chardzhou), penetrated from this area to Unguz at a later time, via the now dry bed. Contrary to our previous assertion (Shcherbak and Golubev 1979), *Ph. rossikowi* is found in various depressions of the Unguz (A.L. Zatoka pers. comm.).

The northeastern group of populations of *Ph. rossikowi* is separated from the major range and is distributed along one of the youngest beds of the Zhanadarya (a dry bed of the Syrdarya). Absence of this lizard in the Akchadarya corridor (our data, see map, N6), which is the most ancient part of the Amudarya delta, could be caused by multiple breakthroughs of the river along this bed (Kes et al. 1980).

Therefore, we suggest that the two largest rivers of Middle Asia, the Amudarya and Syrdarya, played a major role in the formation of modern range of *Ph. rossikowi*. We treat the central populations of this range as the ancestral "core," as compared to the peripheral ones. We believe that this species differentiated during adaptation to local gravel-pebble semidesert in the West Kizylkum. The most ancient part of its range probably lies within the alluvial fan submontane plain of the Sultanuizdagh Range, which is of early Pleistocene age (Kogai 1957) and the Pitnyak elevation in East Turkmenistan, adjacent to the south. Therefore, our opinion differs from that of Ananjeva and Tunijev (1992), who suggested that *Ph. rossikowi* is a Pliocene species whose range shifted during the Pleistocene from the middle portion of the Amudarya River to its lower part.

22. Formation of the Fish Population in the Artificial Hydrographic Network of Turkmenistan (the Amudarya River Basin)

VLADIMIR B. SALNIKOV

Abstract

Fish population in the artificial water bodies of Turkmenistan (the Amudarya River basin) was studied from 1973 to 1991. Since wide-scale irrigation commenced in the republic in the 1950s, the more than 1,000 km long Karakum Canal, its reservoirs, and drainage lakes dramatically changed fauna and ecology of fishes in Turkmenistan. Both the aboriginal and introduced (primarily from the Far East) ichthyofauna are described; a total list includes 44 fish species. Ecological and faunogenetic trends in the formation of the fish population structure, as well as fishing dynamics and fish productivity, are discussed. Morphoecological and developmental anomalies are revealed in fish inhabiting the water bodies which accumulate collection-drainage waters with high content of various agricultural chemicals.

Introduction

A radical restructuring of the continental water ecosystems is one result of modern industrial activities. Human impact has had a profound effect on aquatic communities, resulting in rapid and often irreversible changes in the structure of fish populations. In the continental waters of Middle Asia, including Turkmenistan, the major transforming factors have been irrigation as well as the introduction of new fish species. Human activity in Middle Asia has effected such wide-ranging ecological changes as a radical rearrangement of the entire hydrography of Middle Asia, the deterioration of the natural ecosystem of the Aral Sea – including its fish population, a dramatic decline in the abundance of many native species of fishes outside the Aral basin, the naturalization of Far East fish species in these waters, and the formation of new ecosystems and fish communities in the numerous and diverse artificial water bodies (e.g., reservoirs and canals). The ichthyofauna of the Aral basin has been radically transformed.

Numerous studies have been conducted on the anthropogenic transformation of fish communities in the Aral basin (Kamilov 1973; Tleulov 1981; Pavlovskaya 1982, 1991; Sagitov 1983; Amanov 1985). In Turkmenistan, such studies have been primarily devoted to the formation of fish communities in new artificial water bodies (Mukhamedieva 1974; Aliev 1976; Mukhamedieva and Salnikov 1983; Pavlovskaya and Salnikov 1990; Salnikov and Reshetnikov 1991). The objective of this review is to examine the process of formation and development of fish populations in the new water bodies of the artificial hydrographic network in Turkmenistan.

Brief Characteristics of the Water Bodies

Wide-scale irrigation in Turkmenistan began in the 1950s. It has been primarily limited to the Amudarya River, which is the main water source in this area and forms a prolonged part of the border between Turkmenistan and Uzbekistan. A substantial network of the artificial water bodies has been built in eastern and southeastern Turkmenistan and is today one of the most significant anthropogenic structures in the Amudarya basin.

The largest structure of this artificial network is the Karakum Canal. Originating at the upper part of the lowland portion of the Amudarya River, it continues westward along the southern part of Turkmenistan for more than 1,000 km. The Karakum Canal has several regulatory reservoirs: the Kelif (now defunct due to silting), Khauzkhan, Kurtli, and Kopetdagh Reservoirs.

In addition to the irrigation system, a network of collection and drainage canals has been built to decrease the level of salt ground waters and to drain these waters out of irrigated areas. These mineralized waters of the collection-drainage system have been disposed either in the Amudarya River or in large natural depressions. The waters in these depressions have accumulated, creating a new type of man-made water bodies, which are commonly called "lakes." These lakes include Lake Kattashor (eastern Turkmenistan), Lake Sarykamysh (northwestern Turkmenistan), and others. Brief characteristic of the reservoirs and accumulating lakes most important for the fishing industry is given in Table 1. As an example of a smaller, less significant accumulating lake, we have selected Lake Ovadan which is located next to the city of Ashkhabad and is a part of the Karakum Canal system.

Materials and Methods

This review describes and discusses the data collected by the author from 1973 to 1991 on fish in the artificial water bodies of Turkmenistan. More than 10,000 individual fish have been studied. Specimens were caught by industrial nets and sweep-nets as well as by stationary research nets. Small non-food fish were obtained from dragnets and fine-meshed nets. Biological analysis of the fish was conducted according to the standard techniques (Pravdin 1966). We also used

Table 1. Characteristics of the artificial water bodies of Turkmenistan in the Amudarya River basin

Characteristic	Khauzkhan Reservoir	Kopetdagh Reservoir	Lake Kattashor	Lake Sarykamysh
Year of filling	1961	1973	1950	1963
Volume (million m^3)	875	220	90	28,000
Area (km^2)	210	33	30	2,500
Maximal depth (m)	10.5	13.0	6.5	40.0
Level fluctuations (m)	7.0	7.0–9.0	1.5	None
Clarity (m)	~ 4.6	~ 4.8	0.5–1	2–5
Mineralization (g/l)	0.6–1.0	0.7–1.0	2–5	12–13
Length of vegetative period (months)	9–10	9–10	9	7–8
Trophic status[a]	Meso	Meso	Eu	Meso
Connection with the accessory system[b]	One-way	One-way	Two-way	One-way

[a] Trophic status: Meso = Mesotrophic, Eu = Eutrophic
[b] Connection with the accessory system: One-way: water flow is directed only from the accessory drainage system to a water body; Two-way: water flow is directed in both directions

special techniques described by Doroshev (1964) and Zakharov (1982) in the analysis of morphoecological features of the fish inhabiting reservoirs with accumulating collection-drainage waters.

For analysis of the formation of fish populations, we used the following concepts and approaches: the zoogeographic analysis of the freshwater ichthyofauna of the USSR conducted by L.S. Berg (1949); the concept of faunistic complexes (Nikolsky 1980); and the theory of reproductive ecological groups of the fishes (Kryzhanovsky 1949; Balon 1975; Soin 1981). The Soerensen-Chekanovsky similarity index for different ichthyofaunas has been calculated and depicted as WPGMA dendrograms (Bailey 1970; Andreev and Reshetnikov 1978).

History of the Formation of the Ichthyofauna in Artificial Water Bodies

The ichthyofauna of the artificial water bodies in Turkmenistan has been formed from two general sources: first, by the aboriginal ichthyofauna of the Amudarya drainage basin; and second, by the introduced (in Russian terminology, "acclimatized") fish species.

Table 2 includes a list of all fish species recorded from all discussed water bodies since their creation as well as a list of the ichthyofauna of the middle portion of the Amudarya River for the same period. We have used our data as well as published observations (Shaposhnikova 1950; Poltoratsky and Lyashenko 1972; Mukhamedieva 1974; Sagitov 1983; Pavlovskaya and Zholdasova 1991).

Colonization of the artificial hydrographic network began befor the natural

Table 2. Ichthyofauna of the artificial water bodies of Turkmenistan in the Amudarya River basin

Species	Amudarya River	Karakum Canal	Reservoirs of Karakum Canal	Lake Kattashor	Lake Sarykamysh
Family Acipenceridae – Sturgeons					
1. *Acipenser nudiventris* Lov. – Ship sturgeon	1	1	1	–	2–
2. *Pseudoscaphirhynchus kaufmanni* (Bogd.) – Amudarya shovelnose sturgeon	1	1	1–	–	2–
3. *P. hermanni* Kessl. – little Amudarya shovelnose sturgeon	1	–	–	–	2–
Family Esocidae – Pikes					
4. *Esox lucius* L. – Pike	–	–	–	–	1
Family Cyprinidae – Carps					
5. *Rutilus rutilus aralensis* Berg – Aral roach	1	1	1	1	1
6. *Mylopharyngodon piceus** (Rich.) – Chinese roach	1	1	1	–	2
7. *Leuciscus latus* (Keys.) – Transcaspian dace	–	–	1–	–	–
8. *L. idus oxianis* (Keys.) – Turkestan ide (orfe)	–	–	–	–	2
9. *Scardinius erythrophthalmus* (L.) – Rudd	–	–	–	1	1
10. *Ctenopharyngodon idella** (Val.) – White amur	1	–	1	1	1
11. *Aspiolucius esocinus* (Kessl.) – Asp-pike	–	–	–	–	–
12. *Aspius aspius iblioides* (Kessl.) – Aral asp	–	–	–	1	1
13. *Pseudorasbora parva** (Schenkel) – Amur chebachek	1	–	1	1	–
14. *Gobio gobio lepidolaemus* Kessl. – Turkestan gudgeon	1	–	1–	–	2–
15. *Pseudogobio rivularis** (Bas.) – Chinese false gudgeon	–	–	1–	–	–
16. *Varicorhinus capoëta heratensis* (Keys.) – Transcaspian khramulya	1	1	1	1	–
17. *Barbus capito conocephalus* Kessl. – Turkestan barbel	1	1	1	1	1
18. *B. brachycephalus* Kessl. – Aral barbel	1	1	1	1	1
19. *Chalcalburnus chalcoides aralensis* – Aral shemaya	2	2	2	2	1

Table 2. Continued

Species	Amudarya River	Karakum Canal	Reservoirs of Karakum Canal	Lake Kattashor	Lake Sarykamysh
20. *Alburnoides taeniatus* (Kessl.) – Striped bystryanka	1	1	1–	–	–
21. *A. bipunctatus eichwaldi* (Filippi) – Eastern bystryanka	1	1–	–	1	1
22. *Abramis brama orientalis* – Eastern bream	2	2	2	2	1
23. *A. sapa bergi natio aralensis* Tjapkin – Aral white-eye bream	–	–	–	–	1
24. *Capoetabrama kuschakewitschi* (Kessl.) – Sharpray	1	1	1–	–	2–
25. *Parabramis pekinensis** (Bas.) – White Amur bream	2	1	2	2	3
26. *Hemiculter eigenmanni** (Jordan et Metz) – Korean sharpbelly					
27. *Pelecus cultratus* (L.) – Sabre carp (chekhon)	1	1	1	1	1
28. *Pseudoperilampis ocellatus** (Kner) – Eyed bitterling	3	3	3	3	1
29. *Carassius auratus gibelio* (Bloch) – Goldfish	1	3	1	3	3
30. *Cyprinus carpio* L. – Carp	1	1	1	1	3
31. *Hypophthalmichthys molitrix** – White thickforehead	1	1	1	1	1
32. *Aristichthys nobilis** – Painted thickforehead	1	1	1	–	3
Family Cobitidae – Loaches					
33. *Noemachilus oxianus* Kessl. – Amudarya loach	1	1	–	–	–
34. *N. amudarjensis* Rass – Bokhara loach	1	–	–	–	–
35. *N. malapterurus longicauda* (Kessl.) – Eastern crested loach	1	1	–	–	1–
36. *Cobitis aurata aralensis* Kessl. – Aral spined loach	1	1	–	–	2–
37. *Misgurnus anguillicaudatus** (Cantor) – Amur weatherfish	–	3	3	–	1
Family Siluridae – Catfish					
38. *Silurus glanis* L. – Wels, European catfish	1	1	1	1	1

Table 2. Continued

Species	Amudarya River	Karakum Canal	Reservoirs of Karakum Canal	Lake Kattashor	Lake Sarykamysh
Family Gasterosteidae – Sticklebacks					
39. *Pungitius platygaster aralensis* – Aral stickleback	–	–	–	–	1
Family Oryziatiidae – Oryzias					
40. *Oryzias latipes** (Temminck et Schlegel) – Japanese oryzia	–	3	–	–	3
Family Poeciliidae – Poecilias					
41. *Gambusia affinis holbrooki** (Gir.) – Eastern gambusia	1	1	1	1	1
Family Ophiocephalidae – Snakeheads					
42. *Ophiocephalus argus warpachowskii** Berg – Amur snakehead	2	2	2	3	3
Family Percidae – Perches					
43. *Stizostedion lucioperca* L. – Pike-perch, zander	2	2	2	2	2
Family Gobiidae – Gobies					
44. *Rhinogobius similis** Gill – Amur goby	1	1	1	1	1
Total number of species	35	36	32	21	35

1 – species recorded in the water body before the 1970s
2 – recorded for the first time from the early 1970s to the early 1980s
3 – recorded only after the early 1980s
* introduced species
sign "–" next to a number indicates that the species is not currently found in this water body

pattern of distribution of the native Amudarya fishes was disturbed. The ichthyofauna of the new water bodies was recruited primarily from those portions of the Amudarya River used to supply water: the middle Amudarya for the Karakum Canal system and Lake Kattashor; and the lower Amudarya for Lake Sarykamysh. During this colonization, the lake-type water bodies (reservoirs and accumulating lakes) were colonized predominantly by limnophylic fish species, whereas Karakum Canal also was colonized by both limnophylic and rheophilic fishes. Due to the relative richness of the lower Amudarya limnophylic ichthyofauna (Shaposhnikova 1950), the new ichthyofauna of Lake Sarykamysh also became diverse.

The distribution of natural ichthyofauna along the Amudarya River underwent considerable changes in the 1970s. Mass migration of the Eastern bream (*Abramis brama orientalis*) and the pike-perch (*Stizostedion lucioperca*) into the middle part of the Amudarya occurred, as well as a less significant migration of the Aral shemaya (*Chalcalburnus chalcoides aralensis*). Sources of these migrations are not known precisely although two possibilities, or a combination of both, exist. One is a migration from the Aral Sea and the lower Amudarya, due to the incipient degradation of the Aral Sea and decreasing of the Amudarya flow. A second possibility is a migration of these fish species from the water bodies located to the east of the Amudarya River: the bream and pike-perch were introduced in the 1960s to the Caspian Sea basin from the Ural River, whereas Aral shemaya was recorded even earlier in the Caspian Sea basin (Kamilov 1973). Also, the pike-perch was introduced in 1970 from Kazakhstan into the Karakum Canal system although the results of this introduction are not known.

A second important component of the Turkmenistan ichthyofauna is the introduced species of fish (Table 2). These belong primarily to the Far East ichthyofauna and have been deliberately or accidentally introduced into Middle Asia from the Amur River or from China. In the late 1950s, experimental introduction of the cultured pond fish from China was initiated in Turkmenistan. These species included the white and the painted thickforeheads (*Hypophthalmichthys molitrix* and *Aristichthys nobilis*), the white amur (*Ctenopharyngodon idella*), the Chinese roach (*Mylopharyngodon piceus*), and the white Amur bream (*Parabramis pekinensis*). The goals of this introduction were the prevention of macrophyte algae proliferation in the Karakum Canal and other water bodies of the irrigation network and the increase of fish productivity by introduction of species which actively consume phytoplankton, zooplankton, and detritus. In 1963, natural reproduction of these species began in the Karakum Canal (Aliev 1976). Several "weed" fish species [e.g., Amur chebachek (*Pseudorasbora parva*), Korean sharpbelly (*Hemiculter eigenmanni*), Amur goby (*Rhinogobius similis*), and Amur false gudgeon (*Pseudogobio rivularis*)] were also accidentally introduced into the Karakum Canal system and rapidly colonized. From the Karakum Canal, the introduced species dispersed to the Amudarya River and the connected water bodies.

The colonization of Turkmenistan waters by introduced fish species has continued and still is an ongoing process. In the 1980s, the Karakum Canal system, Lake Kattashor, and Lake Sarykamysh, as well as their connected water bodies, were colonized by the snakehead (*Ophiocephalus argus*) and eyed bitterling (*Pseudoperilampis ocellatus*); these two species were initially introduced to the ponds of Uzbekistan from China (the latter species was introduced accidentally). In 1991, two more introduced Far East species, oryzia (*Oryzias latipes*) and Amur weatherfish (*Misgurnus anguillicaudatus*), were found in the Karakum Canal. Due to continued fish farming and migrations, the ichthyofauna of the Karakum Canal continues to penetrate into the independent basins of the southern Turkmenistan rivers, Tedzhen and Murghab.

During the entire period of existence of the Karakum Canal system, Lake Kattashor, and Lake Sarykamysh, a total of 43 species of fish have been recorded there. The majority (28 species) belong to the carp family (Cyprinidae). Due to colonization, the diversity of ichthyofauna in some water bodies has been declining. The richest ichthyofaunas are those of the Karakum Canal (36 species) and Lake Sarykamysh (35). The reservoirs of the Karakum Canal system have 32 species, and Lake Kattashor, only 21 species of fish. Some species are extremely rare, such as the ship sturgeon, Amudarya shovelnose sturgeon, pike, rudd, Aral white-eye bream, and Turkestan ide. Several species have become practically extinct, such as the little Amudarya shovelnose sturgeon, Transcaspian dace, and asp-pike.

Ecological and Faunistic Trends in the Formation of the Fish Population Structure

As described above, a new, specific ichthyofauna has been formed in the artificial hydrographic network of Turkmenistan. It is comprised almost exclusively by the combination of native species of the Aral Sea and the Amudarya River (28 species) and Far East introductions (12 species). Dendrograms A and B (Fig. 1) show the individual and group specificity of the modern ichthyofaunas of the artificial water bodies as well as of the Amudarya River, the Aral Sea, and the Tedzhen and Murghab Rivers. These calculations have been based on the data from Table 2 and supplemented by the author's and published data on the Tedzhen, Murghab, and Amudarya Rivers (Shaposhnikova 1950; Sagitov 1983; Lim and Yermakhanov 1986; Aliev et al. 1988; Pavlovskaya and Zholdasova 1991). Dendrogram A includes only native species, whereas dendrogram B accounts for the entire ichthyofauna.

The clustering pattern of both dendrograms (Fig. 1) is very similar. Three general faunistic groups can be separated. The first group includes the ichthyofaunas of river-type water bodies: middle (8) and low (9) Amudarya River, the Karakum Canal (1), and the adjacent foothill portion of the Amudarya (7). The second group includes the ichthyofaunas of lake-type water bodies: Lake Sarykamysh (5), the reservoirs of the Karakum Canal system (2),

Formation of the Fish Population in the Network of Turkmenistan 373

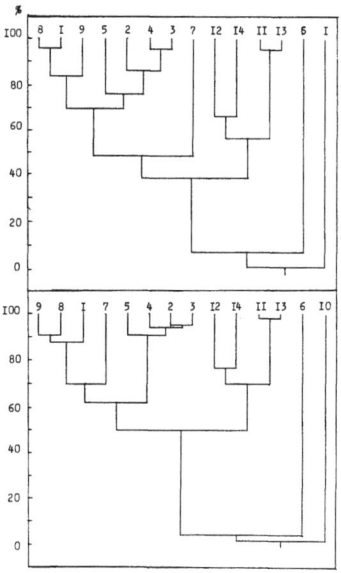

Fig. 1. Dendrograms of similarity (Soerensen-Chekanovsky index) of the ichthyofaunas of the water bodies of Turkmenistan. (A) only native fauna; (B) the entire fauna, including introductions; 1 – the Karakum Canal; 2 – the reservoirs of the Karakum Canal system; 3 – Lake Ovadan; 4 – Lake Kattashor; 5 – Lake Sarykamysh; 6 – the mountain portion of the Amudarya River; 7 – the foothill portion of the Amudarya River; 8 – middle (lowland) portion of the Amudarya River; 9 – lower (lowland) portion of the Amudarya River; 10 – the Aral Sea; 11 – the Murghab River; 12 – the Saryyazy Reservoir; 13 – the Tedzhen River; 14 – the Tedzhen Reservoir.

Lake Kattashor (4), and Lake Ovadan (3). Rheophilic fish species are noticeably absent from the second group. The third group contains ichthyofaunas of the Murghab (11) and Tedzhen (13) Rivers as well as the reservoirs constructed there, Saryyazy (12) and Tedzhen (14) Reservoirs, respectively. The first and the second groups combined form the branch of our dendrogram that characterizes ichthyofaunas of the modern basin of the lower Amudarya River. The third group is well separated from the first two not only because the native ichthyofauna of the Murghab and Tedzhen is different, but also because, for the present, only few Karakum Canal species have colonized these drainages. The most distant from each other are ichthyofaunas of the mountain portion of the Amudarya (6) and of the modern Aral Sea basin (10).

Zoogeographic regionalization of the Aral-Amudarya basin as well as its position in a general zoogeographic scheme for the Holarctic ichthyofauna, was discussed by L.S. Berg (1949), who included this ichthyofauna into the Tibetan province of the Mountain Asian subregion. The native component of the ichthyofauna of reservoirs and accumulating lakes in the Amudarya basin in Turkmenistan is zoogeographically similar to the ichthyofauna of the Aral area of the Ponto-Caspio-Aral province, which is now almost completely degraded due to human industrial activity. The ichthyofauna of these artificial water bodies in the Amudarya basin includes species endemic for the Ponto-Caspian-Aral province and found also in the Caspian Sea, such as *Acipenser nudiventris, Barbus brachycephalus, B. capito, Chalcalburnus chalcoides,* and *Pungitius platygaster*. Some species are represented here by the endemic subspecies of the Aral Sea basin (*Rutilus rutilus aralensis, Barbus brachycephalus brachycephalus, B. capito conocephalus, Chalcalburnus chalcoides aralensis,* and *Pungitius platygaster aralensis*).

The native component of the Karakum Canal ichthyofauna includes forms characteristic for the Aral zoogeographic area as well as representatives of the middle Amudarya ichthyofauna characteristic for the Turkestan zoogeographic province (which covers also the middle portion of the Syrdarya River; Berg 1949). The middle Amudarya forms include two endemic relict genera, *Aspiolucius* and *Pseudoscaphirhynchus* (the latter is close to the North American genus *Scaphirhynchus* from the Mississippi River), the subendemic genus *Capoetobrama*, and the endemic species *Noemachilus oxianus*. The ichthyofauna of the Karakum Canal also includes five taxa shared by the Iranian and Turkestan provinces: *Gobio gobio lepidolaemus, Varicorhinus capoëta heratensis, Alburnoides bipunctatus eichwaldi, Noemachilus malapterurus longicauda,* and *Cobitis aurata*.

The majority of fish species introduced into Turkmenistan originated from the water bodies of the Chinese subregion of the Sino-Indian zoogeographic region (Berg 1949). The most exotic introduction is *Gambusia affinis*, which was introduced into Eurasia from North America.

The ichthyofauna of the artificial water bodies of Turkmenistan is, therefore, a quite heterogeneous one which includes elements of diverse origin. A more precise analysis of the faunogenesis is possible by utilizing the concept of ichthyofaunistic complexes developed by G.V. Nikolsky (1980). In his classification, the ichthyofauna of the discussed water bodies of Turkmenistan includes representatives of six ichthyofaunistic complexes, namely of the Boreal Lowland, the Ancient Upper Tertiary, the Pontic Freshwater, the Southwest Asian, the Chinese Lowland, and the Indian Lowland complexes. From this list, the first four complexes correspond to the natural faunistic composition of the Aral Sea and lowland portion of the Amudarya River, whereas the latter two include the majority of the Far East introductions. The species which are currently ecologically dominant in the lake-type artificial water bodies (reservoirs and accumulating lakes) in Turkmenistan include representatives of the Boreal Lowland (Aral roach, goldfish), the Ancient Upper Tertiary (carp, wels, pike-perch), the Pontic Freshwater (Aral and Turkestan barbels, Eastern bream, sabre carp, Aral asp, Aral shemaya, Aral stickleback), and the Chinese Lowland (white amur, white and painted thickforeheads, white Amur bream, and Korean sharpbelly) ichthyofaunistic complexes.

According to Nikolsky (1980), the faunistic complex is a group of species originating from the same geographic zone and adapted to the abiotic and biotic conditions of this zone. Success in colonization of new water bodies by a certain species is determined by the interaction of its adaptive features with the entire complex of new ecological conditions. The central point of this success is usually successful reproduction.

The most general common feature of the representatives of all faunistic complexes which colonized the new lake-type water bodies in Turkmenistan has been their adaptation to the lowland inland water conditions of Eurasia. The major limiting factors determining the composition and abundance of fish populations have been their reproductive conditions, both in the lake-type water bodies and in the connected hydrographic network (Aliev 1975; Pavlovskaya and Salnikov 1986, 1990; Salnikov and Busheva 1986, 1990; Salnikov 1989, 1990).

The fish population of the water bodies under discussion includes representatives of seven ecological groups as distinguished by the features of their reproductive ecology. The dominant ecological group includes phytophile species (bream, carp, goldfish, roach and other food fishes), which have the most favorable conditions of reproduction in these water bodies. This is true even for the phytophiles in the irrigation reservoirs despite of the fluctuating hydrological regime of these water bodies: the decrease in water level in summer has its positive impact, providing the annual natural melioration of spawning sites.

A second dominant group consists of pelagophile fishes (most of them also food species); among these, sabre carp, Aral barbel, white and painted thickforehead, white amur, and white bream are quite abundant. All these species (excluding sabre carp) reproduce only in rivers and canals.

To reproduce most efficiently, the pelagophile fish species must find a combination of a lake-type water body, for the feeding of the reproductive-age fish (and subsequently of the fry), and an adjacent stretch of a river or canal, long enough to allow for spawning and the development of the floating bathypelagic eggs. The described situation exactly reflects the conditions in the Karakum Canal, with a top portion of the canal connected to the Kelif Reservoir, in the early period of its existence. Later, following the increasing silting of the Kelif Reservoir, the reproductive efficiency of pelagophile fish species here steadily decreased. Presently, the existing artificial hydrographic network in Turkmenistan does not meet the requirements for the reproduction of pelagophile fish species. The majority of reservoirs and accumulating lakes have hydrotechnical constructions or natural barriers which prevent the fish from leaving these water bodies; once a pelagophile species penetrates to these water bodies, it does not participate in reproduction any more.

In the late 1980s, favorable reproductive conditions for pelagophile species were created in the middle portion of the Amudarya River by the construction of the riverbed Tuyamuyun Reservoir. Extensive spawning of white and painted thickforeheads, white amur, and white Amur bream has since been recorded here (Pavlovskaya and Zholdasova 1991).

Of the lithophile ecological group, such species as the Aral shemaya (in Lake Sarykamysh) and the Aral asp reach significant abundance in the artificial water bodies. Species indifferent to egg-laying conditions, such as the Amur goby, are

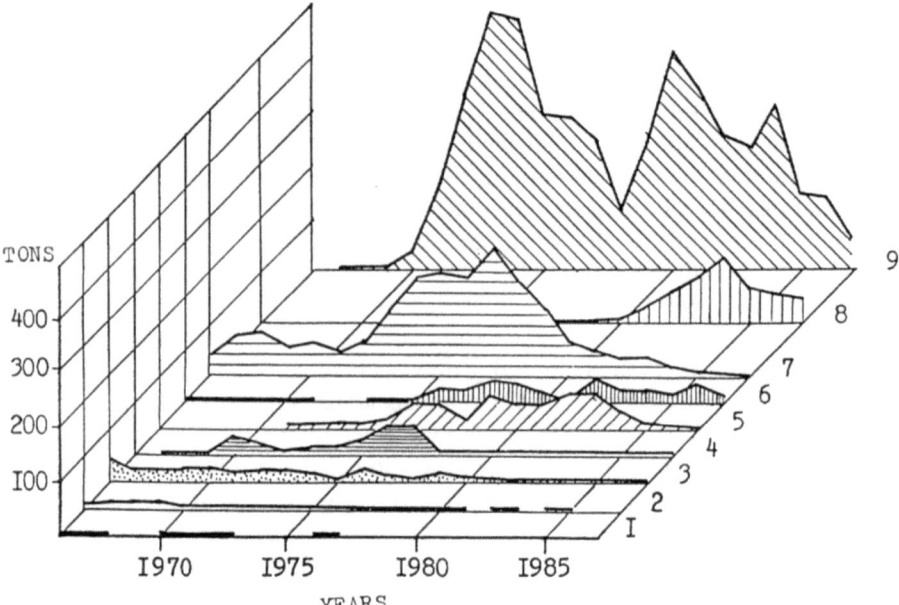

Fig. 2. The fishing dynamics in the Khauzkhan Reservoir. 1 – Aral barbel; 2 – asp; 3 – wels; 4 – white amur; 5 – pike – perch; 6 – others; 7 – carp; 8 – bream; 9 – white and painted thickforeheads.

relatively abundant everywhere. The viviparous gambusia is quite common in all water bodies, and the abundance of the ostracophile bitterling also increasing.

Formation of the Fish Population Structure in Selected Lake-Type Water Bodies

The general ecological and faunistic trends in the formation of the fish population in the artificial water bodies of Turkmenistan have been discussed above. Below, we give the individual characteristics of processes occurring in the selected lake-type water bodies.

Khauzkhan Reservoir. Initial species which have become significantly abundant in the Khauzkhan Reservoir are the carp and the wels (Fig. 2). The asp and the barbel also are abundant and have served as common fishing objects. Simultaneously, the accumulation of Far East introductions has continued through the migration of young fish from the Karakum Canal, where these species first began successful reproduction. In 1972–1974, the catch of thickforeheads species have reached its maximum and fish productivity of the Khauzkhan Reservoir increased four- to fivefold. Both species of thickforehead have dominated the fish community in this reservoir. The abundance and catch of carp have also increased.

In the late 1970s and early 1980s the structure of the fish population in this reservoir substantially changed. Due to the decline in natural reproduction of pelagophile species in the Karakum Canal, the abundance of these fish species in the Khauzkhan Reservoir dramatically decreased.

An increase in the carp catch also had been reversed by the late 1970s and dwindled until it practically ceased in the 1980s. The primary reason for the diminution of this valuable food fish was overfishing without measures taken to preserve carp reproduction.

During this period only the newcomers to the Khauzkhan, the pike-perch and the Eastern bream, maintained or increased their abundance to any extent. Later, their abundance also declined, whereas the abundance of the goldfish and other small and non-valuable fish species began to increase (see Fig. 2, "others"). The total catch in the Khauzkhan Reservoir has presently declined by two to three times its largest yield.

Kopetdagh Reservoir. Located at the end portion of the Karakum Canal, this reservoir was filled later than the Khauzkhan Reservoir, during the time when reproductive efficiency of the Far East fish species in the Karakum Canal was already declining. In 1974, incidental fishing commenced in the Kopetdagh Reservoir, with such species as wels, carp, asp, and pike-perch dominating the catch. Research fishing data demonstrate that in the following years the dominant species in this reservoir were small non-valuable roach and sabre carp, as well as predatory pike-perch and asp (Mukhamedieva and Salnikov 1983). In the 1980s, this list was supplemented by the Eastern bream. This structure of the fish population in the Kopetdagh Reservoir still exists, accounting for its low fish productivity. The total catch here does not exceed 10--14 tons and consists predominantly of the sabre carp (30 to 50%), Eastern bream (20 to 30%), asp (20 to 25%), and pike-perch (3 to 10%).

Lake Kattashor. Fish colonization here began in the 1960s when salt levels of this previously highly mineralized water body decreased. In the 1980s, the lake was a site of intensive eutrophication, silting, and pollution. All these processes, combined with selective fishing, have influenced the dynamics of the fish population. The successive changes in the dominant fish species in Lake Kattashor can be described as follows (species are listed in the order of decreasing dominance): in 1969–1974, carp, amur, wels, and thickforehead; in 1975–1985, bream, carp, thickforehead, and amur; and in 1986–1990, bream, thickforehead, pike-perch, and amur.

Lake Sarykamysh. This lake is the largest water body in Turkmenistan which accumulates collection-drainage waters. The most abundant species here are the catadromous Aral fish species such as carp, bream, pike-perch, sabre carp, and roach, as well as the anadromous Aral barbel (Fig. 3). The dynamics of the fish population in this enclosed brackish lake are defined by its salinity to a significant degree. Salt lakes which existed on the bottom of the former

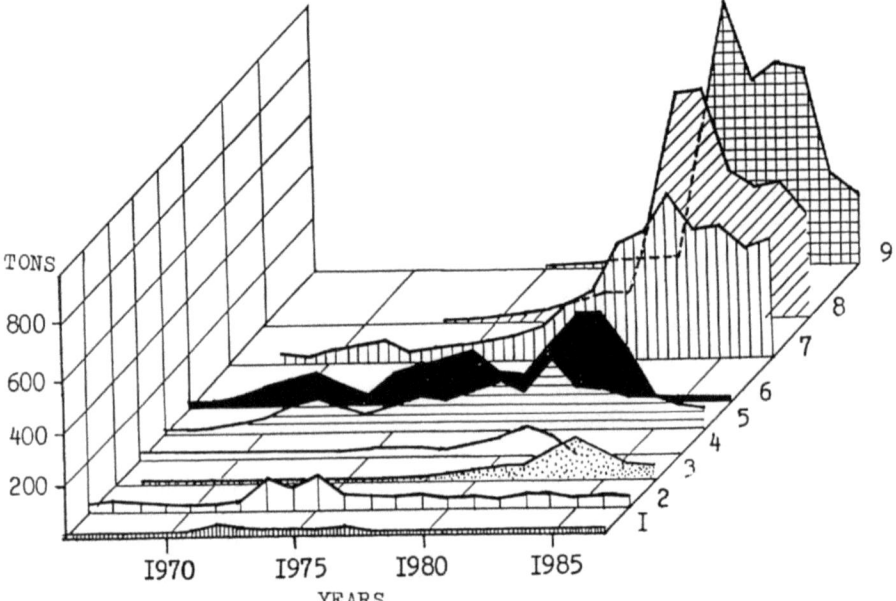

Fig. 3. The fishing dynamics in Lake Sarykamysh. 1 – others; 2 – roach; 3 – wels; 4 – asp; 5 – carp; 6 – Aral barbel; 7 – bream; 8 – pike-perch; 9 – sabre carp.

Sarykamysh Depression contributed to the high salinity of the lake in the initial years of its filling, which started in 1963. Later, the average salinity decreased: from 29 g/l in 1963 to 10–12 g/l in 1969–1985 (Sanin *et al.* 1991). Since the late 1980s, however, the lake has again been subject to rapid salinization.

Generally, these levels of salinity (except for the initial period) have never exceeded the tolerance limit for adult individuals of the dominant species inhabiting Sarykamysh. However, special experimental studies conducted on the young stages of Sarykamysh fish species (Pavlovskaya *et al.* 1986) demonstrated that successful reproduction of such species as the bream and the carp is possible only in the zone of the lake limited by an isohaline of 7–8 g/l and adjacent to the mouth of the freshwater collector canal which carries drainage waters with mineralization 4–5 g/l. Notably, the development of the young stages of fish in Lake Sarykamysh is suppressed under significanly lower salt concentrations than in the Aral Sea.

Due to the rapid increase of the water level, reproduction of phytophile fish species in Lake Sarykamysh was limited until the late 1970s by the lack of the vegetative substrate necessary for spawning. More favorable conditions were created only after the mouth of the collector canal draining into the lake was transformed into a large alluvial delta covered by reeds.

Under these conditions, there have been two sources of fish population growth in Lake Sarykamysh: the colonization by young fish migrating from the lower Amudarya River as well as from the extensive accessory hydromeliorative network connected to the lake, and the efficient reproduction of fish in the lake itself, primarily in the zone next to the delta of the collector canal. The colonization by the young fish was especially significant in the initial period of the formation of Lake Sarykamysh (Poltoratsky and Lyashenko 1972).

Fig. 3 shows that, in the 1980s, numbers of large, valuable species such as the Aral barbel and the carp declined, and the share of small non-valuable (sabre carp) and predatory (pike-perch) species in Lake Sarykamysh increased. This succession effected a decrease of the abundance and catch of valuable food species. Reasons for these changes included increasing salinization as well as excessive and selective fishing: both factors have caused the decline in the fish reproductive efficiency in Lake Sarykamysh and in its accessory hydromeliorative network.

Lake Ovadan. This lake is an example of a small water body formed by the disposal of collection-drainage waters in the Karakum Canal zone. Notably, in this small lake as well as in the numerous similar lakes along the Karakum Canal, many small non-food fish species are preserved which are already extinct or extremely rare in the reservoirs of the Karakum Canal. Among them are a native species, eastern bystryanka (*Alburnoides bipunctatus eichwaldi*), and the Far East introductions, Chinese false gudgeon (*Pseudogobio rivularis*) and Amur chebachek (*Pseudorasbora parva*).

Fish reproduction in the artificial lake-type water bodies of Turkmenistan occurs under the complex ecological conditions. The reproductive stage of their life cycle becomes particularly vulnerable here and can drastically affect the formation and dynamics of the entire fish community. On the other hand, feeding conditions are usually favorable, as has been demonstrated by the rate of linear growth for many fish species (Salnikov and Reshetnikov 1991). In general, morphological indicators of fish biology in these waters, such as meristic and plastic characters, rates of linear growth, fecundity, fatness, and others, though considerably variable, still do not deviate beyond certain species-specific limits of variation. This observation, however, refers primarily to those species inhabiting freshwater reservoirs, since species which live in the accumulative lakes frequently exhibit diverse morphoecological anomalies which are probably due to the toxic pollution.

Morphoecological Anomalies in Fish Inhabiting the Water Bodies Which Accumulate Collection-Drainage Waters

Data on the chemical pollution in the waters of Middle Asia, including Turkmenistan, are scarce and often controversial. The most polluted waters probably are the collection-drainage waters disposed from the agricultural areas. For example, in 1986 the waters of Lake Sarykamysh contained from 0.001 to 0.008 mg/l of g-hexachlorocyclohexane ("Lindane"), and up to 0.04

Table 3. Morphoecological anomalies in fishes inhabiting the water bodies which accumulate collection-drainage waters in Turkmenistan

---

1. Anomalies in the external morphology
    1.1. Bilateral assymmetry of the meristic characters
    1.2. Distorted development of the dorsal and anal fins
    1.3. Curving of the vertebral column and fusion of vertebrae
    1.4. Weight loss
2. Anomalies in the gonad morphology
    2.1. Asymmetry of the gonads following the maturation stages
    2.2. Bilateral size assymetry of the gonads (one smaller than the other)
    2.3. Only one gonad normally developed, the other lacking or manifesting juvenile appearance (in the adults)
    2.4. Gonads of an irregular shape
    2.5. Gonads underdeveloped in large, obese adult fishes
    2.6. Malignant transformation of the gonads
3. Anomalies in the reproductive cycle
    3.1. Gonads of the adult reproductive fishes in the spawning pariod only at second to third stages of maturation (the fish are sick or starved)
    3.2. Sixth stage of gonad maturation prolonged
    3.3. Eggs developed but not spawned, forming a solidified mass
    3.4. Resorbtion of unspawned eggs
4. Low quality of the reproductive material and anomalies in embryogenesis
    4.1. Decreased survivability of spawned eggs in some females
    4.2. Distortion of embryonic development

---

mg/l of 3,4-propionate-dichloranimide ("Propaninide"). The concentration of chemical pollutants in fish tissues probably is also quite high; for example, Sanin *et al.* (1991) detected up to 0.035 mg/kg of chloroorganic pesticides in the eggs of the bream from Lake Sarykamysh.

Table 3 gives a list of the most prominent morphoecological anomalies found in fish species inhabiting the artificial water bodies which accumulate collection-drainage waters in Turkmenistan. In analyzing these anomalies, we were aware that thay could have been caused by stress factors other than or in addition to toxic pollution effects. We also took into account the known data on anomalies detected in experimental studies, and the occurence and degree of expression of such deviations in a normally fluctuating natural environment.

Table 3 reveals that the constant influence of the toxic chemical pollutants on the fish populations first of all affects the reproductive functions: the gonads become distorted and reduced, the quality of eggs declines, the embryogenesis becomes abnormal, and the survivability of eggs and fry decreases. All of these factors cause a decline in abundance of fish populations in the accumulating water bodies.

Fish Productivity

Fish productivity in the artificial water bodies of Turkmenistan depends on a variety of factors other than the biological potential of productivity of the fish species. The waters of the Amudarya River which feed the Karakum Canal and its reservoirs are initially low in productivity because they are formed from the thawing glaciers and mountain snows and thus contain very low amounts of important minerals. Enrichment of the river waters by biogenic minerals occurs primarily through the disposal of drainage waters into the rivers. Bioproduction in the Karakum Canal also is impeded by the high turbidity of the Amudarya waters. Productivity also depends on the contents of initial soils of the construction site; the reservoirs are usually constructed over sands or salty soils with low productivity, and accumulating lakes are formed by filling depressions which contain solonchaks or salt lakes.

Due to the geographic position of Turkmenistan, the vegetative period in the water bodies is protracted (Table 1), especially in the Karakum Canal reservoirs; the water temperature here is further increased by the bottom disposal system, which removes the colder bottom layers of water. The summer temperature of water in these reservoirs reaches 30 °C.

Each year, the ecosystem of every functional irrigation reservoir undergoes a hydrological stress period, with both negative and positive effect. At the very peak of the vegetative period (from June to September), water is drained from a reservoir. The volume, area, and depth of the reservoir gradually decrease, while its turbidity and water exchange intensity increase. A mass of aquatic vegetation is left to dry out. As a result, the feeding grounds and the biomass of the feeding organisms, as well as living space for the fish populations, shrink. Many fish (especially young) are carried out from the reservoir since there is no fish barrier system at the outlet. However, most fish are retained in the reservoir, living in the "dead volume," i.e., the non-disposed of part of the reservoir water (which can be a significant amount if the silting process is not highly developed). At the same time, fishing activities increase resulting in a yield of 50 to 60% of the annual catch. The fish population is replenished throughout the remainder of the year by the water flow to the reservoir from the Karakum Canal.

The summer drying of a significant area of the reservoir bed, the disposal of the bottom water layers which accumulate organic matter and minerals, and the water exchange itself provide a melioration of the reservoir which stabilizes its trophic conditions and prevents eutrophication and pollution. In October, a new cycle of filling begins; the dried areas are again covered by water and bioproduction here is renewed, facilitated by the influx of biogenic minerals from the bed soils which have been mineralized during the summer. In general, however, the seasonal dynamics of the hydrological and biological regimes of an irrigation reservoir are not synchronous with the biological cycles of fish species.

The biological productivity of the accumulating lakes is higher than that of the reservoirs because these former water bodies collect drainage waters enriched by the biogenic minerals from agricultural fields. Productivity is also enhanced by the weak flow (or no flow) of this enriched water through the lakes. These effects are especially pronounced in the water bodies which are medium- and small-sized, shallow, relatively unmineralized, and located further to the

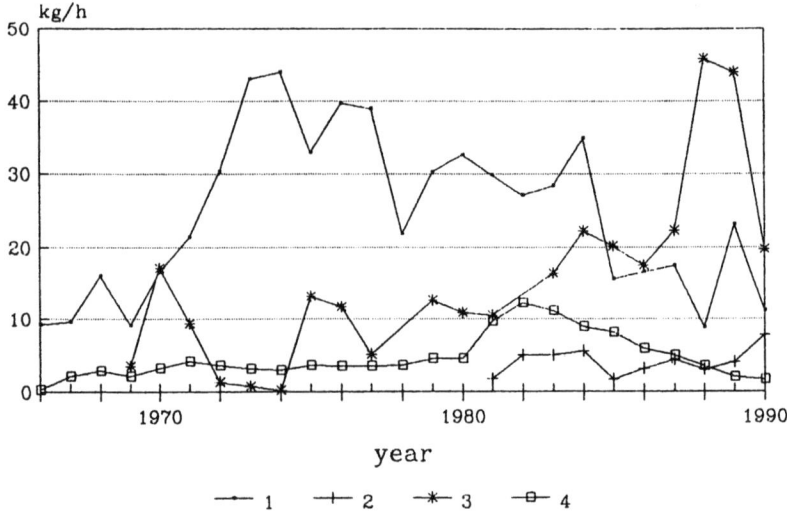

Fig. 4. The food fish productivity of the lake-type water bodies in Turkmenistan. 1 – Khauzkhan Reservoir; 2 – Kopetdagh Reservoir; 3 – Lake Kattashor; 4 – Lake Sarykamysh.

south, as is Lake Kattashor. Such lakes are eventually doomed to eutrophication, pollution, and silting by organic matter.

The quantitative parameters of the described processes are still poorly known. Calculations based on the amount of primary production and phytoplankton biomass (Shkeda 1989) and the analysis of fishing industry data show that, under successful management, natural fish productivity (catch per area unit) can approach, in the mesotrophic filled reservoirs, 60–80 kg/ha; in the eutrophic filled reservoirs, 80–100 kg/ha; in the large mesotrophic accumulating lakes, 80–100 kg/ha; and in the medium and small eutrophic accumulating lakes, 150–200 kg/ha. Fish productivity of the Karakum Canal has not been estimated; it is evidently low since the Canal itself has practically no fishing.

Fig. 4 shows fishing production data (annual catch per area unit; for the reservoirs, unit area refers to the average annual area). These data indicate that only the catch in the Khauzkhan Reservoir (40–45 kg/ha/year) approaches natural levels of fish productivity. Another possible candidate for sufficient fishing production is Lake Kattashor. In both cases, the high productivity is primarily due to the presence of fish species which feed on plants and detritus. In the Khauzkhan Reservoir, these species include very efficient primary consumers, such as painted and white thickforeheads, which feed on detritus and phytoplankton (the painted thickforehead consumes also zooplankton), and, to some extent, the carp, which feeds on plants and zoobenthos. In Lake Kattashor, the productive species include the carp and white thickforehead, but the most abundant species is the small bream. The latter species usually feeds on zoobenthos, but in Lake Kattashor, where zoobenthos always has been poorly developed and sometimes even has disappeared, both carp and bream have switched to consumption of detritus and phytobenthos. This has resulted in

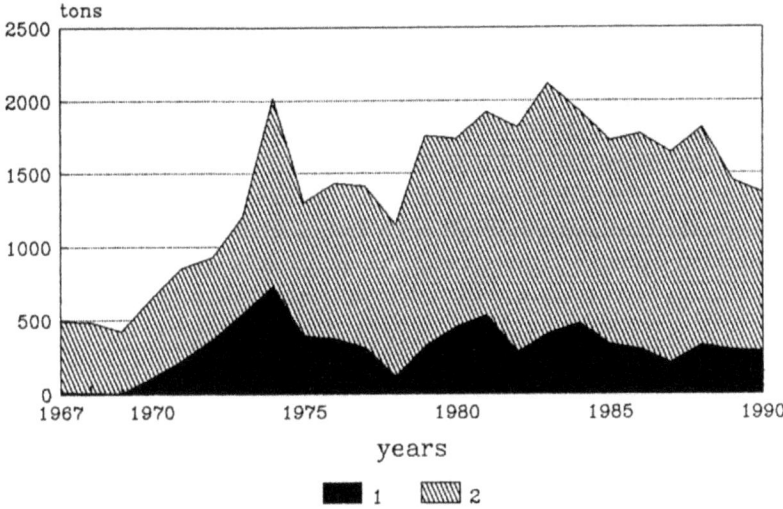

Fig. 5. The fish productivity in the inner water bodies of Turkmenistan by state-owned fishing enterprises: 1 – the Far East introduced species; 2 – native species.

lower growth rates and decreased average size and weight, but, at the same time, the population biomass has increased substantially, especially in the bream population. Such an increase has been made possible by the eutrophication of Lake Kattashor, which increased productivity of aquatic vegetation (phytoplankton and phytobenthos).

Finally, it should be emphasized that, everywhere in the artificial waters of Turkmenistan, the introduced Far East fish species constitute the main component of the fishing industry. The presence of these species increases the fish production of some water bodies by several times. In recent years, however, the catch of both introduced and native fish species has decreased (Fig. 5).

General Features and Trends in the Formation and Dynamics of the Fish Population

Formation of the fish population in artificial water bodies of Turkmenistan bears certain general features which are commonly observed throughout the world during the colonization of new water bodies, for instance, of reservoirs (Nikolsky 1948; Elton 1958; Sharonov 1966; Kamilov 1973; Balon 1974; Reshetnikkov *et al.* 1982). The native ichthyofauna is recruited initially from the original river drainage (the Amudarya River). The lake-type water bodies have lower species diversity due to the absence of rheophile fish species (and, in the accumulative lakes, also of stenohaline species). Due to introduction and fish farming, the ichthyofauna is constantly supplemented by new fish species. Its diversity is constantly changing through the migration of new species (native as well as introduced) from other water bodies and the disappearance of those species that have not found favorable conditions. In lake-type water bodies, the dominant position is occupied by the limnophile species which by their reproductive conditions are either phytophiles (carp, bream, roach, goldfish) or relatively flexible pelagophiles (sabre carp); the typically rheophile species are lacking. The abundance of anadromous (Aral barbel) and Far Eastern pelagophile species (white and painted thickforehead, white amur) depends entirely on their influx from the donor water bodies (the Amudarya River and the Karakum Canal).

The process of formation of the fish population in the artificial hydrographic network of Turkmenistan has not been yet completed. Undoubtedly, the "equalizing" of the species diversity in the Amudarya basin will continue due to fish dispersal (which is facilitated by the construction of the irrigation system), introductions, and fish farming. Potential migrants to the artificial water bodies in Turkmenistan include more than ten native and aboriginal fish species and subspecies from the lower Amudarya, the water bodies of Uzbekistan (where the introduced ichthyofauna is more diverse), and also the Murghab and Tedzhen Rivers.

Fig. 6 gives a general picture of the dynamics of the fish population in the reservoirs and accumulating lakes in Turkmenistan. Initially, such species as carp and wels dominate both population and catch since they are the first colonizers among large valuable fish species of new water bodies. Bream become significantly abundant somewhat later. The replenishment of valuable pelagophile species depends on the influx of young fish from the accessory drainage system; stability of this influx provides a significant increase in fish production (e.g., in production of two thickforehead species in the Khauzkhan Reservoir and Aral barbel in Lake Sarykamysh). Pike-perch usually become abundant subsequent to their migration. Development of such unfavorable processes as changes in the hydrological regime in the accessory drainage system, salinization, eutrophication, pollution, and excessive fishing results in serious structural changes in fish communities. These changes are primarily characterized by the replacement of valuable large species with a long life cycle (and a tendence to be stenobiotic) by small, non-valuable species with a short life cycle (and a tendence to be eurybiotic). The amount of fish catch decreases, or its quality declines. Therefore, successions in these fish communities usually are of anthropogenic nature and follow the patterns common for fish populations in anthropogenic environment (Colby et al. 1972; Leach et al. 1977; Reshetnikov 1986).

In general, the life span of the artificial water bodies, from formation to demise, is quite short. The main factor determining the continuing existence of an irrigation reservoir in the desert zone is the extent of silting by floating particles carried with the turbid river waters (Nikolsky 1948; Kamilov 1973).

Several general points are worth mentioning regarding the succession process in filled reservoirs across the desert zone of Turkmenistan. In the initial stages of the ecosystem development, productivity increases rather gradually. Relative

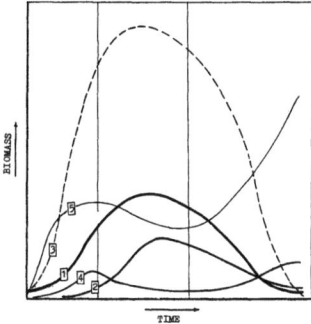

Fig. 6. General changes in the structure of fish population in the artificial water bodies of Turkmenistan (the Amudarya River basin). 1 – carp; 2 – bream; 3 – thickforehead; 4 – pike-perch; 5 – small, realtively short-cycle fish species.

stability is achieved at the oligomesotrophic and mesotrophic stages. This stability is subject to the "impulse" changes (Odum 1986) due to seasonal fluctuations of the hydrological regime. The initially low diversity of the ichthyofauna (as compared to a river or a canal) gradually increases, and the fish biomass may stabilize at a high level. This stage is eventually followed by a regression as a result of increased silting and water exchange. An especially negative effect on fish populations is the silting of a "dead volume" of the reservoir. Predictable trends of the regression stage include a decrease in biogenic minerals, eutrophication, and a decrease in fish abundance. Conditions for fish existence approach those of a river; at this stage, species diversity can even increase, due to the penetration of rheophile fishes. Eventually, however, the ecosystem degrades and ceases to exist when silting of the reservoir is completed and water influx stops.

The Khauzkhan Reservoir is now approaching an initial stage of regression. According to the data provided by the Turkmen State Institute of Planning of Water Industries (*Turkmengiprovodkhoz*), the reservoir's actual capacity has decreased by 30 million cubic meters in 1985, and by additional 45 million cubic meters in 1990. The coefficient of water exchange increased from 3.5–4 to 5–6. A decrease in biogenic minerals and phytoplankton biomass in the reservoir has also been observed.

The main factor for successful development of a lake-type water body accumulating collection-drainage waters is its salinity. A gradual salinization determines the pattern of succession in the accumulating lakes. The first stage in this process is the establishment of an initial ecosystem and formation of an ichthyofauna under brackish conditions. This stage is followed by a period of a relative stabilization when existing salt concentrations are tolerated by the established fauna. At this stage, eutrophication, accompanied by the successional changes in the fish population and an entire ecosystem, can develop. Once the water mineralization exceeds the upper limit of tolerance, the brackish ecosystem degrades and ceases to exist.

Data on salt tolerance have been published for the fish species inhabiting the waters of Turkmenistan (Doroshev 1964; Maceina and Shireman 1979; Nikolsky 1980). A general estimate of the upper limit is approximately 12–15 g/l of mineralization. After this limit is exceeded, negative processes can be expected in the adult fish population; in the accumulating lakes, negative effects can be observed earlier, due to chemical pollution. During the ensuing rapid period of salinization, euryhaline fish species are likely to penetrate and dominate the ecosystem. Theoretically, this development can lead to the formation of a dead salt lake. Sanin *et al.* (1991) predict that, by the year 2010, salinization of Lake Sarykamysh might reach 24–146 g/l, depending on the dimunition of drainage water influx. Of course, all described predictions are only plausible scenarios of development. We are dealing here with artificial ecosystems, and their future can be deliberately changed by human activity.

Conclusions

The irrigation reservoirs and accumulating lakes house temporary ecosystems which experience rapid changes. Their highest fish productivity is achieved under relatively stable abiotic and biotic conditions. Upon spontaneous formation of fish populations, this potential usually is not realized. Productivity is formed, maintained, and increased due to the effect of "coming waves," during which the increase and subsequent decline in the biomass of certain fish species is succeeded by the increase in biomass of other species, either new colonists or previously infrequent ones. One strategy of efficient fish management in the artificial water bodies of Turkmenistan should be directed toward the elimination of this "wave" effect.

The main goal of an efficient fishing industry is the increase in fish productivity. Current environmental conditions in Turkmenistan and in Middle Asia in general, however, dictate that priorities in the management of artificial water bodies be given to the maintenance of clean water and conservation efforts; fish productivity thus is the last priority. From an industrial point of view, those water bodies in Turkmenistan which have an input of collection-drainage waters, are, for the most part, not suitable for ecologically clean fish production, due to the high concentration of pesticides in these waters.

Irrigation reservoirs can play a significant role in fish production, given a necessity of the preservation of native ichthyofauna and the quality of water. The Karakum Canal and its reservoirs are evidently important survival sites for the unique ichthyofauna of the Aral Sea and the Amudarya River.

Fishing industry practices should be directed toward enhancing the natural biological productivity of water bodies. These practices include optimized fishing of the natural populations as well as support for the reproduction of the most productive food species. Fig. 6 shows that, in Turkmenistan, the maximal effect can be achieved by controlled development of fish communities dominated by the Far East introduced species such as white and painted thickforeheads as well as by the native carp, bream, and pike-perch.

Introduction of the Far East species into Turkmenistan has resulted in a great increase in fish production; in the three decades since this introduction began, about 7,000–8,000 tons of introduced fish species were caught in the water bodies, excluding fish farming, of Turkmenistan. One expected biomeliorative effect of the white amur introduction – the prevention of proliferation of water plants and macroalgae in the artificial water bodies in Turkmenistan – is not yet precisely known although such prevention undoubtedly takes place in the irrigation and collection-drainage canals where plant proliferation rate is high. The concepts of biological regulation of the phytomass growth and water quality through the use of fish species with different feeding habits are, however, still a subject of research rather than of industrial practice.

From the conservationist's point of view, the general results of fish introductions into Turkmenistan and Middle Asia are undoubtedly negative: natural ichthyofaunas, populations, and gene pools are mixed, and the original zoogeographic specificity of water ecosystems is lost. This is a price of successful introduction.

Acknowledgements

The author thanks Drs. D.S. Aliev, Yu.S. Reshetnikov, and O.A. Popova for their valuable comments on the manuscript of this article.

23. Arthropods Inhabiting Rodent Burrows in the Karakum Desert

VICTOR A. KRIVOKHATSKY

Abstract

Communities of arthropods were studied in the Repetek Reserve (Karakum Desert) in burrows of desert rodents *Rhombomys opimus, Spermophilopsis leptodactylus, Meriones meridianus*, and *M. libycus*. These burrows differ in complexity and structure of arthropod communities. More than 500 species of burrow-inhabiting arthropods possessing various degrees of ecological specialization were discovered, from obligate bothrobionts to species only accidentally found in burrows. Taxonomic composition of the burrow arthropod communities is discussed as well as their trophic, temporal, and spatial structure.

Introduction

Animal burrows and holes contain peculiar ecosystems. In addition to their hosts, these burrows are often inhabited by many other animal species, predominantly arthropods, which are usually called "bothrobionts" or "nidicoles." Here, we refer to a community of an animal burrow as a "heterotrophic burrow consortium." Its core includes the host(s) of the burrow and its nonliving part, the burrow itself. Other animals in this consortium have ecological connections of various degree with its core.

The term "bothrobiont," currently used in the morphological-ecological classifications of ecosystems, designates any animal that lives in the holes or burrows, digs its own burrow, and has morphological and ecological adaptations to these conditions. We use the following classification system for animals found in burrows: (1) bothrobionts – species with permanent (obligatory) burrow connections; (2) bothrophiles – species with a strong burrow connection; (3) bothroxenes – species with no obligatory connections; and (4) species alien to the burrow, and only accidentally found there.

Studies of burrow ecosystems started in Europe early in this century (Falcoz 1915). In the former USSR, extensive studies were done by Vysotskaya in

Russia (1953–1978) and Chilkilevskaya in Belorussia (1965–1982). In Middle Asian deserts, research on burrow consortia was started by Ya. Vlasov (or Vlassov) (1932–1941, Central Karakum Desert) and continued by E. Nelzina and her colleagues in the Kizylkum (Nelzina 1966, 1971, 1977; Nelzina et al. 1978) and the author (from 1981 to 1989) in the East Karakum.

In the sand desert, many animal species dig burrows as shelters from external conditions and as protection from predators. Mammals which dig only permanent complex burrows include the large gerbil, *Rhombomys opimus* Licht., the ground squirrel, *Spermophylopsis leptodactylus* Licht.; some gerbils of the genus *Meriones* (e.g., *M. meridianus* Pall. and *M. libycus* Licht.); fox (*Vulpes vulpes* L.); and hedgehog (*Hemiechinus auritus* Gmel.). Some mammals do not dig their own holes but rather occupy previously made burrows: these include the active desert predator *Vormela peregusna* Güld., which uses gerbil burrows after eating the host. Even some birds, such as *Oenanthe isabellina* Temm. and *Athene noctua* Scop., build their nests in gerbil burrows. Many lizards, snakes, and tortoises also use previously made burrows as shelters from extreme temperatures in summer and during the winter months. Numerous species of arthropods living in desert conditions strictly inhabit the complex rodent burrows.

In the Karakum Desert, arthropod communities of the burrow were studied in detail in two localities: in the southern part of the Central Karakum near Ashkhabad (Vlasov 1932, 1933, 1937a,b, 1941; Vlasov and Ioff 1937; Vlasov and Miram 1937; Vlasov and Kirichenko 1937; Vlasov and Sychevskaya 1937; Vlasov and Shestoperov 1937; Vlasov and Stackelberg 1937) and the Repetek Reserve, in the East Karakum (Krivokhatsky 1981, 1982a,b, 1983, 1984, 1985a,b, 1987, 1989; Krivokhatsky and Fet 1982; Krivokhatsky and Kashcheev 1983). During these studies, more that 500 species of burrow-inhabiting arthropods possessing various degrees of ecological specialization were discovered.

Materials and Methods

The original research was conducted by the author in the Repetek Reserve, Turkmenistan, from 1978 to 1988. More than 60,000 specimens of Arthropoda, predominantly insects, were studied. Arthropods were collected from rodent burrows (*Rhombomys opimus*, *Spermophilopsis leptodactylus*, and *Meriones meridianus*) using a number of techniques. One of the methods used (total of 5,600 samples) was extraction of material from the holes using a spoon-like shovel with a long handle and a sieve (a modification of the Shiranovich method). Another technique was installation of burrow traps (glass cylinders filled with formalin) inside and outside of each burrow entrance (a modification of the Barber method of pitfall traps). A total of 850 traps were installed, and invertebrates were sampled every three hours. Additional collecting techniques included light trapping and hand collection.

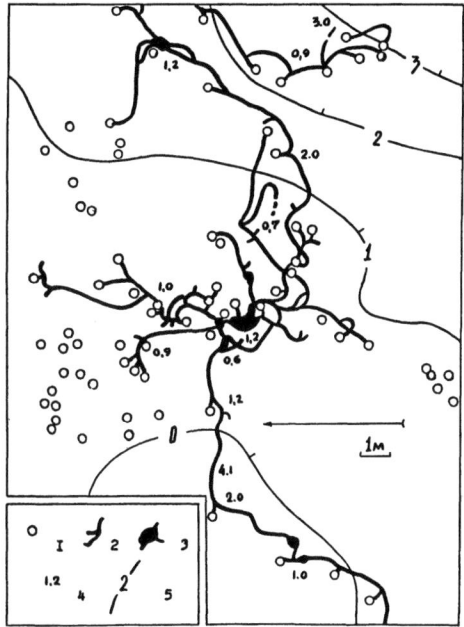

Fig. 1. Part of the large gerbil burrow (Repetek, February 1979): 1 – entrances, 2 – holes, 3 – feeding and nesting chambers, 4 – depth of the holes, 5 – surface.

Environment in the Rodent Burrows

Types of burrows in the Karakum Desert range from the simple and temporary to the complex and permanent. For example, jerboas dig summer burrows which are used only for one day. These holes are destroyed the next day and lack any bothrobionts. Next in complexity follow the burrows of gerbils, *Meriones* spp., and the ground squirrel *Spermophilopsis leptodactylus*. These burrows are permanent, but not complicated. The most complex is the burrow of the large gerbil, *Rhombomys opimus*. It has two to three levels and a number of holes, passages, and chambers (Fig. 1). The microclimate of *R. opimus* burrows has been the object of detailed studies (Shiranovich *et al.* 1965). The temperature in the holes was found to be equal to that of the soil temperature at the same depth. Seasonal and daily fluctuations of the burrow temperatures are less than those on the soil surface. Humidity in the burrow is higher than in the air; it depends on the presence of the animals and decayed food supply, and on the regime of ventilation. Burrows of the large gerbils are peculiar but widespread desert habitats; many desert animals (arthropods as well as vertebrates) cannot exist outside of these burrows.

Table 1. Taxonomic and ecological composition of burrow consortia in the Repetek Reserve

1	2	3	4	5	6	7	8
Isopoda	1	–	–	1	–	–	100
Scorpiones	4	1	2	1	–	–	15
Pseudoscorpiones	4	3	–	1	–	–	20
Solifuga	4	–	1	2	1	–	10
Aranei	40	12	13	8	6	1	1,180
Gamasoidea	3	–	–	–	3	–	1,000
Ixodoidea	6	–	2	2	2	–	1,500
Myriapoda	1	–	–	–	1	–	9
Insecta:	261	106	46	32	58	18	18,730
Collembola	2	–	–	2	–	–	10,000
Protura	1	1	–	–	–	–	6
Thysanura	3	1	2	–	–	–	100
Blattodea	4	1	1	–	2	–	2,240
Orthoptera	6	4	–	1	1	–	212
Psocoptera	1	1	–	–	–	–	10
Homoptera	6	6	–	–	–	–	28
Hemiptera	29	29	4	2	4	–	834
Coleoptera	121	33	31	20	34	3	2,498
Aphaniptera	11	–	–	–	11	–	1,670
Neuroptera	3	–	–	1	2	–	75
Hymenoptera	50	32	3	3	2	10	432
Diptera	8	–	3	1	1	3	500
Lepidoptera	15	8	2	1	2	2	124
Vertebrata:	22	2	3	2	15	–	1,520
Reptilia	10	–	1	–	9	–	500
Mammalia	9	–	2	1	6	–	1,000
Aves	3	1	–	2	–	–	20
Total	345	124	67	49	86	19	25,075

1 – taxonomic groups, 2 – total number of species sampled from the burrows, 3 – alien species, 4 – bothroxenes, 5 – bothrophiles, 6 – bothrobionts, 7 – undetermined species, 8 – total number of animals sampled.

Morphological Structure of the Burrow Arthropod Communities

Morphological structure of a burrow consortium is characterized by classification of arthropod inhabitants according to their taxonomic composition, trophic level, and distribution.

Taxonomic Structure

The list of arthropod groups found in the rodent burrows in the Karakum Desert is given in the Table 1. below, we give a brief review of large systematic groups, which include Crustacea (Isopoda), several classes of Arachnida (Scorpiones, Pseudoscorpiones, Solifuga, Araneï, Acariformes, and Parasitiformes), Myriapoda, and various orders of Insecta.

Isopoda. In the large Asiatic sawbug genus *Protracheoniscus*, only one species, *P. orientalis* Ul., lives in the sandy deserts. This species can dig its own holes only in the moist sand (in Repetek, inside gerbil burrows).

Scorpiones. These are not obligatory bothrobionts and live in natural cavities in the sand desert. Some species, however, such as *Mesobuthus eupeus* (C.L. Koch) and *Orthochirus scrobiculosus* (Grube) are commonly found in gerbil burrows.

Pseudoscorpiones. Among the false scorpions there is an ecological group living in caves, bird nests, and burrows. In the Karakum Desert such species as *Olpium pallipes* Luc. and *Geogarypus shulovi* Beier are obligatory bothrobiontes.

Solifuga. Solpugids *Galeodes fumigatus* Walter and *G. turcomanicus* Birula can use burrows as shelters during the day. This group needs more taxonomic study.

Aranei. Some spider families include many bothrobiont and bothrophile species. In Repetek Reserve, these include *Minosia karakumensis* Spassky, *Berlandina afghana* Denis, and *Minosiella intermedia* Denis (Gnaphosidae); *Theridium varians* Hahn., *Steatoda grossa* (C.L. Koch), and *Latrodectus tredecimguttatus* (Rossi) (Theridiidae); *Artema transcaspica* Spassky (Pholcidae); *Micaria* sp. (Micariidae); *Hersiliola* sp. (Hersiliidae); *Zodarion raddei* Simon (Zodariidae); *Evippa onager* (Simon) (Lycosidae); and *Trachelocamptus asiaticus* Tanasevitsch (Linyphiidae). Some of these spiders in their natural habitats are bothrobionts, but they may also inhabit human habitations (*Artema transcaspica, Steatoda grossa*).

Acariformes. Very small mites belonging to Gamasoidea are common nidicoles on rodents and in rodents' nests. The best known species from the gerbil burrows in the Repetek Reserve are *Eulaelaps stabularis* C.L. Koch and *Androlaelaps angsticutis* Breg.

Parasitiformes. Several tick species (Ixodidae and Argasidae) are rodent parasites during various developmental stages. In *R. opimus* burrows are found larvae of *Haemaphysalis numidiana turanica* Pom., adults and nymphs of *Hyalomma asiaticum* P. Sch. et Sch., and all stages of *Ornithodoros tartakovski* Olen.

Myriapoda, Chilopoda. Several specimens of *Scutigera coleoptrata* Latr. were found only twice, in the outermost holes of the large gerbil burrows in June 1980 and in May 1982, during the days following extraordinary summer rains.

Insecta

Collembola. Willowsia samarkandica Mart. outnumbers all other insects of the burrow communites in the Repetek Reserve. Unlike other desert springtails, this species is an obligatory bothrobiont.

Thysanura. Two species, *Ctenolepisma mauritanica* Luc. and *Apteryskenoma turanica* Kaplin, inhabit soils in Repetek and can be treated as bothroxenes.

Blattodea. Two species of cockroaches from the familiy Polyphagidae (*Polyphaga pellucida* Redt. and *Arenivaga roseni* Br.-W.) are obligatory bothrobionts. *P. pellucida* lives in the burrows as well as in all other desert habitats; *A. roseni* is found only in vegetation-stabilized sands.

Orthoptera. A grasshopper species, *Lesina mutica* Br.-W (Henicidae), and a cricket, *Eremogryllodes vlasovi* (Mir.) (Gryllidae), are obligatory bothrobionts.

Hemiptera. In the arid zone, about 90 species of true bugs are known from the burrows. The majority (Anthocoridae, Stenocephalidae, and Coreidae) sometimes use holes as a wintering place and as a summer shelter from the high daily temperatures. Another family, Reduviidae, has also trophic connections with the burrows. True bugs of the genus *Stirogaster* and *Oncocephalus* are cavity-living bugs, and such species as *Reduvius christophi* Jak., *R. fedtschenkianus* Osh., and *Holotrichius tristis* Jak. are typical bothrobiont predators.

Coleoptera. There are more than 1,000 species of 55 families of beetles known from animal burrows worldwide. About 20 families of beetles have obligatory connections with the desert burrows. More than 300 species of beetles from the rodent burrows are known in the Karakum Desert. Among those, 70 are bothrobionts and bothrophiles: e.g., *Taphoxenus psammophylis* Znoiko and *Anthia mannerheimi* Chd. (Carabidae); *Gnathoncus pygmaeus* Krysh. and *Eremosaprinus vlasovi* Rchdt. (Histeridae); *Cholevinus fuscipes* Men. (Catopidae); *Conosoma lineata* Kasch., *Mycroglotta nidicola* Fairm. and *Aleochara jacobsoni* Kirshbl. (Staphylinidae); *Eremasus cribratus* Sem., *Thynorycter chlamidatus* Sem. et Medv., and *Onthophagus vlasovi* Medv. (Scarabaeidae); *lachnogya squamosa* Men., *Netuschilia hauseri* Rtt., *Cyphogenia gibba* F.-W, *Blaps fausti* Seidl., *B. scutellata* F.-W, *Nanoblaps hiemalis* Sem. et Bog. and *Aphaleria pigmaea* F.-W. (Tenebrionidae); *Attagenus fasciolatus* Sols. (Dermestidae); *Ptinus latro* F. (Ptinidae); *Gronops* sp. (Curculionidae); as well as species of Cucujidae, Cryptophagidae, Lathridiidae, and Anthicidae. Weevils of the genera *Mesostylus* and *Gronops* commonly use burrows as wintering places. Other beetles have trophic connections inside the burrows (they can be saprophages, predators, etc.).

Neuroptera. Ant-lion larvae (Myrmeleontidae) use burrows for making their sandpits inside gerbil holes (*Morter semigriseus* Kriv.) or live under sandy cover in the burrows (*Acanthaclisis pallida* McLachlan and *A. curvispura* Kriv.).

Hymenoptera. Some wasps and bees from the families Mutillidae, Vespidae, Sphecidae, and Apidae build their nests in the rodent burrows. *Vespa orientalis*

F., *Psen pulavskii* Kazenas, and *Nomioides pulverosus* Handl. can either dig their own holes, or use holes of the rodent burrows. Parasitic wasps from the families Ichneumonidae, Braconidae, Eucoliidae, Encyrtidae, Pteromalidae, Torymidae, Ceraphronidae, Platygastridae, and Dryinidae were found in the Repetek Reserve in the burrows of *Rhombomys opimus*. A true trophic connection was discovered between the dauber wasp *Chlorion regale* Sm. (Sphecidae) and the grasshopper *Lesina mutica* Br.-W (Henicidae). Some ants (Formicoidea) commonly use the holes as an underground subway.

Diptera. Many species of flies use burrows as a shelter from the hot weather; some of them have trophic connections with the burrow and its host. Larvae of *Caenophanomyia insignis* Lw. (Therevidae) are very common bothrobionts feeding on live and dead insects. Other flies (e.g., Dolichopodidae, Helomyzidae, and Spheroceridae) also have trophic connections with the burrows. Sand flies (Phlebotomidae) live in the burrows as larvae, and also as adults which suck host's blood.

Aphaniptera. All fleas, as gerbil parasites, use the mammals' nests for larval development and for host availability. In the Repetek Reserve, eleven flea species are found. Most numerous are *Xenopsylla hirtipes* Rorth., *X. conformis* Wagn., *Coptopsylla olgae* Wagn., and *Ceratophyllus turcmenicus* Vlasov et Ioff.

Lepidoptera. Butterflies in the desert can hide themselves during the very hot hours of the day in the shadow of the burrow entrances. Some night moths (Noctuidae) use rodent holes as day shelters, and some Momphidae (*Calycobathra calligoni* Sinev and *Ascelenia decolorella* Sinev), use them obligatorily for shelter during the winter months. Some Tineidae larvae (*Anamallota repetekiella* Zag.) could evolve in the rodent burrows.

Trophic Structure. All burrow arthropods can be classified in several ecological groups characterized by their trophic connections within burrow communities.

Hematophages. – blood-feeding parasites of the burrow hosts (e.g., ticks *Eulaelaps stabularis* and *Ornithodoras tartakovski*, and fleas *Xenopsylla hirtipes*, *X. conformis*, and *Coptopslla olgae*);

Entomophages. – predators and parasitoids of the arthropods (e.g., spiders *Minosiella intermedia* and *Zodarion raddei*, and true bug *Reduvius christophi*);

Saprophages. – arthropods which feed on decaying plant or animal tissues (e.g., springtail *Willowsia samarkandica*, cockroaches *Polyphaga pellucida* and *Arenivaga roseni*, cricket *Eremogryllodes vlasovi*, beetle larvae *Attagenus fasciatus*, and fly larvae *Caenophanomyia insignitis*);

Table 2. Phenology of selected species of burrow arthropods in the Repetek Reserve

Species/stage	Months											
	J	F	M	A	M	J	J	A	S	O	N	D
Xenopsylla conformi (Aphaniptera: Pulicidae)												
ova	+	+	+	+	+	+	+	−	−	−	−	+
larvae	−	+	+	+	+	−	−	−	−	−	−	−
pupa	+	+	+	+	+	+	+	+	+	+	+	+
adult	+	+	+	+	+	+	+	−	−	−	+	+
Reduvius christophi (Hemiptera: Reduviidae)												
ova	−	−	−	+	−	−	−	−	−	−	−	−
larvae, 1 instar	−	−	−	+	+	+	−	−	−	−	−	−
larvae, 2 instar	−	−	−	−	+	+	+	+	+	−	−	−
larvae, 3 instar	−	−	−	−	−	+	+	+	+	+	+	−
larvae, 4 instar	+	−	−	−	−	−	−	+	+	+	+	+
larvae, 5 instar	+	+	+	+	−	−	−	−	+	+	+	+
adult	−	−	−	+	+	+	+	+	−	−	−	−
Caenophamomyia insignis (Diptera: Therevidae)												
ova	−	−	−	−	+	+	+	−	−	−	−	−
larvae	+	+	+	+	+	−	+	+	+	+	+	+
pupa	−	−	−	+	+	−	−	−	−	−	−	−
adult	−	−	−	−	+	+	+	−	−	−	−	−
Attagenus fasciolatus (Coleoptera: Dermestidae)												
ova	−	−	−	−	−	+	+	+	−	−	−	−
larvae	+	+	+	+	+	+	+	+	+	+	+	+
pupa	−	−	−	−	+	+	+	−	−	−	−	−
adult	−	−	−	−	−	+	+	−	−	−	−	−
Minosiella intermedia (Aranei: Gnaphosidae)												
ova	+	+	+	+	+	+	−	−	−	−	−	−
juveniles, young	−	+	+	+	+	+	+	−	−	−	−	−
juveniles, medium	+	−	+	+	+	+	+	+	+	+	+	+
subadults	+	+	−	−	−	+	+	+	+	+	+	+
adult males	+	+	+	+	−	−	−	−	+	+	+	+
adult females	+	+	+	+	+	+	−	−	+	+	+	+

Conventional Aphages. − a group which does not feed in burrows (includes many species that are rarely found in burrows).

Temporal and Spatial Structure of the Burrow Arthropod Communities

Temporal Structure

Changes in the structure of the burrow consortium in time can be studied in terms of daily activities, seasonal dynamics, and multi-year changes and successions. The temporal structure of the arthropod burrow communities depends on the climate and weather changes, and on the availability or absence of the burrow hosts (Krivokhatsky 1981, 1982, 1982b, 1983, 1984, 1985, 1985b).

Seasonal Dynamics

In the East Karakum, four basic seasons can be defined regarding the burrow communities. In April, burrow communities contain high numbers of parasites and saprophages, an increasing number of predators, and the highest abundance of conventional aphages. By July and August, the abundance of parasites in the external holes of rodent burrows reaches its annual minimum; entomophages and saprophages show the highest abundance. Also, this season is characterized by the highest number of the species which hide in burrows in the daytime. In November, the second peak of abundance of all the trophic groups is exhibited. In January, animal abundance is minimal, due to migration from the external holes to the deep burrows. These dynamics are dependent on the time of species development (Table 2) and on their migrations. In general, migrations from the external holes to the deep burrows from spring to summer and from fall to winter are typical for the burrow communities.

Daily Activity

The dynamics of the daily activity are very similar to the temperature dynamics in different seasons. A peak of activity in spring and summer is detected during cool morning hours, but in winter the activity curve closely follows the curve for the daily temperature changes (Fig. 2). Different bothrobiont species exhibit two types of daily activity characterized by single-peak and double-peak curves. In different seasons activity of a species may shift from one type to another.

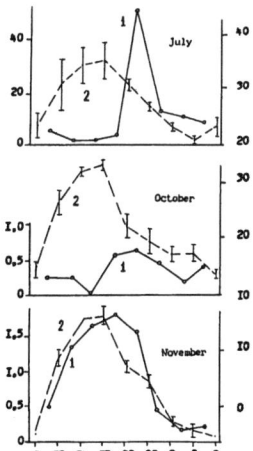

Fig. 2. Temperature dependence of daily activity of the burrow arthropods in Repetek in 1979 (Krivokhatsky 1981): 1 – number of specimens per trap, 2 – soil surface temperature (°C).

One can observe different curves of activity for the same species in the deep burrow and in the external holes. This effect depends on the migratory activity because most arthropods do not stay in burrows day and night; they may spend the day in the burrows and night in the soil surface, or vice versa.

Multi-year Changes and Successions
Climatic changes from year to year can produce delays or accelerations of the individual development of arthropod species, and of burrow communities as a whole.

The population of *Rhombomys opimus* in the East Karakum Desert exhibits a 14-year cycle. In different points of this cycle the abundance of gerbils changes. Consequently, the structure of the burrow arthropod communities also changes. Some stages of the 14-year burrow consortia succession have been simulated (Fig. 3). Following the gerbil depression of 1979-1980, some rodents in the Repetek Reserve left experimental burrows. After short latent period, the structure of burrow consortia started to undergo changes (demutation) toward that of the surrounding soil consortia. The abundance of parasites decreased, and the number of soil saprophages and predators increased. The rate of such demutation varied in different habitats. In sand dunes, demutation proceeded very rapidly and was terminated with the formation of soil communities when the burrow collapsed. In the desert woodlands with stabilized sand where large gerbils live in the depression period, demutation progressed slowly and burrows did not disappear. When numbers of gerbils increased, animals again inhabited their burrows. After the succession toward the burrow consortium climax began, parasites appeared; burrow saprophages and predator species replaced soil arthropods, as shown in the simple model of the successional cycle (Fig. 3). It is interesting to notice that certain burrow consortia located in the desert

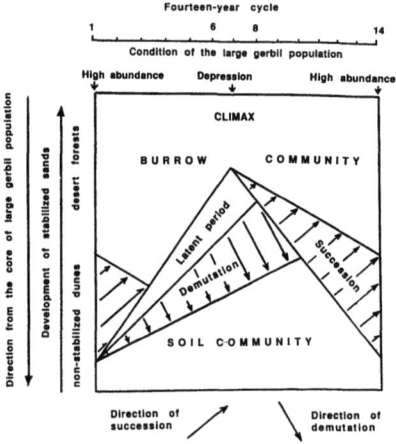

Fig. 3. The 14-year dynamics of the large gerbil population and burrow arthropod communities in Repetek.

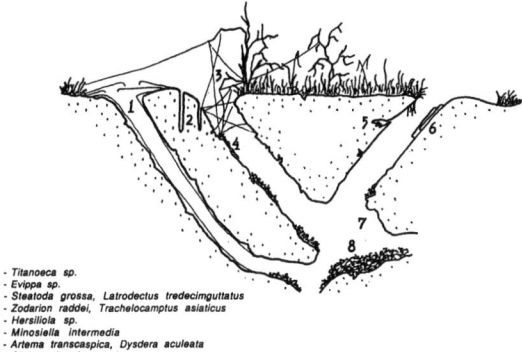

Fig. 4. Distribution of spiders in the rodent burrow in Repetek (Krivokhatsky and Fet 1982).

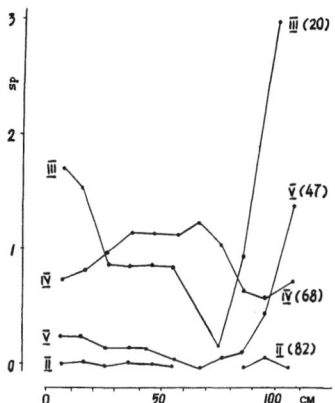

Fig. 5. The distribution of the flea *Xenopsylla hirtipes* (specimens per each 10 cm of a hole) in the external holes of the large gerbil burrows in Repetek, from February (II) to May (V) 1980. Number of experimental burrows is given in parentheses (Krivokhatsky 1984).

woodlands, in the core of the big gerbil population, have exhibited climax indefinitely without demutation.

Spatial Structure
Differences in taxonomic structure of an animal population between different parts of the burrow, and between different burrows, form the spatial structure of the burrow consortium. Some arthropod species can use the entire burrow with all its holes and chambers, whereas other prefer deeper holes, and some live only next to the entrance (Fig. 4). In general, the number of obligatory bothrobionts increases from the entrance toward the depth of the burrow. This pattern, however, may change due to daily and seasonal migrations (Fig. 5).

The taxonomic structure of the burrow communities changes from habitat to habitat, and even among the neighboring burrows. According to these

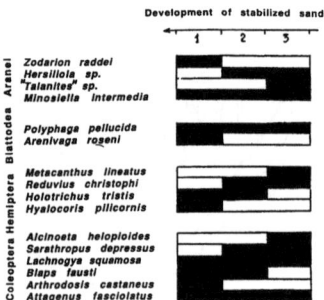

Fig. 6. The habitat distribution of selected burrow arthropods in Repetek. Habitats (plant communities): 1 – *Haloxylon persicum*, 2 – *H. aphyllum*, 3 – *Calligonum* spp.

differences, we can distinguish levels of consortia, which include the following: individual (each burrow as a consortium); habitat (for neighboring burrows in the different habitats); population (for the burrows in the different host populations); geographical (for the burrows in the different regions); host-specific (for the different burrow hosts); and host-animal (for the burrow host-bothrobiont species connection) consortia. As an example of such analysis, the similarity between the neighboring large gerbil burrow consortium and ground squirrel burrow consortium (similarity index of flea fauna in the Repetek Reserve is 73%) is higher than the similarity between large gerbil geographical consortia in the East Karakum and North Kizylkum Deserts (20%), or even in the East and Central Karakum Deserts (13%). Such differences are associated with geographical and habitat distribution of species inhabiting a burrow consortium (Fig. 6).

Functional Structure

Quantitative data on burrow consortia are limited because of the difficulties of sampling of bothrobionts. Our estimate of total abundance of the burrow arthropods in the Repetek Reserve varies from 0 to 20,000 specimens per one hectare in the desert woodland in the spring period. At the same time, we can more precisely describe the quantitative characteristics of the predator-prey model in the burrow consortium. In the burrows of *R. opimus*, the optimal ratio between predators and prey species is estimated as 2.5 predators to 7.5 prey items per one burrow hole. This density, however, is not observed in natural conditions. Dynamics of density correspond to hypothetical models of Leslie and Lotka-Volterra (Fig. 7).

Origin of Burrow Communities

Not less than 4,500 insect species inhabit the Repetek Reserve. Of these, about 3% are bothrobionts which cannot exist in the sand deserts outside of rodent

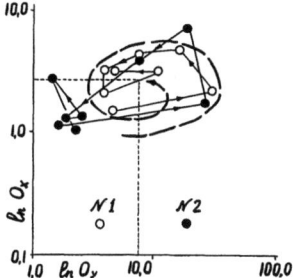

Fig. 7. The quantitative dynamics of predator – prey ratio in two burrow consortia (N_1, N_2) in Repetek (Krivokhatsky 1987). Note similarity to the hypothetical Leslie model. O_x, O_y – number of specimens per one entrance, for predators and prey, correspondingly.

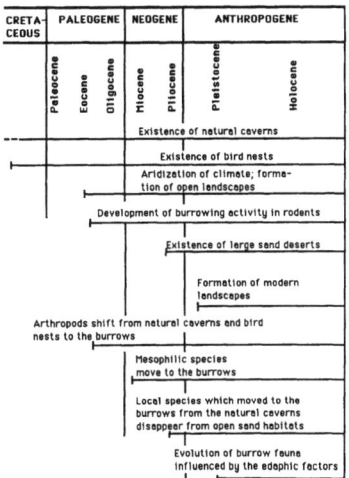

Fig. 8. Origin and evolution of the burrow consortia in the Middle Asian deserts.

burrows. The taxonomic position and geographic distribution of these highly specialized species allow us to trace their possible origin and modes of formation of the burrow communities.

Different characteristics of various arthropod groups have allowed them to adapt to conditions inside burrows: parasites move into the burrows with their host (or change hosts), saprophages and predators search for food, and many arthropods use the burrow as a shelter in the changing environment. First fossils of burrowing rodents in Middle Asia appear during the Eocene and Oligocene. Evolution of burrowing behavior can be attributed to the general aridization and formation of large open areas. Some of the arthropods which lived in the natural cavities and did not construct their own burrows, could have shifted to life in the rodent burrows during this period. Following stages of the the colonization of burrows by the arthropods, and their subsequent adaptive evolution, accompanied by the formation of the large sand deserts lacking natural cavities in the Pliocene and the formation of contemporary landscapes in the Holocene (Fig. 8). Modern burrow communities probably were formed essentially in the Holocene.

24. Zoogeography of Coleoptera in Turkmenistan

OLEG L. KRYZHANOVSKY AND KHABIBULLA I. ATAMURADOV

Abstract

Zoogeographic patterns within the Coleoptera of Turkmenistan (with emphasis on Carabidae, Histeridae, and Tenebrionidae) are discussed. The major zoogeographic complexes are distinguished and described. Peculiarities of the entomofauna are described for the main natural regions of the republic. Special attention is paid to sand desert fauna which is characterized by its extremely high specific and generic endemism. The mountain range of Kopetdagh, in southwest Turkmenistan, represents a transition between two zoogeographic provinces: Turan and Southwest Asia. In these mountains, endemics are fairly numerous but usually have only specific rank. These data correspond to the relatively young geological age of Kopetdagh.

Introduction

The fauna of Turkmenistan is remarkable in many respects. The very peculiar fetures of the animal world of the Turanian desert province are particularly shrply pronounced here; also, elements of other faunistic complexes such as Mediterranean, Middle Eastern, Kazakhstanian, and Afghano-Turkestanian mountainous ones, are prominent in the border areas of Turkmenistan. The Irano-Turanian and the East Mediterranean superprovinces which are two major subdivisions of the immense Ancient Mediterranean biogeographic region, intergrade in the southern Turkmenistan. These features are the result of both the modern physical geography of this country and its Cenozoic geological history at least since the early Neogene.

The term "Central Asia" is used in western European and American literature to describe the entire large arid territory which lies in the center of the Asian continent. Russian authors, however, traditionally make a distinction between two different regions in this area: "Middle Asia" ("Srednyaya Aziya") is used to designate deserts and semideserts of the Aralo-Caspian lowland, Kazakhstan, and the mountainous systems of western Tien Shan, Ghissaro-

Darvaz, and the northern Afghanistan, whereas "Central Asia" ("Tsentral'naya Aziya") is reserved for the Pamir Mountains, Mongolia, and western China (including eastern Tien Shan, Tibet, Kuen Lun, and Nanshan). There are significant differences in the flora and fauna of these two regions in spite of the similarity of their physical geography. These differences are sufficiently ancient to compel us to use the terms "Middle Asia" and "Central Asia" in their restricted senses while describing the biogeography of Turkmenistan and adjacent territories.

This article is a result of long-term studies of both authors. This includes several expeditions by the senior author to Turkmenistan in the 1950s and to the other republics of Middle Asia before and after the 1950s. He elucidated the principal features of the composition and origin of the terrestrial fauna of Middle Asia in a monograph (Kryzhanovsky 1965; English translation: Kryzhanovsky 1980). The junior author carried out intensive faunistical studies in many areas of Turkmenistan in the 1970s and 80s.

Since the publication of the abovementioned monograph, much new data has been accumulated on the fauna of Middle Asia, and Turkmenistan in particular. Here, an attempt is made to formulate the principal patterns of the composition of the Coleopterous fauna in Turkmenistan, and its connections with the faunas of other areas belonging to Middle, Central and Southwest Asia, and to the Mediterranean region.

Three families of beetles, Carabidae, Histeridae and Tenebrionidae, were selected as model groups, although data on other insects is also included. The genus (and sometimes subgenus) was used as a basic unit for zoogeographic analysis. Distributions of species were examined against generic ranges.

To understand the complex composition and origin of the fauna of Turkmenistan, one should have basic knowledge of the geological history of Middle Asia during the Upper Cenozoic. The formation of arid landscapes was the most important factor in the evolution of the terrestrial biota of Turanian lowland, at least from the beginning of the Neogene (and probably since the Upper Oligocene). The formation of high mountain ranges in the system of alpine orogenesis from the Caucasus to Tien Shan and the Himalaya was the main cause of this aridization.

Middle Asia was subdivided quite distinctly into the low western portion and the elevated eastern one as early as the Upper Miocene, but the contrasts in elevations were not so great at that time as they are now. Differentiation of flora and fauna into lowland and mountainous forms was particularly intensive during the Upper Pliocene and Pleistocene, during the rapid increase in mountain elevations and aridization of the lowland climates. Formation of numerous endemic species and probably of some endemic genera among the Coleoptera most likely took place during the Pleistocene (Kryzhanovsky 1965).

There are two general opinions concerning the age of Aralo-Caspian lowland deserts and the origin of their biota. The first of these, promoted by the well-known Russian ornithologist and zoogeographer, M.A. Menzbier (1914), is that these deserts are geologically young formations; their fauna is therefore recent and derived one. This fauna is believed to have been formed primarily after the Pleistocene regression of the Aralo-Caspian sea basin and was the result of dispersal from the most ancient centers of desert fauna, namely, from North Africa and Southwest Asia, and from Central Asia (in the traditional Russian narrow sense of this term, see above). Later, ornithologists B. Stegmann (1938) and G. Dementiev (1952, 1958) and the botanist M.V. Kultiasov (1946, 1952) expresssed similar opinions.

On the other hand, many scholars, including zoologists A. Semenov-Tian-Shansky (1900, 1936) and V. Heptner (1945), and botanists M. Popov (1927, 1931), M. Ilyin (1937, 1946), and E. Lavrenko (1962) considered the desert and mountain biotas of Middle Asia as a result of extensive, predominantly autochthonous evolution. Their opinion was supported by the paleogeographic data. They did not deny the importance of dispersal but also noted the great influence which the Middle Asian center of origin had on the other arid faunas of Asia and North Africa.

Evaluation of the validity of these two opinions is possible by zoogeographic analysis of several groups of animals. Such groups should satisfy the following requirements: they should have numerous genera and species in Middle Asia, sufficiently broad ecological requirements, and a lack of specific limiting factors in their distribution.

The three families of the Coleoptera listed above are very convenient for this purpose. Their distribution in Turkmenistan is briefly described below.

Carabidae

The Middle Asia (including South and Southeast Kazakhstan) has 130 genera and 870 species of ground beetles, including numerous local endemics inhabiting the mountain ranges of Ghissaro-Darvaz and Tien Shan. The carabids are represented in Turkmenistan by 109 genera with ca. 380 species. Among these, 10 genera and ca. 135 species are endemic or subendemic to the Irano-Turanian superprovince (Kryzhanovsky 1965).

A complete list of the ground beetles of Turkmenistan has not yet been published; however, one local list exists for Badghyz (Kryzhanovsky and Atamuradov 1987; 55 genera and 110 species) and another for West Kopetdagh (Kryzhanovsky and Atamuradov 1989; 75 genera and 207 species).

The level of endemism in Turkmenistan carabids is comparatively low compared to the two other studied families. This is probably due to the predominance of mesophilic and hydrophilic forms in this family. Especially scarce are true psammobionts; they include, however, highly specialized forms. The halophiles, as well as specialized dwellers of the cracks in the soil, are rather numerous.

Histeridae

This family is fairly well studied in Middle Asia (Kryzhanovsky and Reichardt 1976; Kryzhanovsky 1985) and the adjacent countries such as Iran (Kryzhanovsky 1975) and Afghanistan (Therond 1969; Kryzhanovsky 1980). Presently known from the Turkmenistan are 27 genera and 106 species. Among histerids, the subfamily Saprininae predominates (19 genera and not less than 82 species). Of these, 5 genera and 38 species are endemic or subendemic to the Irano-Turanian superprovince. The mountain forms in this family are less well represented.

Tenebrionidae

The darkling beetles are especially abundant in Turkmenistan. From this state, 304 species and 105 genera are known (Medvedev and Nepesova 1985; additional species have since been found). Zoogeographic analysis will exclude five genera and ten species from this list which are the synanthropic pests of stored food products distributed throughout the world by man (*Tenebrio molitor*, *Alphitobius diaperinus*, *Tribolium* spp., etc.). The tenebrionids are predominantly xerophilic and live in all arid and semiarid habitats. Curiously, dendrophilous darkling beetles are almost absent from this fauna; only *Eledona agaricola*, *Platydema triste*, *Pentaphyllus chrysomeloides*, and possibly *Metaclisa viridis* are known from the mountain and riparian forests.

The level of generic and especially specific endemism is extremely high in the darkling beetles. At least 35 genera are endemics and 16 more are subendemics of the Irano-Turanian superprovince. Several genera are known only from Turkmenistan. The majority of these genera are highly specialized psammobionts (*Ammozoum*, *Eschatostena*, *Sternodes*, *Meladiesia*, *Remipedella*, *Asiocaedius*), and the genus *Turkmenohelops* is an endemic of the arid mountains of West Turmenistan.

The abundance of the tenebrionid fauna of Turkmenistan is striking. Only 50 genera and 164 species of Tenebrionidae are known from Mongolia (Medvedev 1990), a fairly completely studied arid country of Central Asia. Thus, the fauna of Mongolia is only half as rich as in Turkmenistan, whereas the territory of Mongloia is three times larger. These faunas have 21 common genera and only 10 common free-living species of tenebrionids. These data also illustrate the profound differences between the faunas of Middle and Central Asia.

Regional Review

Turkmenistan is a political state but not a physiogeographic unit. Five main natural regions can be distinguished within this republic:

1. *The plains of Karakum Desert as well as other arid lowland areas in Southwest and East Turkmenistan*. These deserts can be subdivided into three portions which clearly differ in their fauna:
   (a) northern deserts, to the north from Unguz (the dry river bed);
   (b) western deserts (from the Caspian Sea to approximately 57° W;
   (c) central and southeastern deserts.

2. *The plateaus and low mountains in northwestern Turkmenistan* (Ustyurt, Krasnovodsk Plateau, Bolshoi Balkhan Mountains).

3. *The mountain range of Kugitang*. It lies in the extreme east of Turkmenistan and represents the southwestern part of the Ghissaro-Darvaz mountain system.

4. *The hills of Badghyz and Karabil in southern Turkmenistan*, next to the Afghan border. This area includes the northern foothills of the Paropamiz Mountains, which lie in northern Afghanistan.

5. *Kopetdagh Mountains*.
   This regional division includes neither the very peculiar "extrazonal" habitats of the riparian forests (*"tugais,"* or "gallery forests") along the large rivers, nor artificially irrigated agricultural landscapes with their rather uniform and strongly impoverished fauna.

1. Desert Plains

The fauna of the desert plains is especially characteristic for the Turanian province and includes a tremendous number of endemics of this region. The Coleoptera of sand deserts are especially peculiar and deserve more detailed examination.

Although Carabidae are relatively poorly represented in the sands, a few are very numerous here, including species of the endemic genera *Discoptera* and *Eremosphodrus*; a large predator *Scarites bucida* Pall.; *Cymindis capito* Kryzh. et Emetz; and a poorly studied species, *Bleusea ammophila* Tschitsch. The largest beetle of the Turkmenistan fauna, *Anthia (Pachymorpha) mannerheimi* Chaud. lives here and there in the immovable sands. It is the single representative of the mainly Afrotropical tribe in Middle Asia. Some widely distributed small species, e.g., *Amara (Celia) tescicola* Zimm., *A. (Amathitis) fedtschenkoi* Tschitsch., and *Syntomus fuscomaculatus* Motsch. are common near the roots of the desert bushes.

Psammobiotic Histeridae are very characteristic for the sand deserts. The species of endemic genera *Styphrus, Chivaenius, Ammostyphrus*, and especially of the Saharo-Turanian genus *Philothis* are extremely well adapted to digging in the sand. This kind of specialization is also expressed in some *Saprinus* (such as *S. biguttatus* Stev., *S. gilvicornis* Er., *S. intractabilis* Rchdt.), *Hypocacculus* (*H. eremobius* Rchdt., *H. oxytropis* Rchdt., *H. vlasovi* Kryzh.), and even in *Hister megalonyx* Rchdt. Other Histeridae of sand deserts live in the rodent burrows, including *Eremosaprinus vlasovi* Rchdt., *Gnathoncus kiritschenkoi* Rchdt., *G. pygmaeus* Kryzh., and *Pholioxenus orichalceus* Rchdt.

Tenebrionidae are especially numerous and noticeable in the sand deserts. The majority of them are active at night. Some Turkmenian tenebrionid endemics at the generic level were listed above, but the list of Turanian psammobiotic genera is much more extensive. It includes genera *Diaphanidus, Colposphaena, Dengitha, Earophanta, Habrobates, Pisterotarsa, Platyesia, Diesia, Tagona, Aphaleria,* and *Weisea.* Some species of *Sphenaria, Microdera, Trigonoscelis, Trachyderma, Cyphogenia,* and *Blaps* are also abundant in the sands. It is interesting that several psammobiotic genera (e.g., *Sphenaria, Dengitha, Habrobates, Argyrophana*) have strongly depigmented bodies, and that *Sternodes,* which is active partly at daytime, has white waxy stripes on its elythrae.

Other beetle families also have a large number of endemic or subendemic Turanian psammobiotic genera. Especially numerous are chaffers (Scarabaeidae), including such genera as *Thinorycter, Ochranoxia, Dasytrogus, Trigonocnemis, Eremadoretus, Pristadoretus,* and *Eutyctus.* Cerambycidae are represented by a highly specialized genus, *Turkmenigena,* and a very peculiar *Prionus komarovi* Sem., both of which have depigmented bodies and extraordinary sexual dimorphism. Many other endemic genera deserve being mentioned, including *Nyctor* (Elateridae), *Clema* (Buprestidae), *Petria* and *Steneryx* (Alleculidae), *Rhampholyssa* (Meloidae), *Aphilenia* and *Nyctiphantus* (Chrysomelidae), *Mesostylus, Brachycleonus,* and *Leucochromus* (Curculionidae). This list could be extended without difficulty to include many local endemic species which belong to widely distributed genera such as *Aphodius, Mylabris,* and *Cryptocephalus.*

Psammobiotic fauna of Coleoptera in the Karakum Desert is not homogeneous throughout the entire territory. Several genera have allopatric vicariant species in the western and eastern portions of Karakum (Table 1).

L. Arnoldi (1960), in his valuable monograph on the weevil tribe Mesostylini

Table 1. Examples of vicariant beetle species endemic for the Karakum Desert

Family	Genus	Central and Eastern Karakum	Western Karakum
Scarabaeidae	*Achranoxia*	*A. koenigi* Brenske	*A. varentzovi* Sem.
Tenebrionidae	*Ammozoum*	*A. hauseri* Rtt.	*A. bulla* Sem.
	Dengitha	*D. lutea* Rtt.	*D. crystallina* Sem.
	Argyrophana	*A. deserti* Sem.	*A. caspia* Sem.
	Remipedella	*R. deserti* Sem.	*R. semenovi* G. Medv.
Curculionidae	*Mesostylus*	*M. hauseri* Rtt.	*M. uzboicus* L. Arn.

(Curculionidae), proposed a hypothesis explaining the cause of such distribution. He suggested that the isolation of eastern and western species took place during the Pleistocene transgression of the Aralo-Caspian basin when several desert islands remained in modern West Turkmenistan, and the ancestral populations of western species were isolated on these islands from their eastern counterparts. The recent allopatric vicariant species (or subspecies, e.g., in *Discoptera komarovi* Sem., *Sternodes caspicus* Pall., or *Trigonoscelis grandis* Kr.) were allegedly formed due to these events.

The fauna of northern Karakum has some peculiarities as well, but it is studied much less than the fauna of the southern part of this desert. Several species living here are characteristic for the deserts of Kazakhstan, e.g., the tenebrionids *Lasiostola pubescens* Pall. and *Sternoplax deplanata* Kryn. *Mylabris kuzini* Kryzh. is known only from a few places in the northern Karakum Desert. The same region is occupied by a scarab beetle, *Pharaonus semenovi*, whereas the allopatric *Ph. lederi* Rtt. is common in southern Karakum.

The Turkmenistan lowland areas with solid soils differ very significantly from the sand deserts in their fauna. There are few endemic genera here, e.g., *Bedeliolus* (Carabidae), *Axelinus* (Histeridae), and subendemic isolated *Klewaria* (Tenebrionidae). However, endemic Irano-Turanian species are numerous, including *Cicindela jakowlewi* Sem., *Cymbionotum transcaspicum* Sem., *Taphoxenus gracilis* Zubk., *Poecilus* (*Angoleus*) *dissors* Tschitsch., *P.* (*A.*) *warentzowi* Tschitsch. (Carabidae), *Saprinus bimaculatus* Dahlgren, *S. jacobsoni* Rchdt., *Hister semenovi* Rchdt. (Histeridae), and many tenebrionids from the genera *Arthrodosis*, *Calyptopsis*, *Microdera*, and *Hedyphanus*.

Still more abundant in these landscapes are species with extensive Mediterranean or African-Asian ranges. We should mention here, for example, *Paussus turcicus* Friv. (which is found in the nests of *Pheidole* ants), *Siagona europaea* Bon., *Scarites planus* Bon., *Cymbionotum semelederi* Chaud., *Syrdenus grayi* Woll., *Acinopus striolatus* Zoubk., *Chlaeniomimus gracilicollis* B. Jac., *Microlestes plagiatus* Duft., *Zuphium testaceum* Klug (Carabidae), *Saprinus maculatus* Rossi, *S. calatravensis* Fuente, *S. georgicus* Mars., *S. pharao* Mars., *Chalcionellus tyrius* Mars., (Histeridae), *Zophosis punctata* Brullé, *Adesmia servillei* ssp. *schatzmayri* Koch, *Mesostena puncticollis* Sol., *Dilamus fausti* Rtt., *Scleropatrum hirtulum* Baudi, *Gonocephalum setulosum* Fald., *G. rusticum* Ol., *Opatroides punctulatum* Brullé, *Anemia fausti* Sols., and *Cossyphus tauricus* Stev. (Tenebrionidae).

2. Plateaus and Low Mountains of the Northwest Turkmenistan

This geographic area is covered mostly by desert vegetation. Only high elevational belts of Bolshoi Balkhan have small islets of shrubs or trees. The fauna of this district is the poorest one among all natural regions of the republic due to the harsh conditions. This fauna is rather closely related to the fauna of the desert lowlands.

The southern portion of Ustyurt Plateau has many species in common with the deserts of West Kazakhstan, e.g., carabids *Cardioderus chloroticus* Fisch.-W. and *Poecilus (Angoleus) nitens* Chaud., tenebrionids *Gnathosia karelini* Fald., *Philhammus zaitzevi* G. Medv., *Leptodes boisduvali* Sem., *Trigonoscelis muricata* Pall., *Sternoplax echinata* Fisch.-W., and *Lasiostola heterogena* Fisch.-W., and some Curculionidae. Local endemics are very scarce here; among them should be mentioned *Prosodes angustata* Zoubk. (Tenebrionidae).

The fauna of the Krasnovodsk Plateau and the Bolshoi Balkhan Mountains is somewhat richer and includes several local endemics, e.g., tenebrionids *Leptodes zubkovi* Sem., *Trigonoscelis borosi* Kaszab, *Prosodes emiri* Sumakov, *Turkmenohelops balchanicus* G. Medv. et Nepesova, and several species of Buprestidae, Chrysomelidae, and Curculionidae.

3. Kugitang Mountains

This is the highest mountain range (up to 3,137 m) in Turkmenistan. It lies in the extreme east of the republic and stands apart from other geographical districts. This mountain range is the southwestern part of the large Ghissaro-Darvaz mountain system, which lies predominantly within the adjacent republic of Tajikistan.

The beetle fauna of Kugitang consists mainly of the groups characteristic for southeastern Middle Asia. The aridity of this mountain range results in a relative scarcity of mesophilic forms. For example, the genus *Carabus* is absent from here, as well as some forest groups endemic to Ghissaro-Darvaz, e.g., subgenus *Asioplatysma* (genus *Pterostichus*), and the chafers of the genera *Trochalostema* and *Euranoxia*.

Abundant endemic species found in the Kugitang Mountains are typical inhabitants of the arid mountain landscapes of Middle Asia. Among them are *Chilotomus kuhitangi* Kryzh. (Carabidae) and a great number of tenebrionids: *Gnathosia lopatini* G. Medv., *Colposcelis lopatini* G. Medv., *Dichillus dentipes* G. Medv., *Stalagmoptera hybrida* Skopin, *Blaps medvedevi* Bog., *Prosodes kuhitangiana* G. Medv., *P. monticola* G. Medv., *P. subpilosa* G. Medv., and *Cabirutus kuhitangi* G. Medv. Of special interest is the endemic myrmecophylous beetle genus *Kuhitangia* (Colydiidae). The entomological knowledge of Kugitang is still insufficient, and new data may change this brief survey in the future.

4. Badghyz and Karabil Plateaus

This natural district lies in the southernmost part of Turkmenistan and includes foothills and low mountains (not exceeding 1,000 m). Badghyz, the western part of this district, is fairly well studied whereas its eastern part, Karabil, is known only poorly.

The fauna of this region contains many endemic or subendemic forms which are not known from other parts of Turkmenistan but are common in adjacent parts of the northern Afghanistan. A list of such endemics includes at least five species of Carabidae described in the recent decades: *Carabus miles* Sem., *Tachys atamuradovi* Kryzh. and Mich., *Syrdenus debilis* Kryzh. and Mich.,

Chilotomus margianus Kryzh., and *Dromius hiemalis* Kryzh. and Mich. The last species belongs to a forest genus, lives on pistachio trees (*Pistacia vera*), and is active in winter.

Endemic Tenebrionidae are even more numerous here, among them an isolated monotypic genus *Meladiesia* (with its single species, *M. miritarsis* Rtt.), and also *Gnathosia kushkensis* G. Medv., *Microdera badhysi* G. Medv. et Nepesova, *Dichillus ocellaris* Kaszab, *Earophanta pilosissima* Rtt., *Prosodes quadriimpressa* Rtt., *Dissonomus badhysi* G. Medv., *Hedyphanus kushkensis* Kaszab, and *Reitterohelops badhysi* G. Medv.

The largest among the specialized psammobiotic Middle Asian Histeridae, *Reichardtiolus pavlovskii* Kryzh., is known only by two specimens: one from Badghyz and one from the Mary district (which lies to the north of Badghyz).

5. Kopetdagh Mountains

The fauna of this natural district is quite different from the rest of Turkmenistan in its composition and zoogeographic relationships. Kopetdagh forms the northern portion of the Turkmeno-Khorassan Mountains, which connects the large North Iranian range of Elburz and the ranges of northwestern Afghanistan. Kopetdagh, comparatively young geologically, was uplifted during the Neogene alpine orogenesis in the territory previously covered by the Tethys Sea.

The largest part of the Kopetdagh Mountains lies in Iran and is very poorly investigated by entomologists. However, within Turkmenistan this mountain range is quite well studied (except for its eastern part).

The fauna of Kopetdagh differs strikingly from the faunas of all districts discussed above although it still shares common features with the other districts. The zoogeographic composition of the Kopetdagh fauna is much more complicated than that in other regions of Turkmenistan. Here, a very sharp, almost linear boundary separates the typically Turanian desert fauna of adjacent plains from the mountainous fauna of Kopetdagh. This mountainous fauna is an mostly East Mediterranean (or, to be more precise, a West Asian) one. Although numerous Turanian genera are present here, they are, to a considerable degree, represented by endemic and subendemic species.

The Kopetdagh fauna of every beetle family should be discussed separately.

Carabidae of Western Kopetdagh were considered in detail in a recent paper (Kryzhanovsky and Atamuradov 1989); they listed more than 200 species. The carabid fauna of Central and East Kopetdagh differs from the West one in the absence of many western species.

The most numerous zoogeographic group of carabids (ca. 60 spp.) is comprised of more or less widely distributed Mediterranean species, among them *Calosoma maderae* F., *Elaphropus haemorrhoidalis* Ponza, *Bembidion atlanticum megaspilum* Woll., *Poecilus puncticollis* Dej., *Atranus collaris* Mén., *Ophonus cribricollis* Dej., and *Cymindis axillaris* F.

The second zoogeographic group includes more than 30 species with East Mediterranean (mostly Southwest Asian) ranges, e.g., *Omophron rotundatum*

Chaud., *Cymbionotum pictulum* Bat., *Bembidion moschatum* Payk., *Zabrus morio* Mén., *Acinopus laevigatus* Mén., *Chlaenius lederi* Rtt., *Lebia circumducta* Rtt., and *Brachinus brevicollis* Motsch.

The third large and important group (more than 50 species) consists of species with Middle Asian ranges, either ones from Turanian lowland or from the mountainous districts of eastern Middle Asia. Among them are *Cicindela deserticola* Fald., *C. decempustulata* Mén., *Calosoma reitteri* Roeschke, *Notiophilus sublaevis* Sols., *Broscus asiaticus* Ball., three species of *Taphoxenus*, *Machozetus concinnus* Dohrn, *Liochirus cycloderus* Sols., *Chlaenius tenuilimbatus* Ball., and *Lebia turkestanica* Jedl.

Endemic species of the Turkmeno-Khorassan Mountains are generally related to the latter two groups. A small but interesting group of endemics contains about 12 species. Some of them have West Asian relationships, e.g., *Carabus roseni* Rtt., *Amara astrabadensis* Lutschn., *Cymindis walteri* Rtt., and probably *Bembidion (Philochthus) tichomirovae* sp. n. (in litt.), and *Apristus turcmenicus* Kirschenhofer. Others are of Turanian origin, such as *Bembidion hiekei* M.-Motzt., *Taphoxenus humeralis* Sem., *Chilotomus chalybaeus* Fald., *Harpalus (Brachypangus) antonovi* Tschitsch., and possibly *Trichocellus arnoldii* sp. n. (in litt.).

Only about 30 species belong to a group with wide Palearctic ranges, mostly the West Palearctic ones (e.g., *Cicindela germanica* L., *Trechus quadristriatus* Schrank, *Poecilus cupreus* L., *Calathus ambiguus* Payk., *C. melanocephalus* L., and *Chlaenius vestitus* Payk.), and rarely Transpalearctic ranges (e.g., *Amara ovata* F., *Harpalus rubripes* Duft., and *Pseudoophonus griseus* Panz.). Only one species, *Agonum thoeryi* Dej., has a Holarctic distribution.

A small but interesting group of seven to eight species widely distributed beyond the Mediterranean region into the large territories of tropical Africa and Asia includes *Calosoma imbricatum* Klug, *Scarites eurytus* Fisch.-W., *S. planus* Bon., *Siagona europaea* Bon., and *Bembidion niloticum* Dej.

The tropical relationships of Turkmenian Carabidae are manifested by the presence of some primarily tropical genera (*Paussus, Egadroma, Tetragonoderus*) and by the abundance of genera belonging to principally tropical tribes such as Tachyini and Callistini.

The Kopetdagh Mountains serve as an important zoogeographical boundary for many Carabidae. A number of species with Euro-Mediterranean or East Mediterranean distribution have here their eastern limit of distribution, e.g., *Cicindela germanica* L., *Calosoma inquisitor* L., *C. maderae* F., *Leistus lencoranus* Rtt., *Broscus karelini* Zoubk., *Bembidion tetrasemum* Chd., *Elaphropus diabrachys* Kol., *Laemostenus caspius* Mén., *Atranus collaris* Mén., and *Acinopus ammophilus* Dej. In contrast, some Middle Asian taxa have here their western limits, e.g., the genera *Liochirus, Carenochyrus, Machozetus*, and also *Notiophilus sublaevis* Sols., *Perileptus mesasiaticus* Uéno, *Asaphidion transcaspicum* Sem., and *Poecilus liosomus* Chaud.

Some negative characteristics of the carabid fauna of Kopetdagh are very significant, e.g., the absence of Turanian psammobiotic groups as well as genera such as *Callisthenes, Brinislavia, Colposoma* and many subgenera of *Carabus*. The representatives of boreal mesophilic groups are either absent not only from Kopetdagh but from all of Turkmenistan (e.g., such taxa as tribes Elaphrini and Loricerini) or are represented here very poorly (e.g., genera *Carabus, Leistus, Nebria, Pterostichus*).

Histeridae are represented in Kopetdagh by about 50 species. In the foothills, principally the desert species which have Irano-Turanian ranges are found (e.g., *Gnathoncus pygmaeus* Kryzh., *Eremosaprinus vlasovi* Richt., *Saprinus lateralis* Motsch., *S. niger* Motsch., *S. therondianus* Dahlgren, *S. viridicatus* Schm., *Pholioxenus phoenix* Rchdt., and *Paravolvulus binaevulus* Rtt.). Other histerids possess wider distribution (e.g., *Saprinus maculatus* Rossi, *S. pharao* Mars., *S. calatravensis* Fuente, *Chalcionellus tyrius* Mars., and *Dendrophilus sulcatus* Motsch.).

Many of the above species are found also in the higher elevations of Kopetdagh, in the zones of steppe and mountain xerophytic vegetation. However, other widely distributed species are more characteristic there, e.g., *Saprinus planiusculus* Motsch., *S. subnitescens* Bickh., *S. subvirescens* Mén., *S. tenuistrius* Mars., *Chalcionellus blanchei* Mars., *Ch. turcicus* Mars., *Ch. decemstriatus* Rossi, *Hister quadrinotatus* Scr., *H. uncinatus* Ill., and *Eudiplister peyroni* Mars.

Several species with Mediterranean and West Asian ranges have their eastern boundaries in Kopetdagh, e.g., *Saprinus steppensis* Mars., *Pholioxenus shatzmayri* J. Müll., *Onthophilus sulcatus caucasicus* Rtt. (the only representative of this genus in Middle Asia), *Margarinotus graecus* Brullé, *Platysoma simeani* Muls. et God., *Satrapes talyschensis* Rtt. In contrast, few other species inhabiting the eastern parts of Middle Asia here reach their western boundaries, e.g., *Margarinotus solskyi* Schm., *M. bickhardti* Rtt., and *Atholus rudesculptus* Rchdt. Finally, Kopetdagh is the southern limit of range for some Turanian species listed above. Only one species of Histeridae, *Margarinotus atamuradovi* sp. n. (in litt.) can be considered an endemic of Kopetdagh; it is a sister species of the widely distributed West Palearctic *M. stercorarius* Hoffm.

Tenebrionidae of Kopetdagh have been studied by many specialists. Summaries of these studies by G. Medvedev and Nepesova (1985, 1990) indicate that Kopetdagh fauna of tenebrionids includes 96 species from 53 genera. Of this list, 37 species (39%) are endemics or subendemics of these mountains. Seventeen species of these 37 are widely distributed through Kopetdagh, e.g., *Gnathosia hydrobiforfis* Rtt., *G. skobelevi* Stark, *Tentyria robustoides* Rtt., *Leptodes solieri* Sem., *Microblemma simplex* Sem., *Lasiostola grandis* Rtt., *Stalagmoptera ruginota* Rtt., *Prosodes solskyi* Faust, *P. jakovlewi* Sem., *P. laticauda* Rtt., and *Dendarus armeniacus* Baudi. Thirteen other species inhabit only West Kopetdagh: *Gnathosia sublaevigata* Bog. et Kryzh., *Blaps balashovi* G. Medv., *Dendarus transcaspicus* Brancsik, *Penthicus fartilis* Rtt., *Catomus karakalensis* G. Medv., and *Probaticus zoroaster* Seidl. Six to seven species are known only from Central Kopetdagh: *Gnathosia compressa* Schust., *Penthicus turkmenicus* G. Medv., and probably *Somocoelia* sp.

Another group of 29 species has Kopetdagh-Turanian ranges of different types. These are principally the desert species which live in the lower belt of the mountains, e.g., *Arthrodosis schusteri* Rtt., *Microdera globulicollis* Mén., *Eutagenia turcomana* Rtt., *Platamodes dentipes* Mén., *Dichillus tenebrosus* Rtt., *Adesmia karelini* Fisch.-W., *Pimelia cephalotes* Pall., *Trigonoscelis apicalis* Rtt., *Blaps tutanus* Mén., *Penthicus pinguis* Fald., and three species of *Dissonomus*, etc. Thirteen species are inhabiting Irano-Kopetdagh area (sometimes including East Transcaucasia): *Gnathosia modesta* Fald., *Trachyderma christophi* Faust, *Pelorocnemis darwini* Faust, *Pachyscelis gemmans* Baudi, *Blaps dehaani* Baudi, *Prosodes cribrella* Baudi, and others. About 10 species have Mediterranean-Turanian ranges of diffferent types, among them *Zophosis punctata* Brullé, *Dilamus fausti* Rtt., *Gonocephalum pubiferum* Rtt., and *G. rusticum* Baudi. A few species, such as *Penthicus dilectans* Fald. and *Boromorphus opaculus* Rtt., are distributed from Kopetdagh to Central Asia.

Only three species of tenebrionids (among them *Blaps mortisaga* L.) found in Kopetdagh also live in the forest regions of Europe; two of them are dendrobiotic (*Eledona agaricola* Hbst. and *Platydema triste* Lap.-Cast.). A cosmopolitan synanthropic species, *Alphitophagus bifasciatus* Say, also inhabits Kopetdagh.

Many species of Tenebrionidae endemic to the Kopetdagh Mountains belong to genera with Irano-Turanian or Irano-Turan-Central Asian ranges, e.g., *Gnathosia, Lasiostola, Stalagmoptera, Prosodes, Penthicus,* and *Reitterohelops*). Several other genera have mainly Mediterranean connections, such as *Calyptopsis, Tentyria, Pimelia, Cabirutus* (subg. *Cabirutus* s. str.), *Dendarus,* and *Probaticus*. Genera *Zophosis* and *Adesmia* are distributed mainly in the Afrotropical realm. Finally, many genera have wide ranges of different types, e.g., *Dichillus, Blaps, Gonocephalum,* and *Platydema*.

The fauna of Kopetdagh, therefore, should be regarded as a transitional one between the Irano-Turanian fauna of adjacent desert plains and the East Mediterranean fauna of Southwest Asia. Endemic species are rather numerous in these mountains, but endemic genera are practically absent. Some species among these endemics are also found within the ranges of North Iran (Elburz) or North Afghanistan (Paropamiz). The majority of Kopetdagh endemics are probably young forms derived mainly during the Pleistocene or Late Pliocene.

6. Riparian Forests, or "Tugais"

It is also necessary to examine a remarkable fauna of an intrazonal landscape of Turkmenistan, the so-called "*tugais*," or wet riparian (gallery) forests along the large rivers (Amu Darya, Murghab, Tedzhen, and others). Unfortunately, the tugais have been nearly destroyed by the activities of man. Only several portions have escaped destruction; the peculiar fauna living in these areas includes numerous species absent from other parts of Turkmenistan. Some representatives of the tugai fauna have been able to resettle along the edges of irrigation canals.

Carabidae are fairly numerous in tugais and are represented by some endemic species here. Histeridae are represented by the widely distributed taxa although there are few peculiar species. Tenebrionidae in tugais are primarily widespread, actively flying species of the genera *Gonocephalum, Scleropatrum, Opatroides, Penthicus*, and *Belopus*. A mycetophagous Euro-Mediterranean species, *Pentaphyllus chrysomeloides*, is also known from tugai forests.

Tugais also contains other faunal elements of interest. Several zoogeographic groups can be distinguished among the fauna of tugai. Many tugai inhabitants are related to the Palearctic forest species. These are mainly the members of dendrophilic groups, e.g., three species of Buprestidae living on poplars (*Eurythyrea aurata* Pall., *Ancylocheira salomonis* J. Thoms., and *Melanophila picta* Pall.), Elateridae (several species of *Ampedus*), Cerambycidae (*Xylotrechus grumi* Sem., *X. namanganensis* Heyd.), Chrysomelidae (species of *Cryptocephalus* and *Luperus*), and Curculionidae.

Rather numerous in tugai are hydrophilous and dendrophilous species with wide distribution in the East Mediterranean region and Southwest Asia, e.g., carabids *Chlaenius dimidiatus* Chaud., *Hemiaulax morio* Mén., *Loxoncus procerus* Schaum, *Ophonomimus hirsutulus* Dej., *Lebia holomera* Mén., and *Brachinus bayardi* Dej.; buprestids *Capnodis miliaris* Klug and *Sphenoptera mesopotamica* Mars.; the weevil *Liocleonus clathratus* Ol.; and others.

Some inhabitants of tugais are related to tropical taxa, and some are widespread in the tropics of Africa and Asia, e.g., carabids *Cicindela melanoholica* F., *Scarites eurytus* Fisch.-W., *Bembidion niloticum* Dej., *Tetragonoderus intermedius* Sols., and *Cymindoidea famini* Dej.

Several species endemic or subendemic to the Middle Asian tugai habitats are noteworthy from the zoogeographic point of view. These include carabids *Cicindela illicebrosa* Dokht., *C. nox* Sem., *Agonum punctibase* Rtt., *Panagaeus relictus* Sem. et Bog.; the subendemic genus *Mnuphorus* and some species of *Dyschirius* and *Bembidion*, e.g., *B. amnicola* J. Sahlb. and *B. gassneri* Net. Noteworthy histerids include *Saprinus lopatini* sp. n. (in litt.), which is related to a steppe species, *S. lautus* Er., and lives mainly in mushrooms. A tenebrionid species, *Penthicus semenovi* Rchdt., is very characteristic for tugais of both Middle and Central Asia. Representatives of other beetle families are numerous in this group as well, e.g., chafers *Chioneosoma porosum* Fisch.-W. and *Adoretus pruinosus* Fisch.-W.; blister beetles belonging to genera *Mylabris* and *Ctenopus*; and chrysomelids *Parnops glasunovi* Jacobs., and *Malegia jacobsoni* Sum. The single Middle Asian member of the primarily tropical family Brentidae, *Eremoxenus chan* Sem., deserves a special mention. This beetle inhabits the nests of an ant, *Camponotus turkestanicus*, in tugai along the Amu Darya and Murghab Rivers.

Analysis of Selected Genera

The zoogeographic patterns of the representatives of Carabidae from Turkmenistan which have different types of generic distributions, will be discussed in the following summary.

1. Calosoma Weber (sensu lato). The genus is distributed almost worldwide except in tropical rainforests and cold climatic zones. Nine species are known from Turkmenistan (the richest local fauna in the Palearctic realm). The majority of these species have wide ranges.

Calosoma (Calosoma) inquisitor L. and *Calosoma (Calosoma) sycophanta* L. live in European forests and the Mediterranean region; in Turkmenistan, they are found only in the mountain forests of Kopetdagh, where *C. inquisitor* has its southeastern frontier of distribution. Kopetdagh is the eastern limit also for a Mediterranean species *Calosoma (Campalita) maderae* F. This species is an allopatric one to *Calosoma (Calosoma) auropunctatum* Hbst. which has a wide range in Central and East Europe, Kazakhstan and Middle Asia to Tien Shan, Sinkiang and Pamir (up to 4,000 m high). It is known from the north and the east of Turkmenistan. The most common species of this genus in the plains of Turkmenistan is *Calosoma (Caminara) imbricatum* Klug, represented by the subspecies *C. i. deserticola* Sem.; it has an enormous range in the deserts and savannahs from South Africa through East and North Africa and Southwest Asia to the northwest of Indostan and the deserts north from the Caspian Sea and Mongolia.

Two more species, *Calosoma (Calosoma) olivieri* Dej. and *Calosoma (Campalita) algiricum* Gehin, are distributed throughout North Africa and Southwest Asia to Turkmenistan; both live in deserts. *Calosoma (Caminara) denticolle* Gebl. lives in the steppe and semidesert landscapes of Eurasia from Romania to northwestern China; two records are known from the north of Turkmenistan. Finally, *Calosoma (Calosoma) reitteri* Roeschke is found only in the ephemerous desert landscapes of low mountains of Middle Asia and Kazakhstan; it is fairly common in some localities in Kopetdagh and Badghyz at springtime.

The related genus *Callisthenes* Fisch.-W. is, on the contrary, absent from Turkmenistan although it is found in Armenia, Northwest Iran, Central Afghanistan and the eastern parts of Middle Asia.

2. Carabus (l.) Thomson. This primarily Palearctic genus is extremely species-rich (more than 600), and is usually represented by mesophilic forms. Reduced wings prevent the active dispersal of the majority of species. The genus is extremely poorly represented in Turkmenistan (three or four species) although it has many species in the mountains of eastern Middle Asia.

Carabus (Limnocarabus) clathratus L., one of the few Transpalearctic species of this genus, is known from the Bolshoi Balkhan, and many points in Kazakhstan; it lives on the shores of fresh or salt water bodies. *C. (Mimocarabus) roseni* Rtt. belongs to a rather xerobiotic Middle Eastern subgenus and inhabits both Iranian and Turkmenian Kopetdagh; it is active in the early spring in the mountain steppe and semidesert. *C. (Axinocarabus) miles* Sem. is an even more characteristic spring species limited by the ephemeral landscapes of Badghyz and northern Afghanistan. It belongs to a small subgenus endemic to the low mountains of southeastern Middle Asia. The fourth species, *C. (Cyclocarabus) kuznetzovi* Sem., is poorly studied and known only from the Iranian Kopetdagh Mountains. Several other species of the subgenus *Cyclocarabus* live in Tien Shan and also are spring "ephemerous" species.

3. Chilotomus Chaud. This small Irano-Turanian genus of phytophagous carabids is found in semiarid habitats of the middle mountain belt. Eight allopatric species are known: one in Southeast Transcaucasia and adjacent parts of Iran; four in eastern Middle Asia; and three in Turkmenistan. These three are *Ch. chalybaeus* Fald., in Kopetdagh; *Ch. margianus* Kryzh., in Badghyz; and *Ch. kuhitangi* Kryzh., in the Kugitang Mountains (Kryzhanovsky 1962). The single related genus, *Pachycarus*, is limited by the East Mediterranean (Greece with the Aegean Archipelago, western Turkey, and Syria).

4. Cymindis Latr. This genus presents us with an another example of the complexity of Turkmenian fauna. It includes about 150 species common in the Holarctic, and is found primarily in Eurasia. They are especially numerous in semiarid and arid parts of the Ancient Mediterranean region.

Fifteen species of *Cymindis* belonging to six subgenera are known from Turkmenistan, with ranges of very different types. *C. (Cymindis) axillaris* F. and *C. (C.) lineata* (Quens.) have Mediterranean-type ranges, and Kopetdagh forms the eastern limit for the first species. *C. (C.) picta* Pall. has a large range in the steppe zone, from East Russia and Kazakhstan to the mountain steppes and deserts of Iran and Afghanistan; in Turkmenistan, it inhabits mainly mountain landscapes, but in the north of the republic it is also found on the plains. *C. (C.) andreae* Mén. is widely distributed in the Irano-Turanian region. Three species, *C. (C.) accentifera* Zubk., *C. (C.) quadrisignata* Mén., and *C. (Eremocymindis) pallidula* Chd. live in Turanian solid-soil deserts; psammobiotic *C. (Iscariotes) capito* Kryzh. et Emetz is an endemic of the Karakum sand desert. *C. (Menas) walteri* Rtt. inhabits only mountain steppe and habitats with mountain xerophytic vegetation in Kopetdagh. The single species of a monotypic subgenus, *Pseudomenas*, *C. (P.) antonovi* Sem., lives in East Transcaucasia, North Iran, Kopetdagh, and the mountains of Tadjikistan. Finally, the members of Middle and Central Asian subgenus *Mastus* are numerous in Tien Shan, Ghissaro-Darvaz, and Hindu Kush; four of them reach Kugitang, and one, *C. (M.) rufipes* Gebl., reaches West Kopetdagh. An important negative characteristic of Turkmenistan fauna is the absence of the subgenus *Tarsostinus*, which is common in the steppes and forest-steppes of Europe, Siberia, Kazakhstan, and Mongolia.

Conclusions

The authors hope that the review of Turkmenistan beetle fauna gives a general idea of its peculiarities and zoogeographic composition. The extraordinarily high level of generic endemism of the entomofauna of sand deserts in Middle Asia allows consideration of these deserts as an ancient center of origin for the psammophilous fauna. This center exerted strong influence on the desert faunas of Central and Southwest Asia as well as North Africa.

In contrast, the fauna of the Kopetdagh Mountains is a rather young one and its numerous endemics commonly have species rank. This fauna forms a transition between the faunas of Middle and Southwest Asia and at the same time reflects an important zoogeographic boundary.

25. Buprestid Beetles (Coleoptera: Buprestidae) from Kopetdagh and the Adjacent Regions of Southern Turkmenistan

MARK G. VOLKOVICH AND ANATOLY V. ALEXEEV

Abstract

The buprestid fauna of the Kopetdagh Mountains and adjacent regions of South Turkmenistan is reviewed as a whole as well as separately for Southwest Turkmenistan, Bolshoi Balkhan, Maly Balkhan, West Kopetdagh, Central Kopetdagh, East Kopetdagh, the submontane plain of Kopetdagh, the area between the Tedzhen and Murgab Rivers, and the Badghyz Plateau.

There are 193 buprestid species and subspecies belonging to 27 genera in the studied region; 126 species from 25 genera are found in Kopetdagh. A detailed taxonomic and biogeographic analysis of the buprestid fauna is given. The most diverse genera are *Sphenoptera* and *Acmaeoderella* (more than half of the regional buprestid fauna); their role increases in lowlands and decreases in mountains. Generic and subgeneric faunistic diversity in the lowlands is significantly lower than in the mountains; the same is true of specific diversity in the primarily mesophilous genera (*Capnodis*, *Anthaxia*, *Cratomerus*, and *Agrilus*).

The most numerous biogeographic elements in both the regional and the Kopetdagh faunas are species with Turanian, Southwest Asian, Khorassan, subendemic, and endemic ranges. Widely represented are the Irano-Turanian, Hesperian-Sethian, Turkestanian, West Palearctic subboreal-subtropic, Irano-Turanian-Gobian, and East Iran-Turanian elements; Iranian and Afghanian elements also are present. The regional buprestid fauna is comprised primarily of Irano-Turanian elements of western origin with the important participation of the widely distributed (West Palearctic and Hesperian-Sethian) elements. Thus, it differs from buprestid faunas of other Middle Asian regions in which elements of eastern origin are predominant and the role of widely distributed elements decreases. Fifty-nine (46.8%) of the species are not found in Middle Asia eastward beyond Kopetdagh, whereas 67 species are found in the other parts of Middle Asia. Endemic elements (Khorassan endemics, subendemics, and narrow local endemics) include 24 species and subspecies (12.4%) of the regional fauna and 17 species and subspecies (13.5%) of the Kopetdagh fauna, demonstrating the high level of faunistic isolation. There are no endemics of supraspecific rank.

Nine ecological groups of buprestids are designated according to plant communities where species develop as larvae: psammophilous, halophilous, sagebrush, tugai, steppe, xerophilous (shiblyak), mesophilous, juniper, and tragacanthoid groups. A peculiar character of the Kopetdagh buprestid fauna is expressed in the composition of steppe and xerophilous groups, supported by the mesophilous, juniper, and tragacanthoid groups where the importance of West Palearctic, Hesperian-Sethian, Southwest Asian, Khorassanian, and local endemic elements increases. In general, the faunistic specificity increases with the altitude. Tugai forests may have facilitated the dispersal of numerous mesophilous buprestid species over desert areas.

The list of buprestids from Kopetdagh and surrounding areas is given, with notes on their distribution in the studied region, in Middle Asia, in the former USSR, and in the Palearctic realm, as well as with notes on the larval host plants.

Introduction

The rich and original nature of the Kopetdagh Mountains has attracted investigators' attention for a long time. The recent surveys by Kamelin (1970) on the Kopetdagh flora, Kryzhanovsky (1965) and Medvedev and Nepesova (1990) on beetles, and Fet (1983) on spiders emphasized a rather high level of endemism of Kopetdagh flora and fauna and the presence of numerous species not found in other parts of Middle Asia.

Kopetdagh, which is the northern extremity of the Turkmeno-Khorassan Mountains, is formed within Turkmenistan by several mountain ranges latitudinally oriented. It is usually divided into West, Central, and East parts, which differ in their altitudes, climate, and vegetation. That West Kopetdagh is characterized by more-or-less well-developed communities of deciduous, xeromesophilous trees and shrubs secures the presence of many mesophilic insects there. Formations of juniper, tragacanthoid plants, and steppes are more widely present in the higher Central Kopetdagh. East Kopetdagh, with its low mountains, is characterized by the predominance of halophile vegetation and impoverished communities of xerophilous trees and shrubs.

Various viewpoints exist toward the position and status of Kopetdagh and the surrounding areas in different systems of biogeographic division of the Palaearctic realm. We use the system suggested by Yemelyanov (1974), who separates the Turkmeno-Khorasssan Mountains (including Kopetdagh) into the Khorassan Mountainous Province of the Irano-Turanian Subregion of the Sethian Desert Region. Some authors also attribute to this province low, isolated ranges of Bolshoi Balkhan and Maly Balkhan, which are located farther northwest and are separated from Kopetdagh by narrow desert depressions. In the west and northwest, Kopetdagh is bordered by Transcaspian deserts; in the north, by the Karakum sand desert; and in the east and southeast, by desert areas separating the Tedzhen and Murghab Rivers. All of the mentioned deserts are part of the South Turanian Plain Province (Yemelyanov 1974). The Badghyz Plateau belongs to the Paropamiz Mountains, which are contiguous with the Iranian portion of the Turkmeno-Khorassan Mountains. Fauna of Badghyz contains a number of Khorassan species and subendemic species common to the Kopetdagh fauna. However, many authors regard the Badghyz as part of the South Turanian Province; this problem is discussed below.

The Khorassan Province therefore borders the South Turanian Province to the north, west, and east; the Afghanian Province to the southeast; the Iranian Province to the south; and the Southwest Asian Province to the southwest. As a result of the latitudinal orientation of the mountain chains, the Turkmeno-Khorassan Mountains are connected to the mountain ranges of Southwest Asia (including Middle East and Turkey), the Transcaucasia, Paropamiz, and Hindu Kush, all of which greatly affect composition of Kopetdagh flora and fauna.

This work is based primarily on the abundant material from southern Turkmenistan deposited in the Zoological Institute of the Russian Academy of Sciences (St. Petersburg), including material collected by the authors in 1973 through 1991, as well as collections of other Russian and foreign museums, and private collections. It should be noted that Kopetdagh and the surrounding areas are not equally well studied in detail: the buprestid faunas of West Kopetdagh, Central Kopetdagh, and Badghyz are well known, whereas East Kopetdagh, Bolshoi Balkhan, and Maly Balkhan are far less thoroughly investigated. The buprestid fauna of the submontane plain of Kopetdagh is also not well defined because references to "Ashkhabad" and other submontane settlements in old collections are often uncertain. Also, taxonomic problems exist with the identification of some species (especially in the genus *Sphenoptera*, which is a dominant genus in the studied region).

Taxonomic and Biogeographic Analysis of the Buprestid Fauna of Kopetdagh and Adjacent Areas

In the studied region, 193 buprestid species and subspecies belonging to 27 genera have been found (Table 1), which is about 30 % of the total buprestid fauna of the former Soviet Union. The genera *Sphenoptera* (65 spp., or 33.7%) and *Acmaeoderella* (34 spp., 17.6%) are the predominant taxa and constitute a characteristic feature of the Turanian temperate desert areas. The role of these genera increases in the desert lowlands, e.g., 42.5% and 15%, respectively, in southwest Turkmenistan, 44.8% and 12.1% in the area between the Murgab and Tedzhen Rivers, and 33.3% and 23% in Badghyz; this role decreases in the mountains of Kopetdagh (26.4% and 17.6%). Generic and subgeneric diversity of lowland buprestid fauna is significantly poorer than in the mountains; the same is true for specific diversity in the primarily mesophilous genera (e.g., *Capnodis, Anthaxia, Cratomerus,* and *Agrilus*). The most numerous biogeographic elements belong to the species with Turanian (46 spp., or 23.8%),

Southwest Asian (26 spp., or 13.5%), and Khorassan, endemic, and subendemic (total of 24 spp., or 12.4%) types of ranges. Well represented are widely distributed Irano-Turanian (18 spp., or 9.3%), Hesperian-Sethian (16 spp., or 8.3%), Turkestanian (16 spp., or 8.3%), West Palaearctic subboreal-subtropic (12 spp., or 6.2%), Irano-Turanian-Gobian (10 spp., or 5.2%), and East Iran-Turanian (10 spp., or 5.2%) elements. Iranian (five spp.) and Afghanian (one sp.) elements also occur there. The regional buprestid fauna is formed primarily by the Irano-Turanian elements of western origin, with the important participation of widely distributed (West Palaearctic and Hesperian-Sethian) elements. The fauna thus differs from other Middle Asian regional buprestid faunas in which elements of the eastern origin predominate and the proportion of widely distributed elements is reduced.

In Kopetdagh, 126 buprestid species and subspecies belonging to 25 genera have been found. Southwest Asian (21 spp., or 16.7%), Turanian (20 spp., or 15.9%), Khorassan, endemic, and subendemic (total of 17 spp., or 13.5%), widely Irano-Turanian (13 spp., or 10.3%), and Hesperian-Sethian (13 spp., or 10.3%) elements predominate in the composition of the Kopetdagh fauna. Its basic core is formed by Sethian elements (100 spp.), with the rest comprised of widely distributed ones (trans-Palaearctic, 2 spp.; West Palaearctic subboreal-subtropic, 11 spp.; and Hesperian-Sethian, 13 spp.). Kopetdagh is the eastern boundary of distribution inside the former Soviet Union for the West Palaearctic species *Anthaxia cichorii, Agrilus derasofasciatus, Coroebus rubi,* and *Trachys phlyctaenoides*. The West Palaearctic species *Acmaeoderella mimonti* and *A. gibbulosa* as well as trans-Palaearctic *Agrilus viridis* are absent from Middle Asia, but they occur in the Tarbagatai and Altai Mountains (West Kazakhstan). Some widely distributed species (e.g., *Acmaeoderella mimonti*) form distinct geographic races in Kopetdagh; if in the future these races are formally described as species or subspecies, the number of widely distributed species will be reduced. The Irano-Turanian species of western origin are of considerable importance among the Sethian elements; there are 45 species (35.7%) with West Iran-Turanian, Southwest Asian, Iranian, and Khorassan types of ranges. There are 35 species (27.8%) of Irano-Turanian buprestids of eastern origin (with East Iran-Turanian, Turanian, and Turkestanian types of ranges). The remainder belongs to the more widely distributed Sahar0-Irano-Turanian (2 spp.), Irano-Turanian-Gobian (5 spp.), and Irano-Turanian (13 spp.) elements. Kopetdagh forms the western boundary of distribution for many species of this group, primarily for those of the Turkestanian range. The high number of species of eastern origin (particularly Irano-Turanian-Gobian and Turanian) in the Kopetdagh buprestid fauna is explained by high aridity, presence of profound desert depressions, and low altitudes. These species are primarily concentrated in the foothills and large river valleys (e.g., those of the Sumbar and Arvaz). In contrast, the species of western origin, primarily those originating from Southwest Asia, dominate in higher mountain belts. Kopetdagh is the eastern boundary of distribution (at least within Middle Asia) for many of these species, e.g., *Acmaeodera pilosellae persica, A. chalcithorax,*

Table 1. The list of Buprestidae from Kopetdagh and the adjacent regions

Species	Distribution in Kopetdagh and adjacent regions[a]	General distribution[b]	Range type[c]	Larval habitat/host plants[d]
Subfam. Julodinae Trib. Julodini				
Julodis Eschscholtz, 1829				
J. euphratica Laporte et Gory, 1835	1–9	Tu, Uz, Tj, Tr, Ir, Af, Iq, Sa, Eg	SaIT	Soil – outside plant roots
J. variolaris freygessneri Meyr-Darcis, 1883	1–9	Tu, Uz, Ir, Af	Turan	Soil – outside roots of *Alhagi*[*], *Calligonum*[*]
J. laevicostata Gory, 1840	5, 7	Ir, Iq	Iran	Soil
Julodella Semenov, 1893				
J. shestoperovi Stepanov, 1959	5		Endem	Soil
J. kaufmanni (Ballion, 1870)	7, 9	Tu, Uz, Tj	Turk	Soil
J. brevilata Semenov, 1893	5, 7, 9	Ir	Khor	Soil
Subfam. Polycestinae Trib. Polycestini				
Strigopteroides Cobos, 1981				
S. aegyptiacus (Gmelin, 1788)	7	Tu, Uz, Tr, Ir, Sa, Eg, Sy, Iq	SaIT	Dead wood
Subfam. Acmaeoderinae Trib. Acmaeoderini				
Acmaeodera Eschscholtz, 1829				
A. (Acmaeodera)				
A. babatauensis Obenberger, 1935	9	Tu, Uz, Tj, Ky	Turk	*Pistacia*[*], *Ficus*[*], *Amygdalus*[*]
A. ghilarovi Volkovitsh, 1988	4, 9		Endem	Unknown
A. pilosellae persica Mannerheim, 1837	4, 5	Zk, Tr, Ir	SWA	*Colutea*[*]
A. chalcithorax Obenberger, 1935	4, 5, 7	Zk, Tr, Ir, Iq	SWA	*Astragalus* (*Tragacantha*)[*]

Table 1. Continued

Species	Distribution in Kopetdagh and adjacent regions[a]	General distribution[b]	Range type[c]	Larval habitat/host plants[d]
A. (Acmaeotethya)				
A. pallidepicta Reitter, 1895	4, 5, 7	Ir	Khor	Juglans*, Ficus*, Salix*, Amygdalus*, Colutea, Pistacia Paliurus, Pistacia* (oviposition)
A. instabilis Cobos, 1966	9	Af	Afghan	
A. (Cobosiella)				
A. chotanica Semenov, 1890	2, 5, 7, 9	Tu, Uz, Tj, Ch	ITG	Salix*, Ulmus*, Ficus*, Gleditsia*, Populus, Morus, Pistacia
Xantheremia Volkovitsh, 1978				
X. koenigi (Ganglbauer, 1888)	2-5, 7-9	Tu, Uz, Tj, Ka	eIT	Alhagi*, Lycium*
X. subscalaris (Reitter, 1897)	1, 8	Tu, Uz, Tj, Ka, Af	Turan	Alhagi*, Lycium*, Glycyrrhiza
X. steinbergi (Volkovitsh, 1978)	4, 8	Uz, Tj	Turan	Alhagi, Glycyrrhiza
Acmaeoderella Cobos, 1955				
A. (Acmaeoderella) s.str.				
A. caspica caspica (Ganglbauer, 1888)	3-5, 7		Endem	Centaurea*
A. caspica turkestanica (Obenberger, 1934)	9	Tu, Uz, Tj, Af	Turk	Onopordum
A. caspica suturifera (Reitter, 1904)	1, 3, 4, 7-9	Tu, Uz, Ka, Ir, Af, Sy	Turan	Jurinea*, Cousinia*
A. turanica (Reitter, 1890)	5, 7, 9	Tu, Uz, Tj, Af	Turk	Unknown
A. badhysica Volkovitsch, sp.n.	9		Endem	Unknown
A. plavilscikovi (Obenberger, 1936)	1, 2, 4, 5, 9	Tu, Uz, Ky, Ka, Zk	IT	Atriplex, Salsola*, Limonium, Reaumaria*, Noaea*
A. oresitropha (Obenberger, 1936)	3-5	Uz, Tj, Ky, Zk	IT	Salsola*, Reaumuria, Zygophyllum

Table 1. Continued

Species	Distribution in Kopetdagh and adjacent regions[a]	General distribution[b]	Range type[c]	Larval habitat/host plants[d]
A. cinerea Volkovitsch, 1982	2–5, 9	Tu, Uz, Tj, Ky, Ka, Ir	eIT	Artemisia
A. nivetecta Volkovitsch, 1976	7, 9	Tu, Tj, Ir, Af	eIT	Heliotropium*
A. (Carininota)				
A. flavofasciata chorasanica nom. nov.	4	Ir	Khoras	polyphagous (other subspecies) Prunus, Caragana, Astragalus (Tragacantha)
A. mimonti (Boieldieu, 1865)	4	Ka, Ep, Zk, Tr, Ir, Iq, Eu, eMt	WP	Pistacia*, Astragalus (Tragacantha)*, Celtis, Pyrus, Prunus, Robinia, Amygdalus, Rhus, Colutea, Caragana
A. glasunovi (Semenov, 1895)	9	Tu, Uz, Tj, Ky, Ka	Turk	Ammodendron*, Salsola*
A. repetekensis (Obenberger, 1934)	1	Tu, Uz, Ka	Turan	Zygophyllum (Halimiphyllum)*
A. zarudniana Volkovitsch, 1977	2–5, 7, 9	Ir	Iran	
A. (Euacmaeoderella)				
A. villosula (Steven, 1830)	9	Tu, Td, Zk, Tr, Ir, Af, Iq, eMt	HS	Ferula, Malabaila, Zosima
A. coelestina Volkovitsch, 1977	4	Zk, Tr, Ir, Af, Gr	Endem	Unknown
A. subcyanea (Reitter, 1890)	5	Zk, Tr, Ir, Iq, Sa, eMt	HS	Unknown
A. obscura (Reitter, 1889)	4		HS	Unknown
A. vetusta (Menetriés, 1832)	4–5	Zk, Tr, Ir, eMt	HS	Unknown
A. adamantina (Reitter, 1890)	5, 9	Tu, Uz, Tj, Af, Sy	Turk	Ferula*, Dorema*

Table 1. Continued

Species	Distribution in Kopetdagh and adjacent regions[a]	General distribution[b]	Range type[c]	Larval habitat/host plants[d]
A. canescens (Semenov, 1895)	2–3, 8–9	Tu, Uz, Ka, Ir	Turan	Ferula*, Dorema*
A. semiviolacea (Semenov, 1895)	9	Tu, Uz, Tj, Ky	Turk	Ferula*, Dorema*
A. alepidota Volkovitsch, 1977	4, 8–9		Endem	Ferula*
A. strandi (Obenberger, 1918)	4–5		Endem	Ferula*, Dorema*
A. gibbulosa (Menetriés, 1832)	3–7	Ka, sEp, Zk, Tr, Ir, eMt, Iq	WP	Onopordum*, Prangos, Ferula, Malabaila, Zosima, Chondrilla
A. dubia (Ballion, 1870)	1–9	Tu, Uz, Tj, Ky, Ka	eIT	polyphagous; Ferula*, Dorema*, Atriplex*, Suaeda*, Isatis*, Crambe*, Chrozophora*
A. insueta Volkovitsch, 1977	3, 9	Tu, Uz, Tj	Turan	Ferula*, Dorema*, Isatis*
A. valentinae Volkovitsch, 1977	8	Uz, Tj, Ky, Ka	Turan	Zygophyllum*
A. candens Volkovitsch, 1977	5	Tu, Tj	Turan	Chrozophora*
A. iranica (Obenberger, 1934)	5–7, 9	Tu, Ka, Ir, Af	eIT	Cousinia*
A. ballioni (Ganglbauer, 1888)	1–5, 8–9	Tu, Uz, Tj, Ka, Ir, Af	eIT	Salsola*, Halothamnus*, Anabasis, Convolvulus*
A. solskyi (Obenberger, 1934)	1, 3, 7	Tu, Uz	Turan	Astragalus (Ammodendron)*
A. tragacanthae kopetdaghica Volkovitsch, 1977	4, 5, 9		Endem	Astragalus (Tragacantha)*
A. personata (Semenov, 1896)	8, 9	Tu, Ka, Mg	ITG	Halocnemum*, Halothamnus*, Anabasis*, Salsola*, Ephedra*, Haloxylon

Table 1. Continued

Species	Distribution in Kopetdagh and adjacent regions[a]	General distribution[b]	Range type[c]	Larval habitat/host plants[d]
Subfam. Chalcophorinae Trib. Psilopterini				
Capnodis Eschscholtz, 1829				
C. tenebricosa (Olivier, 1790)	4–5, 7–9	Uz, Tj, Ky, Ka, Ir, Iq, Tr, Ep, Zk, Af, Eu, Mt	WP	*Rumex*
C. tenebrionis (Linnaeus, 1758)	4–5, 7	Uz, KA, Ep, Zk, Ir, Tr, Eu, Mt	WP	*Armeniaca, Prunus, Cerasus, Amygdalus, Persica, Pyrus, Crataegus*
C. miliaris (Klug, 1829)	4–5, 7–9	Tu, Uz, Tj, Ky, Ka, Zk, Tr, Ir, Iq, Af, Ch, eMt nIr	HS	*Salix*[*], *Populus*[*], *Elaeagnus*[*]
C. jacobsoni Richter, 1952	5–6, 9	Uz, Tj, Ky, Af, Pa	Khoras	*Amygdalus*[*]
C. parumstriata Ballion, 1870	7, 9		Turk	*Pistacia*[*]
C. excisa Menetriés, 1848	1, 3–4, 7–9	Tu, Uz, Tj, Ka, Ir, Iq, Zk, Sa, Eg	SaIT	*Calligonum*[*]
C. sexmaculata Ballion, 1870	5, 7, 9	Uz, Tj, Ky, Ka, Af, In	Turk	*Amygdalus*
C. anthracina (Fischer, 1830)	4–5, 7–8	Tj, Zk, Tr, Ir, Af	wIT	*Rheum*
Cyphosoma Mannerheim, 1837				
C. tataricum (Pallas, 1773)	7–8	Tu, Uz, Tj, Ka, seEp, Zk, Ir	IT	Unknown
C. turcomanicum (Kraatz, 1883)	1–5, 7–8	Tu, Uz, Ka, Ir	Turan	Unknown

Table 1. Continued

Species	Distribution in Kopetdagh and adjacent regions[a]	General distribution[b]	Range type[c]	Larval habitat/host plants[d]
Aurigena Spinola, 1837				
A. lugubris longicollis Kraatz, 1881	4–5	Zk, Tr, Ir	SWA	*Rosa, Prunus, Cerasus, Pyrus*
Psiloptera Solier, 1833				
P. (*Lampetis*)				
P. argentata (Mannerheim, 1837)	1–5, 7–9	Tu, Uz, Tj, Ka, Zk, Tr, Ir, Af, Iq	IT	*Haloxylon**, *Kalidium**, *Salsola**, *Juglans**
Subfam. Sphenopterinae				
Trib. Sphenopterini				
Sphenoptera Solier, 1833				
S. (*Sphenoptera*), s.str.				
S. glabrata (Menetriés, 1832)	2, 4–5, 7	Zk, Tr, Ir, Af, Iq	SWA	*Astragalus* (*Tragacantha*)*
S. furva Jakowlew, 1907	4–5	Zk, Ir	SWA	*Astragalus* (*Tragacantha*)*
S. rangnowi Kerremans, 1909	4, 7	Ir, Af	Iran	*Astragalus* (*Tragacantha*)*
S. lia Jakowlew, 1901	4–5, 7, 9	Tu, Uz, Af	Turan	*Astragalus**
S. cyanea Jakowlew, 1899	4, 7	Ir, Af	Iran	Unknown
S. violacea Jakowlew, 1899	7	nIr	Khor	Unknown
S. chalybaea Menetriés, 1849	6–7	Tu, Zk, Af	wIT	*Astragalus**
S. serripes Jakowlew, 1901	1, 7	Tu	Turan	*Astragalus**
S. korshinskii Jakowlew, 1900	4	Uz	Turan	Unknown
S. egregia Jakowlew, 1901	8	Tu	Turan	*Acanthophyllum**
S. lateralis Faldermann, 1836	7–8	Uz, Ky, Ka, Af	eIT	*Anabasis*
S. exarata (Fischer, 1824)	7–9	Tu, Uz, Tj, Ky, Ka, seEp, Zk, Ir, Af, Ch	IT	*Glycyrrhiza**

Table 1. Continued

Species	Distribution in Kopetdagh and adjacent regions[a]	General distribution[b]	Range type[c]	Larval habitat/host plants[d]
S. navicula Jakowlew, 1907	2, 7	Tu	Turan	Unknown
S. repetekensis Obenberger, 1927	5, 7–8	Tu, nIr	Turan	Acanthophyllum*
S. aerata Jakowlew, 1892	9	Tu, Uz, Ka, Zk, Ir	IT	Iris*
S. komarovi Jakowlew, 1886	4, 5, 7	Zk, Ir, Af	SWA	Unknown
S. latesulcata Jakowlew, 1886	9	Zk	SWA	Unknown
S. turcmenica Obenberger, 1927	7		Endem	Unknown
S. (Deudora)				
S. caspica Jakowlew, 1904	1	Tu	Turan	Eremosparton*
S. curta Jakowlew, 1885	1, 3–4	Tu	Turan	Unknown
S. unidentata Jakowlew, 1890	4–5, 8–9	Tu, Uz, Tj, Ka, Zk, Tr, Ir, Af	IT	Unknown
S. allecta Jakowlew, 1900	9	Tu, Uz, Zk	SWA	Unknown
S. subtilis Jakowlew, 1899	4	Ir	Khor	Unknown
S. tenax Jakowlew, 1902	5, 7	Zk	SWA	Unknown
S. captiosa Jakowlew, 1902	5, 7	Zk	SWA	Unknown
S. addenda Jakowlew, 1900	8	Zk, Ir, Tr	SWA	Unknown
S. mitrochinae Alexeev, in litt.	9	Tu, Tj	Turan	Unknown
S. kepelensis Zykov et al.exeev, 1992	9	Uz, Af	Turk	Astragalus*
S. koenigi Jakowlew, 1890	4–5, 7, 9		Endem	Cousinia*
S. afflicta Jakowlew, 1900	7, 9	Uz, Tj, Ka	Turan	Unknown
S. bucharica Jakowlew, 1900	4–5, 7–9	Tu, Uz	Turk	Unknown
S. vestita Jakowlew, 1887	1, 3, 8	Tu, Zk	SWA	Salsola*
S. (Rhaphidochila)				
S. coerulea Jakowlew, 1899	5	Zk, Tr, Ir, Af	wIT	Acanthophyllum*

Table 1. Continued

Species	Distribution in Kopetdagh and adjacent regions[a]	General distribution[b]	Range type[c]	Larval habitat/host plants[d]
S. (Chilostetha)				
S. canescens Motschulsky, 1860	7, 9	Tu, Uz, Tj, Ky, Ka, seEp, Af	IT	Artemisia
S. eximia Jakowlew, 1902	4, 7	Uz	Turan	Unknown
S. cataonia Obenberger, 1926	4, 7, 9	Tu	Turan	Unknown
S. erojlandusica Alexeev et Zykov, in litt.	9		Endem	Artemisia[*]
S. rauda Jakowlew, 1908	1–2, 4–5, 7	Tu, Uz, Tj, Ky, Ka, Zk, Ir	IT	Unknown
S. puberula Jakowlew, 1887	1, 4–5, 7, 9	Uz, Tj	Turan	Unknown
S. (Chrysoblemma)				
S. potanini Jakowlew, 1889	1–2, 7–8	Tu, Uz, Ka, Ir, Mg	ITG	Haloxylon[*]
S. beckeri Dohrn, 1866	1, 4–9	Tu, Uz, Tj, Ka, seEp, Zk, Ir, Af, Mg, Iq	ITG	Climacoptera[*], Horaninovia[*], Salsola[*], Calligonum[*]
S. scovitzi Faldermann, 1835	1, 3–4, 6–9	Tu, Uz, Tj, Ky, Ka, Zk, Ir, Af	IT	Climacoptera[*], Salsola[*]
S. ignita Reitter, 1895	4, 7–8	Tu, Uz, Tj, Ka	Turan	Chenopodiaceae
S. pseudoignita Alexeev, 1978	8–9	Tu, Uz, Tj, Ir	Turan	Salsola[*], Suaeda[*], Climacoptera[*], Nitraria[*]
S. hauseri Reitter, 1895	3, 7–9	Tu, Uz, Tj, Ka, Ir, Af	Turan	Haloxylon[*]
S. orichalcea (Pallas, 1781)	1, 4	Tu, Uz, Ki, Ka, Mg, seEp, wSi, Zk	ITG	Anabasis, Kalidium, Halocnemum, Kochia

Table 1. Continued

Species	Distribution in Kopetdagh and adjacent regions[a]	General distribution[b]	Range type[c]	Larval habitat/host plants[d]
S. amplicollis Jakowlew, 1899	3–5, 8–9	Tu, Uz, Tj, Ir	Turan	Halothamnus*
S. tomentosa Jakowlew, 1886	1, 3–5, 6–9	Tu, Uz, Tj, Ka	Turan	Salsola*
S. punctatissima Reitter, 1895	1, 7–9	Tu, Uz, Tj, Ka, Af	Turan	Haloxylon*
S. pubescens Jakowlew, 1886	1, 8	Tu, Tj, Ka	Turan	Halostachys
S. viridula Jakowlew, 1905	9	Tu, Tj, Ka	Turan	Tamarix
S. bifulgida Reitter, 1898	1, 8	Tu, Uz, Ka	Turan	Salsola*
S. viridiaurea Kraatz, 1882	1, 4–9	Tu, Uz, Tj, Ka, Ir, Af	Turan	Salsola*, Kochia*, Climacoptera*, Suaeda, Alhagi*
S. artemisiae Reitter, 1889	9	Zk, Ir, Af	SWA	Unknown
S. zarudnyi schatinensis Alexeev et Zykov, 1991	4–5	Zk, Ir	SWA	Acanthophyllum*
S. tschitscherini Jakowlew, 1900	1, 8	Tu, Tj, Zk	IT	Halothamnus*, Salsola*, Kochia*, Climacoptera*
S. amoena Jakowlew, 1901	4–5	Zk, Ir	SWA	Unknown
S. apta Jakowlew, 1903	4	Zk	SWA	Unknown
S. (Hoplistura)				
S. mesopotamica Marseul, 1865	4–5, 7–8	Tu, Uz, Tj, Ka, Zk, Tr, Ir, Af, Iq	IT	Tamarix, Trachomitum
S. balassogloi Jakowlew, 1885	1, 4, 8–9	Tu, Uz, Ka, Zk, Ir, Af	IT	Tamarix
S. semenovi Jakowlew, 1889	8–9	Tu, Tj, Ir, Af	ITG	Tamarix

Table 1. Continued

Species	Distribution in Kopetdagh and adjacent regions[a]	General distribution[b]	Range type[c]	Larval habitat/host plants[d]
S. (*Tropeopeltis*)				
S. kaznakovi Jakowlew, 1899	4, 7–8	Uz, Tj, Ir, Af	Turk	*Amygdalus, Persica, Armeniaca, Cerasus, Prunus*
S. kambyses Obenberger, 1930	4–5, 7, 9	Ir, Iq	Iran	*Amygdalus*[*], *Armeniaca*[*], *Prunus*[*]
S. schneideri Reitter, 1898	7	Tu, Uz, Ka	Turan	*Calligonum*[*]
S. mujunkumensis Obenberger, 1928	7, 9	Uz, Ka	Turan	*Calligonum*[*]
Trib. Anthaxiini				
Anthaxia Eschscholtz, 1829				
A. (*Haplanthaxia*)				
A. cichorii (Olivier, 1790)	4	Ep, Zk, Ir, Eu, Mt, Iq	WP	polyphagous; *Juglans*[*], *Paliurus*[*]
A. (*Cryptanthaxia*)				
A. nanissima Alexeev, 1968	8	Tu, Uz, Tj, Ka, Mg	ITG	*Populus*
A. spinosa Abeille de Perrin, 1900	4–5	nIr	Khor	*Astragalus* (*Tragacantha*)[*]
A. badghyzica Bily, 1991	9		Endem	*Astragalus* (*Tragacantha*)[*]
A. lucidiceps Gory, 1841	1–5, 7–9	Tu, Uz, Tj, Ka, Zk, Ir, Af, eMt	HS	*Ferula*[*], *Dorema*[*]
A. discicollis Laporte & Gory, 1839	4–5	Zk, eMt	HS	*Juniperus*[*]
A. (*Melanthaxia*)				
A. hemichrysis Abeille de Perrin, 1900	4–5	Zk, nIr	SWA	*Juniperus*[*]

Table 1. Continued

Species	Distribution in Kopetdagh and adjacent regions[a]	General distribution[b]	Range type[c]	Larval habitat/host plants[d]
A. (Cyclanthaxia)				
A. holoptera Obenberger, 1914	4–5	Zk	SWA	Unknown
A. kreuzbergi Richter, 1944	9		Endem	*Pistacia*[*]
A. (Anthaxia)				
A. bicolor Faldermann, 1835	4	Uz, sEp, Ir, eMt	HS	*Fraxinus*
A. muliebris Obenberger, 1918	4	Zk, Tr, Ir	SWA	*Malus, Cydonia, Punica*
A. (Callanthaxia)				
A. passerinii (Pecchioli, 1837)	4–5	Zk, Tr, Ir, eMt	HS	*Juniperus, Cupressus*
Cratomerus Solier, 1833				
C. (Trichocratomerus)				
C. intermedius (Obenberger, 1913)	4–5, 7	Tu, Uz, Tj, Zk, Ir	IT	*Celtis, Ulmus*[*], *Pyrus, Malus*
C. (Cratomerus)				
C. fariniger (Kraatz, 1882)	4–5, 7–8	Tu, Uz, Tj, Ky, Ka, Af	eIT	*Salix, Populus, Juglans*[*]
C. hungaricus sitta (Kuster, 1852)	4, 7	sEp, Zk, Tr, Ir	SWA	Unknown
C. sponsa (Kiesenwetter, 1857)	4	Zk, Tr, Gr, Sy	HS	Unknown
C. medvedevorum Alexeev, 1978	2		Endem	*Halimodendron*[*]
C. (Cryptocratomerus)				
C. fedtschenkoi (Semenov, 1895)	4–5, 8–9	Tu, Uz, Tj, Ir, Af	Turk	*Celtis, Acer, Halimodendron*

Table 1. Continued

Species	Distribution in Kopetdagh and adjacent regions[a]	General distribution[b]	Range type[c]	Larval habitat/host plants[d]
C. elaeagni Richter, 1945	4, 8	Tu, Uz, Tj, Ir, Af	eIT	Elaeagnus*, Amygdalus, Prunus, Persica, Juglans, Tamarix
C. judinae Stepanov, 1954	9	Uz, Tj	Turk	Pistacia*
C. turanus (Obenberger, 1914)	4–5	nIr	Khoras	Rosa*
Trib. Melanophilini				
Trachypteris Kirby, 1837				
T. (Trachypteris)				
T. picta picta (Pallas, 1782)	4–5, 7–8	Tu, Uz, Tj, Ky, Ka, Ir, Af, Ch, Mg	ITG	Salix*, Populus
T. (Oxypteris)				
T. cuspidata (Klug, 1829)	1, 4, 7–9	Uz, Tj, Ka, Zk, Ir, sEu, Mt	HS	Elaeagnus*, Pistacia, Ficus, Juniperus, Pinus, Populus
Trib. Buprestini				
Buprestis Linnaeus, 1758				
B. (Orthocheira)				
B. salomonii Thomson, 1878	7–8	Tu, Uz, Tj, Zk, Tr, Ir, nCh, Sy	HS	Populus*
Trib. Dicercini				
Poecilonota Eschscholtz, 1829				
P. (Poecilonota)				
P. nadezhdae Semenov, 1909	4	Zk, nIr	SWA	Ulmus
P. dives (Guillebeau, 1889)	4	Eu	WP	Salix

Table 1. Continued

Species	Distribution in Kopetdagh and adjacent regions[a]	General distribution[b]	Range type[c]	Larval habitat/host plants[d]
Dicerca Eschscholtz, 1829				
D. (Dicerca)				
D. aenea validiuscula Semenov, 1909	4–5	Uz, Ky, Ka, nIr, Zk	IT	*Juglans*[*]
D. (Hemidicerca)				
D. fritillum (Menetriés, 1832)	4	Zk, nIr	SWA	Unknown
Subfam. Chrysobothrinae Trib. Chrysobothrini				
Chrysobothris Eschscholtz, 1829				
C. (Chrysobothris)				
C. affinis nevskyi Richter, 1944	1, 4–5, 7–8	Tu, Uz, Tj, nIr	eIT	polyphagous; *Juglans*[*]
C. deserticola Semenov et Richter, 1934	1, 4, 9	Tu, Uz, Tj, Ka	Turan	*Ammodendron*[*], *Pistacia*[*]
C. (Sphaerobothris)				
C. globicollis Reitter, 1895	1, 7, 9	Tu, Uz, Tj, Af	Turan	*Ephedra*[*]
C. (Abothris)				
C. nana Fairmaire, 1892	3, 7, 8	Tu, Uz, Tj, Ka	Turan	*Populus*, *Juglans*, *Cercis*
C. jakovlevi Semenov, 1891	9	Tu	Turan	*Ammodendron*[*], *Pistacia*

Table 1. Continued

Species	Distribution in Kopetdagh and adjacent regions[a]	General distribution[b]	Range type[c]	Larval habitat/host plants[d]
Subfam. Agrilinae Trib. Coroebini				
Clema Semenov, 1900				
C. freudei volkovitschi Alexeev, in litt.	4–5, 9		Endem	*Stipa*[*]
C. deserti Semenov, 1900	1, 3, 7	Tu, Uz, Ka, Ir	Turan	*Aristida*[*]
Coroebus Laporte et Gory, 1839				
C. rubi (Linnaeus, 1767)	5	Ep, Zk, Tr, Ir, Eu, Mt	WP	*Rubus, Rosa*
Meliboeus Deyrolle, 1864				
M. (*Meliboeoides*)				
M. amethystinus (Olivier, 1790)	4–5, 9	Tu, Uz, Tj, Ky, Ka, seEp, Zk, Tr, Ir, sEu, Mt	HS	*Cousinia*[*], *Echinops*
M. robustus (Kuster, 1852)	7	seEp, Zk, Tr, Ir, Iq	HS	*Echinops*
M. cyaneus (Ballion, 1870)	4–5, 7, 9	Tu, Uz, Tj, Ky, Ka, seEp, Zk	IT	*Cirsium*[*], *Cousinia, Prangos*
M. (*Meliboeus*)				
M. staneki Obenberger, 1935	4–5	Zk, Tr, Af	SWA	Unknown
M. reitteri Semenov, 1889	1–5, 7, 9	Tu, Uz, Tj, Ky, lKa, seEp, Zk, Tr, Ir, Af	IT	*Artemisia*[*]
M. caucasicus Abeille de Perrin, 1896	5	seEp, Zk	SWA	Unknown

Table 1. Continued

Species	Distribution in Kopetdagh and adjacent regions[a]	General distribution[b]	Range type[c]	Larval habitat/host plants[d]
Trib. Agrilini				
Agrilus Curtis, 1825				
A. derasofasciatus Lacordaire et Boisduval, 1835	4	seEp, Zk, Tr, Ir, Eu, Mt	WP	*Vitis*
A. pistaciophagus Alexeev et Kulinitsch, 1963	9	Uz, Tj, Ky, Ir	Turk	*Pistacia*[*]
A. sericans Kiesenwetter, 1857	4	Ka, sEp, Zk, Eu, eMt	WP	*Artemisia*
A. albogularis Gory, 1841	1, 4–5	Uz, Tj, Ky, Ka, Ep, Zk, Tr, Ir, Eu, Mt, Iq	WP	*Artemisia*
A. pseudoalbogularis Alexeev, in litt.	4–5	Uz, Tj, Ky, Ka	Turk	*Eurotia*
A. ganglbaueri Semenov, 1891	4, 8–9	Tu, Uz, Tj, Ka, Af, Mg	ITG	*Populus*
A. viridis (Linnaeus, 1758)	5	Ka, Ep, Si, Fe, Mg, Eu, Mt	PP	polyphagous; *Acer*[*]
A. cuprescens Menetriés, 1832	4	Uz, Tj, Ky, Ka, Tr, Ir, Mg, Ep, Zk, Si, Fe	PP	*Rosa*[*], *Rubus*[*]
A. lineola schamyl Obenberger, 1922	4–5	Zk, nIr	SWA	*Salix*[*]
A. vaginalis Abeille de Perrin, 1897	1, 3–4	Tu, Uz, Tj, Ka, Zk, Tr	HS	*Astragalus*[*], *Colutea*
A. validiusculus Semenov, 1891	1, 3–5, 7, 9	Tu, Ka, Ir	Turan	*Salsola*[*], *Haloxylon*, *Halostachys*
A. erojlandusicus Alexeev, in litt.	9	Tu, Uz	Turan	*Salsola*[*]
A. araxenus Khnzorian, 1960	4–5	Zk	SWA	Unknown

Table 1. Continued

Species	Distribution in Kopetdagh and adjacent regions[a]	General distribution[b]	Range type[c]	Larval habitat/host plants[d]
Subfam. Cylindromorphinae Trib. Cylindromorphi				
Cylindromorphus Kiesenwetter, 1857				
C. pubescens Semenov, 1895	1	Tu, Uz, Ka	Turan	*Carex*[*]
C. kopetdagicus Alexeev, in litt.	4–5		Endem	*Elytrigia*[*]
Paracylindromorphus Thery, 1930				
P. lebedevi (Obenberger, 1928)	3–4, 7–9	Tu, Uz, Tj, Ka	Turan	*Aeluropus*[*]
P. transversicollis (Reitter, 1913)	7, 9	Tu, Uz, Tj, Ka, seEp, Zk, Ir, Ch, Mg	ITG	*Phragmites*[*]
P. semenovi Thery, 1937	3	Tj, Ka, seEp	Turan	*Phragmites*[*]
P. subuliformis (Mannerheim, 1837)	9	Tj, Ka, Ep, Si, Fe, Tr, Af, Ch, Mg, Ko	PP	*Agropyron, Glyceria*
Subfam. Trachyinae Trib. Trachyini				
Trachys Fabricius, 1801				
T. phlyctaenoides Kolenati, 1846	4–5	Ep, Zk, Tr, Ir, Gr, Bu	HS	*Phlomis*[*]
Habroloma Thomson, 1846				
H. aurea Thomson, 1864	5, 7	Tj, Ky, Ka, Zk, Ir	IT	Unknown

Table 1. Continued

Species	Distribution in Kopetdagh and adjacent regions[a]	General distribution[b]	Range type[c]	Larval habitat/host plants[d]
Trib. Aphanisticini				
Aphanisticus Latreille, 1829				
A. emarginatus (Olivier, 1790)	4–5	Uz, Ka, Ep, Zk, Tr, Ir, Eu, Mt	WP	*Juncus*
A. pygmaeus Lucas, 1849	1	Uz, Ka, Ep, Eu, Mt	WP	*Juncus*

[a] The following areas of Kopetdagh and the adjacent regions are distinguished: 1 – Southwest Turkmenistan, 2 – Bolshoi Balkhan, 3 – Maly Balkhan, 4 – West Kopetdagh, 5 – Central Kopetdagh, 6 – East Kopetdagh, 7 – Submontane plain of Kopetdagh, 8 – Area between the Tedzhen and Murghab Rivers, 9 – Badghyz

[b] Distribution in the Palearctic is given. The characters "n", "s", "w", "e" stand for northern, southern, western, and eastern; the following abbreviations are accepted. Middle Asia: Ka – Kazakhstan, Ky – Kyrghyzstan, Tj – Tajikistan, Tu – Turkmenistan (outside of the studied areas), Uz – Uzbekistan. Other regions of the former USSR: Ep – European Part, Fe – Far East, Si – Siberia, Zk – Transcaucasia. Adjacent countries: Af – Afghanistan, Ch – China, Ir – Iran, Mg – Mongolia, Tr – Turkey. Other countries and regions: Bu – Bulgaria, Eg – Egypt, Eu – Europe, Gr – Greece, In – India, Iq – Iraq, Ko – Korea, Mt – Mediterranean, Pa – Pakistan, Sa – Saudi Arabia, Sy – Syria

[c] Range types (after Yemelyanov 1974): PP – Pan-Palearctic, WP – West Palearctic, HS – Hesperian- Sethian, SaIT – Saharo-Irano-Turanian, ITG – Irano-Turanian-Gobian, IT – broad Irano-Turanian, Turan – Turanian (sT + nT), Turk – Turkestanian, SWA – Southwest Asian, Khor – Khorassanian, Iran – Iranian, Afghan – Afghanian, Endem – endemic

[d] All data on larval host plants (not only from Turkmenistan) are given; host plants in Turkmenistan are marked with (*). Species with wide selection of host plants are designated as "polyphagous"; in this case, host plants are listed only for Turkmenistan

Anthaxia discicollis, A. holoptera, A. hemichrysis, A. muliebris, Cratomerus hungaricus sitta, Sphenoptera glabrata, S. furva, Poecilonota nadezhdae, Dicerca fritillum, Aurigena lugubris, Agrilus lineola shamyl, A. araxenus, Meliboeus robustus, and *M. staneki.* Some species are reported only from West Kopetdagh. Kopetdagh (and, to some extent, Bolshoi Balkhan, Maly Balkhan, and Badghyz) are known as the northern boundary for such Iranian species as *Julodis laevicostata, Acmaeoderella zarudniana, Sphenoptera rangnowi, S. cyanea,* and *S. kambyses.*

Fifty-nine buprestid species (46.8%) are not found in Middle Asia beyond Kopetdagh. Among them prevail Southwest Asian (21 spp.) Khorassan (17 spp.), some widely distributed West Palaearctic (5 spp.), and Hesperian-Sethian species (7 spp.). Genera and subgenera *Acmaeodera (Acmaeotethya), Aurigena, Sphenoptera (Rhaphidochila), Poecilonota, Dicerca (Hemidicerca),* and *Coroebus* are found in Kopetdagh but are absent not only from the surrounding plains but from all other mountains of Central Asia. Such genera as *Julodis, Julodella, Acmaeodera, Capnodis, Anthaxia, Meliboeus,* and *Agrilus* in Kopetdagh are characterized by marked diversity.

Sixty-seven species from Kopetdagh are also found in other parts of Middle Asia: among these prevail species with Turanian (20 spp.), widely Irano-Turanian (13 spp.), East Iran-Turanian (8 spp.), and Turkestanian (7 spp.) types of ranges. Sixty-four species found in Kopetdagh also inhabit the Transcaucasia, with a prevalence of Southwest Asian (21 spp.), widely Irano-Turanian (13 spp.), Hesperian-Sethian (13 spp.), and West Palaearctic (10 spp.) elements.

There are no endemics of superspecific rank in the buprestid fauna of the Kopetdagh and surrounding areas. Endemic elements belong to the Khorassan endemics (which are also distributed in northern Iran). These include twenty-four species and subspecies (12.4%) and are indicative of the high level of isolation of the regional buprestid fauna. As mentioned above, there are 17 endemics (13.5% of the fauna) of different levels in the Kopetdagh. For comparison, endemics constitute 18% of the higher plants (Kamelin 1970) and 38.5% of the tenebrionid beetles (including subendemics) (Medvedev and Nepesova 1990). The following buprestid species belong to Khorassan endemics: *Julodella brevilata, Acmaeodera pallidepicta, Acmaeoderella flavofasciata chorasanica, Capnodis jacobsoni, Sphenoptera violacea, S. subtilis, Anthaxia spinosa,* and *Cratomerus turanus.* The endemics and subendemics of the Kopetdagh are *Julodella shestoperovi, Acmaeoderella caspica caspica* (which is also found in Maly Balkhan), *A. coelestina, A. strandi,* and *Cylindromorphus kopetdagicus.* Endemics of both Kopetdagh and Badghyz include *Acmaeodera ghilarovi, Acmaeoderella alepidota* (which is also found between the Tedzhen and Murghab Rivers), *Acmaeoderella tragacanthae kopetdagica, Sphenoptera koenigi,* and *Clema freudei volkovitshi.* Endemics of Badghyz are *Acmaeoderella badhvsica, Sphenoptera eroilandusica, Anthaxia badhysica,* and *A. kreuzbergi.* The only known endemic of Bolshoi Balkhan is *Cratomerus medvedevorum* (the type locality for poorly known *Sphenoptera turcmenica* was not correctly indicated).

Ecological Groups of the Buprestids

Many species of buprestids are polyphagous with the wide range of larval host plants, a characteristic which makes difficult their placement in definite ecological groups. Some species (e.g., *Acmaeoderella dubia*, *Capnodis tenebricosa*, *Anthaxia lucidiceps*, and *Trachypteris cuspidata*) can be found within various plant communities, from submontane plain through upper mountain belt. On the other hand, larval host plants of many species are unknown although for some species habitats can be surmised by the presence of adults. Therefore, in considering the composition of ecological groups, we refer only to the most typical and best known species.

1. Buprestids associated with Sand Vegetation (the Psammophilous Group)
This group includes 24 species, among them 20 psammophiles widely distributed over Turanian deserts. The most typical species are *Julodis variolaris freygessneri*, *Acmaeoderella caspica suturifera*, *A. repetekensis*, *A. nivetecta*, *A. canescens*, *A. insueta*, *A. solskyi*, *Capnodis excisa*, *Sphenoptera serripes*, *S. caspica*, *S. potanini*, *S. hauseri*, *S. punctatissima*, *S. schneideri*, *Chrysobothris deserticola*, *C. globicollis*, *C. jakovlevi*, *Clema deserti*, and *Cylindromorphus pubescens*. The representatives of this group occur primarily in the sand deserts of Southwest Turkmenistan, in the area between the Tedzhen and Murghab Rivers, and in Badghyz. They are also widely represented in the submontane plain of Kopetdagh and together with psammophilous plants can penetrate to the foothills and even to higher belts.

2. Buprestids associated with salt desert and "painted rock" communities (the halophilous group)
This group constitutes 26 species, among them eleven Turanian and five widely distributed Irano-Turanian elements. There are Irano-Turanian species of western origin: a Southwest Asian *Sphenoptera latesulcata* (Badghyz) and *S. vestita*. The typical species are *Acmaeoderella plavilscikovi*, *A. ballioni*, *A. personata*, *Sphenoptera lateralis*, *S. vestita*, *S. beckeri*, *S. scovitzi*, *S. pseudoignita*, *S. viridiaurea*, *S. orichalcea*, *S. amplicollis*, *S. tomentosa*, *S. tschitscherini*, and *Agrilus validiusculus*. These species are common in the deserts of Southwest Turkmenistan, in Maly Balkhan, in the submontane plain of Kopetdagh, in the area between the Tedzhen and Murghab rivers, and in Badghyz. Many species occur in the foothills of West and East Kopetdagh and are numerous in "painted rocks" of the Sumbar Valley.

3. Buprestids associated with sagebrush-ephemeroid communities (the sagebrush group)
The 12 species of this group are primarily associated with sagebrush (*Artemisia* spp.). The various biogeographic elements range from the West Palaearctic (two spp.) to endemic (one sp.); widely Irano-Turanian and Turanian (3 spp. each) elements prevail. The group includes numerous species of *Sphenoptera*

(*Chilostetha*). Typical species are *Acmaeoderella cinerea, Sphenoptera canescens, S. rauda, S. puberula, S. erojlandusica, S. artemisiae, Meliboeus reitteri, Agrilus sericans*, and *A. albogularis*. Representatives of the sagebrush group are present almost everywhere from the plains to the lower mountain belts in Kopetdagh and Bolshoi Balkhan.

4. Buprestids associated with tugai vegetation (the tugai group)
Riparian desert forests, "tugais," are widely distributed along Middle Asian rivers, reaching lower mountain zones and turning there into mountain riparian forests. The tugai group includes 25 species trophically associated with trees and shrubs (*Populus, Salix, Elaeagnus*, and *Tamarix*) as well as with the herbaceous (*Alhagi, Glycyrrhiza, Phragmites, Aeluropus,* and *Juncus*) vegetation of tugais. Widely distributed Irano-Turanian-Gobian, as well as Turanian (6 spp. each) and widely Irano-Turanian (4 spp.) elements predominate in this group. Other elements include trans-Palaearctic (1 sp.), West Palaearctic (2 spp.), Hesperian-Sethian (2 spp.), East Iran-Turanian (3 spp.), and endemic (1 species, *Cratomerus medvedevorum*, from Bolshoi Balkhan). The typical species are *Acmaeodera chotanica, Capnodis miliaris, Anthaxia nanissima, Cratomerus fariniger, Trachypteris picta picta, Buprestis salomonii, Chrysobothris nana*, and *Agrilus ganglbaueri* (which live on *Populus, Salix*, and sometimes on other trees and shrubs); *Sphenoptera mesopotamica, S. viridula, S. balassagloi,* and *S. semenovi* (which live on *Tamarix*); *S. exarata* (which lives on *Glycyrrhiza*); *Xantheremia* spp. (which live primarily on *Alhagi*); *Paracylindromorphus* spp. (which live on *Phragmites* and *Aeluropus*); and *Aphanisticus* spp. (which live on *Juncus*). The majority of these species are widely distributed from the plains to the upper mountain belt. Species especially associated with the herbaceous vegetation often occur along rivers and streams, around springs and in moist areas. The high proportion of widely distributed Irano-Turanian-Gobian species in the tugai buprestid fauna may indicate that dispersal of many mesophilic species over desert areas may have been faciltated by the presence of tugai forests. Due to their polyphagy, such species as *Acmaeodera chotanica, Cratomerus fariniger, Dicerca aenea,* and *Chrysobothris nana* may live in various forest communities, including those of upper mountain belts. Special consideration should be given to the fact that the tugai group includes many taxa of tropical and subtropical origin, such as *Acmaeodera (Cobosiella)* (distributed mainly in Southeast Asia), *Xantheremia* (North Africa and Southwest Asia), *Sphenoptera (Hoplistura)* and *Aphanisticus* (Africa and Southeast Asia). Such species as *Anthaxia nanissima, Buprestis salomonii, Chrysobothris nana*, and *Agrilus ganglbaueri* also belong to tropical and subtropical taxa.

5. Buprestids associated with steppes (the steppe group)
This group includes 12 species, most of which are asssociated with different Asteraceae (*Acmaeoderella caspica, A. turanica, A. badhysica, A. iranica, Meliboeus amethystinus,* and *M. cyaneus*), Apiaceae (*Acmaeoderella villosula*),

or Lamiaceae (*Trachys sphylctaenoides*). Steppe-group species often occur within various communities. Only a still undescribed endemic *Clema freudei volkovitshi* and *Cylindromorphus kopetdagicus* associated with wild grasses can be considered typical representatives of this group. The core of the steppe group is formed by Hesperian-Sethian (4 spp.) and endemic (4 spp.) elements; the Turkestanian (2 spp.), widely Irano-Turanian (1 sp.), and East Iran-Turanian (1 sp.) elements are also present. Six species do not occur in Middle Asia beyond Kopetdagh and Badghyz. The species of this group inhabit steppe formations from the lower foothills to the upper mountain belts.

6. Buprestids associated with the formation of xerophilous trees and shrubs (shiblyak) (the shiblyak group)

This is the most numerous buprestid group, including 36 species, 22 of which are not found in Middle Asia beyond the studied region. Turkestanian (12 spp.), Southwest Asian (6 spp.), and endemic (6 spp.) elements predominate; Hesperian-Sethian (3 spp.), Khorassan (3 spp.), Iranian (2 spp.), West Palaearctic (1 sp.), widely Irano-Turanian (1 sp.), East Iran-Turanian (1 sp.), and Afghanian (1 sp.) elements are also present. Representatives of the shiblyak group are associated with both trees and shrubs (*Acmaeodera pilosellae persica*, *A. pallidepicta*, *Capnodis sexmaculata*, *Sphenoptera kaznakovi*, *S. kambyses*, and *Cratomerus fedschenkoi*) as well as with herbaceous vegetation [*Sphenoptera koenigi* and a number of species of *Acmaeoderella (Euacmaeoderella)*].

The buprestid group associated with *Pistacia vera* includes seven species (*Acmaeodera babatauensis*, *Acmaeoderella glasunovi*, *Capnodis parustriata*, *Cratomerus judinae*, *Agrilus pistaciophagus*, *Acmaeodera instabilis*, and *Anthaxia kreuzbergi*, which is endemic to Badghyz). Badghyz is the western boundary of distribution for the majority of the Turkestanian species; only *Capnodis parumstriata* has been recorded from Ashkhabad, but this reference needs to be confirmed. Although pistachio communities were almost completely destroyed in Kopetdagh during the historical period, at least polyphagous *Acmaeodera babatauensis* and *Acmaeoderella glasunovi* have a wide range of larval host plants (*Ficus carica*, *Amygdalus* spp., and *Astragalus* sect. *Tragacantha*), many of which are found in Kopetdagh almost everywhere; this seems to prove that their geographic boundary has a natural character. Replacement of insect species composition in Badghyz and Kopetdagh is also confirmed by the example of the tragacanthoid formation group. Another typical representative of the shiblyak group in the studied region is the Iranian species *Acmaeoderella zarudniana*, which develops on *Zygophyllum*. A group of species associated with the gigantic *Ferula* and *Dorema* plants (Apiaceae) can also be distinguished, including endemics *Acmaeoderella alepidota* and *A. strandi*. Adult *A. alepidota*, whose larvae develop in the stems of *Ferula oopoda*, are usually found inside the large egg-like leaf-sinuses of these plants, where often up to twenty beetles accumulate. Representatives of this group are primarily distributed in the lower and upper foothills (1,000–1,200 m) and in the lower mountain belt (up to 1,500 m) in Kopetdagh, Badghyz, and (with an impoverished composition) in Bolshoi

Balkhan and Maly Balkhan. The shiblyak fauna in Kopetdagh differs significantly from the analogous faunas of other parts of Middle Asia due to the presence of the numerous species of western origin as well as Khorassanian and local endemics. The Iranian species *Sphenoptera kambyses* has been shown to be a serious pest of cultured fruit trees.

7. Buprestids associated with xeromesophilous trees and shrubs (the mesophilous group)
The group includes 23 species, of which 16 species are not found in Middle Asia eastward of the studied area. This group includes Southwest Asian (7 spp.) and West Palaearctic (6 spp.) elements as well as Irano-Turanian (3 spp.), Hesperian-Sethian (2 spp.), Khorassan (2 spp.), and East Iran-Turanian (1 sp.) elements. The majority of species occur in the deciduous forests of the southern European portion of the former USSR, the Caucasus, Southwest Asia, and the Eastern Mediterranean; some species are distributed in Middle Asian mountain forests. Typical species include *Acmaeoderella flavofasciata chorasanica*, *Capnodis tenebrionis*, *Aurigena lugubris*, *Anthaxia cichorii*, *A. bicolor*, *A. muliebris*, *Cratomerus intermedius*, *C. hungaricus sitta*, *Poecilonota nadezhdae*, *Dicerca aenea validiuscula*, *Chrysobothris affinis nevskyi*, *Agrilus derasofasciatus*, *A. viridis*, *A. lineola shamyl*, and *Habroloma aurea*. Almost half of these species are known only from West Kopetdagh but may also be possibly found in Central Kopetdagh. Such species as *Capnodis tenebrionis*, *Aurigena lugubris*, and *Chrysobothris affinis* are serious pests of cultivated crops.

8. Buprestids associated with Juniperus spp. (the juniper group)
This group includes only three species, all of which develop in juniper trees. They are Hesperian-Sethian *Anthaxia discicollis* and *A. passerinii*, and Southwest Asian *A. hemichrysis*. These species are rather common in the Transcaucasia, but within Middle Asia they are only known from Kopetdag. In Bolshoi Balkhan they have not been yet found. Other mountain species of *Anthaxia* (subgenus *Melanthaxia*) are found farther east, including the Kugitangtau Mountains.

9. Buprestids associated with tragacanthoid plants (the tragacanthoid group)
This group includes eight species, of which only *Sphenoptera coerulea* develops in *Acanthophyllum*, while other species are associated with *Astragalus* (section *Tragacantha*). None of these species are found in Middle Asia eastward beyond the studied region. Among them are Southwest Asian (3 spp.), endemic (2 spp.), West Iran-Turanian (1 sp.), Khorassan (1 sp.), and Iranian (1 sp.) elements. *Anthaxia badghyzica* is known only from Badghyz; a very closely related *A. spinosa* is found in Kopetdagh and North Iran while other related species inhabit the Transcaucasia. An endemic subspecies *Acmaeoderella tragacanthae kopetdagica* is described from Kopetdagh and Badghyz; the nominal subspecies is found in the Transcaucasia. *Sphenoptera rangnowi* is known from Kopetdagh, Iran, and Afghanistan, whereas *Acmaeodera chalcithorax*, *Sphenoptera*

glabrata, S. furva, and *S. coerulea* inhabit tragacanths in the Transcaucasia, Southwest Asia, Iran, and Afghanistan. Therefore, the tragacanthoid group is faunistically much closer to similar ecological groups from the Transcaucasia and Southwest Asia than to those of the mountainous portion of Middle Asia eastward of the studied region. In Badghyz, species of *Sphenoptera* and *Acmaeodera chalcithorax*, which are very common in Kopetdagh, are not found; *Anthaxia spinosa* is substituted by a closely related *A. badghyzica*, and a Turkestanian polyphagous species *Acmaeoderella glasunovi* is present. The last species is not found in Kopetdagh but is widely distributed in the Turkestanian Province, where the tragacanthoid group is formed by absolutely different species of buprestids than those mentioned above.

In summary, the peculiar character of the Kopetdagh buprestid fauna is somewhat less strongly expressed in the composition of steppe and shiblyak groups but is strongly expressed in mesophilic, juniper, and tragacanthoid groups, which have heavy impact from West Palaearctic, Hesperian-Sethian, Southwest Asian, Khorassan, and endemic elements. This faunistic specificity increases with altitude from lower to upper mountain belts (800 to 2,200 m).

Regional Buprestid Faunas

1. Southwest Turkmenistan

Forty buprestid species belonging to 15 genera are known from this region. Among these, *Sphenoptera* (17 spp., 42.5%) and *Acmaeoderella* (6 spp., 15.0%) predominate; other genera each include one species (or, rarely, two or three). Turanian (20 spp., 50%) and widely Irano-Turanian (7 spp., 17.5%) elements are most numerous in the fauna composition; Hesperian-Sethian, Irano-Turanian-Gobian, East Iran-Turanian (3 spp. each), Saharo-Irano-Turanian (2 spp.), West Palaearctic, and Southwest Asian (1 sp. each) elements are also present. There are no endemic taxa. Only from this region (Gasan-Kuli District) within Turkmenistan has *Aphanisticus pygmaeus* been found. The representatives of halophilous (16 spp., 40%) and psammophilous (13 spp., 32.5%) groups are most abundant. The sagebrush and tugai groups are represented by three species each.

2. Bolshoi Balkhan

The buprestid fauna of these isolated mountains is poorly studied; only 19 species from 10 genera are known from here, and none of them is markedly predominant. *Acmaeoderella* (6 spp.) and *Sphenoptera* (4 spp.) are the most abundant taxa; the majority of other genera are represented by a single species. Widely Irano-Turanian, East Iran-Turanian (4 spp. each), and Turanian (3 spp.) elements prevail; there are also Irano-Turanian-Gobian (2 spp.), West Palaearctic, Hesperian-Sethian, Saharo-Irano-Turanian, Southwest Asian,

Iranian, and endemic (1 sp. each) elements. Endemic to Bolshoi Balkhan (the northern slope) is *Cratomerus medvedevorum*, which develops in *Halimodendron halodendri*; it possibly may be found in Kopetdagh as well. Representatives of halophilous (4 spp.), sagebrush (4 spp.), psammophilous (3 spp.), tugai (3 spp.), xerophilous (1sp.), and tragacanthoid (1 sp.) groups are present. Of special interest are the northernmost records in Middle Asia of an Iranian *Acmaeoderella zarudniana* (xerophilous group) and Southwest Asian *Sphenoptera glabrata* (tragacanthoid group), common to Kopetdagh.

3. Maly Balkhan

The buprestid fauna here has not been studied in detail. Twenty-nine species from 11 genera are known; *Acmaeoderella* (11 spp., 37.9%) and *Sphenoptera* (6 spp., 20.7%) are the most abundant taxa. The Turanian (13 spp., 44.8%), widely Irano-Turanian, and East Iran-Turanian (4 spp. each, 13.8%) elements markedly predominate; there are also Hesperian-Sethian, Saharo-Irano-Turanian (2 spp. each), West Palaearctic, Southwest Asian, Iranian, and endemic (subendemic) elements (1 sp. each). There are no endemic species. Dominant ecological groups are halophilous (10 spp., 34.5%) and psammophilous (8 spp., 27.6%); other groups include the tugai (3 spp.), sagebrush and xerophilous formation (2 spp. each), and steppe (1 sp.). A rare species within Turkmenistan, *Paracylindromorphus semenovi*, is recorded only from Maly Balkhan. Findings of Iranian *Acmaeoderella zarudniana*, West Palaearctic *A. gibbulosa*, and Kopetdagh endemic *A. caspica caspica* demonstrate faunistic connections with Kopetdagh.

4. West Kopetdagh

This is one of the best studied areas of Kopetdagh, with the buprestid fauna including 106 species from 21 genera. The genera *Sphenoptera* (30 spp., 28.3%) and *Acmaeoderella* (17 spp., 16%) are the most abundant ones; *Agrilus* (10 spp., 9.4%), *Anthaxia* (9 spp., 8.5%), and *Cratomerus* (7 spp., 6.6%) are also widely represented. Species of *Poecilonota* and *Dicerca* (*Hemidicerca*) within the studied region are only known from this area. This fauna includes the Southwest Asian (18 spp., 17.0%), Turanian (17 spp., 16.0%), Hesperian-Sethian and widely Irano-Turanian (12 spp. each, 11.3%), and West Palaearctic (10 spp., 9.4%) elements. Of considerable importance are the Khorassan (4 spp., 3.8%) and endemic (9 spp., 8.5%) elements. East Iran-Turanian (7 spp.), Irano-Turanian-Gobian, Turkestanian, Iranian (each 4 spp.), Saharo-Irano-Turanian (2 spp.), trans-Palaearctic and West Iran-Turanian (1 sp. each) elements are also present. *Acmaeoderella coelestina* is the only species which might be regarded as an endemic of West Kopetdagh. Representatives of xerophilous (23 spp., 21.7%), mesophilous (20 spp., 18.9%), halophilous (15 spp., 14.2%), and tugai (13 spp., 12.3%) groups dominate the buprestid fauna of West Kopetdagh; species of tragacanthoid and sagebrush (7 spp. each), steppe (6 spp.), psammophilous (4 spp.), and juniper (3 spp.) groups are also present. Within the xerophilous formation group, Southwest Asian (6 spp.) and endemic (5 spp.) elements predominate; the Turkestanian elements are represented only by three species. Within the studied region, the following 17 species are found only in West Kopetdagh: a trans-Palaearctic *Agrilus cuprescens*, the West Palaearctic *Acmaeoderella mimonti*, *Anthaxia cichorii*, *Poecilonota dives*, *Agrilus derasofasciatus*, and *A. sericans*; the Hesperian-Sethian *Acmaeoderella obscura*, *Anthaxia bicolor*, and *Cratomerus sponsa*; a Turanian *Sphenoptera korshinskii*; Southwest Asian *Sphenoptera apta*, *Anthaxia muliebris*, *Poecilonota nadezhdae*, and *Dicerca fritillum*; the Khorassan *Acmaeoderella flavofasciata chorasanica* and *Sphenoptera subtilis*; and an endemic *Acmaeoderlla coelestina*.

5. Central Kopetdagh

The fauna of this area includes 86 species and subspecies from 23 genera; *Sphenoptera* (21 spp., 24.4%) and *Acmaeoderella* (16 spp., 18.6%) are the most abundant taxa; *Capnodis*, *Agrilus*, *Anthaxia* (6 spp. each, 7.0%), and *Meliboeus* (5 spp.) should be also mentioned. The only representative of *Coroebus* known from Middle Asia has been found in Central Kopetdagh. Southwest Asian (16 spp., 18.6%), widely Irano-Turanian (11 spp., 12.8%), and Turanian (10 spp., 11.6%) elements dominate the fauna; also present are Hesperian-Sethian (8 spp.), East Iran-Turanian and endemic (7 spp. each), West Palaearctic and Turkestanian (6 spp. each), Khorassan (5 spp.), Irano-Turanian-Gobian and Iranian (3 spp. each), West Iran-Turanian (2 spp.), trans-Palaearctic and Saharo-Irano-Turanian (1 sp. each) elements. *Julodella shestoperovi* is the only known endemic of Central Kopetdagh. The representatives of xerophilous (21 spp., 24.4%), halophilous (12 spp., 13.9%), and mesophilous (11 spp., 12.8%) groups predominate; steppe (8 spp.), tugai (7 spp.), sagebrush and tragacanthoid (6 spp. each), and psammophilous and juniper (3 spp. each) groups are present. The xerophilous group is formed primarily by the Southwest Asian (6 spp.) and Turkestanian (4 spp.) elements. Six species, including a trans-Palaearctic *Agrilus viridis*, West Palaearctic *Coroebus rubi*, Hesperian-Sethian *Acmaeoderella subcyanea*, Turanian *Acmaeoderella candens*, Southwest Asian *Meliboeus caucasicus*, and endemic *Julodella shestoperovi*, within the studied region are found only in Central Kopetdagh; of these, only *Acmaeoderella candens* is found in other regions of Middle Asia.

6. East Kopetdagh

This area of Kopetdagh is very poorly studied, with only nine species from four genera recorded. Of special interest are the findings of the Khorassan species *Capnodis jacobsoni* and a West Palaearctic *Acmaeoderella gibbulosa*. Based on the orographic features of East Kopetdagh, we suggest that the Turanian elements from halophilous and xerophilous formation groups will be found to predominate in the local buprestid fauna. A rather large natural grove of

Pistacia vera exists next to the town of Kaakhka, but we have not succeeded in our search of buprestids associated with this tree; thus, we tentatively consider Badghyz as the western boundary of pistachio-associated species.

7. The Submontane Plain of the Kopetdagh

The buprestid fauna of this area includes 84 species from 20 genera; however, many specimens labeled from such settlements as Ashkhabad or Geok-Tepe probably have been collected in the Kopetdagh foothills; this makes the faunistic analysis of this area difficult. The most abundant taxa are *Sphenoptera* (36 spp., 42.9%), *Acmaeoderella* (10 spp., 11.9%), and *Capnodis* (7 spp., 8.3%). Ashkhabad is the only site within the region studied where *Strigopteroides aegyptiacus* developing in dead dry wood has been recorded; according to our observations in Bokhara (Uzbekistan), this species is a timber pest. Turanian (24 spp., 28.6%) and widely Irano-Turanian (11 spp., 13.1%) elements are predominant here; the East Irano-Turanian (7 spp.), Southwest Asian and Turkestanian (6 spp. each), Hesperian-Sethian, Irano-Turanian-Gobian, and Iranian (5 spp. each), West Palaearctic, Saharo-Irano-Turanian, Khorassan, and endemic (3 spp. each), and West Irano-Turanian (1 sp.) elements are also present. A poorly studied *Sphenoptera turcmenica* is an endemic of this area, and two other species (*Acmaeoderella caspica caspica* and *Sphenoptera koenigi*) are subendemic. Among ecological groups, psammophilous (14 spp., 16.7%), halophilous and tugai (12 spp. each, 14.3%), and xerophilous group (11 spp., 13.1%) predominate; there are also representatives of sagebrush (6 spp.), steppe and mesophilous (5 spp. each) and tragacanthoid (3 spp.) groups. The rather high number of mesophilous buprestid species is due to the large areas occupied by orchards in the Kopetdagh Oasis, where the representatives of this group may develop as larvae in various fruit trees and shrubs.

8. The area between the Tedzhen and Murghab Rivers

From this region, 58 species belonging to 14 genera are known, with the predominance of *Sphenoptera* (27 spp., 46.6%) and *Acmaeoderella* (7 spp., 12.1%). Turanian (20 spp., 34.5%); widely Irano-Turanian, Irano-Turanian-Gobian (8 spp. each, 13.8%); and East Iran-Turanian (7 spp., 12.1%) elements are the most abundant; there are also Hesperian-Sethian (4 spp.); Turkestanian (3 spp.); Saharo-Irano-Turanian, west Iran-Turanian, Southwest Asian (2 spp. each), and West Palearctic and subendemic (1 sp. each) elements. There are no endemics. This area is the only place in Turkmenistan where *Acmaeoderella valentinae* has been collected. Ecological groups are represented by halophilous (19 spp., 32.8%), tugai (17 spp., 29.3%), and psammophilous (10 spp., 17.2%) species; there are also species of xerophilous (6 spp.) and mesophilous (1 sp.) groups. The high role of tugai species is due to the presence of extensive tugais (riparian forests) in the Tedzhen and Murghab Valleys. In general, this regional fauna is characteristic for the South Turanian biogeographical province.

9. The Badghyz Plateau

The buprestid fauna of Badghyz has been recently reviewed (Volkovich and Alexeev 1992); it includes 87 species and subspecies from 17 genera. The dominant taxa are *Sphenoptera* (29 spp., 33.3%) and *Acmaeoderella* (20 spp., 23.0%). Relatively numerous are *Capnodis* (6 spp.), *Agrilus* (5 spp.), and *Acmaeodera* (4 spp.); species of *Julodella, Acmaeodera* (*Acmaeodera*), and *Acmaeodera* (*Acmaeotethya*) are also present. Predominant zoogeographic elements are Turanian (23 spp., 26.4%); Turkestanian (14 spp., 16.1%); widely Irano-Turanian (10 spp., 11.5%); endemic and subendemic (9 spp., 10.3%); Southwest Asian (3 spp.); Saharo-Irano-Turanian, Khorassan, Iranian (2 spp. each); trans-Palearctic, West Palearctic, and Afghanian (1 sp. each).

The endemics of Badghyz are *Sphenoptera eroilandusica, Anthaxia badhyzica,* and *A. kreutzbergi*; Southwest Asian *Sphenoptera latesulcata, S. artemisiae,* and the Afghanian species *Acmaeodera instabilis* are known within Middle Asia only from Badghyz. The Khorassan species *Julodella brevilata* and *Capnodis jacobsoni*; the Iranian *Acmaeoderella zarudniana* and *Sphenoptera kambyses*; the endemic (subendemic) *Acmaeodera ghilarovi, Acmaeoderella alepidota, A. tragacanthae kopetdaghica, Sphenoptera koenigi* and *Clema freudei volkovitshi* are common to Badghyz and Kopetdagh. The representatives of xerophilous (20 spp., 23%), halophilous (15 spp., 17.2%), psammophilous and tugai (12 spp. each, 13.8%) groups are the most abundant; species of steppe (8 spp.), sagebrush (7 spp.), and tragacanthoid (2 spp.) groups are also represented. In the xerophilous group, species associated with *Pistacia vera* (7 spp.) play an important part; these are represented mainly by Turkestanian elements, possibly with western boundary of their range in Badghyz. Volkovitch and Alexeev (1992) described the faunistic differences between the lowland and mountainous parts of Badghyz; these differences are expressed primarily in the role of predominating Turanian and Turkestanian species. Our biogeographic analysis demonstrates that Badghyz, which is affiliated with the South Turanian Province by some authors, rather should be regarded as a transitional territory between the South Turanian, Turkestanian, and Khorassan Provinces.

Acknowledgements

The authors would like to thank the following colleagues for loans of material and their comments: N.A. Alexeev (Klimovsk, Russia); S. Bily (Prague, Czech Republic); M.L. Danilevsky (Moscow, Russia), M.L. Kalashian (Yerevan, Armenia); V. Kuban (Brno, Czech Republic); O. Soyunov (Ashgabat, Turkmenistan); and I.Ye. Zykov (Orekhovo-Zuevo, Russia). We would also like to thank T.N. Platonova (Zoological Institute, Russian Academy of Sciences, St. Petersburg, Russia) for her kind help with the English translation, and V. Fet (Loyola University, New Orleans, Louisiana, USA) for preparing our manuscript for publication.

26. Fauna, Zoogeography, and Ecology of Orthoptera in Turkmenistan

TANGIRIBERDY TOKGAEV

Abstract

Fauna of Orthoptera in Turkmenistan includes 179 species belonging to 93 genera and 12 families. Zoogeographically, it is a combination of local endemics (21%), Turanian (27.5%), Mediterranean (24%), Paleotropical (17%), as well as Iranian and European-Siberian species. Lowland Turkmenistan is occupied by the Turanian zoogeographic province, whereas mountains of Kopetdagh and Kugitang belong to the Turkmeno-Iranian and Ghissaro-Tien-Shan mountain provinces, respectively. A few species of Grylloidea and Acridoidea in Turkmenistan are serious agricultural pests and have potential for mass reproduction (especially the Moroccan locust, Turanian prus, and oasis prus). Although the desert locust (*Schistocerca gregaria*) is not a permanent inhabitant of Turkmenistan, it can invade from Iranian territory. Several species of Tettigonidae and Acrididae are rare endemics of Turkmenistan or are found in threatened habitats and therefore require special protection.

Introduction

Turkmenistan possesses a rich fauna of Orthoptera (crickets, grasshoppers, and locusts) which is comparatively well studied and presents good material for faunogenetic and zoogeographical analysis. Most Orthoptera are primary consumers and thus play an important role in the trophic structure of natural communities. Several species among Grylloidea and Acridoidea can seriously damage crops and pastures and therefore deserve detailed ecological study. There are also rare and endemic species of Orthoptera whose existence in natural habitats may be threatened by human activity.

Table 1. Taxonomic spectrum of the orthoptera in Turkmenistan

Superfamilies and Families	Number of genera	Number of species and subspecies
Tettigonoidea	19	29
Stenopelmatidae	1	1
Tettigonidae	18	28
Grylloidea	16	24
Gryllidae	12	17
Myrmecophilidae	2	3
Oecanthidae	1	1
Gryllotalpidae	1	3
Tridactyloidea	1	1
Tridactylidae	1	1
Tetrigoidea	2	4
Tetrigidae	2	4
Acridoidea	55	4
Pamphagidae	6	13
Pyrgomorphidae	2	2
Acrididae	47	106
Total	93	179

Fauna and its Zoogeographic Analysis

There are 179 species of Orthoptera, belonging to 93 genera and 12 families, recorded from the territory of Turkmenistan. Table 1 gives a taxonomic spectrum of higher-level taxa; the complete list of species is given as Appendix 1.

The zoogeographic analysis of the insect fauna of Middle Asia has been attempted by many researchers. Of special interest to us are the works of Uvarov (1912, 1921), who analyzed the 102 species of Orthoptera known then from Turkmenistan. He separated the following elements of the fauna: widespread non-Palearctic, narrow Palearctic, Mediterranean, Southwest Asian, and Turanian. Later, Mishchenko (1952) performed an analysis of the orthopteran subfamily Catantopinae of the USSR and adjacent countries. Shumakov (1963) reviewed Orthoptera of the adjacent Iran and Afghanistan. More general data on the zoogeography of the Middle Asian insects can be found in the works of Semyonov-Tyan-Shansky (1936) and Kryzhanovsky (1965).

We separate the following seven faunistic elements (groups) of Orthoptera in Turkmenistan:

1. Endemics of Turkmenistan (species whose range lies only within the territory of Turkmenistan). Tettigonoidea: *Phaneroptera bivittata* B.-Bien., *Ammoxenulus Pavlovskii* B.-Bien., *A. desertus* B.-Bien., *Bergiola popovi* B.-Bien., *B. montana* B.-Bien., *Metrioptera crassipes* B.-Bien., and *Squamiana squamiptera* (Uv.); Grylloidea: *Acheta turcomana* Gor., *Cophaphonus riparius* Mistsh.,

Eremogryllodes vlasovi (Mir.), and *E. semenovi* (Mir.); Acridoidea: *Uvarovium desertum* Dirsh., *Bufonacridella sumakovi* Ad., *Diexis varentzovi varentzovi* Zub., *D. v. probus* Mistsh., *D. v. salsolae* Mistsh., *D. v. afffinis* Mistch., *Iranella turcmena* B.-Bien., *Conophyma uvarovi uvarovi* Sem., *C. u. vicinum* Mistsh., *C. zimini* Tarb., *C. bactrianum* Mistsh., *C. turcomanum* Mistsh., *Thrinchus turcmenus* B.-Bien., *Saxetania scutata* Mistsh., *S. cultricollis* (Sauss.), *S. pravdini* Tschern., *Duroniella turcomana* Mistsh., *Mizonocara deserti* Uv., *M. inornata insolita* Mistsh., *Notostaurus albicornis turcmenus* (Uv.), *Eremippus carinatus* Mistsh., *E. foveolatus* Mistsh., *E. onerosus* Mistsh., *Chortippus biguttulus pravdini* Syt., *Sphingonotus turcmenus* B.-Bien., *S. obscuratus transcaspicus* Uv., and *Hyalorrhipis turcmena* Uv.

Kryzhanovsky (1965) pointed out that there are at least three different centers of endemism in the Middle Asian insect fauna: sand deserts, ephemerous foothills, and mountains. Among the orthopteran fauna of Turkmenistan, all these three groups of endemics can be traced. Species adapted to the conditions of sand desert are represented by *Ammoxenulus pavlovskii*, *A. desertus*, *Uvarovium desertorum*, *Bufonacridella sumakovi*, *Diexis varentzovi*, *Thrinchus turcmenus*, *Eremippus carinatus*, and *E. foveolatus* Mistsh. Characteristic orthopteran endemic species connected to the ephemerous foothills are *Iranella turcmena*, *Mizonocara deserti*, *M. inornata insolita*, *S. obscuratus transcaspicus*, and *E. onerosus*. Finally, mountain endemics include *Squamiana squamiptera*, *Metrioptera crassipes*, *Bergiola popovi*, *B. montana*, *Conophyma uvarovi*, *C. zimini*, *C. bactrianum*, *C. turcomanum*, *Saxetania scutata*, *S. cultricollis*, *S. pravdini*, *Chortippus biguttulus pravdini*, and *Sphingonotus turcmenus*.

2. *Turanian* (Uvarov 1912, 1921, 1938; Kryzhanovsky 1965; Pravdin 1980; Sergeev 1991), or *Middle Asian species* (Bei-Bienko 1948, 1949; Mishchenko 1951, 1952; Shumakov 1965). Tettigonoidea: *Magrettia mutica* Br.-W., *Glyphonotus thoracicus semenovi* Mir., *Ceraeocercus fuscipennis* Uv., *Bergiola balchashica* Stshelk., *Decticus verrucivoris gracilis* Uv., *Semenoviana tamerlana* (Sauss.), and *Platycleis fatima* Uv.; Grylloidea: *Grylliscus gussakowskii* Tarb., *Turanogryllus lateralis* (Fieb.), *Gryllopsis bolivari* Uv., *Gryllodinus odicus odicus* (Uv.), *Modicogryllus chivensis* Tarb., *M. frontalis* (Fieb.), *M. pallipalpis* (Tarb.), *Gryllus bimaculatus* De Geer, *Melanogryllus desertus* (Pall.), and *Oecanthus turanicus* (Uv.); Acridoidea: *Atrichotmethis semenovi* (Zub.), *Thrinchus tuberculosus* Tarb., *Th. desertus* B.-Bien., *Asiotmethis muricatus australis* (Tarb.), *Melanotmethis fuscipennis fuscipennis* (Redt.), *Egnatioides desertus* Uv., *Egnatius apicalis* Stål., *Strumiger desertorum desertorum* Zub., *S. d. calcaratum* B.-Bien., *Kazakia tarbinskii* B.-Bien., *Mesasippus kozhevnikovi kozhevnikovi* (Tarb.), *Calliptamus barbarus nanus* Mitstsh., *C. turanicus* Tarb., *C. coelesyriensis carbonarius* Uv., *Diexis chivensis* Um., *Ochrilidia hebetata hebetata* (Uv.), *O. turanica* (B.-Bien.), *O. mistshenkoi* (B.-Bien.), *Duroniella gracilis* Uv., *D. kalmyka* (Ad.), *Aiolopus oxianus* Uv., *Hilethera turanica* Uv., *Heteracris theodori delicatus* (Mistsh.), *Dociostaurus plotnikovi* Uv., *Eremippus miramae* Tarb., *Chorthippus biguttulus meridionalis* Mistsh., *Ch. apricarius*

asiaticus Mistsh., *Ch. macrocerus assimilis* Mistsh., *Sphingonotus halophilus* B.-Bien., *S. maculatus maculatus* Uv., *S. miramae* Mistsh., *S. halocnemi* Uv., *S. elegans* Mistsh., and *S. nebulosus discolor* Uv.

That species of certain Middle Asian genera (*Diexis, Strumiger, Trinchus,* and *Conophyma*) produce numerous subspecies in a variety of desert or mountain areas is a clear indication of an ongoing speciation (Pravdin 1978).

3. Mediterranean Species (Uvarov 1912, 1921, 1938; Mishchenko 1949, 1951; Pravdin 1980), or *Ancient Mediterranean species* (*Kryzhanovsky 1965*). Tettigonoidea: *Medecticus assimilis* (Fieb.), *Decticus verrucivorus annaelisae* Rme., *D. albifrons* (Fabr.), *Phytodrymadusa longipes* B.-W., *Campsocleis shelkovnikovae* Ad., *Platycleis intermedia* (Serv.), *P. affinis* Fieb., *P. escalerai* Bol.; Grylloidea: *Gryllodinus kerkennensis* (Finot), *Modicogryllus burdigalensis* (Latr.), *Tartarogryllus tartarus* Sauss.; Acridoidea: *Dericorys albidula* Serv., *D. tibialis* (Pall.), *D. annulata roseipennis* (Redt.), *Calliptamus barbarus cephalotes* F.-W., *C. italicus italicus* (L.), *Thisoicetrinus pterostichus* (F.-W.), *Heteracris adspersus* (Redt.), *H. littoralis similis* (Br.-W.), *Eremippus persicus* Uv., *Ramburiella turcomana* (F.-W.), *R. bolivari* (Kuthy), *R. foveolata* Tarb., *Dociostaurus maroccanus* (Thnb.), *D. tartarus* (Stshelk.), *D. kraussi nigrogeniculatus* Tarb., *Notostaurus albicornis albicornis* (Ev.), *Pyrgodera armata* F.-W., *Oedipoda miniata miniata* (Pall.), *O. m. atripes* B.-Bien., *Mioscirtus wagneri rogenhoferi* (Sauss.), *Sphingoderus carinatus* (Sauss.), *Sphingonotus savignyi* (Sauss.), *S. rubescens rubescens* (Walk.), *S. eurasius eurasius* Mistsh., *S. octofasciatus* (Serv.), *S. satrapes* Sauss., *S. salinus* (Pall.), *Leptopternis gracilis* (Ev.), *Hyalorrhipis clausi* (Kitt.), and *Helioscirtus moseri* Sauss.

4. Iranian (Uvarov 1921, 1938; Shumakov 1963), or *Southwest Asian Species* (Bei-Bienko 1948; Pravdin 1980). The diversity of landscapes and lengthy autochthonous geological development of the Iranian Plateau facilitated the appearance of centers of speciation of orthopterans (Shumakov 1963). Ranges of these species lie within the Iranian Plateau, with its northern boundary within Turkmenistan. Species include the following: Tettigonoidea: *Calopterusa werneri* (Ad.); Acridoidea: *Iranella eremiaphila* Uv., *Brunnerella mirabilis mirabilis* Sauss., *Sphingonotus obscuratus apicalis* Sauss., *S. pilosus* Sauss., *Pseudoceles persa* (Sauss.), *Calliptamus coelesyriensis intricatus* Mistsh., *Sphodromerus luteipes rubripes* Uv., *Melanotmethis fuscipennis unicolor* (Uv.), and *Strumiger desertorum persa* Uv.

5. Paleotropical (Kryzhanovsky 1965; Pravdin 1980), *Ethiopian* (Uvarov 1938; Mishchenko 1952), *Tropical* (Uvarov 1929; Mishchenko 1951), or *Indian* (Shumakov 1963) *species*. Orthopterans included in this group are currently widely distributed in the tropics of Africa and Southeast Asia. In Turkmenistan, many of these species inhabit riparian habitats (tugais) and can overwinter there as larvae or as adults. These include the following: Tettigonoidea: *Conocephalus*

Table 2. Zoogeographic elements of the orthopteran fauna of Turkmenistan

Zoogeographic group of species	Number of species and subspecies	% of total fauna
Endemics of Turkmenistan	38	21.0
Turanian	50	27.5
Mediterranean	43	24.0
Iranian	10	5.5
Paleotropical	31	17.1
European-Siberian	6	3.3
Cosmopolitan	1	0.6

discolor Thnb., *Homorocoryphus nitidulus* (Scop.), and *Phaneroptera falcata* (Poda); Grylloidea: *Pteronemobius heydeni* (Fisch.), *Stenonemobius gracilis* Jac., *Myrmecophilus oculatus* Mir., *Gryllotalpa gryllotalpa* (L.)., *G. unispina* Sauss., and *G. africana* P.-Beauv.; Tridactyloidea: *Tridactylus tartarus* Sauss.; Acridoidea: *Tetrix tartara tartara* (Bol.), *T. depressa* Bris., *T. bolivari* Saul., *Paratettix uvarovi* Sem., *Anacridium aegyptium* (L.), *Tropidopola turanica* Uv., *Oxya fuscovittata* (Marsch.), *Chrotogonus turanicus* Kuthy, *Acrida oxycephala* (Pall.), *Truxalis eximia* (Eichw.), *Eyprepocnemis plorans* (Charp.), *E. unicolor* Tarb., *Gonista sagitta* (Uv.), *Aiolopus thalassinus* (Fabr.), *A. simulatrix simulatrix* Walk., *A. strepens* (Latr.), *Locusta migratoria migratoria* L., *Acrotylus insubricus inficitus* Walk., and *Pyrgomorpha bispinosa* Walk.

6. *European-siberian* (Bei-bienko 1948; Shumakov 1963), *Boreal* (Mishchenko 1951), *or Angaran* (Uvarov 1929; Pravdin 1980) *species.* These species are those widely distributed in the steppe zone of Europe and Siberia and mountain steppes of Middle Asia. These are *Tettigonia viridissima* L., *T. caudata* (Charp.), *Tessellana vittata* (Charp.), *Eremippus simplex simplex* (Ev.), *Euchortippus transcaucasicus* Tarb., *Chortippus albomarginatus karelini* Uv., and *Pararcyptera microptera turanica* (Uv.).

7. *Cosmopolitan species (i.e., found on all continents).* Only one synanthropic species of cricket (*Acheta domestica* L.) found in Turkmenistan can be characterized as a cosmopolitan.

The zoogeographic spectrum of the orthopteran fauna of Turkmenistan is summarized in Table 2.

Zoogeographic Division of Turkmenistan

The territory of Turkmenistan is included in the Saharo-Gobian subregion of the Palearctic region (Kryzhanovsky 1965; Sergeev 1991). It is divided among three zoogeographic provinces: (1) Turanian (Kryzhanovsky 1965; Sergeev 1991), or south Middle Asian desert province (Pravdin 1978); (2) Ghissaro-Tien-Shan (Sergeev 1991), or South Turanian mountain (Pravdin 1978), or

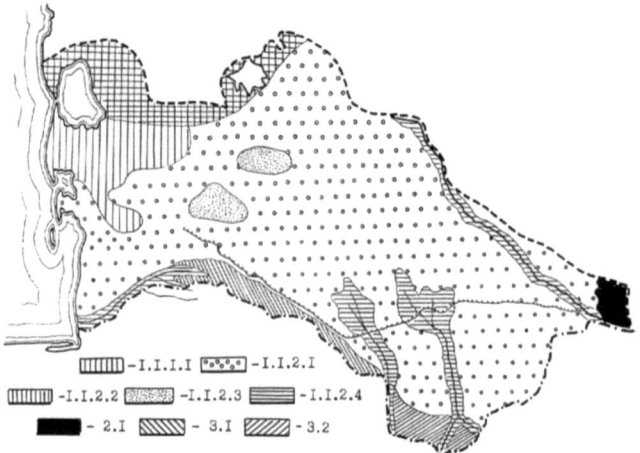

Fig. 1. Zoogeographic regions of Turkmenistan based on the distribution of Orthoptera. 1.1.1.1 – Ustyurt area; 1.1.2.1 – Karakum sand desert area; 1.1.2.2 – West Turkmenistan gravel and gypsum area; 1.1.2.3 – Karakum solonchak area; 1.1.2.4 – Tugai area; 2.1 – Kugitang district; 3.1 – Kopetdagh district; 3.2 – Badghyz district.

Afghano-Turkestan (Kryzhanovsky 1965) province; and (3) Turkmeno-Iranian mountain province (Pravdin 1978; Sergeev 1991). We have attempted to subdivide these areas into smaller zogeographic units according to the distribution of the endemic or subendemic Orthoptera. (Fig. 1).

1. Turanian Province

Most of Turkmenistan is included in the Turanian province, namely the Cis-Caspian Lowland, the southern Ustyurt Plateau, Kaplankyr, Tuarkyr, the Krasnovodsk Plateau, the foothill lowland of Kopetdagh, and the Karakum sand desert. This territory was one of the centers of origin of the psammophilous fauna of Middle Asia. Four genera of the Orthoptera include psammophiles *Ammoxenulus* B.-Bien., *Bufonacridella* Adel., *Diexis* Zub., and *Strumiger* Zub. Within Turkmenistan, the Turanian zoogeographic province includes one subprovince, two areas, and four subareas.

1.1. Karakum Subprovince

Extends from the Caspian Sea to the Amudarya River; occupies sand and clay deserts, the south of the Ustyurt Plateau, plateaus of Kaplankyr, Tuarkyr, and Chelyunkyr, and the riparian tugais of the Amudarya, Murghab, and Tedzhen Rivers, and of the Karakum Canal.

1.1.1. Ustyurt-Cis-Caspian District

Lies mostly within Kazakhstan; in Turkmenistan, includes only the Ustyurt area.

1.1.1.1. Ustyurt Area. Includes deserts with strong soil content of gypsum and gravel in the southern part of the Ustyurt Plateau and the northeastern part of the Krasnovodsk Plateau. Fauna is related to that of semideserts of Kazakhstan. Characteristic species of the Orthoptera are *Kazakia tarbinskii*, *Calliptamus coelesyriensis carbonarius*, *Asiotmethis muricatus australis*, and *Sphingonotus salinus*.

1.1.2. Karakum District

Includes stony deserts of West Turkmenistan and clay and sand deserts of North, Central and Southeast Karakum. This district is divided into four areas.

1.1.2.1. Karakum Sand Desert Area. Includes sand deserts of Northwest Turkmenistan and North, Central and Southeast Karakum desert. Characteristic species of the Orthoptera are *Ammoxenulus pavlovskii*, *A. desertus*, *Diexis varentzovi*, *Strumiger desertorum*, *Ochrilidia hebetata*, *Sphingonotus savignyi*, *Hyalorrhipis clausi*, and *Lepidopternis gracilis*.

1.1.2.2. West Turkmenistan Gravel and Gypsum Desert Area. Occupies the Krasnovodsk Plateau and low plateaus of Kaplankyr and Tuarkyr. Characteristic Orthoptera are *Thrinchus turcmenus*, *Th. desertus*, *Egnatius apicalis*, *Helioscirtus moseri*, and *Sphingonotus satrapes*.

1.1.2.3. Karakum Solonchak Area. Includes takyrs and solonchals mosaically distributed in Central Karakum and the foothills of Kopetdagh. For these areas, characteristic are the halophile species of Orthoptera such as *Gryllodinus odicus*, *Tridactylus tartarus*, *Sphingonotus halocnemi*, *S. halophilus*, and *Mioscirtus wagneri*.

1.1.2.4. Tugai Area. Consists of riparian forests in the river valleys of the Amudarya, Murghab, and Tedzhen Rivers; Karakum Canal; and around lakes and reservoirs of lowland Turkmenistan. Characteristic species of the Orthoptera are *Tropidopola turanica*, *Oxya fuscovittata*, *Gonista sagitta*, *Acrida oxycephala*, *Thisoicetrinus pterostichus*, *Aiolopus oxianus*, *A. thalassinus*, *Duroniella kalmyka*, *Calliptamus italicus*, and *Chrotogonus turanicus*.

2. Ghissaro-Tien-Shan Province

This province includes mountains and foothills of Uzbekistan, Kyrgyzstan, and Tajikistan; within Turkmenistan, its westernmost part occupies the mountains and foothills of the Kugitang Range.

2.1. Kugitang District

Comprises the southwestern part of the Ghissar Range. The fauna is intermediate between the Turkmeno-Khorassan and Pamiro-Alai faunas. The endemic Orthoptera of Kugitang are *Conophyma zimini* and *C. bactrianum*.

3. Turkmeno-Iranian Mountain Province

Occupies mostly mountainous regions of Iran. Within Turkmenistan, two districts are separated: Kopetdagh and Badghyz.

3.1. Kopetdagh District

Occupies the Kopedagh Mountains, which are the northern part of the Turkmeno-Khorassan mountain system; its fauna of the Orthoptera is a mixture of Turanian and Iranian species, with Turanian species predominance. The endemic species of Kopetdagh among orthopterans are: *Squamiana squamiptera, Bergiola montana, B. popovi, Metrioptera crassipes, Conophyma turcomanum, C. uvarovi, Saxetania scutata, Mizonocara deserti, M. inornata insolita, Eremippus onerosus, Chortippus biguttulus pravdini*, and *Sphingonotus turcmenus*.

3.2. Badghyz District

Fauna with North Iranian character; such forms as *Iranella eremiaphila* and *Sphingonotus obscuratus apicalis* penetrate to Badghyz from Iran. The only known endemic species of Orthoptera from Badghyz is *Saxetania pravdini*.

Notes on Ecology of Orthoptera as Agricultural Pests

1. Melanogryllus desertorum (Pall.) – the Desert Cricket. This species of cricket is found everywhere in irrigated areas: along roads and irrigation canals, in alfalfa fields, and grapeyards; in the daytime it hides in soil crevices or burrows. In winter and early spring, *M. desertorum* is usually found at the perimeter of agricultural areas, and it shifts into these areas after the first shoots of crops appear in spring. These insects overwinter as a last instar in burrows 10–20 cm deep, under pieces of soil; sometimes five to ten crickets can be found in one burrow. Crickets appear from their winter shelters from mid-March (in warm years) or April (in cold years). Winged adults appear from late April to late June; they are easily attracted by light. The peak of abundance of winged crickets occurs in late May to early June (Bairamali). In May, females with developed eggs have been recorded. In lab conditions, with a temperature of 29 °C and humidity, 38.8%, eggs of *M. desertorum* have developed in 15 or 16 days.

A new generation of larvae hatches in mid-June. In Turkmenistan, one full and one incomplete generation of this cricket species develop during the year. Appearing from winter shelters, larvae feed on early vegetation such as different species of *Carex*, *Poa*, *Bromus*, *Convolvulus*, and *Plantago*. In agricultural areas, crickets feed on sprouts of different crop plants, and in years of mass reproduction, they present a serious threat to young plants of cotton, cabbage, tomatoes, melons, pumpkins, and watermelons.

2. Gryllotalpa unispina Sauss. – the Mole Cricket. Widespread in Turkmenistan along river valleys, on the shores of lakes, reservoirs, and irrigation canals; often inhabits tree nurseries, greenhouses, and gardens. It also penetrates to the desert (e.g., Repetek: Chernyakhovskii 1972; Inerchaga, north of Ashkhabad: our data).

Larvae of the last instar spend the winter in soil (at 20–90 cm). They are inactive and highly tolerant to cold; in January 1969, temperature at the depth of 15–20 cm fell to -6.2 °C, but there were no deaths of mole crickets. Upon becoming active in late February (in warm winters) to late March (in cold winters), this species makes numerous tunnels in the upper layer of soil. Traces of these elevated tunnels are clearly visible on the soil surface.

In April to May larvae molt into the winged adults; from mid-May, mating takes place at nightime on the soil surface; adult males start chirping and searching for females and readily fly to light. Females deposit developed eggs in underground chambers (150–200 eggs in each chamber) in moist areas at the depth of 10–20 cm. Embryonal development of the eggs lasts for 15–20 days. First larvae have been recorded in mid-June (Sakarchaga District); newly hatched larvae spend some time together in the nest chamber and then disperse. The larvae which hatch early become winged adults before winter, whereas the later generation spends winter in underground shelters.

The mole cricket is an extremely polyphagous insect. In Turkmenistan, it attacks sprouts of tomato and cabbage, young melons, watermelons, and pumpkins, various cereal crops (wheat, barley, and corn), and the roots of young trees in gardens and nurseries. The mole cricket is especially voracious and active in the spring. Not only does it feed on roots and other parts of plants, but it also exposes young roots and seeds by burrowing tunnels in the soil, which leads to the death of annual crops.

3. Dericorys albidula Serv. This species of Acrididae is found in Turkmenistan only in desert landscapes with vegetation dominated by saksaul (*Haloxylon aphyllum* and *H. persicum*) or cherkez (*Salsola richteri*). Mass outbursts of *D. albidula* have been recorded since 1930 or 1932 in all desert areas of Southeast Turkmenistan (especially next to the towns of Kabakly and Deinau, close to the sands along the Amudarya River). In 1937, this species appeared in large numbers in the North Karakum (around Lake Sarykamysh and next to the town of Darganata). After a locust outburst in 1960, anti-locust measures were taken from 1960 to 1963 in an area of about 75,600 ha in the North Karakum. By 1963, the major area of reproduction of *Dericorys albidula* has been restricted to the desert portion of the Lenin District, from the Goshaguyi well to the Sarykamysh Depression.

D. albidula attacks predominantly desert shrubs (such as *Haloxylon aphyllum*, *H. persicum*, *Salsola richteri*, and *S. subaphylla*), thereby constituting a serious pest of sand-stabilizing vegetation. It does not, however, attack agricultural crops (Yenikeyev 1949; our data), and thus the protective measures against this species should be limited to sand desert areas.

Larvae of this species hatch in late April, develop for 45 to 50 days, and molt into the adult stage in early June. Eggs are laid in early July; adult females survive until the end of September and make a total of three to four egg chambers on takyr-like patches and clay areas exposed by wind among *Haloxylon* sand desert communities. Eggs are laid into the chambers during evening hours, from 16:00 or 17:00 to 21:00.

4. Schistocerca gregaria Forsk. – the Desert Locust. This infamous species is not a constant inhabitant of Turkmenistan or any other republic of the former Soviet Union. Its natural range lies in Africa (southern Sahara and Sudan), Pakistan, Baluchistan, and northeastern India. During years of mass outbursts, the desert locust invades Iran and Afghanistan; from these countries it occasionally moves into the former Soviet territory. The largest invasion of the desert locust to Middle Asia was recorded in 1929, when it occupied more than 1,500,000 ha and devastated regional agriculture. The outbursts of 1940 to 1944 and 1957 to 1959 threatened Soviet territory; however, this invasion was prevented by extermination of the locust in Iranian territory by Soviet specialists acting with the support of agricultural aviation. Only single locusts were detected in Kushka, Pulikhatum, Serakhs, and Mary.

Unlike most invasions, which occur in early May, the severe 1962 locust invasion in Turkmenistan occured much earlier (March 24 to April 11). Locust swarms occupied 180,000 ha, and oothecae were laid on 40,000 ha (mostly in Badghyz). Small swarms and single locusts were detected in the Murghab Valley, in the Tedzhen Oasis, and next to Ashkhabad. Eggs were laid in Badghyz during March 27 to 29; since there is no diapause in egg development, hatching of the larvae began May 7, 1962, and the first winged adults emerged on June 10. Locusts occupying irrigated areas yielded two generations; the second generation started to molt into adult stage on September 20, and adults could be found up to the onset of winter. Both larvae and adults of *S. gregaria* are polyphagous. Of wild plants, they consume *Alhagi persarum*, species of *Convolvulus*, *Asparagus*, and *Calligonum*; of crop cultures, they attack leaves of all fruit trees, vegetables, alfalfa, cotton, and cereal crops.

5. Calliptamus turanicus Tarb. – the Turanian Prus. This species is found primarily in the foothills of Kopetdagh and Kugitang. Important areas of outbursts lie in the Kopetdagh foothills south of Kaakhka Station. In 1958, *C. turanicus* gave rise to a mass reproduction which originated in Archinyan and

Novrekchashme (Kaakhka area); together with the Moroccan locust, it invaded crops of melons and watermelons. In the foothills of Badghyz and Karabil, outbursts occured in 1932–1933 and 1956–1958; in Badghyz, density of *C. turanicus* larvae reached 50 to 60 insects/m^2, with average density of winged adults 3 to 5 insects/m^2, and maximal density 30 to 35 adults/m^2. In the foothills of Kugitang (Lelimkam, Malik, and Aktash), density of the Turanian prus during mass reproduction in 1961 reached 20 to 30 insects/m^2. Single individuals were found in the Chandyr and Sumbar Valleys, and along the foothills of Kopetdagh (Kizyl-Arvat, Bakharden, and Geok-Tepe). Along the Karakum Canal, small numbers of *C. turanicus* were detected between Nichka and Lengush wells. This species is completely absent from the Karakum Desert.

Larvae of the Turanian prus in Turkmenistan hatch from early April (in warm years) to April 20 to 25 (in colder years); larval development lasts 40 to 42 days, and first adults can be found in late May to early June. Eggs are laid from late June in soil (typical or grey serozyom) on hilltops as well as in depressions between hills. Each female lays two to five oothecae, with an average of 27 to 28 eggs each (maximum 50 eggs). Adult *C. turanicus* die in August or September.

Natural vegetation in habitats of *Calliptamus turanicus* is dominated by *Poa bulbosa* and *Carex pachystylis*, with participation of *Malcolmia turkestanica*, *Artemisia turanica*, *A. badhysi*, species of *Astragalus*, *Ranunculus*, *Psoralea*, *Papaver*, *Asperula*, *Cousinia*, *Phlomis*, and *Eremopyrum*. Since many of these species are natural food of larvae and adults of the Turanian prus, pasture areas as well as cultured crops such as wheat, barley, corn, watermelons, and melons are often damaged.

6. *Calliptamus italicus* (L.). – the Italian, or Oasis Prus. This species is found in lowland Turkmenistan only in oases, i.e., along the valleys of the Amudarya, Murghab, Tedzhen, and some smaller rivers. It inhabits shores of irrigation canals and ditches, old alfalfa fields, old gardens, vineyards, and roadsides; dominant vegetation here includes *Cynodon dactylon*, *Alhagi persarum*, *Hyosciamus niger*, *Glycyrrhiza glabra*, species of *Mimosa*, *Artemisia*, *Convolvulus*, and *Zygophyllum*. In the mountains of Kopetdagh and Kugitang *C. italicus* occupies both valleys and mountain steppes (dominated by *Elytrigia trichophora*, *Festuca valesiaca*, and species of *Stipa*).

From 1925 to 1933, *C. italicus* considerably damaged crops in the Murghab Valley. Later, the development of agrotechnical culture minimized damage from this locust; however, even now areas of potential mass reproduction still exist in the Tedzhen Oasis. Outbursts of the oasis prus occurred here in 1960 to 1967 and 1974, when density of adults reached a maximum of 25–30 insects/m^2. In years like these, *C. italicus* does considerable damage to alfalfa, cotton, potato, sunflower, and beet fields, and also to tomatoes, onion, cabbage, grapes, dzhidda (*Elaeagnus*), and various fruit trees. This species also readily consumes leaves of such wild plants as *Alhagi*, *Hyosciamus*, *Artemisia*, *Convolvulus*, *Zygophyllum*, and *Portulaca*.

Larvae of *C. italicus* in Turkmenistan hatch from early April to early May;

their development lasts 40 to 45 days and passes through five larval instars. First adults can be found in early to mid-June. Eggs are laid from July to mid-August in soil (typical or grey serozyom) on open places in old alfalfa fields, along irrigation ditches, and in other places with turf formed by *Cynodon dactylon*.

7. Dociostaurus maroccanus – the Moroccan Locust. Within Turkmenistan, this species is found in the foothills of Kopetdagh and Kugitang. We delineate the following twelve areas of potential mass reproduction for this species in Turkmenistan: *Sharlauk* (the lower Sumbar Valley between the villages of Sharlauk and Chat, and foothills of the Songudagh range); *Chandyr* (the Chandyr Valley in West Kopetdagh south from Yartykala, Ak, Kizylimam, and Monzhukly villages); *Kara-Kala* (southern foothills of Mt. Syunt, including Nere village); *Uzyntokai* (eastern portion of the Syunt-Khasardagh range next to Uzyntokai village at 700 m altitude); *Nokhur* [western part of the Nokhur Plateau (Chopirchinar), at 1,300 m altitude]; *Ipaikala* (the Ipaikala Valley in the Bakharden District in small areas of *Poa-Carex* communites; in 1972 to 1973 the abundance of locusts there was low); *Kelyata* (depressions between hills and foothill lowlands to the south of Kelyata Station); *Geok-Tepe* (between the city of Geok-Tepe and Mt. Dushak and the Inzhirli and Aksuv Plateaus; in 1970 to 1971, mass reproduction of the Moroccan locust was observed there); *Ashkhabad* (in valley-like depressions around the villages of Kalininsky, Kuruhaudan, and Gyuaurs; in outburst years swarms of locusts disperse from here northward to the foothill lowlands); *Kaakhka* (hills south from Armansaad Station, around villages of Novrekheshme, Deicha, Kelyata, Sharzha, and Khodzhabulan); *Chaacha* (in Southeast Kopetdagh around the valleys of Kalkhauz, Chilikaman, and Karateken); and *Kugitang* (foothills of the Kugitang range, around the villages of Bazardepe and valleys Myalik, Lelimkam, and Aktash). Chandyr, Kaakhka, and Chaacha areas are connected to Iranian areas of potential outburst of the Moroccan locust.

Usual habitats of the Moroccan locust are slopes of hills, dry valleys, and foothills covered with *Poa bulbosa* and *Carex pachystylis*, which lie at 300 to 700 m elevation. Larvae of *D. maroccanus* in Turkmenistan hatch in late March to early April; their development lasts 31 to 40 days. Adults appear in early May. Eggs are laid from late May to early June in the turf formed by the *Poa – Carex* vegetation. Density of egg chambers in years of mass reproduction reaches $2,000/m^2$.

Dociostaurus maroccanus is a polyphagous species; in Turkmenistan, it feeds on 65 wild plant species, mostly grasses and Asteraceae. Of cultured crops, it damages predominantly wheat, barley, corn, cotton, as well as leaves of fruit and decorative trees.

8. Dociostaurus kraussi nigrogeniculatus Tarb. – Atbasarka. In Turkmenistan, found in the foothills, plateaus, and mountains of Ustyurt, Kyurendagh, Kopetdagh, Badghyz, Karabil, and Kugitang. Its habitats (up to 1,000 m) are dominated by *Poa – Carex* and ephemerous vegetation (in Ustyurt it is also found in sagebrush and halophyte communities), natural food for *D. k. nigrogeniculatus*. Atbasarka significantly damaged crops in Turkmenistan from 1929 to 1932. In 1957, mass reproduction occurred around the village of Sharlauk (West Kopetdagh) where density of larvae reached 20 to 30 insects/m^2. Increase in density was observed also in the Kopetdagh foothills in 1958 and 1959.

Larvae of *D. k. nigrogeniculatus* hatch earlier than those of other pest species of locusts. For instance, after the warm winter and early spring of 1957, hatching of larvae in the Kizyl-Atrek District started on March 16, and the first winged adults were recorded on April 21. Usually, however, hatching takes place in early April, and adults appear in May. This species does not migrate far from its hatching places. Two to three egg chambers are made by females in mid-May in the turf formed by *Carex pachystylis* and *Poa bulbosa*. All adults die off by the end of June.

In its years of mass reproduction, atbasarka, together with other pest species (the Moroccan locust, and the Turanian and oasis prus), can seriously damage crops and pastures.

Rare and Endangered Species of the Orthoptera

1. Magrettia mutica Br.-W. Endemic species of Turkmenistan (Ashkhabad, Kara-Kala, Dushak, Badghyz, and Repetek). This rare nocturnal grasshopper with decreasing abundance inhabits foothills and lowlands with *Poa bulbosa* and *Carex pachystylis*; in the desert it lives in stabilized sands. In daytime, *M. mutica* hides in the burrows belonging to various vertebrates or to beetles of the genus *Lethrus*. Larvae of last instar and adults survive through the winter; from April to June, mostly adult grasshoppers are found. This species requires special protective measures in the Repetek Reserve.

2. Ceraeocercus fuscipennis Uv. This grasshopper is endemic for Middle Asia; in Turkmenistan, it is found only in the foothills of Bolshoi Balkhan, where it is very rarely found on sagebrush vegetation. Winged adults have been recorded in late May, and females with developed eggs, in June.

3. Ammoxenulus pavlovskii B.-Bien. The endemic of the Southeast and Central Karakum Desert; only few specimens are known. Larvae and adults are found on desert shrubs, including *Ammodendron conollyi*, *Salsola richteri*, *Haloxylon persicum*, and *Calligonum* spp. Needs special protective measures in the Repetek Reserve.

4. Saxetania cultricollis (Sauss.). Rare species found in xerophytic stony habitats of Kopetdagh and Kugitang. This curious-looking and conspicuous wingless insect is often collected by amateurs for zoololgical collections. Distribution of *S. cultricollis* is mosaic and depends on a degree of disturbance of vegetative cover by grazing. It spends winter as a larval stage; adults emerge in April. Mating and egg laying occurs from May to June; larval development lasts six months. This species requires special protective measures in the Kopetdagh and Syunt-Khasardagh Reserves.

5. Saxetania pravdini Tschern. The narrow endemic of Badghyz. Solitary insects are found only on stony slopes (*chinks*) in the Yeroyulanduz Depression. In May, both adults and larvae of different age have been recorded. Requires protective measures in the Badghyz Reserve; its collection should be prohibited.

6. Bufonacridella sumakovi Ad. An endemic species of Southeast Turkmenistan (Iolotan, Repetek, Shiramkuyi, and Kerki). Solitary insects are found on barkhans and stabilized sand dunes dominated by such plants as species of *Heliotropium, Tournefortia,* and *Convolvulus*. Larvae hatch in early May; adults are found from early June to mid-August. Density of this species decreases due to intensive sheep grazing and management of sand dunes. Needs special protective measures in the Repetek Reserve.

7. Iranella turcmena B.-Bien. Endemic of Southwest Turkmenistan, where it is known from Malyi Balkhan (Akhchakuima) and Kyurendagh (Danata). Insects are rarely found in stony foothills with sparse vegetation; we collected only three specimens during a three-to-four-hour excursion. Adults are found from mid-June to late July.

8. Sphingonotus turcmenus B.-Bien. The endemic of West Kopetdagh (Yoldere Valley), where it is occasionally found on grey limestone-clay gravels and screes. Requires special protective measures in the Syunt-Khasardagh Reserve.

Appendix: A list of orthoptera recorded from turkmenistan

Superfamily Tettigonoidea

Magrettia mutica Br.-W.
Phaneroptera bivittata B.-Bien.
P. falcata(Poda)
Homorocoryphus nitidulus (Scop.)
Conocephalus discolor Thnb.
Glyphonotus thoracicus semenovi Mir.
Tettigonia viridissima L.
T. caudata (Charp.)
Bergiola balchashica Stshelk.
B. montana B.-Bien.
B. popovi B.-Bien.
Ammoxenulus pavlovskii B.-Bien.

A. desertus B.-Bien.
Campsocleis shelkovnikovae Ad.
Phytodrymadusa longipes B.-W.
Calopterusa werneri (Ad.)
Ceraeocercus fuscipennis Uv.
Semenoviana tamerlana (Sauss.)
Platycleis affinis Fieb.
P. escalerai Bol.
P. fatima Uv.
P. intermedia (Serv.)
Squamiana squamiptera (Uv.)
Tessellana vittata (Charp.)
Metrioptera crassipes B.-Bien.
Decticus albifrons (Fabr.)

D. verrucivorus annaelisae Rme.
D. verrucivoris gracilis Uv.
Medecticus assimilis (Fieb.)

Superfamily Grylloidea

Gryllus bimaculatus De Geer
Acheta domestica L.
A. turcomana Gor.
Melanogryllus desertus (Pall.)
Tartarogryllus tartarus Sauss.
Modicogryllus burdigalensis (Latr.)
M. chivensis (Tarb).
M. frontalis (Fieb.)
M. pallipalpis (Tarb.)
Gryllopsis bolivari Uv.
Gryllodinus kerkennensis (Finot)
G. odicus odicus (Uv.)
Turanogryllus lateralis (Fieb.)
Cophaphonus riparius Mistsh.
Pteronemobius heydeni (Fisch.)
Stenonemobius gracilis Jac.
Grylliscus gussakowskii Tarb.
Oecanthus turanicus (Uv.)
Myrmecophilus oculatus Mir.
Eremogryllodes semenovi (Mir.)
E. vlasovi (Mir.)
Gryllotalpa gryllotalpa (L.).
G. unispina Sauss.
G. africana P.-Beauv.

Superfamily Tridactyloidea

Tridactylus tartarus Sauss.

Superfamily Acridoidea

Fam. Tetrigidae
Tetrix tartara tartara (Bol.)
T. depressa Bris.
T. bolivari Saul.
Paratettix uvarovi Sem.

Fam. Pamphagidae
Melanotmethis fuscipennis fuscipennis
(Redt.)
Melanotmethis fuscipennis unicolor
(Uv.)
Atrichotmethis semenovi (Zub.)
Asiotmethis muricatus australis
(Tarb).
Thrinchus desertus B.-Bien.
Th. tuberculosus Tarb.
Th. turcmenus B.-Bien.
Strumiger desertorum desertorum
Zub.
S. d. calcaratum B.-Bien.
S. d. persa Uv.
Saxetania scutata Mistsh.
S. cultricollis (Sauss.)
S. pravdini Tschern.

Fam. Pyrgomorphidae
Pyrgomorpha bispinosa Walk.
Chrotogonus turanicus Kuthy

Fam. Acrididae
Uvarovium desertum Dirsh.
Dericorys albidula Serv.
D. tibialis (Pall.)
D. annulata roseipennis (Redt.)
Bufonacridella sumakovi Ad.
Diexis chivensis Um.
D. varentzovi varentzovi Zub.
D. v. probus Mistsh.
D. v. salsolae Mistsh.
D. v. afffinis Mistch.
Iranella eremiaphila Uv.
I. turcmena B.-Bien.
Oxya fuscovittata (Marsch.)
Tropidopola turanica Uv.
Conophyma bactrianum Mistsh.
C. turcomanum Mistsh.
C. uvarovi uvarovi Sem.
C. u. vicinum Mistsh.
C. zimini Tarb.
Schistocerca gregaria Forsk.
Anacridium aegyptium (L.)
Calliptamus barbarus cephalotes
F.-W.

C. barbarus nanus Mitstsh.
C. coelesyriensis carbonarius Uv.
C. coelesyriensis intricatus Mistsh.
C. italicus italicus (L.)
C. turanicus Tarb.
Sphodromerus luteipes rubripes Uv.
Thisoicetrinus pterostichus (F.-W.)
Heteracris adspersus (Redt.)
H. littoralis similis (Br.-W.)
H. theodori delicatus (Mistsh.)
Eyprepocnemis plorans (Charp.)
E. unicolor Tarb.
Egnatioides desertus Uv.
Egnatius apicalis Stål.
Acrida oxycephala (Pall.)
Truxalis eximia (Eichw.)
Gonista sagitta (Uv.)
Ochrilidia hebetata hebetata (Uv.)
O. turanica (B.-Bien.)
O. mistshenkoi (B.-Bien.)
Duroniella gracilis Uv.
D. kalmyka (Ad.)
D. turcomana Mistsh.
Pararcyptera microptera turanica (Uv)
Ramburiella turcomana (F.-W.)
R. bolivari (Kuthy)
R. foveolata Tarb.
Dociostaurus kraussi nigrogeniculatus Tarb.
D. maroccanus (Thnb.)
D. plotnikovi Uv.
D. tartarus (Stshelk.)
Notostaurus albicornis albicornis (Ev.)
N. a. turcmenus (Uv.)
Mizonocara deserti Uv.
M. inornata insolita Mistsh.
Kazakia tarbinskii B.-Bien.
Eremippus carinatus Mistsh.
E. foveolatus Mistsh.
E. miramae Tarb.
E. onerosus Mistsh.
E. persicus Uv.
E. simplex simplex (Ev.)
Mesasippus kozhevnikovi

kozhevnikovi (Tarb.)
Chortippus albomarginatus karelini Uv.
Ch. apricarius asiaticus Mistsh.
Ch. biguttulus meridionalis Mistsh.
Ch. b. pravdini Sytshev.
Ch. macrocerus assimilis Mistsh.
Euchortippus transcaucasicus Tarb.
Aiolopus oxianus Uv.
A. simulatrix simulatrix Walk.
A. strepens (Latr.)
A. thalassinus (Fabr.)
Hilethera turanica Uv.
Locusta migratoria migratoria L.
Oedaleus senegalensis (Kr.)
O. decorus (Germ.)
Pyrgodera armata F.-W.
Brunnerella mirabilis mirabilis Sauss.
Mioscirtus wagneri rogenhoferi (Sauss.)
Oedipoda miniata miniata (Pall.)
O. m. atripes B.-Bien.
Acrotylus insubricus inficitus Walk.
Pseudoceles persa (Sauss.)
Sphingoderus carinatus (Sauss.)
Sphingonotus elegans Mistsh.
S. eurasius eurasius Mistsh.
S. halocnemi Uv.
S. halophilus B.-Bien.
S. maculatus maculatus Uv.
S. miramae Mistsh.
S. nebulosus discolor Uv.
S. obscuratus apicalis Sauss.
S. o. transcaspicus Uv.
S. octofasciatus (Serv.)
S. pilosus Sauss.
S. rubescens rubescens (Walk.)
S. salinus (Pall.)
S. satrapes Sauss.
S. savignyi (Sauss.)
S. turcmenus B.-Bien.
Helioscirtus moseri Sauss.
Hyalorrhipis clausi (Kitt.)
H. turcmena Uv.
Leptopternis gracilis (Ev.)

27. Encyrtid Wasps of Turkmenistan (Hymenoptera, Encyrtidae)

SVETLANA N. MYARTSEVA

Abstract

The rich fauna of parasitic wasps Encyrtidae in Turkmenistan includes 254 species belonging to 109 genera. An account of their biology, life cycles, habitat distribution, and zoogeographic connections is given. Encyrtidae parasitize eggs, larvae, and pupae of insects belonging to seven orders: Lepidoptera, Coleoptera, Neuroptera, Diptera, Homoptera, Hemiptera, and Hymenoptera, with a high specificity of parasitism. The majority of Encyrtidae in Turkmenistan are parasites of the Homoptera; of those, 128 species are parasites of the superfamily Coccoidea. The ecological distribution of Encyrtidae is defined by that of both their host insects and the food plants of these hosts. The most diverse in Turkmenistan encyrtid fauna is that of the arid foothills; the second in richness is the tugai fauna; desert fauna (with the exception of sand desert) is rather impoverished. Four groups and thirteen types of zoogeographic ranges were distinguished. The fauna has an Ancient Mediterranean arid character; among encyrtids parasitizing on hosts which feed on shrubs and semishrubs, about 70% are Turanian endemic species.

Introduction

The Encyrtidae are one of the largest families of the parasitic Hymenoptera, including more than 3,000 species worldwide and 1,260 in the Palearctic (Tryapitsyn 1989). Their parasitism on phytophageous insects lends them scientific interest as well as practical importance of this group. Of all Chalcidoidea inhabiting Turkmenistan, the encyrtids are currently the best studied family. Study of the encyrtid fauna, biology, habitats, and geographic distribution has allowed us (Myartseva 1984a, 1986) to reveal their host-parasite relationships as well as other ecological features. We also discuss the mechanisms of the formation of parasitic complexes in natural and agricultural ecosystems as well as the evolution of desert fauna.

Table 1. A list and biological-ecological characteristics of the Encyrtidae in Turkmenistan.

Species	Hosts[a]	Habitat[b]	Range Type[c]
Charitopus trjapitzini Hoff.	–	5, 8	II-3
Ch. fulviventris Fors	–	5	IV-3
Ch. desertus Myarts.	–	1, 2, 4, 5, 7, 8	I-1
Ericydnus tamaricicola Myarts.	Homoptera: 1	1, 5	I-1
E. karakalensis Myarts.	,,	5	I-1
E. sipylus (Walk.)	,,	1	IV-1
E. danatensis Myarts.	,,	5	I-1
E. turkmenicus Myarts.	,,	1, 3, 5, 7, 8	I-1
E. robustior Mercet	,,	2, 5, 8	II-4
E. niger Myarts.	,,	8	I-1
Clausenia mariae Myarts.	,,	5	I-1
Monodiscodes maculipennis Hoff.	–	4, 8	II-4
Tetralophisca dimorpha (Merc.)	–	2, 7	II-3
Metaphaenodiscus bactrianus Trjap.	–	7	I-3
Paratetracnemoidea malenottii (Merc.)	–	5	II-4
Leptomastix fulva (Nik.)	Homoptera: 1	7	I-1
L. flava Merc.	Homoptera: 1, 5	5, 8	II-4
Paranathrix acanthococci (Myarts.)	Homoptera: 5	5	I-1
Anagyrus hammadae Trjap. et Ros.	Homoptera: 1, 5	1, 2, 5	I-1
A. orbitalis (Ruschka)	,,	1, 2, 4, 5, 7	II-4
A. zaitzevi Trjap.	Homoptera: 1	2, 5	II-1
A. diversicornis Merc.	,,	5, 7, 8	IV-1
A. ashkhabadensis Myarts.	,,	3	I-1
A. schoenherri (Westw.)	,,	5	III-2
A. scapularis Myarts.	,,	7	I-1
A. pseudococci (Gir.)	,,	5, 7, 8	IV-4
A. haloxyli Sug.	,,	1, 3, 5, 7, 8	I-1
A. tamaricicola Trjap.	,,	5, 7	I-2
Epidinocarsis archangelskayae (Trjap.)	,,	5, 8	I-1
Dolichoceras amudaryensis Myarts.	,,	7	I-1
D. galinae Myarts.	,,	7	I-1
Rhopus trjapitzini Myarts.	,,	5, 7, 8	I-1
Rh. turanicus Myarts.	Homoptera: 2	7	I-1
Rh. olgae Myarts.	Homoptera: 1	2, 7	I-1
Platyrhopus longicornis Trjap. et Herthevtz.	,,	7, 8	I-2
Leptomastidea enigmatica (Trjap.)	Homoptera: 5	1, 5, 7	II-3
L. turcmenica (Myarts.)	,,	5, 7	I-1
L. acanthococci Myarts.	,,	5	I-1
L. bereketi Myarts.	Homoptera: 1	1, 3, 7	I-1
L. rubra Tach.	,,	7, 8	III-2
L. abnormis (Gir.)	,,	8	IV-4
Callipteroma quinqueguttata Motsch.	–	5, 7	IV-4
Dinocarsis hofferi Grah.	Homoptera: 1	5	III-2
D. hemiptera (Dalm.)	,,	5	III-1
Dicarnosis sugonjaevi Myarts.	,,	2, 5	II-1
D. vicina Trjap. et Hoff.	,,	5	I-3
Dinocarsiella alpina (Gir.)	,,	2, 4, 5	III-2
Tetracnemus diversicornis Westw	,,	5	IV-2
T. peliococci Myarts.	,,	5	I-2

Table 1. Continued

Species	Hosts[a]	Habitat[b]	Range Type[c]
T. phragmitis Myarts.	,,	5, 7	I-1
Phasmopoda loginovae Trjap.	–	1, 5, 8	I-1
Paracopidosoma ephemeri Myarts.	Homoptera: 1	2, 5	II-1
P. parallelum Hoff.	,,	3	II-4
Asencyrtus deserticola Trjap.	–	1, 4	I-1
Mohelencyrtus phenacocci Myarts.	Homoptera: 1	5	I-1
Eremophasma peliococci Myarts.	,,	5, 8	I-1
Microterys sylvius (Dalm.)	Homoptera: 4	6	IV-1
M. darevskii Trjap.	,,	5	I-2
M. praedator Sug.	,,	6, 8	II-2
M. vashlovanicus Jasn.	,,	5	I-2
M. contractus (Hoff.)	,,	5	II-4
M. zygophylli Myarts.	,,	5	I-1
M. ashkhabadensis Myarts.	,,	5	I-1
M. darganatensis Myarts.	,,	7	IV-1
M. chommati Myarts.	,,	2	I-1
M. mesasiaticus Myarts.	,,	1, 5, 8	I-1
M. tricoloricornis (De Stef.)	,,	5	III-2
M. duplicatus (Nees)	,,	6	III-2
Aschitus margaritae Myarts.	Homoptera: Psylloidea	7	I-1
A. naiacocci (Trjap.)	Homoptera: 1, 5	1–5, 7, 8	I-1
A. neoacanthococci Myarts.	Homoptera: 5	5, 7	I-1
A. populi Myarts.	Homoptera: Psylloidea	5, 7	I-1
Hoplopsis minuta (Fabr.)	Homoptera: 4	5	II-4
Neastymachus luteus (Nik.)	Homoptera: 2	5, 7, 8	II-3
N. secundus (Trjap.)	,,	5, 7	I-3
Subprionomitus festucae (Mayr)	Homoptera: 4	5	III-1
Syrphophagus aeruginosus (Dalm.)	Diptera: Syrphidae	1, 5, 8	III-2
S. herbidus (Dalm.)	,,	2, 5, 7	III-2
S. aphidivorus (Mayr)	Homoptera: Aphidoidea	1–5, 7, 8	IV-4
S. orientalis (Myarts.)	,,	8	I-1
Echthrodryinus mesasiaticus Gordh. et Trjap.	Hymenoptera: Braconidae	5, 8	I-1
E. krasilnikovae Myarts.	,,	5, 8	I-1
Ooencyrtus telenomicida (Vass.)	Hemiptera: Pentatomidae, Scutelleridae	5, 7, 8	II-4
O. daritshevae Myarts.	Lepidoptera: Notodontidae	5, 7	I-1
O. minnae Myarts.	Lepidoptera: Attacidae	6	I-1
O. tibialis Myarts.	–	5	I-1
O. indefinitus Myarts.	–	1, 2, 4, 5, 7, 8	I-1
Cerchysius subplanus (Dalm.)	Diptera: Chamaemyiidae	1, 2, 4, 5, 7, 8	IV-2
Helegonatopus coxalis (Myarts.)	Homoptera: Auchenorrhyncha	5, 7	II-1

Table 1. Continued

Species	Hosts[a]	Habitat[b]	Range Type[c]
H. rasnitzyni (Trjap.)	,,	2, 8	II-4
Eremencyrtus unifasciatus Trjap.	–	1, 4, 5	II-1
E. neoptolemus Myarts.	–	1	I-1
Aphyculus slavai Myarts.	Homoptera: 1	5	I-1
A. deynauensis Myarts.	,,	7	I-1
A. mesasiaticus Myarts.	,,	5, 8	I-1
A. frontatus Myarts.	,,	7	I-1
Pseudaphycus malinus Gah.	,,	8	IV-1
A. astanovi Myarts.	,,	7, 8	I-1
A. trabutinae Myarts.	,,	7	I-1
Pseudectroma ciliatum (Myarts.)	,,	7, 8	I-1
P. scenographica Sug.	Homoptera: 4	6, 7, 8	I-1
P. turanica Sug.	,,	5, 6, 8	II-2
Xenaphycus flavovarius (Merc.)	–	5	III-2
X. vigil (Erd.)	–	5	III-2
Metaphycus dispar (Merc.)	Homoptera: 4	5, 7, 8	III-2
M. melanostomatus (Timb.)	,,	5	III-1
M. turanicus Sug.	,,	5, 7, 8	II-2
M. zebratus (Merc.)	Homoptera: 4, Asterolecaniidae	5	IV-2
M. elenae Myarts.	Homoptera: 4	7	I-1
M. hodzhevanishvilii Jasn.	,,	5	I-2
M. turkmenicus Myarts.	,,	1, 3, 4, 5, 7	I-1
M. bogdanovikatkovi Jasn.	,,	1–5	I-2
M. desertus Myarts.	,,	1	I-1
M. acanthococci Myarts.	Homoptera: 5	5	I-1
Prionomitus mitratus (Dalm.)	Homoptera: Psylloidea	5	IV-1
Trechnites flavipes (Merc.)	,,	1, 3, 4, 5, 7	II-4
T. trjapitzini Sug.	,,	1–5	II-1
T. fuscitarsis Thoms.	,,	5	III-2
T. psyllae (Ruschka)	,,	8	III-2
Psyllaephagus caillardiae Sug.	,,	1, 3, 4	I-1
P. longiventris Trjap.	,,	1, 4	II-1
P. cholcinellus Myarts.	,,	1, 7	I-1
P. tyrrheus Myarts.	,,	1	I-1
P. badchysi Myarts.	,,	5	I-2
P. saxaulicus Sug.	,,	1, 3, 4, 5	I-1
P. bachardenicus Myarts.	,,	5, 7	I-1
P. calligonicola Myarts.	,,	1, 3	I-1
P. rubriscutellatus Myarts.	,,	1, 3, 5	I-1
P. colposceniae Trjap.	,,	1, 3, 4, 5, 7	II-4
P. desertus Sug.	,,	1, 4, 5	I-1
P. tamaricicola Myarts.	,,	1, 3–5, 7, 8	I-2
P. populi Trjap.	,,	7	II-1
P. tokgaevi Myarts.	,,	1, 3–5, 7, 8	I-1
P. ogazae Sug.	,,	1, 4	II-1
P. egeirotriozae Trjap.	,,	7	I-1
P. hammadae Myarts.	,,	3	I-1
P. turanicus Myarts.	,,	2–4	I-1

Table 1. Continued

Species	Hosts[a]	Habitat[b]	Range Type[c]
P. obscurus Myarts.	,,	5, 7	I-1
P. arenicola (Trjap.)	,,	5	I-2
P. turkmenicus (Myarts.)	,,	5	I-1
Metapsyllaephagus talassio Myarts.	,,	1	I-1
M. desantisi Myarts.	,,	1	II-1
M. eremita Myarts.	,,	1	I-1
M. tashlievi Myarts.	,,	2, 4, 5	I-2
Discodes obscuriclavus Myarts.	Homoptera: 4	1, 3–5	I-1
D. terebratus Myarts.	–	7	I-1
D. kopetdagicus Myarts.	Homoptera: 5	5	I-1
D. indefinitus Myarts.	,,	4, 5	I-1
D. ipaikalensis Myarts.	,,	5	I-1
D. tugaiensis Myarts.	,,	5	I-1
D. kryzhanovskii Myarts.	,,	1, 4, 5	I-1
D. rhizopulvinariae Myarts.	Homoptera: 4	1, 4, 5	I-1
D. desertus Myarts.	,,	1	I-1
D. acanthopulvinariae Trjap.	,,	1–4	I-2
D. coccophagus (Ratz.)	,,	8	III-1
D. atraphaxidis Myarts.	Homoptera: 1	5	I-1
Paraphaenodiscus murgabicus Myarts.	–	7	I-1
P. sugonjaevi Myarts.	–	7	I-3
Amicroterys asiaticus Myarts.	–	7	I-1
Choreia maculata (Hoff.)	Homoptera: 4	4, 5, 7, 8	II-4
Semen apterum Hoff.	–	2, 5	II-4
Echthroplexiella artemisiae Myarts.	Homoptera: 5	4	I-1
E. popovi Trjap. et Ros.	,,	4	I-1
E. tobiasi Myarts.	,,	7	I-1
E. aeneiventris Mercet	Homoptera: 4	5,7	II-4
Mayridia formosula Mercet	Homoptera: Coccoidea	4, 5	II-4
M. merceti Trjap.	Homoptera: 4	7	II-4
M. pulchra Merc.	Homoptera: Coccoidea	5	IV-2
M. tschairae Myarts.	,,	3	I-1
M. eremobia Myarts.	,,	5	I-1
M. kopetdagica Myarts.	,,	5	I-1
M. murgabensis Myarts.	,,	5	I-1
M. sugonjaevi Trjap.	Homoptera: 2	5, 7	II-1
Adelencyrtus moderatus (How.)	Homoptera: 3	7	IV-4
Habrolepis tergrigorianae Trjap.	,,	5, 8	I-2
Coccidencyrtus artemisiae Myarts.	,,	5	I-1
C. schizotargioniae Myarts.	,,	1, 3, 5	I-1
C. duplachionaspidis Myarts.	,,	5	I-1
Neococcidencyrtus steinbergi Myarts.	,,	5	I-1
Cerapterocerus planus Myarts.	Homoptera: 2	7	I-1
C. pilicornis Thoms.	Homoptera: 5	2	III-2
C. mirabilis Westw.	Homoptera: 4	8	III-2
Cheiloneurus claviger (Thoms.)	,,	1, 3–8	III-2
Ch. elegans (Dalm.)	,,	1–5, 7, 8	IV-1

Table 1. Continued

Species	Hosts[a]	Habitat[b]	Range Type[c]
Ch. yasumatsui Trjap.	Homoptera: Coccoidea	7, 8	IV-2
Ch. paralia (Walk.)	Homoptera: 4, 5	5	III-2
Ch. fulvescens Hoff.	Homoptera: Coccoidea	1, 4, 5, 7	II-4
Ch. flavoscuttatus (Nik.)	Homoptera: 1	7	IV-3
Prochiloneurus pulchellus Silv.	Homoptera: 1, 5	1–3, 5, 7, 8	IV-3
P. bolivari Merc.	,,	1–5, 7, 8	IV-4
P. aegyptiacus (Merc.)	Homoptera: 1	8	IV-4
Tobiasia bifasciata Trjap.	Homoptera: 2	2, 3, 7	I-2
Mahencyrtus comara (Walk.)	Homoptera: 1	5, 7	III-2
Achalcerinys lindus (Merc.)	Homoptera: 1, 5	5, 7, 8	IV-4
Mayrencyrtus maculatus (Hoff.)	–	7	II-4
Bothriothorax aralius (Walk.)	–	2, 7, 8	III-2
Pentacladocerus matranus Erd.	Neuroptera: Hemerobiidae	5	II-4
Trjapitzinellus arboricola Myarts.	–	5	I-1
T. lambeiensis Myarts.	–	7	I-1
Cerchysiella planiscutellum (Merc.)	Coleoptera: Nitidulidae	5	III-1
Copidosoma farabense Myarts.	Lepidoptera	7	I-1
C. clavatum Myarts.	,,	5	I-1
C. nijasovi Myarts.	,,	2, 4, 5	I-1
C. kisilkumense Myarts.	Lepidoptera: Coleophoridae	4, 5	I-1
C. arvense Myarts.	Lepidoptera	8	I-1
C. juliae Myarts.	,,	1	I-1
C. longiventre Myarts.	,,	5	I-1
C. dushakense Myarts.	,,	5	I-1
C. balchanense Myarts.	,,	1, 2, 5	I-1
C. caspicum Myarts.	,,	1, 3	I-1
C. jamansaiense Myarts.	Lepidoptera: Coleophoridae	1, 4	I-1
C. saxaulicum Myarts.	Lepidoptera	1	I-1
C. autumnale Myarts.	,,	5	I-1
C. kushkense Myarts.	,,	5	I-1
C. bucharicum Myarts.	Lepidoptera: Tortricidae	4, 5	I-1
C. dendrophyllum Myarts.	Lepidoptera	5, 6	I-1
C. slavai Myarts.	,,	5	I-1
C. tugaicum Myarts.	,,	7	I-1
C. turanicum Myarts.	,,	1, 5, 7	I-1
C. augasmatis Trjap.	Lepidoptera: Coleophoridae	5	I-2
C. kuhitangense Myarts.	Lepidoptera	5	I-1
C. obscurum (Nik.)	Lepidoptera: Noctuidae	1, 2, 4, 5, 8	I-1
C. agrotis (Fonsc.)	,,	5	III-2

Table 1. Continued

Species	Hosts[a]	Habitat[b]	Range Type[c]
Paralitomastix varicornis (Nees)	Lepidoptera: Tortricidae, Gelechiidae, Pyraustidae	1, 2, 5, 7	IV-4
Ageniaspis fuscicollis (Dalm.)	Lepidoptera: Yponomeutidae, Acrolepiidae	6	IV-4
Holcothorax kopetdagensis Myarts.	Lepidoptera: Lithocolletidae	8	I-1
Parablastothrix reimovi Trjap.	–	1, 3–5, 7	I-1
Homalotylus nigricornis Merc.	Coleoptera: Coccinellidae	1–5, 7, 8	II-4
H. platynaspidis Hoff.	,,	8	II-4
H. flaminius (Dalm.)	,,	5, 7, 8	IV-4
H. balchanensis Myarts.	,,	5	I-1
H. turkmenicus Myarts.	,,	1–5, 7, 8	I-1
H. quaylei Timb.	,,	5, 7, 8	II-4
Isodromus flaviscutum Hoff. et Trjap.	Neuroptera: Chrysopidae	5, 8	II-4
I. luppovae Trjap.	,,	5, 8	I-1
I. ustianae Hoff. et Trjap.	,,	8	II-4
Eupoecilopoda perpunctata (Masi)	,,	1, 2	II-4
Encyrtus lecaniorum (Mayr)	Homoptera: 4	8	IV-4
E. infidus (Rossi)	,,	5	III-2
E. trjapitzini Myarts. et Sug.	,,	1–5	I-3
Astymachus phragmitis Trjap.	Homoptera: 2	5, 7, 8	I-2
Eugahania fumipennis (Ratz.)	–	1, 5	III-2
Anthemus aspidioti Nik.	Homoptera: 3	8	I-4
Ceballosia dusmeti Merc.	–	5	III-2
Coccopilatus zygophylli Myarts. et Sug.	Homoptera: 3	5	I-1
C. babaevi Myarts. et Sug.	,,	5	I-1
Ginsiana carpetana (Merc.)	–	5	III-2
Kurdjumovia plana Trjap.	Homoptera: 2	7	II-1
Lamennaisia ambigua (Nees)	–	5, 7, 8	IV-4
Negeniaspidius nobilis (Nees)	–	1, 2, 5, 7, 8	IV-4
Paraschedius bicolor Myarts.	Homoptera: 3	4, 5	I-1
P. jasnoshae Myarts. et Trjap.	,,	4, 5	I-2
Xerencyrtus compactus Trjap.	–	1, 2, 4, 5	II-3

[a] Within the order Homoptera, host families are: 1 – Pseudococcidae, 2 – Aclerdidae, 3 – Diaspididae, 4 – Coccidae, 5 – Eriococcidae
[b] Habitats: 1 – sand desert, 2 – clay desert, 3 – salt desert, 4 – stony desert, 5 – arid foothills, 6 – forest mountain belt, 7 – tugais, 8 – oases
[c] Types of the zoogeographic range: I-1 – Turanian, I-2 – Kura-Araxes-Turanian, I-3 – Irano-Turanian, II-1 – Turano-Gobian, II-2 – Mountainous Middle Asian, II-3 – Desert Mediterranean, II-4 – Ancient Mediterranean, III-1 – West Palearctic, III-2 – Trans-Palearctic, IV-1 – Holarctic, IV-2 – Palearctic-Indo-Malayan, IV-3 – Palearctic-Ethiopian, IV-4 – Cosmopolitan

Fauna and Biology of Encyrtidae in Turkmenistan

In Middle Asia, there are 314 species of the Encyrtidae belonging to 111 genera; in Turkmenistan, 254 species belonging to 109 genera (Table 1). Of those, we described 3 new genera and 121 new species, especially in genera *Copidosoma* (20 species), *Psyllaephagus* (12), *Discodes* (10) and *Microterys* (%) (Myartseva 1978, 1979, 1980a,b, 1981, 1983a). Fourty-three percent of this family's fauna within the former USSR is comrised of desert and semidesert species, revealing the fauna's expressed arid character.

Encyrtids have diverse life cycles. A majority of species overwinter as larvae (usually non-feeding); very few species overwinter as adults. The encyrtids parasitizing poplar Psylloidea (Homoptera) and Lepidoptera possess only one cycle of seasonal development, which is synchronous with that of their hosts. Those parasitizing Coccidae (Homoptera, Coccoidea) may have two cycles as do their hosts, and encyrtids parasitizing Pseudococcidae (Homoptera, Coccoidea) have many cycles which do not always correspond to those of the host.

In Turkmenistan, encyrtid wasps exhibit several types of seasonality, defined by the period of adult activity. The majority of species (86) are active from spring to fall (adults are found from April to October) or in spring and fall (when adults are rarely found or absent in July and August). These types of seasonal activity are characteristic for encyrtids inhabiting tugais, foothills, and oases. The share of Turanian endemic species in this group is 40%.

Another group (72 species) is active either only in spring, or from late spring to summer; of those, 45 species are active only in April and May. These are species inhabiting sand, salt, and stony deserts as well as xerophilic foothill communities. More than 60% of these encyrtids are endemic or subendemic Turanian species.

Massive swarms of active adults in the spring are characteristic for encyrtids parasitizing Lepidoptera (e.g., species of the genera *Copidosoma*, *Paracopidosoma*, and *Paralitomastix*). Some species of *Copidosoma* are also active as adults from late spring to summer; this type of seasonal activity is also exhibited by the species of *Psyllaephagus*, a parasite of desert Psyllidae (Homoptera), which feed on *Calligonum* and *Haloxylon*. A long period of activity, from spring to fall, is present primarily in widespread species (e.g., those of genera *Ooencyrtus*, *Homalotylus*, and *Prochiloneurus*) which inhabit agricultural landscapes and parasitize eurytopic insects.

The most common form of endoparasitism among encyrtids is the solitary type. Some encyrtids, however, are group, or gregarious, parasites (i.e., more than one female of the same species laying eggs into the same host individual); many of these are small and have flattened bodies, e.g., species of *Aphyculus*, *Astymachus*, *Platyrhopus*, *Pseudectroma*, and *Rhopus*). Encyrtids are the only group among Chalcidoidea in which polyembryonic development is found as a variety of group parasitism; it is characteristic for parasites of the Lepidoptera.

Several types of parasites can be distinguished according to their method of laying eggs and feeding of the larvae: these include egg parasites (e.g., *Ooencyrtus*), egg-larval parasites (e.g., *Holcothorax, Copidosoma,* and *Ageniaspis*), larval parasites (a majority of the genera), and larval-pupal parasites (e.g., *Isodromus*). Some species of the genus *Microterys* also prey on the eggs of Coccoidea.

Most of the studied encyrtids (80 genera) exhibit primary parasitism, in which the host species is a phytophage or predator. Secondary parasitism (hyperparasitism), in which the host species itself is a hymenopteran parasite, is known for 12 genera (e.g., *Aphidencyrtus, Cerapterocerus, Cheiloneurus, Echthrodryinus, Echthroplexiella, Prochiloneururs,* and *Tobiasia*). The secondary parasites always are solitary feeders.

High specificity of parasitism, in which a parasite species invades only a certain taxonomic group of hosts, is very characteristic for the Encyrtidae (Tryapitsyn 1972, 1989; Tachikawa 1974, 1981; Myartseva 1983b). In Turkmenistan, 210 species of Encyrtidae parasitize insects belonging to seven orders: Lepidoptera, Coleoptera, Neuroptera, Diptera, Homoptera, Hemiptera, and Hymenoptera (Myartseva 1984a). The specificity of host utilized by primary parasites is extremely high in Hymenoptera (100%), Homoptera (96.5%), and Lepidoptera (85.7%). The majority of Encyrtidae in Turkmenistan (164 species) are parasites of the Homoptera; of those, 128 species are parasites of the superfamily Coccoidea (Table 1). Also, all four of the largest genera of the Encyrtidae are parasites of the Homoptera; these are *Anagyrus, Metaphycus* and *Microterys* (parasites of Coccoidea) and *Psyllaephagus* (parasites of Psylloidea).

In decreasing order of number of encyrtid genera parasitizing different families of Coccoidea, the distribution of these families is following: Pseudococcidae, Aclerdidae, Diaspididae, Coccidae, and Eriococcidae. The predominance of herbaceous vegetation, shrubs, and semi-shrubs in desert landscapes determines the important role of Eriococcidae as encyrtid hosts as compared to Coccidae, which are mostly dendrophiles: i.e., in Turkmenistan, they feed primarily on trees in tugais and oases. For example, five species of encyrtids belonging to the genus *Discodes*, which primarily parasitize Coccidae, were hatched in Turkmenistan from six species of Eriococcidae. This case can be considered as an example of parasite speciation which followed occupation of a new host taxonomic group.

Also, many of the widespread desert encyrtids which normally feed on Pseudoccocidae can easily shift to Eriococcidae, which are similar to Pseudoccocidae in their morphology (especially in early larval stages) and feed on the same plants. For example, genera *Metaphycus* and *Microterys*, which generally parasitize he dendrophilic Pseudoccocidae, are represented in Turkmenistan by a number of species found in Eriococcidae which feed on herbaceous and shrub vegetation.

The majority of desert encyrtids are monophagous. Of 165 studied species, 124 (76.1%) were monophages, whereas 38 (23.1%) were oligophages, and only 3 species (1.8%), polyphages. Examples of narrow oligophages include

Metapsylaephagus desantisi, which parasitizes four species of *Pachypsylloides* (Psylloidea) feeding exclusively on *Calligonum* in sand desert; or *Copidosoma kisilkumense*, which parasitizes caterpillars of five species of *Coleophora* moths (this is the maximal number of hosts found for an oligophagous encyrtid). Broad oligophages, on the other hand, can utilize taxonomically diverse hosts which feed on a variety of plants. E.g., *Anagyrus haloxyli* in salt desert lives on *Phenacoccus arthrophyti* (Pseudoccocidae), which feeds on *Salsola* spp.; in oases, however, it parasitizes *Trionymus multivorus*, which feeds on herbaceous vegetation, and on *Peliococcus mesasiaticus*, which feeds on mulberry (*Morus* spp.). Many secondary parasites of encyrtids living on Coccoidea have a broad range of hosts; e.g., eight species belonging to seven genera serve as hosts for *Cheiloneurus claviger*; the encyrtid *Prochiloneurus bolivari* was hatched from seven species belonging to six genera of Coccoidea. Among encyrtid genera, broadening of food specialization is determined by the morphoecological similarity between the original and new hosts. In general, the homopteran superfamily Coccoidea (except for the most archaic and some advanced groups) can be considered as a basic host group for Encyrtidae. The evolution of parasitism in encyrtids involved primarily the host families Pseudoccocidae, Coccidae, and Diaspididae (Myartseva 1986).

Ecology of the Encyrtidae in Turkmenistan

Since encyrtid parasites are monophages or narrow oligophages (Myartseva 1983c), their ecological distribution is defined by that of both their host insects (e.g., Coccoidea or Lepidoptera) and the food plants of these hosts (Table 1). In the lowlands and foothills of Turkmenistan, encyrtids are found in the most common plant communities, dominated by *Tamarix* spp., *Salsola* spp., sagebrushes (*Artemisia* spp.), and also in grass-herbaceous communities. However, faunistic composition of encyrtids in certain communities depends on the landscape. For example, *Salsola* communities in all types of lowland deserts and in the foothills contain 10 or 11 species of encyrtids; in tugais, however, *Salsola* communities include only three encyrtid species. The encyrtid fauna of sagebrush communities is the richest in the foothills (28 species), is less rich in stone desert (12 spp.), and includes only from 7 or 8 species in sand and clay deserts. The richest encyrtid fauna on *Tamarix* inhabits tugais (23 species), whereas in sand and salt desert *Tamarix* communities possess 11 and 10 species, respectively, and in stone desert, only four species. The encyrtid complex inhabiting communities of grassy and herbaceous vegetation are richest in tugais (32 species), where large grasses house most of the host insects, and in the foothills (36 species), with their diversity of xerophytic plants. In oases, grassy-herbaceous communities contain only 13 encyrtid species; in clay desert, 11; and in sand and stony desert, 6 or 7 species.

The specificity of encyrtid fauna for each of these landscapes is defined by the specificity of their host phytophagous insects and their food plants. For example, in sand desert with psammophytic vegetation, encyrtid fauna is dominated by species parasitizing Coccoidea and Psylloidea which often lead secretive lives in dominant psammophyte plants such as *Calligonum* or *Haloxylon*. In tugais, most encyrtids feed on phytophages of tall savanna-type grasses or those of *Tamarix*.

The most diverse in Turkmenistan fauna is comprised of encyrtids in the arid foothills (105 species); the second in richness is the tugai fauna (95 species); in oases, 69 species are found. Desert fauna, with the exception of sand desert (68 species), is rather impoverished, with 50 species in stony desert; 41, in clay desert; and only 36, in salt desert. Many encyrtids prefer a specific landscape: their proportion is 61% of the fauna in the arid foothills; 29.3% in both tugais and oases; 14.7% in sand desert; 11.1% in salt desert 5%, in clay desert; and 4% in stony desert.

Zoogeography of Encyrtidae in Turkmenistan

The family Encyrtidae has almost cosmopolitan distribution. We have attempted to classify the encyrtids found in Turkmenistan according to the type of their geographic ranges (Table 1). Range types vary from almost cosmopolitan, in some polyphagous species of the Paleotropic origin, to the limited lowland or mountainous areas in endemic mono- or oligophagous species. Four groups and thirteen types of zoogeographic ranges were distinguished.

Group I: Ranges not extending outside of the Irano-Turanian subprovince of the Saharo-Gobian Desert Province of the Palearctic Realm. Include three types of ranges: I-1. *Turanian* range (127 spp., or 50 % of the Turkmenistan fauna), mostly within the southern desert subzone); I-2. *Kura-Araxes-Turanian* range (17 spp. found in the lowlands and foothills of both Middle Asia and East Transcaucasia); I-3. *Irano-Turanian* range (5 spp.).

Group II: Ranges not extending outside of the Ancient Mediterranean area. Include four types of ranges: II-1. *Turano-Gobian* range (12 species distributed in Middle Asia, Kazakhstan, and Mongolia); II-2. *Mountainous-Middle Asian* range (typical only for three encyrtid spp. of our fauna); II-3. *Desert Mediterranean* range (6 spp.); and II-4. *Ancient Mediterranean* range (26 spp).

Group III: Ranges not extending outside of the Palearctic Realm. Include two types of ranges: III-1. *West Palearctic* range (5 spp., not found eastward from Turkmenistan); and III-2. *Trans-Palearctic* range (24 spp.).

Group IV: Ranges extending outside of the Palearctic Realm. Include four types of ranges: IV-1. *Holarctic* range (6 spp.); IV-2. *Palearctic-Indo-Malayan* range (5 spp.); IV-3. *Palearctic-Ethiopian* range (4 spp.); and IV-4. *Cosmopolitan* range (14 spp.). The latter type is characteristic also for a number of adventive species (e.g., *Encyrtus lecaniorum, Pseudaphicus malinus*, and *Neodiscodes* sp.) which were inadvertently introduced by man or were brought from other regions as effective entomophages.

The encyrtid fauna in Turkmenistan has, expectedly, an arid zoogeographic character. Of this fauna, ranges of 196 species do not extend outside of the Ancient Mediterranean area of the Saharo-Gobian Desert Region as delineated by Lavrenko (1965). Of these, 127 species are endemic for the Turanian Province. The core genera including these endemic species have Paleotropic (ca. 30 genera) or Ancient Mediterranean (ca. 20 genera) origin.

Among encyrtids parasitizing on hosts which feed on shrubs and semishrubs, about 70% are Turanian endemic species; we believe that the major adaptive radiation of the Encyrtidae in desert Turkmenistan has occurred among this group. In contrast, widespread encyrtid species found in Turkmenistan often parasitize on phytophagous insects which feed on trees, large shrubs, and, less often, on herbaceous perennial plants.

The encyrtid complexes in the lowlands and foothills of Turkmenistan have been formed under the progressing aridization and continentalization of climate. They are parasites of insects feeding primarily on dominant arid genera of plants belonging to the families widespread in the desert, such as Chenopodiaceae, Zygophyllaceae, Fabaceae, Tamaricaceae, and sagebrushes *Artemisia* (subgenus *Seriphidium*) (Asteraceae). In general, the encyrtid fauna of Turkmenistan can be characterized as a desert fauna typical for the eastern Saharo-Gobian (or Sethian) Desert Region of the Palearctic. Our data show that Middle Asia is one of the important centers of speciation for a xerophile fauna of parasitic Hymenoptera within the desert zone of the Palearctic.

Practical Importance of Encyrtids

The parasitic Hymenoptera are natural regulators of the density of phytophages. The Encyrtidae feed primarily on Coccoidea (Homoptera), among which several are known to be a number of pests of southern fruit, forest, and decorative tree species. Certain encyrtid species are used for biological control. For example, the encyrtid species *Pseudaphycus malinus* Gah. was introduced in Turkmenistan in 1962 to control the reproduction of *Pseudococcus comstocki* Kuw., which resulted in the decrease of this important pest of mulberry trees. In the 1980s, another species, the native encyrtid *Anagyrus pseudococci* Gir., began active parasitizing of the same pest (Myartseva 1984b). In contrast with a specialized monophage, *Pseudaphycus malinus*, *Anagyrus pseudococci* is able to feed on other Pseudococcidae and inhabits both natural and agricultural habitats in Turkmenistan. The artificial breeding of *Anagyrus pseudococci* and its release into the areas of mass reproduction of Pseudococcidae can be an important method of biological control: the degree of infestation by *Anagyrus pseudococci* can reach 60 to 70%, for example, in the case of *Planococcus ficus* as a host insect (Myartseva 1986).

The Encyrtidae, therefore, are a significant component of natural insect communities. Their conservation is best insured within protected territories such as the Natural Reserves. To protect these and other beneficial insects in agroecosystems, the enforcement of proper pesticide handling is of primary importance.

Acknowledgements

I am grateful to my teacher and colleague, Dr. V.A. Tryapitsyn of the Zoological Institute of the Russian Academy of Sciences, for his constant interest in my work and his important advises and consulting. I also thank my dear colleagues, Drs. E.M. Danzig and M.I. Falkovich, of the same Institute, for their help in identification of encyrtid hosts belonging to Homoptera and Lepidoptera.

28. Zoogeography and Ecological Aspects of the Formation of Horse Fly Fauna (Diptera: Tabanidae) in Turkmenistan

RIMMA V. ANDREEVA

Abstract

The horse fly fauna of Turkmenistan was studied, which includes 44 species and four subspecies belonging to nine genera and two subfamilies. Morphological and ecological analysis of the horse flies is given, and larval stage adaptations are described. Ninety percent of the Turkmenistan horse fly fauna is comprised of species of Southwest Asian or Turanian origin. Ten species are Turanian endemics, and two species are endemics of Turkmenistan. Centers of diversity and abundance of horse flies in Turkmenistan are concentrated in the southern deserts of the Irano-Turanian lowland. This area can be considered a faunogenetic center for the genus *Nanorhynchus*, subgenera *Nemorius* and *Turanochrysops*, and the species groups *Tabanus sabuletorum* and *T. cordiger*.

Introduction

Studies of Turkmenistan horse flies are comparatively recent and not yet complete. The first detailed list of Turkmenistan horse flies, containing 25 species, was published by Olsufyev (1937) in a monographic treatise, "Fauna of the USSR." and the next edition of this monograph (Olsufyev 1977) listed 32 species and 4 subspecies of horse flies for Turkmenistan. Special studies have also been undertaken of the distribution and ecology of the adult (24 species) as well as larval (7 species) stages of horse flies (Babayants 1962; Charykuliev 1967; Mamedniyazov 1971; Charykuliev and Yasakova 1973; Yasakova and Krivosheina 1973; Yasakova 1974). Several papers have addressed rare and understudied species (Paramonov 1961; Olsufyev and Shevchenko 1964; Olsufyev 1966). Some foreign studies have also included characteristics and distribution of several horse fly species in the Turkmenistan fauna (Abbasian-Lintzen 1964; Leclerq 1960, 1966).

Studies by the author (Dolin and Andreeva 1984, 1986; Andreeva 1985, 1990) of the ecology and morphology of larval stages of horse flies in Turkmenistan have supplemented the faunal list, which presently contains 44 species and 4 subspecies. The ecology of 37 species has been studied and their larvae described.

There has been no special zoogeographic or faunogenetic study of Turkmenistan horse flies. The Palearctic fauna of horse flies in general was the subject of arealogical and ecological analysis by Olsufyev (1977, 1980), although Chernov (1984) criticized this treatment of a faunistic complex as a principal zoogeographic unit.

Background of the Research Approach

In the process of evolution, life conditions of species are reflected in their particular morphophysiological adaptations, which thus should not be ruled out when reconstructing the faunogenesis of a group. Given the uncertainty of historical reconstructions based on recent distributions of species, as well as the poor paleontological record for most dipteran familes, the use of ecology in faunogenetic studies can be well justified. Rodendorf (1964) emphasized the importance of correct interpretations of morphoecological changes during phylogenesis. Hennig (1981) held a similar opinion, considering as incomplete those phylogenetic reconstructions which did not account for current knowledge of evolutionary biology and ecology.

In horse flies, morphoecological characteristics are more pronounced in larval forms. As do many other Holometabola, horse flies possess two active life stages, strikingly different in their mobility, life habits, and ecological plasticity. Habitats of larvae in horse flies are much more diverse than those of adult insects, and the larval stage occupies about 85% of the horse fly lifetime. That horse fly species are exposed to ecological conditions primarily during their long larval stages should, therefore, be reflected in larval morphoadaptations. The diversity of larval forms in Tabanidae is one of the highest within the entire order of Diptera. At the same time, variation in adult horse flies is reduced to slight differences in coloration or pubescence due to their highly homogeneous ecological environment.

Data on larval morphology and ecology are now available for about half of the Palearctic horse fly species. These species are classified into different life forms according to larval habitat and morphological adaptations (Andreeva 1982, 1987, 1990a,b; Burger 1988). The adaptive structures serve as indicators of ecological conditions under which larvae undergo their development. The level of morphological specialization of a species determines the limits of its natural distribution, and evolution of life forms can be traced along and characterized from the reconstructed paleoclimatic and paleogeographic changes.

The comparative morphology of larval forms has a high taxonomic value at species, genus, and even tribe levels. Andreeva (1984, 1985, 1986, 1990) demonstrated that species-specific larval morphology in horse flies (especially within a group of closely related species) is taxonomically more informative than the morphology of adult insects. Evolutionary changes in these larval structures can be therefore traced more clearly, allowing phylogenetic reconstructions for closely related species with different geographic ranges.

This approach may be especially useful when discussing historical dispersal within such a wide and complex area as the Ancient Mediterranean. The composition of the horse fly fauna of Turkmenistan has been clearly influenced by species dispersal from Southwest Asia (where elements of the subtropical Mediterranean climate still persist) as well as from the other regions of Middle and Central Asia in the course of their climatic changes. However, the core of this fauna has been formed autochthonously under the changes in landscapes and climates of the Irano-Turanian lowlands, which began no later than in the Paleogene, and have continued since that time with increasing aridization of climate.

Extant classification of many larval life forms of Palearctic horse flies has facilitated our reconstruction of the faunogenesis of these taxa, including those of Turkmenistan (Andreeva 1991).

Morphological and Ecological Analysis of the Turkmenistan Horse Flies

The current distribution of an animal species is determined primarily by the existing ecological conditions. The majority of horse fly larvae are known to be hydrophilous organisms. Of all horse fly species inhabiting arid Turkmenistan, 75% have larvae whose development is directly connected with water (Table 1). About half of these larvae are rheophiles or subrheophiles (Andreeva 1989). Depending on their specialization, these species develop either in a bottom layer of creeks, rivers, or irrigation ditches in low mountains and foothills (ten species) or in the streams and rivers of desert and semidesert lowlands (seven species). As an adaptation to life in streams with different current speeds, larvae of these species possess elongated pseudopodiae equipped with strong chetae, or hooks (Fig. 1).

Table 1. The systematic list of the horse flies of Turkmenistan with habitat distribution, indication of larval life forms, and type of geographic range.

No.	Species	Life Forms	Habitat	Range Type (see Table 2)
	Subfam. Chrysopsinae			
	Tribe Chrysopsini			
	Genus *Chrysops* Meigen			
	Subgenus *Chrysops* Meigen			
1.	Ch. (Ch.) *flavipes punctifer* Lw.	Hhb	De-St	2
2.	Ch. (Ch.) *mlokosiewiczi* Big.	Hhb	De-St	7 (1)
3.	Ch. (Ch.) *oxianus* Pl.	Hhb	De-St	10
	Subgenus *Turanochrysops* Stack.			
4.	Ch. (T.) *hyalipennis* Stack.	Hb-Rh	Ft-St	9 (1)

Table 1. Continued

No.	Species	Life Forms	Habitat	Range Type (see Table 2)
5.	Ch. (T.) stackelbergiellus Ols.	Hb-Rh	Ft-St	9 (1)
	Genus Silvius Mg.			
	Subgenus Nemorius			
6.	S. (N.) vitripennis Mg.	Hb-Rh	Ft-Mt	1
7.	S. (N.) irritans Ric.	Hb-Rh	Ft-St	8
	Subfam. Tabanidae Tribe Diachlorini			
	Genus Nanorhynchus Ols.			
8.	N. crassinervis Villen.	Ed-Pb	De-St	9 (1)
	Genus Dasyrhamphis End.			
9.	D. umbrinus Mg.	Ed-Pb	Ft-St	1 (1)
	Tribe Tabanini			
	Genus Tabanus L.			
	Tabanus sabuletorum group			
10.	T. sabuletorum Lw.	Ed-Pb	De-St	7
11.	T. subsabuletorum Ols.	Ed-Pb	De-St	9
12.	T. accipiter Szil.	Ed-Pb	De-St	9
13.	T. freyi Szil.	?	?	9 (1)
14.	T. ansarii badhysi Ols. et Shevch.	Ed-Pb	Ft-St	9 (1)
15.	T. beschkenticus Bar.	Ed-Pb	De-St	9 (1)
	Tabanus cordiger group			
16.	T. leleani leleani Aust.	Hb-Rh	Ft-St	2
–	T. l. turkestanicus Ols.	Hb-Rh	De-St	9
17.	T. golovi golovi Ols.	Hb-Rh	Ft-St	10
–	T. g. mediasiaticus Ols.	Hb-Rh	De-St	10
18.	T. unifasciatus Lw.	Hb-Rh	Ft-Mt	1 (1)
19.	T. fumidus Aust.	Hb-Rh	Ft-St	8
20.	T. filipjevi Ols.	Hb-Rh	De-St	7
21.	T. zimini Ols.	Hb-Sbrh, Hhb	De-St	6
	Tabanus bromius group			
22.	T. sordes Bog. et Sam.	Hb-Rh	Ft-Mt	6 (1)
23.	T. canipalpis terterjani Dol. et Andr.	Hb-Rh	Ft-Mt	6 (1)
24.	T. regularis Jaenn.	Hb-Rh	Ft-Mt	4 (1)
25.	T. cuculus Szil.	Hb-Sbrh, Hhb	Ft-Mt	8
26.	T. atamuradovi Dol. et Andr.	Hb-Sbrh	De-St	9 (1)
27.	T. laetetinctus Beck.	Hhb	Ft-Mt	6
28.	T. semenovi Ols.	Hhb	Ft-Mt	10
29.	T. bromius L.	Hhb	Ft-Mt	11
–	T. b. flavofemoratus L.	Ed-Cb	Ft-Mt, Ft-St	2
30.	T. indrae Haus.	Ed-Cb	Ft-Mt	4 (1)
31.	T. appendiculatus Szil.	Ed-Cb	Ft-Mt, Ft-St	8
	T. autumnalis group			
32.	T. autumnalis brunnescens L.	Hhb	Ft-St, De-St	2
33.	T. spectabilis Lw.	Hhb	Ft-St, De-St	2
	T. bifarius group			
34.	T. quatuornotatus Mg.	Ed-Cb	Ft-Mt	1 (1)
35.	T. atropathenicus Ols.	Hb-Rh	Ft-Mt	6 (1)
	Genus Hybomitra End.			
	Subgenus Hybomitra s.str.			

Table 1. Continued

No.	Species	Life Forms	Habitat	Range Type (see Table 2)
	H. erberi group			
36.	H. (H.) erberi Br.	Hhb	De-St	3
37.	H. (H.) peculiaris Szil.	Hhb	Ft-St, De-St	5
38.	H. (H.) turanica Ols.	Hhb	De-St	9 (1)
	Subgenus Sipala End.			
39.	H. (S.) acuminata Lw.	Hhb	De-St	5
	Genus Atylotus O.S.			
40.	A. (A.) quadrifarius Lw.	Hhb	De-St	1
41.	A. (A.) pulchellus pulchellus Lw.	Hhb	De-St	2
	A. (A.) p. carybenthinus Szil.	Hhb	De-St	7
42.	A. (A.) sp.	Hhb	De-St	9 (1)
	Genus Therioplectes Zell.			
43.	T. carabaghensis orientalis Ols.	Hhb	Ft-Mt	9 (2)
	Tribe Haematopotini			
	Genus Haematopota Mg.			
44.	H. pallens Lw.	Ed-Cb	Ft-St	4

Life Forms: Hb – hydrobionts; Rh – rheophiles, Sbrh – subrheophiles, Hhb – hemihydrobionts, Eb – edaphobionts, Cb – cespitobionts, Pb – psammobionts
Habitats: Ft-Mt – foothills and mountains; Ft-St – foothill steppes; De-St – lowland deserts and steppes

Among the rheophiles, usually considered to belong to the *Tabanus cordiger* group (Chvala *et al.* 1972; Olsufyev 1977), the larvae of the following two species are distinguished by their peculiar morphology. Larvae of *Tabanus filipjevi* inhabit the bottom layer of slow-current water bodies in the desert regions (Lakes Dashkak and Cheskak); however, this species possesses the longest pseudopodiae, with apices nearly 2.5 times narrower than bases, in the entire *Tabanus cordiger* group. The protruded preanal ridge characteristic to the rest of this species group is replaced in *T. filipjevi* by two protrusions similar to pseudopodia (Fig. 1.2). These very delicate structures differ significantly from analogous structures in the rheophile larvae inhabiting streams and rivers in deserts, and mountain streams of Turkmenistan. These structures probably provide larval attachment to the substrate during the intense water agitations (caused by frequent and strong winds) in coastal zones. The specificity of larval adaptive structures in *T. filipjevi* suggests an early separation of this species from the *T. cordiger* group as well as its long evolution under the constantly changing conditions in the river system of the Turanian lowland, during which small rivers were transformed into flowless lakes in the Paleogene and Neogene (Babaev and Fedorovich 1970).

In the species *Tabanus zimini*, larval pseudopodiae are comparatively poorly developed, and preanal ridges are covered with additional folds (Figs. 1.3 and 1.4). These larvae can be found in great numbers in the bottom layer of streams (e.g., Uzboi and a "salt aryk" in the Yeroyulanduz Depression at Badghyz) as well as in small water bodies with very high mineralization (e.g., filtration basins

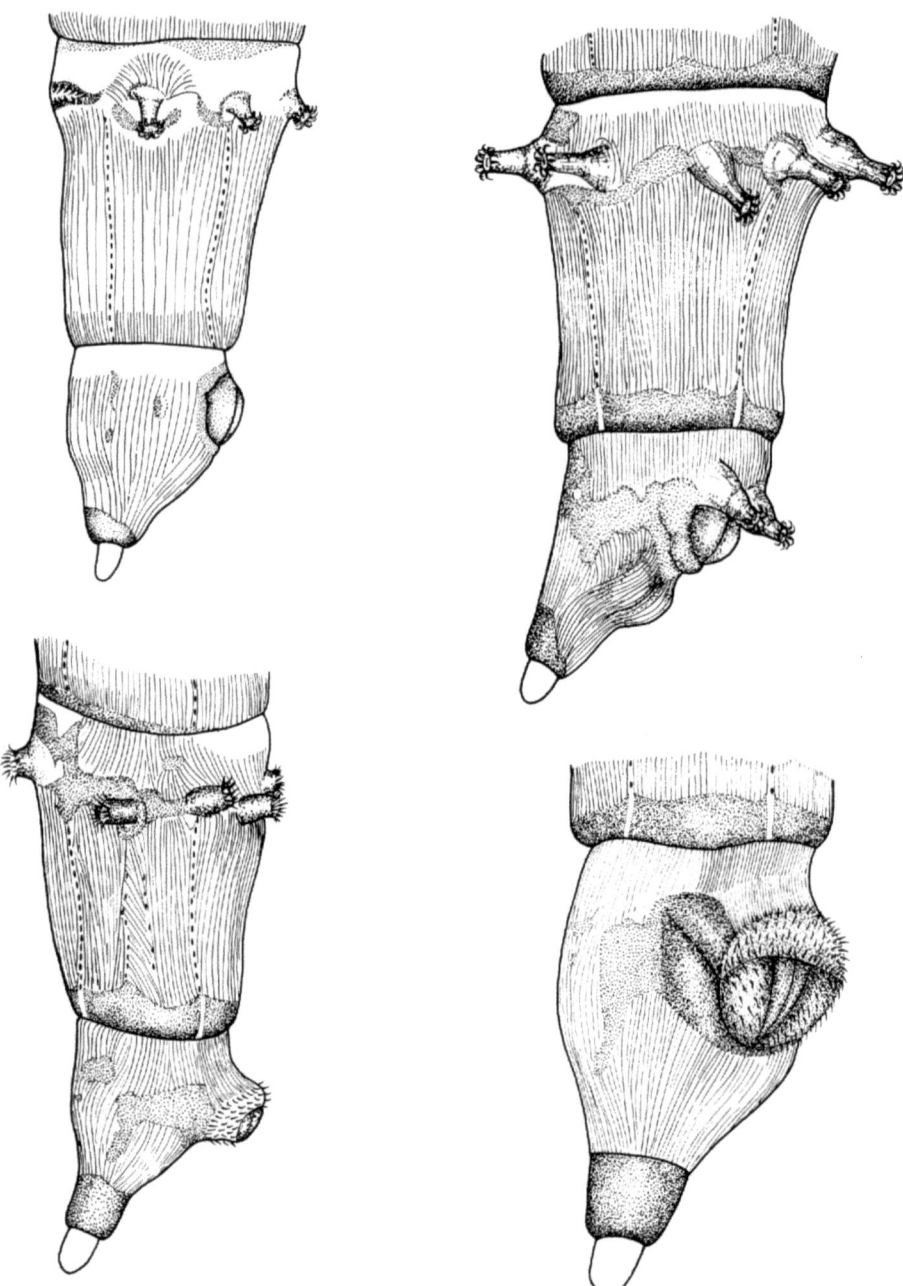

Fig. 1. Morphoecological adaptations of posterior segments in rheophile horse fly larvae: 1 – *Chrysops (Turanochrysops) hyalipennis* Stack.; 2 – *Tabanus filipjevi* Ols.; 3, 4 – *T. zimini* Ols.

in the Karakum Desert). The additional folds at the anal tuber possibly protect the insect from dehydration in a medium with high salt concentration. We (Andreeva 1991) have doubted that *T. zimini* belongs to the *T. cordiger* group because of its significant morphological differences.

Larvae of other sixteen species whose development is connected with water lead a semiaquatic way of life. They live in the moist banks of streams and rivers and in the vicinity of various water bodies and wetland areas. Larval habitats in different species of this ecological group are concentrated either in mountains and foothills or in lowland deserts (Table 1).

Larvae of the remaining species (27% of the total list) are edaphobionts. Such prominent representation of soil dwellers in an arid territory is explained by specific morphological and physiological adaptations, especially by those related to a psammobiotic way of life (Andreeva 1990). Larvae of six species belonging to the *T. sabuletorum* group [which only recently has been separated from the *T. cordiger* group on the basis of principal differences in larval morphology and ecology (Andreeva 1987, 1990)] develop in desert habitats within sand or sand-loess soils moistened either directly from adjacent water bodies or by the condensation of atmospheric moisture.

The highest degree of specialization to desert conditions is demonstrated by the larvae of *Dasyrhamphis umbrinus* and larvae of a monotypic genus, *Nanorhynchus* (Figs. 2.1 and 2.2). Larvae of the first species have been found in the sand-loess soil of the foothills of West Kopetdagh (Sumbar and Chandyr Valleys), which are covered by a sparse grass and shrub vegetation. Larvae of *Nanorhynchus* inhabit sand dunes in the Karakum Desert (Repetek National Park) at a depth of 0.8 to 1.4 m, under the roots of *Haloxylon* spp. and *Ammodendron conollyi*. Larvae of both species are quite similar in body shape, character of the tegument, and presence of protective structures (which are, however, morphologically different; Fig. 2.4). Detailed study shows that the shape of larval perianal structures, especially of the adaptive ones, can be quite variable even within a single genus of the horse flies, such as *Tabanus*.

The taxonomic position of the genus *Dasyrhamphis* is not clear. Our study of the head capsule in *D. umbrinus* larvae has revealed the absence of a connection between the ends of tentorial rods and the walls of the head capsule. This feature has not been previously discovered in horse fly larvae; phylogenetically, the closest group with similar anatomy is the family Rhagionidae. The significance of the discovery of this (ancestral?) feature in the evolution of head anatomy in the lower Diptera has yet to be studied; it will be necessary to compare larvae of different Diachlorinae and other tropical groups of Tabanidae.

Larvae of three more species belong to the ecological group of edaphobionts: *Tabanus quatuornotatus*, *T. indrae indrae*, and *T. appendiculatus*. These species are found in foothills and mountains; their larvae develop in the soils of riparian forests. *T. appendiculatus*, which was previously considered a synonym of *T. bromius flavofemoratus* (Olsufyev 1977), has been revalidated after the morphological differences between larvae of these two species was demonstrated (Andreeva 1985).

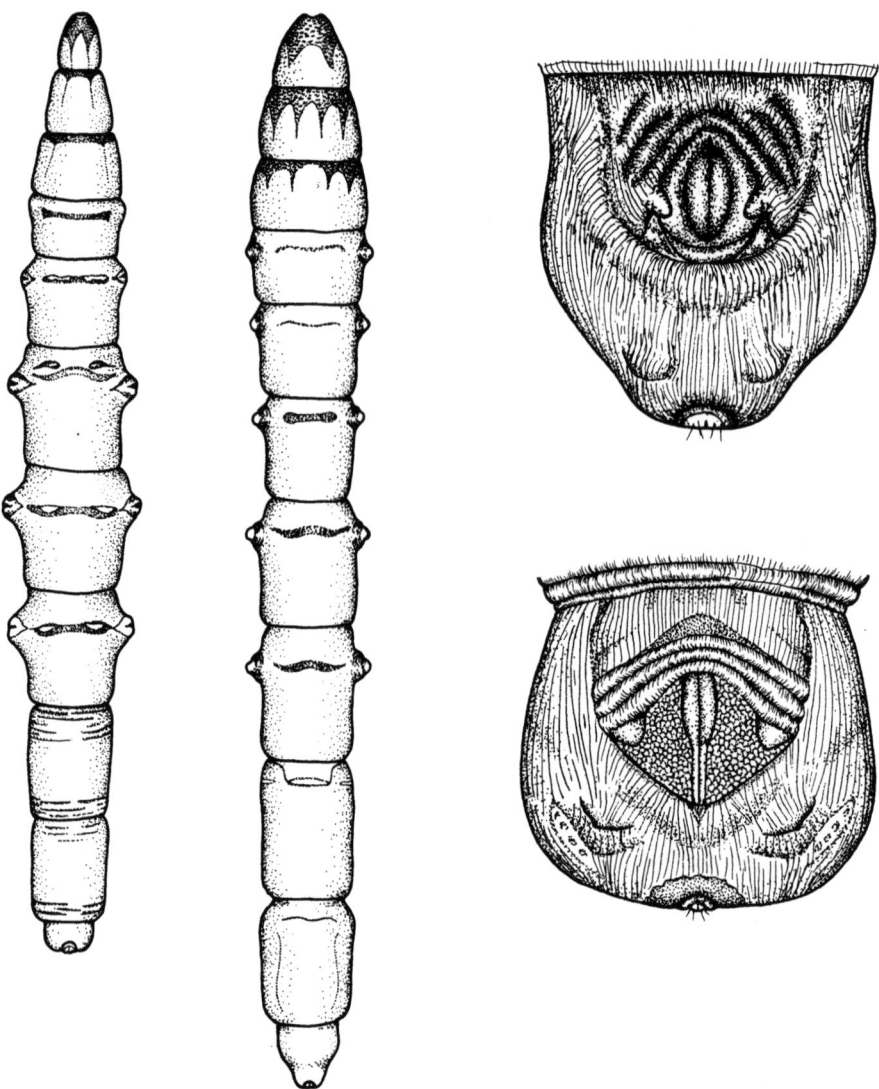

Fig. 2. Edaphobiont horse fly larvae: 1, 4 – *Nanorhynchus crassinervis* Villen.; 2, 3 – *Dasyrhamphis umbrinus* Mg.

Distribution and Faunogenesis in the Horse Flies of Turkmenistan

The core fauna of horse flies in Turkmenistan is comprised of the Turanian endemics, including the endemics of Turkmenistan. These species evolved under the constantly increasing aridization so characterizistic for Middle Asia. In the Upper Cretaceous, aridity in the central parts of Middle Asia was softened by the influence of the Tethys and Middle Asian Seas. Starting in the Miocene, dry steppes, semideserts, and deserts have developed throughout the arid territories of Asia (Sinitsyn 1965). The palynological data confirm the existence of *Haloxylon* plant communities in Middle Asian lowlands in the Pliocene. Naturally, autochthonous endemics of Middle Asia are adapted to the arid conditions of deserts and semideserts. These characteristic morphological adaptations can be found within the genus *Nanorhynchus* and the *Tabanus sabuletorum* species group.

Nanorhynchus crassinervis, an endemic of Middle Asia, is a relatively rare species which is known from the southern Karakum and the Kizylkum Deserts. Its ecology suggests that the ancestors of this peculiar species could have evolved on the arid southern shores of the Middle Asian Sea which existed in the Upper Cretaceous. Representatives of the same tribe (subgenus *Aegialomyia* Phil.), morphologically similar to *Nanorhynchus*, are found in the sand dunes along the Atlantic coast of western Florida (Olsufyev 1966; Goodwin 1974).

It is difficult to determine accurately the composition of the *Tabanus sabuletorum* species group because the horse fly fauna of Iran and Afghanistan is not sufficiently studied. At present, this group includes seven Middle Asian and one Central Asian species; it probably also includes four Iranian species (Jezek 1981, 1990) for which only adults are described; new species may be still discovered in Afghanistan and Central Asia. The ranges of species are usually localized and do not extend outside the desert landscapes of southern Karakum, Kizylkum, and Iran. *Tabanus stackelbergiellus* Ols. inhabits Inner Mongolia and the adjacent parts of China. *T. sabuletorum* (including its northern subspecies, *T. s. gerkei* Br.) has the largest range extending from the steppes of the southern Ukraine through the arid southeastern Transcaucasia, deserts of Southwest and Middle Asia, Tuva and Transbaikalia, and to the mountain deserts of Central Asia.

There are six species belonging to this group in Turkmenistan. *Tabanus accipiter*, *T. freyi*, and *T. ansarii badhysi* are found principally in southern parts of the republic. In the Southeast (Kelifsky Uzboi) *T. beschkenticus* and *T. subsabuletorum* (Olsufyev 1977) have been found although our records do not reveal the presence of the latter species to the south of Farab. The fauna of southeastern Turkmenistan is not sufficiently studied, and additional species of this group may be discovered there.

Whereas the survival of some ancestral forms has been characterized by the adaptation of their larvae to the life in arid soils, another ecological group of horse flies has evolved towards larval adaptation to aquatic life. These include the previuosly discussed *Tabanus cordiger* species group, whose larvae are rheophiles, and species of the subgenera *Turanochrysops* and *Nemorius*.

Turanian endemics belonging to the subgenus *Turanochrysops* are the only representatives of the genus *Chrysops* which exhibit the explicit morphoadaptations to the rheophilous mode of life (Fig. 1.1). They are distributed along the river valleys of Syrdarya, Zeravshan, Amudarya and its present (Vakhsh and Pyandzh) and former (Murghab and Tedzhen) tributaries.

Seven species of the subgenus *Nemorius* are known (Leclercq 1985; Jezek

1990), primarily from Southwest Asia, the Caucasus, and southern Middle Asia. Only *Silvius* (*Nemorius*) *vitripennis* has a wide range, from Spain and southern Europe to Kyrghyzstan. Five species are known from Iran, and only two from Turkmenistan: *S.* (*N.*) *vitripennis*, which is common in the foothills, and *S.* (*N.*) *irritatus*, which is found only in the southernmost part of the republic.

The species *Tabanus atropathenicus* inhabits certain regions of Armenia and Azerbaijan as well as Turkey and Iran; in Turkmenistan it is common in the foothills of West and Central Kopetdagh.

Of species belonging to the *Tabanus cordiger* group, *Tabanus cordiger* itself has Trans-European distribution, whereas *T. unifasciatus* and *T. leleani* can be found in southern Europe. The latter two, together with six more species, are common in the Irano-Turanian province; of these eight species, only three (*T. leleani*, *T. golovi*, and *T. filipjevi*) extend their range to Central Asia. One additional species, *T. kinoshitai* Kono et Tak., is described from East Asia. Larvae of this species group develop in the different types of deserts, steppes, foothills, and mountain streams and rivers. The observed diversity of larval habitats and species-specific habitat distribution might be explained, in part, by the historical instability of river drainages in the deserts as well as by insect migration to the foothills. Following the retreat of the Middle Oligocene sea from the Middle Asian lowlands, the exposed sea bottom was occupied by early rivers which, through their constant wandering, created alluvial deposits by the Lower Miocene. It is known that the Syr Darya River, now the second largest river of Middle Asia, was in the Neogene yet nonexistent as a single artery and repeatedly chaning its bed, while Murghab and Tedzhen were its tributaries (Babaev and Fedorovich 1970).

Among the Palearctic representatives of the genus *Tabanus*, the *T. bromius* group is the largest one, including about 20 species. Ranges of the majority of these species lie within Southeast and Middle Asia. Ten species of this group inhabit Turkmenistan; for six of them (*T. indrae*, *T. regularis*, *T. canipalpis terterjani*, *T. appendiculatus*, *T. cuculus*, and *T. sordes*) Central Kopetdagh is the eastern boundary of distribution. *T. atamuradovi* is described from the desert areas of the utmost southeast of Turkmenistan. Ranges of the remaining three species extend further to the east.

The *T. autumnalis* group is represented in Turkmenistan by two wide-ranging species common in the lowland deserts and foothills.

Two genera, *Atylotus* and *Hybomitra*, which are related to the genus *Tabanus*, are comprised exclusively of desert and steppe species; the majority of these species are quite common and widespread. Only the range of the Turanian endemic *Hybomitra turanica* does not extend beyond the lower Amudarya River; its southern boundary of distribution runs through Chardzhou. The representatives of the genus *Hybomitra* live predominantly in the temperate and cold zones; a complex of highland species is known from Middle and Central Asia (Andreeva 1991). Only the members of the *H. erberi* group and the subgenus *Sipala* inhabit steppe and desert landscapes in the Palearctic region. Larvae of these species possess hemihydrobiotic adaptations which probably evolved slowly under the influence of the increasing aridization, early separating these taxa from the other groups of the genus. Curiously, the squamous, flattened shape of chetoids in *H. (S.) acuminata* is a preserved ancestral feature which is absent from all other species of the genus *Hybomitra*.

The genera *Chrysops* and *Haematopota* are poorly represented in Turkmenistan. The range of *Ch. oxianus* lies primarily within the Turanian lowland; two other speces of *Chrysops* are comparatively widespread throughout Middle Asia, and parts of Southwest and Central Asia. The single species of the genus *Haematopota*, also found in steppes and deserts, is distributed from the northwestern shores of the Black Sea to the Syrdarya River.

Among other distinguishing features of the Turkmenistan horse fly fauna, a low representation of typical Mediterranean and Central Asian elements is noticeable. Only single species belonging to the *T. bifarius* group is found here, and the members of the relatively large *T. bovinus* group are lacking, as well as representatives of genera *Silvius* s.str. and *Philipomyia* Ols. The genera *Dasyrhamphis* and *Therioplectes* are represented only by a single species each; Central Kopetdagh is the eastern boundary of their ranges, being in the same time the eastern boundary of both generic ranges. The species *Dasyrhamphis umbrinum* is common in Southwest Asia and probably dispersed to Kopetdagh from there. The subspecies *Therioplectes carabaghensis orientalis* is known only from western Kopetdagh whereas the nominal subspecies is found in the Transcaucasia, Turkey, and Iran.

Faunogenetic Trends

The major transformations of landscapes and climate, beginning from the end of the Paleogene, have drastically changed the Ancient Mediterranean biotas. However, these changes were of rather gradual nature in the western part of the Ancient Mediterranean, and a certain share of the ancestral flora and fauna has been preserved there. The appearance of for the eastern part of this area, however, has been radically changed with the regression of the Tethys, mountain uplift, and formation of deserts. The simultaneous formation of the deserts and mountain uplift have facilitated the evolution of an original and diverse authochthonous desert biota in Middle Asia (Sushkin 1925; Kryzhanovsky 1965).

Among these authochthonous elements in the horse fly fauna can be undoubtedly listed the psammobiotic species (the genus *Nanorhynchus* and the *Tabanus sabuletorum* group) as well as the rheophilous species of desert lowlands and foothills (the species of the subgenera *Turanochrysops* and *Nemorius*, and the species of the *T. cordiger* group, as well as *T. zimini* and *T. atropathenicus*). All these taxa are adapted to life under arid conditions. Since their centers of diversity and abundance are located in the Irano-Turanian province, it would be more natural therefore when to referto these taxa as Irano-

Turanian faunistic elements rather than Mediterranean or Afro-Eurasian ones (Olsufyev 1977).

Considering the similarity in arid landscapes and the Quaternary connection between Arabia and Africa, dispersal of the Irano-Turanian species toward southern Europe and northern Africa is logical. Four species of horse flies are known from the desert landscapes of Mali and Sudan; their larvae are rheophiles. Of these, the larvae of *T. kingi* Aust. (King 1910) exhibit a number of morphological characters similar to those in the Turanian species *T. fumidus*, which suggests a possibility of comparatively early connection between the Ethiopian and Irano-Turanian faunas of horse flies.

Within the genus *Hybomitra*, the *H. erberi* group and the subgenus *Sipala* are distributed through a latitudinally extended zone between 30 and 50 degrees north, from southern Europe to the Pacific coast (the only exception is a disjunctive range of *H. expollicata* Pand.). These two groups of taxa also are separated from the rest of the genus by features of their morphology and ecology. These significant differences, as well as the endemism of *H. turanica*, suggest the existence of at least two Palearctic centers of speciation in the genus *Hybomitra*: the Irano-Turanian (center of origin of steppe, desert and foothill species) and the Central Asian (center of origin of the mountainous complex).

A diversity of life forms within the larvae of the *T. bromius* group clearly distinguishes these species from other taxonomic groups within the genus *Tabanus* as well as from other tabanid genera. Larvae of many species in this group can live in different habitats, and the duration of their development depends on temperature conditions. It can be suggested that these species evolved under constantly changing climatic conditions of the Quaternary period. Given the recent distribution of the *T. bromius* group, its ecological preferences and paleoclimatic considerations, the center of speciation of this group can be placed in Irano-Turanian province. It is likely that the *T. bromius* group evolved in the desert and foothill landscapes of southwestern Turan and originated from species that inhabited early riparian forests formed after a regression of the Tethys Sea in the Miocene-Pliocene.

The Irano-Turanian species discussed above represent more than 90% of the horse fly fauna in Turkmenistan. The remainder are the species of Mediterranean origin. Evidently, the Mediterranean species which evolved in relatively mild humid conditions have not been able to disperse through Turkey and Iran further into the arid zone of Middle Asia than the valleys of Central Kopetdagh. Similarly, the species which evolved in the relatively humid mountain habitats of eastern Middle Asia and Central Asia (the *Tabanus argenteomaculatus* Kröb. group and a complex of highland species of the genus *Hybomitra*) have not been capable to occupy the arid, relatively low mountains of Turkmenistan.

Therefore, the core of the Turkmenistan horse fly fauna consists of species of the Irano-Turanian or Southwest Asian origin. The types of their ranges are still quite diverse; these differences are determined by ecological adaptations as well as by historical changes in landscapes and climate under which these species

Table 2. The Arealogical Structure of the Horse Fly Fauna of Turkmenistan

No.	Type of a geographic range	Number of species
1.	Mediterraneo-Southwest Asian-Turanian	2
1 (1).	Kopetdagh subtype	3
2.	Mediterraneo-Middle Asian	6
3.	Mediterraneo-Central Asian	1
4.	East Mediterraneo-Turanian	1
4 (1).	Kopetdagh subtype	2
5.	East Mediterraneo-Central Asian	2
6.	Southwest Asian-Transcaucasian-Middle Asian	2
6 (1).	Kopetdagh subtype	3
7.	Transcaucasian-Turanian-Central Asian	3
7 (1).	Turanian-Central Asian subtype	1
8.	Southwest Asian-Turanian	4
9.	Turanian	3
9 (1).	Local desert-steppe subtype	9
9 (2).	Kopetdagh subtype	1
10.	Turan-Middle Asian	4
11.	European-West Siberian	1

evolved. The arealogical analysis reveals eleven types of geographic ranges, with five types containing also subtypes (Table 2). A comparison of the ranges shows that distribution of the majority of species is confined to Southwest or Middle Asia, which corresponds to the concepts of autochthonous origin of the core Irano-Turanian horse fly fauna that have been discussed above.

The importance of the Turanian deserts and the mountains of Middle and Central Asia as the centers of speciation for many specialized insect groups has been discussed previously (Yablokov-Khnzoryan 1961; Kryzhanovsky 1965; Mikhailov 1988). However, a detailed zoogeographic and faunogenetic analysis of the horse fly fauna of the Irano-Turanian province is presented here for the first time. This analysis provides instructive comparisons between the faunogenetic features in this dipteran family and those of certain families of beetles (Kryzhanovsky 1965; Mikhailov 1988; Kryzhanovsky and Atamuradov this volume), where the differences observed would be the result of the specific biology, ecology, and dispersal abilities of each of these insect groups.

29. Ant-lions (Neuroptera, Myrmeleontidae) in Turkmenistan

VICTOR A. KRIVOKHATSKY

Abstract

A list of the known ant-lion fauna (Neuroptera: Myrmeleontidae) of Turkmenistan is given. It includes 25 genera and 53 species, among which the most common are *Acanthaclistis pallida, Myrmecaelurus varians, Lopezus fedtschenkoi,* and *Maracanda amoena.* Distribution of ant-lion species in the deserts, mountains, and river valleys of Turkmenistan is discussed.

Ant-lions are commonly found among insects attracted by light sources at night in Turkmenistan. The ant-lion fauna of this repubic has not been adequately explored, although some details about it can be found in faunistic papers concerning wide zoogeographical areas and neighboring regions (McLachlan 1875; Luppova 1961; Hölzel 1972), and in systematic works (Navas 1913, 1925, 1932; Luppova 1966, 1971, 1979, 1987; Hölzel 1969; Zakharenko 1983; Krivokhatsky 1990a,b, 1991, 1992a,b).

We used the collections of the Zoological Institute in St. Petersburg, Russia, and original materials collected and observations made from 1978 to 1983 and from 1989 to 1992 in different places in Turkmenistan (Kopetdagh, Badghyz, Karakum Desert, Tedzhen-Murghab area, and Amudarya Valley) to provide this current list of the known ant-lion fauna of Turkmenistan. It includes 25 genera and 53 species (Table 1), and notes on their distribution are also given.

Four common ant-lion species are numerous throughout the territory of Turkmenistan: *Acanthaclistis pallida, Myrmecaelurus varians, Lopezus fedtschenkoi,* and *Maracanda amoena.* Other species are specific for certain habitats and distributed differentially in plains and mountains.

Table 1. Ant-lions (Neuroptera, Myrmeleontidae) of Turkmenistan and their distribution.

Species	Areas											
	1	2	3	4	5	6	7	8	9	10	11	12
Palpares Rambur, 1842												
P. solidus Gerstaecker, 1893	+	+	−	−	−	−	−	−	−	−	−	−
Acanthaclisis Rambur, 1842												
A. pallida McLachlan, 1887	−	+	*	−	+	*	*	*	+	+	+	+
A. obscura Hölzel, 1972	−	+	+	−	−	−	−	+	−	+	+	−
A. curvispura Krivokhatsky, 1990a	−	−	+	−	−	−	−	−	−	+	−	−
Myrmecaelurus Costa, 1855												
M. trigrammus (Pallas, 1781)	−	?	−	−	−	−	−	−	−	−	−	−
M. major McLachlan, 1875	+	−	−	+	−	−	−	•	−	−	+	−
M. paghmanus Hölzel, 1969	−	*	*	−	−	−	−	−	−	+	−	−
M. varians Navás, 1913	+	+	−	−	−	−	−	−	−	−	−	−
M. badghisi Krivokhatsky 1992b	+	+	+	+	+	?	+	*	*	*	+	+
M. sp.n.	−	+	+	−	−	−	−	−	−	−	−	−
Nohoveus Navás, 1919												
N. zigan Aspöck, Aspöck et Hölzel, 1969	−	−	−	−	−	−	−	−	−	−	−	+
N. crucifer (Navás, 1913)	−	+	−	−	+	−	−	*	−	−	+	−
Aspoeckiana Hölzel, 1969												
A. caudata (Navás, 1913)	−	+	−	−	+	−	−	+	+	−	+	+
A. carlic Krivokhatsky, 1992a	−	−	−	−	−	−	−	−	•	−	−	−
A. longiventris Zakharenko, 1983	−	−	+	−	−	−	−	−	+	−	−	−
Holzelus Krivokhatsky, 1992a												
H. compactus Krivokhatsky, 1992a	−	−	−	−	−	−	−	−	•	−	−	−
Lopezus Navás, 1913												
L. fedtschenkoi (McLachlan, 1875)	+	+	+	+	*	*	*	*	+	+	+	+
L. autumnalis Krivokhatsky, 1990b	−	−	+	−	−	−	−	+	−	−	+	−
L. karakumicus Krivokhatsky, 1990b	−	−	−	+	−	−	−	+	−	−	+	−
L. nanus Krivokhatsky, 1990b	−	−	−	−	−	−	−	+	−	−	−	−
Solter Navás, 1912												
S. hardei Hölzel, 1968	−	+	+	−	−	−	−	+	+	+	−	−
S. amseli Hölzel, 1967	−	+	+	−	−	−	−	−	−	−	−	−
S. felderi Navás, 1912	−	+	−	−	−	−	−	−	−	−	−	−
S. iranensis Hölzel, 1967	−	+	−	−	−	−	−	−	−	−	−	−
Gepella Hölzel, 1968												
G. modesta Hölzel, 1968	−	+	+	−	−	−	−	−	−	−	−	−
Cueta Navás, 1911												
C. lineosa (Rambur, 1842)	−	*	*	−	+	−	−	•	*	*	*	*
C. kasyi Hölzel, 1969	−	+	−	−	−	−	−	−	−	−	−	−
Maracanda McLachlan, 1875												
M. amoena McLachlan, 1875	+	+	−	+	+	*	*	*	−	−	+	−
M. talitzkii Luppova, 1979	−	+	−	−	−	−	+	+	−	−	+	−
Morter Navás, 1911												
M. hyalinus (Olivier, 1811)	−	+	*	+	−	−	−	*	+	−	*	+
M. semigriseus Krivokhatsky, 1990b	−	−	+	−	−	−	−	+	−	−	+	−
Euroleon Esben-Petersen, 1918												
E. parvus Hölzel, 1967	−	+	−	−	−	−	−	−	−	−	−	−
Geyria Esben-Petersen, 1918												
G. lepidula (Navás, 1912)	−	−	−	−	−	−	−	−	*	−	−	−
Mesonemurus Navás, 1919												

Table 1. Continued.

Species	Areas											
	1	2	3	4	5	6	7	8	9	10	11	12
M. paulus McLachlan, 1875	–	–	–	–	–	–	–	+	–	+	–	–
M. clarus McLachlan, 1875	+	+	+	–	–	+	–	–	–	–	+	–
M. sp. n. –	–	+	–	–	–	–	–	–	–	–	–	–
Macronemurus Costa, 1855												
M. persicus (Navás, 1915)	–	+	*	–	–	–	–	*	+	+	–	–
Quinemurus Kimmins, 1943												
Q. metamerus Krivokhatsky, 1992a	–	–	–	–	–	–	–	*	–	–	+	–
Distonemurus Krivokhatsky, 1992a												
D. desertus Krivokhatsky, 1992a	–	+	–	–	–	–	–	+	–	–	–	–
Delfimeus Navas, 1912												
D. intricatus (Hölzel, 1972)	–	*	–	–	–	–	–	–	–	–	–	–
Deutoleon Navás, 1927												
D. lineatus (Fabricius, 1798)	–	.	–	–	–	–	–	–	–	–	–	–
Creoleon Tilljard, 1918												
C. elegans Hölzel, 1968	–	–	–	–	–	+	+	–	+	+	–	+
C. griseus (Klug, 1834)	–	+	+	+	–	–	–	.	+	+	–	–
C. plumbeus (Olivier, 1811)	–	+	+	–	–	–	–	.	+	+	–	+
C. aegytiacus (Rambur, 1842)	–	–	–	–	–	–	–	–	+	–	+	–
Neuroleon Navás, 1909												
N. nigriventris Navás, 1913	–	?	–	–	–	–	–	–	–	–	–	–
N. jucundus Navás, 1921	–	?	–	–	–	–	–	–	–	–	–	–
N. tenellus (Klug, 1834)	–	+	+	+	–	–	.	.	+	+	–	–
N. leptaleus (Navás, 1912)	–	+	+	–	+	–	–	*	+	+	–	*
N. sp. n.	–	+	–	–	–	–	–	–	–	–	–	–
Nicarinus Navás, 1914												
N. poecilopterus (Stein, 1863)	.	–	–	–	–	–	–	–	–	–	–	–
Nedroleon Navás, 1914												
N. maculatus Zakharenko, 1990	–	+	–	–	–	–	–	–	–	–	–	–
Megistopus Rambur, 1842												
M. flavicornis (Rossi, 1790)	–	+	–	–	–	–	–	–	–	–	–	–

Areas:
Mountains and foothills: 1 – Balkhans, Krasnovodsk Plateau, 2 – Kyurendagh and Kopetdagh, 3 – Badghyz, 4 – Kugitang. Deserts and plains: 5 – Ustyurt, 6 – Trans-Unguz Karakum; 7 – Central Kakakum; 8 – East Karakum.
River valleys and seashore: 9 – Tedzhen, 10 – Murghab, 11 – Amudarya, 12 – Caspian shore).

+ – species is known from this area and confirmed by the author;
? – species is known from the literature but not confirmed by the author;
* – abundant species,
. – single records.

Deserts and Plains

Ant-lion fauna of sand desert (Karakum) and clay plains (Ustyurt) consists primarily of species with Irano-Turanian ranges. Among these are the widespread species *Holzelus compactus*, *Morter semigriseus*, and *Mesonemurus paulus*, which are found in plains from the Caspian Sea to Mongolia; another group includes species that inhabit only Middle Asian deserts: *Acanthaclisis curvispura, Nohoveus crucifer, Aspoeckiana caudata, A. longiventris, Lopezus autumnalis, L. karakumicus, Maracanda talitzkii*, and *Quinemurus metamerus*. Two species, *Aspoeckiana carlic* and *Lopezus nanus*, are endemic for the East Karakum Desert.

A representative East Karakum ant-lion complex from the Repetek Reserve includes fifteen species. Of those, *Acanthaclisis pallida, Lopezus fedtschenkoi, Aspoeckiana longiventris, Maracanda amoena*, and *Neuroleon leptaleus* are common species attracted to the light at night from May to August. From August to October, *Aspoeckiana caudata, Lopezus autumnalis, Geyria lepidula*, and *Quinemurus metamerus* can be found. *Myrmecaelurus varians* and *Nohoveus crucifer* have evening activity and fly above the sand at low altitude. In the sand, one can find numerous pits of ant-lion larvae of *Morter hyalinus*. Other pits made by *Morter semigriseus* larvae are located inside rodent holes but also can be found in human settlements, in cellars and under sheds. Larvae of *Acanthaclisis pallida, A. curvispura, Creoleon elegans*, and *Neuroleon leptaleus* are found in rodent burrows.

Mountains and Foothills
The distinctive ant-lion fauna of the mountains of Turkmenistan falls into two groups: Irano-Turanian species (*Palpares solidus, Myrmecaelurus paghmanus, Solter spp., Gepella modesta, Cueta kasyi, Euroleon parvus, Mesonemurus clarus, M. sp., Delfimeus intricatus*, and species of *Neuroleon*) and Ancient Mediterranian species (*Cueta lineosa, Deutoleon lineatus, Neuroleon tenellus, Nicarinus poecilopterus, Nedroleon maculatus*, and *Megistopus flavicornis*).

Numerous flying *Myrmecaelurus paghmanus* and *Cueta lineosa* have been recorded during the daytime in the Kopetdagh Reserve. Flying species of *Mesonemurus* and *Creoleon* spp. appear during the afternoon. Pits of the *Euroleon parvus* larvae are found in the sand under the rocks.

River Valleys and Seashore
Because sands encroach upon river banks, many ant-lion species that are specific for the deserts (such as species of *Lopezus* and *Quinemurus metamerus*) are found in river valleys. Certain mountain species (e.g., *Cueta lineosa* and species of *Neuroleon* spp. etc.) can disperse from the mountains along the river valleys.

Complexes of ant-lions in river valleys of Turkmenistan are different from those in both desert and mountains. In the sandy banks of the Amudarya River near Chardzhou fly *Creoleon plumbeus* and *Myrmecaelurus major*; in the sands under the bushes of *Tamarix* spp. are found pits of larvae of *Cueta lineosa* and *Morter hyalinus*. In the evening fly *Myrmecaelurus varians* and *Creoleon* spp. fly over the sand dunes in the Tedzhen River valley near Serakhs, whereas *Myrmecaelurus paghmanus* and *Cueta lineosa* are active at dawn. Finally, only on the Caspian Sea shore and islands are found numerous *Nohoveus zigan*, which are widespread in the North Palearctic.

30. Fauna and Zoogeography of Spiders (Aranei) of Turkmenistan

KIRILL G. MIKHAILOV AND VICTOR FET

Abstract

In Turkmenistan, 335 species of spiders (Aranei) belonging to 162 genera and 38 families are found. Data on their distribution are given for Turkmenistan and worldwide. The most diverse spider families are Salticidae (62 species), Gnaphosidae (52 spp.), Linyphiidae (33 spp.), Thomisidae (28 spp.), Lycosidae (26 spp.), and Araneidae (24 spp.). Two new combinations are proposed: *Orthobula charitonovi* (Mikhailov, 1986), *comb. nov.* (= *Trachelas charitonovi* Mikhailov, 1986) (Liocranidae), and *Psammitis turanicus* (Charitonov, 1969), *comb. nov.* (= *Xysticus turanicus* Charitonov, 1969) (Thomisidae). Distribution of spider species within the republic is influenced primarily by the diverse landscape structure. A specific and impoverished xerophilic fauna of lowland deserts, including the great sand desert of Karakum, is almost emtirely different from the rich mesophile spider fauna found in the Kopetdagh Mountains. Zoogeographic analysis reveals details of origin and possible directions of dispersal of spider fauna. The unique historical role of the Kopetdagh region is discussed; this area could have served as an important island/peninsular corridor before the Tethys Sea recession (Oligocene). With mountain uplift and aridization in the Pliocene, Kopetdagh became a sublatitudinal dispersal pathway for mesophilic spiders. About half of araneofauna is represented by widely distributed species; of the other half, such zoogeographic groups as Iranian, Iranian-Turkestanian, European, European-Caucasian, and European-Mediterranean comprise the majority of the mountain araneofauna. Turanian desert species are predominant in the lowland deserts. A number of spider species are currently known only from Turkmenistan, but the degree of local endemism cannot be estimated since the araneofaunas of the adjacent Middle Asian republics, Iran, and Afghanistan are poorly known.

Introduction

In his 1878 treatment of arachnids from the Caucasus, Ludwig Koch described several forms from the other side of the Caspian Sea – namely from Krasnovodsk, then a new Russian colonial settlement in the deserts of Transcaspia (Koch 1878). The first detailed study of spiders from Turkmenistan was begun by the famous French arachnologist Eugene Simon, who described a number of new species collected in 1886 by G. Radde, A. Walter, and A. Konchin during one of the first zoological expeditions to the former Transcaspian Region (Simon 1889, 1899). Part of this material is deposited in Simon's collection in Museum National d'Histoire Naturelle, Paris, France; several specimens are also deposited in the State Museum of Georgia (former Caucasian Museum), Tbilisi, Republic of Georgia.

Turkmenian spider species known by the 1930s were included in the only catalogue of USSR spiders compiled by D.E. Kharitonov (1932, 1936). In the 1930s extensive spider collections were organized by Ya. Vlasov and identified by V. Sytshevskaya (Pereleshina) (Vlasov 1937a,b,c; Vlasov and Sytschevskaya 1937). Materials of V. Sytschevskaya's expedition to Turkmenistan in 1929, deposited in the Zoological Museum of Moscow State University (Moscow, Russia), remained unidentified until the 1980s.

A number of new species from Turkmenistan were described by S. Spassky (1934, 1936, 1937, 1939, 1941). Some unpublished data based on Spassky's personal collection (now in Zoological Institute, St. Petersburg) are included in this paper. Spassky (1952) included all existing data on spiders of Turkmenistan in his zoogeographic analysis of the Middle Asian ("Turanian Zoogeographic Province") araneofauna. Fragmentary faunistic data from Turkmenistan were published by the arachnologists of Perm University (Kharitonov 1955; Utochkin 1956, 1960a,b,c, 1964, 1968) and, later, by Bakhvalov (1978) and Sternbergs (1979). More regular collections in this republic started in the 1970s under the direction of the Zoological Institute (St. Petersburg, then Leningrad, Russia). Faunistic data and descriptions of new species were published specifically on Turkmenistan (Ovtsharenko and Fet 1980; Fet 1982, 1983, 1984b,c,e, 1985a, 1986, 1993; Kuznetsov and Fet 1982, 1986; Dunin and Fet 1985; Nenilin and Fet 1985; Zonshtein and Fet 1985; Mikhailov and Fet 1986; Tanasevitch and Fet 1986) or were incorporated into broader faunistic publications (Dunin 1985, 1990, 1992; Nenilin 1984a,b, 1985; Zonshtein 1985, 1987; Zyuzin 1985; Nenilin and Pestova 1986; Mikhailov 1986, 1987, 1992; Tanasevitch 1989; Platnick and Ovtsharenko 1991; Marusik and Logunov 1990; Ovtsharenko, Platnick, and Song 1992). A number of ecological works also have been recently produced specifically focused on the spider species of this area (Kaplin 1975; Fet 1984d, 1985b; Atamuradov and Sukh 1985; Krivokhatsky and Fet 1981, 1982; Kuznetsov 1985a,b). Other works mention certain spider species and give notes on their ecology (Sabirova 1975, 1977, 1981, 1985, 1986, 1989; Kaplin 1978; Soyunov 1979; Krivokhatsky 1981, 1982a,b, 1983, 1985a,b).

This work is designed to serve as a catalog of spiders from Turkmenistan as well as a review of their distribution with some zoogeographic observations. We incorporated all known literature, as well as some unpublished data from collections of the Zoological Museum of Moscow State University (Moscow, Russia) and the Zoological Institute of the Russian Academy of Sciences (St. Petersburg, Russia) (all spider collections from Turkmenistan compiled by V. Fet from 1975 to 1987 are deposited in these two institutions). During the many years of faunistic and taxonomic studies, numerous type specimens have been analyzed and revised (including those of E. Simon, O. Pickard-Cambridge, S. Spassky, D. Kharitonov, and V. Sytshevskaya). Identification was conducted using modern taxonomic revisions and consultations with available experts and included analysis of external and internal genitalia of adult males and females. Of significant help have been works of Denis (1958) and Roewer (1955, 1960, 1961a,b) on Afghanistan and Iran, and Andreeva (1976) on Tajikistan, all of which include many spiders found also in Turkmenistan. In placement, synonymy, and taxonomy of spiders we followed recent catalogs by Brignoli (1983) and Platnick (1987) as well as classic works of Bonnet (1937-1961) and Roewer (1942, 1954, 1955). Bibliographic references for this paper were compiled using the database "Spider Literature: A Computerized Bibliography Version 1.0," edited by J.A. Coddington (Smithsonian Institution, Washington, DC, USA). Both authors contributed to this project by submitting all known Russian references.

The list below includes names and data on distribution (within Turkmenistan and worldwide) of 343 species of spiders belonging to 164 genera and 38 families. This list undoubtedly will be extended in the future. The most diverse spider families in the Turkmenistan fauna are Salticidae (62 species), Gnaphosidae (54 spp.), Linyphiidae (33 spp.), Thomisidae (28 spp.), Lycosidae (26 spp.), and Araneidae (24 spp.).

A List of Spiders of Turkmenistan

Fam. Atypidae

Atypus muralis Bertkau, 1890. Kopetdagh (Southwest and Central). Mediterranean species.

Fam. Dipluridae

Phyxioschema raddei Simon, 1889. Krasnovodsk, Kopetdagh (Southwest and Central), Serakhs, Badghyz, Repetek. Iranian species.

Fam. Nemesiidae

Nemesia birulai (Spassky, 1937). Kopetdagh (Southwest, Central, and East), Serakhs, Badghyz. Iranian species.
Raveniola fedotovi (Charitonov, 1946). Kopetdagh. Iranian species.
R. kopetdaghensis (Fet, 1984). Kopetdagh (Southwest and Central). Iranian species.
R. redikorzevi (Spassky, 1937). Badghyz, Serakhs. Iranian species.

Fam. Filistatidae

Filistata insidiatrix (Forskål, 1775). Kopetdagh (Southwest, Bakharden Cave). Mediterranean and Afrotropical species.
F. sp. Badkhyz. Iranian species (?).
Pritha crosbyi (Spassky, 1938). Kopetdagh (Southwest). Irano-Turkestanian species.

Fam. Scytodidae

Scytodes bertheloti Lucas, 1839. Madau, Serakhs District, Repetek. Mediterranean species.
S. strandi Spassky, 1941. Kopetdagh (Southwest), Ashkhabad, Badghyz. Irano-Turkestanian species.
S. univittatus Simon, 1882. Aidere. Paleotropical species.

Fam. Loxoscelidae

Loxosceles rufescens (Dufour, 1820). Madau, Bugdaily, Kopetdagh (Southwest and Central), Bakharden, Ashkhabad. Cosmopolitan species.

Fam. Pholcidae

Artema transcaspica Spassky, 1934. Krasnovodsk, Kopetdagh (Southwest and Central), Ashkhabad, Serakhs, Badghyz, Repetek, Farab. Turanian desert species (?).
Ceratopholcus maculipes Spassky, 1934. Krasnovodsk, Ashkhabad, Badghyz, Repetek. Iranian species.
Hoplopholcus forskali (Thorell, 1871). Ashkhabad. Mediterranean-Turkestanian species.
Pholcus nenjukovi Spassky, 1936. Kopetdagh (Southwest). Iranian species.

Fam. Dysderidae

Dysdera aculeata Croneberg, 1875. Krasnovodsk, Kurum, Tekke, Kopetdagh (Southwest and Central), Ashkhabad, Badghyz, Repetek. Turanian desert species (?).
D. limitanea Dunin, 1985. Known only from Badkhyz. Turanian desert species.
D. pococki Dunin, 1985. Known only from Kizyl-Atrek (may also be present in northeastern Iran). Turanian desert species.
D. kugitangica Dunin, 1992. Known only from Kugitang. Turkestanian species.
D. transcaspica Dunin et Fet, 1985. Known only from southern Turkmenistan (Southwest Kopetdagh, Ashkhabad, Badghyz). Iranian species.

Fam. Oonopidae

Dysderina loricata (Simon, 1873). Ashkhabad, Repetek. European-Mediterranean species.

Fam. Palpimanidae

Palpimanus sogdianus Charitonov, 1946. Kopetdagh (Southwest), Ashkhabad, Badghyz, Repetek. Iranian species.

Fam. Mimetidae

Ero furcata (Villers, 1789). Kopetdagh (Southwest). Holarctic species.
Mimetus laevigatus (Keyserling, 1863). Kopetdagh (Southwest), Repetek. Ancient Mediterranean species.

Fam. Eresidae

Eresus cinnaberinus (Olivier, 1789).
E. niger Petagna, 1787, nomen praeoccupatum (Merett & Millidge 1992).
E. niger rotundiceps Simon, 1873.
  Kopetdagh (Southwest, Central, and East), Badghyz. Trans-Palearctic species (eastward to South Korea).
Stegodyphus lineatus (Latreille, 1817). Krasnovodsk, Kopetdagh, Akhal-Teke, Ashkhabad, Badghyz. Mediterranean species (eastward to Tajikistan).

Fam. Oecobiidae (= Urocteidae)

Oecobius nadiae (Spassky, 1936). Krasnovodsk, Kopetdagh (Southwest, Central, and East), Ashkhabad, Badghyz, Repetek. Irano-Turkestanian species.
O. tadzhicus Andreeva et Tystshenko, 1969. Kopetdagh (Southwest, Central, and East). Irano-Turkestanian species.

Uroctea limbata (C.L. Koch, 1843). Kopetdagh (Southwest and Central), Badghyz. Ancient Mediterranean species.

Fam. Hersiliidae

Hersiliola afghanica Roewer, 1962. Kopetdagh (Southwest and Central), Badghyz. Iranian species.
H. sp. n. 1. Kopetdagh (Southwest; foothills). A new species which has not been yet described. Iranian species.
H. sp. n. 2 Tuarkyr, Ashkhabad, Repetek, Badghyz. A new species previously listed as *H. maculata* (Dufour, 1831) (Fet 1983, Krivokhatsky and Fet 1982). Turanian desert species.

Fam. Uloboridae

Uloborus plumipes Lucas, 1846. Kopetdagh (Southwest). Mediterranean-Himalayan species.
Uloborus walckenaerius (Latreille, 1806). Kopetdagh (Southwest and Central), Ashkhabad, Mary (= New Merv), Badghyz. Trans-Palearctic species.

Fam. Theridiidae

Enoplognatha testacea Simon, 1884. Kopetdagh (Southwest and Central). Mediterranean species.
E. thoracica (Hahn, 1831). Kopetdagh (Southwest). European-Mediterranean species.
Euryopis laeta (Westring, 1861). Kopetdagh (Southwest and Central), Badghyz. European-Ancient Mediterranean species.
E. quinqueguttata Thorell, 1875. Kopetdagh (Southwest). European-Mediterranean species.
Latrodectus tredecimguttatus (Rossi, 1790). Meshed-Messerian Plain, Kopetdagh (Southwest and Central), Ashkhabad, Gyuaurs District, Badghyz, Repetek, Kunya-Urgench, Tashauz Oasis. Ancient Mediterranean species.
L. pallidus O. Pickard-Cambridge, 1872. Chilmamedkum Sands, Kopetdagh (Southwest and Central), Kizyl-Arvat District, Geok-Tepe District, Ashkhabad, Gyuaurs District. East Mediterranean species.
Steatoda albomaculata (De Geer, 1778). Krasnovodsk, Kopetdagh (Central), Ashkhabad, Badghyz, Repetek. Holarctic species.
S. bipunctata (L., 1758). Kopetdagh (Central). Holarctic species.
S. castanea (Clerck, 1757). Kopetdagh (East). European-Siberian species.
S. grossa (C.L. Koch, 1838). Gasan-Kuli, Sharlouk, Kopetdagh (Southwest and Central), Ashkhabad, Badghyz, Repetek. Cosmopolitan species.
S. triangulosa (Walckenaer, 1802). Kopetdagh (Southwest and Central), Badghyz, Repetek. Holarctic species.

Theridion pictum (Walckenaer, 1802). Kopetdagh (Southwest). European-Siberian species.
T. simile C.L. Koch, 1836. Repetek. Ancient Mediterranean (trans-Palearctic?) species.
T. sisyphium (Clerck, 1757). Kopetdagh (Southwest). Trans-Palearctic species.
T. varians (Hahn, 1831). Ashkhabad, Repetek. Trans-Palearctic species.

Fam. Linyphiidae (= Erigonidae)

Agyneta fuscipalpis (C.L. Koch, 1836). Kugitang (1,200–1,300 m), Repetek. European-Siberian species.
A. kopetdaghensis Tanasevitch, 1989. Kopetdagh (Southwest: Aidere). Iranian species.
A. ressli (Wunderlich, 1973). Kopetdagh (Southwest). Mediterranean species.
A. rurestris (C.L. Koch, 1836). Ashkhabad. European-Siberian species.
Alioranus avanturus Andreeva et Tystshenko, 1970. Kugitang (1,400 m). Irano-Turkestanian species.
Ceratinella brevis (Wider, 1834). Kopetdagh (Southwest). European-Siberian species.
Diplocephalus bifurcatus Tanasevitch, 1989. Kopetdagh (Southwest). Iranian species.
Donacochara speciosa (Thorell, 1875). Badkhyz. European species.
Erigone dentipalpis (Wider, 1834). Kugitang (1200–1300 m). European-Mediterranean-Siberian species.
Erigonoplus ninae Tanasevitch et Fet, 1986. Kopetdagh (Southwest). Iranian species.
Frontinellina frutetorum (C.L. Koch, 1834). Kopetdagh (Southwest). European-Siberian species (eastward to East Kazakhstan Region).
Gongylidiellum murcidum Simon, 1884. Kopetdagh (Southwest). European species.
Janetschekia necessaria Tanasevitch, 1985. East Karakum (Farab). Turanian desert species.
Lepthyphantes badkhyzensis Tanasevitch, 1986. Badkhyz. Turanian desert species.
L. escapus Tanasevitch, 1989. Kugitang (1,200 m). Turkestanian species.
L. kuhitangensis Tanasevitch, 1989. Kugitang (1,200–1,400 m). Turkestanian species.
L. nebulosoides Wunderlich, 1977. Kopetdagh (Southwest), Kugitang. Ancient Mediterranean species.
L. pinicola Simon, 1884. Kopetdagh (Southwest). European-Mediterranean species.
L. tenuis Blackwall, 1852. Kopetdagh (Southwest and Central). European-Mediterranean species (eastward to Tajikistan).
L. turanicus Tanasevitch et Fet, 1986. Tuarkyr Plateau (Kafigshem Mts.), Kopetdagh (Southwest and Central), Kugitang (1,300 m). Iranian species.

L. turkestanicus Tanasevitch, 1989. East Karakum (Farab). Turanian desert species.
Mecopisthes orientalis Tanasevitch et Fet, 1986. Kopetdagh (Southwest). Iranian species.
Mesasigone mira Tanasevitch, 1989. Kugitang (1300 m). Iranian species.
Microctenonyx subitaneus (O. Pickard-Cambridge, 1875). Kopetdagh (Southwest and Central), Farab. Holarctic species (with a range disjunction in Siberia).
Microlinyphia pusilla (Sundevall 1830). Kopetdagh (Southwest), Ashkhabad, Badghyz, Farab. Holarctic species.
Oedothorax apicatus (Blackwall, 1850). Kopetdagh (Southwest), Kugitang. European species.
Pelecopsis laptevi Tanasevitch et Fet, 1986. Badghyz. Turanian desert species.
P. paralleloides Tanasevitch et Fet, 1986. Kopetdagh (Southwest). Iranian species.
Prinerigone vagans (Audoin, 1826). Kopetdagh (Southwest and Central), Bakharden District, Ashkhabad District, Farab. European-Mediterranean (steppe?) species.
Sphecozone romana (O. Pickard-Cambridge, 1872).
  Ceratinopsis romana O. Pickard-Cambridge, 1872.
  Sphecozone asiatica (Andreeva et Tyschchenko, 1970) (Tanasevitch 1983: 1786).
Repetek. Mediterranean species.
Trachelocamptus asiaticus Tanasevitch, 1989. Repetek. Turanian desert species.
Trichoncoides piscator Simon, 1884. Kopetdagh (Southwest). Mediterranean species.
Walckenaeria monoceros (Wider, 1834). Kopetdagh (Southwest). European species.

Fam. Tetragnathidae

Metellina kirgisica (Bachvalov, 1974).
  Meta kirgisica Bachvalov 1974) (Marusik 1989).
Kugitang. Irano-Turkestanian species.
Tetragnatha extensa (L., 1758). Kopetdagh (Southwest), "Tekke". Holarctic species.
Zygiella caspica (Simon, 1889). Imambaba. Turanian desert species.

Fam. Araneidae

Aculepeira sogdiana (Charitonov, 1969). Kopetdagh (Southwest and Central), Badghyz. Turanian desert (or Iranian?) species.
Agalenatea redii (Scopoli, 1763). Kopetdagh (Southwest). European-Mediterranean species (from North Africa to Middle Asia).
Araneus angulatus Clerck, 1757. Kopetdagh (Southwest). European-

Mediterranean forest species. Records from the New World are erroneous!
A. armida (Auduin, 1825). Ashkhabad. European-Mediterranean species (present in the Caucasus).
A. bituberculatus (Walckenaer, 1802). Kopetdagh (Southwest and Central), Badghyz. European-Ancient Mediterranean species.
A. pallasi (Thorell, 1875). Kyzylsu (Krasnovodsk Region), next to the water. Turkestanian-Turanian desert species (eastward to Mongolia and Tuva).
A. repetecus Bachvalov, 1978. Repetek. Turanian desert species.
A. spasskyi Brignoli, 1983.
  A. cruciferoides Spassky 1952.
Kopetdagh (Southwest). Irano-Turkestanian species.
A. tartaricus (Kroneberg, 1875). Ashkhabad, Badghyz. Irano-Turkestanian species.
A. tedgenicus Bachvalov, 1978. Tedzhen? (place of collection is absent from the original description). Turanian desert species.
Araniella inconspicua (Simon, 1874). Kopetdagh (Southwest). European-Caucasian (European-East Mediterranean?) meadow species.
Argiope ahngeri Spassky, 1932. Akhal-Teke. Irano-Turkestanian species.
A. bruennichi (Scopoli, 1772). Kopetdagh (Southwest). Trans-Palearctic meadow-steppe species.
A. lobata (Pallas, 1772). Krasnovodsk, Kopetdagh (Southwest and Central), Bakharden, Ashkhabad, Badghyz, Murghab, Takhta-Bazar. Paleotropical and Palearctic species (in Palearctic, occupies steppe habitats).
Cyclosa conica (Pallas, 1772). Kopetdagh (Southwest). Holarctic forest species.
Hypsosinga albovittata (Westring, 1851). Kopetdagh (Southwest). Trans-Palearctic species.
H. turkmenica Bachvalov, 1978. Repetek. Turanian desert species.
Larinia nenilini Marusik, 1986. Repetek, Chardzhou. Turanian desert species.
  L. pubiventris Simon, 1889.
  L. turkmenica Spassky 1939.
  Krasnovodsk, Tedzhen, Kalaimor (= Mor-Kala), Imambaba, Repetek. Turanian desert species.
Larinioides folium (Schränk, 1803). Gasan-Kuli, Ashkhabad, "Tekke," Amudarya. European-Mediterranean steppe-desert species, found next to the water (eastward to Middle Asia).
Mangora acalypha (Walckenaer, 1802). Kopetdagh (Southwest). European-Ancient Mediterranean species; eastward to southern Urals, Tajikistan, Kyrghyzstan, and East Kazakhstan Region.
Neoscona adianta (Walckenaer, 1802). Kopetdagh (Southwest), Karakum District (Karakumkanal State Farm). Trans-Palearctic meadow-steppe species.
N. subfusca (C.L. Koch, 1837).
  Araneus dalmaticus Doleschall 1852.
Kopetdagh (Southwest). Mediterranean species (the easternmost record!).

Zilla diodia (Walckenaer, 1802). Kopetdagh (Southwest). European-Caucasian (meadow?) species.

Fam. Lycosidae

Arctosa soror (Simon, 1889). Mary ("New Merv"). Turanian desert species.
A. variana (C.L. Koch, 1848). Badghyz. European-Mediterranean species.
"*Arctosa*" *cereipes* (L. Koch, 1878). Krasnovodsk. Turanian desert species.
Aulonia kratochvili Dunin, Buchar, et Absolon, 1986. Kopetdagh. Caucasian-Iranian species.
Evippa badchysica Sternbergs, 1979. Badghyz. Turanian desert species.
E. onager Simon, 1895. Repetek. Sindian-Turanian desert (steppe) species (?).
E. schenkeli Sternbergs, 1979. Badghyz. Turanian desert species.
E. turkmenica Sternbergs, 1979. Badghyz. Turanian desert species.
Hippasa partita (O.Pickard-Cambridge, 1876).
  H. deserticola Simon, 1889.
  Murghab, Imambaba. Saharo-Sindian species.
Lycosa alticeps Kroneberg, 1875. Krasnovodsk, Ashkhabad, Badghyz. Turanian desert species.
L. nordmanni (Thorell, 1875). Ashkhabad. Mediterranean species.
L. radiata (Latreille, 1819). Kopetdagh (Southwest and Central). Mediterranean species.
Pardosa italica Tongiorgi, 1966. Kopetdagh (Southwest and Central). Mediterranean species.
P. morosa (L. Koch, 1870). Kopetdagh (Southwest and Central). Mediterranean species.
P. nebulosa (Thorell, 1892). Kopetdagh (Southwest), Badghyz, Sultanbent, Iolatan, Mary (= New Merv). Mediterranean species.
P. pontica (Thorell, 1875). Kopetdagh (Southwest and Central). Caucasian-Iranian species (?).
P. proxima C. L. Koch, 1847. Kopetdagh (Southwest). European-Mediterranean species.
Tarentula albofasciata (Brullé, 1832). Kopetdagh (Southwest), Ashkhabad. Mediterranean species.
T. bergsoei (Thorell, 1875). Ashkhabad. Caucasian-Iranian species (?).
T. cronebergi (Thorell, 1875). Murghab District. Ancient Mediterranean species.
T. cursor (Hahn, 1831) *cursorioides* Charitonov, 1969. Kopetdagh (Southwest and Central). Iranian subpecies of an European-Ancient Mediterranean species.
T. raddei (Simon, 1889). Mary ("New Merv"), Amudarya. Turanian desert species.
Trochosa ruricola (De Geer, 1778). Kopetdagh (Southwest). Trans-Palearctic species.
T. terricola Thorell, 1856. Kopetdagh (Southwest). Trans-Palearctic species.

Wadicosa commoventa Zyuzin, 1985. Badghyz (Kushka). Turanian desert species.
Xerolycosa brunneopicta Loksa, 1965. "Duldsch (Sudüfer d. Balchanbühels)", Krasnovodsk. Mongolian desert (steppe) species.

Fam. Pisauridae

Pisaura mirabilis (Clerck, 1757). Kopetdagh (Southwest). European-Mediterranean species.
P. novicia (L. Koch, 1878). Ashkhabad. Caucasian-Iranian species (?). Taxonomic status in relation to *P. mirabilis* is not clear.

Fam. Agelenidae

Agelena labyrinthica (Clerck, 1757). Kopetdagh (Southwest and Central). Trans-Palearctic species (eastward to Manchuria and Japan; southward to the Himalayas).
Tegenaria domestica (Clerck, 1757). Kopetdagh (Southwest and Central). Cosmopolitan species.

Fam. Desidae

Cedicus ephthalitus Fet, 1993. Known only from Southwest Kopetdagh. Iranian species.
C. gennadii Fet, 1993. Kopetdagh (Southwest and Central), Tuarkyr (Kafigshem Mountains). Iranian species.
C. maerens Simon, 1889. Known only from the foothills of Southwest Kopetdagh and Bolshoi Balkhan. Iranian species.
C. parthus Fet, 1993. Known only from Kopetdagh (Southwest and Central). Iranian species.

Fam. Hahniidae

Hahnia sp. Kopetdagh (Southwest and Central). Iranian species.

Fam. Dictynidae

Archaeodictyna ammophila (Menge, 1871). Kopetdagh. European-Caucasian meadow species.
A. consecuta (O. Pickard-Cambridge, 1872). Kopetdagh (Central). East Mediterranean speccies (from Middle East to Turkestan).
Brigittea latens (Fabricius, 1775). Kopetdagh (Southwest). European-Caucasian meadow species, eastward to Tajikistan and Kyrghyzstan.
Devade tenella (Tystshenko, 1965). Badghyz. Turanian desert (or steppe) species.

Dictyna cronebergi Simon, 1889. Mary (= New Merv). Turanian desert species.
D. pusilla Thorell, 1856. Kopetdagh (Southwest). Trans-Palearctic species (eastward to Kamchatka).
D. uncinata Thorell, 1856. Ashkhabad. Trans-Palearctic species (eastward to Kamchatka).
Dictynomorpha strandi Spassky, 1939. Meshed-Messerian Plain, Ashkhabad. Turanian desert species.

Fam. Amaurobiidae

Coelotes charitonovi Spassky, 1939. Krasnovodsk, Badghyz. Turanian desert species.

Fam. Titanoecidae

Titanoeca albomaculata (Lucas, 1846). Kopetdagh (Southwest). Mediterranean species.
T. lehtineni Fet, 1986. Kopetdagh (Southwest, Central, and East), Badghyz. Iranian species.
T. tristis (L. Koch, 1872). Kopetdagh. European-Caucasian meadow species (the easternmost record!).
T. veteranica (Herman, 1879). Kopetdagh (Southwest). Mediterranean steppe species.

Fam. Oxyopidae

Oxyopes badhyzicus Mikhailov et Fet, 1986. Known only from Badghyz (Lake Yeroyulanduz). Turanian desert species.
O. heterophthalmus Latreille, 1804. Ashkhabad, Bagir. European-Ancient Mediterranean species.
O. lineatus Latreille, 1806. Kopetdagh (Southwest), Badghyz. European-Mediterranean species.
O. maracandensis Charitonov, 1946. Kopetdagh (Southwest and Central), Badghyz, Sultanbent, Murghab District, Karakum District, Repetek. Turanian desert species.
O. takobius Andreeva et Tystshenko, 1969. Kopetdagh (Southwest and Central), Badghyz. Iranian-Turkestanian species.

Fam. Anyphaenidae

Anyphaena accentuata (Walckenaer, 1802). Kopetdagh (Southwest). European-Mediterranean species.

Fam. Liocranidae

Agroeca pullata Thorell, 1875. Kopetdagh (Southwest and Central). European-Mediterranean-Siberian species.

Mesiotelus kulczynskii Charitonov, 1946. Kopetdagh (Southwest and Central). Iranian species.

M. tenuissimus (L. Koch, 1866). Kopetdagh (Southwest and Central). Mediterranean species.

Orthobula charitonovi (Mikhailov, 1986) *comb. novi* by K. Mikhailov.
Trachelas charitonovi Mikhailov, 1986.
Kopetdagh (Southwest and Central). Eastern Mediterranean species (from the Caucasus to Kyrghyzstan).

Phrurolithus pullatus Kulczynski, 1897. Kopetdagh (Southwest). Mediterranean species.

Fam. Clubionidae

Cheiracanthium erraticum (Walckenaer, 1802). Kopetdagh (Southwest). Trans-Palearctic species.

C. mildei L. Koch, 1866. Kopetdagh (Southwest). European-Mediterranean species (introduced to the USA).

C. seidlitzi L. Koch, 1864. Kopetdagh (Southwest), Badghyz, Kalaimor (Mor-Kala). European-Mediterranean species.

Clubiona alpicola Kulczynski, 1882. Kopetdagh (Southwest). European meadow species (the easternmost record!).

C. genevensis L. Koch, 1866. Kopetdagh (Southwest). Ancient Mediterranean species.

Fam. Zodariidae

Zodarion proszynskii Nenilin et Fet, 1985. Kopetdagh (Southwest and Central). Iranian species.

Z. raddei Simon, 1889.
  Z. vlasovi Sytshevskaya in Vlasov et Sytshevskaya, 1937.
  Z. denisi: Ovtsharenko and Fet 1980, Krivokhatsky and Fet 1981; misidentification, non *Z. denisi* Spassky, 1938.
  Z. raddei: Fet 1985b.
Kopetdagh (Southwest and Central), Archman, Ashkhabad, Badghyz, Repetek. Iranian species.

Z. sytchevskayae Nenilin et Fet, 1985. Kopetdagh (East), Babadurmaz, Badghyz, Repetek. Turanian desert species.

Fam. Gnaphosidae

Aphantaulax seminigra Simon, 1878. Kopetdagh (Southwest and Central). Mediterranean species (eastward to Kyrghyzstan).
Asiabadus asiaticus (Charitonov, 1946). Badghyz. Turanian desert species.
Berlandina afghana Denis, 1958. Repetek. Turanian desert species.
B. caspica Ponomarev, 1979. Murghab District. Turanian desert species.
B. plumalis (O. Pickard-Cambridge, 1872). Badghyz. Mediterranean and Paleotropical (Burma, Himalayas, China) species.
B. sp. Kopetdagh (Central).
Drassodes jakkabagensis Charitonov, 1946. Badghyz. Turanian desert species.
D. lapidosus (Walckenaer, 1802). Kopetdagh (Southwest and Central). Trans-Palearctic species.
D. proximus (Denis, 1958). Krasnovodsk, Kopetdagh (Central). Iranian species.
D. sp. Kopetdagh (Southwest and Central). Iranian species.
"D." flavomaculatus (L. Koch, 1878). Krasnovodsk. Turanian desert species. Does not belong to the genus *Drassodes* (V. Ovtsharenko, pers. comm.).
"D." thimei (L. Koch, 1878). Krasnovodsk. Turanian desert species. Does not belong to the genus *Drassodes* (V. Ovtsharenko, pers. comm.).
Echemus angustifrons (Westring, 1861). Kopetdagh. European species.
Gnaphosa haarlovi Denis, 1958.
G. ajdahania Roewer, 1961 (Ovtscharenko, Platnick, et Song, 1992.). Ashkhabad Region (Gyuaurs). Irano-Turkestanian species.
G. kuldzha Ovtscharenko, Platnick, et Song, 1992. Murghab District. Irano-Turkestanian species.
G. leporina (L. Koch, 1866). Kopetdagh (Southwest and Central). Ancient Mediterranean species.
G. turkmenica Ovtscharenko, Platnick, et Song 1992. Known only from Badghyz (Lake Eroyulanduz).
G. lugubris (C.L. Koch, 1839). Record for Repetek (Kaplin 1978) is not confirmed (Ovtsharenko *et al.* 1992).
Haplodrassus dalmatensis (L. Koch, 1866). Kopetdagh (Southwest and Central). European-Mediterranean species.
H. signifer (C.L. Koch, 1839). Kopetdagh (Southwest and Central), Ashkhabad, Badghyz. Holarctic species.
Micaria albimana O. Pickard-Cambridge, 1872. Kopetdagh (Central and East), Badghyz, Repetek. East Mediterranean species.
M. kopetdaghensis Michailov in Mikhailov et Fet, 1986. Kopetdagh (Southwest). Caucasian-Iranian species.
M. lenzi Bösenberg, 1899. Sarykamysh. Trans-Palearctic species.
M. pygmaea Kroneberg, 1875. Sharlouk. Ancient Mediterranean species (from Canary Islands to Tajikistan).
M. romana L. Koch, 1866. Kopetdagh (Southwest). European-Mediterranean species.
M. rossica Thorell, 1875. Kopetdagh (Southwest, Central, and East), Dushak,

Badghyz, Repetek, Farab, Tashauz Region (Chirishli, Kankakyr). Holarctic species.
M. septempunctata O. Pickard-Cambridge, 1872. Gasan-Kuli, Kopetdagh (Central), Bairam-Ali, Repetek, Farab. East Mediterranean species.
Minosia karakumensis (Spassky, 1939). Kopetdagh (Southwest and Central), Ashkhabad, Badghyz, Repetek. Turanian desert species.
Minosiella intermedia Denis, 1958. Kopetdagh (Southwest), Serakhs District, Badghyz, Repetek. Iranian species (?).
Nomisia aussereri (L. Koch, 1872). Kopetdagh (Southwest and Central), Ashkhabad. Mediterranean species.
N. conigera (Spassky, 1941). Kopetdagh (Southwest, Central, and East), Badghyz. Irano-Turkestanian species.
N. exornata (C.L. Koch, 1839). Kopetdagh (Southwest, Central, and East), Badghyz. Ancient Mediterranean species.
Poecilochroa conspicua (L. Koch, 1866). Ashkhabad, Repetek. European-Ancient Mediterranean species.
Prodidomus redikorzevi Spassky, 1940. Kopetdagh (Southwest), Serakhs, Badghyz. Iranian species.
Pterotricha strandi Spassky, 1936. Kopetdagh (Southwest and Central), Akhal-Teke, Serakhs District, Badghyz. Iranian species.
Scotophaeus scutulatus (L. Koch, 1866). Kopetdagh. European species.
Synaphosus see Addendum (p. 524).
S. sp. 2. Badghyz. Iranian species.
Talanites dunini Platnick et Ovtsharenko, 1991. Kopetdagh (Southwest and East). Iranian species.
T. fagei Spassky, 1938. Kopetdagh (Central), Serakhs. Iranian species.
"*Talanites*" *aculeatus* Charitonov, 1946. Kopetdagh (Southwest and Central), Repetek. Turanian desert species. Does not belong to the genus *Talanites* (V. Ovtsharenko, pers. comm.).
"*T.*" *sp.* Repetek. Turanian desert species. Belongs to the same genus as "*Talanites*" *aculeatus* Charitonov.
Theuma walteri (Simon, 1889). Kalaimor (= Mor-Kala). Turanian desert species.
Trachyzelotes jaxartensis (Kroneberg, 1875). Kopetdagh (Southwest), Badghyz. Ancient Mediterranean species (introduced to the USA, Mexico, India, China, Senegal, South Africa, and Hawaii).
Zelotes aerosus Charitonov, 1946. Kopetdagh (Southwest and Central). Turanian desert species.
Z. arnoldii Charitonov, 1946. Ashkhabad, Meshed-Messerian Plain. Turanian desert species.
Z. bucharensis Charitonov, 1946. Kopetdagh (Central), Badghyz, Repetek. Turanian desert species.
Z. caucasius (L. Koch, 1866). Kopetdagh (Central). Mediterranean species.
Z. longipes (L. Koch, 1866). Kopetdagh (Southwest). European-Siberian species.

Z. praeficus (L. Koch, 1866). Kopetdagh (Southwest and Central). European-West Siberian species.
Z. pumilus (C.L. Koch, 1839). Kopetdagh (Southwest). European-Mediterranean species.
Z. pusillus (C.L. Koch, 1833). Kopetdagh (Southwest and Central). Trans-Palearctic species.
Z. subterraneus (C.L. Koch, 1833). Kopetdagh (Central), Repetek. Holarctic species.

Fam. Zoridae

Zora nemoralis (Blackwall, 1861). Kopetdagh (Southwest and Central). European species.
Z. silvestris Kulczynski in Chyzer et Kulczynski, 1897. Kopetdagh (Southwest). European species.

Fam. Heteropodidae (= Sparassidae, = Eusparassidae)

Cebrennus sp. Repetek. Turanian species.
Eusparassus oculatus (Kroneberg, 1875). Kopetdagh (Southwest and Central), Archman, Ashkhabad, Badghyz, Repetek. Irano-Turkestanian species.
Micrommata ligurinum C.L. Koch, 1845. Kopetdagh (Southwest and Central). Mediterranean species.
Olios sericeus (Kroneberg, 1875). Kopetdagh (Southwest and Central), Ashkhabad, Badghyz, Kushka, Repetek. Irano-Turkestanian species.

Fam. Philodromidae

Paratibellus oblongiusculus (Lucas, 1846). Kopetdagh (Southwest), Bagir. Mediterranean species.
Philodromus aureolus (Clerck, 1757). Kopetdagh (Southwest). Trans-Palearctic species [? - many records probably belong to *P. cespitum* (Walckenaer 1802)].
P. fallax Sundevall, 1832. Kopetdagh (Southwest). European species.
P. lepidus Blackwall, 1870. Mary (= New Merv), Uch-Adzhi. Mediterranean species.
P. rufus Walckenaer, 1825. Kopetdagh (Southwest). Holarctic species.
Thanatus formicinus (Clerck, 1757). Kopetdagh (Southwest and Central). Holarctic species.
T. imbecillus L. Koch, 1878. Kopetdagh (Southwest), Archman. Caucasian-Turkestanian species.
T. vulgaris Simon, 1870. Kopetdagh (Southwest and Central). Holarctic species.
Tibellus oblongus (Walckenaer, 1802). Kopetdagh (Southwest), Ashkhabad. Holarctic species.

Fam. Thomisidae

Diaea dorsata (Fabricius 1777). Kopetdagh (Central) (the record is dubious). European-Ancient Mediterranean species.
Heriaeus buffonopsis Loerbroks, 1983. Krasnovodsk. Turanian desert species.
H. fedotovi Charitonov, 1946. Kopetdagh (Southwest). Turanian desert species.
H. mellottei Simon, 1886.
H. oblongus Simon, 1918.
Kopetdagh (Southwest). European-Ancient Mediterranean species.
H. spinipalpus Loerbroks, 1983. Krasnovodsk, Firyuza. Caucasian-Turkestanian species.
Monaeses israilensis Levy, 1973. Kopetdagh (Southwest). East Mediterranean species.
M. paradoxus (Lucas, 1846). Kopetdagh (Southwest), Geok-Tepe, Repetek. Mediterranean-Afrotropical species.
Oxyptila baudueri cribrata (Simon, 1885).
  Xysticus cribratus Simon, 1885.
  Psammitis cribratus (Simon, 1885).
  Kopetdagh (Southwest and Central). Synonymy of this Mediterranean form is not clear. Kritscher (1961) placed it as a subspecies into *O. baudueri* Simon, 1877.
O. sanctuaria (O. Pickard-Cambridge, 1871). Kopetdagh (Southwest). European species.
O. tricoloripes Strand, 1913. Kopetdagh (Southwest and Central). East Mediterranean species.
"O". lugibris (Kroneberg, 1875). Kopetdagh (Southwest and Central), Ashkhabad, Repetek. Irano-Turkestanian species; belongs neither to the genus *Oxyptila* nor to *Xysticus* (Marusik and Logunov, 1990: 52).
Psammitis tristrami (O. Pickard-Cambridge, 1872). Durun, Bairam-Ali. Mediterranean species.
P. turanicus (Charitonov, 1969), *comb. novi.*
  Xysticus turanicus Charitonov, 1969.
  Kizyl-Atrek, Sharlouk, Kopetdagh (Southwest and Central), Bakharden District, Badghyz. Irano-Turkestanian species.
Runcinia lateralis (C.L. Koch, 1838). Kopetdagh (Southwest). European-Mediterranean species.
Stiphropus strandi Spassky, 1938. Kopetdagh (Southwest and Central), Badghyz, Repetek. Irano-Turkestanian species.
Synema globosum (Fabricius, 1775). Kopetdagh (Southwest). Trans-Palearctic species.
S. plorator (O. Pickard-Cambridge, 1872). Kopetdagh (Southwest). Mediterranean species.
Thomisus onustus Walckenaer, 1805. Krasnovodsk, Kopetdagh (Southwest and Central), Ashkhabad, Gaudan, Bagir, Badghyz, Islimcheshme, Imambaba. Mediterranean-Paleotropical species.

Xysticus acerbus Thorell, 1872. Kopetdagh (Southwest and Central), Ashkhabad, Repetek. Trans-Palearctic species.
X. caperatus Simon, 1875. Badghyz. Mediterranean species.
X. caspicus Utotchkin, 1968. Kopetdagh (Southwest), Repetek. Turanian desert species.
X. concinnus Kroneberg, 1875. Badghyz, Kopetdagh (Central and East). Turanian desert species.
X. kaznakovi Utotchkin, 1968. Kopetdagh (Southwest). Irano-Turkestanian species.
X. kochi Thorell, 1872. Kopetdagh (Southwest). European-Mediterranean species.
X. lapidarius Utotchkin, 1968. Iolatan. Irano-Turkestanian species.
X. marmoratus Thorell, 1875. Kopetdagh (Central). Mediterranean species.
X. minor Charitonov, 1946. Kopetdagh (Southwest and Central), Ashkhabad, Badghyz, Repetek. Irano-Turkestanian species.
X. ninnii Thorell, 1872. Kopetdagh (Southwest). European-Mediterranean species.

Fam. Salticidae

Aelurillus andreevae Nenilin, 1984; *nomen novum* for specimens misidentified as *A. variegatus* (Kroneberg, 1875) by Andreeva (1976). Ashkhabad. Irano-Turkestanian species.
A. affinis (Lucas, 1846). Ashkhabad. Ancient Mediterranean species.
A. ater (Kroneberg, 1875). Ashkhabad, Imambaba, Kaplankyr. Turanian desert species.
A. concolor Kulczynski, 1901.
  A. iranus Roewer, 1955 (Nenilin 1984).
  Kopetdagh (Southwest and Central), Murghab District. Caucasian-Turkestanian (East Mediterranean?) species.
A. m-nigrum (Kulczynski in Chyzer et Kulczynski, 1891). Gasan-Kuli, Kopetdagh (Southwest and Central), Badghyz. Ancient Mediterranean species (eastward to Sinkiang).
A. variegatus (Kroneberg, 1875). Badghyz. Turanian desert species.
A. v-insignitus (Clerck, 1757). Kopetdagh (Southwest). European-Siberian species.
Ballus chalybeius (Walckenaer, 1802). Kopetdagh (Southwest). European-Mediterranean species (eastward to Tajikistan).
Bianor albomaculatus (Lucas, 1846). Bakharden. Ancient Mediterranean species.
Chalcoscirtus infimus (Simon, 1868). Kopetdagh (Southwest and Central). European-Mediterranean species.
C. martensi parvus Marusik, 1991. Kopetdagh (Southwest). Irano-Turkestanian subspecies.

Cyrba algerina (Lucas, 1846). Kopetdagh (Southwest). Mediterranean and Paleotropical species.
C. ocellata (Kroneberg, 1875). Kopetdagh (Southwest and Central), Repetek. Mediterranean and Paleotropical species.
Eris nidicolens (Walckenaer, 1802). Kopetdagh (Southwest). European-Mediterranean species.
Euophrys frontalis (Walckenaer, 1802). Kopetdagh (Southwest). Trans-Palearctic species.
Evarcha arcuata (Clerck, 1757). Kopetdagh (Southwest), Repetek. Trans-Palearctic species.
Heliophanus auratus C. L. Koch, 1835. Lower Murghab. European-Siberian species.
H. curvidens (O. Pickard-Cambridge, 1872). Bakharden. East Mediterranean species.
H. flavipes (Hahn, 1832). Kopetdagh (Southwest). European-Siberian species.
H. lineiventris Simon, 1868. Badghyz. Trans-Palearctic species.
H. melinus L. Koch, 1867. Kopetdagh (Southwest). Ancient Mediterranean species.
H. niveivestis Simon, 1889. Bairam-Ali (= Old Merv). Turanian desert species. Taxonomic status is not clear.
H. patagiatus Thorell, 1875. Exact locality in Turkmenistan is not published (A. Nenilin, personal notes). European-Siberian species.
H. turanicus Charitonov, 1969. Kopetdagh (Southwest), Khozli-ogly-olum (collected by K. Ahnger, November 1 to 11, 1903). Turanian desert species.
Langona tartarica (Charitonov, 1946). Kopetdagh (Central). Irano-Turkestanian species.
Leptorchestes berolinensis (C.L. Koch, 1846). Kopetdagh (Southwest). European-Mediterranean species.
Menemerops afghanus (Roewer, 1961). Kopetdagh (Southwest), Badghyz. Iranian species.
Menemerus marginatus (Kroneberg, 1875). Sultanbent, Repetek. Irano-Turkestanian species (northwestward to Nakhichevan).
Mogrus antoninus Andreeva, 1976. Kopetdagh (Southwest and Central), Ashkhabad, Badghyz. Irano-Turkestanian species.
M. valerii Kononenko in Andreeva, Kononenko, et Proczynski, 1981. Repetek. Turanian desert species.
Pellenes epularis (O. Pickard-Cambridge, 1872). Exact locality in Turkmenistan is not published (A. Nenilin, personal notes). East Mediterranean species.
P. limbatus Kulczynski, 1895. Ashkhabad (Berzengi), Badghyz. Turkestan-Siberian species.
P. kulabicus Andreeva, 1976. Ashkhabad, Badghyz. Irano-Turkestanian species.
P. nigrociliatus (L. Koch, 1875). Kopetdagh (Southwest). European-Mediterranean and Japanese (possibly Trans-Palearctic) species.

P. simoni (O. Pickard-Cambridge, 1872). Exact locality in Turkmenistan is not published (A. Nenilin, personal notes). East Mediterranean species.

P. tripunctatus (Walckenaer, 1802). Kopetdagh (Southwest). European-Siberian species.

Philaeus chrysops (Poda, 1761). Kopetdagh (Southwest and Central), Ashkhabad, Serakhs, Badghyz. European-Ancient Mediterranean species (eastward to Sinkiang).

Phlegra fasciata (Hahn, 1826). Kopetdagh (Southwest). Ancient Mediterranean species (eastward to Sinkiang).

P. sogdiana Charitonov, 1946. Kopetdagh, Kugitang. Irano-Turkestanian species.

Plexippoides starmuehlneri (Roewer, 1955). Murghab Oasis. Iranian species.

Plexippus coccineus Simon, 1902. Kopetdagh (Southwest and Central), Badghyz. Ancient Mediterranean (Caucasian-Iranian?) species.

P. setipes (Karsch, 1879). Krasnovodsk. Paleotropical species (from Turkestan to Vietnam).

Pseudicius cinctus (O. Pickard-Cambridge, 1872).

P. vittatus Simon, 1889.
    Krasnovodsk, Kopetdagh (Southwest), Badghyz, Murghab Oasis, Sary-Yazy, Repetek. Irano-Turkestanian species (eastward to Sinkiang).

P. spasskyi Andreeva, Heçiak et Proczynski, 1984. Krasnovodsk, Murghab District. Irano-Turkestanian species.

Salticus tricinctus (C.L. Koch, 1846). Krasnovodsk, Badghyz (Kushka). East Mediterranean species.

Sitticus caricis (Westring, 1861). Farab. European-Siberian (?) species.

S. distinguendus (Simon, 1868). Krasnovodsk, Kopetdagh (Central), Repetek. European species.

S. karakumensis Logunov, 1992. Badghyz, Repetek. Turanian desert species.

S. terebratus (Clerck, 1757). Locality in Turkmenistan is not published (Proczynski 1991). European-Siberian (?) species.

S. zimmermanni (Simon, 1877). Kopetdagh (Southwest). European species.

Synageles charitonovi Andreeva, 1976. Murghab Oasis. Irano-Turkestanian species.

S. hilarulus (C. L. Koch, 1846). Exact locality in Turkmenistan is not published (Azheganova 1968). European species (Proczynski 1991).

S. ramitus Andreeva, 1976. Badghyz. Irano-Turkestanian species (eastward to Mongolia).

Thyene imperialis (Rossi, 1846). Kopetdagh (Southwest), Sharlouk, Sultanbent, Gasan-Kuli. Mediterranean-Paleotropical species.

Yllenus albocinctus (Kroneberg, 1875). Deinau. Caucasian-Mongolian species.

Y. auspex (O. Pickard-Cambridge, 1885). Repetek. Turan-Mongolian desert species.

Y. bajan Proczynski, 1968. Krasnovodsk, Repetek. Turan-Mongolian desert and steppe species.

Y. hamifer Simon 1895.

Y. flavociliatus Simon, 1895. "Southern Turkmenistan," Repetek. Turan-Mongolian desert and steppe species.
Y. mongolicus Proczynski, 1968. Repetek. Turan-Mongolian desert and steppe species.
Y. univittatus Simon, 1871. Sultanbent, Repetek. Ancient Mediterranean species.
Y. validus Simon, 1889. Badghyz, Bairam-Ali (= Old Merv), "Bewüste Murghab." Irano-Turkestanian species.
Y. vittatus Thorell, 1875. Kaplankyr. Mediterranean species.

Zoogeographic Connections of the Spider Fauna of Turkmenistan

The spider species listed above belong to a wide assemblage of genera and families with various ecological requirements. Due to their adept dispersal abilities (including the well-known "ballooning" of young spiderlings on silk threads), spiders commonly are not believed to be a very useful group of animals for a zoogeographic analysis. Nevertheless, we can detect certain patterns in their distribution within Turkmenistan and thus, will attempt to reveal zoogeographic connections of the local araneofauna. Of course, these conclusions will be subject to change when more knowledge is gained about distribution of spider species.

More than half of the entire faunal list (184 species, or 54.9%) comprises spiders whose range within the Palearctic Realm is limited to one or a few biogeographic provinces. These species can be classified into several faunistic complexes which show current species distribution (and may also reflect centers of origin).

Iranian and Caucasian-Iranian species. Within Turkmenistan, these species are found in Kopetdagh (including Badghyz). Iranian species (38 spp., or 11.3% of the fauna) are *Phyxioschema raddei, Nemesia birulai, Raveniola fedotovi, R. kopetdaghensis, R. redikorzevi, Filistata sp., Ceratopholcus maculipes, Pholcus nenjukovi, Dysdera transcaspica, Palpimanus sogdianus, Hersiliola afghanica, H. sp., Pelecopsis paralleloides, Cedicus ephthalitus, C. gennadii, C. parthus, C. maerens, Agyneta kopetdaghensis, Diplocephalus bifurcatus, Erigonoplus ninae, Lepthyphantes turanicus, Mecopisthes orientalis, Mesasigone mira, Hahnia sp., Titanoeca lehtineni, Zodarion proszynskii, Z. raddei, Drassodes proximus, D. sp., Berlandina sp., Minosiella intermedia, Prodidomus redikorzevi, Pterotricha strandi, Synaphosus sp., Talanites dunini, T. fagei, Menemerops afghanus,* and *Plexippoides starmuehlneri.* The five Caucasian-Iranian species are *Aulonia kratochvili, Tarentula bergsoei, Pardosa pontica, Pisaura novicia,* and *Micaria kopetdaghensis.*

Irano-Turkestanian Species. These spiders are more widely found than Iranian ones and inhabit mountains and foothills of Middle Asia (including Chinese Turkestan), Afghanistan, and Iran. This group includes 36 species (10.7% of the fauna): *Pritha crosbyi, Scytodes strandi, Oecobius nadiae, O. tadzhicus, Oxyopes takobius, Eusparassus oculatus, Olios sericeus, Alioranus avanturus, Metellina kirgisica, Araneus spasskyi, A. tartaricus, Argiope ahngeri, Gnaphosa haarlovi, G. kuldzha, Nomisia conigera, Tarentula cursor cursorioides, "Oxyptila" lugibris, Psammitis turanicus, Stiphropus strandi, Xysticus kaznakovi, X. lapidarius, X. minor, Mesiotelus kulczynskii, Aelurillus andreevae, Chalcoscirtus martensi parvus, Langona tartarica, Menemerus marginatus, Mogrus antoninus, Pellenes kulabicus, Phlegra sogdiana, Pseudicius cinctus, P. spasskyi, Synageles charitonovi, S. ramitus,* and *Yllenus validus*. Some Turkestanian species, such as *Dysdera kugitangica, Metellina kirghizica, Lepthyphanthes escapus, L. kuhitangensis,* and *Mesasigone mira*, are found in Turkmenistan only in the Kugitang mountains (an offspur of the Pamiro-Alai range). Three species, *Thanatus imbecillus, Heriaeus spinipalpus,* and *Aelurillus concolor*, are found from the Caucasus to Turkestan.

European and European-Caucasian Species. These are spiders of mesophilic habitats; in Turkmenistan, they are found only in the mountains of Kopetdagh, which often represent the easternmost limit of their distribution. The European group includes 16 species (4.8% of the fauna): *Enoplognatha thoracica, Donacochara speciosa, Gongylidiellum murcidum, Oedothorax apicatus, Walckenaeria monoceros, Anyphaena accentuata, Echemus angustifrons, Scotophaeus scutulatus, Clubiona alpicola, Zora nemoralis, Z. silvestris, Philodromus fallax, Oxyptila sanctuaria, Sitticus distinguendus, S. zimmermanni,* and *Synageles hilarulus*. Spiders of the European-Caucasian group include five species: *Araniella inconcpicua, Zilla diodia, Archaeodictyna ammophila, Brigittea latens,* and *Titanoeca tristis*.

East Mediterranean Species. This small group includes 11 species: *Latrodectus pallidus, Micaria albimana, Archaeodictyna consecuta, Orthobula charitonovi, Micaria septempunctata, Oxyptila tricoloripes, Monaeses israilensis, Heliophanus curvidens, Pellenes epularis, Pellenes simoni,* and *Salticus tricinctus*. Their range extends from the Middle East to Kopetdagh but not further eastward.

Turanian Desert Species. These species inhabit the lowland deserts of Karakum and Kizylkum in Middle Asia, may penetrate into semidesert foothills but are not found in mountainous habitats. Some of these species are adopted to sand desert habitats and are specialized psammophiles (e.g., *Sitticus karakumensis*); many inhabit rodent burrows (e.g., *Minosiella intermedia*, Krivokhatsky and Fet 1982). The Turanian desert group, which includes 58 spp., or 17.3% of the total fauna, comprises the following: *Artema transcaspica, Dysdera aculeata, D. limitanea, D. pococki, Hersiliola* sp., *Janetschekia necessaria, Lepthyphantes badkhyzensis, L. turkestanicus, Pelecopsis laptevi, Trachelocamptus asiaticus,*

Aculepeira sogdiana, Araneus repetecus, A. tedgenicus, Hypsosinga turkmenica, Larinia nenilini, L. pubiventris, Zygiella caspica, "Arctosa" cereipes, Tarentula raddei, Arctosa soror, Evippa badchysica, E. schenkeli, E. turkmenica, Lycosa alticeps, Wadicosa commoventa, Devade tenella, Dictyna cronebergi, Dictynomorpha strandi, Coelotes charitonovi, Oxyopes badhyzicus, O. maracandensis, Zodarion sytchevskayae, Asiabadus asiaticus, Berlandina afghana, B. caspica, Drassodes jakkabagensis, "D. "flavomaculatus, "D." thimei, Gnaphosa turkmenica, Minosia karakumensis, Synaphosus sp., "Talanites" aculeatus, "T." sp., Theuma walteri, Zelotes aerosus, Z. arnoldii, Z. bucharensis, Heriaeus buffonopsis, H. fedotovi, Xysticus caspicus, X. concinnus, Aelurillus ater, Heliophanus niveivestis, H. turanicus, Mogrus valerii, and *Sitticus karakumensis.* Related to this group are Turanian-Mongolian spiders that are found also farther to the east, in Mongolian deserts (*Xerolycosa brunneopicta, Yllenus auspex, Y. bajan, Y. hamifer,* and *Y. mongolicus*), as well as *Hippasa partita*, which ranges from the Sahara Desert to the Sind, and *Evippa onager*, which is found from the Sind to Turan.

Our approximation shows that faunistic complexes of spiders found in Turkmenistan are roughly divided into mountain semiarid and humid fauna (species with European, European-Mediterranean, Iranian, Irano-Turkestanian and related types of ranges) and arid fauna of lowland deserts (species with Turanian and related types of ranges).

The remainder of the araneofauna (150 spp., or 44.8%) includes spider species which are broadly distributed within the Palearctic Realm or even more broadly. Several Holarctic species, spiders with Paleotropic connections, and a few cosmopolitan species are found in Turkmenistan. The majority of broadly distributed species are, however, those of the southern Palearctic, or so-called Ancient Mediterranean area, with all kinds of ranges, often including Europe or its southern part. Most of these species in Turkmenistan are found only in the mountains but not in the Karakum Desert. These include, for example, Mediterranean or European-Mediterranean species *Atypus muralis, Scytodes bertheloti, Dysderina loricata, Enoplognatha testacea, Agyneta ressli, Lepthyphantes pinicola, Sphecozone romana, Trichoncoides piscator, Agalenatea redii, Araneus angulatus, A. armida, Larinioides folium, Mangora acalypha, Neoscona subfusca, Titanoeca albomaculata, T. veteranica, Pisaura mirabilis, Tarentula albofasciata, Arctosa variana, Lycosa nordmanni, L. radiata, Pardosa italica, P. morosa, P. nebulosa, Cheiracanthium mildei, C. seidlitzi, Mesiotelus tenuissimus, Phrurolithus pullatus, Haplodrassus dalmatensis, Aphantaulax seminigra, Micaria romana, Nomisia aussereri, Zelotes caucasius, Z. pumilus, Paratibellus oblongiusculus, Philodromus lepidus, Runcinia lateralis, Synema plorator, Xysticus caperatus, X. kochi, X. marmoratus, X. ninnii, Oxyopes lineatus, Aelurillus affinis, A. m-nigrum, Ballus chalybeius, Chalcoscirtus infimus, Eris nidicolens, Leptorchestes berolinensis,* and *Yllenus vittatus.* All these spiders are mesophilic or xeromesopilic, but definitely not adapted to extreme desert conditions. Their dispersal must have been prevented by deserts in at least the last two or three million years. Thus, whatever the origin of a broadly distributed species, it must have migrated through the mountains of Middle Asia in order to disperse along the southern Palearctic region.

In Turkmenistan, mountains and desert are divided only by a narrow (10 to 20 km) belt of the arid foothills. The biogeographic border between the Mediterranean-type mountains of Kopetdagh and the lowland continental desert of Karakum is one of the best expressed (and least studied) ecological and biogeographic boundaries that exists on the earth (Kryzhanovsky 1965). Over its geological history, Kopetdagh could have served as a biogeographic corridor for spider dispersal. In the Late Oligocene-Miocene (25 to 10 million years ago), reduction of the ancient Tethys Sea revealed an island/peninsular chain from the modern Balkans to the Armeno-Iranian Plateau. This chain was a natural corridor for dispersal (Kryzhanovsky 1965), as many island chains are today (e.g., the Antilles or Sunda Islands). With aridization continuing from the Eocene through the Oligocene, landscapes gradually changed. By the Miocene-Pliocene (from 5 to 2 million years ago), the mountain uplift and receding Tethys Sea aridization separated deserts of Middle and Central Asia from those of Sind, the Middle East, and North Africa and promoted the vicariant speciation on desert lowlands. However, the mountain chains of Zagroz, Elburz, and Kopetdagh still could have served as effective dispersal routes. The sublatitudinal position of this uplifting mountain chain suggests that migrations of animals through it eastward or westward was not limited by latitudinal climatic changes (a common problem in such well-known dispersal cases as, for example, North/South American exchange). Moreover, the altitudinal zonation in mountains allowed dispersal of ecologically different animals migrating within specific mountain belts. Located between northern and southern deserts of Asia, these mountains could house mesophile fauna which could not survive aridization of adjacent lowlands.

In the Pleistocene (less than one million years ago), the Iranian mountain corridor undoubtedly was a site of constant dispersal and, probably, also of local speciation. Many local endemic species probably are of this age, especially plant species (Kamelin 1970, 1979). Transgressions of the Proto-Caspian Sea periodically returned these desert mountains to the island condition.

During the recent glaciation (16,000 to 10,000 years ago) this corridor could have been invaded by almost modern European and Asian "refugee" species. Then, a new aridization disrupted many ranges and effectively isolated European forest and meadow species in mountain valleys such as the Aidere Valley in Southwest Kopetdagh. Spider fauna in walnut and elm forests of this magnificent (ca. 30 km long) gorge bears strong European features and includes mesophile species of such genera as *Anyphaena*, *Echemus*, *Aulonia*, *Mesiotelus*, *Walckenaeria*, *Zora*, *Zelotes*, and *Clubiona*.

Extensive local vicariant speciation (well demonstrated, for example, in numerous plant genera) along mountain systems from the Caucasus to Turkestan is observed also in spiders. For instance, a mygalomorph genus *Raveniola* Zonshtein, 1987 inhabits mountain forests at elevations above 1,000 m and has several vicariant species from Georgia to the Himalayas, including *R*.

kopetdaghensis from Southwest and Central Kopetdagh (Fet 1984). Another example is spider genus *Zodarion*, which has numerous endemic species throughout the Ancient Mediterranean area, with three species found only in Turkmenistan. Of these, *Z. raddei* is a very common ant-eating spider in all arid habitats, including sand desert (where it is one of the common inhabitants of rodent burrows; Krivokhatsky and Fet 1982), foothills (Krivokhatsky and Fet 1981; Fet 1983), and mountains (Fet 1985b). Two other species, *Zodarion proszynskii* and *Z. sytchevskayae* (Nenilin and Fet 1985), are vicariant in mountain valleys and desert foothills of Kopetdagh respectively. A third example of local vicariant speciation in Kopetdagh is a recently described (Fet 1993) subgenus *Paracedicus* belonging to the genus *Cedicus*. It includes two closely related species, *Cedicus (Paracedicus) ephthalitus*, which is found in the mountain valleys and humid meadows, and *C. (P.) gennadii*, found in semiarid mountain habitats.

Addressing the problem of local endemics (those of the Kopetdagh Mountains or Karakum Desert), we should remember that southern Palearctic spiders are still poorly studied. Within the Middle Asian republics of the former USSR, special araneological studies have been conducted only in Tajikistan and Turkmenistan; araneofauna of Uzbekistan, Kyrghyzstan, and Kazakhstan is poorly known, as are faunas of adjacent Afghanistan and, especially, Iran. In other areas of the Middle East and Central Asia, only faunas of Israel and India are known to some extent; no serious studies of spiders from Syria, Iraq, Pakistan, or Mongolia exist. Therefore, it is not practical to treat most spider species known presently only from Turkmenistan as endemics of local level. Mountain species may, in the future, be found within the Iranian or Turkestanian mountain systems, and desert spiders may turn out to be widespread Turanian desert elements. Nevertheless, examples from many other groups of animals and plants, some with good dispersal abilities, suggest that local endemism may be found among Turkmenistan spiders. As possible candidates for Kopetdagh endemics, we can list several species of the Iranian faunistic complex which are found so far only in the humid mountain forests of Kopetdagh: *Raveniola kopetdaghensis*, *Cedicus ephthalitus*, *Hahnia sp.*, *Zodarion proszynskii*, *Erigonoplus ninae*, *Mecopisthes orientaliis*, and *Drassodes sp.* A few spider species, discovered so far only in the sand desert of Karakum (e.g., *Sitticus karakumensis*, *Mogrus valerii*, *Cebrennus sp.*, and *"Talanites" sp.*), may belong to the Karakum psammophile endemics, which are quite common in most groups of animals and plants. The narrow belt of sagebrush-covered, loess foothills of Turkmenistan that separates desert from the mountains may also have been a site of speciation in spiders. Among species found only in these habitats (including the hilly plateau of Badghyz) are, e.g., *Nemesia birulai*, *Raveniola redikorzevi*, *Hersiliola sp.*, *Pelecopsis paralleloides*, *Cedicus maerens*, *Titanoeca lehtineni*, *Minosia karakumensis*, *Prodidomus redikorzevi*, and *Zodarion sytchevskayae*. Only from the salty depressions of Baghyz are known *Oxyopes badhyzicus* and *Gnaphosa turkmenica*. The long and complicated geological history of landscapes of Turkmenistan has provided a diverse arena for differentiation of all ecological forms of spiders, from psammophiles of sand desert to mesophiles of mountain meadows.

Acknowledgements

We thank all the people who contributed to spider collection and helped in the field trips in Turkmenistan, and, first of all, we give thanks to Gennady T. Kuznetsov, Svyatoslav I. Zabelin, Nina S. Ustinova, Lyudmila A. Mitroshina, Tamara M. Telyushenko, and Galina N. Fet. Rostislav A. Danov (1941–1993) and Yuri K. Gorelov were always helpful with their advice on natural history and biogeography as well as with their aid in field logistics. Identifications of many spider species were conducted or checked at different times by our colleagues Andrei B. Nenilin (1960–1986), Vladimir I. Ovtsharenko, Sergei L. Zonshtein, Alexei A. Zyuzin, Andrei V. Tanasevich, Yuri M. Marusik, Kirill Yu. Eskov, and Pyotr M. Dunin. Vladimir I. Ovtsharenko kindly helped to clarify recent synonymy of Gnaphosidae from Turkmenistan. We also thank for their constant support of our araneological studies Norman I. Platnick, Jerzy Proszynski, Gershom Levy, Pekka T. Lehtinen, Seppo Koponen, Alexander B. Lange, Yuri S. Balashov, Mikhail V. Heptner, Yaroslav I. Starobogatov, and Gary A. Polis.

Addendum to the List of Spiders of Turkmenistan

Fam. Dysderidae
Dysdera brignolii Dunin, 1989. Repetek. Turanian desert species.
D. kusnetsovi Dunin, 1989. Central Kopetdagh. Iranian species.
D. nenilini Dunin, 1989. Tuarkyr, Kaplankyr. Turanian desert species.
D. tystshenkoi Dunin, 1989. Southwest Kopetdagh. Iranian species.
Harpactea parthica Brignoli, 1980. Krasnovodsk. Iranian species.

Fam. Gnaphosidae
Synaphosus karakumensis Ovtsharenko, Levy et Platnick, 1994, Repetek. Turanian desert species.
S. palearcticus Ovtsharenko, Levy et Platnick, 1994. Kopetdagh (Central), Ashkhabad, Badghyz, Mary, Chardzhou, Amudarya Valley. Ancient Mediterranean species.
S. soyunovi Ovtsharenko, Levy et Platnick, 1994. Sarykamysh, Tashauz Region. Turanian desert species.
S. turanicus Ovtsharenko, Levy et Platnick, 1994. Kopetdagh (Central), Tashauz Region, Ashkhabad, Badghyz, Mary, Repetek, Amudarya Valley. Iranian-Turanian species.

31. Fauna and Zoogeography of Scorpions (Arachnida: Scorpions) in Turkmenistan

VICTOR FET

Abstract

Scorpion fauna in Turkmenistan includes six genera and seven species of Buthidae, all belonging to the Saharo-Gobian (desert Palearctic) genera. *Mesobuthus eupeus*, *M. caucasicus*, and *Orthochirus scrobiculosus* are nearly ubiquitous and exhibit intraspecific variation; they are also widespread beyond the boundaries of Turkmenistan. Specialized Turanian sand desert genera include *Liobuthus kessleri* and *Anomalobuthus rickmersi*, and the endemic *Pectinibuthus birulai*. The last species, *Kraepelinia palpator*, is found in Turkmenistan only at salt lake shores (Yeroyulanduz Depression in the Badghyz Reserve); it is also known from Iran. Scorpion fauna of lowland Turkmenistan is a combination of widespread Asian and endemic Turanian and Iranian desert elements. There are no endemic species in the mountains; only one species, *Mesobuthus eupeus*, is commonly found in Kopetdagh.

Introduction

The fauna of scorpions of the former Russian Empire/USSR (i.e., primarily that of Caucasus and Middle Asia) was reviewed by Birula (1911, 1917a,b) and Fet (1990). In preparing this paper, we used data on scorpion fauna and ecology of Turkmenistan and adjacent countries published by the author (Fet 1980, 1984, 1987a,b, 1990) and other researchers (Simon 1889; Pocock 1899; Radde 1899; Morits 1922; Shestoperov 1934, 1935; Pavlovsky 1934; Vlasov 1937a,b; Vachon 1958a,b, 1966, 1974; Sabirova 1977, 1981, 1986). The largest museum collections reviewed by the author included those of the Zoological Institute of the Russian Academy of Sciences (St. Petersburg, Russia) and the Zoological Museum of Moscow State University, Moscow, Russia). Detailed data on museum collections, complete synonymy of Turkmen species and subspecies of scorpions, and complete references have been published (Fet 1990).

Fauna of Scorpions in Turkmenistan

Six genera and seven species of scorpions belonging to the family Buthidae are found in Turkmenistan. Below, we give general distribution and a list of all known localities within Turkmenistan for every species, with remarks on taxonomic status, intraspecific variability, and ecology.

1. *Mesobuthus eupeus* (C.L. Koch, 1838)

Distribution. Afghanistan, Armenia, Azerbaijhan, China, Georgia, Iran, Iraq, Kazakhstan, Kyrghyzstan, Mongolia, Pakistan, Russia (Astrakhan Region), Tajikistan, Turkey, Turkmenistan, Uzbekistan.
In Turkmenistan, three subspecies are found.

1a. *M. e. thersites* (C.L. Koch, 1839)
Distribution. China (northwest), Iran (northeast), Kazakhstan, Kyrghyzstan, Russia (Astrakhan Region), Tajikistan, Turkmenistan, Uzbekistan.

Localities in Turkmenistan. Ashkhabad Region: Akhalteke, Akdepe, Anau, Archman, Ashkhabad, Bami, Berzengi, Bikrova, Birleshik-3, Chuli, Dushak Mt., Firyuza, Gaudan, Germab, Imarat (20 km north of Saivan), Ipaikala, Kaakhka, Kelyata, Kheirabad, Lake Kopetdaghskoe, Kovata, Lake Kurtli, Mollakurban, Nisa (ruins), Nokhur, Pervomaisky, Serakhs, Tedzhen, Tedzhenstroi, Verkhnee Skobelevo (= Zinovyevka). *Chardzhou Region*: Char-Charagasy, Darganata, Khodzhafilata, Repetek. *Krasnovodsk Region*: Adzhidere, Aidere, Akhchakuima, Arpaklen, Bolshoi Balkhan (including northern slope, Uzunsu or Uzunakar Spring, northern slope of Shakhiburun Mt., Bashmygur, Koshagoi Spring), Chandyr Valley, Chandyr-Sumbar Mountains, Danata, Lake Delili, Dzhebel, Eishem (80 km southwest of Iskander), Kara-Kala, Karateniz (or Karabugaz, a solonchak 40 km north of Kizyl-Arvat), Kazandzhik, Khodzhakala, Kizyl-Arvat, Kizylkaya, Krasnovodsk, Lappi, Madau, Mardagh, Meshed-Misrian, Mollakara, Nebit-Dagh, Pereval (= Perevalnaya), Sharlouk, Syunt-Khasardagh Range (and Reserve), Uzunada (on Krasnovodsk Bay), Yaskhan. *Mary Region*: Badghyz Reserve (including Kepele, Kizyldzhar, Naredevanly, Gyazgyadyk, Pinkhancheshme, Pulikhatum, and Yeroyulanduz), Bairam-Ali, Iolotan, Kushka, Mary, Morgunovka, "Murghab Imperial Estate" (near Bairam-Ali). *Tashauz Region*: Burkhliburun, Chirishly, Gangalykyr, Kankakyr, Kaplankyr Reserve, Shakhsenem.

1b. M. e. philippovitschi (Birula, 1905)
Distribution. Northern Iran, southwestern Turkmenistan.

Localities in Turkmenistan. Krasnovodsk Region: Atrek River.

1c. M. e. afghanus (Pocock, 1889)

Distribution. Northern and eastern Iran, northern Afghanistan, southern Turkmenistan.

Localities in Turkmenistan. Mary Region: Kushka, area between the Tedzhen and Murghab Rivers (Birula 1904, 1905).

Variability. Within its wide range, *M. eupeus* is extremely polymorphic and is subdivided into a number of subspecies distinguished primarily by morphometry, morphosculpture, and coloration. Fourteen subspecies are recognized besides the three forms listed above: these are *M. e. eupeus* (C.L. Koch 1838) (Caucasus, Iran, Turkey); *M. e. bogdoensis* (Birula 1896) (Kazakhstan); *M. e. barszczewskii* (Birula 1904) (Tajikistan); *M. e. haarlovi* Vachon 1958 (Afghanistan); *M. e. iranus* (Birula 1917) (Iran); *M. e. kirmanensis* (Birula 1900) (Iran); *M. e. mesopotamicus* (Penther 1912) (Iraq), *M. e. mongolicus* (Birula 1911) (China, Mongolia); *M. e. pachysoma* (Birula, 1900) (Iran, Pakistan); *M. e. persicus* (Pocock 1899) (Iran, Pakistan); and *M. e. phillipsi* (Pocock 1889) (Iran). Validity of many of these geographic forms of *M. eupeus* will have to be revised.

Our preliminary data on geographic variability in *M. eupeus* show the existence of a possibly undescribed race within *M. e. thersites* in lowlands of western Turkmenistan (Krasnovodsk Region, Madau, and Tashauz Region, Kaplankyr Reserve), which differs in its meristic characters (number of plates on pectinal organs) from *M. e. thersites* inhabiting Kopetdagh as well as from other studied mountainous populations of this subspecies (samples from Tajikistan and Kyrghyzstan). It is interesting that A.A. Birula, according to his unpublished records, intended to describe a separate form from lowland Turkmenistan, *M. e. caspius* (Pavlovsky 1934). With this name (nomen nudum) he had labeled a number of specimens in the collections of the Zoological Institute in St. Petersburg.

This possible evidence of genetic differentiation between lowland and foothill/ mountainous *M. eupeus* in Middle Asia needs to be examined by genetic analysis of these populations.

Ecology. Mesobuthus eupeus is the most common, abundant, and ecologically ubiquitous scorpion species throughout its entire range in Asia, excluding only sand deserts, where it is present but not dominant. It lives in a variety of desert and semi-desert habitats. The boundary of its range in the north (Astrakhan Region of Russia) corresponds to the boundary of semi-desert landscapes. *M.*

e. thersites is the only scorpion commonly found in the mountains of Turkmenistan (Kopetdagh and Bolshoi Balkhan), where it can be found as high as 2,500–2,800 m. It ives in holes and crevices under roots of desert plants and have often been observed in night time (under UV light) in sit-and-wait posture on vegetation (*Artemisia* spp.) rather than on soil surface (Kizyl-Arvat and Ashkhabad Regions, our data). It is not common in pitfall traps (Fet 1980) or buildings.

2. *Mesobuthus caucasicus* (Nordmann, 1840)

Distribution. Afghanistan, Armenia, Azerbaijan, China, Georgia, Iran, Iraq (?), Kazakhstan, Kyrghyzstan, Mongolia, Russia (Chechnya, Daghestan, North Ossetia), Tajikistan, Turkey, Turkmenistan, Ukraine (Odessa and Kherson Regions), Uzbekistan.

In Turkmenistan, two subspecies are found.

2a. *M. c. parthorum* (Pocock, 1889)
Distribution. Afghanistan, Iran, Kazakhstan, Kyrghyzstan, Tajikistan, Turkmenistan, Uzbekistan.

Localities in Turkmenistan. Ashkhabad Region: Anau, Archman, Ashkhabad, Bakharden, "sands of the Central Karakum," Gaudan, Gyaurs, Firyuza, Kaakhka, Lake Kopetdaghskoe, Lake Kurtli, Serakhs, Serakhsky District, Lake Sportivnoe, Sultandepe, Tedzhen, Tedzhenstroi. *Chardzhou Region*: Repetek. *Krasnovodsk Region*: Akhchakuima, Cheleken, Dardzha Peninsula, Dzhebel, Karabugaz (a solonchak 40 km north of Kizyl-Arvat), Kara-Kala, Karatogolok (on the Uzboi), Krasnovodsk, Mollakara, Yaskhan. *Mary Region*: Badghyz Reserve (including Akarcheshme, Kepele, Kizyldzhar, Nardevanly, Pinkhancheshme, and Yeroyulanduz), Bairam-Ali, Chemenibid, Dortkuyu, Iolotan, Kushka, Mary, Morgunovka, "Murghab Imperial Estate" (near Bairam-Ali). *Tashauz Region*: Chirishly, Kaplankyr Reserve, Kunya-Urgench, Shakhsenem.

2b. *M. caucasicus Intermedius* (Birula, 1897)
Distribution. China, Iran, Kazakhstan, Kyrghyzstan, Tajikistan, Turkmenistan, Uzbekistan.

Localities in Turkmenistan. Chardzhou Region: Char-Charagasy, Kelif, Khodzhafilata, Repetek, Saltyrykh, Svintsovy Rudnik. *Krasnovodsk Region*: Krasnovodsk.

Variability. Within its wide range, *M. caucasicus* is subdivided into six subspecies, including the two listed above as well as the following: *M. c. caucasicus* (Nordmann, 1840) (Armenia, Azerbaijan, Georgia, Iran, Russia, Turkey, Ukraine); *M. c. przewalskii* (Birula, 1897) (China, Mongolia,

Tajikistan, Uzbekistan); *M. c. fuscus* (Birula, 1897) (Tajikistan); and *M. c. kaznakovi* (Birula, 1904) (Tajikistan). Validity of all these forms has yet to be confirmed.

Ecology. M. caucasicus lives in a variety of ecological conditions. The nominal form, *M. c. caucasicus*, although limited to arid and semi-arid regions, survives in sandy habitats of southeastern European steppes (Odessa and Kherson Regions of Ukraine) and mountains of East Caucasus (Daghestan and Chechnya). It is also very common in houses and buildings. On the other hand, *M. c. parthorum* from Turkmenistan is a lowland desert psammophile and differs in morphology from both *M. c. caucasicus* and eastern *M. c. intermedius* (the latter inhabits lowlands, foothills, and mountains up to 3,000 m). The largest of all scorpions found in Turkmenistan, *M. c. parthorum*, is very common in sand desert (Repetek), where on moonless nights it can be found by UV-light search (our data) waiting for prey almost under every bush of *Haloxylon persicum*; it does not climb on vegetation or run on sand surfaces. *M. c. parthorum* is one of the most abundant scorpion species in pitfall traps in sand desert (Fet 1980; G.T. Kuznetsov pers. comm.).

3. Orthochirus scrobiculosus (Grube, 1873)

Distribution. Afghanistan, India, Iran, Iraq, Israel, Jordan, Kazakhstan, Pakistan (?), Tajikistan, Turkmenistan, Uzbekistan.

Localities in Turkmenistan. Ashkhabad Region: Akhalteke, Akdepe, Anau, Ashkhabad, Bagir, Bakharden, Chuli, Gaudan, Germab, Gyaurs, Kaakhka, Lake Kurtli, Serakhs, Lake Sportivnoye, Sultandepe. *Chardzhou Region*: Repetek. *Krasnovodsk Region*: Aidere (near Saparbakhar ruins), Akhchakuima, Chandyr Valley, Dzhebel, Kara-Bogaz-Gol Bay, Kara-Kala, Khodzhakala, Kizyl-Arvat, Krasnovodsk, Meshed-Misrian, Mollakara, Nebit-Dagh. *Mary Region*: Badghyz Reserve (including Adamulen, Kizyldzhar, and Yeroyulanduz), Bairam-Ali, Iolotan, Kushka, Mary, Morgunovka, Sultanbent, Takhtabazar. *Tashauz Region*: Chirishly, Sarykamysh, Shakhsenem.

Variability. In Turkmenistan, two subspecies are found: *O. s. scrobiculosus* (Grube 1873) and *O. s. melanurus* (Kessler 1874). Morphological difference between these two forms is not firmly established. Above, we listed all known localities for the entire species in Turkmenistan, without their affiliation to one of the two forms. However, in old literature, and in collections identified and labeled by A.A. Birula (the Zoological Institute in St. Petersburg and the Zoological Museum of Moscow State University), this distinction between forms was made. These two subspecies were recorded by A. A. Birula within Turkmenistan in the following localities: *O. s. scrobiculosus* (= *O. conchini* Simon, 1889) from Akdepe (= Bely Bugor), Ashkhabad, Dzhebel, Kara-Bogaz-Gol Bay, Kizyl-Arvat, and Krasnovodsk; and *O. s. melanurus* (Kessler 1874) from Akhalteke, Repetek, Mollakara, Bairam-Ali, Iolotan, Kushka, and Takhtabazar. This distribution shows that the two forms, as they were treated by A.A. Birula, may form vicariant clusters. The westernmost part of the republic (Kara-Bogaz-Gol Bay and Krasnovodsk) is characterized only by the nominal subspecies, whereas *O. scrobiculosus* from south and southeast (Kushka, Bairam-Ali, Iolotan, Takhtabazar, and Repetek) belong to the subspecies *O. s. melanurus*. Beyond the borders of Turkmenistan, the nominal subspecies is recognized only to the north (Kazakhstan: Ustyurt), whereas *O. s. melanurus* occupies larger range, including Kazakhstan to the north as well as Uzbekistan and Tajikistan to the east. It is possible that, under thorough morphological and genetic examination, these two forms would present valid subspecific complexes.

Other subspecies recognized within this species are *O. s. concolor* (Birula, 1898) (Tajikistan, Uzbekistan), *O. s. negebensis* Shulov et Amitai, 1960 (Israel, Jordan), *O. s. mesopotamicus* (Birula, 1918) (Iraq), *O. s. persa* (Birula, 1900) (Iran), and possibly *O. s. dentatus* (Birula, 1900) (Iran). Status of populations from Afghanistan, India, and Pakistan is not clear.

Ecology. *O. scrobiculosus* is nearly ubiquitous in all lowland deserts although in sand desert it prefers areas with thick vegetation (*Haloxylon persicum* and *H. aphyllum*) and is not common in naked sand dunes (Fet 1980). It can penetrate into the relatively humid mountain valleys as high as 800 m (Saparbakhar in Southwest Kopetdagh, Aidere Valley), where it lives on dry rocky slopes.

4. Liobuthus kessleri Birula, 1898

Distribution. Iran (northeast; Sarakhs), Kazakhstan, Turkmenistan, Uzbekistan.

Localities in Turkmenistan. *Ashkhabad Region*: Anau District, Archman, Ashkhabad, Geok-Tepe, Lake Sportivnoe. *Chardzhou Region*: Amudarya Valley, Chardzhou, Repetek. *Krasnovodsk Region*: Akhchakuima, Dzhebel, Mollakara (= "Muttakary"). *Mary Region*: Bairam-Ali. *Tashauz Region*: Chirishly, Shakhsenem.

Ecology. A psammophile monotypic genus; common in sand dunes in pitfall traps and under UV-light search (Repetek Reserve), where it prefers dunes with vegetation (*Haloxylon persicum*). In sand desert, it is the most abundant scorpion species in pitfall traps (Fet 1980). *L. kessleri* has been observed while quickly digging into the sand (Pavlovsky 1917; our data).

5. *Anomalobuthus rickmersi* Kraepelin, 1900

Distribution. Kazakhstan, Turkmenistan, Uzbekistan.

Localities in Turkmenistan. Ashkhabad Region: Anau District, Lake Sportivnoye, Tedzhen. *Chardzhou Region*: Repetek. *Krasnovodsk Region*: Kheles. *Tashauz Region*: Kunya-Urgench, Shakhsenem.

Ecology. A psammophile monotypic genus; common in sand dunes under UV-light search (Repetek Reserve), where it prefers dune areas without thick vegetation.

6. *Pectinibuthus birulai* Fet, 1987

Distribution. Turkmenistan (endemic).

Localities in Turkmenistan. Chardzhou Region: Repetek.

Ecology. A psammophile monotypic genus. So far, it has been found only in Repetek Reserve (Fet 1987a) in sand dunes without vegetation, where it was captured in pitfall traps and detected under UV-light. It is much less common on the surface than other psammophile scorpions of Repetek. *P. birulai* is able to move very rapidly and dig ("dive") into sand. It has been observed feeding on a tenebrionid, *Habrobates veisovi* (V.I. Kuznetsov pers. comm.).

7. *Kraepelinia palpator* Vachon, 1974

Distribution. Iran (east), Turkmenistan.

Localities in Turkmenistan. Mary Region: Badghyz Reserve (shores of the salt lake in the Yeroyulanduz Depression, Bashchenko Spring).

Ecology. This species has not been observed in natural conditions. Its habitat (a salt lake shore) and relatively small size suggest that it can inhabit crevices in dry salt pans, a very specific niche shared in Yeroyulanduz by insects such as ground beetles of the genus *Syrdenus* (Kryzhanovsky and Mikhailov 1971). Details of the landscape where *Kraepelinia palpator* is found in eastern and southeastern Iran are discussed by Fet (1984).

Zoogeographic Connections

All seven species of Buthidae found in Turkmenistan belong to the Saharo-Gobian (desert Palearctic) genera (Birula 1917; Fet 1984, 1990; Nenilin and Fet 1992). High diversity of desert scorpions in all world deserts (Polis 1990) is a result of authochthonous evolution; e.g., desert scorpion faunas of North America, Australia, and South Africa have nothing in common with Palearctic desert faunas of North Africa and Middle/Central Asia. The latter two have only a few genera of scorpions in common, including *Androctonus, Orthochirus,* and *Hottentotta*. Several genera of desert scorpions found predominantly in Asia have quite wide ranges and exhibit intensive speciation (e.g., *Mesobuthus* and *Compsobuthus*). The most characteristic feature for Palearctic desert scorpions, however, is its high endemism at generic level. Not only vicariant species, but numerous vicariant genera of Buthidae – usually monotypic – occupy desert habitats from Morocco to China; their range in many cases is limited (Vachon 1952, 1966; Levy and Amitai 1980; Kinzelbach 1985; Vachon and Kinzelbach 1987; Nenilin and Fet 1992). Examples of such genera are *Buthiscus, Cicileus,* and *Lissothus* (North Africa), *Leiurus* (North Africa and Middle East), *Vachoniolus, Buthacus, Birulatus, Butheolus,* and *Apisthobuthus* (Middle East), *Odonthobuthus* and *Plesiobuthus* (Iran and Baluchistan), and *Liobuthus, Anomalobuthus, Pectinibuthus,* and *Psammobuthus* (Middle Asia). Thus, there is a high level of speciation among Buthidae in the arid regions of the Palearctic realm.

Several of the listed genera are highly adapted to the life in various sand deserts and are true psammophiles (*Buthiscus, Plesiobuthus, Liobuthus, Anomalobuthus, Pectinibuthus,* and *Psammobuthus*); their ranges are limited to different sand deserts. Within Turkmenistan, the great sand desert of Karakum houses three psammophile monotypic genera of scorpions: *Liobuthus, Anomalobuthus,* and *Pectinibuthus*. The first two are found also in the Kizylkum Desert while the latter is endemic to the East Karakum. In general, the scorpion fauna of lowland Turkmenistan is a combination of endemic Turanian psammophiles (*Liobuthus, Anomalobuthus,* and *Pectinibuthus*), Iranian salt desert elements (*Kraepelinia*), and widespread Asian desert species (*Mesobuthus*).

The genus *Mesobuthus* includes a number of widespread species which form vicariant ranges along the Ancient Mediterranean area from the Balkan and Anatolian Peninsulas (*M. gibbosus*) to the Far East (*M. martensi* in China and Korea). The variety of subspecific forms of *M. eupeus* and *M. caucasicus* was described above. Most of these forms, however, were separated on the basis of vague characters such as coloration and morphosculpture; a complete revision of *Mesobuthus* has never been undertaken.

We have found a difference between lowland, western Turkmenistan populations and populations from mountains of Middle Asia (including Turkmenistan mountains) of a wide-ranging subspecies *Mesobuthus eupeus thersites*, which suggests the presence of two evolutionary lineages of this form within the region. This differentiation corresponds to recent biogeographic events such as the Pleistocene development of the Karakum Desert and western desert massifs due to the last regression of the Proto-Caspian Sea. This hypothesis should be tested using genetic techniques capable of distinguishing population-level differences (through allozyme or mitochondrial DNA variation).

There are no endemic scorpions in the mountains of Turkmenistan; only *Mesobuthus eupeus* is commonly found in Kopetdagh. This situation may seem unusual; a number of other groups, such as beetles (Kryzhanovsky 1965), reptiles (Shcherbak this volume), or spiders (Fet 1983; Mikhailov and Fet this volume), are represented by numerous endemic species in the Kopetdagh Mountains. A possible explanation is that most of endemic mountain species in Turkmenistan are not arid forms, or have evolved from more mesophilic ancestors, primarily of western (Mediterranean) origin. Looking at a map of scorpion distribution (Nenilin and Fet 1992), one can find several taxonomic groups which could have been candidates for mountain endemic speciation but probably never have had a chance to disperse to Kopetdagh from either east or west. A desert genus, *Hottentotta* (= *Buthothus*) (Buthidae), reaches the mountains of Iraq and northwestern Iran in the west (*H. saulcyi*) and the mountains of Tajikistan, Afghanistan, and India in the east (*H. alticola*). There are also two mesophilic, non-buthid scorpion genera, common in the adjacent mountains: *Euscorpius* (Euscorpiinae) in the Mediterranean area, which ranges eastward to northern Turkey and Georgia; and *Scorpiops* (Scorpiopsinae) in India and Afghanistan (Kabul). It is probable that constant aridization and the island regime of Kopetdagh have given no opportunity to any of the above-mentioned mountain elements to disperse to Kopetdagh. Today, the Mediterranean-type communities here are fragmented, and dispersal of mesophilic mountain scorpions is unlikely.

Finally, the salt lake/riparian monotypic genus *Kraepelinia* deserves special attention as a separate zoogeographic element. An ecologically similar specific genus of scorpions, *Australobuthus*, inhabiting salt lake shores has been recently described from Australia (Locket 1990). Many researchers believe that such habitats in the Ancient Mediterranean region originate from shore (littoral) landscapes of the Tethys (Ilyin 1947; Kryzhanovsky 1965). Intensive speciation in littoral habitats, as well as evolution of many species of plants and animals adapted to and tolerant of salt desert conditions, was probably affected by the well-known Messinian salinity crisis in the Mediterranean (Hsu 1983). Many species which originated in the harsh conditions of the Mediterranean salt deserts could have dispersed eastward, from the shores of Dead Sea to the salt lakes of Kazakhstan and Central Asia. Such depressions as Yeroyulanduz in southern Turkmenistan, where *Kraepelinia palpator* is found, house vast salt lakes and could serve as refugia for a peculiar complex of halophile species. We should also note that a number of scorpion species exist that effectively use and prefer littoral habitats. These include *Microbuthus fagei*, *Mesobuthus martensi* (= *M. confucius*) (Buthidae), and *Euscorpius flavicaudis* (Euscorpiinae) in the Palearctic (Vachon 1951) and, especially, *Serradigitus littoralis* (Vaejovidae) in Baja California, Mexico (Due and Polis 1985).

Acknowledgements

I thank Vladimir I. Ovcharenko (Zoological Institute of Russian Academy of Sciences, St. Petersburg, Russia), and Kirill G. Mikhailov (Zoological Museum, Moscow State University, Moscow, Russia) for their help during my work in these museums. I also thank Gary A. Polis, W. David Sissom, and Matt E. Braunwalder for their constant support, encouragement, and advise in my studies of scorpions; the administrations of Repetek, Badghyz, Kopetdagh, and Syunt-Khasardagh Reserves for their support of field research; and many colleagues who collected scorpions in Turkmenistan and other areas of Middle Asia, among them G.T. Kuznetsov, Yu.S. Balashov, K.I. Atamuradov, L.A. Mitroshina, T.A. Pavlenko, K.G. Mikhailov, Yu.K. Gorelov, V.I. Kuznetsov, A.Yu. Tsellarius, S.I. Sukh, and V.B. Kurilenko.

32. Fauna and Zoogeography of Molluscs of Turkmenistan

YAROSLAV I. STAROBOGATOV

Abstract

Zoogeography of terrestrial and inland water malacofauna of Turkmenistan is discussed. Fauna of the Kopetdagh (especially the terrestrial one) is intermediate between the faunas of Southwest Asia and Middle Asia. Freshwater fauna of molluscs in Kopetdagh is related to the Irano-Anatolian fauna. Freshwater fauna of the Amudarya River, excepting its lower portion but including its recent and former tributaries, is an original one and can be placed in the Sogdian-Tibetan superprovince. The mollusc fauna of the lower Amudarya includes several Balkhash-Tarim species as well as an endemic subspecies of *Dreissena polymorpha*. The Caspian aquatories near the shore of Turkmenistan include nine biogeographic provinces belonging to the Ponto-Caspian region; these provinces are characterized here by their complexes of molluscs.

Introduction

Turkmenistan is located completely in the arid zone; the major part of its territory is occupied by the Karakum Desert. This desert is bordered from the south by the Kopetdagh Mountains; the extreme east of the republic is occupied by spurs of the Kugitang range, which belongs to the Ghissaro-Darvaz mountain system. Two isolated mountain masses, Bolshoi Balkhan and Maly Balkhan, are located in the western part of Turkmenistan.

The desert is unfavorable for molluscs and they are absent from the Karakum. The Kopetdagh foothills are covered with desert which changes into grasslands at an altitude of about 350 m; arboreal vegetation in Kopetdagh is present mostly in river and stream valleys. The plateaus of Kopetdagh and Bolshoy and Maly Balkhan are covered with steppe and sparse juniper woodland.

The hydrographic net of Turkmenistan is disrupted and poorly developed due not solely to the current arid climate but also to a peculiar geological history. The part of Turkmenistan occupied presently by the Karakum was a great bay of the Caspian during the Late Pliocene and some periods in the Pleistocene. The sand in this desert contains crushed calcareous thalloms of the alga, *Melobesia* sp., which is still living in the Caspian. The Kopetdagh range formed the southern shore of this ancient bay. The position of the eastern shore has varied in different periods, with the Amudarya, Tedzhen, and Murghab rivers entering the bay from the east. The Murghab and, possibly, the Tedzhen, became tributaries of the Amudarya during the regressions of the Caspian. The Amudarya River (called by ancient authors Οξοσ, or Oxus, which is the Greek spelling of the old Iranian name Vakhsh) flowed into the Caspian for nearly the entire Pleistocene and Holocene, sometimes following the Unguz bed (i.e., turned westward from its recent valley at a latitude 39° 40 to 39° 50 N), and later, following the Uzboi bed (i.e., turned westward from about 42° N) and filling the Sarykamysh Depression. At times, the flow of the Amudarya was directed to the Aral but its current direction has been formed only in the early 1200s. The Caspian was connected with the Aral Sea during its maximal transgressions; the last such connection is believed to have taken place in the early Pleistocene (the Baku Age). The northeastern shores of the ancient bay were formed by the southern cliffs (*chinks*) of the Ustyurt Plateau. The Bolshoi and Maly Balkhans were marine islands in the same manner that they are today isolated within the desert.

Land Molluscs

Due to the arid climate, land molluscs are not diverse in Turkmenistan.
Only about twenty species are known (Likharev and Rammelmeyer 1952; Shileiko 1978, 1984; Likharev and Wiktor 1980; Muratov 1992). The majority of these species are widely distributed in western Middle Asia. Mollusc fauna of the Kopetdagh Mountains is comparatively well studied. The species which are the most resistant to dry climate are found in the steppe and semidesert habitats, e.g., *Gibbulinopsis* (*Primipupilla*) *signata* (Mouss.), *Geminula isseliana* (Bgt.), *Xeropicta candacharica* (L. Pfr.), and *X. krynickii* (Kryn.). Interestingly, the latter is frequent in the Caucasus but in Turkmenistan found only in West Kopetdagh, whereas a congeneric *X. candacharica* is found in East Kopetdagh and is widespread in Middle Asia.

Molluscs in Kopetdagh frequent rocks (especially limestone) in which crevices provide shelter. Representative rock dwellers include *Truncatellina callicratis* (Scacchi), *Pyramidula rupestris* (Drap.), *Chondrina* (*Granopupa*) *granum* (Drap.), and *Helicopsis likharevi* Schil. Species living in gravel and rocky screes are more diverse and include *Sphyradium doliolum* (Brug.), *Pupilla triplicata* (Stud.), *P. turcmenica* (O. Boettg.), *Pseudonapaeus potaninianus* (Anc.), *Ps. sogdianus* (Mts.), *Ps. eremita* (Bens.), *Ps. guttula* Murat., *Geminula continens* (Ros.), and *Ottorosenia varenzovi* (Ros.). Some of these species, together with more hygrophylous molluscs such as *Lauria cylindracea* (da

Costa), *Hesseola transcaspica* (O. Boettg.), *Helicolimax annularis* (Stud.), and *Lytopelte maculata* (Koch et Heyn.), can be found in grass vegetation and leaf litter in riparian forests along mountain valleys.

The mixed zoogeographical character of the land malacofauna in Kopetdagh is evident. A group of widely distributed species ranging from West Europe or from its Mediterranean portion to Central Asia includes *Pyramidula rupestris* (Drap.), *Sphyradium doliolum* (Brug.), *Pupilla triplicata* (Stud.), *Truncatellina callicratis* (Scacchi), and *Helicolimax annularis* (Stud.). The group of European or Mediterranean species of land molluscs which does not penetrate eastward farther than Kopetdagh includes *Lauria cylindracea* (da Costa), *Chondrina granum* (Drap.), and *X. krynickii* (Kryn.). The species *Geminula isseliana* (Bgt.) is found in both Transcaucasia and Kopetdagh. Another group of Kopetdagh land mollusc species includes those distributed mostly in the Middle and Central Asian mountains (or at least in the mountains of South Kazakhstan, Uzbekistan and Tadzhikistan). Many of these species do not penetrate westward farther than Kopetdagh; these include *Pupilla turcmenica* (O. Boettg.), *Pseudonapaeus potaninianus* (Anc.), *Ps. sogdianus* (Mts.), *Ps. eremita* (Bens.), *Geminula continens* (Ros.), and *Xeropicta candacharica* (L. Pfr.); only two species, *Gibbulinopsis signata* (Mouss.) and *Lytopelte maculata* (Koch et Heyn.), penetrate as far westward as Transcaucasia. Finally, four species can be considered endemics of the Kopetdagh Mountains: *Pseudonapaeus guttula* Murat., *Ottorosenia varenzovi* (Ros.), *Helicopsis likharevi* Schil., and *Hesseola transcaspica* (O. Boettg.). Thus, Kopetdagh (probably together with the Iranian portion of the Turkmeno-Khorassan Mountains) forms a bridge connecting mollusc faunas of Southwest Asia and Middle Asia. This agrees with Kryzhanovskii's (1965; present volume) conclusions based on the distribution of Coleoptera.

Only three species of land molluscs, all living in rock gravel, are found in Bolshoi Balkhan: *Chondrina (Granopupa) granum* (Drap.), *Pseudonapaeus sogdianus* (Mts.), and *Geminula continens* (Ros.).

The malacofauna of Turkmen portions of the Kugitang range is very poorly known. Based on data for the adjacent parts of Uzbekistan, we can surmise that a majority of species there is characteristic for Middle Asia; for example, such molluscs as *Pseudonapaeus kaznakovi* (West.), *P. otostomus* (West.), *P. miser* (Mts.), and species of the genus *Leuconozella* such as *L. rubens* (Mts.) and *L. rufispira* (Mts.) can be found there. Among hygrophilous land molluscs, species of *Macrochlamys* as well as *Candaharia (Candaharia) aethiops* (West.) and *C. (Levanderia) levanderi* (Simr.) can be found near water bodies and in irrigated fields.

In summary, the land malacofauna of Turkmenistan (which is found only in mountain areas) clearly demonstrates a gradual transition from that of Southwest Asia and Middle East toward the fauna of Middle Asia; the farther eastward, the more Middle Asian is the representation of land molluscs.

Molluscs of Fresh and Salt Inland Water Bodies

The geological history discussed above explains very well the presence of four separate faunas of molluscs in the inland water bodies of Turkmenistan. The river drainage systems of this republic can be divided into four separate zoogeographivcal provinces (Izzatulaev and Starobogatov 1985; Starobogatov 1986). The term "province" is used here in the sense of an elementary unit or parcel (Russ. *minimalnyi vydel*) defined as an aggregate of aquatories (or territories on the land) which cannot be further subdivided by methods of faunistic zoogeography (or, in a wider sense, of systematic biogeography).

The Kopetdagh province includes water bodies of this mountain range. Their diversity is rather low; there are springs, streams, and *kyarizes* (i.e., underground irrigational constructions). Springs are inhabited by small hydrobiotic forms of moluscs which are not yet sufficiently studied; this fauna includes some species of the genus *Turkmenamnicola* (Fam. Sadlerianidae) such as *T. smaragdovae* (Abr. et Tzw.) and *Valvatamnicola prasina* (Ros.) (Fam. Horatiidae). Molluscs living in streams and kyarizes are better known. Among them are the familiar species of the genus *Melanopsis*, such as *M. transcaspica* Izz. et Star., *M. turkmenica* Izz. et Star., *M. zarudnyi* Izz. et Star., *M. roseni* Izz. et Star., *M. starostini* Izz. et Star., and *M. ashkhabadensis* Izz. et Star. (Izzatulaev and Starobogatov 1984), *Corbicula fluminalis* Müll., and *C. cor* (Lam.). All pectinibranchous molluscs mentioned above are endemics of Kopetdagh but are also related to certain western Iranian species. The pulmonate molluscs in Kopetdagh are represented by *Lymnaea* (*Radix*) *rectilabrum* An. et Pr., *L.* (*R.*) *bactriana* Hutt., *L.* (*R.*) *subdisjuncta* Nev., *Costatella acuta* (Drap.), *Planorbis sieversi* Mouss., *Anisus* (*Gyraulus*) *euphraticus* (Mouss.), and several forms living in springs, such as *Lymnaea* (*Galba*) *schirazensis* (Küst.), *L.* (*G.*) *subangulata* Roff. (= *L. truncatula* var. *ventricosa* Moq.-Tand.), and *L.* (*G.*) *thiesseae* Cl. All these pulmonate molluscs except *Lymnaea subdisjuncta* are present also in the western part of Iran (the Irano-Anatolian province). These data allow us to unite the Kopetdagh province with the Irano-Anatolian one into the Southwest Asian superprovince of the European-Central Asian subregion of the Palearctic region.

Eastern Turkmenistan (the Tedzhen, Murghab, and Amudarya drainages up to approximately 40° N) belongs to the Sogdian zoogeographical province, which we include in the Sogdian-Tibetan superprovince. Water bodies are more diverse here: springs (including thermal ones), streams, cave and underground waters, rivers, and various small water bodies in river valleys are all present. Small hydrobiotic forms belonging to the genera *Martensamnicola* (Fam. Belgrandiellidae), *Turkmenamnicola* (Fam. Sadlerianidae), *Sogdamnicola* and *Valvatamnicola* (Fam. Horatiidae) are especially abundant in springs. Thermal springs (especially the well-known Hodzha-Kainar Spring in the Kugitang range) are inhabited by a peculiar malacofauna (Starobogatov 1972; Starobogatov and Izzatulaev 1980). It includes *Melanoides kainarensis* Star. et Izz., which is known also from one spring in Uzbekistan and two springs in northern Afghanistan and is included in the Red Data Books, and also *Kainarella minima* Star., *K.* sp., *Pseudocaspia kainarensis* Star., and *Ps. starostini* Star. The subterranean brackish lake in the Kaptarkhana Cave (literally, "Pigeon House") in Kugitang is inhabited by *Pseudocaspia ljovuschkini* Star. Species of *Corbicula* are abundant in rivers; *C. purpurea* Prime is found in Kugitang in addition to the two species mentioned above. Also present here are species of the family Unionidae: *Colleptopterum cyreum sogdianum* (Kob.) and *C. bactrianum* (Rolle). Pulmonate molluscs are abundant in the water bodies of river valleys and in irrigation canals: *Lymnaea (Cerasina) impura* Trosch., *L. (Radix) tenera* (Küst.), *L. (R.) subdisjuncta* Nev., *L. (R.) bactriana* Hutt., *L. (Galba) subangulata* Roff., *L. (G.) thiesseae* Cl., *L. (G.) schirazensis* (Küst.), *Costatella acuta* (Drap.), a recently introduced *C. integra* (Hald.), *Planorbis sieversi* Mouss., *Anisus (Gyraulus) convexiusculus* (Hutt.), *A. (G.) euphraticus* (Mouss.), *A. (G.) ladacensis* (Nev.), and several other Planorbidae. The influence of the Irano-Anatolian faunistic center is also evident here although not as significant as in the Kopetdagh province; it is expressed in the list of pulmonate and large bivalvian molluscs. Small Bivalvia of the Amudarya drainage within Turkmenistan are studied rather poorly compared to those within Tadzhikistan and Uzbekistan, but the fauna of the Amudarya is probably rather uniform.

The lower part of the Amudarya drainage is inhabited by a peculiar mixture of molluscs which allows us to separate it in a separate Cis-Aral province which belongs to the Central Asian superprovince. Several species present here are related directly or indirectly to the fauna of Lake Balkhash and the Tarim drainages; these are *Boreoelona moltschanovi* (Ldh.), *Corbiculina tibetensis* (Prash.), and *C. ferghanensis* (Kurs. et Star.). Besides these, an endemic subspecies *Dreissena polymorpha aralensis* (Andr.) and several species of *Caspiohydrobia* are present. The fauna includes also such Sogdian forms as *Colletopterum cyreum sogdianum* (Kob.), *C. bactrianum* (Rolle), and *Corbicula purpurea* Prime. Pulmonate molluscs include *Lymnaea subdisjuncta* Nev., *L. bactriana* Hutt., and *Planorbis planorbis* (L.).

The aquatic malacofauna of Bolshoi Balkhan is peculiar. The only water bodies present here are springs with a debit significantly varying with the seasons. Although the fauna of these springs is practically unknown, two endemic species of aquatic animals can be identified: a pectinibranchiate mollusc *Allocinma caspica* (West.) and a crab, *Potamon zarudnyi* Star. et Vas. Thus, we can include this fauna into the Bolshoi Balkhan province, tentatively belonging to the Southwest Asian superprovince.

The Karakum desert has no permanent water bodies. Molluscs are not found in wells (although protozoan foraminiferans are present there). Temporary water bodies filled by spring rains are inhabited only by *Lymnaea subdisjuncta* Nev. Only a few individuals of the extant thousands can survive the dry period; the surfaces of dried water bodies are covered by scores of white empty shells. There are lake-like water bodies in the Uzboi dry bed, also poorly inhabited by molluscs. Such species are found there as *Dreissena polymorpha aralensis*

(Andr.), *Lymnaea subdisjuncta* Nev., and a species of the genus *Theodoxus*, which inhabits the Amudarya and Aral drainages and is commonly called "*T. pallasi* Ldh.*"* but in fact is different from this species. Finally, Holocene deposits near the eastern shore of the Caspian contain shells of such European pond molluscs as *Lymnaea* (*Stagnicola*) *atra* (Gm.), *L.* (*S.*) *danubialis* (Schr.), *L.* (*S.*) *berlani* Bgt., and *L.* (*Peregriana*) *balthica* (L.). The rare and exteremely mollusc-poor water bodies of the Karakum desert can be included in any of adjacent zoogeographic provinces although they are closest to the Cis-Aral province.

Ecologo-biogeographical division of inland water bodies of Turkmenistan very nearly coincides with the division based on results of systematic biogeography (Starobogatov 1986), which includes only one extra province. The Turan-Cis-Aral desert province which covers the most of Turkmenistan belongs to the North European-West Siberian lowland subregion; it includes water bodies of river valleys and temporary desert water bodies. The Kopetdagh mountainous province includes water bodies of springs and streams and belongs to the Mediterranean subtropical subregion. Finally, the Sogdian oasis province, which belongs to the Middle Asian mountain subregion, contains a significantly diverse number of water bodies in the areas of ancient agriculture. The boundary between the Kopetdagh mountainous and the Sogdian oasis provinces in this ecologo-biogeographical division coincides with that between the Kopetdagh and Sogdian provinces in the systematic- biogeographical scheme.

Molluscs of the Southeastern Caspian

A survey of molluscs of Turkmenistan would remain incomplete without discussion of the fauna of the giant Lake Caspian, commonly called the Caspian Sea due to its enormous size and brackish water. The Caspian malacofauna is very rich and diverse. It consists of 124 species of which 119 are endemics or subendemics (Logvinenko and Starobogatov 1969; Tadjalli-Pour 1977, 1980; Alexenko and Starobogatov 1987; Antsulevich and Starobogatov 1990). Endemic and subendemic species belong to two families of Bivalvia (Dreissenidae and Lymnocardiidae) and to seven families of Gastropoda (Neritidae, Hydrobiidae, Pyrgulidae, Sadlerianidae, Horatiidae, and Planorbidae). Two species (one endemic and one subendemic) became completely extinct during the 1940s. Two other species, *Cerastoderma isthmicum* (Iss.) and *C. rhomboides* (Lam.) (= *Cardium lamarcki* Rve.) (Lymnocardiidae) (the latter species is rare) are not endemic, having dispersed into the Caspian at the boundary of the Pleistocene and Holocene. There are also three recent invaders: the bivalvians *Mytilaster lineatus* (Gm.) (Mytilidae) and *Abra segmentum* (Recl.) (Fam. Scrobiculariidae) and a nudibranch gastropod *Tenellia adspersa* (Nordm.) (= *Eolis pallidus* Ald. et Hanc) (Fam. Tergipedidae).

Subendemic species can be divided in four groups. The first one includes four

Caspian species of the genus *Turricaspia* (subgenus *Clessiniola*), found also in brackish waters of the Black Sea estuaries (*limans*). The second group is also represented in the Black Sea basin but only in fresh waters of the lowest parts of rivers flowing into the Azov and Black Seas. These are bivalvians *Adacna* (*Monodacna*) *angusticostata* (Borc.), *A.* (*M.*) *caspia* (Eichw.), *A.* (*Adacna*) *laeviuscula* (Eichw.), and *Hypanis plicata* (Eichw.) and gastropods *Turricaspia* (*Caspiella*) *derbentina* (Logv. et Star.), *T.* (*C.*) *conus* (Eichw.), *T.* (*C.*) *boltovskoji* (Gol. et Star.), *Caspia gmelini* Cl. et Dyb., and *C. knipowitschi* Mak. These species inhabit fresh waters in the Azov and Black Sea basins, thus being sharply different in salinity preference from their populations in the Caspian. Gastropods are also different in shell proportions. These forms are commonly recognized as subspecies but, given that these and the Caspian forms cannot multiply in the same salinity of water and, thus, cannot interbreed, they may be considered separate species. The third group of subendemics consists of species common for the Caspian and the Aral: *Dreissena polymorpha* (Pall.), *D. caspia* (Eichw.) (recently become extinct), *Adacna (Adacna) vitrea* (Eichw.), and *A.* (*A.*) *minima* Ostr. These are represented by different subspecies in these two giant brackish lakes. The fourth group includes fourteen species of the genus *Caspiohydrobia*, which also live in the salt water bodies of semideserts and deserts of Middle Asia and southern parts of Central Europe, and which were probably introduced to the Aral and Caspian.

The Caspian, together with estuarine parts of the rivers of the Azov and Black Sea drainage, should be considered an independent zoogeographical region due to the diversity, peculiarity, and endemism of its fauna, which exceed those of the Lake Baikal fauna (Starobogatov 1970). Lake Caspian is topographically divided into northern, middle, and southern parts; subdivision of fauna, however, does not completely follow this demarkation because it depends also on vertical subdivision of the lake and salinity. The North Caspian is separated from freshwater aquatories adjacent to the Volga River delta by its 5% isohaline, and from the Middle Caspian by a line which extends from the western shore at Chechen' Island (43° 57' N), along the isobath 5 m to Kulaly Island (45° 00' N) at the eastern shore. The Middle Caspian is separated from the South Caspian by a line which extends from Kilazi (40° 55' N) at the western shore and ends near the eastern shore at Cape Kuuli (40° 10' N). There are also vertical patterns in the distribution of faunistic groups which are different near the western and eastern shores (especially closer to the surface) due to the upwelling at the eastern shore. The shallow water (upper sublittoral) zone is limited by the isobathe 30 to 35 m at the western shore and 20 to 25 m at the eastern shore. The middle sublittoral zone is limited by the isobathe 70 to 80 m, and profundal zone, by the isobathe 500 m. Molluscs are absent below 500 m (pseudoabyssal zone) although other animals exist there.

Such a combination of horizontal and vertical subdivisions of the fauna allow us to divide the Ponto-Caspian zoogeographic region into twelve provinces, ten of which are limited by Lake Caspian proper and two of which include also the estuarine areas of the Azov-Black Sea drainage area (Logvinenko and Starobogatov 1962; Starobogatov 1970).

The Caspian waters adjacent to the shores of Turkmenistan lie between Cape Sue (40° 50' N) and Gasankuli (37° 25' N). The northern part of this aquatory (northward from 40° 10' N) belongs to the Middle Caspian, and the rest, to the South Caspian. Therefore, the aquatory adjacent to Turkmenistan is divided into nine zoogeographical provinces: the Middle-South Caspian upper sublittoral province, the Middle Caspian middle sublittoral province, the South Caspian middle sublittoral province, two respective lower sublittoral provinces, two respective profundal provinces, and two respective pseudoabyssal provinces (devoid of molluscs). The aquatory under discussion is rather poorly studied; the part of the Caspian between Cape Kuuli and the Ulski Bank (38° 30' N) is known fairly well, but northward from Cape Kuuli only two samples have been obtained, and there are no samples known southward from the Ulski Bank. Thus, we are able to present only general characteristics of this aquatory.

Between Cape Kuuli and the Ulski Bank, 90 mollusc species have been recorded (including one that recently became extinct and three which were recently introduced). Few of these species can be considereded eurybathic. Only six species of molluscs are found both in the Middle and South Caspian along the shores of Turkmenistan, from the shoreline to a 50 m depth: *Didacna baeri* (Grimm), *Adacna (Monodacna) caspia filatovae* (Logv. et Star.), *A. (M.) albida* (Logv. et Star.), *A. (Adacna) laeviuscula* (Eichw.), *Abra segmentum* (Recl.), and *Turricaspia (Laevicaspia) pulla* (Dyb. et Gr.). A number of species and subspecies (also, as a rule, from both the Middle and South Caspian) is found between 25 and 80 m and sometimes down to 500 m: *Didacna pyramidata* (Grimm), *Theodoxus schultzi* (Grimm), *Th. jukovi* Kol., *Turricaspia (Trachycaspia) eucalia* (Dyb. et Gr.), *T. (Eurycaspia) pseudobakuana* (Logv. et Star.; this species is found only in the Turkmenistan waters), *Turricaspia (Turricaspia) andrusovi* (Dyb. et Gr.), *T. (Oxypyrgula) pullula* (Dyb. et Gr.) (= *Pyrgula pseudospica* Logv. et Star.), *T. (Laevicaspia) brunnea* (Dyb. et Gr.), *T. (Caspiella) kolesnikoviana* (Logv. et Star.), *T. (C.) abichi* (Logv. et Star.), *T. (C.) conus conus* (Eichw.), *T. (C.) marginata* (West.), *T. (C.) derbentina derbentina* (Logv. et Star.), *T. (C.) trivialis* (Logv. et Star.) (= *Caspia grimmi* Cl. et Dyb., non *Micromelania grimmi* Cl. et Dyb.), *Caspia (Clathrocaspia) gmelini gmelini* Cl. et Dyb., *C. (Clathrocaspia) isseli* (Logv. et Star.), *Caspiohoratia marina* (Logv. et Star.), and *Anisus eichwaldi eichwaldi* Cl. et Dyb.

As can be seen from this short list, the upper sublittoral Caspian fauna is more strongly separated from the faunas of lower horizons than the latter ones are from each other. However, each of the provinces of South Caspian has its own endemic forms of molluscs. The endemics of the Middle-South Caspian upper sublittoral province include *Dreissena elata* (recently extinct), *Theodoxus pallasi* Ldh., *Th. nalivkini* Kol., several species of *Caspiophydrobia*, *Turricaspia (Laevicaspia) caspia* (Eichw.), *T. (L.) meneghiniana* (Issel), and *Abeskunus sphaerion* (Mouss.) (= *Lithogliphus caspius* Cl.; limited by the South Caspian). The endemics of the South Caspian middle sublittoral province include

Theodoxus gaillardi Tadj.-Pour, *Turricaspia (Caspiella) kowalewskii* (Cl. et Dyb.), and *T. (C.) cincta* (Abich). The endemics of the South Caspian lower sublittoral province include *Turricaspia (Turricaspia) turricula* (Cl. et Dyb.), (= *Pyrgula rudis* Logv. et Star.), *T. (T.) elegantula* (Cl. et Dyb.) (the species characteristic for the eastern part of the South Caspian), *T. (Caspiella) orthii* (Cl. et Dyb.), *Caspia (Caspia) baerii* Cl. et Dyb., and *C. (C.) derzhavini* (Logv. et Star.). Finally, the endemics of South Caspian profundal province include *Turricaspia (Caspiopyrgula) nossovi* Kol., *T. (Oxypyrgula) turkmenica* (Logv. et Star.), and *T. (Turricaspia) grimmi* (Cl. et Dyb.); the two former species are found only in the eastern part of the aquatory.

Due to the very scarce data available, it is difficult to characterize the Middle Caspian aquatories adjacent to Turkmenistan in equal detail. Judging from the knowledge of fauna of the Caspian aquatories lying eastward and northward, however, we may conclude that the endemics of the Middle Caspian middle sublittoral province are *Dreissena rostriformis distincta* (Andr.) and *Turricaspia (Turricaspia) lirata lirata* (Dyb. et Gr.); the endemics of the Middle Caspian lower sublittoral province are *Dreissena rostriformis grimmi* (Andr.) and *Turricaspia (Turricaspia) lirata brusinai* (Dyb. et Gr.); and the endemics of the Middle Caspian profundal province are *Turricaspia (Turricaspia) eulimellula* (Dyb. et Gr.), *T. (T.) dubia* (Logv. et Star.), and *T. (T.) fedorovi* (Logv. et Star.).

In summary of the presented data, we should once more state that the upper sublittoral mollusc faunas of Middle and South Caspian are almost identical and at the same time separate from the faunas of deeper layers. That the latter ones are more closely related to each other, may be explained by the fact that the colonization of the deeper horizons historically begins from the shallow parts of the lake. The formation of endemic species is connected either to vertical differentiation of eurybathic species or to geographic differentiation of species widespread within the lake. The peculiarity of the Caspian fauna inhabiting the shallow zone down to 25–35 m is explained by the significant historical fluctuations of the lake level (Starobogatov 1970).

Bibliography

Abbasian-Lintzen, R. 1964. Tabanidae (Diptera) of Iran. X. List, keys, and distribution of species occurring in Iran. Ann. Parasitol. (Paris) 39(3): 258–327.
Abuzyarova, R.Ya. 1954. Tertiary palynological complexes of the Turgai and Pavlodar Irtysh area. Abstr. Ph.D. Thesis. Alma-Ata, 16 pp (in Russian).
Abuzyarova, R.Ya. 1956. Palynological complexes of the Eocene deposits of Turkmenia (Badghyz). Botan. zhurn. [Bot. Journal], 41(9): 1339–1345 (in Russian).
Adamovic, J. 1909. Die Vegetationsverhältnisse der Balkanländer. W. Engelmann, Leipzig, 567 pp.
Adamovic, J. 1902. Die Siblijak-Formation, ein wenig bekanntes Buschwerk der Balkanländer. Engler's Bot. Jahrb. 31: 1–2.
Agakhanyants, O.E. 1981. Aridnye gory SSSR [Arid Mountains of the USSR]. Mysl', Moscow, 270 pp. (in Russian).
Agroklimaticheskie resursy Turkmenskoi SSR [Agroclimatic Resources of the Turkmen SSR]. 1974. Gidrometeoizdat, Leningrad, 244 pp. (in Russian).
Akhmedov, S.B. and N.N. Shcherbak. 1987. Geographical variation and subspecific taxonomy of the golden mabuya (*Mabuya aurata* L.). Vestn. zool. [Mess. Zool.] 5: 20–24 (in Russian).
Alekperov, A.M. 1978. Zemnovodnye i presmykayushchiesya Azerbaidzhana [Amphibians and Reptiles of Azerbaijan]. Elym, Baku, 263 pp (in Russian).
Aleksenko, T.L. and Ya.I. Starobogatov. 1987. Species of *Caspia* and *Turricaspia* (Gastropoda, Pectinibranchia, Pyrgulidae) of the Azov-Black Sea basin. Vestn. zool. [Mess. Zool.] 3: 32–39 (in Russian).
Ali-Zade, A.A. 1961. Akchagyl Turkmenistana [The Akchagylian Age in Turkmenistan]. Gosgeoltekhizdat, Moscow, 1, 300 pp (in Russian).
Ali-Zade, A.A. 1967. Akchagyl Turkmenistana [The Akchagylian Age in Turkmenistan]. Gosgeoltekhizdat, Moscow, 2, 408 pp (in Russian).
Aliev, D.S. 1976. The role of herbivorous fish in a reconstruction of food ichthyofauna and biological melioration of water bodies. Voprosy ikhtiologii [Problems in Ichthyology] 16(2): 247–262 (in Russian).
Aliev, D.S., A.I. Sukhanova, and F.M. Shakirova. 1988. Ryby vnutrennikh vodoyomov Turkmenistana [Fishes of Inland Water Bodies of Turkmenistan]. Ylym, Ashkhabad, 156 pp (in Russian).
Amanniyazov, K., O. Nigarov, and O. Uzakov. 1979. Petrified animal footprints in an ancient desert. Probl. Osvoyen. Pustyn [Problems of Desert Development] 5: 80–82 (in Russian).
Amanov, A.A. 1985. Ekologiya ryb vodoyomov yuga Uzbekistana i sopredel'nykh respublik [Ecology of Fishes in the Water Bodies of Southern Uzbekistan and Adjacent Republics]. Fan, Tashkent, 160 pp (in Russian).
Amanova, M.A. 1968. Ekologiya vodopoinogo rezhima i nekotorye osobennosti vodnogo obmena ptits v pustyne [Ecology of the Drinking Regime and Some Characteristics of Water Balance in Desert Birds]. Ph.D. Thesis, Ashkhabad, (in Russian).

Amanova, M.A. 1979. Ekologo-morfologicheskie i fiziologicheskie adaptatsii vodnogo obmena ptits Karakumov [Ecologo-morphological and Physiological Adaptations of Water Balance in Birds of the Karakum]. D.Sc. Thesis, Ashkhabad, 57 pp (in Russian).

Amanova, M.A., O. Goncharevskaya, and V. Bakhteeva. 1978. Characteristics of the functional organization of the renal cortex in desert birds. Fiziol. zhurn. SSSR [USSR Physiol. Journal] 14 (3): 405–410 (in Russian).

Amursky, G.I. 1961. On the tooth of *Equus caballus* found in the deposits of the Karakum stratum. Izv. AN TSSR, Ser. Geol. Nauk. [Proc. Acad. Sci. Turkm. SSR, Ser. Geol. Sci.] 1: 100–101 (in Russian).

Ananjeva, N.B. and B.S. Tunijev. 1992. Historical biogeography of the *Phrynocephalus* species of the USSR. Asiat. Herp. Res. 4: 76–98.

Ananyeva, N.B. 1986. On the generic status of *Megalochillus mystaceus* (Pall., 1776). Sistematika i ekologiya amfibii i reptilii/Trudy Zool. in-ta AN SSSR [Systematics and Ecology of Amphibians and Reptiles/Trans. Zool. Inst. Acad. Sci. USSR], Leningrad 157: 4–13 (in Russian).

Ananyeva, N.B. and Ch. Ataev. 1984. *Stellio caucasicus triannulatus* ssp. nov. – a new subspecies of an agama from southwestern Turkmenia. Ekologiya i faunistika amfibii i reptilii SSSR i sopredel'nykh stran/Trudy Zool. in-ta AN SSSR [Ecology and Faunistics of Amphibians and Reptiles of the USSR and Adjacent Countries/Trans. Zool. Inst. Acad. Sci. USSR] Leningrad, 124: 4–11 (in Russian).

Ananyeva, N.B. and Yu.K. Gorelov. 1981. On the teeth of the Pliocene lizards found in Baghyz. In: Tez. dokl. V Vsesoyuzn. gerpetol. konf. [Abstr. Rep. 5th All-Union Herpetol. Confer.], Nauka, Leningrad: 8 (in Russian).

Ananyeva, N.B. and N.L. Orlov. 1977. A record of the olive poloz (*Coluber najadum* Eichw., 1831) in Southwest Turkmenia. In: Gerpetologichesky sbornik/Trudy Zool. Inst. AN SSSR [Herpetological Papers/Trans. Zool. Inst. Acad. Sci. USSR], Leningrad 74: 14–16 (in Russian).

Ananyeva, N.B. and O.I. Tsaruk. 1986. On the systematic position of the steppe agama from Cis-Caucasia. In: Gerpetologicheskie issledovaniya na Kavkaze/Trudy Zool. Inst. AN SSSR [Herpetological Studies in the Caucasus/Trans. Zool. Inst. Acad. Sci. USSR], Leningrad 158: 39–46 (in Russian).

Anderson, S.C. 1963. Amphibians and reptiles from Iran. Proc. Calif. Acad. Sci. 31: 417–498.

Anderson, S.C. 1968. Zoogeographic analysis of the lizard fauna of Iran. The Cambridge History of Iran. Vol. 1. The Land of Iran. 305–317.

Anderson, S.C. 1974. Preliminary key to the turtles, lizards, and amphisbaenids of Iran. Field. Zool. 65(4): 27–44.

Anderson, S.C. and A.E. Leviton. 1965. Amphibians and reptiles collected by the Street Expedition to Afghanistan. Proc. Calif. Acad. Sci., 4th Ser. 37(2): 25–56.

Andreev, V.L. and Yu.S. Reshetnikov. 1978. An analysis based on set theory of the freshwater ichthyofauna of the northeastern USSR. Zool. zhurn. [Zool. Journ.] 57(2): 165–175 (in Russian).

Andreeva, E.M. 1968. Contributions to the fauna of spiders of Tajikistan. III. Mygalomorphae. Trudy Akad. nauk Tadzh. SSR [Trans. Acad. Sci. Tadjik SSR] 11: 68–71 (in Russian).

Andreeva, E.M. 1969. Contributions to the fauna of spiders of Tajikistan. V. Salticidae. Izv. Akad. nauk Tadzh. SSR, otdel. biol. [Proc. Acad. Sci. Tajik. SSR, Biol. Div.] 37(4): 89–93 (in Russian).

Andreeva, E.M. 1975a. Zoogeographical characteristics of the spider fauna of Tajikistan. In: Proc. 6th Int. Arachn. Congr. (Amsterdam IV. 1974): 214–215.

Andreeva, E.M. 1975b. Distribution and ecology of spiders (Aranei) in Tajikistan. Fragm. Faunist. 20(19): 323–352.

Andreeva, E.M. 1976. Spiders of Tajikistan. Donish, Dushanbe, 195 pp (in Russian).

Andreeva, E.M. and V.P. Tystchenko. 1968. Contributions to the fauna of spiders (Aranei) of Tajikistan. II. Zodariidae. Zool. zhurn. [Zool. Journ.] 47(5): 684–689.

Andreeva, E.M. and V.P. Tystchenko. 1969. Contributions to the fauna of spiders of Tajikistan. 1. Haplogynae, Cribellatae, Ecribellatae Trionychae (Pholcidae, Palpimanidae, Hersiliidae, Oxyopidae). Entom. Obozr. [Entomol. Rev.] 48(2): 373–384 (in Russian).

Andreeva, E.M. and V.P. Tystchenko. 1970. Contributions to the fauna of spiders (Aranei) of Tajikistan. VI. – Micryphantidae. Zool. zhurn. [Zool. Journ.] 49(1): 38–44 (in Russian).

Andreeva, E.M., S. Heciak, and J. Proszynski. 1984. Remarks on *Icius* and *Pseudicius* (Araneae, Salticidae) mainly from Central Asia. Ann. Zool. (Warsaw) 37(13–18): 349–376.
Andreeva, E.M., A.P. Kononenko, and J. Proszynski. 1981. Remarks on the genus *Mogrus* Simon, 1882 (Araneae, Salticidae). Ann. Zool. (Warsaw) 36(4): 85–104.
Andreeva, R.V. 1982. On ecological and morphological types of horse fly larvae (Diptera, Tabanidae). Entomol. obozr. [Entom. Rev.] 61(1): 43–49 (in Russian).
Andreeva, R.V. 1984. Ekologiya lichinok slepnei i ikh parazitosy [Ecology of Horse Fly Larvae and Their Parasitic Diseases]. Naukova dumka, Kiev, 171 pp (in Russian).
Andreeva, R.V. 1985. Morphology and taxonomy of horse fly larvae belonging to the *Tabanus bromius* L. group (Tabanidae). Parazitologiya [Parasitology] 19(2): 128–133 (in Russian).
Andreeva, R.V. 1986. Development of horse fly larvae belonging to the *Tabanus bromius* L. group (Tabanidae). Parazitologiya [Parasitology] 20(4): 265–271 (in Russian).
Andreeva, R.V. 1987. Life forms of the edaphobiotic larvae of horse flies and their distribution. Vestn. zool. [Mess. Zool.] 2: 42–45 (in Russian).
Andreeva, R.V. 1989. The morphological adaptations of horse fly larvae (Diptera, Tabanidae) to developmental sites in the Palearctic region and their relationship to the evolution and distribution of the family. Can. J. Zool. 67(9): 2286–2293.
Andreeva, R.V. 1990. Opredelitel' lichinok slepnei [A Key to Horse Fly Larvae]. Naukova dumka, Kiev, 169 pp (in Russian).
Andrushko, A.M. and N.E. Mikkau. 1964. Distribution and life habits of the Afghanian lytorhynch (*Lytorhynchus ridgewayi* Boulanger, 1887), with an ecological-geographic review of the genus *Lytorhynchus* Peters, 1862. Vestnik Leningr. Univ., Ser. Biol. Nauk. [Mess. Leningrad Univ., Ser. Biol. Sci.] 9: 5–18 (in Russian).
Andrushko, A.M., N.O. Lange, and Ye.N. Yemelyanova. 1939. Ecological observations on the reptiles in the Kizyl-Arvat area, Iskander Station and Krasnovodsk area (Turkmenia). Voprosy ekologii i biotsenologii [Problems of Ecology and Biocoenology], Leningrad 4: 207–252 (in Russian).
Antsulevich, A.Ye. and Ya.I. Starobogatov. 1990. First record of a nudibranch mollusk (order Tritoniiformes) in the Caspian Sea. Zool. zhurn. [Zool. J.] 69(11): 138–140 (in Russian).
Aranbaev, M.P. 1969. Serozyomy i svetlo-korichnevye sukhostepnye pochvy Tsentralnogo Kopetdaga [Serozems and Light Brown, Dry Steppe Soils of Central Kopetdagh]. Ylym, Ashkhabad, 175 pp (in Russian).
Arnold, E.N. 1980. The reptiles and amphibians of Dhofar, Southern Arabia. J. Oman Stud. Spec. Rep. 3: 273–332.
Arnoldi, K.V. 1946. Insects of Southern Kirghizian walnut and fruit forests and the significance of entomological data for general biological classification and the problem of origin of these forests. Dokl. AN SSSR [Rep. Acad. Sci. USSR], New Ser. 53(9): 345–348 (in Russian).
Arnoldi, L.V. 1960. On the tribe Mesostylini (Coleoptera, Curculionidae) and the problem of the origin of sand desert fauna of Middle Asia. Trudy Zool. inst. AN SSR [Trans. Zool. Inst. Acad. Sci. USSR] 27: 276–292 (in Russian).
Ataev, Ch. 1966a. On the distribution and ecology of some snake species in Central Kopetdagh. Izv. AN TSSR, Ser. Biol. Nauk. [Proc. Acad. Sci. Turkmen SSR, Ser. Biol. Sci.] 2: 97–98 (in Russian).
Ataev, Ch. 1966b. New data on the density of reptiles in Central Kopetdagh. Izv. AN TSSR, Ser. Biol. Nauk. [Proc. Acad. Sci. Turkmen SSR, Ser. Biol. Sci.] 4: 78–80 (in Russian).
Ataev, Ch. 1970. New data on the ecology of shielded skink in Kopetdagh. In: N.A. Gladkov (ed.), Zhivotnyi mir Turkmenii [Animal Life of Turkmenia], Ylym, Ashkhabad: 144–149 (in Russian).
Ataev, Ch. 1973. On the rewintering of the Caucasian agama in Turkmenia. In: I.S. Darevsky (ed.), Voprosy gerpetologii [Problems of Herpetology], Nauka, Leningrad: 19–20 (in Russian).
Ataev, Ch. 1974a. Features of the rewintering of the Caucasian agama in Kopetdagh. Ekologiya [Ecology] 2: 74–78 (in Russian).
Ataev, Ch. 1974b. Dispersal of mountainous reptiles along river valleys in Turkmenia. Izv. AN TSSR, Ser. Biol. Nauk. [Proc. Acad. Sci. Turkmen SSR, Ser. Biol. Sci.] 3: 38–42 (in Russian).
Ataev, Ch. 1975. On winter activity of reptiles in Kopetdagh and Badghyz. Izv. AN TSSR, Ser. Biol. Nauk. [Proc. Acad. Sci. Turkmen SSR, Ser. Biol. Sci.] 4: 63–67 (in Russian).

Ataev, Ch. 1976a. On rare species of reptiles in Turkmenistan. Izv. AN TSSR, Ser. Biol. Nauk. [Proc. Acad. Sci. Turkmen SSR, Ser. Biol. Sci.] 3: 9–14 (in Russian).
Ataev, Ch. 1976b. On the rock lizard from Kopetdagh. In: Gerpetologiya [Herpetology], Krasnodar: 97–99 (in Russian).
Ataev, Ch. 1977a. On the distribution and ecology of some reptile species in Turkmenistan. Izv. AN TSSR, Ser. Biol. Nauk. [Proc. Acad. Sci. Turkmen SSR, Ser. Biol. Sci.] 1: 80–82 (in Russian).
Ataev, Ch. 1977b. Rewintering of some reptiles in Kopetdagh. In: Voprosy gerpetologii [Problems of Herpetology]. Nauka, Leningrad: 17–18 (in Russian).
Ataev, Ch. 1977c. New data on the rewintering of some reptiles in Kopetdagh. In: Voprosy ekologii mlekopitayushchikh i presmykayushchikhsya Turkmenistana [Problems of the Ecology of Mammals and Reptiles of Turkmenistan]. Ylym, Ashkhabad: 156–163 (in Russian).
Ataev, Ch. 1979. New data on the ecology of the Middle Asian tortoise in Kopetdagh. Okhrana prirody Turkmenistana [Natural Conservation in Turkmenistan] 5: 161–167 (in Russian).
Ataev, Ch. 1985. Presmykayushchiyesya gor Turkmenistana [Reptiles of the Mountains of Turkmenistan]. Ylym, Ashkhabad, 344 pp (in Russian).
Ataev, Ch., O.P. Bogdanov, and S. Shammakov. 1968. On a record of the spine-tailed gecko *Alsophylax spinicauda* (Squamata, Gekkonidae) in the USSR. Zool. zhurn. [Zool. Journ.] 48(9): 1420–1421 (in Russian).
Ataev, Ch., Yu.K. Gorelov, and S. Shammakov. 1978. Data on rare and endangered species of reptiles of the Turkmenistan fauna. Izv. AN TSSR, Ser. Biol. Nauk. [Proc. Acad. Sci. Turkmen SSR, Ser. Biol. Sci.] 4: 81–83 (in Russian).
Ataev, Ch., Yu.D. Khomustenko, and S. Shammakov. 1991. New data on the distribution and density of some rare snake species in Southwest Kopetdagh. Gerpetologicheskie issledovaniya [Studies in Herpetology], Leningrad 1: 51–53 (in Russian).
Ataev, Ch., A.K. Rustamov, and S. Shammakov. 1978. Reptiles. In: A.G. Babaev et al. (ed.), Krasnaya KnigaTurkm. SSSR [Red Data Book of the Turkmen SSR]. Turkmenistan, Ashkhabad: 210–276 (in Russian).
Ataev, Ch. and S. Shammakov. 1990. New data on distribution and density of some narrow-ranging or peripheral species of reptiles of Turkmenistan. Izv. AN TSSR, Ser. Biol. Nauk. [Proc. Acad. Sci. Turkmen SSR, Ser. Biol. Sci.] 3: 69–71 (in Russian).
Ataev, Ch., B.S. Tuniev, and S. Shammakov. 1991. New records of the Khorassan stellion in East Kopetdagh. Izv. AN TSSR, Ser. Biol. Nauk. [Proc. Acad. Sci. Turkmen SSR, Ser. Biol. Sci.] 3: 77–78 (in Russian).
Ataev, K. 1974. Some data on the ecology of a Turkmen serpent eagle in the eastern Karakum. Materialy VI Vsesoyuzn. ornitol. konf. [Materials of the 6th All-Union Ornitological Conference], Moscow State Univ., Moscow 2: 19–20 (in Russian).
Ataev, K. 1976. Seasonal dynamics of the bird population in sand desert ecosystems. Materialy sovetsko-amerikanskogo simpoziuma po biosfernym zapovednikam [Materials of the Soviet-American Symposium on Biosphere Reserves]. Moscow 1: 39–50 (in Russian).
Ataev, K., V.I. Vasilyev, R.I. Gorelova, A.A. Karavaev, A.F. Kekilova, O. Sopyev, and A. Eminov. 1978. Data on rare and endangered bird species of Turkmenistan. Izv. AN TSSR, Ser. Biol. Nauk. [Proc. Acad. Sci. Turkmen SSR, Ser. Biol. Sci.] 4: 70–80 (in Russian).
Atamuradov, K.I. 1988. Endemic and little known invertebrates of Badghyz. In: V.E. Flint (ed.), Redkie i maloizuchennye zhivotnye Turkmenistana [Rare and Poorly Known Animals of Turkmenistan], Ylym, Ashkhabad: 161–176 (in Russian).
Atamuradov, K.I. and S.I. Sukh. 1985. Contributions to the study of venomous spiders of the genus *Latrodectus* Walck. in Turkmenistan. Izv. Akad. Nauk. Turkm. SSR. Ser. Biol. Nauk. [Proc. Acad. Sci. Turkmen. SSR. Ser. Biol. Sci.] 5: 29–34 (in Russian).
Babaev, A.G. 1963. Pustynya Karakum [The Karakum Desert]. Ed. Acad. Sci. Turkm. SSR, Ashkhabad, 89 pp (in Russian).
Babaev, A.G. and Kh.D. Durdyev. 1982. A brief outline of physical-geographical characteristics of West Kopetdagh. In: N.T. Nechayeva (ed.), Priroda Zapadnogo Kopetdaga [Nature in West Kopetdagh]. Ylym, Ashkhabad: 7–19 (in Russian).

Babaev, A.G. and B.A. Fedorovich. 1970. Basic stages of the formation of the Karakum relief. Probl. Osvoen. Pustyn [Problems of Desert Development] 5: 3-11 (in Russian).

Babaev, A.G., V.B. German, and I.S. Sukh. 1986. A brief outline of physical-geographical characteristics of Central Kopetdagh. In: N.N. Shcherbak (ed.), Priroda Tsentralnogo Kopetdaga [Nature of Central Kopetdagh]. Ylym, Ashkhabad: 191 pp. (in Russian).

Babaev, A.G. and S.K. Gorelov. 1990. Problemy geomorfologii pustyn' [Problems of Desert Geomorphology]. Ylym, Ashkhabad, 155 pp (in Russian).

Babaev, A.G., N.G. Kharin, and N.S. Orlovsky. 1992. Assessment and Mapping of Desertification. A Methodological Guide. Ylym, Ashkhabad, 125 pp.

Babaev, A.G. and N.G. Orlovsky. 1981. Ecological types of deserts in Middle Asia and Kazakhstan, and the perspectives of their development. In: Aktualnye voprosy osvoeniya i preobrazovaniya pustyn SSSR [Current Problems of the Development and Transformation of Deserts in the USSR]. Ylym, Ashkhabad: 8-37 (in Russian).

Babaev, A.G. and V.S. Zaletaev. 1990. Standard objects of ecological monitoring in arid zones. Probl. osv. pustyn' [Probl. Desert Development] 5: 3-8.

Babaev, Kh. 1974. On certain factors which control density of Chiroptera in Turkmenia. In: Materialy Pervogo Vsesoyuznogo soveshchaniya po rukokrylym [Materials of the First All-Union Conference on Chiroptera). 78-81 (in Russian).

Babaev, Kh. and Ch. Ataev, Ch. 1966. On a record of the Turkmen jerboa (*Jaculus turkmenicus*) in southern Turkmenia. Izv. AN TSSR, Ser. Biol. [Proc. Acad. Sci. Turkmen SSR, Ser. Biol.] 3: (in Russian).

Babaev, Kh. and V.P. Dmitrieva. 1966. *Tadarida teniotis* and *Plecotus auritus*, species of bats new for the fauna of Turkmenia. Zool. Zhurn. [Zool. Journal] 45: 779-780 (in Russian).

Babaev, Kh., N. Ishadov, and Ye.I. Shcherbina. 1978. Data on rare and endangered species of mammals of Turkmenistan. Izv. AN TSSR, Ser. Biol. [Proc. Acad. Sci. Turkmen SSR, Ser. Biol.] 4: 56-69 (in Russian).

Babayants, G.A. 1962. Contributions to the fauna of horse flies (Diptera, Tabanidae) of Turkmenistan. Trudy Ashkhabadsk. nauchn.-issled. in-ta epidemiol. i gigieny [Trans. Ashkhabad Sci. Res. Inst. Epidemiol. and Hygiene] 5: 243-248 (in Russian).

Babushkin, L.N. 1964. Problems of agroclimatic regionalization of Middle Asia. Trudy TashGU [Trans. Tashkent State Univ.] 236: 5-185 (in Russian).

Babushkin, L.N. 1971. On a coefficient of use of thermal resources. Trudy Sredneaz. gidrometeorol. in-ta [Trans. Middle Asian Hydrometeorol. Inst.] 66(81): 27-33 (in Russian).

Babushkin, L.N. 1970. Agroklimaticheskoe raionirovanie khlopkovoi zony Srednei Azii [Agroclimatic Regionalization of the Cotton Zone of Middle Asia]. Gidrometeoizdat, Leningrad, 135 pp (in Russian).

Babushkin, L.N. 1971. Klimatografiya Srednei Azii [Climatography of Middle Asia]. Izd. TashGU, Tahskent, 91 pp (in Russian).

Babushkin, L.N. and N.A. Kogai. 1971. Fiziko-geograficheskoe rayonirovaniye Turkmenskoi SSR [Physical-geographical Regionalization of Turkmen SSR]. Fan, Tashkent, (in Russian).

Babushkin, L.N. and N.S. Orlovsky. 1973. Agroclimatic regionalization of Turkmenistan. Probl. Osvoyen. Pustyn' [Problems Desert Development] 1: 8-16 (in Russian).

Bailey, N. 1970. Mathemathical Methods in Biology and Medicine. Mir, Moscow, 268 pp (a Russian translation).

Bakhvalov, V.F. 1978. New species of spiders from Turkmenia. Zool. zhurn. [Zool. Journ.] 57(5): 790-793 (in Russian).

Balaev, G.I. 1940. On the biology of the striped hyena (*Hyaena hyaena*). Trudy Uzbek. zool. sada [Transactions of the Uzbek Zoological Garden] 2: (in Russian).

Balakirev, Ye.K. 1972. Osobo vrednye klimaticheskie usloviya zimnego vypasa ovets na pastbishchakh Turkmenii [Especially Harmful Climatic Conditions of Sheep Grazing in the Pastures of Turkmenia], Turkmenistan, Ashkhabad, (in Russian).

Balashova, Ye.N., O.M. Zhitomirskaya, and O.A. Semenova. 1960. Klimaticheskoe opisanie respublik Srednei Azii [Climatic Description of the Republics of Middle Asia]. Gidrometeoizdat, Leningrad, 242 pp (in Russian).
Balon, E. 1974. Fish production of a tropical ecosystem. In: E.K. Balon and A.G. Coche (ed.), Lake Kariba. A man-made tropical ecosystem in Central Africa. W. Junk, The Hague: 250–676.
Balon, E. 1975. Terminology in intervals in fish development. J. Fish Res. Board Canada 32(9): 1663–1670.
Bannikov, A.G. 1972. Kulan (The Onager). Lesnaya promyshlennost', Moscow, 120 pp (in Russian).
Bannikov, A.G. 1979. The current status of the [population of] Bokhara deer. Okhrana prirody Turkmenistana [Natural Conservation in Turkmenistan] 5: 19–20 (in Russian).
Bannikov, A.G. 1981. The first record of a clutch of courser (*Cursorius cursor* Latham) in the USSR. Ornitologiya [Ornithology], Moscow State Univ. 10: 326–327 (in Russian).
Bannikov, A.G. and V.B. German. 1981. Kopetdagh State Reserve. Okhota i okhotnichye khozyaistvo [Hunting and Game Management] 6: 12–14 (in Russian).
Bannikov, A.G. and V.Ye. Flint. 1982. My dolzhny ikh spasti (ocherki o zhivotnykh iz Krasnoi Knigi) [We Must Save Them: Studies on Animals of the Red Data Book]. Mysl', Moscow, 174 pp (in Russian).
Bannikov, A.G. and A.I. Sokov. 1972. The tiger in Taijkistan. Priroda [Nature] 1: 121 (in Russian).
Bannikov, A.G., I.S. Darevsky, V.G. Ishchenko, A.K. Rustamov, and N.N. Shcherbak. 1977. Opredelitel' zemnovodnykh i presmykayushchikhsya fauny SSSR [A Key to the Amphibians and Reptiles of the USSR Fauna]. Prosveshcheniye, Moscow, 415 pp (in Russian).
Baran, J. 1976. Türkiye yilanlarinin taksonomik revizyonu ve cografi dagilislari [Taxonomic revision of Turkish snakes]. Türk. Bilims. Arast. Kr., Ankara, 177 pp (in Turkish).
Batenko, L.I. 1972. Some morphological characteristics of kidneys in yellow and steppe lemmings. Izv. Sib. otdel. AN SSSR. [Proc. Siberian Branch Acad. Sci. USSR], Ser. Biol. 15(3): 39–41 (in Russian).
Batenko, L.I. 1973. Morphohistology of kidneys in a narrow-skulled vole. Izv. Sib. otdel. AN SSSR. [Proc. Siberian Branch Acad. Sci. USSR], Ser. Biol. 10(2): 131–138 (in Russian).
Bauer, A.M. 1987. The Gekkonidae fauna of the USSR and adjacent countries by N. N. Szczerbak and M. L. Golubev, 1986 (A Review). Copeia 2: 525–527.
Bazykin, A.D., N.N. Vorontsov, and Ye.A. Lyapunova. 1968. In the underground caves of Turkmenia. Priroda [Nature] 3: 92–96 (in Russian).
Bei-Bienko, G.Ya. 1948. Orthoptera and Dermaptera. Zhivotnyi mir SSSR (zona pustyn') [The Animal Life of the USSR (desert zone)] 2: 270–291 (in Russian).
Bei-Bienko, G.Ya. 1949. On some communities of Orthoptera in Middle Asian deserts of the northern type. Trudy Zool. in-ta AN SSSR [Trans. Zool. Inst. Acad. Sci. USSR] 8 (4): 720–734 (in Russian).
Berdyev, A. 1980. Ekologiya iksodovykh kleshchei Turkmenistana i ikh rol v epizootologii prirodno-ochagovykh boleznei [Ecology of Ixodid Ticks in Turkmenistan and Their Role in Natural Epizootics]. Ylym, Ashkhabad, 280 pp (in Russian).
Berdyev, A.S. 1982. Chto nado znat o sarkotsystoze [What We Should Know About Sarcocystosis]. Ylym, Ashkhabad, 44 pp (in Russian).
Berezin, L.A. 1929. Natural areas of Kopetdagh and their importance for agriculture. Turkmenovedeniye [Turkmen Studies] 4: 9–12; 5: 7–11 (in Russian).
Berg, L.S. 1905. Ryby Turkestana [Fishes of Turkestan]. St. Petersburg, 260 pp (in Russian).
Berg, L.S. 1933. Ryby presnykh vod SSSR i sopredel'nykh stran [Freshwater Fishes of the USSR and Adjacent Countries]. Izd. AN SSSR, Moscow-Leningrad 2: 904 pp (in Russian).
Berg, L.S. 1938. Priroda SSSR [Nature in the USSR]. Izd. AN SSSR, Moscow Gos. Ped. Ucheb. Izdat, 311 p. (in Russian).
Berg, L.S. 1949. Ryby presnykh vod SSSR [Freshwater fishes of the USSR]. Izd. AN SSSR, Moscow-Leningrad 3: 468–925 (in Russian).

Berger, C. 1966. Mikroskopische und histochemische Untersuchungen an der Niere von *Columba livia aberratio domestica* L. Z. Mikrosk.-anat. Forsch. 74: 436–456.

Bidos, V.S. 1985. Ekologiya tadzhikskogo cherno-zolotogo fazana [Ecology of the Tajik Black-Gold Pheasant]. Abstr. Ph.D. Thesis, Moscow, 24 pp (in Russian).

Birula, A.A. 1904a. Miscellanea scorpiologica. VI. Ueber einige *Buthus*-Arten Centralasiens nebst ihrer geographischen Verbreitung. Annu. Mus. Zool. Acad. Imp. Sci., St-Petersbourg 9: 20–27.

Birula, A.A. 1904b. Miscellanea scorpiologica. VII. Synopsis der russischen Skorpione. Annu. Mus. Zool. Acad. Imp. Sci., St-Petersbourg 9: 28–38.

Birula, A.A. 1905a. Beiträge zur Kenntniss der Scorpionenfauna Persiens. (3. Beitrag). Bull. Acad. Imp. Sci. St.-Pétersb., Aer. 5, 23(1–2): 119–148.

Birula, A.A. 1905b. Skorpiologische Beiträge, 1–3. *Microbuthus littoralis* (Pav.), *Anomalobuthus rickmersi* Krpl. und *Buthus zarudnianus* n. nom. Zool. Anz. 29(14): 445–450.

Birula, A.A. 1905c. Skorpiologische Beiträge. Zool. Anz., 29(19): 621–624.

Birula, A.A. 1911. Miscellanea scorpiologica. IX. Ein Beitrag zur Kenntniss der Skorpionenfauna des Russischen Reiches und der angrenzender Länder. Annu. Mus. Zool. Acad. Imp. Sci., St-Petersbourg 16: 161–179.

Birula, A.A. 1917a. Chlenistobryukhie paukoobraznye Kavkazskogo kraya (Arthrogastric Arachnids of the Causasus). Part I. Scorpiones. Yezheg. Kavkazsk. Muzeya [Ann. Caucas. Mus.], Tiflis, Ser. A, 5: 1–253 (in Russian).

Birula, A.A. 1917b. Arachnoidea. Vol. I. Scorpiones. Fauna Rossii i sopredel'nykh stran. [Fauna of Russia and Adjacent Countries]. Petrograd, XX, 224 pp (in Russian).

Blatter, E. 1921. Flora Arabica. Calcutta 8(2): 1–281.

Blyumental, I.Kh. 1979. On principles and methods of classification of vegetation. Bot. zhurn. [Bot. Journal] 64(4): 488–499 (in Russian).

Bobrov, Ye.G. 1931. Vegetation of the Bolshoi Balkhan Mountains. Trudy Bot. sada AN SSSR [Trans. Botanical Garden Acad. Sci. USSR] 44: 3–91 (in Russian).

Bobrov, V.V. 1986. Lizard populations (Squamata, Sauria) in Southwest Kopetdagh. Zool. zhurn. [Zool. Journ.] 65(2): 308–310 (in Russian).

Boettger, O. 1890. Die Reptilien und Batrachier Transcaspiens. In: G. Radde and A. Walter (eds.), Nauchnye rezul'taty ekspeditsii, sovershennoi v Zakaspiiskii krai po Vysochaishemu poveleniyu. Tom 1. Zoologiya [Scientific Results of the Expedition to the Transcaspian Region Undertaken by the Imperial Command. Vol. 1. Zoology]. Tiflis, 3, 102 pp.

Bogdanov, O.P. 1960. Fauna Uzbekskoi SSR, Ch.1. Zemnovodnye i presmykayushchiyesya [Fauna of Uzbek SSR, Part 1. Amphibians and Reptiles]. Izd. ANUzbSSR, Tashkent, 254 pp (in Russian).

Bogdanov, O.P. 1962. Presmykayushchiyesya Turkmenii [Reptiles of Turkmenia]. Izd. AN TSSR, Ashkhabad, 233 pp (in Russian).

Bogdanov, O.P. 1965a. Ekologiya presmykayushchikhsya Srednei Azii [Ecology of the Reptiles of Middle Asia]. Nauka, Tashkent, 258 pp (in Russian).

Bogdanov, O.P. 1965b. Data on the distribution and ecology of the reptiles in the Murghab Valley, Badghyz, and Kopetdagh In: Ya.K. Turakulov (ed.), Gerpetologiya [Herpetology]. Tashkent, zd. ANU, bSSR: 23–25 (in Russian).

Bogdanov, O.P. 1970. Addition to the list of reptiles of the Atrek area. Izv. AN TSSR, Ser. Biol. Nauk. [Proc. Acad. Sci. Turkmen SSR, Ser. Biol. Sci.] 3: 88–89 (in Russian).

Bogdanov, O.P. 1971. The spring ecology of the smooth gecko. Ekologiya [Ecology] 1: 105–107 (in Russian).

Bogdanov, O.P. 1972. New data on the distribution of the shielded gecko. Izv. AN TSSR, Ser. Biol. Nauk. [Proc. Acad. Sci. Turkmen SSR, Ser. Biol. Sci.] 1: 92 (in Russian).

Bogdanov, O.P., Ch. Ataev, and S. Shammakov. 1974. On a record of the spotted toad agama (*Phrynocephalus maculatus*) from the USSR. Zool. zhurn. [Zool. Journ.] 53(2): 304–305 (in Russian).

Bogdanov, O.P. and O.N. Sudarev. 1988. The distribution of the spine-tailed gecko in the Murghab Valley In: V.Ye. Flint (ed.), Redkie i maloizuchennye zhivotnye Turkmenistana [Rare and Poorly Known Animals of Turkmenistan]. Ylym, Ashkhabad: 148–149 (in Russian).

Bogdanov, O.P. and V.D. Potopolsky. 1956. *Telescopus rhinopoma* Blanford, a snake species new for the USSR. Doklady AN SSSR [Rep. Acad. Sci. USSR] 109(4): 877–878 (in Russian).

Bogushevsky, P.N. 1932. Fruit trees of West Kopetdagh. Trudy po prikl. bot., genet., and selektsii rastenii [Studies in Applied Botany, Genetics and Selection of Plants] 8(1): 3–161 (in Russian).

Bondar, Ye.P. and I.V. Zhernovov. 1960. Ecologo-faunistic survey of rodents of western Turkmenia. In: Voprosy prirodnoi ochagovosti i epizootii chumy v Turkmenii [Problems of Nidal Diseases and Plague Epizootics in Turkmenia]. Ylym, Ashkhabad: 291–319 (in Russian).

Bondarenko, D.A. 1982. On the ecology of the Khentau toad agama in summer and fall. Vestn. zool. [Mess. Zool.] 5: 55–59 (in Russian).

Bondarenko, D.A. 1984. The Khentau toad agama. Priroda [Nature] 4: 48–49 (in Russian).

Bondartsev, A.S. 1954. Shkala tsvetov [The Color Scale]. Izd. AN SSSR, Moscow, 28 pp (in Russian)

Bonnet, P. 1945–1961. Bibliographia Araneorum. Analyse metodique de toute la literature araneologique jusqu'en 1939. Toulouse. Vol. 1 (1945), 832 pp; Vol. 2 (1955–1959), 5,058 pp; Vol. 3 (1961), 591 pp.

Borisova, A.G. 1938. Tragacanths of the Kopetdagh Range. Trudy In-ta botan. AN SSSR [Trans. Bot. Inst. Acad. Sci. USSR] 5(1): 19–26 (in Russian).

Borkin, L.Ya. 1977. On a new record and the taxonomic position of brown frogs from Kopetdagh (Turkmenia). Trudy Zool. Instituta [Trans. Zool. Inst.], Leningrad 74: 24–31 (in Russian).

Borkin, L.Ya. and I.S. Darevsky. 1987. A list of amphibians and reptiles in the USSR fauna. In: A.M. Amirkhanov (ed.), Amfibii i reptilii zapovednykh territorii [Amphibians and Reptiles of the Protected Areas]. TsNIILOKhIZ, Moscow: 128–141 (in Russian).

Botchantsev, V.P. 1969a. The genus *Salsola* L., with a brief history of its evolution and dispersal. Bot. zhurn. [Botan. J.] 54(7): 989–1001 (in Russian).

Botchantsev, V.P. 1969b. The genus *Salsola* L. (its composition, history of evolution, and dispersal). Abstr. Report Presented in Lieu of D.Sc. Thesis, Leningrad, 45 pp (in Russian).

Botchantsev, V.P. 1980. Species of the section *Belanthera* Iljin of the genus *Salsola* L. Novosti sist. vysshikh rast. [News in Systematics of Vascular Plants] 17: 112–135 (in Russian).

Brignoli, P.M. 1983. Catalogue of the Araneae described between 1940 and 1981. Manchester Univ. Press., Manchester, 755 pp.

Brown, E. and W.H. Dantzler. 1972. Function of mammalian-type and reptilian-type nephrons in kidneys of desert quail. Amer. J. Physiol. 222(3): 617–629.

Bugaev, V.A. *et al.* 1957. Sinopticheskie protsessy v Srednei Azii [Synoptic Processes in Middle Asia]. IZd. ANUzbSSR, Tashkent, 230 pp (in Russian).

Burachinsky, M.T. 1970. Comparative anatomical data on the arterial bed of the vertebrate kidney. Vestn. zool. [Mess. Zool.] 1: 12–17 (in Russian).

Burger, J.F. and R.V. Andreeva. 1988. Rheophilic larvae of horse flies (Diptera, Tabanidae): ecology, morphology, and distribution. Vestn. zool. [Mess. Zool.] 2: 17–23 (in Russian).

Butyev, V.T. 1976. On the existence of the scaled woodpecker in the USSR. Trudy Okskogo gos. zapov. [Transactions of the Oka State Reserve] 13: 191–192 (in Russian).

Butyev, V.T., O.A. Volkova, and N.N. Drozdov. 1965. The effect of the harsh winter of 1963/1964 on the density of waterfowl in the Kizyl-Agach Reserve. Geografiya resursov vodoplavayushchilkh ptits v SSSR [Geography of Waterfowl Resources in the USSR] 1: 149–150 (in Russian).

Bykov, G.I. and A.F. Milovanov. 1961. On the geographic distribution of the Turkmen jerboa (*Jaculus turkmenicus*). Izv. AN TSSR, Ser. Biol. Nauk. [Proc. Acad. Sci. Turkmen SSR, Ser. Biol. Sci.] 4: 72–73 (in Russian).

Caughley, G. 1966. Mortality patterns in mammals. Ecology 47(6): 906–918.

Charykuliev, D. 1967. On the studies of horse flies (Diptera, Tabanidae) of Turkmenia. Izv. AN Turkm. SSR, Ser. Biol. Nauk. [Proc. Acad. Sci. Turkm. SSR, Ser. Biol. Sci.] 4: 73–79 (in Russian).

Charykuliev, D. and E.I. Yasakova. 1973. New and poorly studied species of horse flies (Diptera, Tabanidae) from Turkmenia. Izv. AN Turkm. SSR, Ser. Biol. Nauk. [Proc. Acad. Sci. Turkm. SSR, Ser. Biol. Sci.] 2: 86–88 (in Russian).

Chelpanova, O.M. 1963. Klimat SSSR. Ch. 3. Srednyaya Aziya [Climate of the USSR. Part 3. Middle Asia]. Gidrometeoizdat, Leningrad, 447 pp (in Russian).
Cheltsov-Bebutov, A.M. 1964. Some problems of zoogeographic mapping (using the example of the map of Kustanai Region). In: Biogeographicheskie ocherki Kustanaiskoi oblasti [Biogeographical Studies of the Kustanai Region]. Moscow State Univ., Moscow: 5–24 (in Russian).
Cheltsov-Bebutov, A.M. 1966. On the principles of zoogeographic mapping. In: Voprosy geografii. Organizm i sreda [Problems of Geography. Organism and Environment]. Mysl', Moscow: 150–157 (in Russian).
Cheltsov-Bebutov, A.M. 1970. Zoogeographical mapping and landscape science. In: Landshaftnyi sbornik [Landscape Papers]. Moscow State Univ., Moscow: 49–94 (in Russian).
Cheltsov-Bebutov, A.M. 1973. Zoogeographical mapping of birds and mammals and its application in complex regional atlases (with the examples of the atlases of the Kustanai Region, North Kazakhstan, and Altai Region). Abstr. D.Sc. Thesis, Moscow, 43 pp (in Russian).
Cheltsov-Bebutov, A.M. 1976. Zoogeographical mapping: basic principles and concepts. Vestnik MGU, Ser. V. Geografiya [Mess. Moscow State Univ., Ser. 5, Geography] 2: 50–56 (in Russian).
Cherlin, V.A. 1981. A new species *Echis multisquamatus*, sp. nov. from Southwest and Middle Asia. Fauna i ekologiya amfibii i reptilii Palearkticheskoi Azii. Trudy Zool. in-ta AN SSSR [Fauna and Ecology of Amphibians and Reptiles of the Palearctic Asia. Trans. Zool. Inst. Acad. Sci. USSR] 101: 92–95 (in Russian).
Cherlin, V.A. 1983. New data on the systematics of the snake genus *Echis*. Vestn. zool. [Mess. Zool.] 2: 42–46 (in Russian).
Cherlin, V.A. 1990. Taxonomic revision of the snake genus *Echis* (Viperidae). II. Analysis of taxonomy and description of new forms. Reptilii gronykh i aridnykh territorii: sistematika i rasprostranenie/Trudy Zool. in-ta AN SSSR [Reptiles of Mountainous and Arid Territories: Systematics and Distribution/Trans. Zool. Inst. Acad. Sci. USSR] 207: 193–223 (in Russian).
Cherlin, V.A. and L.Ya. Borkin. 1990. Taxonomic revision of a snake genus *Echis* (Viperidae). I. Reptilii gronykh i aridnykh territorii: sistematika i rasprostranenie/Trudy Zool. in-ta AN SSSR [Reptiles of Mountainous and Arid Territories: Systematics and Distribution/Trans. Zool. Inst. Acad. Sci. USSR] 207: 175–192 (in Russian).
Chernov, S.A. 1934. Reptiles of Turkmenia. Trudy soveta po izucheniyu proizvopditel'nykh sil (SOPS). Ser. turkmenskaya [Trans. SOPS. Turkmen Series], Leningrad 6: 255–289 (in Russian).
Chernov, S.A. 1959. Fauna of Tajikistan. Reptiles. Trudy In-ta zool. i parazitol. AN Tadzh. SSR [Trans. Inst. Zool. and Parasitol. Acad. Sci. Tajik SSR], Dushanbe 98: 3–202 (in Russian).
Chernov, V.Yu. 1990. Rare bywater birds of Lake Sarykamysh. Okhrana prirody Turkmenistana [Natural Conservation in Turkmenistan] 8: 102–114 (in Russian).
Chernov, Yu.I. 1984. Flora and fauna, vegetation and animal population. Zhurn. obshch. biol. [Journal of General Biol.] 45(6): 732–748 (in Russian).
Chernyakhovsky, M.Ye. 1972. Ecological distribution of orthopteroid insects in the Repetek Reserve. In: Opyt izucheniya i osvoyeniya Vostochnykh Karakumov [Results of Exploration and Development of the East Karakum]. Ylym, Ashkhabad: 79–86 (in Russian).
Chernyakhovsky, M.Ye. 1985. The orthopteroid insects of West Kopetdagh. In: N.T. Nechayeva (ed.), Rastitelnost' i zhivotnyi mir Zapadnogo Kopetdaga [The Vegetation and Animal Life of West Kopetdagh]. Ylym, Ashkhabad: 262–271 (in Russian).
Chernyakovskaya, Ye.G. 1924. Spring vegetation of the Kara-Kala District of the Transcaspian Region. Izv. Glavnogo bot. sada [Proc. Central Botanical Garden] 23(2): 181–191 (in Russian).
Chernyakovskaya, Ye.G. 1927. An outline of the vegetation of Kopetdagh. Izv. Glavnogo bot. sada [Proc. Central Botanical Garden] 27(3): 253–267 (in Russian).
Chopanov, P.Ch., Ye.M. Gudkova, and E.M. Seifulin. 1980. Opredelitel' rastenii Turkmenistana [A Key to the Plants of Turkmenistan]. Ylym, Ashkhabad, Vol. 2, 136 pp (in Russian).
Chopanov, P.Ch., Ye.M. Gudkova, E.M. Seifulin, and S.N. Abramova. 1978. Opredelitel' khvoshcheobraznykh, paporortnikoobraznykh, golosemennykh i odnodolnykh rastenii Turkmenistana [A Key to the Horsetails, Ferns, Gymnosperms, and Monocots of

Turkmenistan]. Ylym, Ashkhabad, 328 pp (in Russian). (This volume represents A Key to the Plants of Turkmenistan, Vol. 1 – Eds.).

Chopanov, P.Ch., Ye.M. Gudkova, E.M. Seifulin, A.D. Ataeva, and A.S. Mishchenko. 1985. Opredelitel' rastenii Turkmenistana [A Key to the Plants of Turkmenistan]. Ylym, Ashkhabad, Vol. 3, 204 pp (in Russian).

Chunikhin, S.P. 1968. New records of *Lanius nubicus* and *Hypocolius ampelinus* in the USSR. Ornitologiya [Ornithology], Moscow State Univ. 9: 377–378 (in Russian).

Chkhikvadze, V.M., N.G. Amiranashvili, and Ch. Ataev. 1990. A new species of tortoise from Southwest Turkmenistan. Izv. AN TSSR, Ser. Biol. Nauk. [Proc. Acad. Sci. Turkmen SSR, Ser. Biol. Sci.] 1: 72–74 (in Russian).

Chvala, M., L. Lyneborg, and J. Moucha. 1972. The horse flies of Europe (Diptera, Tabanidae). Entomol. Soc. Copenh., Copenhagen, 498 pp.

Colby, P.J., G.R. Spangler, D.A. Hurley, and A.M. McCombie. 1972. Effects of eutrophication on salmonid communities in oligotrophic lakes. J. Fish Res. Board Canada 29(6): 975–983.

Dal', S.K. 1936. Vertebrates of the lower Zeravshan River. Trudy Uzbek. gos. un-ta [Trans. Uzbek State Univ.] 7: (in Russian).

Danilenko, A.K. 1974. Kartograficheskii analiz structury arealov ptits otkrytykh landshaftov (na primere zhavoronkov Severnogo Kazakhstana) [Cartographic Analysis of the Structure of Geographic Ranges of the Birds Inhabiting Open Landscapes (with Larks of the North Kazakhstan as an Example)]. Abstr. Ph.D. Thesis, Moscow, 25 pp (in Russian).

Danilenko, A.K., E.A. Danilenko, A.V. Kuprina, and V.Yu. Rumyantsev. 1991. Ecological aspects of the mapping of animal populations. In: Printsipy i metody ekologicheskogo kartografirovaniya/Tez. dokl. mezhdunar. soveshch. [Principles and Methods of Ecological Mapping/Abstr. Intern. Confer.], Pushchino: 37–38 (in Russian).

Danilenko, A.K. and M.V. Mirutenko. 1986. Mapping animal habitats. Mapping Sciences and Remote Sensing 23(1): 36–42.

Danilenko, A.K., M.V. Mirutenko, and E.A. Danilenko. 1991. The role of a habitat map in zoogeographical mapping. In: E.M. Pospelov et al. (ed.), Problemy zoogeograficheskogo kartografirovaniya [Problems in Zoogeographical Mapping]. ANSSSR, Moscow: 10–23 (in Russian).

Danov, R.A. 1984. Some data on the biology of a Middle Asian cobra – *Naja oxiana* (Eichw., 1831) in Turkmenistan. Izv. AN TSSR, Ser. Biol. Nauk. [Proc. Acad. Sci. Turkmen SSR, Ser. Biol. Sci.] 1: 69–72 (in Russian).

Danov, R.A. 1985a. Effect of weather factors on density and age structure of a population of gyurza [*Vipera lebetina*] in Southwest Kopetdagh. In: Voprosy gerpetologii [Problems of Herpetology]. Leningrad: 68–69 (in Russian).

Danov, R.A. 1985b. The life and death of leopards (*Panthera pardus* L.) in the Aidere Area of the Syunt-Khassardagh Reserve. In: N.T. Nechayeva (ed.), Rastitelnost' i zhivotnyi mir Zapadnogo Kopetdaga [The Vegetation and Animal Life of West Kopetdagh]. Ylym, Ashkhabad: 95–99 (in Russian).

Danov, R.A. and S.V. Pereladov. 1985. The list of sphingid moths (Lepidoptera, Sphingidae) of Southwestern Kopetdagh. In: N.T.Nechayeva (ed.), Rastitelnost' i zhivotnyi mir Zapadnogo Kopetdaga [The Vegetation and Animal Life of West Kopetdagh]. Ylym, Ashkhabad: 246–249 (in Russian).

Dantzler, W.H. 1966. Renal response of chickens to infusion of hyperosmotic sodium chloride solution. Amer. J. Physiol. 210: 640–646.

Darevsky, I.S. 1955. Record of *Ablepharus deserti* Strauch in the Murghab Valley and possible zoogeographic reasons for this finding. Izv. AN TSSR, Ser. Biol. Nauk. [Proc. Acad. Sci. Turkmen SSR, Ser. Biol. Sci.] 2: 72–74 (in Russian).

Darevsky, I.S. 1957. Turanian elements in the herpetofauna of Transcaucasia and possible directions of their dispersal from Middle Asia. Izv. AN Arm. SSR [Proc. Acad. Sci. Armenian SSR] 10(12) (in Russian).

Darevsky, I.S. 1978. Which species of *Eublepharis* (Sauria, Gekkonidae) is found in Middle Asia? In: G.S. Medvedev (ed.), Novie vidy zhivotnykh [New Species of Animals]. Nauka, Leningrad: 204–209 (in Russian).
Darevsky, I.S. 1981a. A genus and species of skink lizards, *Chalcides ocellatus* (Forskal) new to the USSR fauna. Fauna i ekologiya amfibii i reptilii Palearkticheskoi Azii. Trudy Zool. in-ta AN SSSR [Fauna and Ecology of Amphibians and Reptiles of the Palearctic Asia. Trans. Zool. Inst. Acad. Sci. USSR] 101: 49–51 (in Russian).
Darevsky, I.S. 1981b. Kopetdagh as the center of endemic herpetofauna and possible reasons for its formation. In: I.S. Darevsky (ed.), Voprosy gerpetologii [Problems of Herpetology]. Nauka, Leningrad: 47–48 (in Russian).
Davies, S.J. 1984. Nomadism as a response to desert conditions in Australia. J. Arid Envir. 7(2): 183–195.
Davis, P.H. 1965–1985. Flora of Turkey and the East Aegean Islands, Vols. 1–9. Edinburgh Univ.Press, Edinburgh.
Dawson, W. 1984. Physiological studies of desert birds: present and future considerations. J. Arid Envir. 7(2): 133–155.
Delacour, J. 1951. The pheasants of the world. Country Life, London, 347 pp.
Dementyev, G.P. 1947. Geographic variation in falcons. Trudy Tsentralnogo Byuro Koltsevaniya [Trans. Central Bird Banding Bureau], Vologda 6: 7–67 (in Russian).
Dementyev, G.P. 1948a. Francolin. Okrana prirody [Natural Conservation] 3: 84–90 (in Russian).
Dementyev, G.P. 1948b. Rare birds as natural monuments in the USSR: Observations on the redheaded falcon, or shakhin, in the Transcaspian Region. Okrana prirody [Natural Conservation] 2: 112–118 (in Russian).
Dementyev, G.P. 1951a. Ornithological observations in Southwest Turkmenia. Izv. Turkm. FAN SSSR, Ser. Biol. Nauk. [Proc. Turkmen Branch Acad. Sci. USSR, Ser. Biol. Sci.] 2: 51–57 (in Russian).
Dementyev, G.P. 1951b. Notes on mammals of South Turkmenia. Izv. Turkm. FAN SSSR, Ser. Biol. Nauk. [Proc. Turkmen Branch Acad. Sci. USSR, Ser. Biol. Sci.] 1: 58–63 (in Russian).
Dementyev, G.P. 1952. Ptitsy Turkmenistana [Birds of Turkmenistan]. Izd. AN TSSR, Ashkhabad, 547 pp (in Russian).
Dementyev, G.P. 1956. On distribution of the cheetah and on hunting with it. Trudy in-ta biologii AN TSSR. Ser. Zool. [Trans. Biol. Inst. Acad. Sci. Turkmen SSR, Ser. Zool.] 4: 66–76 (in Russian).
Dementyev, G.P. 1958. La faune desertique du Turkestan (I). La Terre et la Vie 1: 3–44.
Dementyev, G.P. 1964. Problems of ecological classification and a concept of the life form. In: Problemy ornitologii/Trudy III Vsesoyuzn. ornitol. konfer. [Problems in Ornithology/Trans. 3d All-Union Ornithol. Confer.], Lvov: 5–17.
Dementyev, G.P. and A.K. Rustamov. 1956. Notes on some species of Felidae from Turkmenia. Izv. AN TSSR, Ser. Biol. Nauk. [Proc. Acad. Sci. Turkmen SSR, Ser. Biol. Sci.] 2: 75–78 (in Russian).
Dementyev, G.P., A.K. Rustamov, and Ye.P. Spangenberg. 1955. Contributions to the fauna of terrestrial vertebrates of southeastern Turkmenia. Trudy Turkm. sel'skokhoz. instituta [Trans. Turkmen Agricultural Institute] 7: 125–183 (in Russian).
Dementyev, G.P., N.N. Kartashov, and A.N. Soldatova. 1953. Feeding and the practical importance of some birds of prey in southwestern Turkmenia. Zool. Zhurn. [Zool. Journal] 32(3): 361–375 (in Russian).
Dementyev, G.P., N.N. Kartashov, and A.O. Tashliev. 1956. Contributions to the fauna of terrestrial vertebrates of northeastern Turkmenia. Trudy in-ta biologii AN TSSR. Ser. Zool. [Trans. Biol. Inst. Acad. Sci. Turkmen SSR, Ser. Zool.] 4: 77–119 (in Russian).
Demurina, Ye.M. 1980. Turanian herbaceous dry steppe. In: K.Z. Zakirov (ed.), Rastitel'nyi pokrov Uzbekistana i puti yego ratsional'nogo ispol'zovaniya [Vegetative Cover of Uzbekistan and Directions of Its Rational Exploitation]. Fan, Tashkent 3: 139–158 (in Russian).
Denis, J. 1955. Contribution á l'étude de l'Air. Araignées. Bull. Inst. Fond. Afr. Noire 17(A): 99–146.

Denis, J. 1958. Araignées de l'Afghanistan. Vidensk. Meddr. Dansk Naturh. Foren. (Copenhagen) 120: 81–120.

Denis, J. 1966. Les araignées du Fezzan. Bull. Soc. Hist. Nat. Afr. Noire 55: 103–144.

Denisov, V.P. 1975. On wild fruit plants of West Kopetdagh. Byull. VIR [Bull. All-Union Plant Breeding Inst.] 54: 78–80 (in Russian).

Dimo, V.N. 1972. Teplovoi rezhim pochv SSSR [Thermal Regime of Soils of the USSR]. Kolos, Moscow (in Russian).

Dlussky, G.M. and S.I. Zabelin. 1985. The fauna of ants (Hymenoptera, Formicidae) in the drainage system of the Sumbar River (Southwestern Kopetdagh). In: N.T. Nechayeva (ed.), Rastitelnost' i zhivotnyi mir Zapadnogo Kopetdaga [The Vegetation and Animal Life of West Kopetdagh]. Ylym, Ashkhabad: 208–246 (in Russian).

Dobrin, L.G. 1964. The formation of a barkhan. Izv. AN TSSR, Ser. Biol. Nauk. [Proc. Acad. Sci. Turkmen SSR, Ser. Biol. Sci.] 6: 18–23 (in Russian).

Dobrokhotov, B.P. 1962. The current condition of bird wintering sites on the southeastern Caspian shore. Ornitologiya [Ornithology], Moscow, Moscow State Univ. 5: 362–367 (in Russian).

Dobrynin, M.I. and V.M. Korshunov. 1991. On the helminth fauna of the bearded goat in Turkmenistan. Izv. AN TSSR, Ser. Biol. [Proc. Acad. Sci. Turkmen SSR, Ser. Biol.] 4: 35–39 (in Russian).

Dolgushin, I.A. 1960. Ptitsy Kazakhstana [Birds of Kazakhstan]. Izd. An Kaz. SSR, 1, Alma-Ata, 470 pp (in Russian).

Dolin, V.G. and R.V. Andreeva. 1984. A larva and pupa of a horse fly *Tabanus atropathenicus* (Diptera, Tabanidae) and this species' place in the system of the genus *Tabanus*. Vestn. zool. [Mess. Zool.] 1: 25–27 (in Russian).

Dolin, V.G. and R.V. Andreeva. 1986. New species of horse flies (Diptera, Tabanidae) from the Caucasus and Middle Asia. Zool. zhurn. [Zool. J.] 65(6): 940–944 (in Russian).

Dormidontov, R.V. and A.Yu. Blokhin. 1977. The bezoar goat. In: R.V. Dormidontov et al. (ed.), Kopytnye zveri [Ungulate Mammals]. Lesnaya Promyshlennost', Moscow: 153–164 (in Russian).

Doroshev, S.I. 1964. The salt tolerance of some fish species recommended for introduction into the Azov Sea. Trudy VNIRO [Trans. All-Union Research Institute of Fish Industry and Oceanography] 55(2): 97–107 (in Russian).

Dotsenko, I.B. 1984. Morphological characters and ecological features of *Oligodon taeniolatus* (Serpentes, Colubridae). Vestn. zool. [Mess. Zool.] 4: 23–26 (in Russian).

Dotsenko, I.B. 1985. Revision of the genus *Eirenis* (Reptilia, Colubridae). Communication I. Reestablishment of the genus *Pseudocyclophis* Boettger, 1888. Vestn. zool. [Mess. Zool.] 4: 23–26 (in Russian).

Dotsenko, I.B. 1986. Snakes of the genus *Eirenis* of the Palearctic Region. Ph.D. Thesis, Kiev, 248 pp.

Dotsenko, I.B. and N.N. Shcherbak. 1985. Altitudinal and habitat distribution of *Eirenis meda* and *E. periscus* (Serpentes, Reptilia) in Soviet Kopetdagh. Vestn. zool. [Mess. Zool.] 2: 40–42 (in Russian).

Drozdov, N.N. 1977. Comparative analysis of the ornithofauna and bird population in the arid regions of Middle Asia, Northern Africa, and Australia. In: Adaptivnye osobennosti evolyutsii ptits [Adaptive Features in Bird Evolution]. Nauka, Moscow: 40–48.

Dubrovo, I.A. 1960. Ancient elephants of the USSR. Trudy PIN AN SSSR [Trans. Paleontol. Inst. Acad. Sci. USSR] 85(1): 1–83 (in Russian).

Dubrovo, I.A. and A.N. Nigarov. 1990. Plio-Pleistocene fossil vertebrate localities of South-Western Turkmenia, U.S.S.R. Quartarpaleontologie (Berlin) 8: 35–45.

Dubrovsky, V.P. and N.S. Nardina. 1963. Effect of late spring freezes on tree and shrub vegetation of southeastern Turkmenia. Izv. AN TSSR, Ser. Biol. Nauk. [Proc. Acad. Sci. Turkmen SSR, Ser. Biol. Sci.] 1: (in Russian).

Due, A.D. and G.A. Polis. 1985. Biology of the intertidal scorpion, *Vaejovis littoralis*. Journal of Zoology (London) 207: 563–580.

Dunin, P.M. 1982. New data on *Dysdera concinna* Poc. (Aranei, Dysderidae). Zool. zhurn. [Zool. Journal] 61(4): 605–606 (in Russian).

Dunin, P.M. 1985. The spider family Dysderidae (Aranei, Haplogynae) in Soviet Central Asia. Fauna i ekologiya paukov SSSR/Trudy Zool. in-ta AN SSSR [Fauna and Ecology of Spiders of the USSR/Trans. Zool. Inst. Acad. Sci. USSR] 139: 114–120 (in Russian).
Dunin, P.M. 1989. Data on the fauna of spiders of the family Dysderidae (Aranei) in the Turkmen SSR. Entomol. obozr. [Entomol. Rev.] 68(4): 865–874 (in Russian).
Dunin, P.M. 1990. The spider family Scytodidae (Aranei, Haplogynae) of the USSR fauna. Trudy Zool. in-ta AN SSSR/Trans. Zool. Inst. Acad. Sci. USSR] 226: 74–82 (in Russian).
Dunin, P.M. 1992. New species of *Dysdera* spiders (Aranei, Haplogynae, Dysderidae) from Middle Asia. Zool. zhurn. [Zool. Journal] 71(2): 136–140 (in Russian).
Dunin, P.M. and V.Ya. Fet. 1985. *Dysdera transcaspica* n.sp. (Aranei, Dysderidae) from Turkmenia. Zool. zhurn. [Zool. Journal] 64(2): 298–300 (in Russian).
Durdyev, A.M. and N.S. Orlovsky. 1984. Vaporability in the lowlands of Middle Asia. Probl. Osvoyen. Pustyn' [Problems Desert Development] 1: 18–25 (in Russian).
Dzhuraeva, Z.D. 1978. Likhenoflora Tsentral'nogo Kopetdaga (Turkmenistan) [Lichen Flora of Central Kopetdagh (Turkmenistan)]. Ylym, Ashkhabad, 163 pp (in Russian).
Elpatjewsky, W.S. and L. Sabanejew. 1906. Ergänzungen zur herpetologischen Fauna des Russischen Reiches. Zool. Jahrbuch, Abt. Syst. 24: 247–264.
Elton, Ch.S. 1958. The ecology of invasions by animals and plants. Methuen, London, 181 pp.
Eminov, A. 1981. Rare birds of the Ashkhabad Fishery Enterprise and their conservation. In: Tez. dokl. 2 nauchn. konfer, po okhrane prirody Turkm. SSR [Abstr. 2nd Sci. Conf. on Natural Conservation in the Turkmen SSR], Ashkhabad: 148–149 (in Russian).
Eminov, A. 1982. Observations on *Hypocolius ampelinus* in the Tedzhen Valley. Izv. Akad. Nauk. Turkm. SSR. Ser. Biol. Nauk. [Proc. Acad. Sci. Turkm. SSR, Ser. Biol. Sci.] 5: 75–77 (in Russian).
Eminov, A. and A.F. Kekilova. 1976. Preliminary data on spring bird migration in the Murghab River Valley. In: E.L. Gavrilov et al. (ed.), Migratsii ptits v Azii [Bird Migrations in Asia]. Nauka Kaz. SSR, Alma-Ata: 217–222 (in Russian).
Etchécopar, R.D. and F. Hüe. 1957. Donnees écologiques sur l'avifaune de la zone désertique arabosaharienne. In: Ecologie Humaine et Animale. UNESCO Publ.: 49–61
Falcoz, L. 1914 (1915). Contribution a l'etude de la fauna des microcavernes, fauna des terriers et des nids. Ann. Soc. Linn. Lyon 61: 59–245.
Fedchenko, B.A. 1935. Natural flora of the Turkmen SSR. In: Problemy Turkmenii: Trudy konf. po nauchno-proizv. silam TSSR [Problems of Turkmenia. Mat. Confer. Sci. Industr. Resources of the Turkmen SSR] 2: 3–15 (in Russian).
Fedorovich, B.A. 1946. Problems of the paleogeography of the plains of Middle Asia. Trudy Inst. geograf. AN SSSR [Trans. Geogr. Inst. Acad. Sci. USSR] 37: 152–174 (in Russian).
Fedorovich, B.A. 1952. Ancient rivers in the Turanian desert. Materialy po chetvertichnomu periodu SSSR [Materials on the Quaternary Period in the USSR]. Izd. AN SSSR, Moscow 3: 204–213 (in Russian).
Fedorovich, B.A. 1960. Basic features of the relief of the Karakum sands. In: Origin of sand relief and loess/Trudy Inst. geograf. AN SSSR [Trans. Geogr. Inst. Acad. Sci. USSR] 80: 30–59 (in Russian).
Fedorovich, B.A. and A.S. Kes. 1934. Subaerial delta of the Murghab. Trudy Geomorfol. inst. AN SSSR [Trans. Geomorphol. Inst. Acad. Sci. USSR] 12: 5–20 (in Russian).
Fedorovich, B.A. 1946. Problems of paleogeography of Middle Asian plains. Trudy in-ta geogr. AN SSSR [Trans. Inst. Geogr. Acad. Sci. USSR] 37: 152–174 (in Russian).
Feinbrun, N.D. 1978. Flora Palaestina, 3. Israel Acad. Sci. Hum., Jerusalem, 484 pp.
Fenyuk, B.K., I.V. Zhernovov, P.I. Kamnev, and G.N. Skvortsov. 1955. New records of the Turkmen jerboa (*Jaculus turkmenicus*). In: Gryzuny i bor'ba s nimi [Rodents and the Struggle Against Them]. Saratov Book Publishers, Saratov: 4 (in Russian).
Fet, V.Ya. 1980. On the ecology of scorpions (Arachnida, Scorpiones) of Southeast Karakum. Entomol. obozr. [Entomol. Rev.] 59(1): 223–228 (in Russian).

Fet, V. Ya.1982. A spider species from Kopetdagh, *Oxyptila tricoloripes* Strand 1913 (= *O. pickardi* Levy 1975, *syn.nov.*) (Aranei, Thomisidae), new to the USSR. Izv. Akad. Nauk. Turkm. SSR. Ser. Biol. Nauk. [Proc. Acad. Sci. Turkmen. SSR, Ser. Biol. Sci.]. 5: 74–75 (in Russian).

Fet, V.Ya. 1983. The fauna of spiders (Aranei) of Southwest Kopetdagh. Kopetdagh. Entomol. obozr. [Entomological Review] 62(4): 835–845 (in Russian).

Fet, V.Ya. 1984a. The zoogeographical features of Kopetdagh spider fauna (Aranei). In: Problemy pochvennoi zoologii. Tez. dokl. 8 Vsesoyuzn. konfer. [Problems of Soil Zoology. Abstr. VIII All-Union Conference], Ashkhabad 2: 135–136 (in Russian).

Fet, V.Ya. 1984b. The zoogeographical features of Kopetdagh spider fauna. In: Tez. dokl. 8 Vsesoyuzn. zoogeograficheskoi konfer. [Abstr. VIII All-Union Zoogeographical Conference], Leningrad: 258–259 (in Russian).

Fet, V.Ya. 1984c. A new species of spiders from Turkmenian SSR: *Brachythele kopetdaghensis* sp.n. (Dipluridae). In: A.S. Utochkin et al. (eds.), Fauna i ekologiya paukoobraznykh [Fauna and Ecology of Arachnids]. Perm Univ., Perm: 37–41 (in Russian).

Fet, V.Ya. 1984d. The seasonal changes in spider fauna of Southwest Kopetdagh. In: VI nauchnaya konfer. molodykh uchyonykh Turkmenistana [VI Scientific Conference of Young Scientists of Turkmenistan], Ashkhabad: 221–222 (in Russian).

Fet, V.Ya. 1984e. A new genus and species of scorpions for the USSR from Badghyz: *Kraepelinia palpator* Birula (Scorpiones, Buthidae). Izv. Akad. Nauk. Turkm. SSR. Ser. Biol. Nauk. [Proc. Acad. Sci. Turkmen. SSR, Ser. Biol. Sci.] 4: 37–45 (in Russian).

Fet, V.Ya. 1985a. A zoogeographical analysis of spider fauna of Southwest Kopetdagh. Fauna i ekologiya paukov SSSR/Trudy Zool. in-ta AN SSSR [Fauna and Ecology of Spiders of the USSR/Trans. Zool. Inst. Acad. Sci. USSR] 139: 72–77 (in Russian).

Fet, V.Ya. 1985b. The ecological distribution of spiders in the Syunt-Khasardagh Reserve. In: N.T. Nechayeva (ed.), Rastitelnost' i zhivotnyi mir Zapadnogo Kopetdaga [The Vegetation and Animal Life of West Kopetdagh]. Ylym, Ashkhabad: 271–277 (in Russian).

Fet, V.Ya. 1986. Contributions to the spider fauna of Turkmenistan, IV. A new species of titanoecid spiders, *Titanoeca lehtineni* n.sp. (Aranei, Titanoecidae). Izv. Akad. Nauk. Turkm. SSR. Ser. Biol. Nauk. [Proc. Acad. Sci. Turkmen. SSR, Ser. Biol. Sci.] 6: 65 (in Russian).

Fet, V.Ya. 1987a. A new genus and species of scorpion from East Karakum: *Pectinibuthus birulai* Fet n.g. n.sp. (Scorpiones, Buthidae). Entomol. Obozrenie [Entomol. Rev. (USSR)] 66(2): 443–446 (in Russian).

Fet, V.Ya. 1987b. On the finding of *Kraepelinia palpator* (Birula) (Scorpiones, Buthidae) in East Iran. Izv. Akad. Nauk. Turkm. SSR. Ser. Biol. Nauk. [Proc. Acad. Sci. Turkmen. SSR, Ser. Biol. Sci.] 1: 98 (in Russian).

Fet, V. 1989. A catalogue of the scorpions (Chelicerata: Scorpionida) of the USSR. Rivista di Museo di Scienzi Naturali E. Caffi, Bergamo 13: 73–171 (the 1988 issue, published January 1989).

Fet, V. 1993. The spider genus *Cedicus* Simon 1875 (Arachnida, Aranei, Agelenidae) in Middle Asia. Arthropoda Selecta 2(1): 69–75.

Fet, V.Ya. and G.L. Kamakhina. 1982. The vegetation of West Kopetdagh. In: N.T. Nechayeva (ed.), Priroda Zapadnogo Kopetdaga [Nature in West Kopetdagh]. Ylym, Ashkhabad: 32–37 (in Russian).

Filatov, A.K. 1985. Contributions to the ecology of birds of prey in Western Kopetdagh. In: N.T. Nechayeva (ed.), Rastitelnost' i zhivotnyi mir Zapadnogo Kopetdaga [The Vegetation and Animal Life of West Kopetdagh]. Ylym, Ashkhabad: 155–161 (in Russian).

Filippov, M.N. 1947. On zoogeography of the TSSR. Trudy Ashkhabadsk. gos. ped. in-ta [Trans. Ashkhabad State Pedagogical Inst.] 2: 95–97 (in Russian).

Flerov, K.K. and I.M. Gromov. 1934. Mammals of the Sumbar and Chandyr Valleys. Trudy soveta po izucheniyu proizvopditel'nykh sil (SOPS). Ser. turkmenskaya [Trans. SOPS. Turkmen Series], Leningrad 6: 291–372 (in Russian).

Flint, V.Ye. 1961. *Hypocolius ampelinus*, a new genus and species of birds in the USSR fauna. Byull. MOIP, Otd. biol. [Bull. Moscow Soc. Natur., Biol. Div.] 66(1): 127–129 (in Russian).

Flint, V.Ye. 1962. *Hypocolius ampelinus* in the USSR. Ornitologiya [Ornithology], Moscow State Univ. 4: 186-189 (in Russian).
Flint, V.Ye. 1991. The problem of biodiversity and goals of the ornithologists. In: Mater. X Vsesoyuzn. ornitol. konfer. [Abstr. 10th All-Union Ornithol. Conf.]. Nauka i tekhnika, Minsk 1: 7-8 (in Russian).
Flint, V.Ye. and A.A. Kishinsky. 1975. Sterkh (*Grus leucogeranus* Pallas, 1773) in Yakutia. Zool. zhurn. [Zool. J.] 54(8): 1197-1212 (in Russian).
Flora Turkmenii [Flora of Turkmenia]. 1932-1960. Izd. AN TSSR, Leningrad-Ashkhabad, vol. 1-7 (in Russian).
Fokanov, V.A. 1961. Contributions to the history of formation of the Holocene rodent fauna in Badghyz (South Turkmenia). Byull. MOIP, Otd. biol. [Bull. Moscow Soc. Natur., Biol. Div.] 69(4): 51-56 (in Russian).
Ford, C.E. and J.L. Hamerton. 1956. A colchicine hypotonic citrate squash-sequence for mammalian chromosomes. Stain Technol. 31: 247-251.
Frantskevich, N.A. 1978. Wild relatives of cultural plants and their conservation in the Aidere Valley (Kara-Kala District of Turkmen SSR). Byull. VIR [Bull. All-Union Plant Breeding Inst.] 81: 86-91 (in Russian).
Fyodorov, P.V. 1946. On a mammalian skeleton found in the lower Uzboi. Trudy Turkm. geolog. upravl. [Trans. Turkmen Geol. Survey] 1: 84-85 (in Russian).
Fyodorov, P.V. 1972. General characteristics and paleogeography (Quaternary System). Geologiya SSSR [Geology of the USSR]. Nedra, Moscow 22: 432-442 (in Russian).
Gaiser, M. 1982. Flora of Pakistan (Tamaricaceae). Karachi 141: 1-80.
Gauzer, M.Ye. 1977. The density dynamics of seagull nesting sites on the eastern Caspian. In: Tez. dokl. VIII Vsesoyuzn. ornitol. konfer. [Abstr. 8th All-Union Ornithol. Conf.], Kiev 1: 226-227 (in Russian).
Gauzer, M.Ye., V.I. Vasilyev, and A.A. Karavaev. 1976. On the conservation of colonial Laridae on the Caspian shore within the Turkmen SSR. In: Tez. dokl. I nauchn. konf. po okhhrane prirody Turkm. SSR [Abstr. First Scientific Conference on Natural Conservation in the Turkmen SSR], Ashkhabad: 51-52 (in Russian).
Gazanchyan, M.K. 1965. Density, distribution, and condition of rewintering sites in waterfowl of the Kizyl Agach Reserve. In: Z.I. Ivanova (ed.), Novosti ornitologii [News in Ornithology]. Nauka, Alma-Ata: 150-151 (in Russian).
Geologiya SSSR. Tom 22. Turkmenskaya SSR [The Geology of the USSR. Vol. 22. Turkmen SSR] 1957. Nedra, Moscow, 658 pp (in Russian).
Geptner, V.G. 1945a. Desert and steppe fauna of the Palaearctic region and centers of its origin. Byull. MOIP, Otd. Biol., Nov. Ser. [Bull. Soc. Nat. Mosc. (Biol.), New Ser.] 50(1-2): 17-38 (in Russian).
Geptner, V.G. 1945b. Gady Badghyza [Herpetofauna of Badghyz]. Uchen. zap. MGU, Biol. [Sci. Mem. Moscow State Univ., Biology] 83: 95-126 (in Russian).
Geptner, V.G. 1948. Kulan and perspectives of its existence in the USSR. In: Okhrana prirody [Natural Conservation]. Moscow: 2 (in Russian).
Geptner, V.G. 1949. New data on distribution of some vertebrates of Turkmenia and their zoogeographical importance. Vestn. MGU [Messenger of Moscow State Univ.] 6: 151-158 (in Russian).
Geptner, V.G. 1956. Fauna nazemnykh pozvonochnykh Badkhyza [Fauna of terrestrial vertebrates of Badghyz]. Izd. AN TSSR, Ashkhabad, 334 pp (in Russian).
Geptner, V.G., A.A. Nasimovich, and A.G. Bannikov. 1961. Mlekopitayushchie Sovetskogo Soyuza [Mammals of the Soviet Union]. Vysshaya Shkola, Moscow, 1, 775 pp (in Russian).
Geptner, V.G., N.P. Naumov, P.B. Yurgenson, A.A. Sludsky, A.F. Chirkova, and A.G. Bannikov. 1967. Mlekopitayushchie Sovetskogo Soyuza. Morskie korovy i khishchnye [Mammals of the Soviet Union. Sea cows and carnivores]. Vysshaya Shkola, Moscow 2(1): 1,004 pp (in Russian).
Geptner, V.G. and A.A. Sludsky. 1967. Mlekopitayushchie Sovetskogo Soyuza [Mammals of the Soviet Union]. Vysshaya Shkola, Moscow, 2 (2): 549 pp (in Russian).

Ginetsinsky, A.G. 1961. Role of hialuronic structures in the evolution of osmoregulatory function of the kidney. Obsh. biol. [General Biology] 22(1): 31-42 (in Russian).
Gladkov, N.A. 1951. Order Charadriiformes. Ptitsy Sovetskogo Soyuza [Birds of the Soviet Union]. Sovetskaya Nauka, Moscow 3: 3-369 (in Russian).
Gladkov, N.A., G.P. Dementyev, Ye.S. Ptushenko, and A.M. Sudilovskaya. 1964. Opredelitel' ptits SSSR [A Key to the Birds of the USSR]. Vysshaya shkola, Yaroslavl, 536 pp (in Russian).
Gladkov, N.A. and V.B. Grinberg. 1932. Contributions to the ornithofauna of the Chu River. Byull. MOIP, Nov. Ser., Otdel. Biol. [Bull. Moscow Soc. Natur., New Series, Section of Biology] 41(34): 303-319 (in Russian).
Gladkova, A.N. 1957. Some new data on the flora of the Pliocene deposits of the Cheleken Peninsula. Dokl. AN SSR [Reports Acad. Sci. USSR], 117(1) 121-123 (in Russian).
Gladkova, A.N. 1962. Remnants of the palm *Nipa* Thumb. in the Eocene deposits of Turkmenia. Trudy VNIGRI [Trans. All-Union Sci. Res. Geol. Inst.] 196: 579-600 (in Russian).
Glazovsky, N.F. 1990. The Aral crisis. Priroda [Nature] 10: 14-23.
Gloyd, H.K. and R. Conant. 1990. Snakes of the *Agkistrodon* complex. Society for the Studies of Amphibians and Reptiles, 614 pp.
Godina, A.Ya. and A.A. Dubyansky. 1963. The first record of a fossil giraffe in Turkmenia. Byull. MOIP, Otd. Geol. [Bull. Moscow Soc. Natur., Geol. Div.] 38(1): 155-157.
Golubev, M.L. 1979. On geographic variation and taxonomy of the smooth gecko [*Alsophylax laevis* Nikolsky, 1905 (Sauria, Gekkonidae)]. In: N.B. Ananyeva and L.Ya. Borkin (ed.), Ekologiya i sistematika amfibii i reptilii [Ecology and Systematics of Amphibians and Reptiles]. Nauka, Leningrad: 55-64 (in Russian).
Golubev, M.L. 1989. Three arguable problems in systematics and nomenclature of toad agamas in the USSR fauna (*Phrynocephalus*, Agamidae). In: Voprosy gerpetologii/7 Vsesoyuzn. gerpetol. konf. [Problems of Herpetology/7th All-Union Herpetological Conference], Kiev: 64-65.
Golubev, M.L. 1991. On the name *Agama ocellata* Lichtenstein in Eversmann, 1823 (Reptilia: Agamidae), with redescription of the type specimens. Gerpetologicheskie issledovaniya [Studies in Herpetology] 1: 12-17.
Golubev, M.L. and T. Sattorov. 1979. On the subspecies of shielded gecko *Alsophylax brietus* Strauch, 1887 (Reptilia, Sauria, Gekkonidae). Vestn. zool. [Mess. Zool.] 5: 18-24 (in Russian).
Golubev, M.L. and A.B. Streltsov. 1989. On the distribution of two species of geckos (Reptilia, Gekkonidae) along the middle portion of the Amudarya River. Vestn. zool. [Mess. Zool.] 2: 78-79 (in Russian).
Goncharevskaya, O.A. 1976. Structural organization of nephrons in the kidney and proximal reabsorption in various classes of vertebrates: Nicrodissection and micropuncture data. Zh. evol. biokhim. i Fiziol. [J. Evol. Biochem. Physiol.] 12(2): 113-119 (in Russian).
Goodwin, J. 1974. Immature stages of some Eastern Nearctic Tabanidae (Diptera). V. *Stenotabanus (Aegialomyia) magnicallus* Stone. J. Tenn. Acad. Sci. 48: 14-15.
Gorelov, S.K., A.P. Zhumashov, and L.S. Mokrushina. 1989. Zaunguzskie Karakumy [Trans-Unguz Karakum]. Izd. AN SSSR, Moscow, 120 pp (in Russian).
Gorelov, Yu.K. 1959a. Effect of the snowy winter of 1956-1957 on the ungulate animals of Baghyz. Izv. AN TSSR, Ser. Biol. Nauk. [Proc. Acad. Sci. Turkmen SSR, Ser. Biol. Sci.] 2: 71-73 (in Russian).
Gorelov, Yu.K. 1959b. On the distribution of bezoar goat in Badghyz. Izv. AN TSSR, Ser. Biol. Nauk. [Proc. Acad. Sci. Turkmen SSR, Ser. Biol. Sci.] 4: 62-64 (in Russian).
Gorelov, Yu.K. 1965. Changes in distribution and density of the ungulate animals of Baghyz under anthropogenic pressure. Izv. AN TSSR, Ser. Biol. Nauk. [Proc. Acad. Sci. Turkmen SSR, Ser. Biol. Sci.] 3: 55-61 (in Russian).
Gorelov, Yu.K. 1969. Ostriches in the Neogene of Badghyz. In: Ornitologiya v SSSR/Mater. V Vsesoyuzn. ornitol. konf. [Ornithology in the USSR/Abstr. Fifth All-Union Ornithol. Conf.], Ashkhabad 2: 178-180 (in Russian).
Gorelov, Yu.K. 1970. Factors determining territorial groups of ungulates in Badghyz. In: Populyatsionnaya struktura vida u mlekopitayushchikh [Population Structure of a Species in Mammals]. Nauka, Moscow: 31-33 (in Russian).

Gorelov, Yu.K. 1972a. Reproduction of the dzheiran (*Gazella subgutturosa*) and the problem of its restoration in Badghyz. In: D.I. Bibikov et al. (ed.), Teriologiya [Theriology]. Nauka, Novosibirsk 1: 420–424 (in Russian).

Gorelov, Yu.K. 1972b. On the formation of fauna of terrestrial vertebrates of Turkmenistan. Byull. MOIP, Otd. biol. [Bull. Moscow Soc. Natur., Biol. Div.] 77(6): 27–37 (in Russian).

Gorelov, Yu.K. 1973a. A problem of gene pool conservation in large predators in Badghyz and adjacent regions of Turkmenia. In: V.E. Sokolov (ed.), Redkie vidy mlekopitayushchikh fauny SSSR i ikh okhrana [Rare Mammalian Species of the USSR Fauna and Their Conservation]. Nauka, Moscow: 81–82 (in Russian).

Gorelov, Yu.K. 1973b. Conservation and restoration of the ungulate density in Badghyz. In: V.E. Sokolov (ed.), Redkie vidy mlekopitayushchikh fauny SSSR i ikh okhrana [Rare Mammalian Species of the USSR Fauna and Their Conservation]. Nauka, Moscow: 117–118 (in Russian).

Gorelov, Yu.K. 1976. Badghyz State Reserve and the problem of preservation of its natural complex. In: Tez. dokl. I nauchn. konf. po okhhrane prirody Turkm. SSR [Abstr. First Scientific Conference on Natural Conservation in the Turkmen SSR], Ashkhabad: 23–25 (in Russian).

Gorelov, Yu.K. 1977a. The current condition of kulan [population] in Badghyz. In: V.E. Sokolov (ed.), Redkie vidy mlekopitayushchikh fauny SSSR i ikh okhrana [Rare Mammalian Species of the USSR Fauna and Their Conservation]. Nauka, Moscow: 199–200 (in Russian).

Gorelov, Yu.K. 1977b. Rare species of bats from Badghyz. In: V.E. Sokolov (ed.), Redkie vidy mlekopitayushchikh fauny SSSR i ikh okhrana [Rare Mammalian Species of the USSR Fauna and Their Conservation]. Nauka, Moscow: 52–53 (in Russian).

Gorelov, Yu.K. 1980a. On the possibility of management of territorial distribution of certain ungulates. In: V.E. Sokolov (ed.), Kopytnye fauny SSSR [Ungulates of the USSR Fauna]. Nauka, Moscow: 11–12 (in Russian).

Gorelov, Yu.K. 1980b. [*Ptyas mucosus*]. Priroda [Nature] 11: 100–102 (in Russian).

Gorelov, Yu.K. and R.I. Gorelova. 1976. On the measures toward restoration of tugai landscape areas in the Murghab River basin. In: Tez. dokl. I nauchn. konf. po okhhrane prirody Turkm. SSR [Abstr. First Scientific Conference on Natural Conservation in the Turkmen SSR], Ashkhabad: 98–99 (in Russian).

Gorelov, Yu.K., R.I. Gorelova, I.V. Zhernovov, and V.P. Sosnovtseva. 1978. The honey badger in Turkmenia. Priroda [Nature (Moscow)] 6: 123–125 (in Russian).

Gorelov, Yu.K., I.S. Darevsky, and N.N. Shcherbak. 1974. Two species of geckonid lizards new for the USSR. Vestn. zool. [Mess. zool.] 4: 33–39 (in Russian).

Gorelov, Yu.K. and A.V. Kochkareva. 1975. Data on ticks and parasites of wild sheep and other wild ungulates of Badghyz. Izv. AN TSSR, Ser. Biol. Nauk. [Proc. Acad. Sci. Turkmen SSR, Ser. Biol. Sci.] 3: 81–84 (in Russian).

Gorelov, Yu.K. and V.S. Lukarevsky. 1990. On *Stellio erythrogaster* Nikolsky, Agamidae, Sauria, found in the Soviet portion of East Kopetdagh. Izv. AN TSSR, Ser. Biol. Nauk. [Proc. Acad. Sci. Turkmen SSR, Ser. Biol. Sci.] 5: 63 (in Russian).

Gorelov, Yu.K. and Yu.A. Orlov. 1965. On color forms in *Ptyas mucosus* L. Izv. AN TSSR, Ser. Biol. Nauk. [Proc. Acad. Sci. Turkmen SSR, Ser. Biol. Sci.] 4: 94–95 (in Russian).

Gorelov, Yu.K. and Ye.I. Shcherbina. 1971. The leopard in Turkmenia. Okhota i okhotnichye khozyaistvo [Hunting and Game Management] 2: 26–27 (in Russian).

Gorelova, T.G. and R.V. Kamelin. 1978. On the current condition of rare, endemic, and endangered plant species of Baghyz. Izv. AN TSSR, Ser. Biol. [Proc. Acad. Sci. Turkmen SSR, Ser. Biol.] 4: 26–33 (in Russian).

Granitov, I.I. 1964. Rastitel'nyi pokrov Yugo-Zapadnykh Kizyl-Kumov [Vegetative Cover of the Southwest Kizyl-Kum]. Tashkent, 1, 332 pp (in Russian).

Granitov, I.I. 1967. Rastitel'nyi pokrov Yugo-Zapadnykh Kizyl-Kumov [Vegetative Cover of the Southwest Kizyl-Kum]. Tashkent, 2, 417 pp (in Russian).

Gratsiansky, A.N. 1971. Priroda Sredizemnomorya [Nature in the Mediterranean]. Mysl', Moscow, 570 pp (in Russian).

Grekov, V.S. 1965. Wintering waterfowl in the Kizyl-Agach Reserve from 1955 to 1958. Geografiya resursov vodoplavayushchilkh ptits v SSSR [The geography of waterfowl rersources in the USSR] 1: 145-148 (in Russian).

Gringof, I.G. 1967. Pastbishchnye rasteniya Karakumov i pogoda [Pasture Vegetation of the Karakum Desert and Weather], Gidrometeoizdat, Moscow, 127 pp (in Russian).

Gromov, I.M. and V.A. Fokanov. 1961. On the history of formation of the fauna of the Mangyshlak and Emba deserts. Trudy Rostovskogo-na-Donu Protivoch. in-ta [Trans. Rostov-on-Don Anti-Plague Inst.] 18: 7-24 (in Russian).

Gromov, I.M., A.A. Gureev, G.A. Novikov., I.I. Sokolov, P.P. Strelkov, and K.K. Chapsky. 1963. Mlekopitayushchie Sovetskogo Soyuza [Mammals of the Soviet Union]. Izd. AN SSSR, Moscow-Leningrad 2: 642-2,000 (in Russian).

Grossgeim, A.A. 1926a. Contributions to the knowledge of vegetative formations of northwestern Persia. Zhurn. Russk. bot. obshch. [J. Russ. Bot. Soc.] 10(3-4): 251-278 (in Russian).

Grossgeim, A.A. 1926b. Flora Talysha [Flora of Talysh]. Narkomzem Azerb.SSR, Tiflis, 274 pp (in Russian).

Grossgeim, A.A. 1936. Analiz flory Kavkaza [Analysis of the Flora of the Caucasus. Izd. AN Azerb. SSR, Baku, 259 pp (in Russian).

Grubov, V.I. 1959. Opyt botaniko-geograficheskogo raionirovaniya Tsentral'noi Azii [An Attempt toward the Botanical-Geographical Regionalization of Central Asia]. Izd. AN SSSR, Leningrad, 78 pp (in Russian).

Grubov, V.I. 1966. Chenopodiaceae. Rasteniya Tsentralnoi Azii [Plants of Central Asia]. Nauka, Leningrad 2: 1-135 (in Russian).

Gudkova, Ye.P., E.M. Seifulin, and P.Ch. Chopanov. 1982. Conspectus of the flora of West Kopetdagh. In: N.T. Nechayeva (ed.), Priroda Zapadnogo Kopetdaga [Nature in West Kopetdagh]. Ylym, Ashkhabad: 38-119 (in Russian).

Gursky, A.V. 1932. Walnuts of West Kopetdagh. Trudy po prikl. bot., genet., and selektsii rastenii [Studies in Applied Botany, Genetics, and Selection of Plants] 8(1): 173-200 (in Russian).

Hennig, W. 1981. Insect Phylogeny. Chichester, New York etc., 541 pp.

Hölzel, H. 1969. Beitrag zur Systematik der Myrmeleoniden (Neuroptera - Planipennia, Myrmeleontidae). Ann. Naturhistor. Mus. Wien 73: 275-320.

Hölzel, H. 1972. Die Neuropteren Vorderasiens. IV. Myrmeleontidae. Beitr. Naturkund. Forsch. Südwestdeutsch. 37: 3-103.

Hsü, K.J. 1983. The Mediterranean was a Desert. Princeton Univ. Press, Princeton, 197 pp.

Ilyin, M.M. 1937. On the origin of the desert flora of Middle Asia. Sov. botanika [Soviet Botany] 6: 95-109 (in Russian).

Ilyin, M.M. 1946. Some results of study of deserts of Middle Asia. Materialy po istorii flory i rastitel'nosti SSSR [Materials on the History of the Flora and Vegetation of the USSR]. Moscow-Leningrad, Izd. AN SSSR 2: 197-256 (in Russian).

Ilyin, M.M. 1947. Floras of littorals and deserts and their interconnections. Sov. botanika [Soviet Botany] 15(5): 249-269 (in Russian).

Ilyin, M.M. 1950. The nature of a desert plant (eremophyte) from a viewpoint of plant-breeding knowledge about deserts. E.N. Pavlovsky and M. M. Ilyin (eds.) Pustyni SSSR i ikh osvoenie [Deserts of the USSR and Their Development]. Moscow-Leningrad, Izd. AN SSSR (in Russian).

Ilyin, M.M. 1958. Flora of the deserts of Central Asia, its origin and stages of development. Materialy po istorii flory i rastitel'nosti SSSR [Materials on the History of the Flora and Vegetation of the USSR]. Moscow- Leningrad, Izd. AN SSSR 3: 199-229 (in Russian).

Isakov, Yu.A. 1940. Ecology of waterfowl of the southern Caspian. In: Trudy Vsesoyuzn. ornitol. zapovednika Gasan-Kuli [Trans. All-Union Ornithol. Reserve Gasan-Kuli] 1: 160-313 (in Russian).

Isakov, Yu.A. 1948a. Notes on the biology of flamingo. In: Chaika-khokhotunya i flamingo na Kaspiiskom more [Seagulls and Flamingo on the Caspian Sea]. Moscow (in Russian).

Isakov, Yu.A. 1948b. On the distribution of flamingo in the USSR. In: Chaika-khokhotunya i flamingo na Kaspiiskom more [Seagulls and Flamingo on the Caspian Sea]. Moscow (in Russian).

Isakov, Yu.A. and K.A. Vorobyev. 1940. A survey of bird wintering and migration in the southern Caspian. Trudy Vsesoyuzn. ornitol. zapovednika Gasan-Kuli [Trans. All-Union Ornithol. Reserve Gasan-Kuli] 1: 5-159 (in Russian).

Ishadov, N. 1964. Distribution and biology of the bearded (bezoar) goat in Turkmenia. In: Tez. dokl. 1-i respubl. konf. molod. zoologov Turkmenistana [Abstr. Rep. First Republican Conf. Young Zoologists of Turkmenistan], Ashkhabad: 28-30 (in Russian).

Ishadov, N. 1965a. Ekologo-geograficheskie osobennosti nekotorykh kopytnykh Turkmenii [Ecological-geographical Features of Selected Ungulates of Turkmenia]. Autoref. diss. kand. biol. nauk [Abstr. Ph.D. Thesis], Ashkhabad (in Russian).

Ishadov, N. 1965b. Fauna of the ungulates of Maly Balkhan (West Turkmenia). Zool. zhurn. [Zool. Journ.] 40(2): 1739-1740 (in Russian).

Ishadov, N. 1972. Distribution and current level of density of ungulates in Northwest Turkmenia. In: D.I. Bibikov et al. (ed.), Teriologiya [Theriology]. Nauka, Novosibirsk 1: 417-419 (in Russian).

Ishadov, N. 1973a. Hyena in the Central Karakum. Okhota i okhotnichye khozyaistvo [Hunting and Game Management] 7: 21 (in Russian).

Ishadov, N. 1973b. New data on otter distribution in Turkmenia. Okhota i okhotnichye khozyaistvo [Hunting and Game Management]: 1 (in Russian).

Ishadov, N. 1979. Effect of a harsh and snowy winter on birds in Turkmenistan. In: A.K. Rustamov (ed.), Okhrana prirody Turkmenistana [Natural Conservation in Turkmenistan]. Ylym, Ashkhabad: 103-111 (in Russian).

Ishadov, N. and G.I. Ishunin. 1975. Changes in distribution and habitats of the otter in the Amudarya Basin due to the anthropogenic factor. In: Aktual'nye voprosy zoogeografii. VI Vsesoyuznaya zoogeogr. konfer. [Current Problems of Zoogeography: 6th All-Union Zoogeographical Confer.], Kishinev: 104-105 (in Russian).

Ishadov, N. and G.I. Ishunin. 1976. Taxonomic position, range, density, and ecology of *Lutra lutra seistanica* Birula, 1912. Biologiya zhivotnykh i rastenii Turkmenistana/Trudy Turkm. gos. un-ta [Biology of Animals and Plants in Turkmenistan/Trans. Turkmen State Univ.] 3: 34-47 (in Russian).

Ishadov, N. and Ye.A. Klyushkin. 1975. Ungulates of Kugitang and the adjacent plain. In: Kopytnye fauny SSSR: ekologiya, morfologiya, ispol'zovaniye, okhrana/Tez. dokl. Vsesoyuzn. soveshch. po kopytnym zhivotnym [Ungulates of the USSR Fauna: Ecology, Morphology, Exploitation, and Conservation/ Abstr. All-Union Confer. on Ungulate Animals], Moscow: 98-99 (in Russian).

Ishadov, N. and Ye.A. Klyushkin. 1978. Wild game animals of Kugitang and the adjacent plain (Turkmenia). Voprosy biologiyi zhivotnykh i rastenii Turkmenistana/Trudy Turkm. gos. un-ta [Problems of Biology of Animals and Plants in Turkmenistan/Trans. Turkmen State Univ.] 4: 114-126 (in Russian).

Ishadov, N. and Ye.A. Klyushkin. 1979. On some carnivores of the Karakum Desert. In: V.E. Sokolov (ed.), Ekologicheskie osnovy okhrany i ratsional'nogo ispol'zovaniya khishchnykh mlekopitayushchikh [Ecological Principles of Conservation and Rational Exploitation of Carnivorous Mammals]. Nauka, Moscow: 41-43 (in Russian).

Ishunin, G.I. 1961. Fauna Uzbekskoi SSR. Mlekopitayushchie (khishchnye i kopytnye) [Fauna of the Uzbek SSR. Mammals: Ungulates and Carnivores], Izd. AN Uzb. SSR, Tashkent, 231 pp (in Russian).

Ishunin, G.I. and G.F. Tetyukhin. 1989. Veroyatnye puti obrazovaniya fauny mlekopitayushchikh na territorii Uzbekistana [Possible Directions of Mammalian Fauna Formation in the Territory of Uzbekistan]. Fan, Tashkent, 80 pp (in Russian).

Ivanov, A.P. 1972. Fizicheskiye osnovy deflyatsii peskov pustyn [Physical Principles of the Deflation of Desert Sands]. Ylym, Ashkhabad. 117 pp (in Russian).

Izzatulaev, Z.I., and Ya.I. Starobogatov. 1984. Genus *Melanopsis* (Gastropoda, Pectinibranchia) and its representatives in the water bodies of the USSR. Zool. zhurn. [Zool. J.] 63(10): 1471-1483 (in Russian).

Izzatulaev, Z.I. and Ya.I. Starobogatov. 1985. Zoogeographical characteristics of freshwater mollusks of Middle Asia and the problem of the existence of the Mountainous Asian subregion of the Palearctic. Zool. zhurn. [Zool. J.] 64(4): 506–517 (in Russian).

Jezek, J. 1981. Results of the Czechoslovak-Iranian entomological expeditions to Iran, 1970, 1973, and 1977. Larvae and pupae of two Persian *Tabanus* species. Acta Entomol. Mus. Nat. Pragae 40: 45–55.

Jezek, J. 1990. Results of the Czechoslovak-Iranian entomological expeditions to Iran, 1970, 1973, and 1977. Thee new species of horse flies (Diptera, Tabanidae) from the Palearctic region. Acta Entomol. Mus. Nat. Pragae 43: 119–127.

Johnson, O.W. 1968. Some morphological features of avian kidneys. Auk 85: 218–228.

Johnson, O.W. and J.M. Mugaas. 1970. Some histological features of avian renal medulla. Condor 72: 288–292.

Johnson, O.W., Phipps, G.L., and J.M. Mugaas. 1972. S Injection studies of cortical and medullary organization in the avian kidney. J. Morph. 136(2): 181–190.

Johnson, O.W. and E. Skadhauge. 1975. Structural-functional correlations in the kidneys and observations of color and cloacal morphology in certain Australian birds. J. Anat. 120(3): 195–505.

Kalugin, P.I. 1977. Yuzhny Kopetdagh [South Kopetdagh]. Ylym, Ashkhabad, 216 pp (in Russian).

Kalyonov, G.S. 1977. Ecologo-geographical analysis of the distribution of *Tortula desertorum* Broth. in the deserts of Middle Asia. Bot. Zhurnal [Bot. Journal] 62: 1015–1022 (in Russian).

Kamakhina, G.L. 1986. Species composition of natural pastures of the Kurtusu-Gaudan floristic area. Izv. AN TSSR, Ser. Biol. Nauk. [Proc. Acad. Sci. Turkmen SSR, Ser. Biol. Sci.] 6: 7–14 (in Russian).

Kamakhina, G.L. 1987. Monitoring of the dynamics of the Central Kopetdagh flora. Izv. AN TSSR, Ser. Biol. Nauk. [Proc. Acad. Sci. Turkmen SSR, Ser. Biol. Sci.] 6: 10–16 (in Russian).

Kamakhina, G.L. 1989. Flora of the Kuruhaudan area and its features. Izv. AN TSSR, Ser. Biol. Nauk. [Proc. Acad. Sci. Turkmen SSR, Ser. Biol. Sci.] 2: 21–28 (in Russian).

Kamakhina, G.L. 1991. Rare species in the Central Kopetdagh flora and their selected ecological-biological features. Izv. AN TSSR, Ser. Biol. Nauk. [Proc. Acad. Sci. Turkmen SSR, Ser. Biol. Sci.] 6: 3–9 (in Russian).

Kamalova, Z.Ya. 1969. On the ecology of the Caucasian agama in Southwest Kopetdagh. Uzbek. Biol. zhurnal [Uzbek Biol. Journal] 4: 59–62 (in Russian).

Kamelin, R.V. 1965. On the generic endemism of Middle Asian flora. Bot. zhurnal [Bot. J.] 50(12): 1702–1710 (in Russian).

Kamelin, R.V. 1965. On some outstanding anomalies in the flora of the Mountainous Middle Asian Province. Bot. zhurn. [Bot. J.] 52(4): 447–460 (in Russian).

Kamelin, R.V. 1970. Botanical-geographical features of the flora of the Soviet portion of Kopetdagh. Bot. zhurn. [Bot. J.] 55(10): 1451–1463 (in Russian).

Kamelin, R.V. 1971. Flora reki Varzob i ee analiz [Flora of the Varzob River and its Analysis]. Manuscript. Leningrad, 790 pp (in Russian).

Kamelin, R.V. 1973. Florogeneticheskyi analiz yestestvennoi flory gornoi Srednei Azii [Florogenetic Analysis of the Natural Flora of Mountainous Middle Asia]. Nauka, Leningrad, 356 pp (in Russian).

Kamelin, R.V. 1979. Kukhistanskii okrug gornoi Srednei Azii [The Kuhistan District of Mountainous Middle Asia]. Nauka, Leningrad, 117 pp (in Russian).

Kamelin, R.V. 1990a. Flora Syrdaryinskogo Karatau [Flora of Syrdarya Karatau]. Nauka, Leningrad, 117 pp (in Russian).

Kamelin, R.V. (ed.). 1990b. Fistashka v Badkhyze [The Pistachio in Badghyz]. Nauka, Leningrad, 231 pp (in Russian).

Kamelin, R.V., V.P. Bochantsev, and T.G. Gorelova. 1989. Vascular plants. In: R.V. Kamelin and L.E. Rodin (eds.), Fistashniki Badkhyza [Pistachio Woodlands of Badghyz]. Nauka, Leningrad: 7–21 (in Russian).

Kamelin, R.V. and F.O. Khasanov. 1987. The altitudinal zonation of vegetation of the Kugitang Range (southwestern Pamiro-Alai). Bot. zhurn. [Bot. J.] 72 (in Russian).

Kamelin, R.V. and Dz. Kurbanov. 1985. Endangered and rare plants of Northwest Kopetdagh (Kyurendagh). Bot. zhurn. [Bot. Journal] 70(3): 409–418 (in Russian).

Kamelin, R.V. and Dz. Kurbanov. 1987. On some endangered and rare plants of the western hills of Turkmenistan. Bot. zhurn. [Bot. J.] 72(3): 397–402 (in Russian).

Kamelin, R.V. and L.E. Rodin (eds.). 1989. Fistashniki Badkhyza [Pistachio Woodlands of Badghyz]. Nauka, Leningrad, 245 pp (in Russian).

Kamelin, R.V. and N.I. Zabelina. 1987. *Periploca graeca* (Asclepiadaceae), a new genus and species in the flora of Middle Asia. Bot. zhurn. [Bot. J.] 72(4): 528–529(in Russian).

Kamilov, G.K. 1973. Ryby i biologicheskie osnovy rybokhozyaistvennogo osvoyeniya vodokhranilishch Uzbekistana [Fishes and Biological Principles of the Fishing Industry in the Reservoirs of Uzbekistan]. Fan, Tashkent, 220 pp (in Russian).

Kamnev, P.I., I.V. Zhernovov, and G.N. Skvortsov. 1962. A new record of mouselike dormouse in West Kopetdagh. Zool. zhurn. [Zool. J.] 11(2): 297 (in Russian).

Kaplin, V.G. 1975. The seasonal dynamics of spider communities in the crowns of trees and shrubs in East Karakum. Probl. Osv. Pustyn. [Probl. Desert Development]: 3 (in Russian).

Kaplin, V. G. 1978. The complexes of soil invertebrates in sand deserts of the southern subzone (with East Karakum as an example). Ashkhabad, Ylym, 161 pp (in Russian).

Kara-Murza, E.I., S.N. Kolyadny, and N.I. Forsh. 1955. Flora of the red-painted deposits of Cheleken. Dokl. Akad. Nauk. SSSR [Reports Acad. Sci. USSR] 102(1): 137–139 (in Russian).

Karavaev, A.A. 1979a. Data on the nesting fauna of water and swamp birds of the lower Atrek River. In: Prirodnaya sreda i ptitsy poberezhii Kaspiiskogo morya i prilezhashchikh nizmennostei/Trudy Kizyl-Agachskogo gos. zapovednika [Environment and Birds of the Caspian Sea Shores and Adjacent Lowlands/Trans. Kizyl-Agach State Reserve]. Azgosizdat, Baku 1: 62–82 (in Russian).

Karavaev, A.A. 1979b. On a record of black stork on the eastern shore of the Caspian. In: Prirodnaya sreda i ptitsy poberezhii Kaspiiskogo morya i prilezhashchikh nizmennostei/Trudy Kizyl-Agachskogo gos. zapovednika [Environment and Birds of the Caspian Sea Shores and Adjacent Lowlands/Trans. Kizyl-Agach State Reserve]. Azgosizdat, Baku 1: 248–249 (in Russian).

Karavaev, A.A. 1981. On rare birds of water and swamp complexes in the lower Atrek River and the adjacent shore of the Caspian. In: Tez. dokl. 2 nauchn. konfer, po okhrane prirody Turkm. SSR [Abstr. 2nd Sci. Conf. on Natural Conservation in the Turkmen SSR], Ashkhabad: 108–110 (in Russian).

Kartashov, N.N. 1955. Data on amphibians and reptiles of Southwest Turkmenia. Uch. zap. MGU, Biologiya [Sci. Mem. Moscow State Univ., Biology] 171: 173–202 (in Russian).

Kartashov, N.N. and A.N. Soldatova. 1953. A new record of the Turkmen jerboa (*Jaculus turkmenicus*) in Turkmenia. Byull. MOIP, Otd. Biol. [Bull. Moscow Soc. Natur., Div. Biol.] 58(1): 11–12 (in Russian).

Kashkarov, D.N. 1925. Additions to the knowledge of rodents in Turkmenistan. Trudy Turkestan. nauchn. ob-va Sredneaz. un-ta [Trans. Turkestan Sci. Soc. of Middle Asian State Univ.] 2: 43–47 (in Russian).

Kashkarov, D.N. 1932. Zhivotnye Turkestana [The Animals of Turkestan]. Uzgiz, Tashkent 1, 62 pp (in Russian).

Kashkarov, D.N. and V.P. Kurbatov. 1929. Ekologicheskii obzor fauny pozvonochnykh Tsentralnykh Karakumov/Trudy Srednaz. gos. un-ta, Ser. 12a, Geografiya [Ecological Survey of the Vertebrate Fauna of the Central Karakum/Trans. Middle Asian State Univ., Ser. 12a, Geography] 7: 1–68 (in Russian).

Kes', A.S., B.V. Andrianov, and M.A. Itina. 1980. Dynamics of hydrographic network and fluctuations in the level of the Aral Sea. In: Kolebaniya uvlazhnennosti Aralo-Kaspiiskogo regiona v golotsene [Fluctuations of Water Regime in the Aralo-Caspiuan Region in the Holocene]. Nauka, Moscow: 185–197 (in Russian).

Khakyev, A. 1972. Some data on game animals in the tugais of the Amudarya, their exploitation and conservation. In: Rustamov, A.K. (ed.), Okhrana prirody v Turkmenii [Natural Conservation in Turkmenia]. Ylym, Ashkhabad: 58–63 (in Russian).

Khanmamedov, A.I. 1956. Francolin in Azerbaijan, its distribution, ecology and practical importance. Trudy In-ta zool. AN Az. SSR [Trans. Inst. Zool. Acad. Sci. Azerbaijan SSR], Baku 19: 97–182 (in Russian).
Khanmamedov, A.I. 1966. Distribution of gallinaceous birds in Azerbaijan. In: Chetvertaya mezhvuzovskaya zoogeograficheskaya konferentsiya [4th Intercollegial Zoogeographical Conf.], Odessa: 296–297 (in Russian).
Kharin, N.G. 1975. Distantsionnye metody izucheniya rastitel'nosti [Remote Sense Techniques of Vegetation Studies]. Nauka, Moscow, 132 pp (in Russian).
Kharin, N.G. et al. 1988. Explanatory note to the map of man-made desertification of USSR arid lands, scale 1: 2,500,000. Ylym, Ashkhabad, 21 pp.
Kharin, N.G. et al. 1990. Aridnyi tsentr sbora i obnrabotki kosmicheskoi informatsii [The Arid Center of Collection and Processing of the Satellite Remote Sensing Information]. Ylym, Ashkhabad, 112 pp (in Russian).
Kharin, N.G., N.T. Nechaeva, V.N. Nikolaev et al. 1983. Metodicheskie osnovy izucheniya i kartografirovaniya protsessov opustynivaniya (Na primere aridnykh territorii Turkmenistana) [Methodical Principles of the Study and Mapping of Desertification Processes: Arid Territories of Turkmenistan as an Example]. Ylym, Ashkhabad, 97 pp (in Russian).
Kharin, N.G., V.N. Nikolaev, I.P. Sorokina, and A.A. Kiriltseva. 1989. On the possibility of modeling of desertification processes involving anthropogenic factors. Problemy osvoeniya pustyn [Problems of Desert Development] 1: 64–67 (in Russian).
Kharitonov, D.Ye. 1932. Katalog der Russischen Spinnen. Yezheg. Zool. Muz. [Ann. Mus. Zool.] (Leningrad) 32 (Beilage), 206 pp.
Kharitonov, D.Ye. 1936. Nachtrag zum Katalog der Russischen Spinnen. Uchen. Zap. Permsk. Gos. un-ta [Sci. Mem. Perm. State Univ.] 2(1): 167–225 (in Russian).
Kharitonov, D.Ye. 1946. New forms of spiders in the USSR fauna. Izv. Estestvennonauchn. in-ta Molotovsk. gos. un-ta [Proc. Nat. Sci. Inst. Molotov State Univ.] 12: 19–32 (in Russian).
Kharitonov, D.Ye. 1948. Spiders – Araneina. Zhivotnyi mir SSSR (zona pustyn') [The Animal Life of the USSR (desert zone)] 2: 297–304 (in Russian).
Kharitonov, D.Ye. 1954. A new representative of the genus *Latrodectus* from Turkmenia (*Latrodectus pallidus* O.P. Cambridge subsp. *pavlovskii* n.). Zool. zhurn. [Zool. Journal] 33: 480–485 (in Russian).
Kharitonov, D.Ye. 1955. A rare representative of spiders of the genus *Latrodectus* Walck. (Araneae) from Turkmenia. Trudy Zool. in-ta AN SSSR [Proc. Zool. Inst. Acad. Sci. USSR] 18: 243–247 (in Russian).
Kharitonov, D.Ye. 1956. A review of the spider family Dysderidae in the USSR fauna. Uchen. zapiski Molotovsk. gos. un-ta [Sci. Mem. Molotov State Univ.] 10: 17–39 (in Russian).
Kharitonov, D.Ye. 1969. Contributions to the fauna of spiders of the USSR. Uchen. zapiski Permsk. un-ta [Sci. Mem. Perm Univ., Biol.] 179: 59–133 (in Russian).
Khomustenko, Yu.D. 1981. New data on the herpetofauna of Central Kopetdagh. In: I.S. Darevsky (ed.), Voprosy gerpetologii [Problems of Herpetology]. Nauka, Leningrad: 142 (in Russian).
Khomustenko, Yu.D. 1982. On distribution of *Agkistrodon halys caucasica* in Central Kopetdagh. Izv. AN TSSR, ser. biol. nauk [Proc. Acad. Sci. Turkmen SSR, Ser. Biol. Sci.] 1: 64–65 (in Russian).
Khomustenko, Yu.D. and Ch. Ataev. 1979. On a finding of the Azerbaijan lizard (*Lacerta raddei* Boett.) in Turkmenistan. Izv. AN TSSR, Ser. Biol. Nauk. [Proc. Acad. Sci. Turkmen SSR, Ser. Biol. Sci.] 6: 72–73 (in Russian).
King, H.H. 1910. Some observations on the bionomics of *Tabanus ditaeniatus* Mac. and *Tabanus kingi* Aust. Bull. Entomol. Res. 1: 265–274.
King, J.R. and D.S. Farner. 1961. Energy metabolism, thermoregulation, and body temperature. In: A.L. Marshall (ed.). Biology and Comparative Physiology of Birds. Acad. Press, New York 2: 215–268.
Kinzelbach, R. 1985. Skorpione (Arachnida: Scorpiones). In: Tübinger Atlas der Vorderen Orients (TAVO). Wiesbaden, L. Reichert, A VI 14.2: I–IV.

Kirkley, J.S. and J.A. Lessaman. 1990. Water economy of nesting Swainson's hawks. Condor 92(1): 29-44.
Kirikov, S.V. 1959. Izmeneniya zhivotnogo mira v prirodnykh zonakh SSSR (XIII-XIX vv.). Stepaya zona i lesostep'. [Changes in the Animal Life in Natural Zones of the USSR. Steppe and Forest-Steppe Zones]. Moscow, Izd. AN SSSR, 175 pp (in Russian).
Kirikov, S.V. 1960. Izmeneniya zhivotnogo mira v prirodnykh zonakh SSSR (XIII-XIX vv.). Lesnaya zona i lesotundra. [Changes in the Animal Life in Natural Zones of the USSR. Forest and Forest-Tundra Zones]. Moscow, Izd. AN SSSR, 157 pp (in Russian).
Kleopov, Yu.D. 1938. A project of classification of geographical elements for the analysis of the Ukr. SSR flora. Zhurn. Bot. inst. AN USSR [J. Bot. Inst. Acad,. Sci. Ukrain. SSR], Kiev 17(25): 209-219 (in Russian).
Klyushkin, Ye.A. 1954. Notes on the tugai deer. Izv. AN TSSR, Ser. Biol. Nauk. [Proc. Acad. Sci. Turkmen SSR, Ser. Biol. Sci.] 2: 72-74 (in Russian).
Klychev, A. 1928. [no title] Okhotnik [The Hunter]: 7 (in Russian).
Knyazev, A.V. 1976. On the Holocene history of biogeocoenoses of Badghyz. In: L.G. Dinesman (ed.), Istoriya biogeotsenozov SSSR v golotsene [History of the USSR Biogeocoenoses in the Holocene]. Nauka, Moscow: 229-244 (in Russian).
Koch, L. 1878. Kaukasische Arachniden. In: O. Schneider (ed.), Naturwissen-schaftliche Beiträge zur Kenntnis der Kaukasuslande. Verl. Burdach. Hofbuch, Dresden 3: 36-71.
Kogai, N.A. 1957. Some aspects of the geomorphological analysis of the Sultan-uiz-dagh range and adjacent lowlands. Trudy SAGU [Trans. Middle Asian State Univ.], Tashkent 99(10): 78-102 (in Russian).
Kogan, S.I. 1954. Vegetation of southern Ustyurt. Trudy In-ta biol. AN TSSR [Trans. Inst. Biol. Acad. Sci. Turkmen SSR] 2: 45-115 (in Russian).
Kogan, S.I., Yu.E. Lyubeznov, Kh.S. Sadykov et al. 1985. Vodoyomy Yuzhnoi Turkmenii [The Water Bodies of the Southern Turkmenia]. Ylym, Ashkhabad, 222 pp (in Russian).
Koksharova, N.Ye. 1970. Forests of Turkmenia. In: A.B. Zhukov (ed.), Lesa SSSR [Forests of the USSR]. Nauka, Moscow 5: 184-246 (in Russian).
Kolesnikov, I.I. 1956. Fauna of the terrestrial vertebrates of Kyurendagh. Trudy Sredneaz. gos. un-ta [Trans. Middle Asian State Univ] 86: 151-214 (in Russian).
Kolesnikov, I.I. and K. Ashirov. 1953. On new records of the Turkmen jerboa (*Jaculus turkmenicus*) in Turkmenia. Dokl. AN Uz. SSR [Rep. Acad. Sci. Uzbek SSR]: 11 (in Russian).
Kolodenko, A.I. and O.N. Nurgeldyev. 1977. On death of wild animals on highways. In: Voprosy ekologii mlekopitayushchikh i presmykayushchikhsya Turkmenistana [Problems of Ecology of Mammals and Reptiles of Turkmenistan]. Ylym, Ashkhabad: 136-145 (in Russian).
Kornilova, V.S. 1960. Nizhnepliotsenovaya flora Kushuka (Turgaiskii progib) [The Lower Pliocene Flora of Kushuk (Turgai Deflection)]. Izd. AN Kaz. SSR, Alma-Ata, 112 pp (in Russian).
Kornilova, V.S. 1966. A survey of the history of flora and vegetation of Kazakhstan. In: B.A. Bykov (ed.), Rastitel'nyi pokrov Kazakhstana [Vegetative Cover of Kazakhstan]. Nauka, Alma-Ata 1: 37-190 (in Russian).
Korovin, Ye.P. 1927. Basic features of the structure of vegetative cover of mountainous and foothill portions of Kopetdagh. Izv. In-ta pochvoved. i geobot. SAGU [Proc. Inst. Soil Sci. and Geobotany of Middle Asian State Univ.], Tashkent 3: 71-128 (in Russian).
Korovin, Ye.P. 1934a. On paleoecological changes in Middle Asia. In: Voprosy ekologii i biotsenologii [Problems of Ecology and Biocoenology]. Izd. AN SSSR, Moscow-Leningrad: 16-42 (in Russian).
Korovin, Ye.P. 1934b. Rastitelnost' Srednei Azii i Yuzhnogo Kazakhstana [Vegetation of Middle Asia and South Kazakhstan]. SAGIZ, Tashkent, 480 pp (in Russian).
Korovin, Ye.P. 1935. Studies on the development of the vegetation of Middle Asia. Byull. SAGU [Bull. Middle Asian State Univ.] 20(4): 183-218 (in Russian).
Korovin, Ye.P. 1958. A historical outline of the development of the vegetation of Middle Asia. In: Srednyaya Asia [Middle Asia]. Izd. AN SSSR, Moscow: 277-294 (in Russian).
Korovin, Ye.P. 1961. Rastitelnost' Srednei Azii i Yuzhnogo Kazakhstana [Vegetation of Middle Asia and South Kazakhstan], Tashkent, Vol. 1, 452 pp; Vol. 2, 549 pp (in Russian).

Korovin, Ye.P. and I.I. Granitov. 1949. Vegetative cover. In: Ust'-Urt (Kara-Kalpaksky), yego priroda i khozyaistvo [Kara-Kalpak Ustyurt: Its Nature and Industry]. Izd.ANUzb.SSR, Tashkent, 229 pp (in Russian).

Korovin, Ye.P. and B.A. Mironov. 1935. Major plant associations of the eastern Bet-Pak-Dala and their distribution depending on the relief and soils. Trudy SAGU [Trans. Central Asian State Univ., Tashkent], Ser. 8v (Botany) 21: 62–87 (in Russian).

Korshunov, V.M. 1983. The current condition of rare ungulate species of Central Kopetdagh. In: Biologicheskie osnovy ispolzovaniya i okhrana dikikh zhivotnykh [Biological Principles of Exploitation and Conservation of Wild Animals]. Moscow Veterinary Academy, Moscow: 26–33 (in Russian).

Korshunov, V.M. 1986. Density and habitat distribution of the Turkmen mountain sheep and Turkmen bearded goat in Central Kopetdagh. In: Tez. dokl. IV Syezda Vsesoyuzn. teriol. obshch. [Abstr. Rep. 4th Congr. All-Union Theriol. Soc.], Moscow 1: 249–250 (in Russian).

Korshunov, V.M. 1988. Sravnitelnaya ekologiya turkmenskogo gornogo barana i turkmenskogo borodatogo kozla [Comparative Ecology of the Turkmen Mountain Sheep and Turkmen Bearded Goat]. Autoref. diss. kand. biol. nauk [Abstr. Ph.D. Thesis], Moscow (in Russian).

Korzhinsky, S.I. 1886. Studies on the vegetation of Turkestan. Zap. Imper. Akad. nauk [Mem. Imp. Acad. Sci. St.-Petersb.], Ser. 8, 4(4): 1–74 (in Russian).

Kozlova, Ye.V. Dispersal of the pheasant, *Phasianus colchicus* L., to the deserts of Central Asia. Ornitologicheskii sbornik k 100-letiyu so dnya rozhdeniya P. P. Sushkina/Trudy Zool. in-ta AN SSSR [Ornithological Papers Honoring the Centennial of P. P. Sushkin/Trans. Zool. Inst. Acad. Sci. USSR] 163: 71–76 (in Russian).

Krasheninnikov, I.M. 1921. On the systematics of the genus *Artemisia*. Bot. mat. [Botanical Materials], Leningrad 2(45): 177–191 (in Russian).

Krasheninnikov, I.M. 1946. An attempt toward the phylogenetic analysis of some Eurasian groups of the genus *Artemisia* as related to the features of Eurasian paleogeography. Materialy po istorii flory i rastitel'nosti SSSR [Materials on the History of the Flora and Vegetation of the USSR]. Moscow – Leningrad, Izd. AN SSSR 2: 87–196 (in Russian).

Krasnaya Kniga Kazakhskoi SSR. Pozvonochnye zhivotnye. 1978. [Red Data Book of the Kazakh SSR. Vertebrates], Alma-Ata, Kainar, 1, 204 pp (in Russian).

Krasnaya Kniga SSSR: Redkie i nakhodyashchiesya pod ugrozoi ischeznoveniya vidy zhivotnykh i rastenii [Red Data Book of the USSR: Rare and Endangered Species of Animals and Plants]. 1978. Lesnaya promyshlennost', Moscow, Vol. 1, 459 pp (in Russian).

Krasnaya Kniga SSSR: Redkie i nakhodyashchiesya pod ugrozoi ischeznoveniya vidy zhivotnykh i rastenii [Red Data Book of the USSR: Rare and Endangered Species of Animals and Plants]. 1984. Lesnaya promyshlennost', Moscow, Vol. 2, 478 pp (in Russian).

Krasovskaya, L.S. and I.G. Levichev. 1986. Flora Chatkal'skogo zapovednika [Flora of the Chatkal Reserve]. Fan, Tashkent, 171 pp (in Russian).

Kreitsberg, A.V.-A. and R.A. Danov. 1985. The fauna of papilionids (Lepidoptera, Papilionidae) of Turkmenistan. In: N.T. Nechayeva (ed.), Rastitel'nost' i zhivotnyi mir Zapadnogo Kopetdaga [The Vegetation and Animal Life of West Kopetdagh]. Ylym, Ashkhabad: 249–261 (in Russian).

Krishtofovich, A.N. 1936a. Development of botanical-geographical provinces in the Northern Hemisphere since the end of the Cretaceous Period. Sov. botanika [Soviet Botany] 3: 9–24 (in Russian).

Krishtofovich, A.N. 1936b. Major directions of the development of the flora of Asia. Uchen. zap. LGU [Sci. Mem. Leningrad State Univ.] 2(9): 95113 (in Russian).

Krivokhatsky, V.A. 1981. Seasonal and daily dynamics of abundance of invertebrates in the burrows of gerbils, *Rhombomys opimus* and *Meriones meridionalis* in the East Karakum. Izv. AN TSSR, Ser. Biol. Nauk. [Proc. Acad. Sci. Turkmen SSR, Ser. Biol. Sci.] 4: 32–39 (in Russian).

Krivokhatsky, V.A. 1982a. Connections of arthropods from the burrows of *Rhombomys opimus* to the surrounding ecosystems. Entom. obozr. [Entomol. Rev.] 61(4) : 779–785 (in Russian).

Krivokhatsky, V.A. 1982b. Arthropods from the burrows of gerbils in East Karakum). Ekologiya [Ecology] 2: 60–64 (in Russian).

Krivokhatsky, V.A. 1982c. Fleas from the rodent burrows in the Repetek Reserve. Izv. AN TSSR, Ser. Biol. Nauk. [Proc. Acad. Sci. Turkmen SSR, Ser. Biol. Sci.] 4: 80–83 (in Russian).
Krivokhatsky, V.A. 1983. A case study of *Rhombomys opimus* burrow consortia. Problemy Osvoeniya Pustyn' [Problems of Desert Development] 1: 59–68 (in Russian).
Krivokhatsky, V.A. 1984. Seasonal changes in the distribution of fleas in external burrow passages as an index of their migrational activity. Parazitologiya [Parasitology] 43(2): 150–153 (in Russian).
Krivokhatsky, V.A. 1985a. Insects of gerbil burrow consortia in the deserts of Central Asia. Abstr. Ph.D. Thesis. Leningrad, 20 pp (in Russian).
Krivokhatsky, V.A. 1985b. The results of monitoring of *Rhombomys opimus* burrow consortia in the Repetek Biosphere Reserve. Izv. AN TSSR, Ser. Biol. Nauk. [Proc. Acad. Sci. Turkmen SSR, Ser. Biol. Sci.] 1: 27–32 (in Russian).
Krivokhatsky, V.A. 1985c. On the historical formation of burrow entomofauna in the sand deserts. Entomol. obozr. [Entomol. Rev.] 64(4): 696–704 (in Russian).
Krivokhatsky, V.A. 1987. On the optimal density of predators and prey in the arthropod communities of *Rhombomys opimus* burrows. Vestn. LGU [Mess. Leningrad State Univ.], Ser. 3, 1(3): 95–98 (in Russian).
Krivokhatsky, V.A. 1989. Studies on the inhabitants of mammalian burrows in the USSR. Vestn. LGU [Mess. Leningrad State Univ.], Ser. 3, 4(24): 13–18 (in Russian).
Krivokhatsky, V.A. 1990a. A new species of an ant-lion (Neuroptera, Myrmeleontidae) from Middle Asia. In: Novosti faunistiki i sistematiki [News in Faunistics and Systematics]. Kiev: 61–63 (in Russian).
Krivokhatsky, V.A. 1990b. A revision of ant-lions of the genus *Lopezus* Navás, 1913 (Neuroptera, Myrmeleontidae). Entomol. obozr. [Entomol. Rev.] 69(4): 893–904 (in Russian).
Krivokhatsky, V.A. 1991. A new species of the ant-lion of the genus *Myrmeleon* (Neuroptera, Myrmeleontidae) from Turkmenia. Zool. zhurn. [Zool. J.] 70(2): 147–149 (in Russian).
Krivokhatsky, V.A. 1992a. New taxa of the Asian ant-lions (Neuroptera, Myrmeleontidae). Entomol. obozr. [Entomol. Rev.] 71(2): 405–413 (in Russian).
Krivokhatsky, V.A. 1992b. A new ant-lion from Turkmenia, Middle Asia (Insecta, (Neuroptera, Myrmeleontidae). Reichenbachia 29(14): 77–80.
Krivokhatsky, V.A. and V.Ya. Fet. 1981. The spider distribution in Badghyz during the springtime. Izv. AN TSSR, Ser. Biol. Nauk. [Proc. Acad. Sci. Turkmen SSR, Ser. Biol. Sci.] 1: 45–51 (in Russian).
Krivokhatsky, V.A. and V.Ya. Fet. 1982. The spiders (Aranei) from rodent burrows in Eastern Karakum. Problemy Osvoeniya Pustyn' [Problems of Desert Development] 4: 68–75 (in Russian).
Krivokhatsky, V.A. and V.A. Kashcheev. 1983. Staphylinidae (Coleoptera) from rodent burrows and other habitats of the Repetek Reserve. Izv. AN TSSR, Ser. Biol. Nauk. [Proc. Acad. Sci. Turkmen SSR, Ser. Biol. Sci.] 3: 26–31 (in Russian).
Krivosheev, V.G. 1956. Transcaspian desert sparrow in Kizylkum. Bull. MOIP. Otd. biol. [Bull. Moscow Soc. Natur., Div. Biol.] 61(4) (in Russian).
Krivosheev, V.G. 1961. The Turkmen jerboa (*Jaculus turkmenicus*) in West Kizylkum. Bull. MOIP. Otd. biol. [Bull. Moscow Soc. Natur., Div. Biol.] 66(6): 145–146 (in Russian).
Kroneberg, A. 1875. Spiders. Fedchenko's expedition to Turkestan. Izv. Imper. ob-va lyubit. estestvozn., antropologii i etnografii [Proc. Imp. Soc. Amat. Nat. Hist. Anthropol. Ethnogr.] 29(3): 1–58 (in Russian).
Kryzhanovsky, O.L. 1961. On the zoogeographical features of coleopterous fauna of the deserts of Turkmen SSR. Beitr. zur Entomol. 11(3): 426–445.
Kryzhanovsky, O.L. 1962. The revision of the genus *Chilotomus* Chaud. (Coleoptera, Carabidae) and its geographical distribution. Zool. zhurn. [Zool. J.] 41(4): 539–544 (in Russian).
Kryzhanovsky, O.L. 1965. Sostav i proiskhozhdenie nazemnoi fauny Srednei Azii (glavnym obrazom na materiale po zhestkokrylym [The Composition and Origin of the Terrestraial Fauna of Middle Asia (Based Primarily on Coleopteran Material)]. Moscow-Leningrad. 420 pp (in Russian) (English translation: New Delhi, India, 1980).

Kryzhanovsky, O.L. 1976. Results of the Czechoslovak-Iranian entomological expedition to Iran, 1970. No. 11. Coleoptera: Histeridae. Acta entom. Mus. Nation. Prague, Suppl. 6: 109–118.
Kryzhanovsky, O.L. 1980a. On the fauna of Histeridae (Coleoptera) of Afghanistan. Entomol. obozr. [Entomol. Rev.] 59(1): 135–147 (in Russian).
Kryzhanovsky, O.L. 1980b. On the scope and zoogeographical division of the Paleotropic dominion. In: A.G. Voronov and N.N. Drozdov (eds.), Sovremennye problemy zoogeografii [Current Problems in Zoogeography]. Nauka, Moscow: 61–81 (in Russian).
Kryzhanovsky, O.L. 1985. The fauna of histerid beetles (Coleoptera, Histeridae) of Middle Asia and its zoogeographical features. Zool. zhurn. [Zool. J.] 64(8): 1179–1190 (in Russian).
Kryzhanovsky, O.L. and Kh.I. Atamuradov. 1987. Contributions to the fauna and ecology of ground beetles of Badghyz (Coleoptera, Carabidae). Izv. AN TSSR, Ser. Biol. Nauk. [Proc. Acad. Sci. Turkmen SSR, Ser. Biol. Sci.] 5 : 26–36 (in Russian).
Kryzhanovsky, O.L. and Kh.I. Atamuradov. 1989. Review of the fauna of ground beetles (Coleoptera, Carabidae) of West Kopetdagh and its zoogeographical features. I–II. Izv. AN TSSR, Ser. Biol. Nauk. [Proc. Acad. Sci. Turkmen SSR, Ser. Biol. Sci.] 1: 8–17, 6: 24–34 (in Russian).
Kryzhanovsky, O.L. and V.A. Mikhailov. 1971. New and little known ground beetles (Coleoptera, Carabidae) from Middle Asia. Entomol. Obozr. [Entomol. Rev.] 50(3): 632–640 (in Russian).
Kryzhanovsky, O.L. and A.I. Reichardt. 1976. Fauna of the USSR. Coleoptera. V, 4. Superfamily Histeroidea (families Sphaeritidae, Histeridae, Synteliidae). Nauka, Leningrad, 434 pp (in Russian).
Kryzhanovsky, S.G. 1949. Ecomorphological patterns in the development of carps, weatherfish, and catfish (Cyprinoidae and Siluroidei). Trudy Instituta morphologii zhivotnykh [Trans. Inst. Animal Morphology] 1: 5–333 (in Russian).
Kubanskaya, Z.V. 1956. Rastitel'nost' i kormovye resursy pustyni Bet-Pak-Daly [Vegetation and Feed Resources of the Bet-Pak-Dala Desert]. Izd. ANKaz.SSR, Alma-Ata, 263 pp (in Russian).
Kucheruk, V.V. 1972. An attempt to analyse the development of the views of Russian zoogeographers concerning the regionalization of the Palearctic Region. Trudy MOIP [Trans. Moscow Soc. Natur.] 1972, 48: 150–176 (in Russian).
Kulibaba, V.V. and V.Ya. Fet. 1985. Long-term fluctuations of precipitation and landscape dynamics in West Kopetdagh. In: N.T. Nechayeva (ed.), Rastitelnost' i zhivotnyi mir Zapadnogo Kopetdaga [The Vegetation and Animal Life of West Kopetdagh]. Ylym, Ashkhabad: 6–13 (in Russian).
Kulibaba, V.V. and V.A. Pakulin. 1985. Data on wild boar ecology in the drainage system of the Sumbar River. In: N.T. Nechayeva (ed.), Rastitelnost' i zhivotnyi mir Zapadnogo Kopetdaga [The Vegetation and Animal Life of West Kopetdagh]. Ylym, Ashkhabad: 100–107 (in Russian).
Kullmann, E. and W. Zimmermann. 1976. Beschreibung der neuen Spinnenart *Oecobius afghanicus* mit ergänzenden Angaben zu *Oecobius putus* und *Oecobius annulipes* (Arachnida, Araneae, Oecobiidae). Ent. Germ. 3(1-2): 41–50.
Kultiasov, M.V. 1946. Studies on the formation of vegetation of hot deserts and steppes of Middle Asia. Materialy po istorii flory i rastitelnosti SSSR [Materials on the History of Flora and Vegetation of the USSR], Izd. AN SSSR, Moscow-Leningrad 2: 257–282 (in Russian).
Kultiasov, M.V. 1960. Formation of mountainous forest flora of Middle Asia during the Quaternary. Materialy po chetvertichnomu periodu v SSSR [Materials on the Quaternary Period in the USSR]. Izd. AN SSSR, Moscow 3: 87–98 (in Russian).
Kurbanov, Dz.K. 1981. New data on *Homalodiscus ochradeni* (Boiss.) Boiss. (Resedaceae). Izv. Akad. Nauk. Turkm. SSR. Ser. Biol. Nauk. [Proc. Acad. Sci. Turkmen. SSR, Ser. Biol. Sci.] 4: 93–95 in Russian).
Kurbanov, Dz.K. 1988. Konspekt flory zapadnykh nizkogorii i srednegorii Kopetdaga [Conspectus of the Flora of Foothills and Middle Mountain Belt of West Kopetdagh]. Ylym, Ashkhabad, 257 pp (in Russian).
Kurbanov, Dz.K. 1992. Analiz flory Severo-Zapadnogo Kopetdaga [Analysis of the Flora of Northwest Kopetdagh]. Ylym, Ashgabat, 247 pp (in Russian).
Kurilenko, V.E. 1985. Contributions to the biology of the spine-tailed gecko – *Gymnodactylus*

spinicauda (Str. 1887). In: N.T. Nechayeva (ed.), Rastitelnost' i zhivotnyi mir Zapadnogo Kopetdaga [The Vegetation and Animal Life of West Kopetdagh]. Ylym, Ashkhabad: 205–208 (in Russian).
Kurilenko, V.E. 1987. Contributions to the ecology and reproductive behavior of *Pseudocyclophis persicus*, Serpentes, Colubridae. Izv. Akad. Nauk. Turkm. SSR., Ser. Biol. Nauk. [Proc. Acad. Sci. Turkmen. SSR, Ser. Biol. Sci.] 6: 65–67 (in Russian).
Kurochkina, L.Ya. 1966. Vegetation of the sand deserts of Kazakhstan. In: B.A. Bykov (ed.), Rastitel'nyi pokrov Kazakhstana [Vegetative Cover of Kazakhstan]. Nauka, Alma-Ata 1: 191–292 (in Russian).
Kuvshinova, K.V. 1972. Climatic regions of southwestern Turkmenistan. Problemy osvoeniya pustyn [Problems of Desert Development] 4: 17–23 (in Russian).
Kuznetsov, G.T. 1984a. The common arachnids of Kopetdagh foothills. In: Problemy pochvennoi zoologii [Problems of Soil Zoology, VIII All-Union Confer., Abstr.], Ashkhabad 1: 163–164 (in Russian).
Kuznetsov, G.T. 1984b. On the poisonous arthropods management in Turkmenia. In: IX Syezd Vsesoyuzn. Entomol. Obshch. [9th Congr. All-Union Entomol. Soc.], Kiev 1: 264–265 (in Russian).
Kuznetsov, G.T. 1985a. *Latrodectus pallidus pavlovskyi* Charitonov protection in Turkmenistan. Izv. Akad. Nauk. Turkm. SSR, Ser. Biol. Nauk. [Proc. Acad. Sci. Turkmen. SSR, Ser. Biol. Sci.] 4: 71–72 (in Russian).
Kuznetsov, G.T. 1985b. Ecology of the spiders *Eresus niger* and *Lithyphantes paykullianus* (Aranei, Eresidae, Theridiidae) in southern Turkmen SSR, USSR) . Izv. Akad. Nauk. Turkm. SSR, Ser. Biol. Nauk. [Proc. Acad. Sci. Turkmen. SSR, Ser. Biol. Sci.] 6: 70–72 (in Russian).
Kuznetsov, G.T. and V.Ya. Fet. 1982. On the zoogeographical connections of spider fauna of Kopetdagh. In: Tez. dokl. Nauchn. konfer. molodykh uchyonykh i spetsialistov Tadzhikistana [Abstr. Scientific Conference of Young Scientists and Specialists of Tajikistan], Dushanbe: 59–61 (in Russian).
Kuznetsov, G.T. and V.Ya. Fet. 1986. Contributions to the spider fauna of Kopetdagh. In: N.N. Shcherbak (ed.), Priroda Tsentralnogo Kopetdaga [Nature in Central Kopetdagh]. Ylym, Ashkhabad: 48–67 (in Russian).
Kuznetsov, L.A. 1959. Contributions to the geobotanical characteristics of *Anabaseta salsae* formation. Uchen. zap. Leningradsk. gos. ped. in-ta [Sci. Mem. Leningrad State Ped. Inst.] 178: 38–52 (in Russian).
Kuznetsov, L.A. 1966. Biyurgun communties of the northern Aral area. In: XIX Gertsenskie chteniya [19th Herzen Memorial Lectures]. Leningradsk. gos. ped. in-t., Leningrad: 13–17 (in Russian).
Kuznetsov, V.I., N.V. Skalon, V.V. Kulibaba, and A.K. Filatov. 1981. Problems of conservation of wild gallinaceous fowl in the Sumbar River Basin. In: Tez. dokl. 2 nauchn. konfer, po okhrane prirody Turkm. SSR [Abstr. 2nd Sci. Conf. on Natural Conservation in the Turkmen SSR], Ashkhabad: 120–122 (in Russian).
Kuznetsova, L.P. 1983. Atmosfernyi vlagooborot nad territoriei SSSR [Atmospheric Water Exchange over the Territory of the USSR]. Nauka, Moscow, 173 pp (in Russian).
Kuznetsova, N.V. 1989. From the history of conservation and botanical studies of Kopetdagh. Izv. Akad. Nauk. Turkm. SSR. Ser. Biol. Nauk. [Proc. Acad. Sci. Turkmen. SSR, Ser. Biol. Sci.] 2: 71–75 (in Russian).
Kuzyakin, A.P., 1950. Letuchiye myshi [Bats]. Sovetskaya Nauka, Moscow, 443 pp (in Russian).
Laptev, M.K. 1934. Additions to the knowledge of vertebrate fauna of Turkmenistan (Bolshoi Balkhan and West Kopetdagh). Izv. Turkmen. mezhduved. komit. okhr. prir. i razvit. prir. bogatstv [Proc. Turkmen Interdepartm. Committee Natural Conserv. and Developm. Natural Resources], Ashkhabad 1: 115–195 (in Russian).
Laptev, M.K. 1935. Vertebrates of Turkmenistan and their exploitation. Izv. Turkmen. mezhduved. komit. okhr. prir. i razvit. prir. bogatstv [Proc. Turkmen Interdepartm. Committee Natural Conserv. and Developm. Natural Resources], Ashkhabad 2: 87–114 (in Russian).

Laptev, M.K. 1936. Results of a biological survey of game animals in Central Kopetdagh (Turkmen SSR). Byull. Turkm. zool. stantsii [Bull. Turkm. Zool. Station] 1: 43–54 (in Russian).
Laptev, M.K. 1937. Dikie zhivotnye Kopet-Daga i predgornoi ravniny [Wild Animals of Kopetdagh and its Submontane Plain]. Turkmengosizdat, Ashkhabad-Baku, 50 pp (in Russian).
Laptev, M.K., V.I. Sulima, and L.R. Freiberg. 1934. The All-Union Reserve in Gasan-Kuli. Izv. Turkmen. mezhduved. komit. okhr. prir. i razvit. prir. bogatstv [Proc. Turkmen Interdepartm. Committee Natural Conserv. and Developm. Natural Resources], Ashkhabad 1: 44–111 (in Russian).
Latifi, M. 1991. The snakes of Iran. Soc. Study of Amphibians and Reptiles, Oxford, Ohio, 159 pp.
Lavrenko, Ye.M. 1960. On the Saharo-Gobian desert biogeographic region and its subdivision. Dokl. AN SSSR [Rep. Acad. Sci. USSR] 134(1) (in Russian).
Lavrenko, Ye.M. 1962. Basic features of botanical geography of the deserts of Eurasia and North Africa. Komarovskie Chteniya [Komarov Memorial Lectures]. Izd. AN SSSR, Moscow-Leningrad, 15, 169 pp (in Russian).
Lavrenko, Ye.M. 1965. The provincial delineation of the Central Asian and Irano-Turanian Subregions of the Afro-Asian Desert Region. Bot. zhurn. [Bot. J.] 50(1): 3–15 (in Russian).
Lavrenko, Ye.M. 1982. Plant communities and their classification. Bot. zhurn. [Bot. J.] 62(5): 572–580 (in Russian).
Leach, J., M. Johnson, J. Kelson et al. 1977. Responses of percid fishes and their habitats to eutrophication. J. Fish Res. Board Canada 34(10): 1964–1971.
Leclercq, M. 1960. Revision systematique et biogeographique des Tabanidae (Diptera) Palearctiques. I. Pangoniinae et Chrysopsinae. Mem. Inst. Roy. Sci. Nat. Belgique. Deuxième Ser. 63: 1–77.
Leclercq, M. 1966. Revision systematique et biogeographique des Tabanidae (Diptera) Palearctiques. II. Tabaninae. Mem. Inst. Roy. Sci. Nat. Belgique. Deuxième Ser. 80: 1–235.
Leclercq, M. 1967. Tabanidae (Diptera) de Turquie III. Bull. Res. Agron. Gembloux, N.S. 2: 707–710.
Leclercq, M. 1985. Recent additions and synonymy in Palearctic Tabanidae (Diptera). Myia 3: 341–345.
Levan, A., K. Fredga, and A.A. Sandberg. 1964. Nomenclature for centromeric position on chromosomes. Hereditas 52: 201–220.
Levin, G.M. 1985. The wild pomegranate of Kopetdagh as a source of valuable characters and features for selection. In: N.T. Nechayeva (ed.), Rastitelnost' i zhivotnyi mir Zapadnogo Kopetdaga [The Vegetation and Animal Life of West Kopetdagh]. Ylym, Ashkhabad: 61–77 (in Russian).
Leviton, A.E. and S.C. Anderson. 1963. Third contribution to the herpetology of Afghanistan. Proc. Calif. Acad. Sci., 4th Ser. 31(12): 329–339.
Leviton, A.E. and S.C. Anderson. 1970. The amphibians and reptiles of Afghanistan: a checklist and key to the herpetofauna. Proc. Calif. Acad. Sci., 4th Ser. 38(10): 163–206.
Levy, G. and P. Amitai. 1980. Scorpiones (Fauna Palaestina, Arachnida I). Israel Academy of Sciences and Humanities, Jerusalem, 132 pp.
Leyssac, P.P. 1966. The regulation of proximal tubular reabsorption in the mammalian kidney. Acta Physiol. Scand. 70 (Suppl. 291): 13–151.
Likharev, I.M. and Ye.S. Rammelmeyer. 1952. Nazemnye mollyuski fauny SSSR [Terrestrial Mollusks of the Fauna of the USSR]. Opredeliteli po faune SSSR [Keys to the USSR Fauna], Izd. AN SSSR, Moscow-Leningrad 43, 511 pp (in Russian).
Likharev, I.M. and A.J. Wiktor. 1980. The fauna of slugs of the USSR and adjacent countries (Gastropoda terrestria nuda). Fauna SSSR [Fauna of the USSR], New Ser., Vol. 122. Mollusca 3(5). Nauka, Leningrad, 437 pp (in Russian).
Lim, R.M. and E. Yermakhanov. 1986. A modern hydrobiological regime of the Aral Sea and the perspectives of the fishing industry in its basin under the conditions of sea regression. In: 5 Syezd Vsesoyuznogo Gidrobiologicheskogo Obshchestva [5th Congress of All-Union Hydrobiological Society], Kuibyshev, Part 2: 93–94 (in Russian).

Linchevsky, I.A. 1935a. Vegetation of Badghyz. In: R.I. Abolin (ed.), Rastitel'nye resursy Turkmenskoi SSR [Plant Resources of Turkmen SSR]. Izd. VASKHNIL, Leningrad: 185-291 (in Russian).

Linchevsky, I.A. 1935b. Vegetation of West Kopetdagh. In: R.I. Abolin (ed.), Rastitel'nye resursy Turkmenskoi SSR [Plant Resources of Turkmen SSR]. Izd. VASKhNIL, Leningrad: 15-78 (in Russian).

Lipsky, V.I. 1915. Botanical studies in the Transcaspian Region. In: Trudy pochvenno-botanicheskikh ekspeditsii po issledovaniyu kolonizatsionnykh raionov Aziatskoi Rossii. Chast' 2. Botanicheskie isledovaniya 1912 g. [Results of Soil and Botanical Expeditions Exploring the Colonial Parts of Asian Russia. Part 2. Botanical Studies in 1912]. Petrograd: 1-78 (in Russian).

Lisitsina, G.N. and K.P. Popov. 1988. On the woody vegetation in the lowlands of South Turkmenistan during the Neolite, Eneolite, Bronze, and early Iron Ages. Izv. Akad. Nauk. Turkm. SSR, Ser. Biol. Nauk. [Proc. Acad. Sci. Turkm. SSR, Ser. Biol. Sci.] 1: 68-70 (in Russian).

Livanov, N.A. 1951. Planarians of Kopetdagh and related species in the Crimea, Caucasus, and Transcaucasia. Trudy Murgabsk. gidrobiol. stants. [Trans. Murghab Hydrobiol. Station] 1: 103-114 (in Russian).

Lobachev, V.S. and G.I. Shenbrot. 1973. Comparative analysis of various techniques of jerboa surveys. Byull. MOIP [Bull. Moscow Soc. Natur.] 78(2): 47-57 (in Russian).

Locket, N.A. 1990. A new genus and species of scorpion from South Australia (Buthidae: Buthinae). Trans. R. Soc. S. Aust. 114(1/2): 67-80.

Logvinenko, B.M. and Ya.I. Starobogatov. 1962. Caspian malacofauna and its zoogeographical connections. Byull. MOIP [Bull. Moscow Soc. Natur.] 67(1): 153-154 (in Russian).

Logvinenko, B.M. and Ya.I. Starobogatov. 1969. Phylum Mollusca. In: Ya.A. Birshtein et al. (ed.), Atlas bespozvonochnykh Kaspiiskogo morya [Atlas of Invertebrates of the Caspian Sea]. Pishchevaya promyshlennost', Moscow: 308-375; 407-410 (in Russian).

Loskot, V.M. 1971. Notes on some birds of Badghyz. Zb. prats' Zool. Muz. AN URSR [Trans. zool. Mus. Ukrainian SSR] 34: 97-99 (in Ukranian).

Loudon, H. 1910 (1911). Meine vierte Reise nach Zentral-Asien und Talysch. In: Verhandl. V Intern. Ornitol. Kongr., Berlin: 335-369.

Loudon, H. 1901. Ergebnisse einer ornithologischen Sammelreise nach Zentral-Asien (1901). Ornith. Jb. 13(3/4): 81-106.

Loudon, H. 1902. Ergebnisse einer ornithologischen Sammelreise nach Zentral-Asien (1901). Ornith. Jb. 13(5/6): 190-233.

Luppov, N.M. 1931. Geological structure of the notheastern portion of the Krasnovodsk district of the Turkmen SSR. Izv. Glavn. geologorazved. upravl. [Proc. Central Geol. Surveying Dept.] 50(54): 3-21 (in Russian).

Luppov, N.M. 1932. A geological survey of the East Karabugaz area based on the studies in 1929 and 1930. Trudy Vses. geologorazved. obyedin. [Trans. All-Union Geol. Surveying Trust] 269: 3-35 (in Russian).

Luppov, N.M. 1956. History of the geological development of the territory of Turkmen SSR. Trudy Geol. inst. AN SSSR [Trans. Geol. Inst. Acad. Sci. USSR] 1: 5-42 (in Russian).

Luppov, N.M. (ed.). 1972. Geologiya SSSR: Turkmenskaya SSR. Geologicheskoe opisanie [Geology of the USSR: Turkmen SSR, Geological Description]. Nedra, Moscow 22: 1-768 (in Russian).

Luppova, Ye.P. 1961. On ant-lions (Neuroptera, Myrmeleontidae) of Middle Asia. Trudy In-ta zool. i parazitol. AN Tadzh. SSR [Trans. Inst. Zool. and Parasitol. Acad. Sci. Tajik SSR] 20: 193--210 (in Russian).

Luppova, Ye.P. 1966. Results of the study of Neuroptera in Middle Asia. In: M.N. Narzikulov and E.P. Luppova (ed.), Fauna i zoogeografiya nasekomykh Srednei Azii [Fauna and Zoogeography of Insects of Middle Asia]. Donish, Dushanbe: 245-252 (in Russian).

Luppova, Ye.P. 1971. New data on the ant-lion *Acanthaclisis pallida* McLachlan (Neuroptera, Myrmeleontidae) from Middle Asia. Izv. AN Tadzh. SSR, Otd. Biol. Nauk. [Proc. Acad. Sci. Tajik SSR, Div. Biol. Sci.] 45(4): 82-88 (in Russian).

Luppova, Ye.P. 1979. New species of the ant-lion of the genus *Maracanda* McL. (Neuroptera, Myrmeleontidae) in the USSR fauna. Novye vidy nasekomykh/ Trudy Zool. in-ta AN SSSR [New Species of Insects/Trans. zool. Inst. Acad. Sci. USSR] 61: 90-92 (in Russian).

Luppova, Ye.P. 1979. Superfamily Myrmeleontoidea. Opredelitel' nasekomykh evropeiskoi chasti SSSR [A Key to the Insects of the European Portion of the USSR]. Nauka, Leningrad 4(6): 73-96 (in Russian).

Maceina, M. and J. Shireman. 1979. Grass carp: effects of salinity on survival, weight loss and muscle tissue water conntent. The Progressive Fish Culturist 41(2): 69-73.

McGregor, G. and J. Varley. 1986. Techniques of Chromosome Studies. Mir, Moscow, 262 pp (translated into Russian).

McLachlan, R. 1875. Neuroptera. Puteshestvie v Turkestan A.P. Fedchenko [A.P. Fedchenko's expedition to Turkestan] 2(8): 1-24 (in Russian).

MacMillen, R.E. and C.H. Trost. 1967. Thermoregulation and water loss in the inca dove. Comp. Biochem. Physiol. 20(1): 71-82.

McNabb, E.N.A. 1969. A comparative study of water balance in three species of quail. 1. Water turnover in the absence of temperature stress. Comp. Biochem. Physiol. 28(38): 35-42.

Makeev, V.M. 1969. On the comparative ecology of gyurza (*Vipera lebetina*) and cobra (*Naja oxiana*). Zool. zhurn. (Zool. J.) 48(12): 1832-1837 (in Russian).

Makeev, V.M. 1972. Characteristics of the behavior of the Caucasian agama. In: Povedeniye zhivotnykh [Animal Behavior]. Nauka, Moscow: 119-120 (in Russian).

Makeev, V.M. 1978. The Middle Asian cobra (*Naja oxiana* Eichw.) and its conservation in the USSR. In: Okhrana i ratsional'noye ispol'zovaniye reptilii [Conservation and Rational Exploitation of Reptiles]. TsLOP, Moscow 1: 59-103 (in Russian).

Makeev, V.M. 1979. Density and biomass of reptiles in the Southeast Karakum. Zool. zhurn (Zool. J.) 58(1): 59-103 (in Russian).

Makeev, V.M., A.T. Bozhansky, and Yu.D. Khomustenko. 1983. Distribution and density of reptiles in Central Kopetdagh. Zool. zhurn. (Zool. J.) 62: 1122-1125 (in Russian).

Makeev, V.M., A.T. Bozhansky, and Yu.D. Khomustenko. 1986. Density and ecology of the Middle Asian gyurza in Central Kopetdagh. Ekologiya [Ecology] 5: 82-84 (in Russian).

Makeev, V.M., A.T. Bozhansky, S.V. Kudryavtsev, V.Ye. Frolov, and Yu.D. Khomustenko. 1988. Some results of the heropetological survey of East Turkmenia. In: V.Ye. Flint (ed.), Redkie i maloizuchennye zhivotnye Turkmenistana [Rare and Poorly Known Animals of Turkmenistan]. Ylym, Ashkhabad: 127-143 (in Russian).

Makeev, V.M. and N.A. Zemlyakova. 1968. On the reproduction of the Middle Asian cobra (*Naja oxiana* Eichw.). Zool. zhurn. (Zool. J.) 47(12): 1872-1874 (in Russian).

Malgina, E.A. 1958. Palynological spectra of the Quaternary and Upper Pliocene deposits of West Turkmenia. Dokl. AN SSR [Rep. Acad. Sci. USSR] 120(3): 633-636 (in Russian).

Malgina, E.A. 1961. Results of palynological analysis of the Quaternary and Upper Pliocene deposits of the Balkhan area of West Turkmenia. In: Mater. vses. soveshch. po izuch. chetvert. perioda [Mat. All-Union Confer. on Studies of Quaternary Period], Izd. AN SSSR, Moscow 1: 296-303 (in Russian).

Mamedkuliev, I.D. 1990. Structure and dynamics of the plant organic matter in steppe communities of Central Kopetdagh. Izv. AN TSSR, Ser. Biol. Nauk. [Proc. Acad. Sci. Turkmen SSR, Ser. Biol. Sci.] 3: 9-14 (in Russian).

Mamedniyazov, O. 1971. Krovososushchie dvukrylye nasekomye v rayonakh 4-i ocheredi Karakumskogo kanala i meropriyatiya po bor'be s nimi [Bloodsucking Dipterous Insects in the Area of the 4th Section of the Karakum Canal and the Struggle Against Them], Abstr. Ph.D. Thesis, 24 pp (in Russian).

Manilo, V.V. 1986. Karyotypes of geckos of the genera *Alsophylax* and *Crossobamon*. Vestn. zool. [Mess. Zool.] 5: 46-54 (in Russian).

Markov, K.K. 1960. Paleogeografiya Azii [Paleogeography of Asia]. Izd. MGU, Moscow, 268 pp (in Russian).

Marusik, Yu.M. and D.V. Logunov. 1990. The crab spiders of Middle Asia, USSR (Aranei,

Thomisidae). I. Descriptions and notes on distribution of some species. Korean Arachnology 6(1): 31-62.
Matthews, G.V.T. and M.E. Evans. 1974. On the behavior of the Whiteheaded Duck with special reference to breeding. Wildfowl 25: 56-60.
Medvedev G.S. 1990. Key to the darkling beetles of Mongolia. Trudy Zool. in-ta AN SSSR [Proc. Zool. Inst. Acad. Sci. USSR] 220: 1-253 (in Russian).
Medvedev G.S. and M.G. Nepesova. 1985. Opredelitel' zhukov-chernotelok Turkmenistana [Key to the Darkling Beetles of Turkmenistan]. Ashkhabad, Ylym, 180 pp (in Russian).
Medvedev, G.S. and M.G. Nepesova. 1990. Composition and geographic distribution of the fauna of tenebrionid beetles (Coleoptera, Tenebrionidae) of Kopetdagh. Entomol. obozr. [Entom. Rev.] 69(4): 879-893 (in Russian).
Meikle, P.D. 1977. Flora of Cyprus. Royal Bot. Gardens, Kew, Vol. 1, 833 pp.
Meikle, P.D. 1985. Flora of Cyprus. Royal Bot. Gardens, Kew, Vol. 2, 1969 pp.
Menzbier, M.A. 1914. Zoologicheskie uchastki Turkestanskogo kraya i vozmozhnoye proiskhozhdenie yego fauny [Zoological Districts of the Turkestan Region and the Possible Origin of its Fauna]. Moscow, 144 pp (in Russian).
Merrett, P. and A.F. Millidge. 1992. Amendments to the checklist of British spiders. Bull. British Arachnol. Soc. 9(1): 4-9.
Mezhzherin, S.V. and M.V. Golubev. 1992. Allozyme variation and genetic differentiation of toad agamas belonging to the group *Phrynocephalus ocellatus* (Agamidae). Gerpetol. issled. [Herpetolological Studies]: 1 (in Russian).
Mikeshin, G.V. 1946a. Sagebrush deserts of Kopetagh. Byull. MOIP [Bull. Moscow Soc. Natur.] 51(6) (in Russian).
Mikeshin, G.V. 1946b. *Stipa-Festuca* steppes of Kopetagh. Byull. MOIP [Bull. Moscow Soc. Natur.] 52(3) (in Russian).
Mikeshin, G.V. and V.V. Alyokhin. 1945. An outline of altitudinal belts of vegetation in Kopetdagh. Uchen. zap. MGU [Sci. Mem. Moscow State Univ.] (in Russian).
Mikhailov, K.G. 1986. New spider species of the families Clubionidae and Liocranidae from Central Asia and the Caucasus Mountains, USSR. Zool. zhurn. [Zool. Journal] 65(5): 798-802 (in Russian).
Mikhailov, K.G. 1987a. Redescription of a spider, *Trachelas maculatus* (Aranei, Corinnidae). Zool. zhurn. [Zool. Journal] 66: 1583-1586 (in Russian).
Mikhailov, K.G. 1987b. Contribution to the USSR spider fauna of the genus *Micaria* Westring, 1851. I. (Aranei, Gnaphosidae). Spixiana 10(3): 319-334.
Mikhailov, K.G. 1992. The genus *Clubiona* Latreille, 1804 (Arachnida, Aranei, Clubionidae) in the USSR fauna: a critical review with taxonomical remarks. Arthropoda Selecta 1(3): 3-34.
Mikhailov, K.G. and V.Ya. Fet. 1986. Contributions to the spider fauna of Turkmenia, I. Families Clubionidae, Sparassidae, Oxyopidae, Micariidae, Anyphaenidae, Zoridae. Trudy Zool. Muzeya MGU [Trans. Zool. Mus. Moscow State Univ.] 24: 168-186 (in Russian).
Mikhailov, V.A. 1988. Zhuki-zhuzhelitsy (Coleoptera, Carabidae) yugo-vostoka Srednei Azii [Ground-beetles (Coleoptera, Carabidae) of Southeastern Middle Asia]. Abstr. D.Sc. Thesis, Leningrad, 35 pp (in Russian).
Mikhailovsky, M. 1904. On the herpetofauna of the Transcaspian Region. Ezheg. zool. muzeya Akademii nauk [Annu. Zool. Mus. Acad. Sci.] 9: 2-6 (in Russian).
Mikhel', N.M. 1935. Preliminary results of the work of the ecological team of the South Ustuyrt expeditional division of the USSR Academy of Sciences. In: Predvaritel'nye itogi ekspeditsionnykh rabot v Turkm. SSR za 1934 g. SOPS AN SSSR [Preliminary Results of the Expeditional Work in the Turkmen SSR in 1934]. Moscow: 72-80 (in Russian).
Mikhelson, Ye.G. 1955. Dynamics of vegetaton in Southeast Karakum (Repetek). Trudy Repeteksk. peschano-pust. stantsii AN Turkm. SSR [Trans. Repetek Sand Desert Station Acad. Sci. Turkm. SSR] 3: 141-175 (in Russian).
Minton, S.A. 1966. A contribution to the herpetology of West Pakistan. Bull. Amer. Mus. Nat. Hist. 134: 29-184.

Mirkin, B.M. 1985. Teoreticheskie osnovy sovremennoi fitotsenologii [Theoretical Principles of Modern Phytocoenology]. Nauka, Moscow, 136 pp (in Russian).
Mishchenko, L.L. 1949. Dermaptera, Blattodea, Mantoptera, Phasmidea, and Orthoptera of the Ghissar Valley (Tadjik SSR). Trudy Zool. in-ta AN SSSR [Trans. Zool. Inst. Acad. Sci. USSR] 8: 735–749 (in Russian).
Mishchenko, L.L. 1951. On the fauna of Dermaptera, Blattodea, Mantoptera, Phasmidea, and Orthoptera on the southern slope of the Ghissar Range. In: E.N. Pavlovsky and V.I. Zhadin (eds.), Ushchelye Kondara [The Kondara Valley], Izd. AN SSSR, Moscow-Leningrad: 198–205 (in Russian).
Mishchenko, L.L. 1952. Nasekomye pryamokrylye [Orthoptera]. Catantopinae. Izd. AN SSSR, Moscow-Leningrad, Fauna SSSR [Fauna of the USSR] 4(2), 610 pp (in Russian).
Mishchenko, Yu.V. and N.N. Shcherbak. 1980. On new records of rare and little known birds from Turkmenia. Vestn. zool. [Mess. Zool.] 1: 13–17 (in Russian).
Momotov, I.F. 1953. Rastitel'nye kompleksy Ust'-Urta [Complexes of Vegetation of Ustyurt]. Izd.ANUzb.SSR, Tashkent, 124 pp (in Russian).
Mordvilko, A. 1935. Die Blattläuse mit unfollständigen Generationszyklus und ihre Ersteunge. Ergebnisse und Fortschritte der Zoologie 8: 37–43.
Morits, L.D. 1922. A report on a trip to Turkestan in Summer 1921. Trudy Stavropol'sk. sel'skokhoz. in-ta (Trans. Stavropol Agric. Inst.), Vol. 1 (Zoology) 18: 209–226 (in Russian).
Morits, L.D. 1929. Reptiles of Turkmenia and adjacent Persia. Turkmenovedeniye [Turkmen Studies] 4/7: 17–35 (in Russian).
Morits, L.D. 1930. The bezoar goat in Turkmenia. Okhotnik [The Hunter]. 7 (in Russian).
Mukhamedieva, F.D. 1974. Changes in the fish faunIstic complexes of inland water bodies of Turkmenia under the influence of anthropogenic factors. Uchyonye zapiski TGU [Sci. Mem. Turkmen State Univ.] 1: 80–85 (in Russian).
Mukhamedieva, F.D. and V.B. Salnikov. 1983. The formation of ichthyofauna and the perspectives of the fishing industry in the Kopetdagh Reservoir. Izv. AN TSSR, Ser. Biol. Nauk. [Proc. Acad. Sci. Turkmen SSR, Ser. Biol. Sci.] 4: 45–50 (in Russian).
Mukhammedov, G.M. 1972. Khvoinik shishkonosnyi v pustyne Karakum [*Ephedra strobilacea* in the Karakum Desert]. Ylym, Ashkhabad, 92 pp (in Russian).
Mukhammedov, G.M. 1979. Uluchsheniye pastbishch v Tsentral'nykh Karakumakh [Enhancement of the Pastures in Central Karakum]. Ylym, Ashkhabad, 215 pp (in Russian).
Muratov, I.V. 1992. New taxa of Pseudonapaeinae (Gastropoda, Pulmonata, Enidae). Ruthenica 2(1): 37–44.
Murzaev, E.M. 1966. Priroda Sin'tszyana i formirovanie pustyn' Tsentral'noi Azii [The Nature of Sinkyang and Formation of the Deserts of Central Asia]. Izd.ANSSSR, Moscow, 382 pp (in Russian).
Myartseva, S.N. 1978. New species of the genus *Psyllaephagus* Ashmead (Chalcidoidea, Encyrtidae), the parasites of psyllids (Homoptera, Psylloidea) on *Calligonum* in Turkmenia. Izv. AN TSSR, Ser. Biol. Nauk. [Proc. Acad. Sci. Turkmen SSR, Ser. Biol. Sci.] 6: 3–9 (in Russian).
Myartseva, S.N. 1979. The encyrtids of the genus *Psyllaephagus* (Hymenoptera, Chalcidoidea) from *Tamarix* in Turkmenia. Izv. AN TSSR, Ser. Biol. Nauk. [Proc. Acad. Sci. Turkmen SSR, Ser. Biol. Sci.] 5: 27–33 (in Russian).
Myartseva, S.N. 1980a. On the fauna of *Psyllaephagus* (Hymenoptera, Encyrtidae) of South Turkmenia. Izv. AN TSSR, Ser. Biol. Nauk. [Proc. Acad. Sci. Turkmen SSR, Ser. Biol. Sci.] 1: 47–54 (in Russian).
Myartseva, S.N. 1980b. The species of the genus *Microterys* Thomson (Hymenoptera, Encyrtidae) in the fauna of Turkmenia. Izv. AN TSSR, Ser. Biol. Nauk. [Proc. Acad. Sci. Turkmen SSR, Ser. Biol. Sci.] 4: 43–53 (in Russian).
Myartseva, S.N. 1981. A survey of the encyrtids belonging to the genus *Discodes* Förster (Hymenoptera, Encyrtidae) of the USSR fauna, with a description of new species from Turkmenia. Entomol. Obozr. [Entomol. Rev.] 61(2): 364–379 (in Russian).
Myartseva, S.N. 1983a. Family Encyrtidae – Encyrtids. In: M.A. Daricheva, G.A. Krasilnikova, S.N. Myartseva, O.R. Sabirova, T.G. Sidyak, and D. Charykuliev (eds.), Fauna i ekologiya

nasekomykh doliny srednego techeniya Amudaryi [Fauna and Ecology of Insects of the Middle Amudarya]. Ylym, Ashkabad: 40-110 (in Russian).
Myartseva, S.N. 1983b. The specificity of parasitism in encyrtids (Hymenoptera, Chalcidoidea, Encyrtidae) in Middle Asia. Izv. AN TSSR, Ser. Biol. Nauk. [Proc. Acad. Sci. Turkmen SSR, Ser. Biol. Sci.] 4: 39-44 (in Russian).
Myartseva, S.N. 1983c. The landscape distribution of encyrtids (Hymenoptera, Encyrtidae) in arid regions of Middle Asia. Izv. AN TSSR, Ser. Biol. Nauk. [Proc. Acad. Sci. Turkmen SSR, Ser. Biol. Sci.] 5: 22-31 (in Russian).
Myartseva, S.N. 1984a. Paraziticheskie pereponchatokrylye semeistva Encyrtidae (Hymenoptera, Chalcidoidea) Turkmenistana i sopredelnykh raionov Srednei Azii [Parasitic Hymenopterans of the Family Encyrtidae (Hymenoptera, Chalcidoidea) of Turkmenistan and A Regions of Middle Asia]. Ylym, Ashkabad, 304 pp (in Russian).
Myartseva, S.N. 1984b. Parazity chervetsa Komstoka [Parasites of *Pseudococcus comstocki*]. Ylym, Ashkabad, 56 pp (in Russian).
Myartseva, S.N. 1986. Encirtidy (Hymenoptera, Chalcidoidea, Encyrtidae) pustyn i polupustyn Srednei Azii (fauna, biologiya, ekologiya, rasprostraneniye, khozyaistvennoye znacheniye) [Encyrtids (Hymenoptera, Chalcidoidea, Encyrtidae) of the Deserts and Semideserts of Middle Asia (Fauna, Biology, Ecology, Distribution, Practical Importance]. Ylym, Ashkabad, 304 pp (in Russian).
Myartseva, S.N. and O.D. Niyazov. 1986. Vinogradny muchnisty chervets i yego yestestvennye vragi v Turkmenistanae [The Grape Scale Insect and Its Natural Enemies in Turkmenistan]. Ylym, Ashkabad, 93 pp (in Russian).
Nabokov, V.V. 1952. Dar [The Gift]. New York. Reprint: Ardis, Ann Arbor, 1975 (in Russian).
Nasimovich, A.A. 1950. Reasons for the dynamics of the range boundaries and abundance of pheasant in the USSR. Okhrana prirody [Natural Conservation] 12: 94-106 (in Russian).
Nastyukov, N.Z. 1976. Fauna of small mammals of Ustyurt and its changes in the late Anthropogene. In: L.G. Dinesman (ed.), Istoriya biogeotsenozov SSSR v golotsene [History of the USSR Biogeocoenoses in the Holocene]. Nauka, Moscow: 270-272 (in Russian).
Nauchno-prikladnoi spravochnik po klimatu SSSR. Turkmenskaya SSR [Scientific and Applied Reference Book on the USSR Climate. Turkmen SSR]. 1989. Gidrometeoizdat, Leningrad. Ser. 3, 1-6(30): 1-501 (in Russian).
Navás, L. 1913. Neuroptera asiatica. I Series. Russk. entomol. obozr. [Russ. Entomol. Review] 13(2): 271-284.
Navás, L. 1925. Insectos exoticos nuevos o poco conocidos. II. Mem. R. Acad. Ci. Barcelona 19: 181-203.
Navás, L. 1932. Decadas de insectos nuevos. Decadas 19, 20. Broteria Ci. Nat. 5(1): 62-85.
Nechayeva, N.T. 1956. *Sagebrush-Salsola* pastures of northwestern Turkmenistan. Trudy in-ta zhivotnovodtsva AN TSSR [Trans. Inst. Animal Breeding Acad. Sci. Turkm. SSR] 1: 18-127 (in Russian).
Nechayeva, N.T. 1958. Dinamika pastbishchnoi rastitel'nosti Karakumov pod vliyaniyem meteorologicheskikh uslovii [Dynamics of the Pasture Vegetation of the Karakum Desert Depending on Meteorological Conditions]. Izd. AN TSSR, Ashkhabad, (in Russian).
Nechayeva, N.T. 1973. A problem of development of the indicators of desertification. Problemy osvoeniya pustyn [Problems of Desert Development] 4: 16-23 (in Russian).
Nechayeva, N.T. (ed.). 1982. Priroda Zapadnogo Kopetdaga [Nature in West Kopetdagh]. Ylym, Ashkhabad, 229 pp (in Russian).
Nechayeva, N.T. (ed.). 1985. Rastitelnost' i zhivotnyi mir Zapadnogo Kopetdaga [The Vegetation and Animal Life of West Kopetdagh]. Ylym, Ashkhabad, 278 pp (in Russian).
Nechayeva, N.T., V.K. Vasilevskaya, and K.G. Antonova. 1973. Zhiznennye formy rastenii pustyni Karakum [Life Forms of the Plants of the Karakum Desert]. Nauka, Moscow, 241 pp (in Russian).
Nelzina, Ye.N. 1966. Summer burrows of ground squirrels and their role in the formation of microbiocoenoses. Zool. zhurn. [Zool. J.] 45(8): 1235-1240 (in Russian).
Nelzina, Ye.N. 1971. The structure of burrow microbiocoenoses of the little ground squirrel and some species of gerbils. Parazitologiya [Parasitology] 5(3): 266-273 (in Russian).

Nelzina, Ye.N. 1977. Basic taxonomic groups of the organisms which form nidal and burrow microbiocoenoses. Parazitologiya [Parasitology] 11(4): 326–331 (in Russian).
Nelzina, Ye.N., Z.I. Klimova, and M.G. Protopopyan. 1978. Structure and level of organization of burrow microbiocoenoses of *Rhombomys opimus* in different parts of its range. In: M.S. Gilyarov (ed.), Itogi i perspektivy issledovanii po parazitotsenologii v SSSR [Results and Perspectives of Studies in Parasitocoenology in the USSR]. Nauka, Moscow: 87–97 (in Russian).
Nenilin, A.B. 1984a. On the taxonomy of spiders of the family Salticidae in the fauna of the USSR and adjacent countries. Zool. zhurn. [Zool. J.] 63(8): 1175–1180 (in Russian).
Nenilin, A.B. 1984b. Contributions to the fauna of the spider family Salticidae in the USSR. I. Catalog of the Salticidae of Middle Asia. In: A.S. Utochkin et al. (eds.) Fauna i ekologiyua paukoobraznykh [Fauna and Ecology of Arachnids]. Univ. of Perm, Perm: 6–37 (in Russian).
Nenilin, A.B. 1985. Contributions to the fauna of the spider family Salticidae in the USSR. II. Results of the study in the USSR. Fauna i ekologiya paukov SSSR/Trudy Zool. in-ta AN SSSR [Fauna and Ecology of Spiders of the USSR/Trans. Zool. Inst. Acad. Sci. USSR] 139: 129–134 (in Russian).
Nenilin, A.B. and M.V. Pestova. 1986. Spiders of the family Eresidae in the fauna of the USSR. Zool. zhurn. [Zool. J.] 65(11): 1734–1736 (in Russian).
Nenilin, A.B. and V.Ya. Fet. 1985. New species of spiders of the genus *Zodarion* (Aranei, Zodariidae) from Turkmenia. Zool. zhurn. [Zool. J.] 64(4): 618–620 (in Russian).
Nenilin, A.B. and V. Fet. 1992. Zoogeographical analysis of the world scorpion fauna (Arachnida: Scorpiones). Arthropoda Selecta (Moscow) 1(2): 3–31 (in Russian, with extended English summary).
Nepesova, M.G. 1975. Zoogeographical analysis of the fauna of darkling beetles (Coleoptera, Tenebrionidae) of Turkmenistan. Izv. AN TSSR, Ser. Biol. Nauk. [Proc. Acad. Sci. Turkmen SSR, Ser. Biol. Sci.] 5: 65–69 (in Russian).
Neronov, V.M. 1976. Zoogeographical analysis of the rodent fauna of Iran. Byull. MOIP [Bull. Moscow Soc. Natur.] 81(2): 32–47 (in Russian).
Neronov, V.M. and L.P. Arsenyeva. 1980. Zoogeographical analysis of the rodent fauna of Afhganistan. In: A.G. Voronov and N.N. Drozdov (ed.), Sovremennye problemy zoogeografii [Current Problems in Zoogeography]. Nauka, Moscow: 254–271 (in Russian).
Neshataeva, G.Yu. 1985. The classification of vegetation of Southwest Kopetdagh. In: N.T. Nechayeva (ed.), Rastitelnost' i zhivotnyi mir Zapadnogo Kopetdaga [The Vegetation and Animal Life of West Kopetdagh]. Ylym, Ashkhabad: 24–38 (in Russian).
Neshatayeva, G.Yu. and V.Ya. Fet. 1984. Current conditions and succession trends in plant communities in arid mountains of Southwest Kopetdagh. In: Problemy sokhraneniya genofonda i upravleniyua ekosistemami v stepnykh i pustynnykh zapovednikakh [Problems of Gene Pool Conservation and Ecosystem Management in Steppe and Desert Reserves]. Moscow: 248–250 (in Russian).
Nikitin, V.V. 1965. Illyustrirovannyi opredelitel' rastenii okrestnostei Ashkhabada [An Illustrated Key to the Plants of the Environs of Ashkhabad]. Nauka, Moscow-Leningrad, 457 pp (in Russian).
Nikitin, V.V. 1978. Endangered rare plants of Kopetdagh and the problem of restoration of their range. Izv. AN TSSR, Ser. Biol. Nauk. [Proc. Acad. Sci. Turkmen SSR, Ser. Biol. Sci.] 2: 3–9 (in Russian).
Nikitin, V.V. and A.M. Geldykhanov. 1988. Opredelitel' rastenii Turkmenistana [A Key to the Plants of Turkmenistan]. Nauka, Leningrad, 680 pp (in Russian).
Nikitin, V.V. and Ye.A. Klyushkin. 1971. On rare and endangered species of the flora of Turkmenia. Izv. AN TSSR, Ser. Biol. Nauk. [Proc. Acad. Sci. Turkmen SSR, Ser. Biol. Sci.] 5: 3–5 (in Russian).
Nikitin, V.V. and N.S. Krasikova. 1978. *Homalodiscus ochradeni*, an endangered rare plant requiring protection. Izv. AN TSSR, Ser. Biol. Nauk. [Proc. Acad. Sci. Turkmen SSR, Ser. Biol. Sci.] 6: 71–73 (in Russian).
Nikitina, V.N. 1954. Vegetation of East Kopetdagh in relation to its altitudinal zonation. Trudy Inst. biol. AN TSSR [Trans. Inst. Biol. Acad. Sci. Turkmen SSR] 1: 132–209 (in Russian).

Nikitina, V.N. 1956. On conditions of life forms in the vegetation of East Kopetdagh. Trudy TSKhI [Trans. Turkmen Agricult. Inst.] 8: 261–266 (in Russian).
Nikolsky, A.M. 1898. Two new lizard species from Russia. Yezheg. zool. muzeya Akademii nauk [Annu. Zool. Mus. Acad. Sci.] 3: 284–288 (in Russian).
Nikolsky, A.M. 1899. Herpetologia Turanica. Reptiles and amphibians of the Turkestan Gouvernor-Generalship. A. fedtshcenko, Reise in Turkestan, Zool. 2 (7). Izv. Ob-va lyubit. yestestvozn, antropol. i etnografii [Proc. Soc. Amat. Natur., Anthropol., and Ethnography], 84 pp (in Russian).
Nikolsky, A.M. 1903. *Contia transcaspica* n. sp. (Ophidia, Colubridae). Yezheg. zool. muzeya Akademii nauk [Annu. Zool. Mus. Acad. Sci.] 8: 11–13 (in Russian).
Nikolsky, A.M. 1905. Herpetologia Rossica. Reptiles and amphibians of the Russian Empire. Zap. Akad. nauk [Mem. Acad. Sci. St.-Petersburg], Ser. 8, 1: 1–518 (in Russian).
Nikolsky, A.M. 1908. Contributions to the herpetology of Russian Turkestan. Yezheg. zool. muzeya Akademii nauk [Annu. Zool. Mus. Acad. Sci.] 13: 336–344 (in Russian).
Nikolsky, A.M. 1911. Contributions to the herpetology of eastern Bokhara. Yezheg. zool. muzeya Akademii nauk [Annu. Zool. Mus. Acad. Sci.] 16(3): 272–284 (in Russian).
Nikolsky, A.M. 1915. Fauna Rossii i sopredel'nykh stran. Presmykayushchiyesya. [Fauna of Russia and Adjacent Countries. Reptiles]. Vol. 1. Chelonia, Sauria. Petrograd, 532 pp (in Russian).
Nikolsky, A.M. 1916. Fauna Rossii i sopredel'nykh stran. Presmykayushchiyesya. [Fauna of Russia and Adjacent Countries. Reptiles]. Vol. 2. Ophidia. Petrograd, 349 pp (in Russian).
Nikolsky, V.S. and M.I. Molyukov. 1975. A new record of mouselike dormouse (*Myomimus personatus* Ognev) in West Kopetdagh. Zool. zhurn. [Zool. J.] 54(10): 1583–1585 (in Russian).
Nikolsky, G.V. 1938. Ryby Tadzhikistana [Fishes of Tajikistan]. Izd. AN SSSR, Moscow-Leningrad, 228 pp (in Russian).
Nikolsky, G.V. 1945. Contributions to the systematics and biology of *Discognathichthys rossicus* (Nik.). Uchen. zap. MGU [Sci. Mem. Moscow State Univ], pp 127–131 (in Russian).
Nikolsky, G.V. 1947. Loaches of closed water bodies of Turkmenia. Byull. MOIP [Bull. Moscow Soc. Natur.] 52(3): 29–34 (in Russian).
Nikolsky, G.V. 1948. Contributions to the knowledge of the formation and development of reservoir ichthyofauna in some geographical zones of the Soviet Union. Zool. zhurn. [Zool. Journal] 27(2): 149–158 (in Russian).
Nikolsky, G.V. 1980. Structura vida i zakonomernosti izmenchivosti ryb [Species Structure and Patterns of Variation in Fishes]. Moscow, Pishchevaya Promyshlennost, 184 pp (in Russian).
Novikov, G.A. 1956. Khishchnye mlekopitayushchiye fauny SSSR [Carnivorous Mammals of the USSR Fauna]. Izd. AN SSR, Moscow-Leningrad, 294 pp (in Russian).
Nuratdinov, T. and R. Reimov. 1979. Effect of the anthropogenic factor on changes in the density of carnivorous mammals in the southern Aral area. In: V.E. Sokolov (ed.), Ekologicheskie osnovy okhrany i ratsionalnogo ispol'zovaniya khishchnykh mlekopitayushchikh [Ecological Principles of Conservation and Rational Exploitation of Carnivorous Mammals]. Nauka, Moscow: 50–51 (in Russian).
Nurgeldyev, O.N. 1960. Materialy po faune i ekologii mlekopitayushchikh trassy Karakumskogo kanala pervoi ochgeredi i ikh prakticheskoye znacheniye [Materials on the Fauna and Ecology of Mammals of the Karakum Canal and Their Practical Importance]. Ylym, Ashkhabad, 287 pp (in Russian).
Nurgeldyev, O.N., S. Shammakov, and Ch. Ataev. 1970. On distribution of some snake species in Turkmenia. In: N.A. Gladkov (ed.), Zhivotnyi mir Turkmenii [Animal Life of Turkmenia]. Ylym, Ashkhabad: 187–190 (in Russian).
Odingo, R.S. 1990. Review of UNEP's definition of desertification and iots programmatic implication. In: Desertification Revisited. Ed. UNEP-DC/PAC, Nairobi: 7–44.
Odum, E.P. 1971. Fundamentals of Ecology. 3rd Ed. Saunders, Philadelphia, 574 pp.
Odum, E.P. 1983. Basic Ecology. Saunders College Publ., Philadelphia, 613 pp.
Ognyev, S.I. 1924. A wonderful animal. Priroda i okhota na Ukraine [Nature and Hunting in Ukraine] 1: 115–116 (in Russian).

Ognyev, S.I. 1947. Zveri SSSR i prilezhashchykh stran: Gryzuny [Mammals of the USSR and Adjacent Countries: Rodents]. Izd. AN SSSR, Moscow-Leningrad 5, 809 pp (in Russian).
Ognyev, S.I. and V.G. Geptner. 1929. Mlekopitayushchie Srednego Kopetdaga i prilegayushchei ravniny [Mammals of Central Kopetdagh and Adjacent Plain]. Trudy NII zoologii MGU [Trans. Sci. Res. Inst. Zoology of Moscow State Univ.] 3(1): 47-171 (in Russian).
Okladnikov, A.P. 1956. The ancient past of the Turkmenistan. Trudy Inst. istorii, arkheol. i etnograf. AN TSSR [Trans. Inst. Hist., Archaeol., and Ethnogr. Acad. Sci. Turkmen SSR]: 181-221 (in Russian).
Olsufyev, N.G. 1937. Slepni - Tabanidae. Fauna SSSR. Dvukrylye nasekomye. [Horse Flies - Tabanidae. Fauna of the USSR. Diptera, Insecta]. Izd AN SSSR, Leningrad 7(2), 433 pp (in Russian).
Olsufyev, N.G. 1966. On ancient Atlantic connections of the southern part of North America and Middle Asia, with horse flies as an example. Byull. MOIP, otd. biol [Bull. Mosc. Soc. Natur., Div. Biol.] 69(3): 73-76 (in Russian).
Olsufyev, N.G. 1977. Fauna SSSR. Nasekomye dvukrylye [Fauna of the USSR. Diptera, Insecta]. Izd AN SSSR, Leningrad 7(2), 434 pp (in Russian).
Olsufyev, N.G. 1980. Types of horse fly faunas and zoogeographical regionalization of the USSR. In: A.G. Voronov and N.N. Drozdov (ed.), Sovremennye problemy zoogeografii [Current Problems in Zoogeography]. Nauka, Moscow: 81-115 (in Russian).
Olsufyev, N.G. and V.V. Shevchenko. 1964. A species of horse fly, *Tabanus ansarii badhysi* subsp. nova, new for the USSR fauna. Trudy in-ta zool. AN Kazakh. SSR [Trans. Inst. Zool. Acad. Sci. Kaz. SSR] 22: 197-198 (in Russian).
Opredelitel' rastenii Srednei Azii [A Key to the Plants of Middle Asia]. 1968-1987. Fan, Tashkent, Vols. 1-9, (in Russian).
Orlov, V.I. 1970. On rewintering of birds on the Turkmen shore of the Caspian Sea. In: Voprosy popul. ekologii i geografii zhivotnykh [Problems of Population Ecology and Geography of Animals]. Moscow: 171-183 (in Russian).
Orlovsky, N.S. 1962. Some data on dust storms in Turkmenia. Sb. rabot Ashkhabadsk. gidrometeorol. observat. [Collected Papers of Ashkhabad Hydrometeorol. Observatory] 3: 17-42 (in Russian).
Orlovsky, N.S. 1976. On the thermal balance in Turkmenistan. In: Gidrometeorologicheskii rezhim Turkmenii [Hydrometeorological Regime of Turkmenia]. Ylym, Ashkhabad: 48-66 (in Russian).
Orlovsky, N.S. 1981. Pogoda i tonkovoloknistyi khlopchatnik Turkmenistana [Weather and Thin-fiber Cotton in Turkmenistan]. Ylym, Ashkhabad, 163 pp (in Russian).
Orlovsky, N.S., A.G. Nilova, and G.A. Orlovskaya. 1975. An attempt to categorize the agroclimatic zonation of Kopetdagh. In: Gidrometeorlogicheskii rezhim Turkmenii [Hydrometeorological Regime of Turkmenia]. Ylym, Ashkhabad, (in Russian).
Orlovsky, N.S. and L.P. Shlikhter. 1981. On the sun radiation regime in the lowland Turkmen SSR. In: Gidrometeorologicheskii rezhim Turkmenii [Hydrometeorological Regime of Turkmenia]. Ylym, Ashkhabad: 90-103 (in Russian).
Orlovsky, N.S. and Z.I. Volosyuk. 1981. Pogoda i otgonno-pastbishchnoe zhivotnovodstvo Turkmenistana [Weather and the Distant Pasture Livestock Industry in Turkmenistan]. Ylym, Ashkhabad, 104 pp (in Russian).
Ovchinnikov, P.N. 1940a. Major directions in speciation trends related to the origin of vegetation types in Middle Asia. Trudy Tadzh. fil. AN SSSR [Trans. Tajik Branch Acad. Sci. USSR] 3: 114-133 (in Russian).
Ovchinnikov, P.N. 1940b. On the history of vegetation of southern Middle Asia. Sov. bot. [Soviet Botany] 3: 23-48 (in Russian).
Ovchinnikov, P.N. 1948a. Major features of the origin of mountain steppes. Soobshch. Tadzh, fil. AN SSSR [Comm. Tajik Branch Acad. Sci. USSR] 3: 18-20 (in Russian).
Ovchinnikov, P.N. 1948b. On the development of phytocoenological classification of woody vegetation of Tajikistan. Soobsh. Tadzhik. Fil. AN SSSR [Reports of Tajik Branch of the Academy of Sciences of the USSR] 7: 3-12 (in Russian).

Ovchinnikov, P.N. 1955. Major directions in speciation related to the origin of vegetation types. Trudy AN Tadzh. SSR [Trans. Acad. Sci. Tajik SSR] 31: 107–140 (in Russian).

Ovchinnikov, P.N. 1957a. Major features of vegetation and floristic regions of Tajikistan. Flora Tadzhikskoi SSR [Flora of the Tajik SSR], Leningrad 1: 9–20 (in Russian).

Ovchinnikov, P.N. 1957b. On major directions in the classification of the vegetation of Middle Asia. Izv. Otd. estestv. nauk AN Tadzh. SSR [Proc. Div. Natural Sci. Acad. Sci. Tajik SSR] 18: 49–65 (in Russian).

Ovchinnikov, P.N. 1971. The Varzob River Valley as one of the areas of the Ancient Mediterranean botanical-geographical region. In: P.N. Ovchinnikov (ed.), Flora i rastitel'nost' ushchelya reki Varzob [Flora and Vegetation of the Varzob River Valley]. Nauka, Leningrad: 396–447 (in Russian).

Ovtsharenko, V.I. and V.Ya. Fet. 1980. The fauna and ecology of spiders (Aranei) in Badghyz (Turkmen SSR). Entomol. obozr. [Entom. Rev.] 59(2): 442–447 (in Russian).

Ovtsharenko, V.I., G. Levy, and N.I. Platnick. 1994, A review of the ground spider genus *Synaphosus* (Araneae, Gnaphosidae). Amer. Mus. Novit., 3095: 1–27.

Ovtsharenko, V.I., N.I. Platnick, and D.X. Song. 1992. A review of the North Asian ground spiders of the genus *Gnaphosa* (Araneae, Gnaphosidae). Bull. Amer. Mus. Nat. Hist. 212: 1–88.

Ozorovsky, V.V. 1968. *Coluber jugularis* and *Ptyas mucosus* in Turkmenia. Priroda [Nature] 5: 91–92 (in Russian).

Panov, E.N. and L.Yu. Zykova. 1985. The comparative biology of steppe and Caucasian agamas (*Agama sanguinolenta, A. caucasica*) in the Sumbar River basin (West Kopetdagh). In: N.T. Nechayeva (ed.), Rastitelnost' i zhivotnyi mir Zapadnogo Kopetdaga [The Vegetation and Animal Life of West Kopetdagh]. Ylym, Ashkhabad: 185–204 (in Russian).

Paramonow, S.J. 1961. Dipterologische Fragmente XXXXVII–XL. Rev. Espan. Entomol. 37(1): 71–90.

Parin, N.V. and V.Yu. Dolzhansky. 1982. *Nemochilus starostini*, the first cave fish in the USSR fauna. Priroda [Nature] 8: 29–32 (in Russian).

Pashchenko, Yu.I. 1964. New data on the distribution of some reptiles of Turkmenia. In: I.S. Darevsky (ed.), Voprosy gerpetologii [Problems of Herpetology]. Nauka, Leningrad: 16 (in Russian).

Paulson, T.L. 1965. Countercurrent multipliers in avian kidney. Science 148(3668): 389–391.

Paulson, T.L. and G.A. Bartholomew. 1962. Salt utilization in the House Finch. Condor 64: 245–252.

Pavlenko, T.A. and E.I. Zinchenko. 1981. Food objects and their role in rodent feeding in natural desert pastures of southern Kizylkum. In: Ekologiya nekotorykh vidov mlekopitayushchikh i ptits ravnin i gor Uzbekistana [Ecology of Some Species of Mammals and Birds in Lowlands and Mountains of Uzbekistan]. Fan, Tashkent: 39–52 (in Russian).

Pavlovskaya, L.P. 1982. Promyslovye ryby nizhnego techceniya Amudaryi i gidrostroitel'stvo [Food Fishes of the Lower Amudarya and Hydrotechnic Construction]. Fan, Tashkent, 97 pp (in Russian).

Pavlovskaya, L.P. 1990. Struktura rybnogo naseleniya v kontsevykh sbrosakh orositel'nykh sistem [Structure of the Fish Population in the Distal Ends of the Irrigation Disposal System]. Fan, Tashkent, 110 pp (in Russian).

Pavlovskaya, L.P., M.A. Kunin, and V.B. Salnikov. 1986. Egg incubation in some fish species from Lake Sarykamysh. Manuscript deposited at TsNIITERKh, No. 806, 24 pp (in Russian).

Pavlovskaya, L.P. and V.B. Salnikov. 1986. On the reproduction and feeding of the carp in Lake Sarykamysh. Vestnik Kara-Kalpaksgogo filiala Akademii Nauk. Uzbekskoi SSR [Mess. Kara Kalpak Branch Acad. Sci. Uzbek SSR] 4: 44–52 (in Russian).

Pavlovskaya, L.P. and V.B. Salnikov. 1990. The patterns of formation and development of the fish population in enclosed water bodies which accumulate drainage waters (utilizing Lake Sarykamysh as an example). Gidrobiol. zhurn. [Hydrobiol. J.] 26(1): 39–47 (in Russian).

Pavlovskaya, L.P. and I.M. Zholdasova. 1991. The anthropogenic changes in the fish population of the Amudarya River (based on egg and fry studies). Voprosy ikhtiologii [Problems of Ichthyology] 31(4): 585–595 (in Russian).

Pavlovsky, E.N. 1934. Some data on the venomous animals of Turkmenia. In: Trudy Kara-Kalinskoi i Kzyl-Atrekskoi parazitl. ekspeditsii 1931 g., s materialami po faune Turkmenii [Transactions of the Kara Kala and Kzyl Atrek Parasitology Expedition in 1931, with Contributions to the Fauna of Turkmenia]. Izd. AN SSSR i Narkomzdrava Turkmenii, Leningrad: 191–204 (in Russian).

Peklo, A.M. and O.S. Sopyev. 1980. *Hypocolius ampelinus* (Aves, Bombycillidae) nesting in the USSR. Vestn. zool. [Mess. Zool.] 3: 47–52 (in Russian).

Pelt, N.N. and V.F. Chervinsky. 1956. Directions of the agricultural development of the Ustyurt. Tr. Aralo-Kaspiiskoi kompleksnoi eksped. AN SSSR [Trans. Aralo-Caspian Complex Expedition Acad. Sci. USSR] 5: 47–62 (in Russian).

Perskaya, A.D. 1955. Fruits and seeds of trees and shrubs stabilizing sand in the Middle Asian deserts. Trudy Repeteksk. peschano-pust. stantsii AN Turkm. SSR [Trans. Repetek Sand Desert Station Acad. Sci. Turkm. SSR] 3: 235–293 (in Russian).

Pestinsky, B.V. 1939. Data on the biology of poisonous snakes of Middle Asia, their capture, and containment in captivity. Trudy Uzbek. zool. sada [Trans. Uzbek Zoo] 1: 4–62 (in Russian).

Petrosyants, M.A. 1956. On the Akchagyl paleogeography of the Central and Southeast Karakum. Izv. AN SSSR, Ser. geograf. [Proc. Acad. Sci. USSR, ser. geograph.] 5: 79–86 (in Russian).

Petrosyants, M.A. 1961. Palynological spectra of the Neogene deposits of the Central and Southeast Karakum. In: Novye danye po geologii i neftegazonosnosti Srednei Azii [New Data on the Geology and Oil and Gas Deposits in Middle Asia]. Gostoptekhizdat, Moscow: 17–22 (in Russian).

Petrov, M.P. 1933. Root sytems of plants of the Karakum sand desert, and their distribution according to ecological conditions. Trudy po prikl. bot., genet., and selektsii rastenii [Studies in Applied Botany, Genetics and Selection of Plants] 1(1): 113–208 (in Russian).

Petrov, M.P. 1935. Development of root systems of shrubs in the Karakum sand desert. Problemy rastenievodcheskogo osvoyeniya pustyn' [Problems of Plant Breeding in Desert Development] 4: 67–98 (in Russian).

Petrov, M.P. 1945. Tree and shrub vegetation of Southwest Kopetdagh and its connection with the forest vegetation of northern Iran. Izv. Turkm. FAN SSSR, Ser. Biol. Nauk. [Proc. Turkmen Branch Acad. Sci. USSR, Ser. Biol. Sci.] 1: 25–31 (in Russian).

Petrov, M.P. 1973. Pustyni zemnogo shara [Deserts of the World]. Nauka, Leningrad, 435 pp (in Russian).

Petrova, Ye.F. 1975. Wild figs of West Kopetdagh. Trudy po prikl. bot., genet., and selektsii [Studies in Applied Botany, Genetics and Selection] 54(1): 199–204 (in Russian).

Platnick, N.I. and V.I. Ovtsharenko. 1991. On Eurasian and American *Talanites* (Araneae, Gnaphosidae). J. Arachnol. 19(2): 115–121.

Platnick, N.I. and M.U. Shadab. 1980. A revision of the North American spider genera *Nodocion, Litopyllus* and *Synaphosus* (Araneae, Gnaphosidae). Amer. Mus. Novitates 2691: 1–26.

Polozov, S.A. 1980. New data about the francolin in the USSR: data on the population in West Kopetdagh. In: Fauna i ekologiya nazemnykh pozvonochnykh. Sborn. nauchn. trudov MGPI [Fauna and Ecology of Terrestrial Vertebrates/Sci. Trans. Moscow State Pedagogical Institute], Moscow: 27–39 (in Russian).

Polozov, S.A. 1981. New data about the francolin distribution in Southwest Kopetdagh. Khozyaistvennaya deyatel'nost' i okhotnichya fauna [Industrial Activities and Game Fauna], Kirov 2: 174–175 (in Russian).

Poltoratsky, S.V. and A.G. Lyashenko. 1972. Formation of the modern ichthyofauna and perspectives of the fishing industry in Lake Sarykamysh. Rybnoye khozyaistvo [Fish Industry] 10: 14–16 (in Russian).

Popov, K.P. 1974. Do the pistachio woodlands of Midle Asia constitute a forest-type vegetation? Bot. zhurn. [Bot. J.] 59(12): 1755–1759 (in Russian).

Popov, K.P. 1979. Fistashka v Srednei Azii [Pistachio in Middle Asia]. Ylym, Ashkhabad, 160 pp (in Russian).

Popov, K.P. 1981. The genus *Amygdalus* in Turkmenia. Izv. Akad. Nauk. Turkm. SSR, Ser. Biol. Nauk. [Proc. Acad. Sci. Turkm. SSR, Ser. Biol. Sci.] 6: 24–29 (in Russian).

Popov, K.P. 1985. The walnut (*Juglans regia* L.) in Kopetdagh. In: N.T. Nechayeva (ed.), Rastitelnost' i zhivotnyi mir Zapadnogo Kopetdaga [Vegetation and Animal life in West Kopetdagh]. Ylym, Ashkhabad: 51–60 (in Russian).
Popov, K.P. and E.M. Seifulin. 1985. Cushion plants of Turkmenistan. Izv. Akad. Nauk. Turkm. SSR, Ser. Biol. Nauk. [Proc. Acad. Sci. Turkm. SSR, Ser. Biol. Sci.] 4: 37–41 (in Russian).
Popov, M.G. 1923. Flora of the "painted rocks" (red sandstone foothills) of Bokhara. Trudy Turkest. nauchn. ob-va [Trans. Turkestan Sci. Soc.], Tashkent 1: 3–42 (in Russian).
Popov, M.G. 1927. Major features of the historical development of flora of Middle Asia. Byull. Sredneaz. dnevnika Sredneaz. gos. un-ta [Bull. Middle Asian Journal Middle Asian State Univ.], Tashkent 15: 239–290 (in Russian).
Popov, M.G. 1929a. The genus *Cicer* and its species. On the problem of the origin of the Mediterranean flora. Trudy po prikl. bot., genet., and selektsii rastenii [Studies in Applied Botany, Genetics and Selection of Plants] 21(1): 234–239 (in Russian).
Popov, M.G. 1929b. The wild fruit trees and shrubs in Central Asia. Trudy po prikl. bot., genet., and selektsii rastenii [Studies in Applied Botany, Genetics and Selection of Plants] 22(3): 241–483 (in Russian).
Popov, M.G. 1931. Between Mongolia and Iran. Trudy po prikl. bot., genet., and selektsii rastenii [Studies in Applied Botany, Genetics and Selection of Plants] 26(3): 45–84 (in Russian).
Popov, M.G. 1938. The principal periods of the formation of and immigrations in the flora of Middle Asia during the era of Anthophyta, and the relict types of this flora. Problema relictov vo flore SSSR [Problem of Relicts in the Flora of the USSR], Izd. AN SSSR, Moscow-Leningrad 1: 10–26. (in Russian).
Popov, M.G. 1940. An attempted monograph of the genus *Eremostachys* Bunge. Nov. nem. MOIP [New Mem. Moscow Soc. Natur.] 19: 7–11 (in Russian).
Popov, M.G. 1941. Geographical-genetic elements of the flora of the Alma-Ata Reserve. In: Materialy issledovanii rastitel'nosti Kazakhstana [Materials on the Studies of the Vegatation of Kazakhstan]. Izd. ANSSSR Moscow-Leningrad 20: 29–34 (in Russian).
Popov, M.G. 1963. Osnovy florogenetiki [Foundations of Florogenetics]. Izd. ANSSSR, Moscow, 135 pp (in Russian).
Popov, M.G. 1983. Filogeniya, florogenetika, florografia, sistematica. Izbrannye trudy [Phylogeny, Florogenetics, Florography, and Systematics. Selected Works]. Naukova dumka, Kiev 1–2, 478 pp (in Russian).
Petrov, M.P. 1939. On the wild fruit trees of mountainous Turkmenia. Priroda [Nature] 5: (in Russian).
Potapov, R.L. 1978. Disappearance of the Turanian tiger. Priroda [Nature] 6: 23–25 (in Russian).
Potapov, R.L. 1987a. On the formation of the range of pheasant (*Phasianus colchicus* L.). Issledovaniya po faune i ekologii ptits Palearktiki/Trudy Zool. in-ta AN SSSR [Studies on Fauna and Ecology of Birds in the Palearctic/Trans. Zool. Inst. Acad. Sci. USSR] 163: 71–76 (in Russian).
Potapov, R.L. 1987b. Order Galliformes. In: R.L. Potapov and V.E. Flint (ed.), Ptitsy SSSR [Birds of the USSR: Galliformes, Gruiformes]. Nauka, Leningrad: 7–260 (in Russian).
Potapov, R.L. and G.N. Sapozhnikov. 1976. The Zeravshan pheasant in Tajikistan. Izv. AN Tadzhik. SSR, otd. biol. nauk [Proc. Acad. Sci. Tajik SSR. Div. biol. sci.] 1: 62–66 (in Russian).
Pottier, G.A. 1981. La petite Flora de la Tunisie. Tunis 2: 1092 pp.
Pratov, U. 1986a. The position of the genus *Nanophyton* in the system of the famliy Chenopodiaceae and its phylogenesis. Bot. zhurn. [Bot. J.] 71(2): 175–185 (in Russian).
Pratov, U. 1986b. Rod *Climacoptera* Botsch. [The genus *Climacoptera* Botsch.]. Fan, Tashkent, 72 pp (in Russian).
Pravdin, I.F. 1966. Rukovodstvo po izucheniyu ryb [A Manual on Fish Studies]. Pishchevaya Promyshlennost', Moscow, 376 pp (in Russian).
Pravdin, F.N. 1978. Ekologicheskaya geografiya nasekomykh Srednei Azii [Ecological Geography of Insects of Middle Asia]. Nauka, Moscow, 272 pp (in Russian).
Pravdin, F.N. and L.L. Mishchenko. 1980. Formirovaniye i evolyutsiya ekologicheskikh faun

nasekomykh Srednei Azii [Formation and Evolution of the Ecological Faunas of Insects of Middle Asia]. Nauka, Moscow, 256 pp (in Russian).

Prikhod'ko, S.Ya. and N.D. Prikhod'ko. 1968. Growth and development of root systems of white and black saksauls. Izv. Akad. Nauk. Turkm. SSR, Ser. Biol. Nauk. [Proc. Acad. Sci. Turkm. SSR, Ser. Biol. Sci.] 4: 22–26 (in Russian).

Proskuryakova, G.M. 1964. Conspectus of the flora of Bolshoi Balkhan. I. Nauchn. dokl. vyssh. shkoly. Biol. nauki. [Sci. Rep. High Educ., Biol. Sciences] 3: 107–118 (in Russian).

Proskuryakova, G.M. 1965. Conspectus of the flora of Bolshoi Balkhan. II. Nauchn. dokl. vyssh. shkoly. Biol. nauki. [Sci. Rep. High Educ., Biol. Sciences] 3: 97–107 (in Russian).

Proskuryakova, G.M. 1966. Conspectus of the flora of Bolshoi Balkhan. III. Nauchn. dokl. vyssh. shkoly. Biol. nauki. [Sci. Rep. High Educ., Biol. Sciences] 2: 134–143 (in Russian).

Proskuryakova, G.M. 1967. Conspectus of the flora of Bolshoi Balkhan. IV. Nauchn. dokl. vyssh. shkoly. Biol. nauki. [Sci. Rep. High Educ., Biol. Sciences] 3: 70–79 (in Russian).

Proskuryakova, G.M. 1978. Juniper woodlands of Turkmenia and problems of their conservation. Izv. AN TSSR, Ser. Biol. Nauk. [Proc. Acad. Sci. Turkmen SSR, Ser. Biol. Sci.] 4: 34–41 (in Russian).

Proskuryakova, G.M. 1985. Conspectus of the flora of Palyzan. In: A.K. Skvortsov (ed.), Floristicheskie issledovaniya v raznykh raionakh SSSR [Floristic Studies in Various Regions of the USSR]. Nauka, Moscow: 90–138 (in Russian).

Prozorovsky, A.V. 1940. Deserts and semideserts of the USSR. Rastitel'nost; SSSR [Vegetation of the USSR]. Izd. AN SSSR, Moscow-Leningrad 2: 267–480 (in Russian).

Pulatova, M.Z. 1971. The Eocene flora of the Tadzhik Depression. Tez. Dokl. III Mezhdunar. palinol. konf. Sekt. VI [Abstr. Rep. 3d Intern. Palinol. Conf., Sect. 6] (in Russian).

Pyatayeva, A.D. 1956. The vegetation of the Kyurendagh Range. Trudy SAGU [Trans. Central Asian State Univ., Tashkent] 86: 9–75 (in Russian).

Quezel, P. and S. Santa. 1962. Nouvelle flora de l'Algerie et des regions desertiques meridionales. Paris 2: 1–1060.

Rachkovskaya, Ye.I. 1957. On the biology of desert semishrubs. Trudy BIN AN SSR [Trans. Bot. Inst. Acad. Sci. USSR], Ser. 3 (Geobotany) 3: 5–87 (in Russian).

Radde, G. 1898. Wissenschaftliche Ergebnisse der in Jahre 1886 allerhochst befohlenen Expedition nach Transcaspiens und Nord-Chorassan von G. Radde. J. Perthes, Gotha, 126: 196 pp.

Radde, G.I. (ed.). 1899. Museum Caucasicum. Collektsii Kavkazskogo Museya [Collections of the Caucasian Museum], Vol. 1. Zoology. Tiflis, 520 pp (in Russian and Latin).

Radde, G. and A. Walter. 1889. Die Vögel Transcaspiens. Ornis 5: 1–128, 165–279.

Raevsky, M.I. 1969. Major features of the paleogeography of East Turkmenia in the Late Pliocene. Sov. geologiya [Soviet Geology] 9: 70–81 (in Russian).

Rechinger, K.H. 1963–1981. Flora Iranica. Graz, Hefte 1–161.

Reshetnikov, Yu.S. 1986. A synecological approach toward the dynamics of fish density. In: Yu.S. Reshetnikov (ed.), Dinamika chislennosti promyslovykh ryb [Dynamics of Density of Food Fish Species]. Nauka, Moscow: 22–36 (in Russian).

Reshetnikov, Yu.S., O.A. Popova, O.P. Sterligova *et al.* 1982. Izmenenie structury rybnogo naseleniya evtrofiruemogo ozera [Changes in the Structure of Fish Populations in a Lake During Eutrophication]. Nauka, Moscow, 248 pp (in Russian).

Reshkevich, N.A. 1958. Reproduction in the Khentau toad agama. Priroda [Nature] 6: 114–115 (in Russian).

Ricklefs, R.E. 1976. The Economy of Nature: A Textbook in Basic Ecology. Portland, Ore., Chiron Press, 455 pp.

Riedl, H. 1968. Cupressaceae. Flora Iranica, Graz 50/28(2): 1–10.

Rodendorf, B.B. 1964. Istoricheskoe razvitie dvukrylykh nasekomykh [Historical Development of Dipterous Insects]. Nauka, Moscow, 310 pp (in Russian).

Rodin, L.Ye. 1940. Major features of the vegetative cover of North (Trans-Unguz) Karakum. Prirodnye resursy Karakumov [Natural Resources of the Karakum Desert]. Izd. AN SSSR, Leningrad 1: 61–106 (in Russian).

Rodin, L.Ye. 1963. Rastitel'nost' pustyn' Zapadnoi Turkmenii [Vegetation of the Deserts of West Turkmenia]. Izd. AN SSSR, Moscow-Leningrad, 294 pp (in Russian).
Rodin, L.Ye. and N.I. Rubtsov. 1948. In: Semishrub sagebrush and *Salsola* deserts. Rastitel'nyi pokrov SSSR [Vegetative Cover of the USSR], Izd. AN SSSR, Moscow-Leningrad 2: 731–796 (in Russian).
Roewer, C.F. 1942. Katalog der Araneae von 1758 bis 1940. Vol. 1. Natura, Bremen, 1040 pp.
Roewer, C.F. 1954. Katalog der Araneae von 1758 bis 1940. Vol. 2a. Inst.Roy. Sci. Natur., Bruxelles, 1–923 pp, Vol. 2b, pp 927–1751.
Roewer, C.F. 1955. Die Araneen der Oesterreichischen Iran-Expedition 1949/50. Sber. öst. Akad. Wiss. (I) 164(9): 751–782.
Roewer, C.F. 1960. Lycosidae aus Afghanistan. Acta Univ. Lund. (N.F.) (2), 56(17): 3–34.
Roewer, C.F. 1961a. Araneae Dionychae aus Afghanistan. I. Acta Univ. Lund. (N.F.) (2), 58(3): 3–33.
Roewer, C.F. 1961b. Araneae Dionychae aus Afghanistan. II. Acta Univ. Lund. (N.F.) (2), 58(4), 3–34.
Rossolimo, O.L. and I.Ya. Pavlinov. 1982. Mlekopitayushchie Zapadnogo Kopetdaga [Mammals of West Kopetdagh]. In: N.T. Nechayeva (ed.), Priroda Zapadnogo Kopetdaga [Nature in West Kopetdagh]. Ylym, Ashkhabad: 203–228 (in Russian).
Rozanov, M.P. 1940. Hunting in Badghyz and in tugais of the Tedzhen. Boyets-okhotnik [The Soldier Hunter] 6 (in Russian).
Rumyantseva, A. 1953. Development of the root system in sarsazan (*Halocnemum strobilaceum*). Izv. AN TSSR, Ser. Biol. Nauk. [Proc. Acad. Sci. Turkmen SSR, Ser. Biol. Sci.] 4: 27–29 (in Russian).
Rustamov, A.K. 1945. European forest "Hyrcanian" fauna in Kopetdagh. Izv. Turkm. FAN SSSR, Ser. Biol. Nauk. [Proc. Turkmen Branch Acad. Sci. USSR, Ser. Biol. Sci.] 3/4: 132–135 (in Russian).
Rustamov, A.K. 1948. On the southern boundary of distribution of psammophile fauna in the Karakum. Byull. MOIP [Bull. Moscow Soc. Natur.] 53(5): 85–91 (in Russian).
Rustamov, A.K. 1951. New data on zoogeography and ornithofauna of southern Ustyurt. Izv. AN Kaz. SSR [Proc. Acad. Sci. Kaz. SSR] 10: 61–71 (in Russian).
Rustamov, A.K. 1954. Ptitsy pustyni Karakum [Birds of the Karakum Desert]. Ashkhabad, Izd. AN TSSR, 344 pp (in Russian).
Rustamov, A.K. 1955a. On the nesting bird fauna and practical importance of terrestrial vertebrates in the Tashauz Region (northeastern Turkmenistan). Uchen. zap. Turkmen. gos. un-ta [Sci. Mem. Turkmen State Univ] 4: 95–130 (in Russian).
Rustamov, A.K. 1955b. On the term "life form" in animal ecology. Zool. zhurn. [Zool. J.] 4: 710–718 (in Russian).
Rustamov, A.K. 1955c. La faune avieenne des déserts de l'URSS. Essai écologique et biogeographique. In: Acta 9th Congr. Intern. Ornithol., Basel-Stuttgart: 510–515.
Rustamov, A.K. 1956. On the fauna of amphibians and reptiles of southeastern Turkmenia. Trudy Turkm. sel'skokhoz. in-ta [Trans. Turkmen Agric. Inst.] 8: 293–306 (in Russian).
Rustamov, A.K. 1958. Ptitsy Turkmenistana [Birds of Turkmenistan]. Ashkhabad, Izd. AN TSSR 2, 333 pp (in Russian).
Rustamov, A.K. 1961. On ornithogeographical connections of Kopetdagh with Hyrcania. Trudy In-ta zool. AN Kaz. SSR [Trans. Inst. Zool. Acad. Sci. Kazakh SSR] 15: 132–138 (in Russian).
Rustamov, A.K. 1964. Further remarks on the term "life form" in animal ecology. In: Problemy ornitologii/Trudy III Vsesoyuzn. ornitol. konfer. [Problems in Ornithology/Trans. 3d All-Union Ornithol. Conf.], Lvov (in Russian).
Rustamov, A.K. 1966. A brief survey of the herpetofauna of Turkmenia and its zoogeographical characteristics. In: Pozvonochnye zhivotnye Srednei Azii [Vertebrates of Middle Asia]. Fan, Tashkent: 158–168 (in Russian).
Rustamov, A.K. 1968. Zoogeographical features of various groups of terrestrial fauna of the Middle Asian deserts. Ornitologiya [Ornithology] 9: 131–136 (in Russian).

Rustamov, A.K. 1973. The francolin is on the verge of extinction. Priroda [Nature] 7: 46–49 (in Russian).
Rustamov, A.K. 1976. Water in the desert as ecological and zoogeographical factors. Teoreticheskie i prikladnye aspekty okhrany prirody i okhotovedeniya [Theor. and Applied Aspects of Natural Conservation and Game Management] 84: 40–44 (in Russian).
Rustamov, A.K. 1978. Rare mammals and birds in Turkmenistan and their protection. In: Some Problems of Wildlife Conservation in the USSR. Moscow: 98–103.
Rustamov, A.K. 1979. Conservation of the fauna. In: Okhrana prirody Turkmenistana [Natural Conservation in Turkmenistan] 5: 10–18 (in Russian).
Rustamov, A.K. 1980. To revive the cheetah in Transcaspia. Priroda [Nature] 7: 46–49 (in Russian).
Rustamov, A.K. 1981a. Strategic tasks of natural conservation in Turkmenistan. In: Tez. dokl. 2 nauchn. konfer, po okhrane prirody Turkm. SSR [Abstr. 2nd Sci. Conf. on Natural Conservation in the Turkmen SSR], Ashkhabad: 126–130 (in Russian).
Rustamov, A.K. 1981b. Zoogeographical connections of the herpetofauna of Middle Asia and the Caucasus. Byull. MOIP [Bull. Moscow Soc. Natur.] 86(4): 31–36 (in Russian).
Rustamov, A.K. 1981c. An attempt to estimate the generic endemism of the herpetofaunas of Iran, Afghanistan, and Middle Asia. In: I.S. Darevsky (ed.), Voprosy gerpetologii [Problems of Herpetology]. Nauka, Leningrad: 118–119 (in Russian).
Rustamov, A.K. 1981d. M.A. Menzbier's views on parallelism. Vestn. zool. [Mess. Zool.] 2: 3–9 (in Russian).
Rustamov, A.K. 1983. Turkmenistan as a key region for preservation of the gene pool of rare and endangered animal species. Izv. Akad. Nauk. Turkm. SSR, Ser. Biol. Nauk. [Proc. Acad. Sci. Turkm. SSR, Ser. Biol. Sci.] 5: 3–11 (in Russian).
Rustamov, A.K. 1987. Ecology and conservation of the desert. Probl. osv. pustyn' [Probl. Desert Development] 5: 19–25 (in Russian).
Rustamov, A.K. 1991a. On the conservation of biodiversity. Izv. Akad. Nauk. Turkm. SSR, Ser. Biol. Nauk. [Proc. Acad. Sci. Turkm. SSR, Ser. Biol. Sci.] 3: 3–8 (in Russian).
Rustamov, A.K. 1991b. On some aspects of conservation of biodiversity under the conditions of anthropogenic transformation of the desert. Probl. osv. pustyn' [Probl. Desert Development] 3/4: 31–36 (in Russian).
Rustamov, A.K. and Ch. Ataev. 1976. New data on the herpetofauna of Turkmenistan. Izv. Akad. Nauk. Turkm. SSR, Ser. Biol. Nauk. [Proc. Acad. Sci. Turkm. SSR, Ser. Biol. Sci.] 5: 47–53 (in Russian).
Rustamov, A.K. and Ch. Ataev. 1990. Distribution and ecology of *Eirenis meda* and *Pseudocyclophis persicus* in Turkmenistan. In: Ekolog. aspekty okhrany i retsional'noye ispolzovaniye dikikh zhivotnykh [Ecolog. aspects of conservation and rational exploitation of wildlife]. Moscow: 29–35 (in Russian).
Rustamov, A.K., Ch.A. Ataev, O.S. Sopyev, and A.N. Makarov. 1985. On the ecology of *Eublepharis turcmenicus*. Izv. Akad. Nauk. Turkm. SSR, Ser. Biol. Nauk. [Proc. Acad. Sci. Turkm. SSR, Ser. Biol. Sci.] 1: 3–6 (in Russian).
Rustamov, A.K. and N.N. Drozdov. 1984. Parallelism and convergent evolution in the adaptations of birds of the arid ecosystems. Ornitologiya [Ornithology] 19: 64–67 (in Russian).
Rustamov, A.K. and D.N. Kashkarov. 1990. On the applicability of the Le Chatelier principle in biology. Izv. Akad. Nauk. Turkm. SSR, Ser. Biol. Nauk. [Proc. Acad. Sci. Turkm. SSR, Ser. Biol. Sci.] 5: 3–9 (in Russian).
Rustamov, A.K., A. Kurbanov, and O. Sopyev. 1962. On the fauna of amphibians and reptiles of the Atrek area. Trudy Turkm. sel'skokhoz. in-ta [Trans. Turkmen Inst. Agricult.] 11(2): 95–107 (in Russian).
Rustamov, A.K. and V.M. Makeev. 1981. Conservation of rare and endangered species of herpetofauna in Turkmenistan and rational exploitation of poisonous snakes. Izv. Akad. Nauk. Turkm. SSR, Ser. Biol. Nauk. [Proc. Acad. Sci. Turkm. SSR, Ser. Biol. Sci.] 4: 10–17 (in Russian).
Rustamov, A.K., V.M. Makeev, O.S. Sopyev, and S. Shammakov. 1988. Problems of conservation

of reptiles in Turkmenistan and work with Red Data Books. In: Redkie i maloizuchennye zhivotnye Turkmenistana [Rare and Poorly Known Animals of Turkmenistan]. Ashkhabad, Ylym: 101–118 (in Russian).
Rustamov, A.K. and Ye.S. Ptushenko. 1948. Caravan paths in the Karakum desert as elements of a cultured landscape. Trudy Tsentralnogo Byuro Koltsevaniya [Trans. Central Bird Banding Bureau], Vologda 7: 68–73 (in Russian).
Rustamov, A.K. and S. Shammakov. 1977. Ecology of the spotted toad agama (*Phrynocephalus maculatus*). Zool. zhurn. [Zool. J.] 56(9): 1351–1356 (in Russian).
Rustamov, A.K. and S. Shammakov. 1979. Rare and endangered species of reptiles of Turkmenistan. Okhrana prirody Turkmenistana [Natural Conservation in Turkmenistan] 5: 47–53 (in Russian).
Rustamov, A.K. and S. Shammakov. 1982. On the herpetofauna of Turkmenistan. Vertebrata Hungarica (Budapest) 21: 215–226.
Rustamov, A.K. and N.N. Shcherbak. 1985. Herpetogeographical regions of Middle Asia. In: I.S. Darevsky (ed.), Voprosy gerpetologii [Problems of Herpetology]. Nauka, Leningrad: 181–182 (in Russian).
Rustamov, A.K. and N.N. Shcherbak. 1989. Herpetology of Middle Asia. In: I.S. Darevsky (ed.), Voprosy gerpetologii [Problems of Herpetology]. Naukova dumka, Kiev: 215–216 (in Russian).
Rustamov, A.K. and N.N. Shcherbak. 1990. Herpetology of Middle Asia (a brief survey). Povyshenie effektivnosti sel'skokhoz. proizvodstva [Increasing Efficiency of the Agricultural Production] 34: 161–188 (in Russian).
Rustamov, A.K. and Ye.I. Shcherbina. 1957. On the Felidae of Badghyz. Izv. Akad. Nauk. Turkm. SSR, Ser. Biol. Nauk. [Proc. Acad. Sci. Turkm. SSR, Ser. Biol. Sci.] 3: 119–120 (in Russian).
Rustamov, A.K., O. Sopyev, and M. Amanova. 1974. Behavioural and ecological adaptations of birds in Middle Asia. In: Abstr. 16th Intern. Ornithol. Congress, Canberra: 34–36.
Rustamov, A.K. and A.N. Sukhinin. 1957. New data on ornithofauna of southern Turkmenia. Izv. Akad. Nauk. Turkm. SSR, Ser. Biol. Nauk. [Proc. Acad. Sci. Turkm. SSR, Ser. Biol. Sci.] 4: 70–76 (in Russian).
Rustamov, A.K. and V.I. Vasilyev. 1976. Prirodnye zapovedniki SSSR: Vsesoyuznyi ornitologicheskii Krasnovodskii zapovednik [Natural Reserves of the USSR: All-Union Ornithological Krasnovodsk Reserve], Moscow, Znanie, 47 pp (in Russian).
Rustamov, E.A. 1978. Use of a meridional automobile survey for study of bird migration in the Karakum Desert. In: Vtoraya Vsesoyuzn. konfer. po migratsii ptits.Tez. soobshch. [Abstr. 2nd All-Union Confer. on Bird Migrations]. Alma-Ata, Nauka 2: 201–202 (in Russian).
Rustamov, E.A. 1981. On the conservation of the dzhek in southeastern Turkmenistan. In: Tez. dokl. 2 nauchn. konfer, po okhrane prirody Turkm. SSR [Abstr. 2nd Sci. Conf. on Natural Conservation in the Turkmen SSR], Ashkhabad: 130–132 (in Russian).
Rustamov, E.A. 1988. Naseleniye ptits Tedzheno-Murgabskogo mezhdurechya i prilegayushchikh pustyn' (kartograficheskii analiz) [Bird Population of the Tedzhen-Murghab Interfluvial Area and Adjacent Deserts (a cartographical analysis)]. Ylym, Ashkhabad, 209 pp (in Russian).
Rustamov, E.A. 1991. Features of the habitat map in the mapping of seasonal aspects of bird population in arid regions. In: E.M. Pospelov et al. (ed.), Problemy zoogeograficheskogo kartografirovaniya [Problems in Zoogeographical Mapping]. Izd. AN SSSR, Moscow: 71–75 (in Russian).
Rustamov, E.A. 1991. Ekologicheskaya struktura naseleniya ptits aridnykh regionov (na primere Srednei i Tsentral'noi Azii) [Ecological Structure of Bird Population in Arid Regions (of Middle and Central Asia as examples)]. Abstr. D.Sc. Thesis, Moscow, 56 pp (in Russian).
Rustamov, E.A. and A. Khakyev. 1978. On wintering of the waterfowl on the Kelif Lakes. Izv. Akad. Nauk. Turkm. SSR, Ser. Biol. Nauk. [Proc. Acad. Sci. Turkm. SSR, Ser. Biol. Sci.] 4: 96–99 (in Russian).
Rustamov, E.A., A.N. Poslavsky, A.A. Karavaev, V.D. Myatlev, and V.Yu. Chernov. 1990. Geography, ecology, and conservation of the wintering sites of waterfowl in Turkmenistan. Okhrana prirody Turkmenistana [Natural Conservation in Turkmenistan] 8: 56–100 (in Russian).

Rustamov, I.G. 1962. Vegetation of the middle and lower portions of the West Uzboi. Uchen. zap. Turkm. gos. un-ta [Sci. Mem. Turkmen State Univ.] 9(4).
Rustamov, I.G. 1970. On the classification of vegetation of the lowland deserts of Middle Asia. Problemy osvoeniya pustyn [Problems of Desert Development] 6: 39–48 (in Russian).
Rustamov, I.G. 1973. Kornevye sistemy i produktivnost' rastitel'nykh soobshchestv Severo-Zapadnoi Turkmenii [Root Systems and Productivity of Plant Communitites of Northwest Turkmenia]. Abstr. D.Sc. Thesis, Ashkhabad, 64 pp (in Russian).
Rustamov, I.G. 1992. Rastitel'nost' pustyn' [Vegetation of the Deserts]. In: Rastitel'nost' Turkmenistana [Vegetation of Turkmenistan]. Ylym, Ashkhabad: 111–176 (in Russian).
Rustamov, O.I. 1991. On the feeding of the Middle Asian cobra. Izv. Akad. Nauk. Turkm. SSR, Ser. Biol. Nauk. [Proc. Acad. Sci. Turkm. SSR, Ser. Biol. Sci.] 5: 59 (in Russian).
Ryumin, A.V. 1968. On the ecology of the grey monitor in southern Turkmenia. In: I.Kh. Turakulov (ed.), Gerpetologiya Srednei Azii [Herpetology in Middle Asia]. Fan, Tashkent: 28–33 (in Russian).
Sabirova, O.R. 1975. Studies on spiders (Araneina) from the Repetek Reserve. Izv. AN TSSR, Ser. Biol. Nauk. [Proc. Acad. Sci. Turkmen SSR, Ser. Biol. Sci.] 6: 79–82 (in Russian).
Sabirova, O.R. 1977. Pochvennaya fauna pod peskoukrepitel'nymi rasteniyami Vostochnykh Karakumov [Soil Fauna Under the Sand-Stabilizing Vegetation of East Karakum]. Ylym, Ashkhabad, 100 pp (in Russian).
Sabirova, O.R. 1981. Soil arthropods of Central Karakum. Izv. AN Turkm. SSR, Ser. Biol. Nauk. [Proc. Acad. Sci. Turkm. SSR, Ser. Biol. Sci.] 2: 28–33 (in Russian).
Sabirova, O.R. 1985. Features of soil mesofauna of southeastern Turkmenistan. Ent. Obozr. [Entomol. Review] 64(3): 525–532 (in Russian).
Sabirova, O.R. 1986. Features of the soil mesoarthropods in East and Southeast Karakum. Izv. AN Turkm. SSR, Ser. Biol. Nauk. [Proc. Acad. Sci. Turkm. SSR, Ser. Biol. Sci.] 2: 53–56 (in Russian).
Sagitov, N.I. 1983. Ryby i kormovye bespozvonochnye srednego i nizhnego techeniya Amudaryi [Fishes and Feed Invertebrates of the Middle and Lower Amudarya]. Fan, Tashkent, 116 pp (in Russian).
Sakharov, A.K. 1982. On the experience in francolin containment in captivity (Syunt Khasardagh Reserve of the Ministry of Forestry of the Turkmen SSR. In: Razvedeniye i sozdanie novykh populyatsii redkikh i tsennykh vidov zhivotnykh/Tez. dokl. 3 soveshch. [Breeding and Creation of New Populations of Rare and Valuable Animal Species/Abstr. 3rd Conf.], Ashkhabad: 41–44 (in Russian).
Salnikov, V.B. 1989. On the biology of the carp, *Cyprinus carpio* L., in the Khauzkhan Reservoir. Izv. AN TSSR, Ser. Biol. Nauk. [Proc. Acad. Sci. Turkmen SSR, Ser. Biol. Sci.] 6: 50–56 (in Russian).
Salnikov, V.B. 1990. The Aral barbel, *Barbus brachycephalus* Kessler, in the water bodies of Turkmenistan: modern conditions and ways of conservation of a species. Izv. AN TSSR, Ser. Biol. Nauk [Proc. Acad. Sci. Turkmen SSR, Ser. Biol. Sci.] 5: 25–31 (in Russian).
Salnikov, V.B. and T.N. Busheva. 1986. Biology and the fishing industry in the Khauzkhan Reservoir. In: Yu.S. Reshetnikov (ed.), Dinamika chislennosti promyslovykh ryb [Dynamics of Abundance of Food Fish Species]. Nauka, Moscow: 204–215 (in Russian).
Salnikov, V.B. and T.N. Busheva. 1990. The biology of the bream, *Abramis brama orientalis* Berg, in the Khauzkhan Reservoir. Izv. AN TSSR, Ser. Biol. Nauk. [Proc. Acad. Sci. Turkmen SSR, Ser. Biol. Sci.] 1: 39–47 (in Russian).
Salnikov, V.B. and L.P. Pavlovskaya. 1987. On the reproduction and feeding of the bream in Lake Sarykamysh. Vestnik Kara-Kalpakskogo filiala Akademii Nauk. Uzbekskoi SSR [Mess. Kara Kalpak Branch Acad. Sci. Uzbek SSR] 4: 32–40 (in Russian).
Salnikov, V.B. and Yu.S. Reshetnikov. 1991. The formation of the fish population in the artificial water bodies of Turkmenistan. Voprosy ikhtiologii [Problems in Ichthyology] 31(4): 565–575 (in Russian).
Salnikov, V.B. and A.M. Shkeda. 1990. The bitterling, *Pseudoperilampus* (= *Rhodeus*) *ocellatus* Kner, 1867 (Fam. Cyprinidae), a new species of ichthyofauna for the water bodies of

Turkmenistan. Izv. AN TSSR, Ser. Biol. Nauk. [Proc. Acad. Sci. Turkmen SSR, Ser. Biol. Sci.] 6: 63–66 (in Russian).
Samorodov, A.V. 1953. Contributions to the mammalian fauna of the lower Atrek River. Izv. AN TSSR, Ser. Biol. Nauk. [Proc. Acad. Sci. Turkmen SSR, Ser. Biol. Sci.] 6: 4–48 (in Russian).
Sanin, M.V., V.I. Kostyukovsky, S.I. Shaporenko *et al.* 1991. Ozero Sarykamysh i vodoyomy-nakopiteli kollektorno-drenazhnykh vod [Lake Sarykamysh and the Water Bodies Accumulating Collection-Drainage Waters]. Nauka, Moscow, 149 pp (in Russian).
Sapozhenkov, Yu.F. 1960. On the distribution of karakal (*Felis caracal* Schr.) in Turkmenia. Zool. zhurn. [Zool. J.] 40(2): 195–197 (in Russian).
Sapozhenkov, Yu.F. 1961. On the distribution and ecology of the barkhan cat (*Felis margarita*) in eastern Turkmenia. Zool. zhurn. [Zool. J.] 40(7): 1086–1089 (in Russian).
Sapozhenkov, Yu.F. 1962. On the ecology of karakal (*Felis caracal*) in the Karakum. Zool. zhurn. [Zool. J.] 41(7): 1111–1112 (in Russian).
Sapozhenkov, Yu.F., Yu.K. Gorelov, I.V. Zhernovov, and A.I. Svyatoi. 1963. On the distribution and ecology of the honey badger in Turkmenia. Zool. zhurn. [Zool. J.] 42(6): 961–963 (in Russian).
Sarianidi, V.I. 1967. Tainy ischeznuvshego iskusstva Karakumov [Mysteries of the Vanished Art of the Karakum]. Ylym, Ashkhabad, 174 pp (in Russian).
Schacht, W. 1983. Eine neue Bremsenart aus der Türkei (Diptera, Tabanidae). Entomofauna 4(27), Suppl.: 483–492.
Schacht, W. 1984. Beitrag zu einigen palearktischen Bremsenarten vornehmlich aus der Türkei (Diptera, Tabanidae). Entomofauna 5(35): 483–498.
Schatti, B. 1988. Systematik und Evolution der Schlangengattung *Hierophis* Fitzinger, 1843 (Reptilia, Serpentes). Inaugural-Dissertation, Zürich, 50 pp.
Schatti, B. and A. Agasian. 1985. Ein neues Konzept für den *Coluber raviergieri* – *C. nummifer* Complex (Reptilia, Serpentes, Colubridae). Zool. Abh. Staat. Mus. f. Tierkunde Dresden 40(9): 109–123.
Schmidt-Nielsen, K. 1972. The desert animals. Nauka, Leningrad, 308 pp (Russian translation).
Seel, P.D. 1976. Flora Europaea. New York, Cambridge, etc. 4: 308 pp.
Selyaninov, G.T. 1961. Perspektivy razvitiya subtropicheskogo khozyaistva SSSR v svyazi s prirodnymi usloviyami [Perspectives on the Development of Subtropical Industry in the USSR Depending on Natural Conditons]. Gidrometeoizdat, Leningrad, 195 pp (in Russian).
Semenov-Tian-Shansky, A.P. 1900. The genus *Pseudobroscus* Sem. (Coleoptera, Carabidae), its genetic connections, and significance in Turanian fauna. Horae Soc. Entom. Ross. 34(1–2): 41–51 (in Russian).
Semenov-Tian-Shansky, A.P. 1936. Predely i zoogeograficheskie podrazdeleniya Palearkticheskoi oblasti dlya nazemnykh sukhoputnykh zhivotnykh na osnovanii geograficheskogo raspredeleniya zhestkokrylykh nasekomykh [Limits and Zoogeographical Subdivision of the Palearctic Region for Terrestrial Animals Based on the Geographical Distribution of the Coleopterous Insects]. Izd. AN SSSR, Moscow-Leningrad, 16 pp, 1 map (in Russian).
Semenova, O.A. 1961. A climatic description of the Karakum Desert. Trudy Sredneaz. gidrometeorol. in-ta [Trans. Middle Asian Hydrometeorol. Inst.] 6(21): 95–151 (in Russian).
Serebrovsky, P. V. 1928 (1929). On the ornithogeography of Southeast Asia. Ezheg. Zool. Muz. AN SSSR [Annu. Zool. Mus. Acad. Sci. USSR] 29: 289–392 (in Russian).
Serebryakov, I.G. 1962. Ecologicheskaya morphologiya rastenii [Ecological Morphology of Plants]. Vysshaya shkola, Moscow, 378 pp (in Russian).
Serebryakov, I.G. 1984. Life forms of plants and their study. Polevaya geobotanika [Field Geobotany]. Nauka, Moscow-Leningrad 3: 146–205 (in Russian).
Sergeev, A.M. and Yu.A. Isakov. 1941. On the feeding of the grey monitor. Priroda [Nature] 6: 75–76 (in Russian).
Sergeev, M.G. 1991. Zakonomernosti rasprostraneniya pryamokrylykh nasekomykh Aziatskoi chasti SSSR [Features of the Distribution of Orthoptera in the Asian Part of the USSR]. Abstr. D.Sc. Thesis, St. Petersburg, 37 pp (in Russian).
Shalna, A.A. and E.Yu. Shalnene. 1981. Breeding of the baloban falcon in captivity at the Kirghiz

State Enterprise for Game Management. In: Ekologiya i okhrana ptits/Tez. dokl. 8 Vsesoyuzn. ornitol. konf. [Ecology and Conservation of Birds/Abstr. All-Union Ornithol. Conf.], Stiinca, Kishinev: 238 (in Russian).
Shammakov, S. 1964a. On the density of reptiles of the Lesser Ranges of western Turkmenia. Izv. Akad. Nauk. Turkm. SSR, Ser. Biol. Nauk. [Proc. Acad. Sci. Turkmen. SSR, Ser. Biol. Sci.] 1: 86–88 (in Russian).
Shammakov, S. 1964b. On some rare species of reptiles from Turkmenia. Izv. Akad. Nauk. Turkm. SSR, Ser. Biol. Nauk. [Proc. Acad. Sci. Turkmen. SSR, Ser. Biol. Sci.] 6: 86–88 (in Russian).
Shammakov, S. 1968. Faunistic data on the reptiles of the Lesser Ranges (Maly Balkhan, Kyurendagh, Karagyoz) of western Turkmenia. In: I.Kh. Turakulov (ed.), Gerpetologiya Srednei Azii [Herpetology of Middle Asia]. Fan, Tashkent: 10–15 (in Russian).
Shammakov, S. 1969. On the ecology of *Eremias velox* in western Turkmenia. Izv. Akad. Nauk. Turkm. SSR, Ser. Biol. Nauk. [Proc. Acad. Sci. Turkmen. SSR, Ser. Biol. Sci.] 4: 37–42 (in Russian).
Shammakov, S. 1974. On the ecology of the shielded gecko (*Alsophylax loricatus* Strauch) in Turkmenia. Izv. Akad. Nauk. Turkm. SSR, Ser. Biol. Nauk. [Proc. Acad. Sci. Turkmen. SSR, Ser. Biol. Sci.] 2: 76–77 (in Russian).
Shammakov, S. 1981. Presmykayushchiesya ravninnogo Turkmenistana [Reptiles of Lowland Turkmenistan]. Ylym, Ashkhabad, 309 pp (in Russian).
Shammakov, S. and Ch. Ataev. 1971. On the smooth gecko (*Alsophylax laevis* Nikolsky) in Turkmenia. Izv. Akad. Nauk. Turkm. SSR, Ser. Biol. Nauk. [Proc. Acad. Sci. Turkmen. SSR, Ser. Biol. Sci.] 3: 65–69 (in Russian).
Shammakov, S. and Ch. Ataev. 1987. On the density and distribution of *Gymnodactylus spinicauda*. Izv. Akad. Nauk. Turkm. SSR, Ser. Biol. Nauk. [Proc. Acad. Sci. Turkmen. SSR, Ser. Biol. Sci.] 5: 70–71 (in Russian).
Shaposhnikova, G.Kh. 1950. Fishes of the Amudarya. Trudy Zoologicheskogo Instituta Akademii Nauk. SSSR [Trans. Zool. Inst. Acad. Sci. USSR] 9(1): 16–54 (in Russian).
Sharonov, I.V. 1966. Formation of the ichthyofauna in reservoirs. In: G.M. Belyaev et al. (ed.), Ekologiya vodnykh organizmov [Ecology of Aquatic Organisms]. Nauka, Moscow: 103–110 (in Russian).
Shcherbak, N.N. 1966. Zemnovodnye i presmykayushchiyesya Kryma [Amphibians and Reptiles of the Crimea]. Naukova dumka, Kiev, 240 pp (in Russian).
Shcherbak, N.N. 1971. Systematics of the genus *Eremias* (Sauria, Reptilia) and its relationship to the centers of origin of desert-steppe fauna of the Palearctic. Vestn. zool. [Mess. Zool.] 2: 48–55 (in Russian).
Shcherbak, N.N. 1972. A new subspecies, *Eremias strauchi kopetdaghica* (Sauria, Reptilia) from Turkmenia. Vestn. zool. [Mess. Zool.] 2: 83–85 (in Russian).
Shcherbak, N.N. 1974. Yashchurki Palearktiki [The Lizard Genus *Eremias* in the Palearctic Region]. Naukova dumka, Kiev, 293 pp (in Russian).
Shcherbak, N.N. 1978. *Gymnodactylus turcmenicus* sp.n., Reptilia, Sauria – a new species of gecko from southern Turkmenia. Vestn. zool. [Mess. Zool.] 3: 39–44 (in Russian).
Shcherbak, N.N. 1979a. The Turkmen eublephare. Priroda [Nature] 6: 83–85.
Shcherbak, N.N. 1979b. New records of lizards and snakes in Middle Asia. Vestn. zool. [Mess. Zool.] 1: 68–70 (in Russian).
Shcherbak, N.N. 1982. Grundzüge einer herpetogeographischen Gliederung der Paläarktis. Vertebrata Hungarica 21: 227–239.
Shcherbak, N.N. (Szczerbak, N.N.) 1986a. Review of Gekkonidae of the USSR and neighbouring countries. In: Z. Rocek (ed.), Studies in Herpetology. Proc. Europ. Herp. Meeting, Prague, 1985: 705–709 (with collaboration from M.L. Golubev).
Shcherbak, N.N. (ed.). 1986b. Priroda Tsentralnogo Kopetdaga [Nature in Central Kopetdagh]. Ylym, Ashkhabad, 190 pp (in Russian).
Shcherbak, N.N. 1988. On the nomenclature of the Palearctic thin-toed geckos. Vestn. zool. [Mess. Zool.] 4: 96–97 (in Russian).

Shcherbak, N.N. 1990. Systematics and geographic variation of *Eumeces taeniolatus* (Blyth, 1854) (Scincidae, Sauria, Reptilia). Vestn. zool. [Mess. Zool.] 3: 33–40 (in Russian).
Shcherbak, N.N. and S.B. Akhmedov. 1990. Systematics and geographic variation of *Eumeces schneideri* (Daud., 1802) (Scincidae, Sauria, Reptilia). Vestn. zool. [Mess. Zool.] 1: 23–28 (in Russian).
Shcherbak, N.N. and M.V. Golubev. 1979. A new subspecies of Khentau toad agama – *Phrynocephalus rossikowi shammakowi* ssp. n. (Reptilia, Sauria, Agamidae) from the Central Karakum. Vestn. zool. [Mess. Zool.] 6: 81–83 (in Russian).
Shcherbak, N.N. and M.V. Golubev. 1981. New records of amphibians and reptiles from Middle Asia and Kazakhstan. Vestn. zool. [Mess. Zool.] 1: 70–72 (in Russian).
Shcherbak, N.N. and M.V. Golubev. 1986. Gekkony fauny SSSR i sopredel'nykh stran [The Gekkonidae Fauna of the USSR and Adjacent Countries]. Naukova dumka, Kiev, 232 pp (in Russian).
Shcherbak, N.N., Yu.D. Khomustenko, and M.V. Golubev. 1986. Amphibians and reptiles of the Kopetdagh State Reserve and adjacent territories. In: N.N. Shcherbak (ed.), 1986. Priroda Tsentral'nogo Kopetdaga [Nature in Central Kopetdagh]. Ylym, Ashkhabad: 76–110 (in Russian).
Shcherbina, Ye.I. 1958. On the ecology (feeding and reproduction) of the fox, *Vulpes vulpes*, in Badghyz. Trudy In-ta zool. i parazitol. [Trans. Inst. Zool. and Parasitol.], Ashkhabad 3: 5–46 (in Russian).
Shcherbina, Ye.I. 1966. On the ecology of the fox in Karabil and Obruchevskaya Steppe. Izv. Akad. Nauk. Turkm. SSR, Ser. Biol. Nauk. [Proc. Acad. Sci. Turkmen. SSR, Ser. Biol. Sci.] 3: 66–74 (in Russian).
Shcherbina, Ye.I. 1970. Okhotnichye-promyslovye zveri Turkmenii [Game Mammals of Turkmenia]. Ylym, Ashkhabad, 66 pp (in Russian).
Shcherbina, Ye.I. and Yu.K. Gorelov. 1980. Rare and endangered mammals of Turkmenistan. Izv. Akad. Nauk. Turkm. SSR, Ser. Biol. Nauk. [Proc. Acad. Sci. Turkmen. SSR, Ser. Biol. Sci.] 4: 80–82 (in Russian).
Shcherbina, Ye.I. and V.I. Kravchenko. 1960. On measures toward conservation and increase of population of onagers in Badghyz. Izv. Akad. Nauk. Turkm. SSR, Ser. Biol. Nauk. [Proc. Acad. Sci. Turkmen. SSR, Ser. Biol. Sci.] 4: 69–74 (in Russian).
Shepelin, M.M. 1972. Reproduction of the karakal in the Tashkent Zoo. In: E.I. Gan (ed.), Ekologiya i biologiya zhivotnykh Uzbekistana [Ecology and Biology of Animals in Uzbekistan]. Fan, Tashkent: 332–337 (in Russian).
Sheroina, A.A. 1979. New data on nesting colonial and bywater birds of the islands of Karabogazgol Bay. Prirodnaya sreda i ptitsy poberezhii Kaspiiskogo morya i prilezhashchikh nizmennostei/Trudy Kizyl-Agachskogo gos. zapovednika [Environment and Birds of the Caspian Sea Shores and Adjacent Lowlands/Trans. Kizyl-Agach State Reserve]. Azgosizdat, Baku 1: 89–190 (in Russian).
Shestoperov, Ye.L. 1928. Notes on the ornithofauna. Turkmenovedeniye [Turkmen Studies] 5/6: 61–75 (in Russian).
Shestoperov, Ye.L. 1934. Fauna of the Repetek Reserve in TSSR. Izv. Turkmen. mezhduved. komit. okhr. prir. i razvit. prir. bogatstv [Proc. Turkmen Interdepartm. Committee Natural Conserv. and Developm. Natural Resources], Ashkhabad 1: 197–232 (in Russian).
Shestoperov, Ye.L. 1935. A preliminary zoological survey of the Akhcha-Kuima Reserve area. Izv. Turkmen. mezhduved. komit. okhr. prir. i razvit. prir. bogatstv [Proc. Turkmen Interdepartm. Committee Natural Conserv. and Developm. Natural Resources], Ashkhabad 2: 161–193 (in Russian).
Shestoperov, Ye.L. 1936a. Data on the game management in the Tashauz Region. Byull. Turkm. zool. stantsii [Bull. Turkm. Zool. Station] 1: 142–155 (in Russian).
Shestoperov, Ye.L. 1936b. Additions to the knowledge of the fauna of the Karluyuk District of TSSR. Byull. Turkm. zool. stantsii [Bull. Turkm. Zool. Station] 1: 157–171 (in Russian).
Shestoperov, Ye.L. 1936c. A faunistic survey of the environs of Tedzhen. Byull. Turkm. zool. stantsii [Bull. Turkm. Zool. Station] 1: 174–189 (in Russian).

Shestoperov, Ye.L. 1937. Opredelitel' pozvonochnykh zhivotnykh Turkmenskoi SSR [A Key to the Vertebrate Animals of the Turkmen SSR: Birds], Ashkhabad-Baku 5, 331 pp (in Russian).
Shileiko, A.A. 1978. Terrestrial mollusks of the superfamily Helicoidea. Fauna SSSR [Fauna of the USSR], New Ser., Vol. 117, Mollusca 3(6). Nauka, Leningrad, 384 pp (in Russian).
Shileiko, A.A. 1984. Terrestrial mollusks of the suborder Pupillina of the fauna of the USSR. Fauna SSSR [Fauna of the USSR], New Ser., Vol. 130, Mollusca 3(3). Nauka, Leningrad, 399 pp (in Russian).
Shiranovich, P.K., A.V. Molodovsky, B.Ye. Osolinker, K.I. Derevyanchenko, and Ye.G. Samarin. 1965. On the microclimate of the burrows of *Rhombomys opimus* Licht. Zool. zhurn. [Zool. J.] 44(8): 1245-1254 (in Russian).
Shishkin, I.B. 1981. U sten velikoi Namazgi [Under the Walls of the Great Namazga]. Nauka, Moscow, 207 pp (in Russian).
Shkeda, R.V. 1989. Vodoyomy Karakumskogo kanala [Water Bodies of the Karakum Canal]. Ylym, Ashkhabad, 132 pp (in Russian).
Shtegman, B.K. 1938. Osnovy ornitogeograficheskogo deleniya Palearktiki. Fauna SSSR. Ptitsy. [Fauna of the USSR. Birds]. Izd. AN SSSR, Moscow-Leningrad 1(2), 157 pp (in Russian).
Shukurov, G.Sh. 1962. Fauna pozvonochnykh zhivotnykh gor Bolshie Balkhany (Yugo-Zapadnyi Turkmenistan) [The Vertebrate Fauna of the Bolshoi Balkhan Mountains (Southwest Turkmenistan)]. Izd. AN TSSR, Ashkhabad, 159 pp (in Russian).
Shukurov, O.Sh. 1965. New records of some species of reptiles from Turkmenia. Zool. zhurn. [Zool. J.] 44(12): 1873-1874 (in Russian).
Shukurov, O.Sh. 1973. Snakes of the middle Amudarya River. In: I.S. Darevsky (ed.), Voprosy gerpetologii [Problems of Herpetology]. Nauka, Leningrad: 216-218 (in Russian).
Shukurov, O.Sh. 1976. Herpetofauna of Kugitang and its foothills within Turkmenia. In: Gerpetologiya [Herpetology]. Krasnodar: 74-76 (in Russian).
Shumakov, Ye.M. 1963. Acridoidea of Afghanistan and Iran. Trudy Vses. entomol. ob-va [Trans. All-Union Entomol. Soc.] 49: 3-248 (in Russian).
Shvarts, S.S. 1954. On the specificity of a species in vertebrates. Zool. zhurn. [Zool. J.] 33(3): 1120-1129 (in Russian).
Shvarts, S.S. 1958. Method of morphophysiological indices in the ecology of terrestrial vertebrates. Zool. zhurn. [Zool. J.] 37(2): 161-173 (in Russian).
Shvarts, S.S. 1960. Principles and methods of modern animal ecology. Trudy in-ta ekol. rast. i zhivotn. [Trans. Inst. Plant and Animal Ecol.]: 21 (in Russian).
Shvarts, S.S. 1969. Evolutionary ecology of animals. Trudy in-ta ekol. rast. i zhivotn. [Trans. Inst. Plant and Animal Ecol.]: 65 (in Russian).
Shvarts, S.S. 1975. Ecological principles of the conservation of the biosphere. Okhota i okhotnichye khozyaistvo [Hunting and Game Management] 8: 8-12 (in Russian).
Sidorenko, A.V. 1952. On the origin of closed depressions (with the example of Badghyz). Izv. VGO [Proc. All-Union Geogr. Soc.] 84(3): 255-264 (in Russian).
Sidorenko, A.V. and O.A. Mikhelson. 1948. On the pistachio in Badghyz found on the Paleogene rocks. Bot. zhurn. [Bot. J.] 33(6): 614-616 (in Russian).
Sikstel, T.A. and R. Khudaiberdyev. 1968. On the ancient floras of Middle Asia. In: Paleobotanika Uzbekistana [Paleobotany of Uzbekistan]. Fan, Tashkent 1: 3-87 (in Russian).
Silantyev, A.A. 1898. Obzor promuslovykh okhot v Rossii [A Survey of the Game Hunting in Russia]. Tipogr. Kirshbauma, St. Petersburg, 234 pp (in Russian).
Simakin, L.V. 1991. A new record of the scaled woodpecker in southern Turkmenia. Vestnik zoologii [Messenger of Zoology] 2: 85 (in Russian).
Simon, E. 1889. Archnidae transcaspicae ab ill. Dr. G. Radde, Dr. A. Walter et A. Conchin inventae (annis 1886-1887). Verh. Zool.-Bot. Ges. Wien 39: 373-386.
Sinev, I.E. 1985. The boundary of dry subtropics in Southwestern Turkmenistan. In: N.T. Nechayeva (ed.), Rastitelnost' i zhivotnyi mir Zapadnogo Kopetdaga [The Vegetation and Animal Life of West Kopetdagh]. Ylym, Ashkhabad: 14-24 (in Russian).
Sinytsin, V.M. 1962. Paleogeografiya Azii [Paleogeography of Asia]. Izd. AN SSSR, Moscow-Leningrad, 268 pp (in Russian).

Sinytsin, V.M. 1965–1966. Drevnie klimaty v Evrazii [Ancient Climates of Eurasia]. Leningrad State Univ., Leningrad, Part 1 (1965), 166 pp; Part 2 (1966), 166 pp (in Russian).
Sinytsin, V.M. 1967. Vvedenie v paleoklimatologiyu [Introduction to Paleoclimatology]. Nedra, Leningrad, 237 pp (in Russian).
Sinytsin, V.M. 1980. Prirodnye usloviya i klimaty territorii SSSR v rannem i srednem kainozoe [Natural Conditions and Climates of the USSR Territory in the Early and Middle Cenozoic Period]. Leningrad State Univ., Leningrad, 104 pp (in Russian).
Sirotina, I.V. 1988. Mkhi Kopetdaga [The Mosses of Kopetdagh]. Abstr. Ph.D. Thesis, Leningrad, 18 pp (in Russian).
Skadhauge, E. 1973. Renal and cloacal salt and water transport in the fowl (*Gallus domesticus*). Danish Med. Bull. 20 (Suppl. 1): 1–82.
Skadhauge, E. and B. Schmiedt-Nielsen. 1967. Renal function in domestic fowl. Amer. J. Physiol. 212: 793–798.
Skalon, N.V. 1982. Amphibians and reptiles of Southwest Kopetdagh. In: N.T. Nechayeva (ed.), Priroda Zapadnogo Kopetdaga [Nature in West Kopetdagh]. Ylym, Ashkhabad: 146–157 (in Russian).
Sludsky, A.A. 1973. Distribution and density of wild cats in the USSR. In: Promyslovye mlekopitayushchie Kazakhstana/Trudy In-ta zoologii [Game Mammals of Kazakhstan/Trans. Zool. Inst.]. Izd. AN Kaz. SSR, Alma-Ata: 5–106 (in Russian).
Smolko, A.I., L.N. Smirnov, and G.I. Popov. 1972. General characteristics and paleogeography (Neogene System). Geologiya SSSR [Geology of the USSR]. Nedra, Moscow 22: 382–401 (in Russian).
Soin, S.G. 1981. On the diversity of the ecological groups in fishes according to their reproductive conditions. In: N.V. Butorin, L.S. Berdichevsky, A.S. Konstantinov, S.G. Soin, and M.I. Shatrunovsky (eds.), Current Problems in Ichthyology. Nauka, Moscow: 124–141 (in Russian).
Sokolov, V.E. 1973. Kozhnyi pokrov mlekopitayushchikh [Mammalian Skin]. Nauka, Moscow, 484 pp (in Russian).
Sokolovsky, V.V. 1974. Comparative karyological study of the lizard family Agamidae. I. Chromosomal sets of eight species of the genus *Phrynocephalus*. Tsitologiya [Cytology] 16(7): 920–925 (in Russian).
Sokolovsky, V.V. 1977. Systematic relationshipa within the family Agamidae according to the karyological data. In: Voprosy gerpetologii. Avtoref. dokl. IV Vsesoyuzn. gerpetol. konf. [Problems of Herpetology/Abstr. 4th All-Union Herpetol. Conf.]. Nauka, Leningrad: 195 (in Russian).
Sokov, A.I. 1976. On rare carnivorous mammals of Tajikistan. In: V.E. Sokolov (ed.), Redkie mlekopitayushchie fauny SSSR [Rare Mammals of the USSR Fauna]. Nauka, Moscow: 97–102 (in Russian).
Solokha, A.V. 1990. On the problem of dispersal of pheasant (*Phasianus colchicus* L.) in Middle Asia and phylogenetic connections among its subspecies. In: Okhrana prirody Turkmenistana [Natural Conservation in Turkmenistan] 8: 192–204 (in Russian).
Solokha, A.V. 1991. Fazan v Turkmenistane (biologiya, razvedeniye i ratsional'noye ispol'zovaniye) [Pheasant in Turkmenistan (biology, breeding, and rational exploitation)]. Abstr. Ph.D., Ashkhabad, 20 pp (in Russian).
Solomatin, A.O. 1973. Kulan [The Onager]. Nauka, Moscow, 145 pp (in Russian).
Sopyev, O. 1962. Some data on the ecology of the golden eagle and *Athene noctua* in eastern Karakum. Izv. Akad. Nauk. Turkm. SSR. Ser. Biol. Nauk. [Proc. Acad. Sci. Turkmen. SSR, Ser. Biol. Sci.] 2: 79–83 (in Russian).
Sopyev, O. 1965. The desert sparrow in the Karakum. Ornitologiya [Ornitology], Moscow State Univ. 1: 134–141 (in Russian).
Sopyev, O. 1979. On distribution and ecology of the desert sparrow in the Central Karakum. In: Okhrana prirody Turkmenistana [Natural Conservation in Turkmenistan] 5: 52–55 (in Russian).
Sopyev, O. 1981a. A new nesting population of *Hypocolius ampellinus*. In: Tezisy dokl. 2 nauchn. konf. po okhrane prirody Turkm. SSR [Natural Conservation in Turkmenistan. Abstracts of the

2nd Scientific Conference on Natural Conservation in Turkmen SSR], Ashkhabad: 132-134 (in Russian).
Sopyev, O. 1981b. First nesting population of *Hypocolius ampellinus* in the USSR. Priroda [Nature] 9: 58 (in Russian).
Soskov, Yu.D. 1974. On the section *Medusa* Sosk. et L. Alexandr. of the genus *Calligonum* L. Novosti sistem. vyssh. rast. [News in Systematics of Higher Plants] 11: 94-109 (in Russian).
Soyunov, O.S. 1979. The termite-hill: a peculiar community of organisms. Izv. AN TSSR, Ser. Biol. Nauk. [Proc. Acad. Sci. Turkmen SSR, Ser. Biol. Sci.] 1: 26-31 (in Russian).
Spanner, R. 1925. Der Pfortaderkreislauf in der Vogelniere. Morphol. Jahrbuch 54: 560-632.
Spangenberg, Ye.P. 1951. Orders Phoenicopteriformes and Anseriformes. Ptitsy Sovetskogo Soyuza [Birds of the Soviet Union]. Sovetskaya Nauka, Moscow 2: 341-475 (in Russian).
Spassky, S. 1934a. Aranearum species novae II. New York Ent. Soc. 42: 1-4 (in Latin).
Spassky, S. 1934b. Aranearum species novae III. Rev. Ent. Franc. 1(2): 135-139 (in Latin).
Spassky, S. 1934c. Araneae palaearcticae novae. Fam. Pholcidae. Bull. Mus. Hist. Nat. Paris, Ser. 2, 6(4): 361-372 (in Latin).
Spassky, S. 1936a. Araneae palaearcticae novae. Festschrift f. Embrik Strand 1: 37-46 (in Latin).
Spassky, S. 1936b. New species and genera of spiders for the USSR. Proc. Zool. Inst. Acad. Sci. USSR 3: 533-536 (in Russian).
Spassky, S. 1937. Araneae palaearcticae novae. Mygalomorphae I. Festschrift f. Embrik Strand 3: 361-368 (in Latin).
Spassky, S. 1938. Araneae palaearcticae novae II. Festschrift f. Embrik Strand 4: 573-582 (in Latin).
Spassky, S. 1939a. Araneae palaearcticae novae III. Festschrift f. Embrik Strand 5: 138-144 (in Latin).
Spassky, S. 1939b. Araneae palaearcticae novae. Folia Zool. Hydrobiol. 9: 299-308.
Spassky, S. 1952. Spiders of the Turanian zoogeographical province. Entom. Obozr. [Entomol. Review] 32: 192-205 (in Russian).
Spassky, S. and E. Luppova. 1945. Spiders of Tajikistan. Entom. Obozr. [Entomol. Review] 28: 43-55 (in Russian).
Sperber, V. 1960. Excretion. In: A.L. Marshall (ed.), Biology and Comparative Physiology of Birds. Vol. 1. Acad. Press, New York: 469-492.
Spravochnik po klimatu SSSR [Reference Book on the USSR Climate]. 1967-1969. Gidrometeoizdat, Leningrad 30(1-5) (in Russian).
Stalmakova, V.A. 1957. On a record of Turkmen jerboa in the North Karakum and some its ecological and morphological features. Zool. zhurn. [Zool. J.] 36(2): 235-279 (in Russian).
Stanyukovich, K.V. 1955. Major types of altitudinal zonation in the mountains of the USSR]. Izv. VGO [Proc. All-Union Geographical Soc.] 87(3) (in Russian).
Stanyukovich, K.V. 1973. Rastitel'nost' gor SSSR [Vegetation of the Mountains of the USSR]. Donish, Dushanbe, 416 pp (in Russian).
Starobogatov, Ya.I. 1970. Fauna mollyuskov i zoogeograficheskoe delenie vnutrennikh vodoyomov zemnogo shara [Molluscan Fauna and Zoogeographical Division of Inland Water Bodies of the Globe]. Nauka, Leningrad, 372 pp (in Russian).
Starobogatov, Ya.I. 1972. New species of gastropod molluscs from springs and underground waters of Soviet Middle Asia. Trudy Zool. in-ta AN SSSR [Trans. Zool. Inst. Acad. Sci. USSR]. Leningrad 51: 165-172 (in Russian).
Starobogatov, Ya.I. 1986. Fauna of lakes as a source of information about their history. In: D.D. Kvasov et al. (ed.), Istoriya ozer SSSR: Obshchie zakonomernosti proiskhozhdeniya i razvitiya ozer. Metody izucheniya istorii ozer [History of Lakes of the USSR: General Patterns of Origin and Development of Lakes; Methods of Study of Lake History]. Nauka, Leningrad: 33-50 (in Russian).
Starobogatov, Ya.I. and Z.I. Izzatulaev. 1980. Molluscs of the family Melanoididae (Gastropoda, Pectinibranchia) of Soviet Middle Asia and adjacent territories. Zool. zhurn. [Zool. J.] 59(1): 21-31 (in Russian).

Starobogatov, Ya.I. 1984. Genus *Melanopsis* (Gastropoda, Pectinibranchia) and its representatives in the water bodies of the USSR. Zool. zhurn. [Zool. J.] 63(10): 1471–1483 (in Russian).
Starostin, I.V. 1936. Ichthyofauna of rivers of the northen slope of Kopetdagh. Byull. Turkm. zool. stantsii [Bull. Turkm. Zool. Station] 1: 89–93 (in Russian).
Stemmler, O. 1969. Die Sandrasselotter aus Pakistan: *Echis carinatus sochureki* subsp. nov. Aquaterra 6(10): 118–125.
Stepanyan, L.S. 1966. New data on the ecology of *Falco peregrinus babylonicus* Sclater. Izv. Akad. Nauk. Turkm. SSR. Ser. Biol. Nauk. [Proc. Acad. Sci. Turkmen. SSR, Ser. Biol. Sci.] 2: 89–91 (in Russian).
Stepanyan, L.S. 1969. Observations on *Falco peregrinus babylonicus* Sclater in Middle Asia. Byull. MOIP [Bull. Moscow Soc. Natur.] 74(6): 37–48 (in Russian).
Stepanyan, L.S. 1975. Sostav i raspredelenie ptits fauny SSSR. Nevorobyinye – Non-Passeriformes [Composition and Distribution of Birds of the USSR Fauna: Non-Passeriformes]. Moscow, Nauka, 370 pp (in Russian).
Stepanyan, L.S. 1978. Sostav i raspredelenie ptits fauny SSSR. Vorobyinoobvraznye – Passeriformes [Composition and Distribution of Birds of the USSR Fauna: Passeriformes]. Moscow, Nauka, 392 pp (in Russian).
Sternbergs, M.T. 1979. New and little known species of spider of the genus *Evippa* (Aranei, Lycosidae) from Turkmenistan. Izv. Akad. Nauk. Turkm. SSR, Ser. Biol. Nauk [Proc. Acad. Sci. Turkmen. SSR, Ser. Biol. Sci.] 5: 65–67 (in Russian).
Stolzmann, J. 1890. Liste des oiseaux d'Askhabad. Mem. Soc. Zol. France 3: 88–96.
Stolzmann, J. 1892. Contribution a l'ornithologie de la Transcaspie, d'apres les recherches faites par Mr. Thomas Bary (1888–1891). Bull. Soc. Imp. Natur. Moscou: 382–487.
Strelkov, P.P., V.P. Sosnovtseva, and Kh. Babaev. 1979. Bats of Turkmenia. Trudy Zool. in-ta AN SSSR [Trans. Zool. Inst. Acad. Sci. USSR] 79: 3–71 (in Russian).
Sudilovskaya, A.M. 1972. A forgotten article by the academician M.A. Menzbier. Byull. MOIP. Otd. biol. [Bull. Moscow Soc. Natur., Div. Biol.] 77(1): 145 (in Russian).
Sultanov, G.S. 1953. Markhor in Uzbekistan. Trudy In-ta zoologii i parazitol. AN Uz. SSR [Trans. Inst. Zool. and Parasitol. Acad. Sci. Uzbek SSR] 79: 3–71 (in Russian).
Sukhinin, A.N. 1955. Black stork in Turkmenia. Priroda [Nature] 4: 117 (in Russian).
Sukhinin, A.N. 1957. Data on the ecology of the Turkestan snake eagle in Badghyz. Izv. Akad. Nauk. Turkm. SSR. Ser. Biol. Nauk. [Proc. Acad. Sci. Turkmen. SSR, Ser. Biol. Sci.] 5: 133–135 (in Russian).
Sukhinin, A.N. 1961. On feeding of the Saker Falcon in Badghyz (southeastern Turkmenia). Izv. Akad. Nauk. Turkm. SSR. Ser. Biol. Nauk. [Proc. Acad. Sci. Turkmen. SSR, Ser. Biol. Sci.] 3: 82–84 (in Russian).
Sukhinin, A.N. 1967. Characteristics of the bird population in the Tedzhen and Murghab River valleys. In: A.O. Tashliev et al. (ed.), Ptitsy kul'turnogo landshafta (Tedzhenskogo i Murgabskogo oazisov) [Birds of a Cultured Landscape (Tedzhen and Murghab Oases)]. Ylym, Ashkhabad: 35–59 (in Russian).
Sukhinin, A.N. 1971. Ekologiya sov i khishchnykh ptits Badkhyza [Ecology of Owls and Birds of Prey in Badghyz]. Ylym, Ashkhabad, 100 pp (in Russian).
Sukhinin, A.N. 1969. On reproduction and feeding of the lammergeier. Izv. Akad. Nauk. Turkm. SSR, Ser. Biol. Nauk. [Proc. Acad. Sci. Turkmen. SSR, Ser. Biol. Sci.] 4: 83–86 (in Russian).
Sukhinin, A.N. and Ye.I. Shcherbina. 1955. Honey badger in Turkmenia. Priroda [Nature] 5: 117–118 (in Russian).
Surova, T.G. 1962. Sporovo-pyltsevye spektry novokaspiiskikh otlozhenii Balkhanskogo shora yugo-zapadnogo Turkmenistana [Palynological Spectra of the New Caspian Deposits from the Balkhan Shor in Southwestern Turkmenistan]. Gostoptekhizdat, Moscow, 218 pp (in Russian).
Sushkin, P.P. 1925. Zoogeographical areas of Central Siberia and the adjacent parts of mountainous Asia, with an attempt to characterize the history of modern fauna of Palearctic Asia. Byull. MOIP, otd. biol [Bull. Mosc. Soc. Natur., Div. Biol.] 69(3): 73–76 (in Russian).
Syroyechkovsky, Ye.Ye. 1958. An ecological-geographical outline of the fauna of the reptiles of the western Kyzyl-Kum Desert. Zool. zhurn. [Zool. J.] 37(2): 240–250 (in Russian).

Tachikawa, T. 1974. Hosts of Encyrtidae (Hymenoptera, Chalcidoidea). Mem. Coll. Agr. Egime Univ. 19: 185-204.
Tachikawa, T. 1981. Hosts of encyrtid genera in the world (Hymenoptera, Chalcidoidea). Mem. Coll. Agr. Egime Univ. 25(2): 85-110.
Tackholm, V. 1974. Student's Flora of Egypt. 2nd Ed. Cairo Univ., Cairo, 888 pp.
Tadjalli-Pour, M. 1977. Les mollusques marins de cote Iraniennes de la mer Caspienne (Astara-Hachtpar). J. de Conchyliol. 114(3-4): 87-117.
Tadjalli-Pour, M. 1980. Contribution a l'etude de la faune macroscopique bentique de la partie ouest de la mer Caspienne-sud. Soc. Nation. Iran. de sci. de la mer, Ahwaz, 126 pp (in French) and 128 pp (in Farsi).
Takhtadzhyan, A.L. 1941. Botaniko-geograficheski ocherk Armenii [Botanical-geographical Survey of Armenia]. Ed. Bot. inst. AN Arm. SSR, Tbilisi-Yerevan 2: 1-180 (in Russian).
Takhtadzhyan, A.L. 1946. On the historical development of the vegetation of Armenia. Trudy Bot. inst. AN Arm. SSR [Trans. Bot. Inst. Acad. Sci. Armenian SSR] 4: 51-107 (in Russian).
Takhtadzhyan, A.L. 1966. Sistema i filogeniya tsvetkovykh rastenii [System and Phylogeny of the Angiosperm Plants]. Nauka, Moscow-Leningrad, 611 pp (in Russian).
Takhtadzhyan, A.L. 1978. Floristicheskie oblasti Zemli [The Floristic Regions of the World]. Leningrad, Nauka, 247 pp (in Russian).
Takhtadzhyan, A.L. (ed.). 1981. Redlkie i ischezayushchiye vidy flory SSSR, nuzhdayushchiyesya v okhrane [Rare and Endangered Species of the USSR Flora Which Require Protection]. Nauka, Leningrad, 263 pp (in Russian).
Takhtadzhyan, A.L. 1987. Sistema magnoliofitov [System of the Magnoliophyta]. Nauka, Leningrad, 438 pp (in Russian).
Tanasevich, A.V. 1988 (1989). The linyphiid spiders of Middle Asia (Arachnida: Araneae: Linyphiidae). Senckenberg. Biol. 69(1-3): 83-176.
Tanasevich, A.V. and V.Ya. Fet. 1986. Contributions to the spider fauna of Turkmenistan, III. Family Linyphiidae. Izv. Akad. Nauk. Turkm. SSR, Ser. Biol. Nauk. [Proc. Acad. Sci. Turkmen. SSR, Ser. Biol. Sci.] 1: 33-42 (in Russian).
Tarasov, R.N. 1954. The vegetation of Maly Balkhan. Trudy In-ta biol. AN TSSR [Trans. Inst. Biol. Acad. Sci. Turkmen SSR] 2: 5-44 (in Russian).
Tashliev, A.O. and S.A. Yermakova. 1961. Data on birds of the areas adjacent to the Karakum Canal. Trudy In-ta zool. i parazitol. AN TSSR [Trans. Inst. Zool. and Parasitol. Acad. Sci. Turkmen SSR] 7: 5-42 (in Russian).
Temple, S.A. 1989. Biodiversity issues: Do they involve Wisconsin birds? Passenger Pigeon 51(3): 285-288.
Terentyev, P.V. and S.A. Chernov. 1949. Opredelitel' presmykayushchikhsya i zemnovodnykh. [A Key to the Reptiles and Amphibians]. Sov. Nauka, Moscow, 340 pp (in Russian).
Tertyshnikov, M.F. and N.N. Shcherbak. 1989. On the systematic position of *Ophisaurus apodus* (Pall., 1775) from the USSR. Vestn. zool. [Mess. Zool.] 5: 35-37 (in Russian).
Therond, J. 1969. Beiträge zur Kenntnis der Fauna Afghanistans. Histeridae. Casop. Moravsk. Mus. Brno, 54, Suppl.: 183-198.
Thomas, D.H. 1984a. Adaptations of desert birds: sandgrouse (Pteroclididae) as highly successful inhabitants of Afro-Asian arid lands. J. Arid Envir. 7(2): 157-181.
Thomas, D.H. 1984b. Sandgrouse as models of avian adaptations to deserts. S. Afr. J. Zool. 19(2): 113-120.
Thomas, D.H., B. Pinshow, and A.A. Degen. 1984. Renal and lower intestinal contributions to the water economy of desert-dwelling phasianid birds: Comparison of free living and captive chuckars and sand partridges. Physiol. Zool. 57(1): 128-136.
Tleulov, R.T. 1981. Novyi rezhim Arala i yego vliyanie na ikhtiofaunu [The New Regime of the Aral Sea and its Effect on the Ichthyofauna]. Fan, Tashkent, 190 pp (in Russian).
Tleulov, R. and N.N. Sagitov. 1973. Osetrovye ryby Amudaryi [Sturgeon Fishes of the Amudarya River]. Fan, Tashkent, 155 pp (in Russian).
Tokar, A.A. 1990. Sistematika zmei roda *Eryx* fauny SSSR i sopredel'nykh stran [Systematics of the

Snakes of the Genus *Eryx* from the USSR and Adjacent Countries]. Ph.D. Thesis, Kiev (in Russian).

Tokgaev, T.T. 1966. Marokkskaya sarancha v Turkmenii [The Moroccan Locust in Turkmenia]. Turkmenistan, Ashkhabad, 127 pp (in Russian).

Tokgaev, T.T. 1968a. Fauna and practical importance of the orthoptherans (Tettigonoidea, Acridoidea) of Badghyz. Izv. AN TSSR, Ser. Biol. Nauk. [Proc. Acad. Sci. Turkmen SSR, Ser. Biol. Sci.] 3: 75–80 (in Russian).

Tokgaev, T.T. 1968b. Fauna and distribution of the orthoptherans (Tettigonoidea, Acridoidea) of West Kopetdagh. Izv. AN TSSR, Ser. Biol. Nauk. [Proc. Acad. Sci. Turkmen SSR, Ser. Biol. Sci.] 5: 68–75 (in Russian).

Tokgaev, T.T. 1972. Fauna and ecological complexes of the orthoptherans in Central Kopetdagh. In: T. Tokgaev and S.N. Myartseva (ed.), Nasekomye Yuzhnoi Turkmenii [Insects of Southern Turkmenia]. Ashkhabad, Ylym: 18–27 (in Russian).

Tokgaev, T.T. 1973. Fauna i ekologiya saranchovykh Turkmenii [Fauna and Ecology of the Acridoidea in Turkmenia]. Ylym, Ashkhabad, 222 pp (in Russian).

Tokgaev, T.T. 1974. Fauna and ecology of the tettigonoids (Orthoptera, Tettigonoidea) in Turkmenia. Trudy Vsesoyuzn. Entomol. Obshchestva [Transactions of All-Union Entomological Society] 57: 123–131 (in Russian).

Tolmachev, A.I. 1974. Vvedenie v geografiyu rastenii [Introduction to Plant Geography]. Izd. LGU, Leningrad, 244 pp (in Russian).

Tolmachev, A.I. 1986. Metody sravnitel'noi floristiki i problemy florogeneza [Methods of Comparative Floristics and Problems of Florogenesis]. Nauka, Novosibirsk, 195 pp (in Russian).

Tolstov, S.P. 1964. Po drevnim del'tam Oksa i Yaksarta [Along the Ancient Deltas of the Oxus and the Yaxart]. Vostochnaya literatura, Moscow, 324 pp (in Russian).

Treus, V.D. 1968. Akklimatizatsiya i gibridizatsiya zhivotnykh v Askanii – Nova [Acclimatization and Hybridization of Animals in Askania Nova]. Urozhai, Kiev, 316 pp (in Russian).

Tryapitsyn, V.A. 1972. Trophic relationships in the family Encyrtidae (Hymenoptera, Chalcidoidea). In: V.A. Zaslavsky (ed.), Khozyainno-parazitnyye otnosheniya u nasekomykh [Host-Parasite Relationships in Insects]. Nauka, Leningrad: 31–47 (in Russian).

Tryapitsyn, V.A. 1989. Nayezdniki-encirtidy (Hymenoptera, Encyrtidae) Palearktiki [The Encyrtid Wasps (Hymenoptera, Encyrtidae) of the Palearctic]. Nauka, Leningrad, 488 pp (in Russian).

Tsalkin, V.I. 1948. Wild sheep of Turkmenia. Byull. MOIP [Bull. Moscow Soc. Natur.] 53(1): 31–48 (in Russian).

Tsalkin, V.I. 1950. A new subspecies of bezoar goat from Turkmenia. Dokl. AN SSSR [Rep. Acad. Sci. USSR] 70(2): 323–325 (in Russian).

Tsvelev, N.N. 1976. Zlaki SSSR [The Grasses of the USSR]. Nauka, Leningrad, 788 pp (in Russian).

Tuniev, B.S., Ch. Ataev, and S. Shammakov. 1991. *Stellio erythrogaster nurgeldievi* spp. nov. (Aganidae, Suaria), a new subspecies of the Khorassan stellion from East Kopetdagh. Izv. AN TSSR, Ser. Biol. Nauk. [Proc. Acad. Sci. Turkmen SSR, Ser. Biol. Sci.] 6: 50–56 (in Russian).

Tupikova, N.V. 1969. Zoologicheskoe kartografirovanie [Zoological Mapping]. Moscow State Univ., Moscow, 249 pp (in Russian).

Tupikova, N.V. 1976. Mapping of animal populations. Itogi nauki. Biogeografiya [Scientific Results: Biogeography]. VINITI, Moscow 1: 98–219 (in Russian).

Tupikova, N.V. and L.V. Komarova. 1979. Printsipy i metody zoologicheskogo kartografirovaniya [Principles and Methods of Zoological Mapping]. Moscow State Univ., Moscow, 189 pp (in Russian).

Ushko, K.A. and L.S. Isaeva-Petrova. 1959. New data on the Pliocene flora of West Turkmenia. Dokl. AN SSSR [Rep. Acad. Sci. USSR] 126(2): 392–395 (in Russian).

Utochkin, A.S. 1956. Contributions to the study of the venomous spider *Latrodectus pallidus* O.P. Cambr. subsp. *pavlovskyi* Charit.). Zool. zhurn. [Zool. J.] 35: 1657–1660 (in Russian).

Utochkin, A.S. 1960a. Spiders belonging to the genus *Synaema*, the group *plorator*. Zool. zhurn. [Zool. J.] 39: 375–380 (in Russian).

Utochkin, A.S. 1960b. Spiders of the genus *Synaema*, the group *globosum*. Zool. zhurn. [Zool. J.] 39: 1018–1024 (in Russian).
Utochkin, A.S. 1960c. Contributions to the fauna of spiders of the genus *Oxyptila* Sim. in the USSR. Uchen. zap. Permsk. gos. un-ta [Sci. Mem. Perm. State Univ.] 13: 47–61 (in Russian).
Utochkin, A.S. 1968. Oprdelitel' paukov roda *Xysticus* fauny SSSR [A Key to the Spiders of the Genus *Xysticus* in the USSR Fauna]. Perm State Univ., Perm, 73 pp (in Russian).
Utochkin, A.S. 1981. A contribution to the systematics of the spider genus *Tibellus* in the fauna of the USSR. In: Fauna i ekologiya nasekomykh [Fauna and Ecology of Insects]. Perm State Univ., Perm: 8–20 (in Russian).
Utochkin, A.S. 1985. Contributions to the [knowledge of] spider genus *Heriaeus* (Aranei, Thomisidae) of the USSR. Fauna i ekologiya paukov SSSR/Trudy Zool. in-ta AN SSSR [Fauna and Ecology of Spiders of the USSR/Trans. Zool. Inst. Acad. Sci. USSR] 139: 105–113 (in Russian).
Uvarov, B.P. 1912. Ueber die Orthopterenfauna Transcaspiens. Trudy Russk. entomol. ob-va [Trans. Russian Entomol. Soc.] 15(3): 54.
Uvarov, B.P. 1921. The geographical distribution of orthopterous insects in the Caucasus and in Western Asia. Proc. Zool. Soc. London 31: 447–472.
Uvarov, B.P. 1938. Ecological and biogeographical relations of Eremian Acrididae. Mem. Soc. Biogeogr. 6: 231–273.
Vachon, M. 1951. Sur quelques Scorpions "halophiles" (*Microbuthus fagei*, *Mesobuthus confucius* et *Euscorpius flavicaudis*). Bull. Mus. natn. Hist. natur., Sér. 2, 23(3): 256–260.
Vachon, M. 1952. Etudes sur les scorpions. Inst. Pasteur de l' Alger, Alger, 482 pp.
Vachon, M. 1958a. A propos de *Liobuthus kessleri* Birula, Scorpion psammophile nouveau pour le faune iranienne. Bull. Mus. natn. Hist. natur., Sér. 2, 30(5): 422–426.
Vachon, M. 1958b. Scorpionidea (Chelicerata) de l'Afghanistan (The 3rd Danish Expedition to Central Asia. Zool. results 23). Vidensk. Medd. fra Dansk naturh. For. Kobenhavn 120: 121–187.
Vachon, M. 1966. Liste des scorpions connus en Égypte, Arabie, Israel, Liban, Syrie, Jordanie, Turquie, Irak, Iran. Toxicon 4: 209–217.
Vachon, M. 1974. Etude des caractères utilisés pour classer les familles et les genres du Scorpions (Arachnides). I. La trichobothriotaxie en arachnologie. Sigles trichobothriaux et types de trichobothriotaxie chez les Scorpions. Bull. Mus. natn. Hist. natur., Sér. 3, 140 (Zool. 104): 857–958.
Vachon, M. and R. Kinzelbach. 1987. On the taxonomy and distribution of the scorpions of the Middle East. In: F. Krupp, W. Schneider and R. Kinzelbach (eds.), Proc. Symp. on the Fauna and Zoogeography of the Middle East, Mainz 1985. Beihefte zum TAVO A 28. Dr. Ludwig Reichert Verlag, Wiesbaden: 91–103.
Vakhrameev, V.A. 1964. The Jurassic and Early Cretaceous floras of Eurasia and paleofloristic provinces of that time. Trudy Geol. inst. AN SSSR [Trans. Geol. Inst. Acad. Sci. USSR] Nauka, 102 Moskow, 260 pp. (in Russian).
Vashetko, E.V. 1969. On the ecology of *Eremias strauchi* in southwestern Turkmenia. Zool. zhurn. [Zool. J.] 48(12): 1893–1895 (in Russian).
Vashetko, E.V. 1970. On the ecology of *Eremias velox velox* during fall period in southwestern Turkmenia. Uzbek. biol. zhurn. [Uzbek Biol. Journal] 6: 43–45 (in Russian).
Vashetko, E.V. and M.Ye. Chernyakhovsky. 1969. Feeding of *Eremias velox* Pallas in southern Turkmenia. Izv. AN TSSR, Ser. Biol. Nauk. [Proc. Acad. Sci. Turkmen SSR, Ser. Biol. Sci.] 2: 85–87 (in Russian).
Varentsov, P.A. 1894. Nablyudeniya za pozvonochnymi i spiski zhivotnykh, naidennykh v 1890–1892 gg. v Zakaspiiskoi oblasti. Prilozheniye k obzoru Zakaspiiskoi oblasti za 1892 g. [Observations on the vertebrates and lists of animals found in 1890–1892 in the Transcaspian Region. A supplement to the 1892 survey of the Transcaspian Region]. Ashkhabad: 4–38 (in Russian).
Vasilevskaya, N.D. 1949. On the Poltava xerophite flora of Turkmenia. Dokl. AN SSR, Nov. Ser. [Reports Acad. Sci. USSR, New ser.] 18(4): 30–36 (in Russian).

Vasilevskaya, N.D. 1957. The Eocene flora of Badghyz in Turkmenia. In: P.I. Dorofeev (ed.), Sbornik pamyati A. N. Krishtofovicha [A Memorial Issue for A.N. Krishtofovich]. Izd. AN SSSR, Moscow-Leningrad: 103–175 (in Russian).

Vasilyev, I.V. 1905. On a record of the shielded skink in the Transcaspian Region. Yezheg. zool. muz. Akademii nauk. [Annu. Zool. Mus. Acad. Sci. St. Petersb.] 9: 312 (in Russian).

Vasilyev, V.I., V.I. Yefimov, and V.A. Zarkhidze. 1960. New data on distribution of some vertebrates in northwestern Turkmenia. Izv. AN TSSR, Ser. Biol. Nauk. [Proc. Acad. Sci. Turkmen SSR, Ser. Biol. Sci.] 4: 82–83 (in Russian).

Vasilyev, V.I. 1976a. Data on seasonal distribution, density, and ecology of the flamingo in the eastern Caspian. In: Materialy soveshchaniya po promyslovoi ornitologii [Materials of the Conference on Industrial Ornithology], Moscow: 137–138 (in Russian).

Vasilyev, V.I. 1976b. On waterfowl rewintering on the southeastern shore of the Caspian. In: Okhrana prirody v Turkmenistane. Tezisy dokl. 1 nauchn. konf. po okhrane prirody Turkm. SSR [Natural Conservation in Turkmenistan. Abstracts of the 1st Scientific Conference on Natural Conservation in Turkmen SSR], Ashkhabad: 52–55 (in Russian).

Vaurie, C. 1965. The birds of the palearctic fauna. A systematic reference. Non-Passeriformes. Witherby, London, 763 pp.

Vavilov, N.I. 1967. Izbrannye proizvedeniya v dvukh tomakh [Selected Works in Two Volumes]. Nauka, Leningrad 1, 423 pp (in Russian).

Velikanov, V.P. 1977. On new records of the squeaking gecko and *Agkistrodon halys* in Turkmenia. Izv. AN TSSR, Ser. Biol. Nauk. [Proc. Acad. Sci. Turkmen SSR, Ser. Biol. Sci.] 5: 81–82 (in Russian).

Velikanov, V.P. and A.N. Khokhlov. 1979. On the ornithofauna and biology of waterfowl and bywater birds of Lake Sarykamysh. In: Prirodnaya sreda i ptitsy poberezhii Kaspiiskogo morya i prilezhashchikh nizmennostei/Trudy Kizyl-Agachskogo gos. zapovednika [Environment and Birds of the Caspian Sea Shores and Adjacent Lowlands/Trans. Kizyl-Agach State Reserve]. Azgosizdat, Baku 1: 236–240 (in Russian).

Vengerov, M.P. 1975. On colonial nesting birds of the protected Lake Delili. In: Kolonial'no gnezdyashchieyesya okolovodnye ptitsy i ikh okhrana [Colonial, Nesting Bywater Birds and Their Conservation]. Moscow (in Russian).

Vereshchagin, N.K. and B. Batyrov. 1967. Fragments of the history of theriofauna of Middle Asia. Byull. MOIP, Otd. biol. [Bull. Moscow Soc. Natur., Biol. Div.] 72(4): 104–115 (in Russian).

Vernander, T.B., S.V. Viktorov, A.G. Voronov, T.I. Dybskaya, N.P. Osadchaya, and A.M. Cheltsov-Bebutov. 1959. An proposed method for medium-scale biogeographic mapping. In: A.N. Formozov (ed.), Geografiya naseleniya nazemnykh zhivotnykh i metody yego izucheniya [Geography of the Population of Terrestrial Animals and Methods of its Study]. Izd. AN SSSR, Moscow: 33–34 (in Russian).

Vinogradov, B.S., A.I. Argiropulo, and V.G. Geptner. 1936. Gryzuny Srednei Azii [Rodents of Middle Asia]. Izd. AN SSSR, Moscow-Leningrad, 228 pp (in Russian).

Vinogradov, B.S. and Ye.P. Bondar'. 1949. On the record in Turkmenia of a new species of jerboa belonging to the genus *Jaculus*. Dokl. AN SSR [Reports of the Academy of Sciences of the USSR] 15(4): 559–562 (in Russian).

Vinogradov, B.S. and I.M. Gromov. 1952. Gryzuny fauny SSSR [Rodents of the USSR Fauna]. Izd. AN SSSR, Moscow-Leningrad, 297 pp (in Russian).

Vinogradov, V.V. and V.B. Potapova. 1964. Hidden metachromasia: a new method for histochemical detection of sialomucines. Arkhiv anat., gistol. i embriol. [Arch. Anat., Histol., and Embryol.] 47(11): 69–75 (in Russian).

Vlasov, Ya.P. 1932. On a record of mosquitoes in the burrows of rodents (*Rhombomys opimus* and *Spermophilopsis leptodactylus*). Parazitologicheskii sbornik [Parasitol. Papers] 3: 63–72 (in Russian).

Vlasov, Ya.P. 1933 – see Wlassow, J.P.

Vlasov, Ya.P. 1937a. The burrow as a specific habitat in environs of Ashkhabad. Problemy parazitologii i fauny Turkmenii/Trudy SOPS, ser. turkmen. [Problems of Parasitology and Fauna of Turkmenia/Trans. SOPS, Turkmen series], Leningrad 9: 223–240 (in Russian).

Vlasov, Ya.P. 1937b. Arachnids from the burrows in environs of Ashkhabad. Problemy parazitologii i fauny Turkmenii/Trudy SOPS, ser. turkmen. [Problems of Parasitology and Fauna of Turkmenia/Trans. SOPS, Turkmen series], Leningrad 9: 241–245 (in Russian).

Vlasov, Ya.P. 1937c. The fauna of Bakharden (Durun) Cave. Problemy parazitologii i fauny Turkmenii/Trudy SOPS, ser. turkmen. [Problems of Parasitology and Fauna of Turkmenia/Trans. SOPS, Turkmen series], Leningrad 9: 289–296 (in Russian).

Vlasov, Ya.P. 1941. Burrows of *Spermophilopsis leptodactylus* and *Rhombomys opimus* as a specific habitat of *Phlebotomus* in the environs of Ashkhabad. In: Problemy kozhnogo leishmanioza [Problems of Cutaneous Leishmaniasis]. Ashkhabad: 74–89 (in Russian).

Vlasov, Ya.P. and I.G. Ioff. 1937. Fleas from the burrows in the environs of Ashkhabad. Problemy parazitologii i fauny Turkmenii/Trudy SOPS, ser. turkmen. [Problems of Parasitology and Fauna of Turkmenia/Trans. SOPS, Turkmen series], Leningrad 9: 277–282 (in Russian).

Vlasov, Ya.P. and A.N. Kirichenko. 1937. Hemiptera-Heteroptera from the burrows in the environs of Ashkhabad. Problemy parazitologii i fauny Turkmenii/Trudy SOPS, ser. turkmen. [Problems of Parasitology and Fauna of Turkmenia/Trans. SOPS, Turkmen series], Leningrad 9: 263–268 (in Russian).

Vlasov, Ya.P. and E.F. Miram. 1937. Cockroaches and Orthoptera from the burrows in the environs of Ashkhabad. Problemy parazitologii i fauny Turkmenii/Trudy SOPS, ser. turkmen. [Problems of Parasitology and Fauna of Turkmenia/Trans. SOPS, Turkmen series], Leningrad 9: 259–262 (in Russian).

Vlasov, Ya.P. and Ye.L. Shestoperov. 1937. Beetles from the burrows in the environs of Ashkhabad. Problemy parazitologii i fauny Turkmenii/Trudy SOPS, ser. turkmen. [Problems of Parasitology and Fauna of Turkmenia/Trans. SOPS, Turkmen series], Leningrad 9: 269–276 (in Russian).

Vlasov, Ya.P. and A.A. Shtakelberg. 1937. On the dipterofauna from the vertebrate burrows in the environs of Ashkhabad. Problemy parazitologii i fauny Turkmenii/Trudy SOPS, ser. turkmen. [Problems of Parasitology and Fauna of Turkmenia/Trans. SOPS, Turkmen series], Leningrad 9: 283–288 (in Russian).

Vlasov, Ya.P. and V.I. Sytschevskaja. 1937. The spiders of the environs of Ashkabad. Problemy parazitologii i fauny Turkmenii/Trudy SOPS, ser. turkmen. [Problems of Parasitology and Fauna of Turkmenia/Trans. SOPS, Turkmen series], Leningrad 9: 247–258 (in Russian).

Voitsekhovsky, D.P. 1958. An ecologo-faunistic survey of rodents of the Karabil Plateau. Trudy Turkmenskoi protivochumnoi stantsii [Trans. Turkmen Anti-Plague Station] 1: 145–152 (in Russian).

Volkovich, M.G. and A.V. Alexeev. 1990. The buprestid beetles (Coleoptera, Buprestidae) of Badghyz. In: K.I. Atamuradov (ed.), Priroda Badkhyza [Nature of Badghyz]. Ylym, Ashkhabad: 146–170 (in Russian).

Vorobyev, K.A. 1946. Data on the biology of the francolin and its distribution in southern Turkmenia. Byull. MOIP. Otd. biol. [Bull. Moscow Soc. Natur. Div. biol.] 51(1): 62–68 (in Russian).

Vorobyev, K.A. 1955. Contributions to the ornithofauna of Badghyz (Southeast Turkmenia). Zool. Zhurn. [Zool. J.] 34(4): 898–901 (in Russian).

Vorobyev, K.A. 1967. The francolin. Okhota i okhotnichye khozyaistvo [Hunting and Game Management] 5: 18–19 (in Russian).

Vorobyev, K.A. 1968. In the tugais of Tedzhen. Okhota i okhotnichye khozyaistvo [Hunting and Game Management] 6: 40–41 (in Russian).

Voronov, A.G. and A.M. Cheltsov-Bebutov. 1962. On the technique of the biogeographic mapping of open landscapes. In: V.B. Sochava (ed.), Printsipy i metody geobotanicheskogo kartografirovaniya [Principles and Methods of Geobotanical Mapping]. Izd. AN SSSR, Moscow-Leningrad: 186–193 (in Russian).

Vorontsov, N.N., I.V. Kartavtseva, and Ye.G. Potapova. 1979. Systematics of the hamster genus *Calomyscus*. I. Karyological differentiation of sibling species from Transcaucasia and Turkmenia, with a survey of species belonging to the genus *Calomyscus*. Zool. zhurn. [Zool. J.] 58(8): 1213–1224 (in Russian).

Vtorov, P.P. and N.N. Drozdov. 1978. Biogeografiya [Biogeography]. Prosveshcheniye, Moscow, 268 pp (in Russian).
Walter, H. 1927. Einführung in die allgemeine Pflanzengeographie Deutschlands. Jena, G. Fischer, 458 pp.
Walter, H.I. and A.A. Alyokhin. 1936. Osnovy botanicheskoi geografii [Foundations of Botanical Geography]. Biomedgiz, Moscow-Leningrad, 714 pp (in Russian).
Whittaker, R.H. 1960. Vegetation of the Siskiyou Mountains, Oregon and California. Ecol. Monogr. 30: 279–338.
Whittaker, R.H. 1965. Dominance and diversity in land plant communities. Science 147: 250–260.
Willoughby, E.J. 1966. Water requirements of the ground dove. Condor 68(3): 243–248.
Wlassow, J.P. 1933. Die Fauna der Wohnhoglen von *Rhombomys opimus* und *Spermophilopsis leptodactylus* in der Umgebung von Aschhabad. Zool. Anz. 101 (5/6): 143–158.
Wulff, E.V. 1944. Istoricheskaya geografiya rastenii. Istoriya flor zemnogo shara [Historical Geography of Plants. History of the Floras of the Globe]. Izd. AN SSSR, Moscow-Leningrad, 545 pp (in Russian).
Wuster, W. and R.S. Thorpe. 1992. Asiatic cobras: population systematics of the *Naja naja* species complex (Serpentes: Elapidae) in India and Central Asia. Herpetologica 48(1): 69–85.
Yablokov-Khnzoryan, S.M. 1961. Opyt vosstanovleniya genezisa fauny zhestkokrylykh Armenii [An Attempt to Reconstruct the Genesis of the Fauna of Coleoptera in Armenia]. Izd. AN Arm. SSR, Yerevan, 266 pp (in Russian).
Yamnov, A.A. and V.N. Kunin. 1953. Some theoretical results of the newest studies in paleogeography and geomorphology in the area of Uzboi. Izv. AN SSSR. Ser. geograf. [Proc. Acad. Sci. USSR, Ser. Geogr.] 3 (in Russian).
Yasamanov, N.A. 1978. Landshaftno-klimaticheskie usloviya yury, mela i paleogena yuga SSSR [Landscape and Climatic Conditions in the Jurassic, Cretaceous, and Paleogene in Southern USSR]. Nedra, Moscow, 224 pp (in Russian).
Yasakova, E.I. 1974. On the ecology of horse fly larvae (Diptera, Tabanidae) in Turkmenia. Med. parasit. bolezni [Med. Parasit. Diseases] 43(5): 563–568 (in Russian).
Yasakova, E.I. and N.P. Krivosheina. 1973. Notes on the morphology of the larva of a horse fly *Tabanus unifasciatus* L.W. (Diptera, Tabanidae). Izv. AN Turkm. SSR, Ser. Biol. Nauk. [Proc. Acad. Sci. Turkm. SSR, Ser. Biol. Sci.] 1: 68–70 (in Russian).
Yefimenko, N.N. 1989. Rare and poorly studied bird species of the Kopetdagh Reserve. In: Ekolog. aspekty izucheniyua, prakticheskogo ispolzovaniya i okhrany ptits v gornykh ekosistemakh [Ecolog. Aspects of Study, Exploitation, and Conservation of Birds in Mountain Ecosystems]. Frunze: 34–36 (in Russian).
Yefremov, Yu.K. 1956. Peredneaziatskiye nagorya [Plateaus of Southwest Asia]. In: B.F. Dobrynin and E.M. Murzaev (ed.), Zarubezhanya Aziya [Foreign Asia]. Gosuchpedgiz, Moscow: 88–188 (in Russian).
Yemelyanov, A.F. 1972. A review of viewpoints toward the history of development of the biota of Central Asian deserts. In: I.M. Kerzhner (ed.), Nasekomye Mongolii [Insects of Mongolia]. Nauka, Leningrad 1: 11–49 (in Russian).
Yemelyanov, A.F. 1974. Suggestions on classification and nomenclature of geographic ranges. Entomol. obozr. [Entomol. Rev.] 53(3): 497–522 (in Russian).
Yenikov, I.F. 1949. The saksaul locust, *Dericorys albidula* Serv., in Turkmenistan. Izv. AN TSSR, Ser. Biol. Nauk. [Proc. Acad. Sci. Turkmen SSR, Ser. Biol. Sci.] 1: 61–63 (in Russian).
Yeremchenko, V.K. and N.N. Shcherbak. 1986. Ablefaridnye yashcheritsy fauny SSSR i sopredel'nykh stran [Ablepharid Lizards of the USSR and Adjacent Countries]. Ilim, Frunze, 170 pp (in Russian).
Yevgenov, D.N. and A.P. Parinkin. 1955. Data on the geographic distribution of the Khentau toad agama. Uchen. zap. LGU [Sci. Mem. Leningrad State Univ.], Ser. Biol. 181(38): 51–52 (in Russian).
Yurevich, A.A. 1966. Material composition and conditions of the formation of the Akchagyl deposits in the Cis-Balkhan area of Southwest Turkmenia. Trudy Geolog. Inst. AN SSSR [Trans. Geol. Inst. Acad. Sci. USSR] 164 (in Russian).

Yurin, V.A. and N.Ya. Myagkov. 1959. Calculation of sunshine duration for Turkmenistan. Uchen. zap. Turkmen. gos. un-ta [Sci. Mem. Turkmen State Univ.] 15: 170–180 (in Russian).
Zabelina, N.I. 1985. New data on the flora of Southwest Kopetdagh. In: N.T. Nechayeva (ed.), Rastitelnost' i zhivotnyi mir Zapadnogo Kopetdaga [The Vegetation and Animal Life of West Kopetdagh]. Ylym, Ashkhabad: 38–50 (in Russian).
Zakharenko, A.V. 1983. Ant-lions of the genus *Aspoeckiana* Hölzel (Neuroptera, Myrmeleontidae) in the USSR fauna. Entomol. obozr. [Entomol. Rev.] 62 (3): 586–589 (in Russian).
Zakharov, V.M. 1982. A phenogenetic aspect in the studies of natural populations. In: A.V. Yablokov (ed.), Fenetika populyatsii [Population Phenetics]. Nauka, Moscow: 45–55 (in Russian).
Zakhidov, T.Z. 1971. Biotsenozy pustyni Kyzylkum [Biocoenoses of the Kyzylkum Desert]. Fan, Tashkent, 304 pp (in Russian).
Zakhidov, T.Z. and R.N. Meklenburtsev. 1969. Priroda i zhivotnyi mir Srednei Azii. Pozvonochnye zhivotnye [Nature and Animal Life of Middle Asia. Vertebrates]. Ukituvchi, Tashkent 1, 428 pp (in Russian).
Zakirov, K.Z. and P.K. Zakirov. 1978. Opyt tipologii rastitel'nosti zemnogo shara na primere Srednei Azii. [An Attempt to Typologize the World Vegetation usingMiddle Asia as an Example]. Fan, Tashkent, 56 pp (in Russian).
Zaklinskaya, Ye.D. 1953. Palynological spectra of marine Eocene sediments from northern Aral area. Dokl. AN SSR [Rep. Acad. Sci. USSR] 42(5): 1043–1046 (in Russian).
Zaklinskaya, Ye.D. 1958. Types of palynological spectra of the Paleogene sediments from various physical-geographical provinces. Materialy po istorii flory i rastitel'nosti SSSR [Materials on the History of the Flora and Vegetation of the USSR]. Izd. AN SSSR, Moscow- Leningrad 3: 43–61 (in Russian).
Zaktreger, N.I. 1966. *Amygdalus scoparia* Spach. in the Kara-Kala District. Trudy TOS VIR [Trans. Turkmen Experim. Station of All-Union Plant Breeding Inst.] 4: 290–296 (in Russian).
Zapryagayeva, V.I. 1964. Dikorastushchie plodovye Tadzhikistana [Natural Fruit Plants of Tajikistan]. Izd. AN SSSR, Moscow-Leningrad, 695 pp (in Russian).
Zapryagayeva, V.I. 1975. Lesnye resursy Pamiro-Alaya [Forest Resources of Pamiro-Alai]. Nauka, Leningrad, 594 pp (in Russian).
Zarkhidze, V.A. 1979. Density and conservation of predatory mammals of western Turkmenia. In: V.E. Sokolov (ed.), Ekologicheskie osnovy okhrany i ratsionalnogo ispol'zovaniya khishchnykh mlekopitayushchikh [Ecological Principles of Conservation and Rational Exploitation of Carnivorous Mammals]. Nauka, Moscow: 39–40 (in Russian).
Zarkhidze, V.A. 1980a. Ungulates of West Turkmenia. In: V.E. Sokolov (ed.), Kopytnye fauny SSSR [Ungulates of the USSR fauna]. Nauka, Moscow: 88–90 (in Russian).
Zarkhidze, V.A. 1980b. Density of dzheirans in the southern Ustyurt. In: V.E. Sokolov (ed.), Kopytnye fauny SSSR [Ungulates of the USSR Fauna]. Nauka, Moscow: 90–91 (in Russian).
Zarkhidze, V.A. 1981. The grey monitor in western Turkmenia. In: I.S. Darevsky (ed.), Voprosy gerpetologii [Problems of Herpetology]. Nauka, Leningrad: 55–56 (in Russian).
Zarudny, N.A. 1896. Ornithological fauna of the Transcaspian area (North Persia, Transcaspian Region, Khiva Oasis, and lowland Bokhara]. Materialy k poznaniyu fauny i flory Rossiskoi Imperii, Otd. zoologii [Contributions to the Knowledge of Fauna and Flora of the Russian Empire, Div. Zool.] 2: 1–555 (in Russian).
Zarudny, N.A. 1897. Notes on reptiles and amphibians of northeastern Persia. Trudy zool. muz. Imper. Akad. nauk [Annu. Zool. Mus. Imper. Acad. Sci.], St. Petersburgh 2: 349–361 (in Russian).
Zarudny, N.A. 1900. Excursion through northeastern Persia and birds of this country. Zap. Akad. Nauk. po fiz.-mat. otd. [Bull. Acad. Sci., St.-Petersburg, Div. Physico-Math.], Ser. 8, 10(1): 1–262 (in Russian).
Zarudny, N.A. 1903. On amphibians, reptiles, and fishes of eastern Persia. Zap. Imper. Russk. Geogr. obshchestva [Mem. Imper. Russian Geogr. Soc.] 36(3): 1–39 (in Russian).
Zarudny, N.A. 1915. Birds of the Aral Sea. Izv. Turkest. otdel. RGO [Proc. Turkestan Branch Russian Geogr. Soc.] 12(1) (in Russian).

Zarudny, N.A. and S.I. Bilkevich. 1918. A list of birds of the Transcaspian region and their distribution over zoological areas of this country. Izv. Zakaspiiskogo muzeya [Proc. of Transcaspian Museum] 1: 1–56 (in Russian).
Zhirin, V.M. 1974. Juniper woodlands of Turkmenia. Lesnoye khozyaistvo [Forestry]. (in Russian).
Zhirnov, L.V., A.A. Vinokurov, and V.A. Bychkov. 1978. Redkie i ischezayushchie zhivotnye SSSR. Mlekopitayushchie i ptitsy [Rare and Endangered Animals of the USSR: Mammals and Birds]. Lesnaya promyshlennost, Moscow, 303 pp (in Russian).
Zhitnikov, M. 1900. Ornithological observations in the lower Atrek River: Winter 1898 and spring 1899. Psovaya i ruzheinaya okhota [Hunting with Dogs and Guns] 10: 1–57 (in Russian).
Zhitnikov, M. 1901. On the rewintering of *Scolopax* in the Transcaspian Region. Psovaya i ruzheinaya okhota [Hunting with Dogs and Guns] 8 (in Russian).
Zhitnikov, M. 1903. The flamingo in the lower Tedzhen River. Psovaya i ruzheinaya okhota [Hunting with Dogs and Guns] 26: 347–348 (in Russian).
Zinchenko, V.K. and O.Ye. Ritskov. 1989. On the ecology of *Eryx elegans* and *Agkistrodon halys caucasicus* in Central Kopetdagh. In: N.N. Shcherbak (ed.), Voprosy gerpetologii [Problems of Herpetology]. Naukova dumka, Kiev: 93–94 (in Russian).
Znamensky, A.I. 1958. Experimentalnye issledovaniya protsessov vetrovoi erozii peskov i voprosy zashchity ot peschanykh zanosov [Experimental Studies of the Processes of Wind Erosion of Sands and Problems of Protection from Sand Smothering]. Izd. AN TSSR, Ashkhabad, 137 pp (in Russian).
Zohary, M. 1963. On the geobotanical structure of Iran. Bull. Res. Council Israel D11(4), Suppl., 113 pp.
Zohary, M. 1966–1986. Flora Palaestina. Vols. 1–3. Israel Acad. Sci. Human, Jerusalem.
Zohary, M. 1973. Geobotanical foundations of the Middle East. Vols. 1–2. G. Fischer, Stuttgart-Amsterdam, 739 pp.
Zonshtein, S.L. 1985. Preliminary data on the fauna of the spider suborder Mygalomorphae from the USSR. Fauna i ekologiya paukov SSSR/Trudy Zool. in-ta AN SSSR [Fauna and Ecology of Spiders of the USSR/Trans. Zool. Inst. Acad. Sci. USSR] 139: 156–161 (in Russian).
Zonshtein, S.L. 1987. A new genus of mygalomorph spiders of the subfamily Nemesiinae (Aranei, Nemesiidae) in the Palaearctic fauna. Zool. zhurn. [Zool. J.] 66(7): 1013–1019 (in Russian).
Zonshtein, S.L. and V.Ya. Fet. 1985. Contributions on the spider fauna of Turkmenistan, II. Family Atypidae. Izv. Akad. Nauk. Turkm. SSR, Ser. Biol. Nauk. [Proc. Acad. Sci. Turkmen. SSR, Ser. Biol. Sci.] 6: 65–67 (in Russian).
Zoogeography of the Paleogene of Asia. 1974. Nauka, Moscow, 301 pp (in Russian).
Zykova, L.Yu. and Ye.N. Panov. 1989. Long-term studies of growth of the Caucasian agama *Stellio caucasius* Eichwald in individually tagged populations. In: N.N. Shcherbak (ed.), Voprosy gerpetologii [Problems of Herpetology]. Naukova dumka, Kiev: 97–98 (in Russian).
Zyuzin, A.A. 1985. Generic and subfamilial criteria in the systematics of the spider family Lycosidae (Aranei) with the description of a new genus and two new subfamilies. Fauna i ekologiya paukov SSSR/Trudy Zool. in-ta AN SSSR [Fauna and Ecology of Spiders of the USSR/Trans. Zool. Inst. Acad. Sci. USSR] 139: 40–51 (in Russian).

Index of Taxa

Abeskunus sphaerion 542
Ablepharus 318, 319
A. bivittatus 311, 349
A. deserti 311, 325
A. pannonicus 192, 193, 311, 334, 337, 340, 341, 343, 345, 349
Abothris 435
Abra segmentum 540, 542
Abramis brama orientalis 369, 371
A. sapa bergi natio *aralensis* 369
Acanthaclisis curvispura 394, 496, 497
A. obscura 496
A. pallida 394, 496, 497
Acanthis cannabina fringillirostris 263
Acantholimon 59, 113, 127, 132, 144, 166, 181
A. balchanicum 122
A. bracteatum 122
A. bromifolium 122
A. erythraceum 181
A. kjurendaghi 122
A. korovinii 122
A. procumbens 136
A. pterostegium 123, 126
A. stocksii 122
A. tragacanthium 122
Acanthophyllum 93, 113, 127, 132, 136, 166, 181, 428, 429
A. adenophorum 180
A. elatus 98, 99
A. microcephalum 161
A. mikeschianum 144
A. multiflora 155

A. stenostegium 180, 182
Acari 191
Acariformes 393
Accipiter badius cenchroides 258
A. nisus nisus 258
Acer 125, 433, 437
A. cinerescens 123
A. fraxinifolia 138
A. ibericum 123
A. laetum 127
A. monspessulanum 126
A. persicum 126
A. pubescens 179
A. turcomanicum 63, 123, 126, 132, 158, 161, 164, 171, 182
Aceraceae 110
Achalcerinys lindus 472
Acheta domestica 455, 465
A. turcomana 452, 465
Achillea biebersteinii 156, 157
Achranoxia 408
A. koenigi 408
A. varentzovi 412
Acinonyx jubatus 206
A. jubatus raddei 206
Acinopus ammophilus 409
A. laevigatus 412
A. striolatus 409
Acipenser nudiventris 368, 374
Acipenseridae 368
Aclerdidae 473, 475
Acmaeodera 424, 440, 449
A. babatauensis 423, 443
A. chalcithorax 422, 423, 444, 445

A. chotanica 424, 442
A. ghilarovi 423, 440, 449
A. instabilis 424, 449
A. pallidepicta 424, 440, 443
A. pilosellae persica 422, 423, 443
Acmaeoderella 419, 421, 424, 425, 445-449
A. adamantina 425
A. alepidota 426, 440, 443, 449
A. badhysica 424, 440, 442
A. ballioni 426, 441
A. candens 426, 447
A. canescens 426, 441
A. caspica 442
A. c. caspica 424, 440, 446, 448
A. c. suturifera 424, 441
A. c. turkestanica 424
A. cinerea 425, 442
A. coelestina 425, 440, 446, 447
A. dubia 426, 441
A. flavofasciata chorasanica 425, 440, 444, 447
A. gibbulosa 422, 426, 446, 447
A. glasunovi 425, 443, 445
A. insueta 426, 441
A. iranica 426, 442
A. mimonti 422, 425, 447
A. nivetecta 425, 441
A. obscura 425, 447
A. oresitropha 424
A. personata 426, 441
A. plavilscikovi 424, 441

A. repetekensis 425, 441
A. semiviolacea 426
A. solskyi 426, 441
A. strandi 426, 440, 443
A. subcyanea 425, 447
A. tragacanthae kopetdaghica 426, 440, 449
A. turanica 424, 442
A. valentinae 426, 448
A. vetusta 425
A. villosula 425, 442
A. zarudniana 425, 440, 443, 446, 449
Acmaeoderinae 423
Acmaeoderini 423
Acmaeotethya 424, 449
Acrida oxycephala 455, 457, 466
Acrididae 333, 334, 452
Acridoidea 451
Acridotheres tristis tristis 256, 257, 261
Acrocephalus agricola brevipennis 262
A. arundinaceus 262
A. dumetorum 262
A. scirpaceus fuscus 262
A. stentorius brunnescens 262
Acrolepiidae 473
Acrotylus insubricus inficitus 455, 466
Actitis hypoleucos 260
Acucularia italica 59
Aculepeira sogdiana 506, 520
Adacna albida 542
A. angusticostata 541
A. caspia 541
A. c. filatovae 542
A. laeviuscula 541, 542
A. minima 541
A. vitrea 541
Adelencyrtus moderatus 471
Adesmia 414
A. karelini 414
A. servillei ssp. *schatzmayri* 409
Adianthaceae 111
Adonis parviflora 156
A. vernalis 154
Adoretus pruinosus 415
Aegialomyia 489
Aegylops 59
A. cylindrica 156, 158
A. tauschii 159

A. triuncialis 156, 158, 163
Aegypius monachus 259
Aellenia subaphylla 163
Aelurillus affinis 516, 521
A. andreevae 516, 519
A. ater 516, 520
A. concolor 516, 520
A. iranus, see *A. concolor*
A. m-nigrum 516, 521
A. v-insignitus 516
A. variegatus, see also *A. andreevae* 516
Aeluropus 438, 442
A. littoralis 133
Aethionema 114
A. kopetdaghi 133, 124, 144
A. transhyrcanum 156
Agalenatea redii 506, 521
Agama sanguinolenta, see *Trapelas sanguinolentus*
Agamidae 318, 331, 342, 344
Agelena labyrinthica 508
Agelenidae 509
Ageniaspis 475
A. fuscicollis 473
Agkistrodon 313, 314, 320
A. halys 225, 317, 326, 338, 341, 343, 345, 348, 349
A. h. caraganus 225, 312
A. h. caucasicus 225, 312, 317, 338
A. intermedius 317
Agonum punctibase 415
A. thoeryi 412
Agrilinae 436
Agrilini 437
Agrilus 419, 421, 440, 446, 447, 449
A. albogularis 437, 442
A. araxenus 437, 440
A. cuprescens 437, 447
A. derasofasciatus 422, 437, 444, 447
A. erojlandusicus 437
A. ganglbaueri 437, 442
A. lineola schamyl 437, 440, 444
A. pistaciophagus 437, 443
A. pseudoalbogularis 437
A. sericans 437, 442, 447
A. vaginalis 437
A. validiusculus 437, 441
A. viridis 422, 437, 444, 447
Agrionemys 347

A. horsfieldi 192, 331, 340–342, 345, 347
A. h. horsfieldi 331
A. h. rustamovi 310, 331
Agriophyllum latifolium 101, 102
A. minus 98–100
Agroeca pullata 510
Agropyron 438
Agrostis stolonifera 137
Agyneta fuscipalpis 505
A. kopetdaghensis 505, 519
A. ressli 505, 521
A. rurestris 505
Ailanthus 125, 126
A. altissima 133, 138
Aiolopus oxianus 453, 457, 466
A. simulatrix simulatrix 455, 466
A. strepens 455, 466
A. thalassinus 455, 457, 466
Aizoaceae 109, 114
Aizoon 114
Alauda arvensis dulcivox 261
Alaudidae 256
Alburnoides bipunctatus eichvaldi 369
A. taeniatus 369
Alcea 113
A. antoninae 145
A. karakalensis 117
A. kopetdaghensis 145
Alchemilla 113
Alcinoeta helopioides 400
Alectoris chukar 251
A. c. dementievi 259
A. c. laptevi 259
Aleochara jacobsoni 394
Alhagi 59, 423, 424, 431, 432, 442, 461
A. persarum 90, 91, 94, 461
Alioranus avanturus 505, 519
Alismatidae 111, 112
Alleculidae 408
Alliaceae 111, 112, 133
Alliaria alliacea 162, 164
Allium 85, 113, 131, 133, 141, 144
A. akaka 123
A. alxeianum 123
A. bodeanum 162
A. brachyodon 135, 144
A. brachyscapum 145

A. christophii 123
A. derderianum 123
A. dictyoscordum 136
A. fibrosum 82
A. helicophyllum 145
A. iliense 136, 137
A. janischianum 123
A. kirindicum 138, 144
A. kopetdaghense 115
A. monophyllum 136, 145
A. paradoxum 63, 127, 165, 167
A. rubellum 153, 158, 160
A. sabulosum 96, 331
A. scabriscapum 159
A. transvestiens 115
A. vavilovii 133, 138, 144
Allocinma caspica 539
Alnus 53
Alphitobius diaperinus 406
Alphitophagus bifasciatus 414
Alsophylax laevis 222, 310, 318, 322, 325, 327
A. loricatus 221, 229
A. l. szczerbaki 221, 310, 322, 325, 327
A. pipiens 221, 310, 318, 322, 324, 326
Alyssum campestre 159
A. desertorum 157
Amara astrabadensis 412
A. fedtschenkoi 407
A. ovata 412
A. tescicola 407
Amaranthaceae 109
Amaryllidaceae 111
Amathitis 407
Amaurobiidae 510
Amberboa turanica 83, 87, 89, 92, 95, 96
Amicroterys asiaticus 471
Ammodendron 53, 59, 125, 425, 426, 435
A. conollyi 19, 41, 460, 461, 487
Ammomanes deserti parvirostris 251, 261
Ammostyphrus 407
Ammothamnus 53, 59, 125
Ammoxenulus 456
A. desertus 452, 453, 457, 464
A. pavlovskii 452, 453, 457, 463, 464
Ammozoum 406, 408

A. bulla 408
A. hauseri 408
Ampedus 415
Amygdalus 59, 124, 126, 180, 423–425, 427, 432, 434, 443
A. bucharica 179
A. communis 159, 164, 165, 176, 177
A. scoparia 126, 156, 165, 169, 179
A. spinosa 180
A. turcomanica 126, 132, 133, 137, 143, 179
A. vavilovi 177
Anabasis 59, 60, 426, 428, 430
A. brachiata 128
A. eriopoda 87, 128
A. ramosissimum 78
A. salsa 18, 78, 79, 84, 86, 87
A. truncata 128
Anacardiaceae 52, 110, 124
Anacridium aegyptium 455, 465
Anagyrus 475
A. ashkhabadensis 468
A. diversicornis 468
A. haloxyli 468, 478
A. hammadae 468
A. orbitalis 468
A. pseudococci 468, 478
A. scapularis 468
A. schoenherri 468
A. tamaricicola 468
A. zaitzevi 468
Anamallota repetekiella 395
Anas acuta acuta 258
A. angustirostris 214, 258
A. clypeata clypeata 258
A. crecca crecca 257, 258
A. penelope penelope 258
A. platyrhynchos platyrhynchos 257, 258
A. querquedula 258
A. rufina 258
A. strepera 258
Ancylocheira salomonis 415
Andrachne stenophylla 126
Androctonus 532
Androlaelaps angsticutis 393
Andromeda 124
Androsace turczaninovii 154, 159
Anemia fausti 409

Anemone petiolulosa 159
Angoleus 409, 410
Anguidae 332, 343, 344
Anisantha sterilis 158, 164, 165, 167
A. tectorum 81, 83, 92, 96, 99, 101, 102, 153, 159, 163, 165, 331
Anisus convexiusculus 539
A. eichwaldi eichwaldi 542
A. euphraticus 538, 539
A. ladacensis 539
Anogramma 114
Anomalobuthus rickmersi 525, 531
Anser anser 258
A. erythropus 258
Anthaxia 419, 421, 432, 440, 446, 447, 449
A. badghyzica 432, 440, 444, 445, 449
A. bicolor 433, 444, 447
A. cichorii 422, 432, 444, 447
A. discicollis 432, 440, 444
A. hemichrysis 432, 440, 441
A. holoptera 433, 440
A. kreuzbergi 433, 440, 443, 449
A. lucidiceps 432, 441
A. muliebris 433, 440, 444, 447
A. nanissima 432, 442
A. passerinii 433, 444
A. spinosa 432, 440, 444, 445
Anthaxiini 432
Anthemis altissima 156, 157, 162
Anthemus aspidioti 473
Anthia mannerheimi 394, 407
Anthicidae 394
Anthochlamis turcomanica 136
Anthocoridae 393, 394
Anthriscus longirostris 159, 162, 167
Anthropoides virgo 215
Anthus campestris boehmii 261
A. campestris griseus 261
A. cervinus rufogularis 261
A. pratensis 261
A. richardi richardi 261
A. spinoletta blakistoni 261
A. trivialis trivialis 261

Anyphaena 522
A. accentuata 510, 520
Anyphaenidae 510
Apera interrupta 133, 156
Aphaleria 408
A. pigmaea 394
Aphaniptera 392, 395, 396
Aphanisticini 439
Aphanisticus 442
A. emarginatus 438
A. pygmaeus 438, 445
Aphantaulax seminigra 511, 521
Aphidencyrtus 475
Aphidoidea 192, 469
Aphilenia 408
Aphodius 408
Aphyculus 474
A. deynauensis 470
A. frontatus 470
A. mesasiaticus 470
A. slavai 470
Apiaceae 108, 110, 112, 114, 138, 144, 442, 443
Apidae 394
Apisthobuthus 532
Apocrinaceae 110
Apristus turcmenicus 412
Apteryskenoma turanica 394
Apus apus pekinensis 260
A. melba petrensis 260
Aquila chrysaëtos 216
A. c. homeyeri 216, 248–252, 259
A. heliaca 219
A. h. heliaca 219, 259
A. rapax 219, 251
A. r. nipalensis 259
A. r. orientalis 259
Arabidopsis wallichi 136
Arabis auriculata 154
Araceae 111
Araliaceae 52
Aranei 191–193, 332, 334, 392, 393, 396, 400, 499–524
Araneidae 499, 501, 506
Araneus angulatus 506, 521
A. armida 507, 521
A. bituberculatus 507
A. cruciferoides, see *Araneus spasskyi*
A. pallasi 507
A. repetecus 507, 520
A. spasskyi 507, 519

A. tartaricus 507, 519
A. tedgenicus 507, 520
Araniella inconspicua 507, 520
Archaeodictyna ammophila 509, 520
A. consecuta 509, 520
Arctiidae 192
"*Arctosa*" *cereipes* 508, 520
Arctosa soror 508, 520
A. variana 508, 521
Ardea cinerea cinerea 258
A. purpurea purpurea 258
Arecidae 111, 112
Arenaria insignis 136
A. serpyllifolia 161
Arenivaga roseni 394, 395, 400
Argasidae 393
Argiope ahngeri 507, 519
A. bruennichi 507
A. lobata 507
Argusia sibirica 128
Argyrophana 408
A. caspia 408
A. deserti 408
Arhrophytum 60
Aristichthys nobilis 369, 371
Aristida 60, 436
Aristida pennata, see *Stipagrostis pennatus*
Armeniaca 427, 432
Arnebia decumbens 83, 85, 87–89, 92, 95, 96, 102, 156
Arrenatherum 114
A. elatius 135
Artema transcaspica 393, 399, 502, 520
Artemisia 18, 45, 59, 78, 112, 113, 131, 152, 153, 425, 430, 436, 437, 461, 476, 478, 528
A. badhysi 79, 82, 94, 101, 137, 146, 152, 461
A. balchanorum 63, 127
A. ciniformis 133, 137, 146, 158, 163
A. deserti 63, 127
A. diffusa 128
A. dimcana 99
A. gypsacea 131, 137, 146
A. halophida 82
A. herbaalbae 152
A. kelleri 93
A. kemrudica 79, 82–88, 96

A. kopetdaghensis 127, 131, 132
A. kulbadica 152
A. lobulifolia 128
A. olivieriana 132, 163
A. santolina 82, 97, 99–101
A. serotina 156
A. sieberi 128
A. turanica 461
A. turcomanica 63, 127, 137, 146, 152
Arthrodosis 409
A. castaneus 400
A. schusteri 414
Arundo donax 127, 168
Asaphidion transcaspicum 412
Ascelenia decolorella 395
Aschitus margaritae 469
A. naiacocci 469
A. neoacanthococci 469
A. populi 469
Asclepidaceae 110
Asencyrtus deserticola 469
Asiabadus asiaticus 511, 520
Asilidae 191
Asio flammeus flammeus 260
A. otus otus 260
Asiocaedius 406
Asioplatysma 410
Asiotmethis muricatus australis 453, 457, 465
Asparagaceae 53, 111
Asparagus 463
A. brachyphyllus 137
A. verticillatus 154
Asperula 461
A. balchanorum 181
Aspidiaceae 111
Aspiolucius 374
A. esocinus 226, 368
Aspius aspius iblioides 368
Aspleniaceae 111
Aspoeckiana carlic 496
A. caudata 496, 497
A. longiventris 496, 497
Asteracae 53, 105, 110, 112, 114, 133, 135, 138, 139, 144, 442, 478
Asteridae 110, 112
Asterolecaniidae 473
Astragalus 41, 63, 93, 101, 112, 131, 133, 135, 141, 144, 423, 425, 426, 428, 429, 432, 437, 443, 461

A. ackerbergensis 138, 145
A. ammodendron 101, 128, 155
A. ammophilus 128
A. arpilobus 85, 92, 95
A. bachardeni 144
A. brachypetalus 159, 160
A. cerasocrenus 132
A. chivensis 93
A. chrysostachys 134, 144
A. confiniorum 145
A. curvipes 145
A. excedens 94
A. filicaulis 156
A. flexus 92
A. uhsii 138, 145
A. gaudanensis 134, 144
A. jarmolenkoi 144
A. jolderensis 138
A. kucanensis 134, 144
A. kuschkensis 145
A. meschedensis 127
A. nephthonensis 157
A. nigricans 128
A. nigriceps 145
A. oxyglottis 83, 85, 88, 89, 102
A. paucijudus 331
A. piletocladus 127, 132
A. pulvinatus 136
A. raddei 145
A. rawlinsianus 144
A. retamocarpus 132, 159, 162
A. squarrosus 128
A. subdjenarensis 144
A. transcaspicus 101
A. tribuloides 156
A. turcomanicus 83, 128
A. vicarius 159
A. xanthoxiphidium 153, 159
A. xiphidioides 85, 144
Astymachus 474
A. phragmitis 473
Athene noctua 390
A. n. bactriana 248-250, 252, 253, 260
Atholus rudesculptus 413
Athyriaceae 111
Atranus collaris 411, 412
Atraphaxis 180
A. kopetdaghensis 138, 145
A. spinosa 87, 96, 143, 155, 159, 171, 179

Atrichotmethis semenovi 453, 465
Atriplex 424, 426
A. dimorphostegia 83, 95
Atropa 114
A. komarovii 125
Attacidae 469
Attagenus fasciolatus 394-396, 400
Atylotus 485, 490
A. pulchellus 485
A. p. arybenthinus 485
A. p. pulchellus 485
A. quadrifarius 485
Atypidae 501
Atypus muralis 501, 521
Auchenorrhyncha 469
Auchonium 114
Aulonia 522
A. kratochvili 508, 519
Aurigena 428, 440
A. lugubris longicollis 428, 440, 444
Australobuthus 533
Avena barbata 133, 154, 157, 163
A. eriantha 159
Axelinus 409
Axinocarabus 416
Aythya ferina 258
A. fuligula 258
A. marila 258
A. nyroca 258

Ballus chalybeius 516, 521
Barbus brachycephalus 368, 374
B. b. brachycephalus 374
B. capito conocephalus 368, 374
Bedeliolus 409
Belanthera 120
Belgrandiellidae 538
Belopus 415
Bembidion 415
B. amnicola 415
B. atlanticum megaspilum 411
B. gassneri 415
B. hiekei 412
B. moschatum 412
B. niloticum 412, 415
B. tetrasemum 412
B. tichomirovae 412
Berberidaceae 109

Berberis 179, 184
B. densiflora 132
B. interregima 159, 164, 165, 183
B. turcomanica 132, 143, 179
Bergiola balchashica 453, 464
B. montana 452, 458, 464
B. popovi 452, 453, 458, 464
Berlandina 519
B. afghana 393, 511, 520
B. caspica 511, 520
B. plumalis 511, 512
Beta 114
Betula 53
Bianor albomaculatus 516
Bibionidae 193
Bidens tripartita 138
Bieberssteiniaceae 110
Bienertia cycloptera 128
Bilegnum bungei 135, 144
Birulatus 532
Bivalvia 539
Blaps 408, 414
B. balashovi 413
B. dehaani 414
B. fausti 394, 400
B. medvedevi 410
B. mortisaga 414
B. scutellata 394
B. tutanus 414
Blattodea 392, 394, 400
Bleusea ammophila 407
Blysmus compressus 135
Boidae 335
Boiga 320
B. trigonatum 226, 326, 335, 340, 348
B. t. melanocephala 226, 312, 335
Bolboschoenus maritimus 137
Bombycilla garrulus garrulus 262
Bongardia chrysogonum 153, 159
Boraginaceae 105, 108, 110, 112, 131, 134, 139
Boreidae 191
Boreoelona moltschanovi 539
Boromorphus opaculus 414
Botaurus stellaris stellaris 258
Bothriochloa ischaemum 156
Bothriothorax aralius 472
Brachinus bayardi 412
B. brevicollis 412

Brachycleonus 408
Brachypangus 412
Braconidae 395, 469
Brassica tournefortii 128
Brassicaceae 105, 109, 112, 114, 124, 134, 138, 139
Brentidae 415
Brigittea latens 509, 520
Brinislavia 413
Bromopsis inermis 135, 137
Bromus 113, 135, 459
B. danthoniae 156, 158
B. japonicus 135, 153, 163
B. oxyodon 156
B. popovii 135
B. sewerzowii 136
Bromus tectorum, see *Anisantha tectorum*
Broscus asiaticus 412
B. karelini 203, 412
Brunnerella mirabilis mirabilis 454, 466
Bryonia dioica 137
Bubo bubo 253
B. b. omissus 260
B. b. turcomanus 263
Bucanetes githagineus crassirostris 263
Bucephala clangula clangula 258
Bufo viridis 337
Bufonacridella 456
Bufonacridella sumakovi 453, 464, 465
Buhsea coluteoides 137
Bunium korovinii 145
B. longipes 153, 157, 164
Bunopus abudhabi 314
B. blanfordi 314
B. tuberculatus 223, 227, 310, 314, 348
B. t. tuberculatus 223
Bupleurum gerardii 156, 162
Buprestidae 408, 410, 415, 419–449
Buprestini 434
Buprestis salomonii 434, 442
Burhinus oedicnemus harterti 248, 250, 252, 254, 259
Buteo buteo japonicus 259
B. b. vulpinus 259
B. rufinus rufinus 248–250, 252, 253, 259

Buthacus 532
Butheolus 532
Buthidae 531
Buthiscus 532
Buthothus, see *Hottentotta*

Cabirutus 414
C. kuhitangi 410
Caccinia crassiflora 154
Caenophanomyia insignitis 395, 396
Caesalpinaceae 110
Calamagrostis epigeios 135
C. persica 135
Calandrella cinerea longipennis 248–250, 252, 261
C. rufescens heinei 248, 250–252, 261
Calathus ambiguus 412
C. melanocephalus 412
Calendula 114
C. persica 159
Calidris alpina alpina 260
C. ferruginea 260
C. minuta 260
Callanthaxia 433
Callicephalus 114
Calligonum 19, 41, 59, 63, 80, 83, 93, 94, 96–99, 101, 113, 125, 400, 423, 427, 430, 432, 460, 463, 474, 475, 476
C. alatum 82
C. aralense 128
C. arborescens 98
C. caput-medusae 92, 98, 100, 128
C. eriopodum 92, 98
C. leucocladum 98, 128
C. macrocarpum 128
C. microcarpum 98
C. rubens 98
C. setosum 85, 86, 92, 94, 97, 100
C. triste 116, 126
Callipeltis cucullaris 153
Calliptamus barbarus cephalotes 454, 465
C. b. nanus 453, 466
C. coelesyriensis carbonarius 453, 457, 466
C. c. intricatus 454, 466
C. italicus italicus 454, 457, 466, 461, 466

C. turanicus 453, 460, 461, 466
Callipteroma quinqueguttata 468
Callisthenes 413, 416
Callistini 412
Calomyscus mystax 209
Calopterusa werneri 454, 464
Calosoma 416
C. algiricum 416
C. auropunctatum 416
C. denticolle 416
C. deserticola 416
C. imbricatum 412, 416
C. inquisitor 412, 416
C. maderae 411, 412
C. olivieri 416
C. reitteri 412, 416
C. sycophanta 416
Calycobathra calligoni 395
Calyptopsis 409, 414
Camelina rumelica 159
Camelostrongylus mentulatus 245
Caminara 416
Campalita 416
Campanula incanescens 133
C. khorassanica 135, 144
Campanulaceae 110
Camponotus turkestanicus 415
Campsocleis shelkovnikovae 454, 464
Candaharia aethiops 537
C. levanderi 537
Canis aureus 321
C. lupus 321
Capnodis 419, 427, 440, 441, 447–449
C. anthracina 427
C. excisa 427, 441
C. jacobsoni 427, 440, 447, 449
C. miliaris 415, 427, 442
C. parumstriata 427, 443
C. sexmaculata 427, 443
C. tenebricosa 427, 441
C. tenebrionis 427, 444
Capoetabrama 374
C. kuschakewitschi 379
Capparidaceae 53, 109
Capparis spinosa 156
Capra aegagrus 210, 231–246
C. ae. turkmenicus 210, 233

C. falconeri 209
C. f. heptneri 209, 231
C. sibirica 231
Caprifoliaceae 110, 114
Caprimulgus aegyptius arenicolor 248, 249, 252, 254, 255, 260
C. europaeus zarudnyi 260
Carabidae 191, 394, 403–405, 407, 409–412, 414, 415
Carabus 413, 416
C. clathratus 416
C. kuznetzovi 416, 417
C. miles 410, 416
C. roseni 412, 416
Caragana 425
C. grandiflora 127, 179
Carassius auratus gibelio 369
Cardaria draba 159, 162
Cardioderus chloroticus 410
Cardium lamarcki 540
Carduus arabicus 133
Carenochyrus 412
Carex 59, 113, 135, 167, 438, 459
C. diluta 135, 168
C. divisa 168
C. divulsa 135
C. orbicularis 135
C. pachystylis 82, 85, 86, 88, 89, 96, 128, 131, 146, 152, 154, 159, 163, 175, 189, 270, 461–463
C. physodes 19, 43, 45, 78, 82, 92–94, 96, 97, 99, 100, 128
C. polyphylla 162, 168
C. stenophylloides 136
Carininota 425
Carlina 114
Carpoceras 114
Carpodacus erythrinus erythrinus 263
Carum carvi 135
Carya typica 53, 124
Caryophyllaceae 108, 109, 112, 114, 139
Caryophyllidae 108, 109
Caspia baerii 543
C. derzhavini 543
C. gmelini 541
C. g. gmelini 542
C. grimmi 542
C. isseli 542

C. knipowitschi 541
Caspiella 541–543
Caspiohoratia marina 542
Caspiohydrobia 539, 541, 542
Caspiopyrgula 543
Castanea 125, 126
Catabrosa aquatica 135
Catabrosella parviflora 135
Catantopinae 452
Catomus karakalensis 413
Catopidae 394
Ceballosia dusmeti 473
Cebrennus 514, 523
Cedicus ephthalitus 509, 519, 522, 523
C. gennadii 509, 519, 523
C. maerens 509, 519
C. parthus 505
Celasteraceae 110
Celia 407
Celtidaceae 109
Celtis 125, 126, 179, 433
C. caucasica 53, 124, 132, 137, 143, 159, 164, 169, 171, 180
Centaurea 59, 113, 424
C. androssowii 115, 131, 144
C. kopetdaghensis 161
C. sintenisii 159
C. squarrosa 156
C. vulgare 159
Cephalorrhynchus brassicifolius 123
C. polycladus 135
Ceraeocercus fuscipennis 453, 463, 464
Cerambycidae 415
Ceraphronidae 395
Cerapterocerus 475
C. mirabilis 471
C. pilicornis 471
C. planus 471
Cerasina 539
Cerastium glomeratum 156
C. inflatum 159
C. perfoliatum 157
Cerastoderma isthmicum 540
C. rhomboides 540
Cerasus 12, 116, 179, 427, 428, 432
C. blinovskii 125, 126, 165
C. erythrocarpa 165
C. mahaleb 179
C. microcarpa 126, 156, 158,

164, 165, 170, 171, 179
C. pseudoprostrata 126, 132
Ceratinella brevis 505
Ceratinopsis romana, see *Sphecozone romana*
Ceratocarpus utriculosus 81, 82, 87, 88, 89, 101, 102
Ceratocephala falcata 83, 85–88, 92, 95, 96, 102, 155, 159
Ceratopholcus maculipes 502, 519
Ceratophyllus turcmenicus 395
Cerchysiella planiscutellum 472
Cerchysius subplanus 469
Cercis 59, 125, 126, 435
C. griffithi 124, 131, 133, 137, 143
C. siliquastrum 125
Cercotrichas galactotes 193
C. g. familiaris 262
Cervus elaphus 231
C. e. bactrianus 208
Chaerophyllum 114
C. khorassanica 135
Chalcalburnus chalcoides aralensis 368, 371, 374
Chalcides ocellatus 202, 225, 311, 334, 340, 343, 348
C. o. ocellatus 334
Chalcidoidea 467, 474
Chalcionellus blanchei 413
C. decemstriatus 413
C. turcicus 413
C. tyrius 409, 413
Chalcophorinae 427
Chalcoscirtus infimus 516, 521
C. martensi parvus 516, 519
Chamaemyiidae 469
Chara meriane 59
Charadrius alexandrinus alexandrinus 259
C. asiaticus 248, 250–252, 259
C. dubius curonicus 259
C. leschenaultii crassirostris 248, 250–252, 259
Chardinia orientalis 155, 159
Charitopus desertus 468
C. fulviventris 468
C. trjapitzini 468

612 Index of Taxa

Chartoloma platycarpum 98
Cheilanthes persica 133
Cheiloneurus 475
C. claviger 471, 476
C. elegans 471
C. flavoscuttatus 472
C. fulvescens 472
C. paralia 472
C. yasumatsui 472
Cheiracanthium erraticum 511
C. mildei 511, 521
C. seidlitzi 511, 521
Chenopodiaceae 12, 53, 60, 88, 91, 105, 108, 109, 112, 114, 122, 131, 133, 139, 430, 478
Chesneya astragalina 143, 144
C. botschantzevii 131, 144
C. kopetdaghensis 156
Chettusia gregaria 218, 259
Chilopoda 393
Chilostetha 430, 442
Chilotomus 417
C. chalybaeus 412, 417
C. kuhitangi 410, 417
C. margianus 411, 417
Chioneosoma porosum 415
Chivaenius 407
Chlaeniomimus gracilicollis 409
Chlaenius dimidiatus 415
C. lederi 412
C. tenuilimbatus 412
C. vestitus 412
Chlamydotis undulata 217
C. u. macqueeni 248–252, 259
Chlidonias hybrida hybrida 260
Chlorion regale 395
Chloris chloris turkestanicus 263
Cholevinus fuscipes 394
Chondrilla 426
Chondrina granum 536, 537
Choreia maculata 471
Chortippus albomarginatus karelini 455, 466
C. apricarius asiaticus 453, 466
C. biguttulus meridionalis 453, 466
C. b. pravdini 453, 458, 466

C. macrocerus assimilis 454, 466
Chrotogonus turanicus 455, 457, 465
Chrozophora 426
C. gracilis 98
Chrysoblemma 430
Chrysobothrinae 435
Chrysobothrini 435
Chrysobothris affinis nevskyi 435, 444
C. deserticola 435, 441
C. globicollis 435, 441
C. jakovlevi 435, 441
C. nana 435, 442
Chrysomelidae 408, 410, 415
Chrysopidae 473
Chrysops 483, 489, 491
C. flavipes punctifer 483
C. hyalipennis 483, 486
C. mlokosiewiczi 483
C. oxianus 483
C. stackelbergiellus 483, 484
Cicadoidea 192
Cicer kopetdaghensis 145
Cicileus 532
Cicindela decempustulata 412
C. deserticola 412
C. germanica 412
C. illicebrosa 415
C. jakowlewi 409
C. melanoholica 415
C. nox 415
Ciconia nigra 218
Cinnamommum polymorpha 59
Circaetus ferox, see *C. gallicus*
C. gallicus 219, 321
C. g. heptneri 219, 259
Circus aeruginosus aeruginosus 258
C. cyaneus cyaneus 258
C. macrourus 258
C. pygargus 258
Cirsium 436
C. rhizocephalum 135, 136
C. vulgare 156
Cissus 59
Cistaceae 109, 114
Cistanche flava 99
Cithareloma lehmanni 98
C. vernum 98
Cladium martii 137
Clangula hyemalis 258

Clathrocaspia 542
Clausenia mariae 468
Clema 408
C. deserti 436, 441
C. freudei volkovitschi 436, 440, 443, 449
Clematis orientalis 137
Cleome 60
C. gordjaginii 120
C. turkmena 115, 120, 126, 127
Cleomopsis 120
Clessiniola 541
Climacoptera 113, 128, 430, 431
C. brachiata 137
C. czelekenica 123, 128
C. lanata 83, 86–91, 95
C. sukaczevii 123
Clubiona 522
C. alpicola 511, 520
C. genevensis 511
Clubionidae 501
Clypeola jonthlaspi 156, 159
Cnicus benedictus 156
Cobitidae 369
Cobitis aurata aralensis 369, 374
Cobosiella 424, 442
Coccidae 473, 475
Coccidencyrtus artemisiae 471
C. duplachionaspidis 471
C. schizotargioniae 471
Coccinellidae 473
Coccoidea 471, 472, 474–478
Coccopilatus babaevi 473
C. zygophylli 473
Coccothraustes coccothraustes nigricans 263
Codonocephalum 114
C. peacockianum 132
Coelotes charitonovi 510, 520
Colchicum bifolium 137
Coleophora 476
Coleophoridae 472
Coleoptera 59, 191, 192, 199, 320, 332–334, 344, 392, 394, 396, 400, 403–418, 472, 475
Collembola 191, 193, 392, 393
Colleptopterum bactrianum 539
C. cyreum sogdianum 539

Index of Taxa 613

Colposcelis lopatini 410
Colposoma 413
Colposphaena 408
Coluber 320, 325, 326
C. ataevi 224, 329, 336, 339–341, 347–349
C. caspius 316, 335, 340, 345, 348, 349
C. jugularis 224, 316
C. karelini 320, 336, 340, 347
C. k. karelini 311, 336
C. ladacensis 335, 340, 344, 347
C. najadum 224, 311, 336
C. nummifer 311
C. ravergieri 192, 316, 320, 326, 336, 340, 343, 345, 347, 349
C. r. nummifer 316
C. r. ravergieri 311
C. rhodorhachis 192, 316, 320, 326, 336, 340, 343, 345, 347
C. r. ladacensis 316
C. r. rhodorhachis 311
C. schmidti 224, 311, 313, 316, 322, 325, 326
C. spinalis 316
Colubridae 318, 335
Columba eversmanni 260
C. livia neglecta 260
C. oenas oenas 260
Columbidae 256
Colutea 179, 182, 423–425, 437
C. atabajevii 119, 31, 143, 144
C. buhsei 132
C. gracilis 158, 163–165, 169–171
Colydiidae 410
Compsobuthus 532
Conocephalus discolor 464
Conophyma 454
C. bactrianum 453, 458, 465
C. turcomanum 453, 458, 465
C. uvarovi 458, 465
C. u. uvarovi 453, 465
C. u. vicinum 453, 465
C. zimini 453, 458, 465
Conosoma lineata 394
Consolida rugulosa 92, 95
Convolvulaceae 110
Convolvulus 84, 113, 426,

459–461, 464
C. arvensis 162
C. divaricatus 93, 98, 99
C. erinaceus 100
C. fruticosus 137, 153
C. korolkovii 93, 97
C. persicus 128
C. subhirsutus 133, 155, 158, 163
Cophaphonus riparius 452, 465
Copidosoma 474, 475
C. agrotis 472
C. arvense 472
C. augasmatis 472
C. autumnale 472
C. balchanense 472
C. bucharicum 472
C. caspicum 472
C. clavatum 472
C. dendrophyllum 472
C. dushakense 472
C. farabense 472
C. jamansaiense 472
C. juliae 472
C. kisilkumense 472, 476
C. kuhitangense 472
C. kushkense 472
C. longiventre 472
C. nijasovi 472
C. obscurum 472
C. saxaulicum 472
C. slavai 472
C. tugaicum 472
C. turanicum 472
Coptopsylla olgae 395
Coracias garrulus semenovi 260
Corbicula cor 538
C. fluminalis 538
C. purpurea 539
Corbiculina ferghanensis 539
C. tibetensis 539
Coreidae 394
Corispermum lehmannianum 98
C. papillosum 98
Coroebini 436
Coroebus 440, 447
C. rubi 436, 422, 447
Coronilla 114
Corvus cornix sharpii 262
C. corone orientalis 262
C. frugilegus frugilegus 261

C. monedula monedula 261
C. ruficollis ruficollis 248, 250, 252, 253, 262
Corydalis chionophilla 135, 144
C. macrocalyx 144
Cossyphus tauricus 409
Costatella acuta 538, 539
C. integra 539
Cotoneaster 137, 179, 180, 182, 184
C. nummularia 160, 164, 171
C. nummularioides 159
C. ovatus 132
C. turcomanicus 132
Coturnix coturnix coturnix 259
Cousinia 59, 112, 113, 135, 141, 144, 198, 424, 426, 436, 461
C. albiflora 144, 154, 163, 165, 169
C. bipinnata 162
C. chaetocephala 115
C. congesta 132
C. cryptadena 138
C. glochidiata 145
C. hypopodia 163, 145, 154
C. mucida 138, 144
C. oreoxerophila 115, 134, 144
C. oxiana 93
C. schistoptera 101
C. tenella 162
C. umbrosa 159, 160, 167
Crambe 426
C. kotchyana 157
Crassulaceae 110
Crataegus 112, 126, 137, 143, 159, 176, 427
C. androssowi 117
C. melanocarpa 165
C. nikitinii 125, 126
C. pontica 165, 176
C. pseudoambigua 161, 165
C. turkestanica 165
Cratomerus 419, 421, 446
C. elaeagni 434
C. fariniger 433, 442
C. fedtschenkoi 433, 443
C. hungaricus sitta 433, 440, 444
C. intermedius 433, 444
C. judinae 434, 443

C. medvedevorum 433, 440, 442, 446
C. sponsa 433, 447
C. turanus 434, 440
Crematogaster 192
Creoleon 497, 498
C. aegyptiacus 497
C. elegans 497
C. griseus 497
C. plumbeus 497, 498
Crepis 59
C. sancta 159
Crex crex 259
Crocus michelsonii 144
Crossobamon 322, 324
C. eversmanni 310, 317
Crotalidae 338
Crucianella gilanica 156
C. sintenisii 156, 159, 160, 164
Crupina vulgare 155, 157, 162
Cryptanthaxia 432
Cryptocephalus 408, 415
Cryptocratomerus 433
Cryptophagidae 394
Cryptospora falcata 154
C. omissa 159
Ctenolepisma mauritanica 394
Ctenopharyngodon idella 368, 371
Ctenopus 415
Cucujidae 394
Cuculus canorus subtelephonus 260
Cucurbitaceae 109
Cueta kasyi 496, 498
C. lineosa 496, 498
Cuminum setifolum 136
Cupressaceae 54, 111, 124
Cupressus 433
Curculionidae 191, 192, 394, 408, 410
Cursorius cursor 220
C. c. bogolubovi 220, 248–252, 254, 255, 260
Cutandia memphitica 92, 94, 99
Cyclanthaxia 433
Cyclocarabus 417
Cyclosa conica 507
Cydonia 114, 433
C. oblonga 127, 165, 179
Cygnus bewickii 218
Cylindromorphi 438

Cylindromorphinae 438
Cylindromorphus 438
Cylindromorphus kopetdagicus 438, 440, 443
C. pubescens 438, 441
Cymbionotum pictulum 412
C. semelederi 409
C. transcaspicum 409
Cymindis 417
C. accentifera 417
C. andreae 417
C. antonovi 417
C. axillaris 411, 417
C. capito 407
C. lineata 417
C. pallidula 417
C. picta 417
C. quadrisignata 417
C. rufipes 417
C. walteri 412
Cymindoidea famini 415
Cynodon dactylon 156, 461, 462
Cynosurus 114
Cyperaceae 111, 133
Cyperus fuscus 135
C. glaber 135
Cyphogenia 408
C. gibba 394
Cyphosoma tataricum 427
C. turcomanicum 427
Cyprinidae 368
Cyprinus carpio 369
Cyrba algerina 516
C. ocellata 516
Cyrtodactylus 314
Cyrtopodion 314, 328
C. caspius, see *Tenuidactylus caspius*
C. longipes, see *Tenuidactylus longipes*
C. spinicauda, see *Mediodactylus spinicaudus*
C. turcmenicus, see *Tenuidactylus turcmenicus*

Daboia 320
D. lebetina, see *Vipera lebetina*
D. l. turanica 312
Dactylis glomerata 156, 158, 160, 164
Dasyrhamphis 484, 487, 491
D. umbrinus 484, 487, 488, 491
Dasytrogus 408

Daucus carota 159
Decticus albifrons 454, 464
D. verrucivoris gracilis 453, 465
D. v. annaelisae 454, 465
Delfimeus intricatus 498
Delphinium camptocarpum 159
D. semibarbatus 161
D. turcomanicum 156
Dendarus 414
D. armeniacus 413
D. ranscaspicus 413
Dendrocopos leucopterus albipennis 248–250, 252, 261
Dendrophilus sulcatus 413
Dendrostellera turkmenorum 128
Dengitha 408
D. crystallina 408
D. lutea 408
Dericorys albidula 454, 459, 460, 465
D. annulata roseipennis 454, 465
D. tibialis 454, 465
Dermacentor daghestanicus 244
Dermestidae 394, 396
Desidae 509
Deudora 429
Deutoleon lineatus 497, 498
Devade tenella 509, 520
Diaea dorsata 514
Dianthus turcomanica 133
Diaphanidus 408
Diaspididae 473, 475, 476
Dicarnosis sugonjaevi 468
D. vicina 468
Dicerca 435, 446
D. aenea validiuscula 434, 444
D. fritillum 435, 440, 447
Dicercini 434
Dichillus 414
D. dentipes 410
D. ocellaris 411
D. tenebrosus 414
Dictyna cronebergi 509, 520
D. pusilla 510
D. uncinata 510
Dictynidae 509
Dictynomorpha strandi 510,

520
Didacna baeri 541
D. pyramidata 542
Dielsiocharis 114
D. kotschyi 133
Diesia 408
Diexis 454, 456
D. chivensis 453
D. varentzovi afffinis 453, 465
D. v. probus 453, 465
D. v. salsolae 453, 465
D. v. varentzovi 453, 465
Dilamus fausti 409, 414
Dilleniidae 109
Dinocarsiella alpina 468
Dinocarsis hemiptera 468
D. hofferi 468
Dionysia kossinskyi 118, 134, 181
D. tapetoides 133, 181
Diplocephalus bifurcatus 505, 519
Diplomesodon pulchellum 321
Dipluridae 501
Dipsaceae 110
Diptera 59, 191, 244, 392, 395, 396, 469, 475, 481–494
Diptychocarpus strictus 136
Discodes 474, 475
D. acanthopulvinariae 471
D. atraphaxidis 471
D. coccophagus 471
D. desertus 471
D. indefinitus 471
D. ipaikalensis 471
D. kopetdagicus 471
D. kryzhanovskii 471
D. obscuriclavus 471
D. rhizopulvinariae 471
D. terebratus 471
D. tugaiensis 471
Discognathichtys rossicus 227
Discoptera 407
D. komarovi 409
Dissonomus 414
D. badhysi 411
Distonemurus desertus 498
Dociostaurus kraussi nigrogeniculatus 454, 462, 463, 466
D. maroccanus 454, 462, 466
D. plotnikovi 453, 466
D. tartarus 454, 466
Dodartia orientalis 331

Dolichoceras amudaryensis 468
D. galinae 468
Dolichopodidae 59, 395
Donacochara speciosa 505, 520
Dorema 173, 425, 426, 432, 443
D. balchanorum 127
D. hyrcanum 162
D. kopetdaghense 132
Drassodes 512, 519, 523
"*D.* " *flavomaculatus* 512, 520
D. jakkabagensis 512, 520
D. lapidosus 512
D. proximus 512, 519
"*D.* " *thimei* 512, 520
Dreissena caspia 541
D. elata 542
D. polymorpha 535, 541
D. p. aralensis 539
D. rostriformis distincta 543
D. r. grimmi 543
Dreissenidae 540
Dromius hiemalis 411
Dryandra 124
Dryinidae 395
Dugesia 198
Duroniella gracilis 453, 466
D. kalmyka 453, 457, 466
D. turcomana 453, 466
Dyschirius 415
Dysdera aculeata 399, 503, 520
D. brignolii 524
D. concinna 203
D. kugitangica 503, 519
D. kusnetsovi 524
D. limitanea 503, 520
D. nenilini 524
D. pococki 203, 503, 520
D. tartarica 524
D. transcaspica 503, 519
D. tystshenkoi 524
Dysderidae 503
Dysderina loricata 503, 521

Earophanta 408
E. pilosissima 411
Echemus 522
E. angustifrons 522, 520
Echinaria capitata 156
Echinops 436
E. ritro 159

E. transcaspicus 138
Echis 316
E. carinatus 321, 326, 339, 341
E. c. multisquamatus 312, 316, 339, 341, 343, 347
E. c. sochureki 316
E. multisquamatus , see *E. carinatus multisquamatus*
Echthrodryinus 475
E. krasilnikovae 469
E. mesasiaticus 469
Echthroplexiella 476
E. aeneiventris 471
E. artemisiae 471
E. popovi 471
E. tobiasi 471
Egadroma 412
Egnatioides desertus 453, 466
Egnatius apicalis 453, 457, 466
Egretta alba 258
Eirenis collarius 349
E. meda 312, 320, 322, 324, 329, 336, 340, 344, 347, 348
E. modestus 349
Elaeagnaceae 110
Elaeagnus 427, 434, 442, 461
E. orientalis 137, 168, 183
Elaphe 320
E. dione 312, 313, 322, 325, 326
E. quatuorlineata 225, 349
E. q. sauromates 225, 312, 322, 324
Elaphrini 413
Elaphropus diabrachys 412
E. haemorrhoidalis 411
Elapidae 338
Elateridae 408, 415
Eledona agaricola 406, 414
Eleocharis meridionalis 135
E. uniglumis 169
Ellobius fuscocapillus 193
Elymus longearistatus 136
Elytrigia 20, 438
E. caespitosa 138
E. repens 159, 161, 164, 167
E. trichophora 133, 145, 146, 157, 160, 163–165, 461
Emberiza bruniceps 193, 263
E. calandra buturlini 263
E. cia 251
E. citrinella erythrogeni 263

E. hortulana 263
E. leucocephala
    leucocephala 263
E. rustica 263
E. schoeniclus harmsi 263
E. s. pallidior 263
E. s. passerina 263
Emberizidae 193
Emydidae 330
Emys orbicularis 313, 330,
    340–342, 345, 348, 349
E. o. orbicularis 310, 325, 326
Enchytraeidae 191, 193
Encyrtidae 395, 467–480
Encyrtus infidus 473
E. lecaniorum 473, 477
E. trjapitzini 473
Enoplognatha testacea 504,
    521
E. thoracica 504, 520
Eolis pallidus 540
Ephedra 52, 59, 113, 125,
    179, 180, 426, 435
E. distachya 84
E. equisetina 132, 157, 163,
    164
E. intermedia 93, 132, 159,
    163, 164, 171, 179
E. lomatolepis 132
E. strobilacea 19, 80.92, 94,
    97, 98, 100, 101
Ephedraceae 111, 133
Epidinocarsis
    archangelskayae 468
Epilasia hemilasia 85, 89, 92,
    95, 102
Epilobium parviflora 137
E. tetragonum 133
E. velutinum 137
Epipactis veratrifolia 144, 167
Equisetaceae 111
Equisetophyta 111, 140, 142
Equisetum ramosissimum 167
Equus caballus fossilis 60
E. hemionus 213, 231
E. h. onager 3, 193, 213
E. onager, see E. h. onager
Eremadoretus 408
Eremasus cribratus 394
Eremencyrtus
    neoptolemus 470
E. unifasciatus 470
Eremias 308–310, 314, 318,
    320–322, 324

E. acutirostris 309, 310
E. afghanistanica 309
E. aporosceles 309, 310
E. argus 309, 310
E. arguta 223, 309, 310, 313,
    324, 326, 347
E. a. uzbekistanica 223, 311
E. arida 309, 310
E. brenchleyi 309, 310
E. buechneri 309, 310
E. fasciata 309, 310
E. grammica 309–311, 313,
    324, 326
E. intermedia 309–311, 318,
    347
E. lineolata 309–311, 318,
    325, 347
E. multiocellata 309, 310
E. nigrocellata 224, 309–311,
    318, 326, 347
E. nikolskii 309, 310
E. persica 311, 322, 324, 345,
    348
E. pleskei 309, 310
E. przewalskii 309, 310
E. quadrifrons 309, 310
E. regeli 223, 311, 322, 326
E. s. kopetdaghica 311, 313,
    316, 322, 329, 333, 348
E. s. scripta 311
E. scripta 309, 310, 313, 318,
    347
E. strauchi 309, 310, 313,
    316, 323, 333, 340, 341, 343,
    344, 347–349
E. v. velox 311
E. velox 309, 310, 313, 325,
    326, 333, 340, 341, 343, 345,
    347
E. vermiculata 309, 310
Eremippus carinatus 453, 466
E. foveolatus 453, 466
E. miramae 466
E. onerosus 453, 458, 466
E. persicus 454, 466
E. simplex simplex 455, 466
Eremocymindis 417
Eremogryllodes semenovi 453,
    465
E. vlasovi 394, 395, 453, 465
Eremophasma peliococci 469
Eremophila alpestris
    albigula 261
Eremopoa oxyglumis 135

Eremopyrum 461
E. orientale 81, 83, 85–87, 89,
    92, 94, 96, 100–102, 153,
    169, 163
Eremosaprinus vlasovi 394,
    408, 413
Eremosparton 53, 59, 125,
    429
E. flaccidum 98–99
Eremosphodrus 407
Eremostachys
    boissieriana 153, 158, 160
E. labiosiformis 156, 159
Eremoxenus chan 415
Eremurus 112, 113, 143
E. anisopterus 92–94
E. kopetdaghensis 115, 120,
    122, 137, 138, 144
E. luteus 136, 137
E. olgae 157
E. persicus 122
E. stenophyllus 136
Eresidae 503
Eresus cinnaberinus 503
E. niger rotundiceps, see E.
    cinnaberinus
E. niger, see E. cinnaberinus
Erianthus 59
E. ravennae 127, 133, 168,
    173
Ericydnus danatensis 468
E. karakalensis 468
E. niger 468
E. robustior 468
E. sipylus 468
E. tamaricicola 468
E. turkmenicus 468
Erigeron dolichostylus 135
Erigone dentipalpis 505
Erigonoplus ninae 505, 519,
    523
Eriococcidae 473, 475
Eris nidicolens 516, 521
Erithacus rubecula
    rubecula 263
Ero furcata 503
Erodium cicutarium 153, 157
E. oxyrrhynchum 83, 92, 95,
    102
Eruca sativa 156, 164
Eryngium biebersteinii 157,
    161
E. bungei 156
E. badchysi 123

Index of Taxa 617

Erysimum ischnostylum 154, 158, 162
E. kerbabaevii 115, 123, 145
Erythropygia galactotes 255
Eryx 314, 320, 321
E. elegans 224, 311, 318, 324, 329, 335, 340, 341, 345, 347, 348
E. miliaris 192, 224, 311, 313, 316, 321, 335, 340, 341, 343, 347
E. m. miliaris 335
E. m. tataricus 311, 326
Eschatostena 406
Esocidae 368
Esox lucius 368
Euacmaeoderella 425, 443
Eublepharis turcmenicus 223, 227, 329, 333, 340, 342, 344, 346–348, 310, 318, 322, 323
Euchortippus transcaucasicus 455, 466
Eucoliidae 395
Eudiplister peyroni 413
Eugahania fumipennis 473
Eulaelaps stabularis 393, 395
Eumeces 318, 319, 321
E. schneideri 315, 326, 334, 340, 341, 343–347, 349
E. s. princeps 311, 315, 334
E. taeniolatus 315, 334, 340, 341, 343, 344, 346, 347
E. t. parthianicus 311, 315, 322, 334
Euonymus velutina 53, 117, 124, 162, 167, 171, 182
Euophrys frontalis 516
Eupatorium 114
Euphorbia 113, 133, 135, 141
E. bungei 159
E. alcata 162
E. helioscopia 154, 162
E. monostyla 159
E. sclerocyathum 126
E. sequieriana 161
E. turkomanica 99
Euphorbiaceae 109
Eupoecilopoda perpunctata 473
Euranoxia 410
Euroleon parvus 496, 498
Eurotia 437
Eurycaspia 542
Euryopes laeta 504

E. quinqueguttata 504
Eurythyrea aurata 415
Euscorpiinae 533
Euscorpius 533
E. flavicaudis 533
Eusparassus oculatus 514, 519
Eutagenia turcomana 414
Eutyctus 408
Evarcha arcuata 517
Evippa badchysica 508, 520
E. onager 393, 399, 508, 521
E. schenkeli 508, 520
E. turkmenica 508, 520
Eyprepocnemis plorans 455, 466
E. unicolor 455, 466

Fabaceae 53, 105, 108, 110, 112, 114, 131, 133, 135, 139, 144, 181, 478
Falco cherrug 219
F. c. cherrug 219
F. c. coatsi 219, 251, 259
F. columbarius 259
F. naumanni 259, 321
F. pelegrinoides 214, 251
F. p. babylonicus 214
F. peregrinus 229
F. tinnunculus 321
F. t. tinnunculus 259
Felis caracal 212
F. c. michaëlis 212
F. chaus 321
F. lynx 206
F. l. isabellina 206
F. manul 213
F. m. ferrugineus 213
F. lybica 321
Ferula 60, 113, 132, 173, 425, 426, 432, 443
F. assafoetida 96, 98, 101
F. badrakema 127
F. gummosa 127, 153
F. karakalensis 127
F. litwinowiana 98
F. oopoda 154, 164, 443
F. ovina 127, 159, 160
F. plurivittata 115
F. turcomanica 127
Ferulago 114
Festuca 20, 63, 113, 127, 145, 146
Festuca arundinacea 156
F. pratensis 135

F. regeliana 135
F. valesiaca 133, 157, 160, 163, 164, 461
Ficaria 114
Ficedula parva parva 262
Ficus 59, 179, 423, 424, 434
F. afghanistanica 174
F. carica 53, 124, 127, 143, 165, 171, 174, 179, 443
Filistata 502, 519
F. insidiatrix 502
Filistatidae 502
Foeniculum 114
Formicidae 192
Formicoidea 395
Francolinus francolinus 220
Frankenia hirsuta 90, 91, 137
Frankeniaceae 109, 112, 124
Fraxinus 125, 433
F. lanceolata 137
F. sogdiana, see *F. syriaca*
F. syriaca 127, 167, 183
Fringilla coelebs coelebs 263
F. montifringilla 263
Fritillaria raddeana 144, 165
Frontinellina frutetorum 505
Fulica atra atra 257, 259
Fumana 114
Fumariaceae 109
Fungivoridae 59

Gagea 112, 113, 131, 133, 135
G. divaricata 92
G. dubia 137
G. gageoides 153
G. reticulata 82, 85
G. stipitata 159
Gaillonia brugierii 126
Galba 539
Galeodes fumigatus 393
G. turcomanicus 393
Galerida cristata 281, 282
G. c. iwanowi 248–250, 252, 253, 255, 261
Galium 113
G. aparine 158, 161, 164
G. teniussimum 157, 155, 160
G. tricorne 155, 159
G. verticillatum 156, 162
G. verum 158, 160, 164
Gallinago gallinago gallinago 260
Gallinula chloropus chloropus 259

Gamasoidea 392, 393
Gambusia affinis
  holbrooki 370, 374
Garhadiolus hedypnois 156
G. papposus 156
Gasterosteidae 370
Gastropoda 540
Gaudinopsis 114
Gazella subgutturosa 193,
  208, 231
Gekkonidae 318, 332, 342,
  344
Gelechiidae 473
Gelochelidon nilotica
  nilotica 260
Geminula continens 536, 537
G. isseliana 536, 537
Gentianaceae 110
Geogarypus shulovi 393
Geometridae 192
Gepella modesta 496, 498
Geraniaceae 110
Geranium lucidum 159, 162
G. regelii 135
Geyria lepidula 496, 497
Gibbulinopsis signata 536
Giniocarpa 126
Ginsiana carpetana 473
Girgensonnia oppositiflora 88,
  89
Gladiolus italicus 153, 158
Glareola pratincola
  pratincola 260
Glaucium oxylobum 115
Gleditsia 125, 424
G. caspia 127
Glyceria 438
G. plicata 135
Glycyrrhiza 424, 428, 442
G. glabra 133, 159, 461
Glyphonotus thoracicus
  semenovi 453, 464
Gnaphosa ajadahania, see G.
  haarlovi
G. haarlovi 512, 519
G. kuldzha 512, 519
G. leporina 512
G. lugubris 512
G. turkmenica 512, 520, 523
Gnaphosidae 393, 396, 499,
  501, 511
Gnathoncus kiritschenkoi 408
G. pygmaeus 394, 408, 413
Gnathosia 414

G. compressa 413
G. hydrobiforfis 413
G. karelini 410
G. kushkensis 411
G. lopatini 410
G. modesta 414
G. skobelevi 413
G. sublaevigata 413
Gnetopsida 111
Gobiidae 370
Gobio gobio
  lepidolaemus 368, 374
Goldbachia laevigata 83, 87,
  88
Gongylidiellum
  murcidum 505, 520
Gongylonema pulchrum 245
Goniolimon 60
Gonista sagitta 455, 457, 466
G. pubiferum 414
G. rusticum 409, 414
G. setulosum 409
Granopupa 536, 537
Gronops 394
Grossulariaceae 109
Grus grus lilfordi 259
G. leucogeranus 215
Gryllidae 394, 452
Grylliscus gussakowskii 453,
  465
Gryllodinus kerkennensis 454,
  465
G. odicus odicus 453, 457, 465
Grylloidea 451, 452
Gryllopsis bolivari 453, 465
Gryllotalpa africana 455, 465
G. gryllotalpa 455, 465
G. unispina 455, 459, 465
Gryllotalpidae 452
Gryllus bimaculatus 453, 465
Gundelia 114
Gymnodactylus 314
Gypaëtus barbatus 214
G. b. aureus 214
Gyps fulvus fulvus 259
Gypsophila 127
G. antoninae 133, 138, 144
G. aretioides 132, 143, 181
G. bicolor 159, 160
Gyraulus 538, 539

Habrobates 408
H. veisovi 531
Habrolepis tergrigorianae 471

Habroloma aurea 438, 444
Haemaphysalis numidiana
  turanica 393
H. otophila 244
H. pospalovashtromae 244
H. punctata 244
H. sulcata 244
Haematopota 485, 490, 491
H. pallens 485
Haemonchus contortus 245
Hahnia 509, 519
Hahniidae 509
Haliaëtus albicilla 217, 259
H. leucoryphus 214
Halimiphyllum 425
Halimocnemis 128
H. karelinii 86–89
H. longifolia 90, 91, 94
H. villosa 85, 86, 102
Halimodendron 59, 433
H. halodendri 137, 446
Halocnemum 426, 430
H. strobilaceum 78, 79, 89–
  91, 128
Halopeplis pygmaea 128
Halostachys 431, 437
H. belangeriana 128
H. caspica 90, 91, 137
H. europaea 128
Halothamnus 53, 125, 426,
  431
H. glaucus 128, 137
H. hispidulus 128, 137
H. subaphyllus 82, 83, 98,
  101, 128
Haloxylon 19, 39, 53, 59,
  125, 426, 428, 430, 431, 437,
  474, 477, 487, 488
H. aphyllum 18, 41, 45, 78,
  80, 83, 86, 89–91, 94–97,
  128, 400, 459, 460, 530
H. persicum 19, 41, 42, 80,
  91–95, 98, 100, 187, 400,
  459, 460, 463, 529, 530
Hamamelidae 109
Hammada 60
Haplanthaxia 432
Haplodrassus
  dalmatensis 512, 521
H. signifer 512
Haplophyllum 60
H. acutifolium 156
H. obtusifolium 156, 158
H. ramosissimum 84

Index of Taxa

H. versicolor 133
Harpactea parthica 524
Harpalus antonovi 412
H. rubripes 412
Hedyphanus 409
H. kushkensis 411
Hedysarum kopetdaghi 138
Helegonatopus coxalis 469
H. rasnitzyni 470
Helianthemum salicifolium 153, 157, 160, 163
Helichrysum kopetdaghense 145
Helicolimax annularis 537
Helicopsis likharevi 536, 537
Helictotrichon turcomanicum 138, 144
Heliophanus auratus 517
H. curvidens 517, 520
H. flavipes 517
H. lineiventris 517
H. melinus 517
H. niveivestis 517, 520
H. patagiatus 517
H. turanicus 517, 520
Helioscirtus moseri 454, 457, 466
Heliotropium 85, 425, 464
H. argusioides 94, 99, 101
H. dasycarpum 94, 99, 101
Helminthia 114
Helomyzidae 395
Hemerobiidae 472
Hemiaulax morio 415
Hemiculter eigenmanni 369, 371
Hemidicerca 435, 440
Hemiechinus auritus 321, 390
Hemiontidaceae 111, 114
Hemiptera 192, 392, 394, 396, 400, 469, 475
Henicidae 394, 395
Heriaeus buffonopsis 514, 520
H. fedotovi 520
H. mellottei 520
H. oblongus, see *H. mellottei*
H. spinipalpus 515, 520
Hersiliidae 393, 504
Hersiliola 393, 399, 400, 504, 519, 520, 523
H. afghanica 519
H. maculata 504
Hesseola transcaspica 537

Heteracris adspersus 454, 466
H. littoralis similis 454, 466
H. theodori delicatus 453, 466
Heteropappus canescens 159
Heteropodidae 514
Hieracium procerum 156, 159, 160
Hierophis 316
Hilethera turania 453, 466
Himantopus himantopus himantopus 259
Hippasa deserticola, see *H. partita*
H. partita 508, 521
Hippolais caligata caligata 262
H. languida 248–250, 252, 262
H. pallida 193
H. p. elaeica 262
H. rama 248, 250, 252, 262
Hirundo rustica rustica 261
Hister megalonyx 407
H. quadrinotatus 413
H. semenovi 409
H. uncinatus 413
Histeridae 394, 403–404
Holcothorax 475
H. kopetdagensis 473
Holometabola 482
Holosteum glutinosum 156
H. umbellatum 157
Holotrichius tristis 394, 400
Holzezus compactus 496, 497
Homalodiscus 114
H. ochradeni 122, 123, 126
Homalotylus 474
H. balchanensis 473
H. flaminius 473
H. nigricornis 473
H. platynaspidis 473
H. quaylei 473
H. turkmenicus 473
Homoptera 192, 468–475, 478
Homorocoryphus nitidulus 455, 464
Hoplistura 431, 442
Hoplopholcus forskali 502
Hoplopsis minuta 469
Horaninovia 101, 430
H. ulcina 98, 99
Horatiidae 538, 540
Hordeum brevisubulatum 136

H. bulbosum 159, 164
H. leporinum 100
H. murinum 164
H. spontaneum 156
Hottentotta 532, 533
H. alticola 533
H. saulcyi 533
Hulthemia persica 156
Hulthemosa blinovskyana 132, 144
Hyacinthaceae 111, 114
Hyacinthus 114
H. litwinowii 137, 144
H. transcaspica 137, 144
Hyaena hyaena 210, 321
Hyalocoris pilicornis 400
Hyalomma asiaticum 393
H. detritum 244
Hyalorrhipis clausi 454, 457, 466
H. turcmena 453, 466
Hybomitra 484, 490–492
H. acuminata 485
H. erberi (species group) 485, 492
H. erberi 485, 490
H. expollicata 491
H. peculiaris 485
H. turanica 485
Hydrobiidae 540
Hymenocrater bituminosus 131, 155, 158, 163, 164, 171
Hymenoptera 192, 320, 392, 467, 475
Hyosciamus 461
H. niger 461
Hypanis plicata 541
Hypecoaceae 109
Hypecoum pendulum 83, 85, 89, 92, 95, 102
H. trilobum 154
Hypericaceae 109
Hypericum perfoliatum 156
H. scabrum 159, 160, 164
Hypocacculus eremobius 407
H. oxytropis 407
H. vlasovi 407
Hypocolius ampelinus 220
Hypoderma silenus 244
Hypophthalmichthys molitrix 369, 371
Hyppocrepis 114
Hypsosinga albovittata 507

H. turkmenica 507, 520
Hystrix leucura 193

Ichneumonidae 395
Imperata cylindrica 168
I. cordata 159, 161
I. oculuschristii 167
I. rhizocephala 135
Iranella eremiaphila 454, 458, 465
I. turcmena 453, 464, 465
Iridaceae 111
Iris 429
I. ewbankiana 145, 154
I. kopetdaghensis 331
I. longiscapa 82, 155
Isatis 426
I. minima 83, 92, 95
Iscariotes 417
Isodromus 475
I. flaviscutum 473
I. luppovae 473
I. ustianae 473
Isopoda 392, 393
Isoptera 332, 334
Ixioliridaceae 111
Ixiolirion 143
I. tataricum 156
Ixobrychus minutus 258
Ixodidae 244, 393
Iynx torquilla torquilla 261

Jaculus turkmenicus 211
Janetschekia necessaria 505, 520
Jasminum 179
J. fruticans 53, 63, 118, 124, 127, 159, 164, 165, 171, 182
Juglandaceae 109
Juglans 53, 175, 424, 428,432–435
J. regia 63, 127, 137, 143, 167, 203
J. r. ssp. *turcomanica* 183, 184
Julodella 423, 440, 449
J. brevilata 423, 440, 449
J. kaufmanni 423
J. schestoperovi 423, 440, 447
Julodinae 423
Julodini 423
Julodis 423, 440
J. euphratica 423
J. laevicostata 423, 440

J. variolaris freygessneri 423, 441
Juncaceae 111
Juncaginaceae 111
Juncus 113, 439, 442
J. gerardii 169
Juniperus 432–434
J. excelsa 123, 166
J. polycarpos 123, 166, 177
J. seravschanica 123, 166, 178
J. turcomanica 63, 123, 143, 144, 156, 159, 166, 171, 173, 174, 177, 178, 187
Juno fosterana 156
Jurinea 113, 424
J. antonovi 145
J. kultiassovii 145
J. ludmilae 138
J. sintenisii 132

Kainarella minima 539
Kalidium 428, 430
K. caspicum 128
Kazakia tarbinskii 453, 457, 466
Klewaria 409
Kochia 430, 431
K. odontoptera 101
K. prostrata 128
K. schrenkiana 102
Koelpinia linearis 83, 85, 88, 89, 92, 95, 100, 102, 156, 159
Kraepelinia 532, 533
K. palpator 525, 531
Kuhitangia 410
K. popovii 181
Kurdjumovia plana 473

Lacerta agilis 349
L. defilippii 224, 311, 313, 316, 318, 326
L. raddei, see also *L. defilippii* 333, 339–341, 348, 349
L. saxicola defilippii, see *L. defilippii*
L. strigata 311, 313, 326, 333, 340, 341, 348, 349
Lacertidae 318, 333, 343, 344
Lachnogya squamosa 394, 400
Lactuca rosularis 115
L. serriola 156, 159

Laemostenus caspius 412
Laevicaspia 542
Lagochillus balchanicus 123
L. kabulicus 154
Lallemantia royleana 83, 159
L. sulphurea 156
Lamennaisia ambigua 473
Lamiaceae 53, 108, 110, 112, 114, 134, 139, 144, 166, 443
Lamidae 108, 110
Lamium turkestanicum 153, 159, 160, 165
Lampetis 428
Langona tartarica 517, 519
Laniidae 193
Lanius collurio 193, 261
L. excubitor 193, 253, 255
L. e. homeyeri 261
L. e. pallidirostris 248, 250, 252, 261
L. isabellinus phaenicuroides 261
L. minor 193, 260
Lappula 113
L. barbata 153, 158, 161
L. semiglabra 83, 85, 88, 89, 92, 95, 96, 99
Larinia nenilini 507, 520
L. pubiventris 507, 520
Larinioides folium 507, 521
Larus argentatus heuglini 260
L. canus heinei 260
L. genei 260
L. ichthyaetus 215, 260
L. minutus 260
L. ridibundus 260
Lasiopogon 114
Lasiostola 414
L. grandis 413
L. heterogena 410
L. pubescens 409
Lathridiidae 394
Lathyrus inconspicuus 156
L. sphaericus 156
L. tuberosus 138
Latrodectus pallidus 504, 520
L. tredecimguttatus 393, 399, 504
Lauraceae 52, 124
Lauria cylindracea 536, 537
Lebia circumducta 412
L. holomera 415
L. turkestanica 412
Leistus 413

L. lencoranus 412
Leiurus 532
Lemnaceae 111
Leontodon 114
L. asperrimum 161
Lepechiniella persica ssp.
  kopetdaghensis 145
Lephothryx gambelii 284
Lepidium perfoliatum 87, 88
Lepidoptera 191, 192, 332,
  334, 392, 395, 469, 472–476
Leptaleum filifolium 83, 85–
  89, 96, 156
Leptodes boisduvali 410
L. solieri 413
L. zubkovi 410
Leptomastidea abnormis 468
L. acanthococci 468
L. bereketi 468
L. enigmatica 468
L. rubra 468
L. turcmenica 468
Leptomastix flava 468
L. fulva 468
Leptopternis gracilis 454, 457, 466
Leptorchestes
  berolinensis 517, 521
Leptyphantes
  badkhyzensis 505, 520
L. escapus 505, 519
L. kuhitangensis 505
L. nebulosoides 505
L. pinicola 505, 520
L. tenebricola 505
L. tenius 505
L. turanicus 505, 519
L. turkestanicus 506, 520
Lepus tolai 213
Lesina (=Magrettia)
  mutica 394, 395, 453, 463, 464
Lethrus 463
Leuciscus idus oxianus 368
L. latus 368
Leucochromus 408
Leuconozella rubens 537
L. rufispira 537
Levanderia 537
Leymus kopetdaghensis 136
L. nikitinii 131
L. tianschanicus 136
Liliaceae 111, 112, 134, 139
Lilidae 111, 112

Liliopsida 111, 112, 140, 142
Limnocarabus 416
Limnocryptes minimus 260
Limoniaceae 109
Limoniidae 59
Limonium 60, 424
L. reniforme 137
L. subfruticosum 90, 91, 128
Liniaceae 110
Linum corymbulosum 138
L. turcomanicum 138, 144
Linyphiidae 393, 499, 501, 505
Liobuthus 532
L. kessleri 525, 530
Liochirus 412
L. cycloderus 412
Liocleonus clathratus 415
Liocranidae 510
Liquidambar 125, 127
Liriodendron 125
Lissothus 532
Lithocolletidae 473
Lithogliphus caspius 542
Lithospermum arvense 156
L. officinale 156
L. turkestanicum 154
Lobivanellus indicus 220
L. i. aigneri 220
Locusta migratoria
  migratoria 455, 466
Locustella naevia
  straminea 262
Loliolum subulatum 155
Lonicera 126, 137, 179
L. bracteolaris 162, 164, 171
L. floribunda 126, 167, 183
L. nummularifolia 132
L. persica 179
Lopezus 498
L. autumnalis 496, 497
L. fedtschenkoi 496, 497
L. karakumicus 496, 497
L. nanus 496
Loranthaceae 110
Loricerini 413
Loxoncus procerus 415
Loxosceles rufescens 502
Loxoscelidae 502
Luperus 415
Luscinia luscinia 263
L. megarhynchos hafizi 262
L. svecica pallidogularis 263
L. s. svecica 263

Lutra lutra 212
L. l. seistanica 212
Lycium 424
L. kopetdaghi 137
L. ruthenicum 90, 91
L. turcomanicum 156
Lycodon 320
L. striatus 225, 326, 336, 339, 340, 343–345
L. s. bicolor 225, 312, 337
Lycosa alticeps 508, 520
L. nordmanni 508, 521
L. radiata 508, 521
Lycosidae 393, 499, 501, 508
Lygophyllum 113
Lymnaea atra 540
L. bactriana 538, 539
L. balthica 540
L. berlani 540
L. danubialis 540
L. impura 539
L. rectilabrum 538
L. schirazensis 538, 539
L. subangulata 538, 539
L. subdisjuncta 538, 539, 540
L. tenera 539
L. thiesseae 538, 539
L. truncatula var.
  ventricosa 538
Lymnocardiidae 540
Lythraceae 110
Lythrum salicaria 137
Lytopelte maculata 537
Litorhynchus 320
L. ridgewayi 224, 312, 320, 337, 340, 347

Mabuya 319
M. aurata 315, 321, 322, 326, 334, 340, 341, 343–346, 347, 349
M. a. affinis 311, 315, 350
M. a. septemtaeniata 315, 334
Machozetus 412
M. concinnus 412
Macrochlamys 537
Macronemurus persicus 497
Magnoliophyta 140, 142
Magnoliopsida 109, 110, 112, 140, 142
Magrettia mutica, see *Lesina mutica*
Mahencyrtus comara 472
Malabalia 425, 426

Index of Taxa

Malacocarpus
  crithmifolium 137, 183
Malcolmia africana 154, 159
M. grandiflora 156
M. turkestanica 461
Malegia jacobsoni 415
Mallophagus ovinus 244
Malus 433
M. sieversii 137, 143
M. s. var. turkmenorum 138
M. turkmenorum 118, 125,
  165, 171, 183
Malva neglecta 159
Malvaceae 109, 114
Malvalthaea 114
Mandragora 114
M. turcomanica 117, 125
Mangora acalypha 507, 521
Maracanda amoena 496, 497
M. talitzkii 496, 497
Margarinotus
  atamuradovi 413
M. bickhardti 413
M. graecus 413
M. solskyi 413
M. stercorarius 413
Marrubium propinquum 144
Marshallagia marshalli 245
Martensamnicola 538
Martes foina 321
Mastus 417
Matricaria 59
M. lamellata 95
Matthiola farinosa 153, 159
Mauremys caspica 325, 330,
  340, 348, 349
M. c. caspica 310, 330
Mausolea eriocarpa 98, 101,
  128
Mayrencyrtus maculatus 472
Mayridia eremobia 471
M. formosula 471
M. kopetdagica 471
M. merceti 471
M. murgabensis 471
M. pulchra 471
M. sugonjaevi 471
M. tschairae 471
Maytenus 124
Mecopisthes orientalis 506,
  519, 523
Medecticus assimilis 454, 465
Medicago 59, 113
M. falcata 133

M. rigidula 155, 159
M. sativa 159, 160
Mediodactylus 314, 328
M. russowi 314, 317, 320
M. r. russowi 310
M. spinicaudus 223, 310,
  314.323, 332, 340-342, 347
Megalochilus 315
Megistopus flavicornis 498
Meladiesia 406
M. miritarsis 411
Melandrium boissieri 159
Melanocorypha bimaculata
  torquata 251, 261
M. calandra
  psammochroa 261
Melanogryllus desertus 453,
  458, 459, 465
Melanoides kainarensis 538
Melanophila picta 415
Melanophilini 434
Melanopsis
  ashkhabadensis 538
M. roseni 538
M. starostini 538
M. transcaspica 538
M. turkmenica 538
M. zarudnyi 538
Melanotmethis fuscipennis
  fuscipennis 453, 465
M. fuscipennis unicolor 454,
  465
Melanthaxia 432, 444
Melastomaceae 52
Meles meles 321
Meliboeoides 436
Meliboeus 436, 440, 447
M. amethystinus 436, 442
M. caucasicus 436, 447
M. cyaneus 436, 442
M. reitteri 436, 442
M. robustus 436, 440
M. staneki 436, 440
Melica persica 133
Mellivora capensis 212
M. c. indica 212, 321
M. indica, see M. c. indica
Meloidae 408
Menas 417
Menemerops afghanus 517,
  519
Menemerus marginatus 517,
  519
Meniocus linifolius 156, 157

Mentha longifolia 133, 137,
  167
Merendera jolantae 144, 155
M. robusta 155
M. sobolifera 136
Mergus albellus 258
M. merganser merganser 258
M. serrator 258
Meriones 391
M. libycus 193, 389, 390
M. meridianus 389, 390
M. tamariscinus 193
Merops apiaster 261
M. superciliosus persicus 261
Mesalina guttulata 315, 318,
  322, 326, 333, 340, 347, 348
M. g. watsonana 311, 315,
  328, 333, 350
M. watsonana, see M. g.
  watsonana
Mesasigone mira 506, 519,
  523
Mesasippus kozhevnikovi
  kozhevnikovi 453, 466
Mesiotelus 522
M. kulczynskii 511, 519
M. tenuissimus 511, 521
Mesobuthus 321, 532
M. caucasicus 525, 528, 529
M. c. caucasicus 528
M. c. fuscus 529
M. c. intermedius 528, 529
M. c. kaznakovi 529
M. c. parthorum 528, 529
M. c. przewalskii 528
M. confucius, see M. martensi
M. eupeus 393, 525-527, 533
M. e. afghanus 527
M. e. barszczewskii 527
M. e. bogdoensis 527
M. e. caspius (nomen
  nudum) 527
M. e. eupeus 527
M. e. haarlovi 527
M. e. iranus 527
M. e. kirmanensis 527
M. e. mesopotamicus 527
M. e. mongolicus 527
M. e. pachysoma 527
M. e. persicus 527
M. e. philippovitschi 527
M. e. phillipsi 527
M. e. thersites 526, 527, 532
M. gibbosus 532

Index of Taxa 623

M. martensi 532, 533
Mesonemurus 497, 498
M. clarus 498
M. paulus 497
Mesostemma 143
M. kotschyana 133
Mesostena puncticollis 409
Mesostylini 408
Mesostylus 394, 408
M. hauseri 408 M.
 uzboicus 408
Mespilus 114
M. germanica 117, 165, 179
Messor intermedius 192
Meta kirgisica , see Metellina
 kirgisica
Metaclisa viridis 203, 406
Metaphaenodiscus
 bactrianus 468
Metaphycus 475
M. acanthococci 470
M. bogdanovikatkovi 470
M. desertus 470
M. dispar 470
M. elenae 470
M. hodzhevanishvilii 470
M. melanostomatus 470
M. turanicus 470
M. turkmenicus 470
M. zebratus 470
Metapsyllaephagus
 desantisi 471, 475
M. eremita 471
M. talassioi 471
M. tashlievi 471
Metellina kirgisica 506, 519
Metrioptera crassipes 452,
 453, 458, 464
Micaria 393
M. albimana 512, 520
M. kopetdaghensis 512, 519
M. lenzi 512
M. pygmaea 512
M. romana 512, 521
M. rossica 512
M. septempunctata 512, 520
Microblemma simplex 413
Microbuthus fagei 532
Microcephala lamellata 92
Microctenonyx subitaneus 506
Microderus 408, 409
M. badhysi 411
M. globulicollis 414
Microlestes plagiatus 409

Microlinyphia pusilla 506
Micromelania grimmi 542
Micrommata ligurinum 514
Microterys 474, 475
M. ashkhabadensis 469
M. chommati 469
M. contractus 469
M. darevskii 469
M. darganatensis 469
M. duplicatus 469
M. mesasiaticus 469
M. praedator 469
M. sylvius 469
M. tricoloricornis 469
M. vashlovanicus 469
M. zygophylli 469
Microtus afghanus 193
Milium vernale 156, 159, 162
Milvus migrans migrans 258
Mimetidae 503
Mimetus laevigatus 503
Mimocarabus 416
Mimosa 461
Mimosiaceae 110
Miniopterus schreibersi 209
M. schreibersi pallidus 209
Minosia karakumensis 393,
 512, 520, 523
Minosiella intermedia 393,
 395, 396, 400, 513, 519, 520
Minuartia litwinovii 132
Mioscirtus wagneri
 rogenhoferi 454, 457, 466
Miridae 192
Misgurnus
 anguillicaudatus 369, 372
Mizonocara deserti 453, 458,
 466
M. inornata insolita 453, 458,
 466
Mnuphorus 415
Modicogryllus
 burdigalensis 454, 465
M. chivensis 453, 465
M. frontalis 453, 465
M. pallipalpis 453, 465
Mogrus antoninus 517, 519
M. valerii 517, 520, 523
Mohelencyrtus
 phenacocci 469
Molucella 114
Momphidae 395
Monaeses israilensis 515, 520
M. paradoxus 515

Moniezia benedeni 245
Monodacna 541, 542
Monodiscodes
 maculipennis 468
Monticola saxatilis 193
M. s. saxatilis 262
Moraceae 109
Moriera 114
Morter hyalinus 497, 498
M. semigriseus 394, 497
Morus 424, 476
M. alba 165
Motacilla alba
 dukhunensis 261
M. citreola werae 261
M. feldegg melanogrisea 261
M. flava beema 261
M. f. thunbergi 261
M. personata 261
Muscari bucharica 156, 159
Muscicapa striata
 neumanni 262
Mustela eversmanni 321
Mutillidae 394
Mycetobas carnipes 178
Mycroglotta nidicola 394
Mylabris 408, 415
M. kuzini 409
Mylopharyngodon piceus 368,
 371
Myomimus personatus 213
Myosurus 114
Myotis nattereri 213
M. n. tschuliensis 213
Myriapoda 344, 392, 393
Myrica 52
Myricaceae 52
Myrmecaelurus 496
M. badghisi 496
M. major 496, 498
M. paghmanus 496, 498
M. trigrammus 496
M. varians 497, 498
Myrmecophilidae 452
Myrmecophilus oculatus 455,
 465
Myrmecophyes 192
Myrmeleontidae 394,
 495–498
Myrtaceae 52
Mytilaster lineatus 540
Mytilidae 540

Naja oxiana 192, 222, 312,

319, 320, 326, 338–341, 344, 345, 347
Nanoblaps hiemalis 394
Nanophyton 60
N. erinaceum 88
Nanorhynchus 481, 484, 487, 489, 491
N. crassinervis 484, 488, 489
Natrix 320, 321
N. natrix 320, 326, 349
N. n. persa 311
N. tessellata 311, 313, 320, 326, 335, 337, 340, 343, 345, 348, 349
Neastymachus luteus 469
N. secundus 469
Nebria 413
Nedroleon maculatus 498
Negeniaspidius nobilis 473
Nemachilus starostini 227
Nematodirus abnormalis 245
N. davtiani 245
N. dogieli 245
N. filicollis 245
N. helvetianus 245
N. oiratianus 245
N. spathiger 245
Nemesia birulai 502, 519, 523
Nemesiidae 502
Nemorius 484
Neococcidencyrtus steinbergi 471
Neodiscodes 477
Neophron percnopterus percnopterus 259
Neoscona adianta 507
N. subfusca 507, 521
Nepeta sintenisii 160, 164
Neritidae 540
Netta rufina 257
Netuschilia hauseri 394
Neuroleon 496, 498
N. jucundus 496
N. leptaleus 496, 497
N. nigriventris 496
N. tenellus 496, 498
Neuroptera 191, 392, 472, 475, 495–498
Nicarinus poecilopterus 498
Nikitinia 134
N. leptoclada 144
Nitidulidae 472
Nitraria 60, 430
N. komarovii 128

N. schoberi 90, 91, 128
Nitrariaceae 110, 112
Noaea 424
Noctuidae 192, 395, 472
Noemachilus amudarjensis 369
N. malapterurus longicauda 369, 374
N. oxianus 369, 374
Nohoveus crucifer 496, 497
N. zigan 496, 498
Nomioides pulverosus 395
Nomisia aussereri 513, 521
N. conigera 513, 519
N. exornata 513
Nonea caspica 83, 85, 87, 89, 92, 95, 102, 156, 159
N. pulla 135
N. turcomanica 153
Notiophilus sublaevis 412
Notodontidae 469
Notostaurus albicornis albicornis 454, 466
N. albicornis turcmenus 453, 466
Numenius phaeopus phaeopus 260
Nyctalus noctula 211
Nyctea scandiaca 260
Nyctiphantus 408
Nyctor 408

Ochranoxia 408
Ochrilidia hebetata hebetata 453, 457, 466
O. mistshenkoi 453, 466
O. turanica 453, 466
Odonthobuthus 532
Oecanthidae 452
Oecanthus turanicus 453, 465
Oecobiidae 503
Oecobius nadiae 503, 519
O. tadzhicus 503, 519
Oedaleus decorus 466
O. senegalensis 466
Oedipoda miniata atripes 454, 466
O. m. miniata 454, 466
Oedothorax apicatus 506, 520
Oenanthe deserti salina 248–252, 262
O. finschii 262
O. isabellina 248–252, 262, 390

O. oenanthe oenanthe 262
O. picata picata 262
O. pleschanka pleschanka 262
Oesophagostomum columbianum 245
Oestrus ovis 244
Oleaceae 110
Oligochaeta 191
Oligodon taeniolatus 225, 312, 324, 326, 337, 340, 344
Olios sericeus 514, 519
Olpium pallipes 393
Omophron rotundatum 411
Onagraceae 110
Oncocephalus 394
Onobrychis 59, 127, 156
O. cornuta 181
O. echidna 181
O. pulchella 159, 162
Onopordum 424, 426
Onosma angustilobum 122
O. dichroanthum 156, 159
O. longiloba 122
Onthophagus vlasovi 394
Onthophilus sulcatus caucasicus 413
Ooencyrtus 474, 475
O. daritshevae 469
O. indefinitus 469
O. minnae 469
O. telenomicida 469
O. tibialis 469
Oonopidae 503
Opatroides 415
O. punctulatum 409
Ophiocephalidae 370
Ophiocephalus argus warpachowskii 372
Ophioglossaceae 111
Ophioglossum vulgatum 167
Ophiomorus chernovi 225, 229, 311, 324, 348
Ophisaurus 350
O. apodus 311, 313, 319, 320, 321, 326, 332, 340, 341, 343, 344, 346, 348, 349
O. a. apodus 315
O. a. thracicus 332 (as *Pseudopus*)
Ophonomimus hirsutulus 415
Ophonus cribricollis 411
Ophrys 114
O. transhyrcana 144, 167
Orchidaceae 111, 114, 133

Orchis pseudolaxiflora 137
Oriolus oriolus 193
O. o. kundoo 261
O. o. oriolus 261
Ornithodoros tartakovskii 393
Ornithogalum arianum 144
Orthantha 114
Orthobula charitonovi 499, 511, 520
Orthocheira 434
Orthochirus 532
O. conchini, see *O. scrobiculosus scrobiculosus*
O. scrobiculosus 393, 525, 529
O. s. concolor 530
O. s. dentatus 530
O. s. melanurus 529, 530
O. s. mesopotamicus 530
O. s. negebensis 530
O. s. persa 530
O. s. scrobiculosus 529, 530
Orthoptera 392, 451–466
Orthurus heterocarpus 161, 165
Oryzias latipes 370, 372
Oryziatidae 370
Ostertagiella circumcinta 245
O. occidentalis 245
O. trifida 245
O. trifurcata 245
Otis tarda 217
O. t. tarda 217, 259
O. tetrax 217, 259
O. t. orientalis 217
O. undulata macqueenii, see *Chlamydotis undulata macqueenii*
Ottorosenia varenzovi 536, 537
Otus brucei 260
O. scops scops 260
Ovis ammon 193, 210, 231
O. a. bucharensis 211
O. a. cycloceros 210
Ovulites renata 59
Oxya fuscovittata 455, 457, 465
Oxyopes badhyzicus 510, 520, 523
O. heterophthalmus 510, 520
O. lineatus 510, 521
O. marakandensis 510, 520
O. takobius 510, 519

Oxyopidae 510
Oxypteris 434
"Oxyptila" lugubris 515, 519
Oxyptila baudueri cribrata 515
O. sanctuaria 515, 520
O. tricoloripes 515, 520
Oxypyrgula 542, 543
Oxytropis czapandaghi 136, 144
Oxyura leucocephala 216, 258

Pachycarus 417
Pachymorpha 407
Pachypsylloides 476
Pachyscelis gemmans 414
Palaeoloxodon turkmenicus 61
Palandra 59
Palibinia 124
Paliurus 179, 424, 432
P. spinachristi 164–165, 170–172, 177, 179, 182
Pallenis 114
Palpares solidus 496, 498
Palpimanidae 503
Palpimanus sogdianus 503, 519
Pamphagidae 452
Panagaeus relictus 415
Pandion haliaëtus 218
Panthera pardus 207
P. p. tullianus 207, 243
P. tigris virgata 208
Panurus biarmicus russicus 263
Papaver 461
P. hybridum 155, 159
P. pavoninum 95, 331
Papaveracae 109
Parablastothrix reimovi 473
Parabramis pekinensis 369, 371
Paracaryum turcomanicum 156
Paracedicus 522
Paracopidosoma 474
P. eparallelum 469
P. ephemeri 469
Paracylindromorphus 438, 442
P. lebedevi 438
P. semenovi 438, 446
P. subuliformis 438

P. transversicollis 438
Paraechinus hypomelas 321
Paralitomastix 474
P. varicornis 473
Paranathrix acanthococci 468
Paraphaenodiscus murgabicus 471
P. sugonjaevi 471
Pararcyptera microptera turanica 455, 466
Paraschedius bicolor 473
P. jasnoshae 473
Parasitiformes 393
Paratetracnemoidea malenottii 468
Paratettix uvarovi 455, 465
Paratibellus oblongiusculus 514, 521
Paravolvulus binaevulus 413
Pardosa italica 508, 521
P. morosa 508, 521
P. nebulosa 508, 521
P. pontica 508, 519
P. proxima 508
Parnops glasunovi 415
Parrotia 125
P. persica 127, 203
Parus bokharensis bokharensis 248–250, 252, 253, 263
Passer 256
P. ammodendri 281, 282, 284–286, 288, 289
P. a. ammodendri 248–250, 252, 255, 263
P. domesticus 193, 281, 282, 284–286
P. hispaniolensis transcaspius 263
P. indicus bactrianus 255, 263
P. montanus 281, 282, 284–286, 289–293
P. m. dilutus 255, 263
P. simplex 220, 281, 282, 284–286
P. s. zarudnyi 220, 248–252, 254, 255, 263
Pastor roseus 261
Paussus 412
Paussus turcicus 409
Pectinibuthus 532
P. birulai 525, 531
Pedaliaceae 53
Peganaceae 110, 184

Peganum harmala 80, 100, 101, 137, 159
Pelecanus crispus 218, 257, 258
P. onocrotalus 215, 257, 258
Pelecopsis laptevi 506, 520
P. paralleloides 506, 519, 523
Pelecus cultratus 369
Peliococcus mesasiaticus 476
Pellenes epularis 517, 520
P. kulabicus 517, 519
P. limbatus 517
P. nigrociliatus 517
P. simoni 517, 520
P. tripunctatus 517
Pelorocnemis darwini 414
Peltaria 114
P. turkmena 162
Peltariopsis 114
Pennisetum orientale 133
Pentacladocerus matranus 472
Pentaphyllus chrysomeloides 406, 415
Pentatomidae 469
Penthicus 414, 415
P. dilectans 414
P. fartilis 413
P. pinguis 414
P. semenovi 415
P. turkmenicus 413
Percidae 370
Peregriana 540
Perileptus mesasiaticus 412
Periploca 59, 114
P. graeca 125, 179, 183
Perovskia abrotanoides 132, 158, 164-166
Perplexia 114
P. microcephala 135, 144
Persica 427, 432.434
Petria 408
Petrosimonia 128
P. glauca 90, 91
Peucedanum sintenisii 162
Phagnalon androssovii 115
Phalacrocorax carbo sinensis 258
Ph. pygmaeus 258
Phalaris minor 154
Phalaropus lobatus 260
Phaneroptera bivittata 454
P. falcata 455, 464
Pharaonus lederi 409
P. semenovi 409

Phasianus colchicus 295-306
Ph. c. bergii 299, 300, 304
Ph. c. bianchii 295, 297-299, 300, 303, 304
Ph. c. chrysomelas 295, 297-299, 300, 304, 305
Ph. c. colchicus 300-304
Ph. c. hagenbecki 299, 300
Ph. c. lorensi 300, 301, 303, 304
Ph. c. mongolicus 295, 297, 299, 300, 303, 304
Ph. c. pallasii 295
Ph. c. persicus 295, 297, 299-301, 303, 304
Ph. c. principalis 295, 297-301, 304, 305
Ph. c. satscheuensis 299, 300
Ph. c. septentrionalis 300, 302-304
Ph. c. shawi 299, 300, 304
Ph. c. talischensis 295, 299-304
Ph. c. tarimensis 295, 299, 300, 304
Ph. c. torquatus 295
Ph. c. turkestanicus 295, 297, 299, 300, 303, 304
Ph. c. vlangalii 304
Ph. c. zarudnyi 259, 295, 297--301, 304
Ph. c. zerafschanicus 295, 297-304
Phasmopoda loginovae 469
Pheidole 409
Phenacoccus arthrophyti 476
Philaeus chrysops 517, 517
Philhammus zaitzevi 410
Philipomyia 491
Philochthus 412
Philodromidae 514
Philodromus aureolus 514
Ph. cespitum 514
Ph. fallax 514, 520
Ph. lepidus 514, 521
Ph. rufus 514
Philomachus pugnax 260
Philothis 407
Phlebotomidae 395
Phlegra fasciata 517
Ph. sogdiana 518, 519
Phleum paniculatum 156, 159
Phlomis 438, 461
Ph. cancellata 153, 163

Ph. kopetdaghensis 157, 160, 165, 169
Phoenicopterus roseus 216, 258
Phoenicurus erythronota 262
Ph. ochruros rufiventris 262
Ph. phoenicurus phoenicurus 262
Pholcidae 393, 502
Pholcus nenjukovi 502, 519
Pholioxenus orichalceus 408
Ph. phoenix 413
Ph. shatzmayri 413
Phragmites 59, 125, 438, 442
Ph. australis 127, 133, 156, 167-169, 173
Ph. communis, see *Phragmites australis*
Phrurolithus pullatus 511, 521
Phrynocephalus 362
Ph. golubevi 328
Ph. guttatus 313, 318
Ph. g. guttatus 311
Ph. helioscopus 310, 318, 325, 347
Ph. interscapularis 318, 324, 347
Ph. i. interscapularis 311
Ph. maculatus 221, 311-313, 318, 322, 327, 328
Ph. mystaceus 313, 315, 317, 324
Ph. m. galli 311, 315
Ph. ocellatus 315, 318, 362
Ph. o. bannikovi 311, 315, 322, 324
Ph. o. ocellatus 310
Ph. raddei 318, 324
Ph. r. raddei 311
Ph. reticulatus 310, 315, 322, 324
Ph. rossikowi 223, 318, 324, 327, 347, 351-364
Ph. r. schammakowi 223, 310, 322, 324, 352, 353, 355, 356, 361
Ph. r. rossikowi 223, 310, 325, 353-356
Phyllitis scolopendrium 135
Phylloscopus collybita fulvescens 262
Ph. trochiloides viridanus 262
Ph. trochilus acredula 262
Physocaulis nodosus 167
Phythodrymadusa

longipes 454, 464
Phyxioschaema raddei 501, 519
Pica pica bactriana 261
Picus squamatus 215
Pimelia 414
P. cephalotes 414
Pinophyta 111, 140, 142
Pinus 434
Piptatherum vicarium 136
Pisaura mirabilis 508, 521
P. novicia 508, 519
Pisauridae 509
Pistacia 59, 124, 126, 423–425, 427, 433–435, 437
P. chinensis 175
P. mutica 175
P. terebinthus 175
P. vera 131, 133, 137, 143, 173–176, 187, 188, 270, 411, 448, 449
Pisterotarsa 408
Planococcus ficus 478
Planorbidae 539, 540
Planorbis planorbis 539
P. sieversi 538, 539
Plantaginaceae 110
Plantago 459
Platalea leucorodia 216, 258
Platamodes dentipes 414
Platanaceae 109
Platanus 59, 125
P. orientalis 137, 143, 167, 168, 177, 183, 185, 203
Platycleis affinis 454, 464
P. escalerai 454, 464
P. fatima 453, 464
P. intermedia 454, 464
Platydema 414
P. triste 406, 414
Platyesia 408
Platygastridae 395
Platyrhopus 474
P. longicornis 468
Platysoma simeani 413
Plegadis falcinellus 258
Plesiobuthus 532
Plexippoides starmuehlneri 518, 519
Plexippus coccineus 518
P. setipes 518
Plumbaginaceae 53, 109, 114, 181
Plumbago 114

Pluvialis dominica 259
P. squatarola 259
Poa 63, 113, 127, 135, 459
P. angustifolia 156
P. bulbosa 131, 133, 146, 152.153, 157, 160, 163, 165, 175, 189, 270, 461–463
P. nemoralis 156
P. pratensis 135
P. trivialis 135, 137, 159, 161
Poaceae 105, 111, 112, 114, 133, 135, 139
Podiceps auritus auritus 258
P. cristatus cristatus 258
P. grisegena grisegena 258
P. nigricollis nigricollis 258
P. ruficollis capensis 258
Podoces panderi 248–250, 252, 254, 255, 261, 281, 282, 284–292
Poeciliidae 370
Poecilochroa conspicua 513
Poecilonota 434, 440, 446
P. dives 434, 447
P. nadezhdae 434, 440, 444, 447
Poecilus cupreus 412
P. dissors 409
P. liosomus 412
P. nitens 410
P. puncticollis 411
P. warentzowi 409
Polycestinae 423
Polycestini 423
Polygonaceae 109, 134
Polygonum 113
Polyphaga pellucida 394, 395, 400
Polyphagidae 393
Polypodiophyta 111, 140, 142
Popoviolimon turcomanicum 131, 143, 144
Populus 59, 125–127, 137, 143, 424, 427, 433–435, 437, 442
P. alba 137
P. diversifolia 125
P. euphratica 137, 168, 183
P. pruinosa 125, 137, 183
Porphyrio porphyrio 219
P. p. seistanicus 219
Portulaca 461

Portulacaceae 109
Porzana porzana 259
Potamogeton 137
Potamogetonaceae 111
Potamon zarudnyi 539
Potentilla 113
P. botschanzeviana 135, 144
P. transcaspica 159, 160
Poterium sanguisorba 156, 159
Prangos 173, 426, 436
P. diduma 99
P. latiloba 155, 159
Primipupilla 536
Primulaceae 109
Prinerigone vagans 506
Prinoidae 59
Prionomitus mitratus 470
Prionotrichon 134
P. gaudanense 144
Prionus komarovi 408
Pristadoretus 408
Pritha crosbyi 502, 519
Probaticus 414
P. zoroaster 413
Prochiloneurus 475, 475, 476
P. aegyptiacus 472
P. bolivari 472, 476
P. pulchellus 472
Prodidomus redikorzevi 513, 519, 523
Prosodes 414
P. angustata 410
P. cribrella 414
P. emiri 410
P. jakovlewi 413
P. kuhitangiana 410
P. laticauda 413
P. monticola 410
P. quadriimpressa 411
P. solskyi 413
P. subpilosa 410
Proteaceae 52, 124
Protozoa 244
Protracheoniscus orientalis 393
Protura 392
Prunella atrogularis atrogularis 262
Prunus 184, 425, 427, 428, 432, 434
P. divaricata 165, 167, 171, 183
P. domestica 183

Psammitis cribratus, see *Oxyptila baudueri cribrata*
P. tristrami 515
P. turanicus 499, 515, 519
Psammobuthus 532
Psammogeton setifolium 159, 162
Psammophis lineolatus 312, 320, 326, 337, 340, 345, 347, 349
P. schokari 312, 326, 337, 341, 345, 348
P. s. schokari 337
Psen pulavskii 395
Pseudaphycus astanovi 470
P. malinus 470, 477, 478
P. trabutinae 470
Pseudectroma 474
P. ciliatum 470
P. scenographica 470
P. turanica 470
Pseudicius cinctus 518, 519
P. spasskyi 518, 519
P. vittatus, see *P. cinctus*
Pseudocaspia kainarensis 539
P. ljovuschkini 539
P. starostini 539
Pseudoceles persa 454, 466
Pseudococcidae 473–476
Pseudococcus comstocki 478
Pseudocyclophis 318
P. persicus 316, 320, 324, 337, 341, 344, 347
P. p. persicus 312, 316, 338
P. p. walteri 316
Pseudogobio rivularis 368, 371, 379
Pseudolinosyris sintenisii 138, 144
Pseudonapaeus eremita 536, 537
P. guttula 536, 537
P. kaznakovi 537
P. miser 537
P. otostomus 537
P. potaninianus 536, 537
P. sogdianus 536, 537
Pseudoophonus griseus 412
Pseudoperilampis ocellatus 369, 372
Pseudopus 350
P. apodus, see *Ophisaurus apodus*
Pseudorasbora parva 368,
371, 379
Pseudoscaphirhynchus 374
P. hermanni 226, 227, 368
P. kaufmanni 226, 227, 368
Pseudoscorpiones 392, 393
Psiloptera argentata 428
Psilopterini 427
Psocoptera 392
Psoralea 461
Psyllaephagus 474
P. arenicola 471
P. bachardenicus 470
P. badchysi 470
P. caillardiae 470
P. calligonicola 470
P. cholcinellus 470
P. colposceniae 470
P. desertus 470
P. egeirotriozae 470
P. hammadae 470
P. longiventris 470
P. obscurus 471
P. ogazae 470
P. populi 470
P. rubriscutellatus 470
P. saxaulicus 470
P. tamaricicola 470
P. tokgaevi 470
P. turanicus 470
P. turkmenicus 471
P. tyrrheus 470
Psyllidae 474
Psylliostachys spicata 156
Psylloidea 469, 470, 474–477
Pterocephalus khorassanicus 135
Pterocles 255, 256
P. alchata caudacutus 248–250, 252, 254, 260
P. orientalis arenarius 248–250, 252, 254, 260
Pteromalidae 395
Pteronemobius heydeni 455, 465
Pteropyrum aucheri 115, 116
Pterostichus 410, 413
Pterotricha strandi 513, 519
Ptinidae 394
Ptinus latro 394
Ptyas 320
P. mucosus 221, 229, 312, 320, 322, 325–327
Puccinellia distans 136
P. tenuissima 135, 136

Pulicidae 392
Pungitius platygaster aralensis 370, 374
Punica 179, 433
P. granatum 53, 124, 126, 164, 165, 179
Punicaceae 110
Pupilla triplicata 536, 537
P. turcmenica 536, 537
Pyramidula rupestris 536, 537
Pyraustidae 473
Pyrgodera armata 454, 466
Pyrgomorpha bispinosa 455, 465
Pyrgomorphidae 452
Pyrgula pseudospica 542
P. rudis 543
Pyrgulidae 540
Pyrus 126, 425, 427, 428, 433
P. boissieriana 125, 126, 165, 183
P. communis 183
P. regelii 180
P. turcomanica 125, 138, 143

Quercus 53, 124, 125, 127
Q. balloot 125
Q. castaneifolia 125, 203
Q. ilex 125
Queria hispanica 158
Quinemurus metamerus 497, 498

Radix 538, 539
Rallus aquaticus aquaticus 259
R. a. korejewi 259
Ramburiella bolivari 454, 466
R. foveolata 454, 466
R. turcomana 454, 466
Rana macrocnemis 226
R. ridibunda 330, 331, 337
Ranunculaceae 109, 112, 114
Ranunculidae 109, 112
Ranunculus 113, 135, 461
R. meyeranus 135
R. muricatus 156
R. pinnatisectus 159
R. trichocarpus 136, 144
Raveniola 522
R. fedotovi 502, 519
R. kopetdaghensis 502, 519, 522, 523
R. redikorzevi 502, 519, 523

Index of Taxa 629

Reaumuria 60, 79, 113, 424
R. botschantzevii 115, 118, 123
R. cistoides 136
R. fruticosa 90, 96, 128
R. oxiana 96
R. pangjgurica 123
R. reflexa 156
R. tatarica 128
R. turkestanica 131, 136
Reduviidae 394, 396
Reduvius christophi 394–396, 400
R. fedtschenkianus 394
Regulus regulus tristis 262
Reichardtiolus pavlovskii 411
Reitterohelops 414
R. badhysi 411
Remipedella 406, 408
R. deserti 408
R. semenovi 408
Remiz pendulinus coronatus 263
R. p. jaxarticus 263
Reseda dschebeli 123
Resedaceae 109, 114
Reuteria 114
Rhagionidae 487
Rhamnaceae 52, 110
Rhamnus 59, 179
R. sintenisii 126, 132, 156, 159, 165, 169, 171
Rhampholyssa 408
Rhaphidochila 429, 440
Rheum 427
R. turkestanicum 92, 93
Rhinogobius similis 370, 371
Rhinolophus blasii 209
R. euryale 211
R. hypposideros 211
Rhinopetalum gibbosum 154
Rhipidura leucophrus 255
Rhodopechys sanguinea sanguinea 263
Rhodospiza obsoleta 255, 263
Rhombomys opimus 193, 389--391, 395, 398–400
Rhopus 474
R. olgae 468
R. trjapitzini 468
Rhopus turanicus 468
Rhus 52, 175, 425
R. coriaria 53, 117, 124, 177, 179

R. turkomanica 124
Ribes melananthum 132, 144
Rindera coechinata 144
Riparia riparia riparia 261
Robinia 425
Rochelia retorta 159
Roemeria hybrida 83, 85, 92, 95
R. refracta 156
Roeperia 120
Rosa 112, 132, 137, 179, 182, 184, 428, 434, 436, 437
R. fertilis 134, 144
R. karakalensis 144
R. lacerans 167
Rosaceae 110, 112, 114, 134, 139
Rosidae 108, 109
Rostraria cristata 154
Rosularia elymaitica 133
Rubia florida 131, 132, 155, 159, 165, 171
Rubiaceae 110, 114, 133
Rubus 436, 437
R. anatolicus 183
R. caesius 137
R. sanguineus 167, 171
Rufibrenta ruficollis 218
Rumex 427
R. tuberosus 159, 160
Runcinia lateralis 515, 521
Rutaceae 53, 110
Rutilus rutilus aralensis 368, 374

Sadlerianidae 538, 540
Sageretia 59
Saiga tatarica 231
Salicaceae 109
Salicornia europaea 128
Salix 113, 126, 137, 183, 424, 427, 433, 434, 437, 442
S. acmophylla 137
S. persa 168
Salsola 18, 53, 59, 60, 113, 125, 126, 128, 131, 424–426, 428–431, 437, 476
S. arbuscula 45, 79, 80, 82–86, 88–90, 94, 96, 100, 101, 123, 128
S. aucheri 154
S. botschantzevii 59, 115, 122, 126, 127
S. buhseana 122

S. bungeana 59, 126
S. carinata 102
S. dendroides 137, 163
S. gemmascens 78, 79, 82–90, 94, 96, 128, 156, 163
S. glabella 115
S. iljinii 59, 115, 120, 126
S. incanescens 94
S. junatovi 122
S. kopetdaghensis 59, 126
S. leptoclada 99, 100
S. orientalis 18, 78, 79, 82, 84–88, 94, 96, 128
S. paletzkiana 99
S. paulsenii 81, 102
S. podlechii 120
S. richteri 18, 19, 41, 45, 82, 83, 92, 94, 96–101, 459, 460, 463
S. sclerantha 86, 89–91, 94, 102
S. stelullata 122, 126
S. subaphylla 460
S. titovii 120
S. transhyrcanica 123, 128
Salticidae 499, 501, 516
Salticus tricinctus 518, 520
Salvia virgata 158, 160
S. viridis 156
Sambucus 114
Santalaceae 110
Sapindaceae 52
Saprininae 406
Saprinus 407
S. biguttatus 407
S. bimaculatus 409
S. calatravensis 409, 413
S. georgicus 409
S. gilvicornis 407
S. intractabilis 407
S. jacobsoni 409
S. lateralis 413
S. lautus 415
S. lopatini 415
S. maculatus 409, 413
S. niger 413
S. pharao 409, 413
S. planiusculus 413
S. steppensis 413
S. subnitescens 413
S. subvirescens 413
S. tenuistrius 413
S. therondianus 413
S. viridicatus 413

Index of Taxa

Sarcocystis 244
Satrapes talyschensis 413
Satureia 114
Sauria 331
Saxetania 198
S. cultricollis 453, 463
S. pravdini 453, 458, 464, 465
S. scutata 453, 458
Saxicola caprata rossorum 262
S. torquata maura 262
Saxifragaceae 109
Scabiosa micrantha 157, 161
S. rotata 159
Scandix stellata 159
Scaphirhynchus 374
Scarabaeidae 191, 394, 408
Scardinius erythrophthalmus 368
Scariola orientalis 131
Scarites bucida 407
S. eurytus 412, 415
S. planus 409, 412
Schistocerca gregaria 451, 460, 465
Scincidae 202, 318, 334, 343, 344
Scirpus 59
Sclerochloa dura .136
Scleropatrum 415
S. hirtulum 409
Scorpiones 336, 344, 392, 393, 525–534
Scorpiops 533
Scorpiopsinae 533
Scorzonera litvinovii 159
S. pusilla 155
Scotocerca inquieta 281, 282, 292
S. i. platyura 248–250, 252, 254, 255, 262
Scotophaeus scutulatus 513, 520
Scrobiculariidae 540
Scrophularia 112
S. benthamiana 122
S. kjurendaghi 122, 126, 127
S. kurbanovii 122
S. pruinosa 156
S. turcomanica 156, 159, 169
S. variegata 122
Scrophulariaceae 110, 112, 114, 134, 135, 139
Scutellaria litwinowii 122,
126, 155, 169
S. luteo-coerulea 122, 126
S. multicaulis 122
Scutelleridae 469
Scutigera coleoptrata 393
Scyaridae 191, 193
Scytodes bertheloti 502, 521
S. strandi 502, 519
S. univittatus 502
Scytodidae 502
Semen apterum 471
Semenoviana tamerlana 453, 464
Senecio khorossanica 134
S. subdentatus 83, 85, 7, 92, 95, 98, 99
S. vernalis 154
Seriphidium 78, 152, 478
Serpentes 335, 343
Serradigitus littoralis 533
Setaria glauca 156
Sherardia 114
Shumeria 114
Siagona europaea 409, 412
Siebera 114
S. nana 115
Silene 113
S. bupleuroides 160
S. conica 155, 157
S. coniflora 154
S. czopandagensis 135, 144
S. litwinowii 144
S. swertiifolia 155
Siluridae 369
Silurus glanis 321, 369
Silvius 491
S. irritans 484
S. vitripennis 484, 490
Simarubaceae 110
Sinopteridaceae 111
Sipala 485, 490, 492
Sisymbrium loeselii 156
Sitona 192
Sitticus caricis 518
S. distinguendus 518, 520
S. karakumensis 399, 518, 520, 523
S. terebratus 518
S. zimmermanni 518, 520
Skrjabinema ovis 245
Smirnowia 59
S. turkestana 94, 98, 99, 128
Smyrnium 114
Sogdamnicola 538
Solanaceae 110, 114
Solifuga 392, 393
Solter 496, 498
S. amseli 496
S. felderi 496
S. hardei 496
S. iranensis 496
Somocoelia 413
Sonchus palustris 138
Sophora 137
Sorbus 137
S. graeca 114, 179
S. luristanica 138
S. persica 177
S. turkestanica 138, 143
Spalerosophis 320
S. diadema 312, 326, 338, 343, 345, 348
S. d. schiraziana 338, 339, 340
Spermophilopsis leptodactylus 389–391
Sphaerobothris 435
Sphecidae 394, 395
Sphecozone asiatica, see *S. romana*
S. romana 506, 521
Sphenaria 408
Sphenoptera 419, 421, 428, 445–449
S. addenda 429
S. aerata 429
S. afflicta 429
S. allecta 429
S. amoena 431
S. amplicollis 431, 441
S. apta 431, 447
S. artemisiae 431, 449
S. balassogloi 431, 442
S. beckeri 430, 441
S. bifulgida 431
S. bucharica 429
S. canescens 430, 442
S. captiosa 429
S. caspica 429, 441
S. cataonia 430
S. chalybaea 428
S. coerulea 429, 444, 445
S. curta 429
S. cyanea 428, 440
S. egregia 428
S. erojlandusica 430, 440, 442, 449
S. exarata 428, 442

S. eximia 430
S. furva 428, 440, 445
S. glabrata 428, 440, 445, 446
S. hauseri 430, 441
S. ignita 430
S. kambyses 432, 440, 443, 449
S. kaznakovi 432, 443
S. kepelensis 429
S. koenigi 429, 440, 443, 448, 449
S. komarovi 429
S. korshinskii 428, 447
S. lateralis 428, 441
S. latesulcata 429, 441, 449
S. lia 428
S. mesopotamica 415, 431, 442
S. mitrochinae 429
S. mujunkumensis 432
S. navicula 429
S. orichalcea 430, 441
S. potanini 430, 441
S. pseudoignita 430, 441
S. puberula 430, 442
S. pubescens 431
S. punctatissima 431, 441
S. rangnowi 428, 440, 444
S. rauda 430, 442
S. repetekensis 429
S. schneideri 432, 441
S. scovitzi 430, 441
S. semenovi 431, 442
S. serripes 428, 441
S. subtilis 429, 440, 447
S. tenax 429
S. tomentosa 431, 441
S. tschitscherini 431, 444
S. turcmenica 429, 440, 448
S. unidentata 429
S. vestita 429, 441
S. violacea 428, 440
S. viridiaurea 431, 441
S. viridula 431, 442
S. zarudnyi schatinensis 431
Sphenopterinae 428
Sphenopterini 428
Spheroceridae 395
Sphingoderus carinatus 454, 466
Sphingonotus elegans 454, 466
S. eurasius eurasius 454, 466
S. halocnemi 454, 457, 466

S. halophilus 454, 457, 466
S. maculatus maculatus 454, 466
S. miramae 454, 466
S. nebulosus discolor 466
S. obscuratus apicalis 454, 458, 466
S. obscuratus transcaspicus 453, 466
S. octofasciatus 454, 466
S. pilosus 454, 466
S. rubescens rubescens 454, 466
S. salinus 454, 457, 466
S. satrapes 454, 457, 466
S. savignyi 454, 457, 466
S. turcmenus 454, 458, 464, 466
Sphodromerus luteipes rubripes 454, 466
Sphyradium doliolum 536, 537
Spinus spinus 263
Spirorhynchus sabulosa 98
Squamata 331
Squamiana squamiptera 452, 458, 464
Stachys turcomanica 153, 157, 162, 163
Stagnicola 540
Stalagmoptera 414
S. hybrida 410
S. ruginota 413
Staphylinidae 394
Steatoda albomaculata 504
S. bipunctata 504
S. castanea 504
S. grossa 393, 399, 504
S. triangulosa 504
Stegodyphus lineatus 503
Stelleropsis antoninae 138
Stellio 318, 320, 322
S. caucasius 322, 326, 331, 339–342, 344–347, 349
S. c. caucasius 311, 315, 331
S. c. triannulatus 311, 315, 322, 331
S. chernovi 311, 322, 326
S. erythrogaster 311, 322, 324, 331, 339–347
S. e. nurgeldievi 332
S. himalayana 326
S. lehmani 311, 322, 326
Steneryx 408

Stenocephalidae 394
Stenonemobius gracilis 455, 465
Stenopelmatidae 452
Sterna hirundo hirundo 260
Sternodes 406
S. caspicus 409
Sternoplax deplanata 409
S. echinata 410
Stipa 20, 63, 117, 133, 145, 146, 164, 436, 461
S. arabica 154, 164
S. caucasica 156
S. crassiculmis 144
S. hohenackeriana 161, 164
S. kopetdaghensis 144
S. lingua 144
S. turcomanica 158, 162, 163, 164
Stipagrostis 19
S. karelinii 99–101
S. pennata 78, 82, 92–94, 97–101
Stiphropus strandi 515, 521
Stirogaster 394
Stizostedion lucioperca 370, 371
Streptoloma desertorum 92, 95
Streptopelia orientalis meena 260
S. senegalensis 281, 282, 284–286, 291, 292
S. s. ermanni 257, 260
S. turtur 193, 281, 282, 284–291
S. t. arenicola 260
S. t. turtur 260
Streptorhamphus linczevskii 144
Strigopteroides aegyptiacus 423, 448
Strigosella 83, 85, 102, 113
S. africana 87, 88
S. circinnata 88, 89, 92, 95
S. grandiflora 83, 85, 88, 89, 92, 95, 128
Stripiturus ruficeps 255
Strumiger 454, 456
S. desertorum 457
S. d. calcaratum 453, 465
S. d. desertorum 453, 465
S. d. persa 454, 465
Sturnus vulgaris

632 *Index of Taxa*

poltaratskyi 261
S. v. porphyronotus 261
Styphrus 407
Suaeda 60, 426, 430, 431
S. acuminata 137
S. arcuata 87, 88, 137
S. microphylla 128, 137
S. physophora 128
Subprionomitus festucae 469
Suchtelenia acmophylla 156
S. calycina 126
Sus scrofa nigripes 231
*Sylvia communis
   communis* 262
S. c. icterops 262
S. curruca blythi 262
S. c. curruca 262
S. c. halimodendri 262
S. c. jaxartica 248–250, 252, 262
S. hortensis 193
S. mystacea turcmenica 262
S. nana nana 248, 250, 252, 262
Sylviidae 193
Synageles charitonovi 518, 519
S. hilarulus 518, 520
S. ramitus 518, 519
Synaphosus 519, 520
S. karakumensis 524
S. palearcticus 524
S. soyunovi 524
S. turanicus 524
Synema globosum 515
S. plorator 515, 521
Syntomus fuscomaculatus 407
Syrdenus 531
S. debilis 410
S. grayi 409
Syrphidae 469
*Syrphophagus
   aeruginosus* 469
S. aphidivorus 469
S. herbidus 469
S. orientalis 469
Syrrhaptes 255
S. paradoxus 251, 260

Tabanidae 481–494
Tabanus 484, 490, 492
T. accipiter 484, 489
T. ansarii badhysi 484, 489
T. appendiculatus 484, 487, 490
T. atamuradovi 484, 490
T. atropathenicus 484, 490, 491
T. autumnalis (species group)484
*T. autumnalis
   brunnescens* 484
T. beschkenticus 484, 489
T. bifarius (species group)484, 491
T. bovinus (species group)491
T. bromius (species group)484, 490
T. bromius 484
T. b. flavofemoratus 484, 487
T. canipalpis terterjani 484, 490
T. cordiger (species group)484, 485, 487, 489–491
T. cuculus 484, 490
T. filipjevi 484, 485, 486, 490
T. freyi 484, 489
T. fumidus 484, 492
T. golovi 490
T. g. golovi 484
T. g. mediasiaticus 484
T. indrae 484, 490
T. i. indrae 484, 487
T. kinoshitai 490
T. laetetinctus 484
T. leleani 490
T. l. leleani 484
T. l. turkestanicus 484
T. quatuornotatus 484, 487
T. regularis 484, 490
T. sabuletorum (species group)481, 487, 489, 491
T. sabuletorum 484
T. s. gerkei 489
T. semenovi 484
T. sordes 484, 490
T. spectabilis 484
T. stackelbergiellus 489
T. subsabuletorum 484, 489
T. unifasciatus 484, 490
T. zimini 484, 486, 487, 491
Tachyini 412
Tachys atamuradovi 410
Tadarida teniotis 211
Tadorna ferruginea 258
T. tadorna 258
Taenia hydatigena 245
Taeniatherum crinitum 158, 163, 165
Tagona 408
Talanites dunini 513, 519
Talanites fagei 513, 519
"Talanites" 400, 520, 523
"Talanites" aculeatus 513, 520
Tamaricaceae 53, 109, 112, 124, 478
Tamarix 90, 113, 125, 127, 183, 431, 434, 442, 476, 477
T. bungei 137
T. florida 137, 168
T. hispida 91
T. hohenackeri 137
T. meyeri 168
T. passerinoides 128
T. ramosissima 94, 137
Tanacetopsis kjurendaghi 122, 127
T. paropamisica 122
T. platyrachys 122
Taphoxenus 412
T. gracilis 409
T. humeralis 412
T. psammophylis 394
Taraxacum lipskyi 144
T. muricatum 136, 138
T. officinalis 159, 162
Tarentula albofasciata 508, 521
T. bergsoei 508, 519
T. cronebergi 508
T. cursor cursorioides 508, 519
T. raddei 508
Tarsostinus 417
Tartarogryllus tartarus 454, 465
Taxodiaceae 51, 54, 124
Tegenaria domestica 508
Telephium 114
Telescopus rhynopoma 226, 312, 338, 341, 348
Tenebrio molitor 406
Tenebrionidae 199, 394, 403–406, 408–411, 413, 414
Tenellia adspersa 540
Tentyria 414
T. robustoides 413
Tenuidactylus 314, 322
T. caspius 314, 317, 332, 340––342, 345, 347
T. c. caspius 310

T. fedtschenkoi 310, 314, 318, 320, 322
T. longipes 223, 314, 322
T. l. microlepis 223, 310, 324
T. turcmenicus 223, 348, 308, 310, 314, 318, 322, 324
Teratoscincus 322, 324
T. scincus 317, 324
T. s. scincus 310
Tergipedidae 540
Tessellana vittata 455, 464
Testudines 330
Testudinidae 331, 342
Testudo horsfieldi, see *Agrionemys horsfieldi*
Tetracme quadricornis 83, 87, 89
T. recurvata 92, 95
Tetracnemus diversicornis 468
T. peliococci 468
T. phragmitis 469
Tetradiclidaceae 110
Tetragnatha extensa 506
Tetragnathidae 506
Tetragonoderus 412
T. intermedius 415
Tetralophisca dimorpha 468
Tetraogallus caspius 220
T. c. caspius 220
Tetrapogon 114
Tetrigidae 452
Tetrigoidea 452
Tetrix bolivari 455, 465
T. depressa 455, 465
T. tartara tartara 455, 465
Tettigonia caudata 455, 464
T. viridissima 455, 464
Tettigonidae 452
Tettigonoidea 452
Teucrium polium 155, 158
Thalictrum isopyroides 156
T. minus 156
Thanatus formicinus 514
T. imbecillus 514, 520
T. vulgaris 514
Thelycrania meyeri 161, 167, 171, 177, 183, 185
Theodoxus 540
T. gaillardi 543
T. jukovi 542
T. nalivkini 542
T. pallasi 540, 542
T. schultzi 542
Therevidae 395, 396

Theridiidae 393, 504
Theridion pictum 505
T. simile 505
T. sisyphium 505
T. varians 393, 505
Therioplectes 485, 491
T. carabaghensis orientalis 485, 491
Thesium 60
T. arvense 156
Theuma walteri 513, 520
Thinorycter 408
Thisoicetrinus pterostichus 454, 457, 466
Thlaspi perfoliatum 156, 162
Thomisidae 499, 501, 514
Thomisus onustus 515
Thrinchus 454
T. desertus 453, 457, 465
T. tuberculosus 453, 465
T. turcmenus 453, 457, 465
Thyene imperialis 518
Thymeleaceae 109
Thymus transcaspicus 159, 161, 164–166
Thynorycter chlamidatus 394
Thysanura 392, 394
Tibellus oblongus 514
Tineidae 395
Tipulidae 191, 193
Titanoeca 399
T. albomaculata 510, 521
T. lehtineni 510, 519, 523
T. tristis 510, 520
T. veteranica 510, 521
Titanoecidae 510
Tobiasia 475
T. bifasciata 472
Tordylium 114
Tortricidae 472, 473
Tortula desertorum 68, 96
Torymidae 395
Tournefortia 464
T. sibirica 93
Trachelas charitonovi, see *Orthobula charitonovi*
Trachelocamptus asiaticus 393, 399, 506, 520
Trachomitum 431
Trachycaspia 542
Trachyderma 408
T. christophi 414
Trachyinae 438
Trachyini 438

Trachypteris cuspidata 434, 441
T. picta picta 434, 442
Trachys phlyctaenoides 422, 438, 443
Trachyzelotes jaxartensis 513
Tragacantha 59, 156, 159, 166, 181, 423, 425, 426, 428, 432, 443
T. marshalli 181
Tragopogon 156, 159
T. capitatus 161
Tragus 114
Trapelas sanguinolentus 192, 193, 311, 313–315, 319–321, 325, 328, 332, 340–343, 345, 347
T. s. aralensis 315, 332
Trechnites flavipes 470
T. fuscitarsis 470
T. psyllae 470
T. trjapitzini 470
Trechus quadristriatus 412
Tribolium 406
Trichanthemis 60
Trichocellus arnoldii 412
Trichocratomerus 433
Trichoncoides piscator 506, 521
Trichostrongylis axei 245
T. probolurus 245
T. vitrinus 245
Tridactylidae 452
Tridactyloidea 452
Tridactylus tartarus 455, 457, 465
Trigonella 113
T. orthoceras 156, 159
Trigonocnemis 408
Trigonoscelis 408
T. apicalis 414
T. borosi 410
T. grandis 409
T. muricata 410
Tringa glareola 259
T. nebularia 259
T. ochropus 259
T. stagnatilis 260
T. totanus totanus 260
Trionymus multivorus 476
Tripleurospermum disciforme 156
Trisetum flavescens 162, 164, 165

Index of Taxa

Trjapitzinellus arboricola 472
T. lambeiensis 472
Trochalostema 410
Trochosa ruricola 508
T. terricola 508
Tropeopeltis 432
Tropidopola turanica 455, 457, 465
Truncatellina callicratis 536, 537
Truxalis eximia 455, 466
Trypaniedae 59
Tulipa 112, 143
T. botschantzevae 144
T. hoogiana 144, 159
T. micheliana 153
T. sogdiana 82, 85, 92, 94, 96
Turaniphytum kopetdaghense 134
Turanochrysops 481, 483, 489, 491
Turanogryllus lateralis 453, 465
Turanophyllum 114
Turdus atrogularis 263
T. pilaris 263
T. ruficollis 263
T. viscivorus bonapartei 263
Turgenia latifolia 156
Turkmenamnicola 538
T. smaragdovae 538
Turkmenigena 410
Turkmenohelops 406
T. balchanicus 410
Turricaspia 541
T. abichi 542
T. andrusovi 542
T. boltovskoji 541
T. brunnea 542
T. caspia 542
T. cincta 543
T. conus 541
T. c. conus 542
T. derbentina 541
T. d. derbentina 542
T. dubia 543
T. elegantula 543
T. eucalia 542
T. eulimellula 543
T. fedorovi 543
T. grimmi 543
T. kolesnikoviana 542
T. kowalewskii 543
T. lirata brusinai 543

T. l. lirata 543
T. marginata 542
T. meneghiniana 542
T. nossovi 543
T. orthii 543
T. pseudobakuana 542
T. pulla 542
T. pullula 542
T. trivialis 542
T. turkmenica 543
T. turricula 543
Typha 137
Typhaceae 111
Typhlopidae 338
Typhlops vermicularis 312, 318, 320, 338, 339, 341, 343, 344, 347, 349

Ulmaceae 109
Ulmus 125, 126, 424, 433, 434
U. carpinifolia 167, 171, 183
Uloboridae 504
Uloborus plumipes 504
U. walckenaerius 504
Ulotricha 123
Umbelliferae 181
Unionidae 539
Upupa epops 193
U. e. epops 261
Uroctea limbata 504
Urospermum 114
Ursus arctos 206
U. a. syriacus 206
Urticaceae 109
Uvarovium desertum 453, 465

Vachoniolus 532
Vaejovidae 533
Valerianaceae 110
Valerianella 113
V. capitata 159
V. coronata 156, 159
V. dufresnia 156
V. platycarpa 156
V. turkestanica 162
Valvatamnicola 538
V. prasina 538
Vanellochettusia leucura 259
Vanellus vanellus 259
Varanidae 334
Varanus griseus 192, 193, 222, 319, 320, 321, 325, 326, 334, 340, 341, 343, 345–347

V. g. caspius 311
Varicorhinus capoëta heratensis 368, 374
Varthemia 113
Velezia rigida 159
Vencetoxicum pumilum 156
Verbascum 59, 113
V. songoricum 159, 160
Verbenaceae 110
Veronica 112, 113, 141
V. arvensis 156
V. gaubae 135
V. kopetdaghensis 135
Vespa orientalis 394
Vespidae 394
Vexibia pachycarpa 137
Vicia 113
V. angustifolia 154, 157
V. venulosa 145
Viola karakalensis 156
V. occulta 156
V. suavis 162
Violaceae 109
Vipera kaznakovi 349, 350
V. lebetina 192, 319, 320, 325, 326, 339–341, 343, 345, 347–349
V. ursini 349
V. xanthina 349
Viperidae 339
Vitaceae 110
Vitis 126, 437
V. sylvestris 127, 143, 165, 171, 183
V. vinifera 165
Vormela peregusna 211, 321, 390
V. p. koshevnikovi 211
Vulpes corsac 321
V. vulpes 193, 321, 390
Vulpia myuros 156, 159
V. persica 156

Wadicosa commoventa 508, 520
Walckenaeria 522
W. monoceros 506, 520
Weisea 408
Willowsia samarkandica 393

Xantheremia 424, 442
X. koenigi 424
X. steinbergi 424
X. subscalaris 424

Xanthium spinosum 162
Xenaphycus flavovarius 470
X. vigil 470
Xenopsylla conformis 395, 396, 399
X. hirtipes 395
Xerencyrtus compactus 473
Xerolycosa brunneopicta 509, 521
Xeropicta candacharica 536, 537
X. krynickii 536, 537
Xylotrechus grumi 415
X. namanganensis 415
Xysticus acerbus 515
X. caperatus 515, 521
X. caspicus 515, 520
X. concinnus 515, 520
X. cribratus, see *Oxyptila baudueri cribrata*
X. kaznakovi 516, 519
X. kochi 516, 521
X. lapidarius 516, 519
X. marmoratus 516, 521
X. minor 516, 519
X. ninnii 516, 521
X. turanicus, see *Psammitis turanicus*

Yllenus albocinctus 518

Y. auspex 518, 521
Y. bajan 518, 521
Y. flavociliatus, see *Y. hamifer*
Y. hamifer 518, 521
Y. mongolicus 518, 521
Y. univittatus 518
Y. validus 518, 519
Y. vittatus 518, 521
Yponomeutidae 473

Zabrus morio 412
Zannicheliaceae 111
Zelkovia carpinifolia 127
Zelotes 522
Z. aerosus 513, 520
Z. arnoldii 513, 520
Z. bucharensis 513, 520
Z. caucasius 513, 521
Z. longipes 513
Z. praeficus 513
Z. pumilus 513, 521
Z. pusillus 513
Z. subterraneus 514
Zilla diodia 507, 520
Ziziphora capitata 155
Z. clinopodioides 159, 161, 166
Z. tenuior 153, 157
Zizyphus 179
Z. jujuba 53, 124, 164, 165, 179
Zodariidae 393, 511
Zodarion 522
Z. denisi, see *Z. raddei*
Z. proszynskii 511, 519, 522, 523
Z. raddei 393, 395, 399, 400, 511, 519, 522
Z. sytchevskayae 511, 520, 522, 523
Z. vlasovi, see *Z. raddei*
Zophosis 414
Z. punctata 409, 414
Zora 522
Z. nemoralis 514, 520
Z. silvestris 514, 520
Zoridae 514
Zosima 425, 426
Z. absinthifolia 155, 161
Zuphium testaceum 409
Zygiella caspica 506, 520
Zygophyllaceae 53, 110, 478
Zygophyllum 60, 113, 125, 424–426, 443, 461
Z. atriplicoides 131, 133, 153, 163, 179
Z. eichwaldii 128
Z. jaxarticum 128
Z. macrophyllum 126
Z. turcomanicum 128

Index of Subjects

Adamulen 529
Adzhidere 127, 526
Adzhikuyu 107
Aegean Sea 233
Afghanian (range) 419, 421, 422, 443, 446-449
Afghanistan 1, 5, 12, 106, 113, 120, 174, 202,207, 213, 215, 223, 229, 233, 303, 309, 322,444, 445, 452, 460, 499, 501, 523, 526-530, 532, 539
Afghano-Turkestan province 456
Africa 454, 460, 492
Afrotropical (range) 502
Agama 193
Agamids 60
Aidere 127, 164, 166, 167, 177, 183, 184, 206, 210, 225, 228, 502, 505, 522, 526, 529, 530
Aidere Area, of Syunt-Khasardagh Reserve 170
Aidere Refuge 228
Aidin 24
Air humidity 36, 37
Air temperature 30-32
Ak 462
Akarcheshme 40, 53, 106, 528
Akchadarya 364
Akchagylian 57, 59, 126
Akchagylian Sea 49, 57
Akchakaya 6, 279
Akdarya 18
Akdepe 526, 529
Akhal-Teke 503, 507, 513, 526, 529, 530

Akhchakuima 116, 464, 526, 528, 529, 530
Akibai 221
Akkyr 9
Akmolla 24, 26, 33
Akoba 107, 118
Aksuv 462
Aktam 10
Aktash 461, 462
Aktash Well 60
Aladag 107, 134
Alai Valley 325
Alashan Sand Desert, center of origin 309, 310
Albian 51, 107
Alder 127
Algae 81
Alluvial fan deposits 7
Alma 234
Almadzhik 137
Almond 165, 176, 177, 179
Alps 124
Altai 422
Altitudinal belts, in Southwest Kopetdagh 150-151
Altybai 117
Amu-Bokhara Canal 304
Amudarya Reserve 3, 208, 216, 218, 219, 227, 228, 279
Amudarya River 5, 10-15, 18-19, 25, 26, 34-38, 43, 45, 49, 51, 61, 62, 82, 87, 95, 97, 99, 208, 210, 212, 214-217, 219, 221, 223-226, 229, 255, 256, 273, 274 , 298, 300, 301, 303, 305, 312, 318, 319, 324, 325, 351, 359, 362-364, 365-375, 379, 381, 383, 384,

386, 414, 415, 456, 457, 459, 489, 495, 498, 507, 508, 524, 530, 535, 538-540
Amudarya Tugai Area 325
Amur River 371
Anatolia 126, 200, 532
Anau 526, 528, 529
Anau District 530, 531
Ancient Mediterranean (range) 473, 477, 478, 503-505, 508, 511-513, 516-518, 522, 524
Ancient Mediterranean 60, 105, 112, 113, 115, 125, 129, 131, 141, 149, 199, 203, 313, 403, 454, 478, 483, 532, 497
Annakara 167, 183
Ant-lions 495-498
Antilles 522
Ants 192
Aphids 192
Apple 2, 165, 183
Apsheron Peninsula 56, 302, 309
Apsheronian 107
Apsheronian Sea 58
Aptian 51, 107
Arabia 492
Aral Sea 58, 62, 257, 298, 299, 304, 312, 365, 366, 371--374, 378, 536, 540, 541
Aral-Sarykamysh Depression 303
Aralo-Caspian Lowland 403, 404
Aralo-Caspian Sea 405
Ararat 233
Archinyan 460

Archinyansu River 14
Archman 24
Archman 184, 225, 511, 514, 526, 528, 530
Aridization 127, 182
Arkhar, see Wild sheep
Armansaad 462
Armenia 233, 416, 490, 526, 528
Armenian (range) 133
Armenian Plateau 106
Armenian province 201
Armeno-Iranian Plateau 521
Armeno-Iranian subregion 115
Arpagan 79, 101
Arpaklen 226, 235, 335, 526
Artificial water bodies, of Turkmenistan 365–388
Artlan 126
Artyk River 14, 213
Arvaz River 14, 166, 208, 337, 422
Asassin flies 191
Aselma, Mt. 115, 132, 146, 234
Ash 167, 183
Ashgabat, see Ashkhabad
Ashkhabad 2, 26, 33, 36, 46, 44, 48, 66, 108, 215, 224, 266, 279, 292, 317, 318, 349, 366, 421, 448, 460, 462, 463, 502–517, 524, 526, 528–530
Ashkhabad Region 526–531
Asia Minor 27, 187, 233
Asian Mountainous (range) 307, 323, 326
Assake-Audan Depression 62, 94
Astrakhan Region 526
Ata 24
Atbasarka 462, 463
Atmospheric circulation 26–30
Atmospheric events 38
Atrek River 8, 13, 15, 19, 90, 130, 144, 183, 202, 208, 212–216, 218–220, 222, 224, 257, 325, 330, 527
Atrek Tugai Area 325
Atropatene 106
Atropatenian province 201
Australia 255, 532, 533
Azerbaijan 199, 210, 233, 490, 526, 528
Azov Sea 541, 542

Babadurmaz 235, 511
Babazo 208, 234, 335
Badaitugai Reserve 279
Badger 321
Badghyz 5, 6, 12, 14, 20–21, 37, 39, 40–42, 45, 46, 51–53, 57, 60, 62, 63, 106–108, 116, 123, 124, 139, 144, 174–176, 181, 182, 207, 210–220, 222–225, 234, 235, 266, 308, 309, 317, 318, 331, 347, 348, 405, 407, 410, 416, 417, 419–421, 440, 441, 443, 444, 456, 458, 460–464, 485, 495, 501–519, 523, 524
Badghyz Reserve 3, 116, 187––196, 208, 210, 211, 213, 219, 221–229, 279, 327, 526, 528, 529, 531
Badghyz, the Eocene flora of 52–54
Badghyz, volcanic activity in Paleogene 52
Badghyz-Karabil Area 325
Bagir 510, 514, 515, 529
Baikal, Lake 541
Bairam-Ali (=Old Merv) 24, 33, 35, 36, 39, 44, 46–48, 221, 458, 512, 515, 517, 518, 526, 528–530
Bairamali Spring 120
Bairs (hills) 12
Baja California 533
Bakharden 24, 36, 212, 461, 502, 507, 516, 517, 528, 529
Bakharden Cave 209, 211, 225, 502
Bakharden District 206, 462, 506, 515
Bakhardok 24, 36, 220
Baku 61
Baku Age 536
Baku transgression 61
Balakhanian 49, 56, 57
Balkan Peninsula 532
Balkans 349, 521
Balkh 303
Balkhan Bay 8, 10
Balkhan Corridor 10, 59
Balkhash, Lake 295, 312, 539
Balkhash-Tarim (range) 535

Baloban, see Saker falcon
Baluchistan 201, 233, 460, 532
Bami 221, 312, 313, 322, 526
Bami River 335
Barley 2
Barremian 51, 107
Barsovoe 234
Bashchenko Spring 531
Bashmygur 526
Bats 209, 211, 213
Bazardepe 462
Bear 60, 206
Bearded goat 210, 231–246
Bearded vulture, see Lammergeier
Begarslan 9, 235
Begi-Arslan, see Begarslan
Bekdash 24
Bekibent 24, 33, 36, 218
Belize 3
Beltau Plateau 83
Berkut, see Golden eagle
Berzengi 225, 517, 524, 526
Betpakdlala Desert 88
Bikrova 526
Biological stations, in Turkmenistan 2, 3
Biomass 189–192
Birds, of the Karakum Desert 247–264
Birleshik-3 526
Biyurgun 78, 79, 84, 86–88
Black bear, see Bear
"Black forest" 132
Black Sea 55, 57, 303, 312, 491, 541, 542
Black stork 218
Blackberry 167, 183
Bogara (non-irrigated cultivation) 21
Bokhara1
Bokhara deer 208, 228
Bokhara goat, see Markhor
Boktybulak Well 352, 354, 356, 362, 368
Bolshie Karanki 224, 225, 333
Bolshoi Balkhan 5, 8–10, 49, 50, 52, 55, 56, 62, 105–107, 114, 126, 127, 144, 149, 166, 176–178, 207, 210, 213, 214, 216, 222, 234–236, 313, 318, 407, 409, 419, 420, 440,

442–446, 463, 509, 526, 528, 535, 537, 539
Bordzhok 79, 97, 100
Bot fly 244
Boyadagh 8, 59, 106
Boyalych 79
British Empire 1
Broad-leaf forests 60
Brontotherium fauna 54
Bronze Age 176, 183
Bugdaily 24, 218, 502
Bukan Well 352, 354
Burdalyk 24
Burkhliburun 526
Burrows, rodent 389–402
Butentau 88, 221
Byukabaicheshme 225

California 25
Camels 60
Canary Islands 512
Caracal (sand lynx) 212, 213
Carpathians 124
Cartographic methods, in zoogeography 265–280
Caspian (West Turkmen) Lowland 6, 8, 10, 13, 15, 55, 56, 61
Caspian Depression 55
Caspian Lake, see Caspian Sea
Caspian Sea 1, 6, 8, 10, 18, 26, 27, 30, 31, 34–38, 40, 43, 55, 61, 62, 68, 73, 100, 106, 108, 117, 122, 124–128, 202, 203, 209, 210, 214–219, 225, 233, 257, 302, 303, 309, 312, 313, 317, 324, 349, 364, 371, 374, 407, 416, 456, 497, 498, 500, 535, 536, 540–543
Caucasian-Anatolian (range) 177, 307, 329, 348
Caucasian-Iranian (range) 508, 509, 512, 519
Caucasian-Mongolian (range) 518
Caucasian-Turkestanian (range) 514–516
Caucasus 1, 54, 124, 126, 185, 203, 233, 295, 296, 299, 303, 309, 312, 313, 315, 349, 404, 490, 500, 507, 511, 520, 522, 525, 527
Cenomanian 51, 107
Cenozoic 106, 403, 404

Central Asia 23, 50, 54, 59, 62, 124, 178, 295, 308, 309, 404–406, 490–492, 532, 533, 537
Central Asian superprovince 539
Central Europe 416
Central Iranian province 202
Chaacha 234, 462
Chaacha River, see Chaachachai River
Chaachachai River 14
Chaata 130
Chafers 191
Chagyl 24, 33, 36, 46
Chaili 206
Chakankala 127, 210
Chalayuk 219, 220
Chaldzha 107
Chalsu 127
Chamchakly 250
Chandyr River 13, 127, 206, 208, 209, 211, 220, 224, 228, 330, 335, 339, 462, 487, 526, 529
Chandyr-Sumbar Mountains 526
Char-Charagasy 526, 528
Chardzhou 12, 24, 26, 33, 36, 44, 47, 48, 66, 225, 226, 229, 298, 300, 319, 351, 354, 356, 364, 490, 498, 507, 524, 530
Chardzhou District 225
Chardzhou Region 526, 528–531
Chardzhou–Ashkhabad Railroad 12, 14
Charshanga 24, 36, 227
Chashdepe 225
Chashkent 24
Chaskak, see Cheskak 24
Chat 24, 462
Chechen' Island 541
Chechnya 528, 529
Cheetah 2, 60, 206, 207
Cheleken 24, 44, 46–48, 106, 115, 528
Cheloyuk, see Chalayuk
Chelyungkyr 9, 456
Chemenibid, see Cheminibit
Chemenebit, see Cheminibit
Cheminibit 528, 221, 223
Cheminibit Refuge 228
Chengurek, Mts. 207

Cherkez 79, 97, 99
Cherkezli 106
Cherries 165, 170, 171
Cherry 179, 180, 183
Cheshme 24, 36
Cheskak, Lake 485
Chetli 107
Chewing lice 244
Chikishlyar 215, 217
Childukhtor 221
Chilikaman 462
Chilmamed, see Chilmamedkum
Chilmamedkum 5, 9, 100, 106, 251, 266, 504
China 54, 371, 372, 404, 416, 489, 512, 526–528, 532
Chinar, see Plane
Chinks (escarpments) 10
Chirishli 512, 524, 526, 528–530
Chokalaks 90–91
Chokhagach 117
Chopandagh, Mt. 6, 114, 134, 146, 225
Chopirchinar 456
Chortobionts 191
Christ's thorn 164, 165, 170–172, 179
Chuli 106, 136, 337, 526, 529
Circum-Euxinian (range) 313, 348
Cis-Aral (Priaralye) area 51, 52, 212
Cis-Aral province 539, 540
Cis-Caspian (Prikaspiiskaya) Lowland 5, 56, 456
Cis-Ustyurt area 15
Climate, of Turkmenistan 23–48
Climatic rhythms 46–48
Climatic seasons, in Turkmenistan 39–46
Cloudiness 34
Cobra 192, 222, 227, 320, 338
Common Syrt 56
Cosmopolitan (range) 473
Costa Rica 3
Cotton 3, 20
Crab 539
Crane flies 191
Cream-colored courser 220
Cretaceous 9, 12, 49, 51, 107,

Index of Subjects

177, 488, 489
Crimea 349
Curly pelican 218
Cushion plants
  (tragacanthoids) 166, 180–182

Dagada 106
Dagdanli 117
Daghestan 233, 528, 529
Dagish 335
Daina 235
Danagermab 207, 223
Danata 107, 108, 464, 526
Danian 51
Danisher, Lake 223, 354
Danisher-Kala 24, 36
Danishor, Lake, see Danisher, Lake
Dardzha 115, 123, 212
Dardzha Peninsula 100, 528
Darganata 24, 221, 298, 354, 459, 526
Darganata District 208
Darvaza 24, 36, 40, 211, 352
Daryalyk 18, 217
Dashkak, Lake 485
Date palm 20
Daudan 18
Dauletabad 220
Davali 24
Dead Sea 533
Deer 231
Deforestation 3
Deicha 462
Deinau 11, 24, 298, 300, 352, 354, 459, 518
Deinau District 208
Deindere 167
Delili, Lake 526
Demoiselle crane 215
Dendrobionts 191, 192
Dengizkul, Lake 214
Desert birds, kidney structure 281–294
Desert hare (tolai) 213
Desert Mediterranean (range) 473, 477
Desert sparrow 220
Desertification 3, 65–76
Desertification, in Central Kopetdagh 145–147
Desertification, maps of 73
Desertification, scale of 67, 68
Desertification, technogenic 68
Deserts, clay 15, 16
Deserts, gypsum 60
Deserts, loess 15, 16
Deserts, salt 15, 16, 441
Deserts, sand 15, 16
Deserts, stony (gypsum) 15, 16
Deserts, vegetation 77–104
Digermendzhik 234
Dortkuyu 528
Droughts 43
Duedzhidagh 235
Durun 515
Dushak, Mt. 14, 24, 126, 136, 207, 210, 224, 344, 462, 463, 512, 526
Dushakerekdagh 134, 136, 146, 147, 234–235
Dust storms 38
Duyeboyun 223, 354
Dzhebel 24, 33, 526, 528–530
Dzheiran (goiter gazelle) 2, 193, 208, 231
Dzhek 217
Dzhungar Desert 303
Dzhuzgun 79, 97–99

Eagle, imperial 219
Eagle, short-toed (serpent) 219
Eagle, steppe 219
Eagles 321
East Africa 416
East Asia 490
East Europe 416
East Iran-Turanian (range) 419, 422, 440, 442–448
East Kazakhstan Region 505, 507
East Mediterranean (range) 129, 141, 177, 188, 198, 411, 412, 444, 504, 509, 511, 512, 515, 517, 518, 520
East Mediterranean superprovince 403, 404
East Mediterranean-Central Asian (range) 493
East Mediterranean-Turanian (range) 493
East Palearctic (range) 307, 326, 329, 348
Ebelek 79, 102
Ecological regions, of Turkmenistan 17–21
Ecotourism 3
Edikhauz 88
Egrigek River 220
Eishem 107, 526
Ekechinar 117
Elburz 55, 56, 144, 177, 180, 185, 202, 224, 303, 313, 349, 411, 522
Eldzhik 229
Elephant 61
Elm 167, 183
Endangered species, animals 3
Endangered species, plants 3
Endemism, in Kopetdagh flora 121–124
England 54
Environmental problems in Turkmenistan 3
Eocene 49, 52, 54, 105, 401, 522
Ephemerous deserts 163
Erne, see White-tailed sea eagle
Eroilanduz, see Yeroyulanduz
Esenbai stratum 60
Eshek-Ankren-Kyr 97
Ethiopian (range) 454, 492
Eureshdagh 115, 120, 126
Europe 54, 417, 490, 492
European (range) 329, 348, 499, 505, 506, 511–515, 518, 520, 521
European-Caucasian (range) 499, 507, 509, 510, 521
European-Central Asian subregion 538
European-Mediterranean (range) 198, 307, 326, 412, 499, 503–517, 521
European-Mediterranean-Siberian (range) 505, 510
European-Siberian (range) 451, 455, 504, 505, 513, 516–518
European-West Siberian (range) 493
Evaporation 37
Ezetdagh 115

Falcons 321
Fall, in Turkmenistan 45-46
Far East 54, 317, 365, 371, 372, 376, 377, 387, 532
Farab 489, 502, 505, 506, 512
Farab District 225
Ferghana Valley 54, 325
Ferns 50, 108, 167
Fig 20, 164, 174, 179
Fires 195
Firyuza 24, 46, 106, 130, 134--136, 208, 210, 220, 224-226, 234, 338, 344, 515, 524, 526, 528
Firyuza River 14
Fistashkovy (Pistachio) Range 218
Flamingo 216
Flora, elementary 130
Florida 489
Fogs 38
Fox 193, 243, 321
France 54
Francolin 220

Gadameli 234
Gangalykyr 526
Gasan-Kuli 24, 26, 36, 43, 215-217, 220, 504, 507, 512, 516, 518, 542
Gasan-Kuli District 445
Gatygyzyl 279
Gaudan 24, 46, 106, 207, 220, 225, 515, 526, 528, 529
Gaudan Pass 132
Gaurdak 24
Gaurdak-Kugitang area 12, 50-52, 58, 62
Gaurdak-Kugitang region 12
Gazelle, see Dzheiran
Gazelles 60
Geabyuldag 134
Geldere 229
Genghis Khan 1
Geok-Tepe 24, 209, 211, 448, 461, 462, 515, 530
Geok-Tepe District 206
Georgia 301, 522, 526, 528, 533
Gerat Valley 14
Gerbils 193, 389-402
Germab 24, 46, 210, 336, 33, 526, 529
Germab Depression 136

Ghilan 202
Ghissar 12, 176, 180, 455
Ghissaro-Alai 325
Ghissaro-Alai District 325
Ghissaro-Darvaz 403, 404, 417, 535
Ghissaro-Tien-Shan province 455
Ghissaro-Turkestan Foothill Area 325
Ghissaro-Turkestan Highland Area 326
Ghissaro-Turkestan Juniper Area 326
Gindere, Mt. 164
Giraffe 60
Glazed frost 38
Gobi Desert 309
Goiter gazelle, see Dzheiran
Golden eagle (berkut) 216
Gorgan River 304
Goshaguyi Well 460
Grapes 2, 165, 183
Grazing 66, 195
Great black-headed gull 215
Great bustard 217
"Great Game" 1
Greater Caucasus 49
Greece 417
Greenland Province, paleofloristic 53
Grey monitor 222
Ground squirrel 390
Guen 206
Gyaurs 130, 209, 224, 279, 462, 512, 528, 529
Gyaurs District 504
Gyaurs Valley 305
Gyaursdagh 130, 234
Gyazgyadyk 106, 144, 207, 210, 223, 225, 234, 235, 526
Gyuen 127
Gyuni 234

Hail 38
Halophytes 84-89
Hammada 60
Hare, see Desert hare
Harirud River, see Tedzhen River
Hauterivian 107
Hawaii 513
Hawthorn 165, 176, 177, 180
Hedgehog 321, 390

Heel fly 244
Helminths 244, 245
Hesperian-Sethian (range) 419, 422, 440, 442-448
Hibernation, in reptiles 345
Himalayas 404, 509, 512, 522
Hindu Kush 106, 129, 188, 312, 417, 421
Hipparion 60
Hodzha-Kainar Spring 538
Holarctic (range) 113, 473, 477, 503-507, 512, 514
Holocene 49, 63, 123, 127, 128, 307, 402, 536, 540
Honey badger 212, 321
Honeysuckle 167, 179, 180, 183
Horse 60
Horse flies 481-494
Horsetail 167
Humus 190
Huns 1
Hyena 2, 193, 210, 243, 321
Hyrcania 125, 127, 202, 203
Hyrcanian (range) 197, 199, 200

Ilchik 24
Ili River 312
Imam-Baba 221, 506-508, 515, 516
Imarat 526
India 303, 460, 513, 523, 513, 529, 530, 532
Indian (range) 307, 326, 329, 348, 454
Indostan 23, 416
Indricotherium fauna 54
Introduction, of fish 365-388
Inzhirli 174, 462
Iolatan, see Iolotan
Ioldere, see Yoldere
Iolotan 24, 33, 35, 221, 464, 508, 516, 526, 528-530
Ipaikala 235, 462, 526,
Iran 7, 13, 14, 55, 59, 63, 106, 113, 115, 122, 123, 129, 139, 149, 174, 176, 198, 200, 201, 206-208, 213, 214, 224, 228, 229, 233, 303, 309, 312, 313, 316, 322, 330, 406, 416, 417, 452, 458, 460, 490-492, 499, 501, 503, 519, 523,

525–532, 537, 538
Iranian (range) 105, 129, 133, 422, 443, 446–449, 451, 454, 499, 501–506. 507–519, 521
Iranian Mountainous District 323
Iranian Plateau 55, 182, 198, 303, 454, 455
Iranian-Turkestanian (range) 499, 502, 503, 505–507, 510, 512–519, 521
Irano-Afghanian (Mountain Steppe) center of origin 323
Irano-Afghanian (range) 307, 314, 326, 329, 347, 348
Irano-Afghanian Mountainous Subprovince 323
Irano-Afghanian Transitional Province 323
Irano-Anatolian province 538, 539
Irano-Azerbaijanian province 199
Irano-Turan-Central Asian (range) 414
Irano-Turanian (range) 105, 187, 414, 422, 440–449, 473, 477, 492, 497, 498, 524
Irano-Turanian Lowland 483
Irano-Turanian province 199, 201, 491
Irano-Turanian subprovince 477
Irano-Turanian superprovince 403–406
Irano-Turanian-Gobian (range) 419, 422, 442, 445–448
Iraq 312, 523, 526–530, 532
Irrigation 14, 23, 195, 255–257
Irsarybaba 9
Iskander 7, 226, 526
Islimcheshme 515
Ispas 223
Israel 523, 529, 530
Issyk-Kul 297

Jackal 321
Japan 54, 509
Jasmine 179
Jerboa, see Turkmen jerboa

Jordan 529, 530
Jujube 164, 165, 179
Juniper 1, 20, 127, 128, 143, 166, 177–179, 239, 444
Jurassic 9, 12, 50, 106, 107, 134

Kaakhka 24, 33, 36, 212, 332, 448, 460, 462, 529, 528, 529
Kabakly 229, 459, 524
Kabul 533
Kafigshem 505, 509, 524
Kaifata 106
Kalaimor (=Mor-kala) 213, 218, 507, 511, 513
Kalinin Refuge 213, 228
Kalininsky 462
Kalkhauz 462
Kalymkhoz 166
Kamchatka 510
Kankakyr 512, 526
Kaplankyr 9, 10, 251, 324, 456, 457, 516, 518, 524
Kaplankyr Reserve 3, 207, 213, 227, 228, 279, 327, 526–528
Kaptaringa 332
Kaptarkhana Cave 539
Kara-Bogaz 60
Kara-Bogaz-Gol Bay 8–10, 24, 26, 34, 42, 68, 73, 75, 85, 90, 215–217, 529, 530
Kara-Bogaz sand bank 9
Karabugaz, see Karateniz
Kara-Bulag 235
Karabil 5, 6, 12, 14, 20, 37, 39, 40–42, 45–46, 57, 62, 108, 207, 210, 213, 214, 222, 223, 229, 309, 318, 347, 348, 407, 410, 461, 462
Karadagh 134
Karadzha 127
Karagan 225
Karagyoz 107, 223, 235, 354
Karagysy 235
Kara-Kala 2, 24, 33, 40, 45, 149, 211, 220, 225, 226, 335, 337–339, 462, 463, 526, 528, 529
Kara-Kala District 206
Karakalpakistan 10, 13, 57, 231, 352, 361, 362
Karakorum 302

Karakum Canal 20, 212, 215, 219, 222, 226, 255, 276, 305, 365–377, 379, 381, 382, 384, 386, 456, 457, 461
Karakum Desert 2, 5, 6, 10–12, 14, 15, 18–20, 68, 77–104, 106, 107, 174, 182, 184, 211, 212, 217, 219, 220, 222, 229, 247–264, 266, 277, 281, 303, 309, 312, 324, 389–402, 407–409, 417, 456, 461, 487, 489, 495, 497, 520, 521, 523, 532, 535, 536, 539
Karakum Desert, formation of 49, 51, 52, 62
Karakum District 324, 507, 510
Karakum Plateau 56
Karakum Sand Desert Area 324
Karakum subprovince 324, 456
Karakum Takyr Area 325
Karakum, Central 10, 11, 15, 27, 30–32, 37, 38, 40–42, 45, 59, 98, 103, 210, 211, 223, 224, 250, 251, 317, 352, 457, 463, 528
Karakum, East 250, 251, 317, 497, 505, 532
Karakum, Lowland 10, 19, 25, 34, 46, 62, 94, 99
Karakum, North 457, 459
Karakum, Northwest 251
Karakum, Southeast 10–12, 14, 19, 30–32, 34, 37, 38, 46, 57, 59, 94, 96, 210, 457, 463
Karakum, Trans-Unguz 6, 10, 11, 18, 31, 34, 37, 39, 42, 46, 55, 60, 62, 82, 85, 87, 94, 96, 97, 212, 250
Karakum, West 251
Karakumkanal State Farm 507
Karanki 524
Karashor 9, 10, 15
Karasu 167
Karateken 116, 462
Karateniz 526, 528
Karatogolok 528
Karayalchi 184, 137, 225
Karlyuk 208, 223, 224, 227
Karlyuk Refuge 228
Karst 21

Index of Subjects 643

Karyotype, of toad agama 351–356
Kash River, see Kashan River
Kashan River 14, 211, 212, 229
Kashgarian Mountainous Steppe, center of origin 309
Kashkadarya 304
Kat 235
Kattashor, Lake 366–373, 377, 382, 383
Kaushutbent 221
Kavalerizou 234
Kazakhstan 10, 50, 51, 54, 86, 88, 93, 125, 126, 216, 228, 231, 280, 371, 403, 409, 416, 417, 422, 457, 477, 523, 526–531, 533, 537
Kazakhstan, center of origin 309, 310
Kazakhstan-Mongolian (range) 307, 326
Kazakhstanian province 403
Kazandzhik 7, 24, 36, 46, 106–108, 255, 526
Kazganchai River 14
Kazyklybent 215
Kelat 207
Kelat Cave 209
Kelatachai River 14
Keletchaya 347
Keletkaya 207, 223, 235
Kelif 223, 257, 298, 304, 528
Kelif Refuge 216, 219, 228
Kelif Reservoir 375
Kelif Uzboi 12, 15, 216, 219, 266, 276, 298, 303, 489
Kelkor, see Kyolkor
Keltechinar 130, 210
Kelyata 106, 462, 526
Kelyatkaya 106
Kelykhalyk 206
Kenaf 19
Kepele 189, 526, 528
Kerki 11, 12, 24, 33, 36, 46, 48, 298, 464
Kerlek 106
Kerman 122
Keshi River 14
Ketgen 79, 102
Kevreik 79, 84, 88, 89
Khachmas, Lake 49, 56
Khanbaagykum 100

Khangul, see Bokhara deer
Kharlasang, Mt. 136
Kharvar 130
Khasar, see Khasardagh, Mt. 126
Khasardagh, Mt. 126, 164, 208, 228
Khauzkhan Reservoir 224, 366, 367, 376, 377, 382, 385 Kopetdagh Reservoir 366, 367, 377
Khazaraspa 354
Khazarian 61, 303
Kheirabad 24, 43, 46, 115, 136, 220, 225, 526
Kheles 531
Khen-tau 354
Kherson Region 528, 529
Khiva 1, 10, 14
Khiveabad 126, 332, 336
Khodzhabulan 462
Khodzhaburdzhibelend Refuge 228
Khodzhaeken 127
Khodzhafilata, see Khodzhapil
Khodzhagaskar 209
Khodzhakala 36, 107, 130, 235, 526, 529
Khodzhakala Refuge 228
Khodzhakala Valley 166
Khodzhakul, Lake 354
Khodzhambas District 223
Khodzhapil 209, 225, 526, 528
Khodzheipil, see Khodzhapil
Khorassan (range) 419, 420, 422, 440, 443–449
Khorassan 134, 135, 138, 180, 181, 185, 197
Khorassan province 197, 201, 203
Khorassan-Kopetdagh 105, 113, 114, 197, 323
Khozdere 183
Khozli-ogly-olum 517
Khozly 167, 235
Khudaidagh 61
Khvalynian Sea 62
Khvalynian transgression 61, 62
Khwarazm 1
Khwarazm, Lake 364
Kilyazi 541
Kilyazi Peninsula 55

Kilyazi-Krasnovodsk mountains 49, 56
Kishimiri 235
Kizyl-Arvat 24, 33, 36, 60, 62, 107, 108, 118, 130, 235, 461, 526, 528, 529
Kizyl-Arvat District 504
Kizyl-Atrek 24, 33, 36, 40, 46, 219, 220, 503, 515
Kizyl-Atrek District 218, 463
Kizylcheshme 107
Kizyldagh 211, 234
Kizyldzhar 207, 524, 526, 528, 529
Kizyldzhar Refuge 228
Kizylimam 462
Kizylkaya 526
Kizylkum Desert 19, 52, 62, 88, 91, 93, 97, 100, 222, 247, 251, 312, 317, 351, 362, 363, 489, 520, 532
Kizylkum Reserve 279
Kizyl-Kun 24
Koimat 88
Koine-Kesir 34, 210, 235
Kopetdagh 2, 3, 5–8, 10, 11, 13–18, 20, 27, 30, 38, 39, 42, 45, 46, 49, 51, 56–58, 82, 105–128, 129–148, 197–203, 207–211, 213–217, 222, 224, 309, 312, 313, 318, 323, 325, 329–349, 411–413, 416, 417, 419, 422, 440–444, 446–449, 456, 457, 460–463, 495, 519, 521–523, 528, 533, 535–538
Kopetdagh mountainous province 538, 540
Kopetdagh Oasis 20
Kopetdagh Reserve 3, 116, 143, 178, 231–246, 327, 464, 498
Kopetdagh, Central 105, 108, 114–115, 129–148, 149, 163, 177–179, 181, 183, 206–208, 213, 214, 217, 218, 220, 222–226, 231–246, 299, 312, 329, 333–339, 341, 344, 419–421, 444–447, 490–492, 501–518, 522, 524
Kopetdagh, East 51, 52, 105, 108, 115, 116, 132, 144, 206, 218, 234, 235, 299, 303, 329, 332, 336, 337, 339, 341, 344, 419–421, 447–448, 502–504,

510, 512, 513, 515, 536
Kopetdagh, Northwest 107, 114, 118, 120, 139, 235, 329, 339, 341
Kopetdagh, Southwest 2, 114, 115, 139, 149–172, 174, 176, 183–185, 198, 203, 213, 222, 226, 329, 333–335, 337–339, 341, 342, 464, 501–518, 522, 524, 530
Kopetdagh, West 56, 59, 105, 107, 108, 114, 115, 120, 125, 138, 179, 206, 208, 210, 211, 213, 220, 223–226, 235, 299, 301, 405, 411, 419–421, 440, 444, 446, 447, 462–464, 487, 490, 536
Kopetdagh-Khorassan, see Khorassan-Kopetdagh
Kopetdaghskoe, Lake 526, 528
Korea 532
Kosha 126
Kosha-Seira-Porsukh 126
Koshagoi Spring 526
Koshaseira 235
Koshoba 24
Kovata 526
Krasnovodsk 1, 24, 31, 36, 46, 61, 66, 123, 234, 500–503, 507–510, 512, 514, 515, 518, 524, 526, 528–530
Krasnovodsk Bay 8, 217, 526
Krasnovodsk Mountains 56
Krasnovodsk Peninsula 5, 8, 9, 51, 55, 61, 62, 309
Krasnovodsk Plateau 6, 39, 42, 56, 83, 85, 86, 106, 108, 114, 115, 407, 457
Krasnovodsk Region 507, 524, 526–531
Krasnovodsk Reserve 3, 208, 210–220, 223–228, 327
Krasnovodsk sand bank 9
Krasnoye Znamya 221
Kubadagh 50, 55, 56, 62, 106, 210, 234, 235
Kuchan 13
Kuchan highway 224
Kuen Lun 302, 404
Kugitang 5, 6, 12, 13, 27, 139, 178, 180, 181, 209, 211, 214, 216, 219, 222–225, 227, 229, 318, 325, 326, 347, 407, 410, 417, 444, 457, 458, 460–463, 503, 505, 506, 518, 520, 524, 537, 539
Kugitang Reserve 3, 178, 209, 214, 224, 227, 228
Kugitang-Darya River 12, 13
Kugitangtau, see Kugitang
Kuilyar 107
Kuitan 24
Kulaly Island 541
Kulan, see Onager
Kulandagh 10
Kuldzhuktau 363
Kulmach 107
Kulmachbaba 119
Kumdagh 106, 115, 116
Kumsebshen 15, 94
Kunya-Urgench 24, 33, 40, 221, 321, 327, 504, 528, 531
Kunyadarya River 88, 95, 221
Kura-Araxes-Turanian (range) 473, 477
Kurdistan 122
Kurkulab 136, 208, 225
Kurkulab River 14
Kurtli, Lake 366, 526, 528, 529
Kurtlibil 235
Kurtusu 132
Kurtusu-Gaudan 130, 132–134
Kuruhaudan 130–132, 147, 462
Kurum 503
Kurydere 166
Kushka 24, 33, 36, 44, 48, 174, 221, 321, 460, 508, 514, 518, 526–530
Kushka River 14, 207, 212, 213, 215, 218, 220, 221, 227, 229
Kushkanatau 354
Kutlyubil 126
Kuuli, Cape 541, 542
Kuuli-Mayak 24
Kyolkor 8, 10, 15, 90
Kyr (flat-topped ridge) 11, 18
Kyrghyzstan 54, 316, 457, 490, 507, 509, 511, 523, 526–528
Kyrkoili 223
Kyrs 88
Kyuren 126
Kyurendagh 107, 115, 122, 149, 207, 213, 235, 464
Kyuryanyn-Kyure 106
Kyzyldagh, see Kizyldagh
Kyzylkum, see Kizylkum
Kyzylsu 507

Lainsu River 14
Lammergeier (bearded vulture) 214, 243
Laos 295
Lappi 526
Lebab 351–353, 356, 362, 363
Lekker 24, 33, 36
Lelimkam 461, 462
Lengush 461
Lenin District 460
Leninsk 24, 33
Lenkoran 202
Lenkoran, Lake 49, 56
Leopard 2, 193, 207, 208, 228, 243
Lesser Caucasus 56, 177, 182
Levant 200
Lichens 79, 81, 108, 182
Lithoedaphic types of deserts, in Turkmenistan 15–17
Little bustard 217
Lizards 192
Locust, see Moroccan locust
Logging 1, 195
Ludzha, Mt. 132
Lynx 206, 243

Madau 315, 331, 502, 526, 527
Maestrichtian 51
Mahaleb cherry 183
Makhtumkala 208
Mali 492
Malik 461
Maloe Delili, Lake 212, 216, 219, 224, 341, 342
Maly Balkhan 5, 6, 8, 10, 56, 105–108, 114–116, 120, 125, 176, 177, 207, 210, 216, 222, 234, 235, 313, 318, 419–421, 440, 441, 444, 446, 464
Malye Karanki 225
Manchuria 509
Mandrake 118
Mangyshlak 6, 8, 9, 50–52, 55, 63

Manul 213, 243
Manysh syncline 130
Maple 20, 164, 171, 179, 182, 183, 239
Marbled polecat 211
Marbled teal 214
Mardagh 526
Marghyz 119
Markhor 209, 231
Markou, Mt. 126, 132
Mary (=New Merv) 14, 24, 66, 220, 221, 460, 504, 508, 509, 514, 524, 526, 528, 529
Mary Region 526–530
Mazandaran 202
Meadows 164
Meana 116, 235
Meana River, see Meanachai River
Meana-Chaacha 206, 213, 279
Meana-Chaacha Refuge 213, 228
Meanachai River 14, 332
Mediterranean (range) 177, 198–200, 329, 349, 409, 411, 416, 451, 452, 454, 455, 501--508, 510–516, 491, 492, 533
Mediterranean province 314, 403
Mediterranean Region 23
Mediterranean subtropical subregion 540
Mediterranean-Afrotropical (range) 515
Mediterranean-Asian Arid subregion 322
Mediterranean-Caucasian (range) 183
Mediterranean-Central Asian (range) 493
Mediterranean-Himalayan (range) 504
Mediterranean-Middle Asian (range) 493
Mediterranean-Paleotropical (range) 515, 518
Mediterranean-Southwest Asian-Turanian (range) 493
Mediterranean-Turkestanian (range) 502
Medlar 165, 179
Mekran 55, 201

Melon 19
Mergenchuk 234
Mergenolen, see Mergenulen
Mergenulen 225, 334
Mergenulya 136, 208
Merv, New, see Mary
Merv, Old, see Bairam-Ali
Meshed-Messerian Plain 88, 222, 224, 313, 327, 504, 510, 513, 526, 529
Meshed-Misrian, see Meshed-Messerian
Mesopotamia 55
Mesozoic 10, 106, 136
Messerian Plain, see Meshed-Messerian Plain
Messinian salinity crisis 533
Mexico 513, 533
Middle Asia 5, 50–52, 54, 58, 59, 86, 94, 97, 113, 124, 138, 143, 149, 163, 173–175, 177, 179, 185, 188, 248, 249, 295--306, 365, 371, 403–418, 477, 478, 488–490, 532
Middle Asian cobra, see Cobra
Middle Asian mountain subregion 540
Middle Asian Sea 488, 489
Middle East 421, 509, 520, 522, 523, 532, 537
Middle Eastern province, see Southwest Asian province
Mint 167
Miocene 9, 49, 55, 106, 117, 125, 126, 404, 488, 492, 521
Mirzadagh 137, 146, 225, 234, 235
Missinev 115, 136, 137, 146, 234
Mississippi River 374
Mollakara 526, 528–530
Mollakurban 526
Mongolia 54, 404, 406, 416, 417, 477, 489, 497, 507, 518, 523, 526–528
Mongolian (range) 509
Mongolian Steppe and Forest-Steppe, center of origin 309, 310
Monitor lizard 60, 193
Moths 192
Monzhukly 8, 59, 61, 106, 115, 462

Mor-kala, see Kalaimor
Morgunovka, see Morgunovsky
Morgunovsky (or Morgunovka) 220, 221, 223, 526, 528, 529
Moroccan locust 462
Morocco 532
Morphoecological anomalies, in fish 379–380
Moscow 2, 3
Mosses 81, 108
Mountain sheep. see Wild sheep
Mountainous-Middle Asian (range) 473, 477
Mt. Ararat, see Ararat
Mt. Chopandagh, see Chopandagh
Mt. Dushak, see Dushak
Mt. Khasardagh. see Khasardagh
Mt. Riza, see Riza
Mt. Syunt, see Syunt
Mt. Tagarev, see Tagarev
Mudslide (sel) 7
Mugodzhary 56
Mulberry 165
Murghab cyclone 27
Murghab District 508, 510–512, 516–518
Murghab Imperial Estate 526, 528
Murghab Oasis 20, 37
Murghab River 5, 11, 12, 14, 19, 26, 38, 49, 61, 62, 95, 174, 188, 191, 207, 208, 212, 214–223, 227, 229, 256, 266, 273, 277, 279, 299, 301, 303, 305, 318, 322, 324, 325, 327, 372, 373, 419–421, 440, 441, 448, 456, 457, 460, 461, 489, 490, 507, 508, 518, 527, 536, 538
Muttakary, see Mollakara
Myalik 462

Naivadai 127
Nakhduin 337
Nakhichevan 517
Nanshan 404
Nardevanly 526, 528
Narli 127, 206
Natural reserves

(zapovedniki) 116
Nazarekerem 106
Nebit-Dagh 8, 38, 108, 526, 529
Neftezavodsk 223
Nematodes 191
Neocomian 107
Neogene 49, 54–61, 105, 107, 115, 122, 126, 136, 150, 185, 201, 303, 307, 309, 403, 404, 485, 490
Neolithic 1, 183
Neotropical zoogeographic region 314
Neprokhodimoye 234
Nere 462
New Caspian transgression 61
New Merv, see Mary
New Mexico 25
Nichka 24, 461
Nisa 1, 349, 526
Nitrogen, total 190
Nokhur 462, 526
Nokhurli 183
North Africa 63, 308, 405, 416, 492, 506, 522, 532
North America 374, 532
North Caspian, Lake 55
North European-West Siberian subregion 540
North Ossetia 528
Northern Afghanian District 324
Novaya Nisa 344
Novrekcheshme 461, 462
Novruzabad 220
Nukus 351, 352, 361, 363

Oak 125, 126
Oat 2
Oboi 12, 61, 107
Odessa Region 528, 529
Ogurchinsky Island (and Refuge) 24, 209, 228
Oktum 9
Old Merv, see Bairam-Ali
Oligocene 49, 52, 54, 124–126, 401, 402, 404, 490, 521, 522
Olive 20
Onager (wild ass, kulan) 2, 193, 213, 227
Orchids 167

Oriental zoogeographic region 314
Osmoregulation 286–288
Osprey 218
Ostriches 60
Otter 212
Ovadan, Lake 366, 373, 379
Oxus 536

Pacific Ocean 295, 492
"Painted rocks" (*pyostrotsvety*) 120, 126, 130, 136, 441
Pakistan 54, 201, 229, 460, 523, 526–530
Palearctic 201, 322, 498, 477, 452, 519, 455, 538
Palearctic-Ethiopian (range) 473, 477
Palearctic-Indo-Malayan (range) 473, 477
Paleo-Amudarya River 55, 57
Paleo-Atrek River 56
Paleo-Emba River 56
Paleo-Murghab River 55, 57
Paleo-Tedzhen River 55, 57
Paleo-Ural River 56
Paleo-Uzboi River 56
Paleo-Volga River 56
Paleocene 49, 52
Paleogene 49, 52, 105–107, 115, 122, 130, 131, 136, 189, 201, 483, 485, 491
Paleogeography, of Turkmenistan 49–63
Paleolithic 176
Paleotropical (range) 201, 451, 454, 455, 507, 512, 516, 518
Paleozoic 12
Pallas' sea eagle 214
Palms 51
Pamir 13, 152, 180, 325, 404, 416
Pamiro-Alai 55, 106, 120, 121, 129, 166, 174, 177, 180, 181, 188, 302, 312, 458, 520
Parkhai 225, 345
Paropamiz 10, 12, 55, 122, 188, 303, 407, 414, 421
Parthia 1, 203
Pastures, overgrazing 3
Pear 3, 165, 180

Pedobionts 192
Pelengovali 174, 223
Perednyaya Aziya, see Southwest Asia
Peredovoi 107
Peregrine falcon (sapsan) 217
Pereval 116, 526
Perevalnaya, see Pereval
Pervomaisky 207, 526
Pesticides 3
Petrophytes 169
Pheasant 295–306
Phryganoid vegetation (tomillares) 166
Phytomass 190, 191
Pink pelican 215
Pinkhancheshme 106, 526, 528
Pistachio 1, 174–176, 187–196
Pitnyak 351, 354, 364
Planarians 198
Plane (sycamore) tree 167, 183
Pleistocene 121, 127, 150, 302, 312, 313, 351, 363, 364, 404, 405, 409, 522, 532, 536
Pliocene 55–61, 105, 249, 307, 312, 313, 347, 349, 364, 401, 402, 404, 414, 489, 492, 522, 536
Plum 165, 167
Pobeda Sovkhoz 221
Poltava (paleofloristic province) 53, 124
Poltavka 221
Pomegranate 2, 20, 164, 165, 179
Pontian 57
Pontic Sea 49
Ponto-Caspian region 541
Ponto-Caspio-Aral province 373, 374
Poplar 168, 183
Porcupine 193
Pordere 127, 167
Pordere 183
Porsukh 126
Pre-shiblyak 133
Precipitation 34–36
Priaralye, see Cis-Aral area
Prikaspiiskaya Lowland, see Cis-Caspian Lowland
Productivity, of fish in

Turkmenistan 380–383
Proto-Amudarya River 61, 303, 364
Proto-Atrek River 61
Proto-Caspian Sea 49, 522, 532
Proto-shiblyak 124, 126
Proto-Volga River 312
Proto-Zeravshan 303
Protoshiblyak 53
Psammophytes 19, 84, 98
Pulikhatum 24, 106, 117, 174, 188, 207, 223, 225, 460, 526
Pulikhatum Refuge 228
Pulikhatyn, see Pulikhatum
Purnuar 335
Purple swamphen 219
Pyandzh River 298, 489

Quaternary 20, 55, 61–63, 117, 249, 347
Quince 165, 179

Radiation balance 25
Red Data Book, of the IUCN 116
Red Data Book, of the USSR 143, 167, 327
Red Data Book, of Turkmenistan 143, 205–230, 327, 333–336, 338
Red-breasted goose 218
Red-wattled lapwing 220
Reed 168
Repetek 24, 32, 33, 44, 46, 48, 59, 96, 212, 220, 250, 292, 463, 464, 487, 501–518, 524, 526, 528–531
Repetek Reserve 2, 219, 220, 224, 226–228, 266, 279, 327, 389–402, 497
Repetek Sand Desert Station, see Repetek Reserve
Reptiles, of Kopetdagh 329–350
Reptiles, of Turkmenistan 191–193, 307–328
Research, scientific, in Turkmenistan 2, 3
Reush 235
Rice 19
Rime 38

Riparian forest 167, 168
Riparian forest, mountain 183
Riparian forests, desert, see Tugais
Russian Platform 51
Riza, Mt. (or Rizarash, Mt.) 106, 134
Robergovsky 126
Romania 416
Rose 167
Russia 526–528
Russian Empire 1

Saber-tooth cats 60
Sagebrush 79, 82
Sagebrush-ephemeroid desert 441
Sahara Desert 460, 521
Saharo-Gobian (range) 199–201, 531
Saharo-Gobian Desert Province 477, 478
Saharo-Gobian subregion 455
Saharo-Irano-Turanian (range) 445, 447–449
Saharo-Sindian (range) 307, 326, 329, 508
Saharo-Turanian (range) 407
Saiga 231
Saivan 24, 184, 235, 335, 336, 526
Sakarchaga District 459
Saker falcon (baloban) 219
Saksaul 1, 39, 41
Saksaul, black 78, 79, 94–97
Saksaul, white 79, 91–97
Salinization 73
Saltyrykh 528
Samkum 279
Sand dunes (barkhans) 9–11, 19
Sand sedge 43, 78, 93
Sandagzov 234
Saparbakhar 529, 530
Sapsan, see Peregrine falcon
Sarakhs (in Iran) 530
Sarmatian 55, 107, 125
Sarsazan 79, 89–91
Sary-Chengarak 106
Sarychop 215
Sarykamysh, Lake (Depression) 5, 10, 18, 8,

62, 95, 95, 97, 212, 215–218, 221, 215, 217, 218, 221, 228, 250, 251, 257, 274, 276, 324, 364, 365–373, 375, 377–380, 382, 386, 459, 460, 529, 536
Sarykamysh-Khwarazm Lowland 5, 10
Sarymsakli 117
Saryyazy 24, 221, 518
Saryyazy Reservoir 373
Sarzavu 234
Saudi Arabia 312
Savanna 60
Sawflies 192
Sayat 24
Scaled woodpecker 215
Scorpions 321
Seira 126
Seit-Kerderi 107
Sekidagh 106
Sekizyab 106, 130, 134, 136–138
Selin 79, 100, 101
Semansur 134
Semirechye 297
Semisavanna 163
Semishrub (sagebrush) deserts 152
Senegal 513
Serakhs 14, 24, 33, 460, 498, 501, 502, 513, 517, 526, 528, 529
Serakhs District 220, 502, 513, 528
Serny Zavod 223, 352–354
Serpent eagle, see Eagle, short-toed
Shakhgadam 106
Shakhiburun, Mt. 526
Shakhin falcon 214
Shakhsenem 24, 526, 528–531
Shakhshakh, Mt. 134
Shalcheklen Plateau 228
Shamli 119, 207
Sharlouk 462, 463, 504, 512, 515, 518, 526
Sharzha 462
Sheikharyk 223, 354
Sherlok River 130
Shevlan 117
Shiblyak 53, 63, 124, 127, 164, 175, 179, 180
Shikhindere 225

Shiramkuyi 464
Shor (solonchak) 9, 11, 15, 18
Shorgol 220
Shortepe 213
Shrew 321
Siberia 23, 317, 417, 451, 506
Siberian anticyclone 26
Siberian goat 231
Sibir, Mt. 132
Simpson Desert 255
Sind 201, 233, 521, 522
Sindian-Turanian (range) 507
Sindzhou 107
Sinkiang 416, 516–518
Sino-Indian zoogeographic region 374
Snakes 192
Snow 40
Sociable plover 218
Sogdian (range) 539
Sogdian province 538, 540
Sogdian-Tibetan superprovince 538
Soil deflation 70
Soil erosion 195
Soil, fauna of 191, 192
Soil, swamping of 72
Soil, temperature of 32–34
Solar radiation 25
Solonchak vegetation 60
Solonchak, see Shor
Solpugids 321
Solyukli, see Syulyukli
Songudagh 107, 115, 117, 462
South Africa 416, 513, 532
South Caspian cyclone 27
South Caspian, Lake 55
South Dashtoi, Mt. 132
South Korea 503
South Russian (range) 133
South Tajik Depression 302
South Turanian province 421, 455
Southeast Asia 454
Southwest Asia (Russ. *Perednyaya Aziya*) 403–405, 411, 416, 442, 444, 445, 483, 489–491, 537
Southwest Asian (range) 105, 177, 307, 326, 419–422, 440, 444, 445–449, 452, 454, 492

Southwest Asian province 198, 200, 201, 403
Southwest Asian superprovince 539
Southwest Asian-Transcaucasian-Middle Asian (range) 493
Southwest Asian-Turanian (range) 493
Spain 490
Spiders 191, 192, 203
Sportivnoye, Lake 528–531
Spring frosts 41
Spring, in Turkmenistan 40–42
St. Petersburg 2
Steppe eagle 219
Steppe, mountain 164
Stolovye, Mts. 234
Sudan 460, 492
Sukhaya Balka 234
Sultanbent 221, 508, 510, 518, 529
Sultandagh 279
Sultandepe 528, 529
Sultanuizdagh 50, 351, 354, 363, 364
Sultanuvais 354, 362
Sumbar Refuge 228
Sumbar River 13, 19, 45, 120, 127, 130, 163, 176, 184, 198, 208, 211, 212, 220, 222, 224, 228, 235, 330, 335, 422, 441, 443, 444, 462, 487
Summer, in Turkmenistan 42–45
Sunda Islands 522
Sundukli 12, 19, 212, 220, 229
Sunshine, duration of 25
Surkhan-Darya River 13
Sushanka 333
Svintsovy Rudnik 528
Swan 218
Sycamore, see Plane
Syrdarya River 298, 300, 312, 351, 364, 374, 489–491
Syria 417, 523
Syrtlany 59, 106
Syulyukli (tributary of the Chandyr River) 208
Syulyukli (on Mt. Dushak) 136, 224, 524

Syunt, Mt. 126, 182, 462
Syunt-Khasardagh Mts. 117, 464, 526
Syunt-Khasardagh Reserve 3, 116, 117, 149–172, 182, 327, 464, 526
Syuzen 79, 97, 98

Tagarev, Mt. 146, 206, 234, 235
Taiwan 295
Tajik Depression 51
Tajikistan 50, 152, 188, 208, 210, 224, 228, 417, 457, 501, 503, 505, 507, 509, 526–530, 532, 537, 539
Takhta-Bazar 14, 24, 33, 36, 220, 221, 507, 529, 530
Takyr 9, 11, 18, 78, 87
Talas Alatau 298
Talimardzhan 223
Talkhatanbaba 221
Talysh 56, 202
Tamarisk 168
Tamerlane 1
Tarbagatai 422
Tarim Depression 302
Tarim River 303, 539
Tarimkaya 97
Taschukho Well 352, 354
Tashauz 24, 30, 33, 36, 37, 46, 66, 211
Tashauz Oasis 18, 504
Tashauz Region 88, 512, 524, 526–531
Tashkent 1, 52
Tashkepri 215, 221, 223, 227
Tedzhen 7, 24, 33, 36, 46, 507, 526, 528, 531
Tedzhen Oasis 20, 27, 37, 45
Tedzhen Reservoir 373
Tedzhen River 5, 11, 12, 14, 19, 26, 38, 40, 49, 61, 62, 95, 106, 174, 188, 208, 212–222, 227, 229, 256, 266, 273, 277, 279, 299, 301, 303, 305, 322, 324, 325, 372, 373, 414, 419–421, 440, 441, 448, 456, 457, 460, 461, 489, 490, 498, 527, 536, 538
Tedzhen-Murghab area 495
Tedzhenstroi 24, 526, 528
Tedzhenstroi II 221
Tekechingasy 130, 131, 234,

Index of Subjects 649

235
Tekedzhik 235
Tekke 503, 506, 507
Temperature 17–21
Termez 298, 304
Tersakan River 13, 127, 130
Tertiary 10, 53, 54, 173, 177, 187, 203, 347
Tethys Sea 49, 60, 121, 124, 177, 309, 312, 411, 488, 492, 522, 533
Tetyr 79, 84, 85
Tiamil 127
Tibet 325, 404
Tibetan province 374
Ticks 244
Tien Shan 50, 55, 120, 152, 163, 166, 182, 188, 303, 312, 325, 329, 403–405, 416, 417
Tien Shan Mountainous, center of origin 309, 310
Tiflis 2
Tiger 2, 208
Toad agama, geographic variation 351–364
Togarev, see Tagarev, Mt.
Tortoise 192
Tragacanthoid vegetation 166
Tragacanthoids, see Cushion plants
Trans-Ili Alatau 297
Trans-Palearctic (range) 422, 442, 446–449, 473, 477, 503–505, 507–517
Trans-Unguz 9, 57, 212
Trans-Uzboi 10, 212
Transbaikalia 489
Transcaspian railroad 1
Transcaspian Region 1, 3, 265–280
Transcaucasia 105, 106, 122, 123, 235, 421, 440, 444, 445, 89, 491, 537
Transcaucasia, East 210
Transcaucasia, Southeast 181, 183, 301
Transcaucasian-Iranian (range) 166
Transcaucasian-Turanian-Central Asian (range) 493
Trgoi 107, 115, 119
Tuarkyr 9, 50–52, 456, 457, 504, 505, 509, 524
Tugais (riparian forests) 13, 59, 60, 62, 127, 168, 183, 208, 215, 325, 414, 442, 464
Turan 173, 175, 249
Turan Plateau 51, 59
Turan-Cis-Aral province 540
Turan-Mongolian (range) 518, 520
Turangli 174
Turanian (range) 63, 105, 307, 309, 312, 314. 326, 329, 408, 419–422, 440–442, 445–449, 451–455, 473, 477, 478, 485, 493, 502–518, 520, 524, 532
Turanian Lowland 6
Turanian province 202,323, 324, 403, 455, 456
Turanian tiger, see Tiger
Turanian, center of origin 309
Turanian-Middle Asian (range) 493
Turano-Gobian (range) 473, 477
Turgai (paleofloristic province) 53, 125
Turkestan 509, 518, 520, 522
Turkestan province 374
Turkestanian (range) 419, 422, 440, 445–449, 503, 505, 524
Turkestanian-Siberian (range) 517
Turkestanian-Turanian (range) 507
Turkey 115, 316, 417, 421, 490–492, 526–528, 533
Turkmen Academy of Sciences 3
Turkmen jerboa 211
Turkmen mountain sheep, see Wild sheep
Turkmen-Kala 221
Turkmeno-Iranian province 197, 456
Turkmeno-Khorassan Mountains 6, 58, 106, 129, 139, 176, 180, 203, 209, 303, 412, 420, 537, 188
Turkmeno-Khorassan province 458
Turkmeno-Nishapur Mountains 144
Turonian 51, 107
Turtkul 298, 352
Tutly 135
Tuva 489, 507
Tuyamuyun Reservoir 13, 298, 304, 375

Uch-Adzhi, see Uchadzhi
Uchadzhi 24, 36, 32, 37, 57, 215, 514
Uchgyoz 106
Uchtagan 5, 9, 10, 88, 94, 97, 251
Ufra 106
Uilya 107
Ukraine 528, 529
Ular 220
Ulski Bank 542
Ulyshor 327, 328
Umbelmes 107
Unguz 11, 15, 212, 224, 317, 327, 352, 353, 363, 364, 407, 536
Upper Amudarya cyclone 27
Ural Mts. 56, 507
Ural River 371
USA 511, 513
USSR 115
Ustyurt 5, 6, 9, 10, 15, 18, 51, 55–57, 63, 82, 83, 85–87, 179, 184, 212, 213, 217, 219, 221, 251, 266, 274, 275, 279, 317, 324, 407, 409, 456, 457, 462, 497, 530, 536
Ustyurt Area 324
Ustyurt-Caspian District 324
Ustyurt-Cis-Caspian district 457
Utah 25
Uvun-Kair Island 299
Uyaly Island 299
Uzbekistan 12, 19, 50, 88, 91, 93, 163, 208, 210, 212, 214, 222, 224, 228, 257, 280, 296, 316, 317, 366, 372, 457, 523, 526, 528–531, 537–539
Uzboi 5, 9, 10, 15, 57, 61, 62, 85, 93, 96, 97, 99, 207, 210, 214, 250, 251, 276, 279, 317, 341, 485, 528, 536, 539
Uzunada 526
Uzunakar Spring, see Uzunsu Spring
Uzunsu Spring 526
Uzyntokai 462

Vakhsh River 295, 489, 536
Valanginian 51
Valdai glaciation 304
Vannovsky 134, 226
Vavilov, Nikolai 2
Vekil-Bazar 221
Verkhnee Skobelevo 526
Vietnam 295, 518
Viper 192, 319, 339
Vole 193
Volga Plateau 56
Volga River 295, 541

Walnut 2, 127, 137, 167, 183–185
Water erosion 71, 72
Water resources 3
Watermelon 19
Weevils 191
Wels 321
West Palearctic (range) 419, 420, 422, 440–447, 473, 477
White crane (sterkh) 215
White spoonbill 216
White-headed duck 216

White-tailed sea eagle (erne) 217
Wild ass, see Onager
Wild boar 231
Wild cats 60, 243, 321
Wild sheep (arkhar) 2, 60, 193, 210–211, 243–244
Willow 168, 183
Wind 37–38
Winter, in Turkmenistan 39, 40
Wolf 60, 193, 321

Yablonnaya 207
Yablonovsky 132
Yartykala 462
Yaskhan 24, 26, 33, 59, 526, 528
Yatyk 106
Yekedzhe 24, 33, 46
Yepelek 81, 101
Yeradzhi 220, 250
Yeradzhi Refuge 228
Yerbent 24, 212, 210
Yeroilan, see Yeroyulanduz

Yeroyulanduz 3, 12, 95, 207, 215, 223, 314, 464, 485, 510, 512, 525, 526, 528, 529, 531, 533
Yezd 122
Yoldere 127, 167, 182, 210, 464
Yuzarlik 79, 100

Zagroz 55, 106, 200, 522
Zaisan 303
Zakli 223
Zau 120
Zeagli 24, 33, 39, 40, 46
Zelenaya 234
Zeravshan River 12, 304, 363, 489
Zerkau 234, 235
Zhanadarya 364
Zinovyevka, see Verkhnee Skobelevo
Zirakev 130
Zirik 107
Zoomass 191
Zyulfagar 106, 235

MONOGRAPHIAE BIOLOGICAE

1. *Physiologia Comparata et Oecologia*, Vol. I. 1948   out of print
2. *Physiologia Comparata et Oecologia*, Vol. II. 1952   out of print
3. *Physiologia Comparata et Oecologia*, Vol. III. 1954   out of print
4. *Physiologia Comparata et Oecologia*, Vol. IV. 1957   out of print
5. (1) R. A. Kennedy: *The Metzner Theory of Urine Formation*. 1957
   (2) M. D. Grmek: *On Ageing and Old Age*. Basic Problems and Historic Aspects of Gerontology and Geriatrics. 1958
   (3–4) O. Bassir: *Biochemical Aspects of Human Malnutrition in the Tropics*. 1962
     out of print
6. F. S. Bodenheimer: *Animal Ecology Today*. 1958   out of print
7. R. R. Gates: *Taxonomy and Genetics of Oenothera*. 1958   ISBN 90-6193-061-8
8. A. Keast, R. L. Crocker and C. S. Christian (eds.): *Biogeography and Ecology in Australia*. 1959   out of print
9. S. Stanković: *The Balkan Lake Ohrid and Its Living World*. 1960   out of print
10. E. Rivnay: *Field Crop Pests in the Near East*. 1962   ISBN 90-6193-063-4
11. M. Kozhov: *Lake Baikal and Its Life*. 1962   ISBN 90-6193-064-2
12. M. S. Mani: *The Ecology of Plant Galls*. 1964   out of print
13. E. van Oye (ed.): *The World Problem of Salmonellosis*. 1964
      ISBN 90-6193-066-9
14. D. H. S. Davis (ed.): *Ecological Studies in Southern Africa*. 1964   out of print
15. J. van Mieghem and P. van Oye (eds.): *Biogeography and Ecology in Antarctica*. 1965   ISBN 90-6193-067-7
16. H. Boyko (ed.): *Salinity and Aridity*. New Approaches to Old Problems. 1966
      out of print
17. M. J. Coe: *The Ecology of the Alpine Zone of Mount Kenya*. 1967
      ISBN 90-6193-068-5
18. E. J. Fittkau, J. Illies, H. Klinge, G. H. Schwabe and H. Sioli (eds.): *Biogeography and Ecology in South America*, Vol. I. 1968   out of print
19. E. J. Fittkau, J. Illies, H. Klinge, G. H. Schwabe and H. Stioli (eds.): *Biogeography and Ecology in South America*, Vol. II. 1969   ISBN 90-6193-071-5
20. R. N. Kaul (ed.): *Afforestation in Arid Zones*. 1970   out of print
21. R. Battistini and G. Richard-Vindard (eds.): *Biogeogaphy and Ecology of Madagascar*. 1972   ISBN 90-6193-073-1
22. J. E. Bishop: *Limnology of a Small Malayan River Sungai Gombak*. 1973
      ISBN 90-6193-074-X
23. M. S. Mani (ed.): *Ecology and Biogeography in India*. 1974   ISBN 90-6193-075-8
24. E. K. Balon and A. G. Coche (eds.): *Lake Kariba*. A Man-made Tropical Ecosystem in Central Africa. 1974   ISBN 90-6193-076-6
25. W. D. Williams (ed.): *Biogeography and Ecology in Tasmania*. 1974
      ISBN 90-6193-077-4
26. A. de Vos: *Africa, the Devastated Continent?* Man's Impact on the Ecology of Africa. 1975   ISBN 90-6193-078-2
27. G. Kuschel (ed.): *Biogeography and Ecology in New Zealand*. 1975
      ISBN 90-6193-079-0
28. I. Prakash and P. K. Ghosh (eds.): *Rodents in Desert Environments*. 1975
      ISBN 90-6193-080-4
29. J. Rzóska (ed.): *The Nile*. Biology of an Ancient River. 1976   ISBN 90-6193-081-2
30. G. Kunkel (ed.): *Biogeography and Ecology in the Canary Islands*. 1976
      ISBN 90-6193-082-0

MONOGRAPHIAE BIOLOGICAE

31. M. J. A. Werger (ed.): *Biogeography and Ecology of Southern Africa*, 2 vols. 1978
ISBN 90-6193-083-9
32. C. Serruya (ed.): *Lake Kinneret* [Lake of Tiberias / Sea of Galilee]. 1978
ISBN 90-6193-085-5
33. Ph. D. Mordukhaí-Boltovskoi (ed.): *The River Volga and Its Life.* 1979
ISBN 90-6193-084-7
34. Ch. W. Heckman: *Rice Field Ecology in Northeastern Thailand.* The Effect of Wet and Dry Seasons on a Cultivated Aquatic Ecosystem. 1979   ISBN 90-6193-086-3
35. M. Kalk, A. J. McLachlan and C. Howard-Williams (eds.): *Lake Chilwa.* Studies of Change in a Tropical Ecosystem. 1979   ISBN 90-6193-087-1
36. B. R. Allanson (ed.): *Lake Sibaya* [South Africa]. 1979   ISBN 90-6193-088-X
37. H. Löffler (ed.): *Neusiedlersee.* The Limnology of a Shallow Lake in Central Europe. 1979   ISBN 90-6193-089-8
38. J. Rzóska: *Euphrates and Tigris.* Mesopotamian Ecology and Destiny. With Contributions by J. F. Talling and K. E. Banister. 1980   ISBN 90-6193-090-1
39. K. A. Brodsky: *Mountain Torrent of the Tien Shan.* A Faunistic-Ecology Essay. Translated from Russian by V. V. Golosov. 1980   ISBN 90-6193-091-X
40. M. S. Mani and L. E. Giddings (eds.): *Ecology of Highlands.* 1980
ISBN 90-6193-093-6
41. A. Keast (ed.): *Ecological Biogeography of Australia*, 3 vols. 1981
ISBN 90-6193-092-8
42. J. L. Gressit (ed.): *Biogeography and Ecology of New Guinea*, 2 vols. 1982
ISBN 90-6193-094-4
43. H. Heathwole, T. Done and E. Cameron: *Community Ecology of a Coral Cay.* A Study of One-Tree Island, Great Barrier Reef, Australia. 1981   ISBN 90-6193-096-0
44. P. S. Maitland (ed.): *The Ecology of Scotland's Largest Lochs.* Lomond, Awe, Ness, Morar and Shiel. 1981   ISBN 90-6193-097-9
45. K. Müller (ed.): *Coastal Research in the Gulf of Bothnia.* 1982   ISBN 90-6193-098-7
46. G. K. Rutherford (ed.): *The Physical Environment of the Faeroe Islands.* 1982
ISBN 90-6193-099-5
47. J. I. Furtado and S. Mori (eds.): *Tasek Bera* [Malaysia]. The Ecology of a Feshwater Swamp. 1982   ISBN 90-6193-100-2
48. G. R. South (ed.): *Biogeography and Ecology of the Island of Newfoundland.* 1983
ISBN 90-6193-101-0
49. J. A. Thornton (ed.): *Lake McIlwaine.* The Eutrophication and Recovery of a Tropical African Man-made Lake. 1982   ISBN 90-6193-102-9
50. R. W. Edwards and M. P. Brooker: *The Ecology of the Wye* [U.K.]. 1982
ISBN 90-6193-103-7
51. T. Petr (ed.): *The Purari* [Papua New Guinea]. Tropical Environment of a High Rainfall River Basin. 1983   ISBN 90-6193-104-5
52. H. Kuhbier, J. A. Alcover and C. Guerau d'Arellano Tur (eds.): *Biogeography and Ecology of the Pityusic Islands.* 1984   ISBN 90-6193-105-3
53. J.-P. Carmouze, J.-R. Durand and C. Lévêque (eds.): *Lake Chad.* Ecology and Productivity of a Shallow Tropical Ecosystem. 1983   ISBN 90-6193-106-1
54. S. Horie (ed.): *Lake Biwa* [Japan]. 1984   ISBN 90-6193-095-2
55. D. R. Stoddart (ed.): *Biogeography and Ecology of the Seychelles Islands.* 1984
ISBN 90-6193-107-X
56. H. Sioli (ed.): *The Amazon.* Limnology and Landscape Ecology of a Mighty Tropical River and Its Basin. 1984   ISBN 90-6193-108-8

MONOGRAPHIAE BIOLOGICAE

57. C. H. Fernando (ed.): *Ecology and Biogeography in Sri Lanka.* 1984
 ISBN 90-6193-109-6
58. S. J. Casper (ed.): *Lake Stechlin* [Germany]. A Temperate Oligotrophic Lake. 1985
 ISBN 90-6193-512-1
59. U. Th. Hammer: *Saline Lake Ecosystems of the World.* 1985   ISBN 90-6193-535-0
60. B. R. Davies and K. F. Walker (eds.): *The Ecology of River Systems.* 1986 out of print
61. P. de Deckker and W. D. Williams (eds.): *Limnology in Australia.* 1986
 ISBN 90-6193-578-4
62. Y. Yom-Tov and E. Tchernov (eds.): *The Zoogeography of Israel.* The Distribution and Abundance at a Zoogeographical Crossroad. 1988   ISBN 90-6193-650-0
63. F. D. Por: *The Legacy of Tethys.* An Aquatic Biogeography of the Levant. 1989
 ISBN 0-7923-0189-7
64. B. R. Allanson, R. C. Hart, J. H. O'Keeffe and R. D. Robarts: *Inland Waters of Southern Africa.* An Ecological Perspective. 1990   ISBN 0-7923-0266-4
65. F. di Castri, A. J. Hansen and M. Debussche (eds.): *Biological Invasions in Europe and the Mediterranean Basin.* 1990   ISBN 0-7923-0411-X
66. R. W. Edwards, A. S. Gee and J. H. Stoner (eds.): *Acid Waters in Wales.* 1990
 ISBN 0-7923-0493-4
67. S. K. Jain and L. W. Botsford (eds.): *Applied Populations Biology.* 1992
 ISBN 0-7923-1425-5
68. C. Dejoux and A. Iltis (eds.): *Lake Titicaca* [Peru/Bolivia]. A Synthesis of Limnological Knowledge. 1992   ISBN 0-7923-1663-0
69. R.B. Wood and R.V. Smith (eds.): *Lough Neagh.* The Ecology of a Multipurpose Water Resource. 1993   ISBN 0-7923-2112-X
70. P.E. Ouboter (ed.): *The Freshwater Ecosystems of Suriname.* 1993
 ISBN 0-7923-2408-0
71. M.A. Brunt and J.E. Davies (eds.): *The Cayman Islands.* Natural History and Biogeography. 1994   ISBN 0-7923-2462-5
72. V. Fet and K. Atamuradov (eds.): *Biogeography and Ecology of Turkmenistan.* 1994
 ISBN 0-7923-2738-1

KLUWER ACADEMIC PUBLISHERS – DORDRECHT / BOSTON / LONDON

MIX
Papier aus verantwortungsvollen Quellen
Paper from responsible sources
FSC® C105338

If you have any concerns about our products,
you can contact us on
ProductSafety@springernature.com

In case Publisher is established outside the EU,
the EU authorized representative is:
**Springer Nature Customer Service Center GmbH
Europaplatz 3, 69115 Heidelberg, Germany**

Printed by Libri Plureos GmbH
in Hamburg, Germany